BIOMIMETICS
Biologically Inspired Technologies

BIOMIMETICS
Biologically Inspired Technologies

EDITED BY

Yoseph Bar-Cohen

Jet Propulsion Laboratory (JPL),
California Institute of Technology
Pasadena, California, USA

Taylor & Francis
Taylor & Francis Group
Boca Raton London New York

A CRC title, part of the Taylor & Francis imprint, a member of the
Taylor & Francis Group, the academic division of T&F Informa plc.

Published in 2006 by
CRC Press
Taylor & Francis Group
6000 Broken Sound Parkway NW, Suite 300
Boca Raton, FL 33487-2742

International Standard Book Number-10: 0-8493-3163-3 (Hardcover)
International Standard Book Number-13: 978-0-8493-3163-3 (Hardcover)
Library of Congress Card Number 2005048511

Library of Congress Cataloging-in-Publication Data

Bar-Cohen, Yoseph.
   Biomimetics : biologically inspired technologies / Yoseph Bar-Cohen.
      p. cm.
   Includes index.
   ISBN 0-8493-3163-3 (alk. Paper)
   1. Biomimetics. 2. Bionics I. Title.

QP517.B56B37 2005
600—dc22   2005048511

Taylor & Francis Group
is the Academic Division of Informa plc.

Visit the Taylor & Francis Web site at
http://www.taylorandfrancis.com

**and the CRC Press Web site at
http://www.crcpress.com**

Abstract

Over the 3.8 billion years since life is estimated to have begun to appear on Earth, evolution has resolved many of nature's challenges leading to lasting solutions with maximal performance using minimal resources. Nature's inventions have always inspired human achievement and have led to effective algorithms, methods, materials, processes, structures, tools, mechanisms, and systems. There are numerous examples of biomimetic successes including some that are simple copies of nature, such as the use of fins for swimming. Other examples were inspired by biological capabilities with greater complexity including the mastery of flying that became possible only after the principles of aerodynamics were better understood. Some commercial implementations of biomimetics can be readily found in toy stores, where robotic toys are increasingly appearing and behaving like living creatures. More substantial benefits of biomimetics include the development of prosthetics that closely mimic real limbs as well as sensory-enhancing microchips that are being used to interface with the brain to assist in hearing, seeing, and controlling instruments. In this book, various aspects of the field of biomimetics are reviewed, examples of inspiring biological models and practical applications of biomimetics are described, and challenges and potential directions of the field are discussed.

Editor

Yoseph Bar-Cohen is a physicist who has specialized in electroactive materials and mechanisms, and ultrasonics nondestructive evaluation (NDE). A senior research scientist and group supervisor, Advanced Technologies, at the Jet Propulsion Laboratory (JPL), he is also responsible for the Nondestructive Evaluation and Advanced Actuators (NDEAA) Lab (http://ndeaa.jpl.nasa.gov/). The NDEAA lab established in 1991 is listed on the JPL's Chief Technologies as one of the JPL unique facilities. Bar-Cohen is a fellow of two technical societies: The International Society for Optical Engineering (SPIE) and American Society for Nondestructive Testing (ASNT). He received his PhD in Physics (1979) from the Hebrew University, Jerusalem, Israel. His notable discoveries are the leaky Lamb waves (LLW) and polar backscattering phenomena in composite materials. Bar-Cohen has over 280 publications as author and coauthor, and 16 registered patents. He has made numerous presentations at national and international conferences and chaired and co-chaired 27 conferences. As coauthor, editor, and coeditor, he has written and edited 4 books and 11 conference proceedings. He is the initiator of the SPIE Conference on electroactive polymers (EAP), which he has been chairing since 1999. He challenged engineers worldwide to develop a robotic arm driven by EAP to wrestle with humans and win, and he organized the competition as part of the EAPAD conferences. The first of this competition took place on March 7, 2005. For his contributions to the field of EAP, he has been named one of five technology gurus who are pushing tech's boundaries by *Business Week* in April 2003. His scientific and engineering accomplishments have earned him many honors and awards including the NASA Exceptional Engineering Achievement Medal in 2001, two SPIE awards — the NDE Lifetime Achievement Award (2001) and the Smart Materials and Structures Lifetime Achievement Award (2005).

Acknowledgments

The research at Jet Propulsion Laboratory (JPL), California Institute of Technology was carried out under a contract with National Aeronautics and Space Administration (NASA).

The editor would like to thank David Hanson, University of Texas at Dallas and Hanson Robotics, Inc. for his drawing of the book cover page. Further, the editor would like to acknowledge and express his appreciation to the following individuals who took the time to review various book chapters. Their contribution is highly appreciated and it has helped to make this book of significantly greater value to its readers. The 65 individuals who served as reviewers of chapters in this book are:

Rajat Agrawal	Department of Mechanical Engineering, University of Delaware, Newark, DE
Tony Aponick	Foster-Miller, Inc., Waltham, MA
Limor Bar-Cohen	Department of Public Policy, School of Public Affairs, University of California, Los Angeles, CA
Jon Barnes	Institute of Biomedical and Life Sciences, University of Glasgow, Scotland, U.K.
William Baumgartner	Cardiac Surgery, Johns Hopkins Medical Institutions, Baltimore, MD
Peter J. Bentley	University College London, London, U.K.
Reinhard Blickhan	Science of Motion, Institute of Sportscience, Friedrich Schiller University, Jena, Germany
Paul Calvert	Department of Textile Sciences, Dartmouth, MA
Robert E. Cleland	University of Washington, Seattle, WA
Jack Cohen	Newent, Glasgow, Scotland, U.K.
John Conte	Cardiac Surgery, Johns Hopkins Medical Institutions, Baltimore, MD
Bob Dennis	University of North Carolina, Department of Biomedical Engineering, Chapel Hill, NC
Ezequiel Di Paolo	Sussex University, Brighton, U.K.
Tammy Drezner	California State University, Fullerton, Fullerton, CA
Zvi Drezner	California State University, Fullerton, Fullerton, CA
Wolfgang Fink	Jet Propulsion Laboratory/Caltech, Pasadena, CA
Greg Fischer	Strategic Analysis, Inc., Arlington, VA
Robert A. Freitas Jr.	Institute for Molecular Manufacturing, Los Altos, CA
Udi Gazit	University of Tel Aviv, Israel
Vincent Gott	Cardiac Surgery, Johns Hopkins Medical Institutions, Baltimore, MD

H. Thomas Hahn   Mechanical and Aerospace Engineering, University of California,
                          Los Angeles, CA
Roger T. Hanlon   Marine Resources Center, Marine Biological Laboratory, Woods
                          Hole, MA
Walter Herzog   Human Performance Laboratory, University of Calgary, Calgary,
                          Alberta, Canada
Neville Hogan   Massachusetts Institute of Technology, Cambridge, MA
George Jeronimidis   Centre for Biomimetics, School of Construction Management and
                          Engineering, Reading University, Reading, Berkshire, U.K.
Brett Kennedy   Jet Propulsion Laboratory, Pasadena, CA
Roy Kornbluh   SRI International, Menlo Park, CA
Dimitris C. Lagoudas   Texas Institute for Intelligent Bio-Nano Materials, and Structures for
                          Aerospace Vehicles, Texas A&M University, College Station, TX
Matthias Langer   Universitätsklinikum Ulm, Ulm, Germany
Luke Lee   Berkeley Sensor and Actuator Center, University of California,
                          Berkeley, CA
Cornelius Leondes   University of California, Los Angeles, CA
Hod Lipton   Mechanical and Aerospace Engineering, and Computing and
                          Information Science, Cornell University, Ithaca, NY
Jian R. Lu   Department of Physics, University of Manchester Institute of
                          Science and Technology (UMIST), Manchester, New Hampshire
Michael Lysaght   Biomedical Engineering Artificial Organs Laboratory, Department
                          of Molecular Pharmacology and Biotechnology, Brown University,
                          Providence, RI
John Madden   University of British Columbia, Vancouver, BC, Canada
John Main   DARPA, Washington, DC
Ajit Mal   Mechanical and Aerospace Engineering, University of California,
                          Los Angeles, CA
Paul S. Malchesky   International Center for Artificial Organs and Transplants,
                          Painesville and STERIS Corporation, Mentor, OH
George A. Marcoulides   California State University, Fullerton, CA
Adi Marom   Research Artifacts Center Engineering, The University of Tokyo,
                          Japan
William Megill   University of Bath, Bath, U.K.
Chris Melhuish   Intelligent Autonomous Systems Laboratory, CEMS Faculty,
                          University of the West of England, Bristol, U.K.
Kenneth Meijer   Universiteit Maastricht, The Netherlands
Dharmendra Modha   Computer Science Department, IBM Almaden Research Center,
                          White Plains, New York
Yuki Nose   Department of Surgery, Houston, TX
Michael E. DeBakey   Department of Surgery, Houston, TX
Qibing Pie   Department of Materials Science and Engineering, University of
                          California, Los Angeles, CA
Glen Pennington   East Tennessee State University Medical Center, Johnson City, TN
Gerald H. Pollack   University of Washington, Seattle, WA
Gill Pratt   F.W. Olin College of Engineering, Needham, MA
Sumitra Rajagopalan   Biomedical Engineering Institute, Université de Montréal,
                          Montréal, Canada
Roy E. Ritzmann   Department of Biology, Case Western Reserve University,
                          Cleveland, OH
Nick Rowe   Botanique et Bioinformatique, Montpellier, France

Said Salhi	School of Mathematics and Statistics, University of Birmingham, Birmingham U.K.
Gernot Schmierer	BMW Group, Munich, Germany
Bill Seaman	Digital Media Department, Rhode Island School of Design, Providence, RI
Ji Su	NASA Langley Research Center, Langley, VA
Mikhail Svinin	RIKEN BMC, Nagoya, Japan
Ken Toda	Department of Zoology, Graduate School of Science, Kyoto University, Japan
Tihamer "Tee" Toth-Fejel	General Dynamics Advanced Information Systems, Arlington, VA
Blaire Van Valkenburgh	Department of Ecology and Evolutionary Biology at University of California, Los Angeles, CA
James Weiland	Doheny Eye Institute, USC, Los Angeles, CA
Stefan Wölfl	Klinik für Innere Medizin Friedrich Schiller Universität, Jena, Germany
Julian F.V. Vincent	Department of Mechanical Engineering, Centre for Biomimetics and Natural Technologies, University of Bath, Bath, U.K.
Steven Vogel	Duke University, Durham, NC
Masaki Yamakita	Tokyo Institute of Technology, Tokyo, Japan

Contributors

Rajat N. Agrawal
Doheny Eye Institute
University of Southern Carolina
Los Angeles, California

Pramod Bonde
School of Medicine
The Johns Hopkins University
Baltimore, Maryland

Justin Carlson
Departments of Biomedical Engineering,
    Chemical and Biological Engineering,
    and Biology
Tufts University
Medford, Massachusetts

Robert G. Dennis
Department of Biomedical Engineering
University of Michigan
Ann Arbor, Michigan

Tammy Drezner
California State University-Fullerton
Fullerton, California

Zvi Drezner
California State University-Fullerton
Fullerton, California

Atul Dubey
Department of Mechanical and Industrial
    Engineering
Northeastern University
Boston, Massachusetts

Shail Ghaey
Departments of Biomedical
    Engineering, Chemical and Biological
    Engineering, and Biology
Tufts University
Medford, Massachusetts

Stanislav N. Gorb
Max Planck Institute for Metals
    Research
Stuttgart, Germany

Keyoor Chetan Gosalia
North Carolina State University
Raleigh, North Carolina

David Hanson
University of Texas at Dallas
Hanson Robotics, Inc.
Los Angeles, California

Robert Hecht-Nielsen
Computational Neurobiology
University of California
San Diego, California

Hugh Herr
Massachusetts Institute of
    Technology
Media Laboratory
Cambridge, Massachusetts

Shigeyuki Hosoe
RIKEN BMC
Nagoya, Japan

Mark S. Humayun
Doheny Eye Institute
University of Southern Carolina
Los Angeles, California

Masami Ito
RIKEN BMC
Nagoya, Japan

David L. Kaplan
Departments of Biomedical Engineering,
   Chemical and Biological Engineering,
   and Biology
Tufts University
Medford, Massachusetts

Gianluca Lazzi
North Carolina State University
Raleigh, North Carolina

Luke P. Lee
Berkeley Sensor and Actuator Center
Department of Bioengineering
University of California
Berkeley, California

Hod Lipson
Mechanical and Aerospace Engineering, and
   Computing and Information Science
Cornell University
Ithaca, New York

Zhiwei Luo
RIKEN BMC
Nagoya, Japan

Constantinos Mavroidis
Department of Mechanical and Aerospace
   Engineering
Rutgers University
Newark, New Jersey

Kenneth Meijer
Department of Biomedical Engineering
Technische Universiteit Eindhoven
Eindhoven, The Netherlands

Sean Moran
Departments of Biomedical Engineering,
   Chemical and Biological Engineering,
   and Biology
Tufts University
Medford, Massachusetts

Juan C. Moreno
Instituto de Automática Industrial
Madrid, Spain

Sia Nemat-Nasser
Center of Excellence for Advanced
   Materials, Mechanical and Aerospace
   Engineering
University of California
San Diego, California

Syrus Nemat-Nasser
Center of Excellence for Advanced Materials,
   Mechanical and Aerospace Engineering
University of California,
San Diego, California

Thomas Plaisted
Center of Excellence for Advanced Materials,
   Materials Science and Engineering
University of California,
San Diego, California

Hans H.C.M. Savelberg
Department of Health Sciences
Universiteit Maastricht
Maastricht, The Netherlands

Rainer Stahlberg
Department of Biology
University of Washington
Seattle, Washington

Anthony Starr
Center of Excellence for Advanced Materials,
   Mechanical and Aerospace Engineering
University of California
San Diego, California

Robert Szema
Department of Bioengineering
University of California
Berkeley, California

Minoru Taya
Department of Mechanical Engineering
University of Washington
Seattle, Washington

Cam Anh Tran
Departments of Biomedical Engineering,
    Chemical and Biological Engineering,
    and Biology
Tufts University
Medford, Massachusetts

Ajay Ummat
Department of Mechanical and Industrial
    Engineering
Northeastern University
Boston, Massachusetts

Alireza Vakil Amirkhizi
Center of Excellence for Advanced Materials,
    Mechanical and Aerospace Engineering
University of California
San Diego, California

Julian F.V. Vincent
Department of Mechanical Engineering
The University of Bath
Bath, U.K.

James Weiland
Doheny Eye Institute
University of Southern Carolina
Los Angeles, California

Hidenori Yokoi
Massachusetts Institute of Technology
Cambridge, Massachusetts

Shuguang Zhang
Massachusetts Institute of Technology
Cambridge, Massachusetts

Xiaojun Zhao
Massachusetts Institute of Technology
Cambridge, Massachusetts
and
Institute for Nanobiomedical Technology and
    Membrane Biology
Sichuan University
Sichuan, China

Contents

1

Introduction to Biomimetics: The Wealth of Inventions in Nature as an Inspiration for Human Innovation

Yoseph Bar-Cohen

CONTENTS

1.1 INTRODUCTION

Imagine a smart microchip that is buried in the ground for a long time. Upon certain triggering conditions this chip begins to grow and consume materials from its surroundings, converting them into energy and structural cells. As the chip grows further, it reconfigures its shape to become a mobile robot. Using its recently created mobility, the chip becomes capable of searching and locating critical resources consuming them to grow even more. The type and function of the specific cells that are formed depend on each cell's role within the growing structure. This science-fiction scenario is inspired by true-life biology such as the growth of chicks from an egg or plants from a seed. Yet given all our technological advances, it is still impossible to engineer such a reality.

Bionics as the term for the field of study involving copying, imitating, and learning from biology was coined by Jack Steele of the US Air Force in 1960 at a meeting at Wright–Patterson Air Force Base in Dayton, Ohio (Vincent, 2001). Otto H. Schmitt coined the term *Biomimetics* in 1969 (Schmitt, 1969) and this field is increasingly involved with emerging subjects of science and engineering. The term itself is derived from *bios,* meaning life, and *mimesis,* meaning to imitate. This new science represents the study and imitation of nature's methods, designs, and processes. While some of its basic configurations and designs can be copied, many ideas from nature are best adapted when they serve as inspiration for human-made capabilities. In this book, both biologically inspired and biologically mimicked technologies are discussed, and the terms biology, creatures, and nature are used synonymously.

Nature has always served as a model for mimicking and inspiration for humans in their desire to improve their life. By adapting mechanisms and capabilities from nature, scientific approaches have helped humans understand related phenomena and associated principles in order to engineer novel devices and improve their capability. The cell-based structure, which makes up the majority of biological creatures, offers the ability to grow with fault-tolerance and self-repair, while doing all of the things that characterize biological systems. Biomimetic structures that are made of multiple cells would allow for the design of devices and mechanisms that are impossible with today's capabilities. Emerging nano-technologies are increasingly enabling the potential of such capabilities.

The beak of birds may have served as an inspiring model for the development of the tweezers and the tong. While it is difficult to find evidence that it had inspired early humans, one can argue that since nature invented this device first it was a widely known concept way before humans began making tweezers and tongs. The mimicking of the beak is illustrated graphically on the cover page of this book, where a virtual mirror is drawn to represent the inspiration of adapting nature's capabilities. Although enormous advances have been made in the field of biomimetics, nature is still far superior to what we are capable of making or adapting. Given the limitation of today's technology, copying nature may not be the most effective approach. Many examples exist where humans using nature as inspiration have used its principles to invent far more effective solutions; flying is one such example. This book focuses on the technologies that resulted from both mimicking and being inspired by biology.

Nature evolves by responding to its needs and finding solutions that work, and most importantly, that last through innumerable generations while passing the test of survival to reach its next generation. Geological studies suggest the presence of life on Earth as early as 3.8 billion years ago (Lowman, 2002). Specifically, in Greenland, a series of ancient metamorphosed sediments were found with carbon isotope signatures that appear to have been produced by organisms that lived when the sediments were deposited. Furthermore, fossil evidence indicates that ancient bacteria, Archea (Archaebacteria), have existed on the Earth for at least 3.5 billion years (Schopf, 1993; Petr, 1996). After billions of years of trial and error experiments, which turn failures to fossils, nature has created an enormous pool of effective solutions. It is important to note however that the extinction of a species is not necessarily the result of a failed solution; it can be the result of outside influences, such as significant changes in climate, the impact of asteroids, volcanic activity, and other conditions that seriously affect the ability of specific creatures to survive. The adaptations of nature have led to the evolution of millions of species — each with its own way of meeting its needs in harmony with the environment (Research Report, 1992).

Through evolution, nature has "experimented" with various solutions to challenges and has improved upon successful solutions. Organisms that nature created, which are capable of surviving, are not necessarily optimal for their technical performance. Effectively, all they need to do is to survive long enough to reproduce. Living systems archive the evolved and accumulated information by coding it into the species' genes and passing the information from generation to generation through self-replication. Thus, through evolution, nature or biology has experimented with the principles of physics, chemistry, mechanical engineering, materials science, mobility, control, sensors, and many other fields that we recognize as science and engineering. The process has also involved scaling from nano and macro, as in the case of bacteria and virus, to the macro and mega, including our life scale and the dinosaurs, respectively. Although there is still doubt regarding the reason for the extinction of creatures such as the mammoth, it may be argued that the experiment in the evolution of mega-scale terrestrial biology failed. While marine creatures such as the whales survived, nature's experiment with large size terrestrial biology ended with the extinction of the prehistoric mega-creatures (e.g., dinosaurs and mammoths). Such creatures can now be found only in excavation sites and natural history museums.

As the evolution process continues, biology has created and continues to create effective solutions that offer great models for copying or as inspiration for novel engineering methods, processes, materials, algorithms, etc. Adapting biology can involve copying the complete appearance and function of specific creatures like the many toys found in toy stores, which are increasingly full of simplistic imitations of electro-mechanized toys such as dogs that walk and bark, frogs that swim, and such others. However, while we have copied or adapted many of nature's solutions an enormous number of mysteries remain unravelled. Humans have learned a lot from nature and the results help surviving generations and continue to secure a sustainable future.

This book reviews the various aspects of biomimetics from modeling to applications as well as various scales of the field from cell to macro-structures. Chapter 1 provides an overview of the field of biomimetics addressing technologies that mimic biology versus those that adapt its principles using biology as an inspiring model. Chapter 2 describes biological mechanisms as models for mimicking. Chapter 3 examines the mechanization of cognition and the creation of knowledge, and the various aspects of processing by the brain as a basis for autonomous operation. Another angle of this issue is covered in Chapter 4, where evolutionary robotics and open-ended design automation are described. One of the widely used biologically inspired algorithms, the genetic algorithm, is described in Chapter 5 using a mathematical imitation of evolution and natural selection. Robotics is increasingly inspired by biology and robots that are close imitation of animals and humans are emerging with incredible capability as described in Chapter 6. The details of making a biological system as a model are discussed in the following chapters where biologically inspired molecular machines are described in Chapter 7 and molecular design of biological and nano-materials in Chapter 8. The next two chapters deal with biological and artificial muscles with Chapter 9 describing engineered muscle actuators and Chapter 10 covering the topic of artificial muscles using electroactive polymers (EAP). An important aspect of biology and systems is the use of sensors and Chapter 11 covers the topic of vision as an example of bio-sensors. One of the unique characteristics of biological materials and structures is their multifunctionality and these materials are covered in Chapter 12. Other aspects of biological systems that offer important models for imitation are described in the chapters that follow. Chapter 13 covers defense and attack strategies and mechanisms in biology; Chapter 14 covers biological materials in engineering mechanisms; Chapter 15 describes mechanisms and applications of functional surfaces in biology. One of the critical issues of operating systems is that of control and Chapter 16 examines the issue of biomimetic and biologically inspired control. Interfacing the body with artificial devices is covered in the next two chapters with Chapter 17 describing interfacing microelectronics and the human body and Chapter 18 covering artificial support and replacement of human organs. Plants also serve as a model for inspiration and Chapter 19 describes the topic of nastic structures, which are active materials that enact and mimic plant movements. Chapter 20 of this book includes an overview, description, challenges, and outlook for the field of biomimetics.

This chapter provides an overview of some of the key biology areas that inspired humans to produce an imitation. This includes making artificial, synthesized, inspired, and copied mechanisms, as well as processes, techniques, and other biomimetic aspects. There are many examples but only a select few are given in this chapter to illustrate the successes and the possibilities.

1.2 MIMICKING AND INSPIRATION OF NATURE

Biology offers a great model for imitation, copying and learning, and also as inspiration for new technologies (Benyus, 1998). Flying was inspired by birds using human developed capabilities (Figure 1.1), whereas the design and function of fins, which divers use, was copied from the legs of water creatures such as the seal, goose, and frog. But the distinction between technologies resulting from the various adaptive approaches is not always clear. For instance, studying photosynthesis in a leaf may lead some to argue that the invention of the solar cell is an *imitation,* while others may see it as a biologically *inspired* technology. While both photosynthesis and solar cell use sunlight as a source of energy, they neither perform the same process nor create the same output.

Biologically inspired terms such as *male* and *female* connectors, as well as *teeth* of a saw are common, and it is very clear to us what they mean. Other terms derived from biology the usage of which are clearly understood include the *heart* to suggest the center, the *head* to indicate the beginning, the *foot* or *tail* to imply the end, the *brain* to describe a computing system. Likewise, the

Figure 1.1   The image of the Egyptian God Khensu with wings (left) illustrates the age-old fantasy of humans of being able to fly. (Photographed by the author at the Smithsonian Museum, Washington, DC.) This fantasy turned to reality with the use of aerodynamic principles leading to enormous capabilities such as the supersonic passenger plane, the Concord on the right. (Photographed by the author at the Boeing Aerospace Museum, Seattle, Washington.)

use of the terms *intelligent* or *smart* suggests the emulation of biological capabilities with a certain degree of feedback and decision making. Other terms include *aging, fatigue, death, digestion, life cycle*, and even *"high on the food chain"* (referring to a high management level). In the world of computers and software many biological terms are used to describe aspects of technology including *virus, worm, infection, quarantine, replicate*, and *hibernate*. Other forms of imitating nature comprise virtual reality, simulations and copying of structures and materials. Shapes are also used as recognizable terms where the *dog-bone* provides a clear description of the shape of test coupons that are used to measure the tensile module and strength of materials. Structures are also widely copied, for example the honeycomb. Used for its efficient packing structure by bees (which is different from its use in aerospace — for low weight and high strength), the honeycomb has the same overall shape in both biological and aerospace structures. It could be reasoned that the honeycomb structures, which are used in many of the aircraft structures of today's airplanes, were not copied from the bees (Gordon, 1976). However, since it is a commonly known structure invented by nature many years before humans arrived, no patent can be granted in the "patent court" of nature to the first human who produced this configuration. Generally, biological materials (Chapter 14), including silk and wool that are widely used in clothing, have capabilities that surpass those made by humans. This superb capability of biological materials, structures, and processes has been the subject of imitation in artificial versions of materials.

Plants can also offer a model for imitation (Chapter 19). Besides their familiar characteristics, some plants exhibit actuation capabilities that are expected of biological creatures. Such plants include the *mimosa* and the Sensitive Fern (*Onoclea sensibilis*) that fold or close their leaves when touched (Figure 1.2). There are also bug-eating plants with a leaf derived trap "door" that closes and traps unsuspecting bugs that enter to become prey. Examples of such plants include the Venus Flytrap (*Dionaea muscipula*) and the Pitcher plant (*Sarracenia purpurea*) (Figure 1.3). The sunflower tracks the sun's direction throughout the day to maximize exposure to its light. Plants have evolved in various ways, and some have produced uncommon solutions to their special needs. For example, some desert plants have flowers that produce the malodor of rotten meat, and some even have a brown color that appears very much like decomposing meat. Such characteristics are critical for these plants to attract flies, rather than bees, to pollinate their flowers.

Figure 1.2   The Sensitive Fern (*O. sensibilis* form the Woodsiaceae family of plants) has its leaves open (left) until they are touched (right).

Figure 1.3 (See color insert following page 302)   Bug-eating plants with traps that developed from their leaf.

1.2.1 Synthetic Life

Advances in understanding and unraveling the genetic code, and the ability to manipulate and splice genes have made the possibility of creating synthetic life an increasing reality. Biologists are now able to engineer bacteria and develop drugs that otherwise must be extracted from rare plants at very high costs. Further, bacteria and yeast are being produced to build proteins with synthetic amino acids having novel properties that are impossible to find in nature. Researchers are also working on assembling simple cells from basic components with an ability that is much broader than recombining DNA. The possibility of synthetically producing living cells fromscratch is increasingly becoming a near future potential (http://www.nature.com/cgi-taf/ DynaPage.taf?file = /nature/journal/v431/n7009/full/431624a_fs.html). This subject, however, will not be covered any further in this book since the subject is outside the scope of the book's objective.

1.3 ARTIFICIAL LIFE

The name artificial life (A-Life) suggests the synthesizing of life from nonliving components. A-Life is a technical field that is dedicated to the investigation of scientific, engineering, philosophical, and social issues involved in our rapidly increasing technological capability to synthesize from scratch life-like behaviors using computers, machines, molecules, and other alternative media (Langton, 1995). A-Life focuses on the broad characteristics of biology and contributes to the development of machines that evolve, sociable robots, artificial immune systems that protect computers from malicious viruses, and virtual creatures that learn, breed, age, and die. Moreover, biologists can now study evolution in virtual worlds, and medical students and doctors can study operation mechanisms of various living organs, including the heart with its cells, enabling learning in ways that are impossible with actual living organs.

The field of A-Life consists of a broad range of topics related to the synthesis and simulation of living systems in the form of self-replicating computer code that allows learning about fundamental aspects of evolution and their ecological context (Ray, 1992). The enormous advances of computer capability have led to the creation of an incredible computation and information processing power in support of the analytical development of biologically inspired capabilities. These advances have led to biological concepts and systems that are systematically modeled, copied, or adapted (Chapters 4 and 5; Adami, 1998) enabling predictions of what life can be beyond what we know from empirical research. Some of the topics that are covered under the umbrella of A-Life include origin of life, evolutionary and ecological dynamics, self-assembly, hierarchy of biological organization, growth and development, animal and robot behavior, social organization, and cultural evolution.

A-Life is often described as the effort to understand high-level behavior using low-level rules that are based on the laws of physics. The field itself covers the simulation or emulation of living systems or parts of living systems with the intent to understand their behavior. Another aspect of this field is the attempt to study emergent properties of living populations, usually by making a simulation of many agents and neglecting the precise details of members of an individual population. Adami (1998) approached the field of A-Life from physical sciences with life-like entities taking life as a property of an ensemble of units that share information coded in a physical substrate. In the presence of noise, each unit manages to keep its entropy significantly lower than the maximal entropy of the ensemble. This information is shared on timescales that exceed the "natural" timescale of decay of the information-bearing substrate by many orders of magnitude. For this purpose, he introduced the necessity for a synthetic approach and formulated a principle of living systems based on information and thermodynamic theory.

The founding of the field of A-Life is attributed to John Horton Conway, a mathematician from the University of Cambridge, who in 1968 invented a game called "The Game of Life" (Gardner, 1970). Using a simple system inspired by cell biology, this game exhibits complex, life-like behavior. The rules involve cell patterns that move across the Life universe, simulating life in the form of living and dead objects. After playing the game for a while, Conway discovered an interesting emergence of a pattern of five cells. He named this stable, repeating cell pattern, *glider*. This discovery was followed by R. William Gosper, Jr, who designed a glider gun that fires new gliders every 30 turns. The glider gun proved that it was possible for a single group of living cells to expand into the Life universe without limit (Levy, 1984; and Gardner, 1983). Later, using powerful computers, the study expanded into "organisms" in the Life universe with some starting at random patterns of cells seeking stable repeating patterns, or patterns that move like the gliders. An interesting aspect of this game was that the patterns found by computers were discovered rather than invented.

Some of the benefits of using computers have been the development of the "genetic programming" or "evolutionary programming" (Chapters 4 and 5; Koza, 1992). The "DNA" of genetic programming consists of a set of equations and operations where the computer software measures how well each program solves a particular problem. The programs that fare the worst are eliminated and new strains of program code are bred by recombination, either with or without mutation. The solutions produced by evolutionary programming emulate the solutions in the real world, and it may use functions that seemingly have no logical relevance to the problem that is being solved but it produces effective solutions (Chapters 4 and 5).

1.4 ARTIFICIAL INTELLIGENCE

According to the American Association for Artificial Intelligence (AAAI), artificial intelligence (AI) is, "the scientific understanding of the mechanisms underlying thought and intelligent behavior and their embodiment in machines." AI is a branch of computer science that studies the computational requirements for such tasks as perception, reasoning, and learning, to allow the development of systems that perform these capabilities (Russell and Norvig, 2003). AI researchers are addressing a wide range of problems that include studying the requirements for expert performance of specialized tasks, explaining behaviors in terms of low-level processes, using models inspired by the computation of the brain, and explaining them in terms of higher-level psychological constructs such as plans and goals. The field seeks to advance the understanding of human cognition (Chapter 3), understand the requirements for intelligence in general, and develop artifacts such as intelligent devices, autonomous agents, and systems that cooperate with humans to enhance their abilities. The name AI was coined in 1956, though the roots of the field may be attributed to the efforts in World War II to crack enemy codes by capturing human intelligence in a machine that was called Enigma. This approach eventually led to the 1997 computer success of IBM's Deep Blue in beating the world-champion chess player Garry Kasparov. Even though this was an enormous success for computers, it still does not resemble human intelligence. AI technologies consist of an increasing number of tools, including artificial neural networks, expert systems, fuzzy logic, and genetic algorithms (Luger, 2001; Chapters 4 and 5).

Advances in AI are allowing analysis of complex nonlinear problems that are beyond the capability of conventional methods by using such tools as neural networks (i.e., networks of artificial brain cells) that can learn and recognize patterns and reach solutions. This is providing enormous capabilities in the area of robotics including the ability to operate autonomously. One of the milestones in AI is the development of "Shakey" robot, which was completed by SRI International's Artificial Intelligence Center (AIC) in 1972. This six-foot tall robot (http://www-clmc.usc.edu/~cs545/Lecture_I.pdf) was named for its erratic and jerky movement. Shakey is the first mobile robot to visually interpret its environment, locate items, navigate around them, and reason about its actions. Shakey was equipped with a TV camera, a triangulating range finder, bumpers, and a wireless video system and it has the capability of autonomous decision making.

The subject of AI is widely covered in the literature (e.g., Luger, 2001; Russell and Norvig, 2003). Chapter 3 of this book addresses the topic of modeling computers after the processes in the human brain. One area of AI, which mimics nature, is the swarm intelligence that involves the study of self-organizing processes in artifacts of nature and humans. Algorithms inspired by social insect behavior have been proposed to solve difficult computational problems such as discrete optimization where the ant colony optimization process was followed. Resulting algorithms were used to solve such problems as vehicle routing and routing in telecommunication networks.

Some of the research areas in the field of AI today include web search engines, knowledge capture, representation and reasoning, reasoning under uncertainty, planning, vision, robotics, natural language processing, and machine learning. Increasingly, AI components are embedded in devices and machines that combine case-based reasoning and fuzzy reasoning to operate automatically or even autonomously. AI systems are used for such tasks as identifying credit card fraud, pricing airline tickets, configuring products, aiding complex planning tasks, and advising physicians. AI is also playing an increasing role in corporate knowledge management, facilitating the capture and reuse of expert knowledge. Intelligent tutoring systems make it possible to provide students with more personalized attention or even have computers listen and respond to speech-provided information. Moreover, cognitive models developed by AI tools can suggest principles for effective support for human learning — guiding the design of educational systems (Russell and Norvig, 2003).

1.5 NATURE AS A MODEL FOR STRUCTURES AND TOOLS

Biological creatures can build amazing shapes and structures using materials in their surroundings or the materials that they produce. The shapes and structures produced within a species are very close copies. They are also quite robust, and support the required function of the structure over the duration for which it is needed. Such structures include birds' nests and bees' honeycombs. Often the size of a structure can be significantly larger than the species that built it, as is the case of the spider web. One creature that has a highly impressive engineering skill is the beaver, which constructs dams as its habitat on streams. Other interesting structures include underground tunnels that gophers and rats build. Birds make their nests from twigs and other materials that are secured to various stable objects, such as trees, and their nests are durable throughout the bird's nesting season. Many nests are hemispherical in the area where the eggs are laid. One may wonder how birds have the capability to design and produce the correct shape and size of nests that matches the requirements of allowing eggs that are laid to hatch and grow as chicks until they leave the nest. The nest's size even takes into account the potential number of eggs and chicks, in terms of required space. Even plants offer engineering inspiration. Velcro was invented by mimicking the concept of seeds that adhere to an animal's fur, and has led to an enormous impact in many fields including clothing and electric-wires strapping. Because of their intuitive characteristics, the use of biologically based rules allows for the making of devices and instruments that are user-friendly and humans can figure out how to operate them with minimal instructions. Examples of devices and structures that were most likely initiated from imitation of biological models are listed below. These examples illustrate the diverse and incredible number of possibilities that have already been biomimicked.

1.5.1 Constructing Structures from Cells

Using cells to construct structures is the basis of the majority of animals and plants. Adapting this characteristic offers many advantages including the ability to grow with fault-tolerance and self-repair. Advances in nano- and micro-technologies are allowing the fabrication of minute elements that could become the basis for making artificial cells. Recently, scientists from the University of Washington, Seattle (Morris et al., 2004) reported on the use of guided and unconstrained self-assembled silicon circuits to constructed micro-electro-mechanical systems (MEMS)-based cells that can potentially have this capability. The term *self-assembly* is defined as the spontaneous generation of higher-order structures from lower-order elements. Self-assembly is the basis of the structure for all biological organisms, which exhibit massively parallel fabrication processes that generate three-dimensional structures with nanoscale precision. As a result, many orders of magnitude are spanned from the elemental or device-size scale to the final system level. This

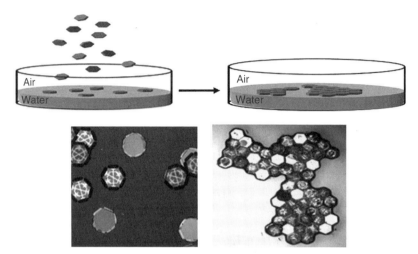

Figure 1.4 Self-assembly of large numbers of MEMS parts into two- and three-dimensional arrays of engineered crystals. (Courtesy of Babak Amir Parviz, University of Washington, Seattle, WA.)

biologically inspired characteristic is pursued through the use of self-assembly towards developing an engineering tool to produce structures, devices, and systems. Progress using self-assembly has allowed for the guided assembly of micro-devices on substrates and self-assembly of large numbers of parts into two- and three-dimensional arrays or engineered crystals (Figure 1.4). These methods are expected to allow the integration of devices from different manufacturing processes (CMOS, MEMS, micro-optics) into one system, addressing some of the main challenges to manufacturing that are foreseen in 21st century.

1.5.2 Biologically Inspired Mechanisms

Many mechanisms are attributed to a biological source for their inspiration. Some of these mechanisms include:

1.5.2.1 Digging as the Gopher and the Crab

Since 1998, the author, his Advanced Technologies Group, and engineers from Cybersonics, Inc., have been involved with research and development of sampling techniques for future *in situ* exploration of planets in the Universe. The investigated techniques are mostly based on the use of piezoelectric actuators that drive a penetrator at the sonic-frequency range. Using the mechanism developed, which they called the Ultrasonic/Sonic Driller/Corer (USDC), deep drills were developed that was inspired by the gopher and sand-crab with respect to penetrating soil and debris removal (Bar-Cohen et al., 2001). A piezoelectric actuator induces vibration in the form of a hammering action and the mechanism consists of a bit that has a diameter that is the same or larger than the actuator. In the device that emulates the gopher, it is lowered into the produced borehole, cores the medium, breaks and holds the core, and finally the core is extracted on the surface. This device can be lowered and raised from the ground surface via cable as shown in Figure 1.5. Analogy to the biological gopher is that the gopher digs into the ground and removes the loose soil out of the underground tunnel that it forms, bringing it to the surface.

Another digging device emulates the sand-crab. Like the sand-crab, this device uses mechanical vibrations on the front surface of the end-effector to travel through particulate media, such as soil and ground. In this configuration, the device digs and propagates itself through the medium. The biological crab shakes its body in the sand and thus inserts itself into the sand, as can commonly be

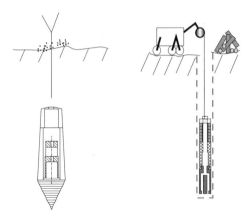

Figure 1.5   Biologically inspired ground penetrators.

seen in sand-crab habitats, such as the beach. While the ultrasonic/sonic gopher was developed to a prototype device and was demonstrated to perform its intended function, the ultrasonic/sonic crab has not yet been produced even though its implementation is not expected to pose any major challenges.

1.5.2.2 Inchworm Motors

The biologic inchworm is a caterpillar of a group of moths called Geomeridae, which has six front legs and four rear legs. Emulating the mobility mechanism of this larva or caterpillar led to the development of motors and linear actuators that are known as *inchworms*. These commercially available motors are driven by piezoelectric actuators (made by Burleigh Instruments) and they are capable of moving at a speed of about 2 mm/sec with a resolution of nanometers while providing hundreds of millimeters of travel. The forces produced by these types of motors can reach over 30 N with zero-backlash and high stability. Their nonmagnetic content offers advantages for applications in test instruments such as Magnetic Resonance Imagers (MRI). As opposed to biological muscles, the piezoelectric actuated inchworms are involved with zero-power dissipation when holding position. One of the limitations of this mechanical inchworm is its inability to operate at extreme temperatures that are as low as cryogenic temperatures and as high as 200°C. The brakes and shaft materials have different thermal expansion coefficients, and as a result, at lower temperatures the shaft–brake fit becomes tighter breaking the ceramic piezoelectric material that is used. At higher temperatures, on the other hand, the shaft–brake fit gets loose and the motor stops operating. Eventually, the curie temperature of the piezoelectric material is exceeded and the motor ceases to work. Using thermally compatible expansion coefficients is broadening the operating range of temperatures in which inchworms can be used.

Inchworm mechanisms have many configurations where the unifying drive principle is the use of two brakes and an extender. An example of the operation of an inchworm is shown in Figure 1.6 where the brakes and clamp are riding linearly on a shaft. These motors perform cyclic steps where the first brake clamps onto the shaft and the extender pushes the second brake forward. Brake no. 2 then clamps the shaft, brake no. 1 is released, and the extender retracts to move brake no. 1 forward. Another example of such a motor can be a modification where the brakes and extender operate inside a tube. The motor elements perform similar travel procedure as shown in Figure 1.6 while gripping the wall of the internal diameter of the tube in which the inchworm travels. This type of motion is performed by geometrid larva worms that move inside the ground. Generally, worms use their head and tail sections as support, similar to the brake in the inchworm, where the legs grab the ground or the two ends expand sequentially to operate as a brake. A simplified view of the movement of the *millipede* (different from that described for the inchworm) is illustrated schematically in Figure 1.7 showing steps that are made while progressing over the surface of objects such

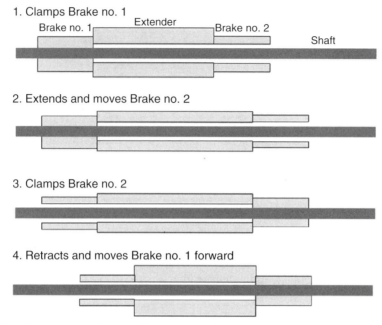

Figure 1.6   Operation sequence of a typical inchworm mechanism.

as the ground or plants. Other creatures that perform worm-like movement but in a different way can be seen in the earthworm, maggot, hornworm, ragworm (swimming, walking, burrowing), eel, geometrid larva, snake, millipede, and centipede.

1.5.2.3 Pumping Mechanisms

Nature uses various pumping mechanisms that are also used in mechanical pumps. The lungs pump air in and out (tidal pumping) via the use of the diaphragm to enable our breathing. Peristaltic pumping is one of the most common forms of pumping in biological systems, where liquids are

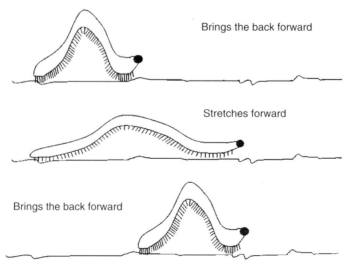

Figure 1.7   One of the forms of mobility seen in worms (the *millipede*).

squeezed in the required direction. Such pumping is common in the digestion system. Pumping via valves and chambers that change volume is found in human and animal hearts, with expansion and contraction of chambers. The use of one-way valves is the key to the blood flow inside the veins, where the pressure is lower.

1.5.2.4 Controlled Adhesion

Controlled adhesion is achieved by many organisms using a highly fibrillated microstructure. The *Hemisphaerota cyanea* (a beetle) uses wet adhesion that is based on capillary interaction (wet adhesion) (Eismer and Aneshansly, 2000). The gecko exhibits remarkable dry adhesion using van der Waals forces. Even though these forces provide low intrinsic energy of approximately 50 mJ/m^2, their effective localized application allows for the remarkable capability (Autumn et al., 2002). Using this adhesion mechanism, the gecko can race up a polished glass at a speed of approximately 1 m/sec and support its body weight from a wall with a single toe. Geckos have millions of 10 to 20 μm long setae, which are microscopic hairs at the bottom of their feet. Each seta ends with about 1000 pads at the tip (called spatulae) that significantly increase the surface density, and allow getting into close contact with the adhered surface. This capability motivated efforts to mimic the gecko adhesion mechanism, and some limited success was reported. Researchers like Autumn and Peattie (2003) sought to develop artificial foot-hair tip model for a dry, self-cleaning adhesive that works under water and in a vacuum. Their limited success effectively created a synthetic gecko adhesive that can potentially operate in vacuum areas of clean rooms as well as outer space.

1.5.2.5 Biological Clock

The body processes are controlled by our biological clock and it is amazing in its precision. It is critical in assuring the timely execution of the genetic code to form the same characteristics for the given creatures at the same sequence of occurrence at about the same age. The cicada matures for 17 years, after which it lives for another 1-week period. During this week, all cicadas mate, the females lay eggs, and then they all die. The hatched cicadas then develop for another 17 years and these synchronized processes are repeated again.

1.5.3 Biologically Inspired Structures and Parts

Parts and structures also have a biological model of inspiration. Some of these are discussed below.

1.5.3.1 Honeycomb as a Strong, Lightweight Structure

Honeycombs consist of perfect hexagonal cellular structures and they offer optimal packing shape. For the honeybees, the geometry meets their need for making a structure that provides the maximum amount of stable containment (honey, larvae) using the minimum amount of material (Figure 1.8). The honeycomb is, for the same reasons, an ideal structure for the construction of control surfaces of an aircraft and it can be found in the wing, elevators, tail, the floor, and many other parts that need strength and large dimensions while maintaining low weight. An example of a control surface part of an aircraft with a honeycomb is shown in Figure 1.9.

1.5.3.2 Hand Fan

Historically, hand fans were one of the most important ways of cooling down during the hot summer months (Figure 1.10). This simple tool used to be made of feathers, which copy the shape

Figure 1.8   The honeycomb (left) and the nest of the wasp (right) are highly effective structures in terms of low weight and high strength.

Figure 1.9   A cross-section of a honeycomb structure that plays an important role in the construction of aircraft control surfaces.

Figure 1.10   The hand fan, which is also produced in a folding form, was probably inspired by the peacock tail and the ability of this bird to open its tail into a wide screen that is shaken to impress the female.

of a bird's wing or the tail of the male peacock. The advantage of using feathers is their lightweight structure and their beauty.

1.5.3.3 *Fishing Nets and Screens*

The fishing net is another of nature's invention that most likely has been imitated by humans after observing the spider's use of its web to catch flies. At an even more basic level, the concept of fiber or string may well have been inspired by the spider. Both the spider web and the fishing net have structural similarities and carry out the same function of trapping creatures passing by. The spider uses a sticky material that helps capture the trapped insects by gluing them onto the web, and the spider knows how to avoid being glued to its own web. Depending on the type of spider, the distance between the fibers in the web can be as large as several centimeters and as small as fractions of a millimeter. Beside the use of nets to catch fish, insects, and animals, humans further expanded the application of the concept of the net to such tools as bags for carrying and storage of objects, protective covers against insects, and mounting stored food while allowing aeration. The screen, mesh, and many other sieving devices that allow separation of various size objects may also be attributed to the evolution of the net. Also, it is possible to attribute the invention of the net configuration to many medical supplies including the bandage and the membranes that are used to cover burns and other wounds.

1.5.3.4 *Fins*

Unlike the failure to fly by copying the flapping of birds' wings, the use of fins to enhance swimming and diving has been highly successful. While it may be arguable whether the fins were a direct biologically inspired invention, it is common knowledge that swimming creatures have legs with gossamer (geese, swans, seagulls, seals, frogs, etc.). Imitating the legs of these creatures offered the inventors of the fins a model that was improved to the point where it resembles the leg of the seal and to a lesser extent that of the frog. This similarity to the latter led to the naming of divers — frogmen, which is clearly a biomimetically inspired name.

1.5.4 Defense and Attack Mechanisms in Biology

A critical aspect of the survival of various species is having effective defense and attack mechanisms to protect against predators, catch prey, secure mating, protect the younger generation, procure and protect food, and other elements that are essential to survival. The following are some of the biologically inspired mechanisms that were adapted by humans. Further details are discussed more extensively in Chapter 13.

1.5.4.1 Camouflage

The chameleon and the octopus are well known for their capability of changing their body color. The octopus matches the shape and texture of its surroundings as well as releases ink to completely mask its location and activity — and yet, the octopus is a color-blind creature (Hanlon et al., 1999). Another aspect of the octopus' behavior is its ability to configure its body to allow traveling through narrow openings and passages. These include tubes, which are significantly smaller than its normal body cross-section. Generally, camouflage is not used solely for concealment alone, it also allows the predator to get close to its prey before charging ahead and capturing it by gaining the element of surprise while minimizing the response time of the prey. In some creatures, camouflage provides deterrence. For instance, some snakes, which are harmless, clone the appearance of highly poisonous snakes. Further, some harmless flies camouflage themselves with bright colors, pretending that they are wasps.

Minutes after birth, a baby deer is already capable of recognizing danger and taking action of passive self-defense. Since oftentimes the baby deer is left alone after birth, while the mother goes off to search for food, the baby has to rely on its ability to hide. It does this by finding shelter and taking advantage of basic camouflage rules. Without training, it is able to recognize which animals pose a threat to its life. Furthermore, the baby deer is equipped with the basic skill of taking advantage of objects in its terrain (e.g., plants), to reduce its body profile by ducking low, and to use a surrounding background that matches its colors in order to minimize its visibility. These skills, which are innate in the baby deer, are taught in human military training as camouflage methods.

While it is impossible for humans to imitate the octopus' ability to squeeze its body through narrow openings (since we have bones and the octopus does not), its camouflage capabilities have been the subject of imitation by all armies. In World War II, the zoologist Hugh Cott (1938) was instrumental in guiding the British army in developing camouflage techniques. Modern military uniforms and weapons are all colored in a way that makes them minimally visible by matching the background colors in the area where the personnel operate. Further, like the use of the ink by the octopus, soldiers in the army and on large naval vessels at sea use a smoke screen when they do not want to be seen. Until recently, camouflage has been used in the form of fixed colors for uniforms, armor and various military vehicles. With advancement in technology, the possibility of using paint that changes color is becoming increasingly feasible, and the use of liquid crystal color displays as a form of external coating are under consideration for active camouflage. Recent efforts are producing colors that can be changed to adapt to the local terrain (http://www.csmonitor.com/2004/0108/p14s01-stct.htm).

1.5.4.2 Body Armor

The shell is another means of protection that some creatures are equipped with, both on Earth and under water, and to a certain extent also in some flying insects. Creatures with body armor include the turtle, snail, and various shelled marine creatures (e.g., mussel, etc.). There are several forms of shells ranging from shelter that is carried on the back (e.g., snails) to those with full body cover in

Figure 1.11 The snail protects its soft body with a hard-shell which it carries on its back when safe.

which creatures can completely close the shell as a means of defense against predators. While the snail is able to emerge from its shell and crawl as it carries the shell on the back (Figure 1.11), the turtle lives inside its "body armor" and is able to use its legs for mobility when it is safe and hide its legs and head when it fears danger. The turtle was probably a good model for human imitation in terms of self-defense. The idea of body protection was adapted by humans many thousands of years ago in the form of hand-carried shields that allowed for defense against sharp objects, such as knives and swords. As the capability to process metals improved, humans developed better weapons to overcome the shield and therefore forced the need for better body armor in order to provide cover for the whole body. The armor that knights wore for defense during the Middle Ages provided metal shield from head to toe. Figure 1.12 shows such an armor for the upper part of the body.

In Japan, a more flexible armor was produced that consisted of thin metal strips connected with flexible leather bands. Relying on such protection led to defeat when faced against soldiers with rifles. As weapon technology in the West evolved, efforts were made to reduce the use of armor on individual soldiers for the sake of increased speed and maneuverability, as well as to lower the cost of fabrication and operation. In parallel, armored vehicles, which included tanks providing mobile shield and weaponry, with both defense and offense capabilities were developed. In nature, the use of shell for body protection is limited mostly to slow moving creatures and nearly all of them are plant-eaters.

1.5.4.3 Hooks, Pins, Sting, Syringe, Barb, and the Spear

Most of us have experienced at least once in our lifetime the pain of being hurt by a prick from plants — sometimes from something as popular and beautiful as the rose bush. Such experience can

Figure 1.12   An armor used as a body protection for knights can be viewed as mimicking the turtle's hard-shell body cover.

also occur when interacting with certain creatures, such as the bee. In the case of the bee, the stinger is left in the penetrated area and does not come out because of its spear shape. Humans adapted and evolved the concept of sharp penetrators in order to create many tools for applications in medicine, sports, and weaponry. These tools include the syringe, spears, fishing hooks, stings, barbs, and many others. Once penetrated, the hook and barb section on the head of a harpoon or an arrow makes it difficult to remove the weapon from the body of fish, animal, or human being.

1.5.4.4 Decoy

The use of decoy is as ancient as the lizard's use of its tail as a method to distract the attention of predators. The lizard autotomizes its tail and the tail moves rapidly, diverting the attention of the suspected predator while the lizard escapes to safety. This method is quite critical to lizard's survival and the tail grows back again without leaving a scar. This capability is not only a great

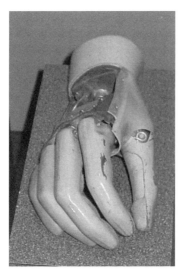

Figure 1.13 A mechanical hand for use as a prosthetic. (Photographed by the author at the Smithsonian Museum in Washington, DC.)

model for military strategies but also offers a model for potential healing of maimed parts of the human body. Success in adapting this capability could help people with disabilities the possibility of regrowing amputated or maimed parts of their body.

1.5.5 Artificial Organs

It is increasingly common to augment body organs with artificial substitutes. This is the result of significant advances in materials that are biocompatible, powerful electronics, and efficient miniature actuators. An artificial hand is shown in Figure 1.13 where a mechanism was designed to allow control of the fingers using a hand that matches the appearance of a human real hand. Artificial organs already include the heart, lung, kidney, liver, hip, and others (Chapter 18). Smart limbs, also known as Cyborgs, are also increasingly being developed with various degrees of sophistication and operation similar to the biological model. Moreover, the possibility of an artificial vision allowing a blind person to see is another growing reality, and a description of the state of the art as well as the expected future of this technology is provided in Chapter 17.

1.6 MATERIALS AND PROCESSES IN BIOLOGY

The body is a chemical laboratory that processes chemicals acquired from nature and turns them to energy, construction materials, waste, and various multifunctional structures (Mann, 1995). Natural materials have been well recognized by humans as sources of food, clothing, comfort, and so on. These include fur, leather, honey, wax, milk, and silk (see Chapter 14). Even though some of the creatures and insects that produce materials are relatively small, they can produce quantities of materials that are sufficient to meet human consumption on a scale of mass production (e.g., honey, silk, and wool). The use of natural materials can be traced back to thousands of years. Silk, which is produced to protect the cocoon of the silkmoth, has great properties that include beauty, strength, and durability. These advantages are well recognized by humans and the need to make them in any desired quantity has led to the production of artificial versions and imitations. Some of the fascinating capabilities of natural materials include self-healing, self-replication, reconfigurability, chemical balance, and multifunctionality. Many man-made materials are processed by heating and pressurizing, and this is in contrast to nature which uses ambient conditions. Materials, such as bone, collagen, or silk, are made inside the organism's body without the harsh treatment that is used to make our materials. The fabrication of biologically derived materials produces minimum waste and no pollution, where the result is biodegradable, and can be recycled by nature. Learning how to process such materials can make our material choices greater, and improve our ability to create recyclable materials that can better protect the environment. There are also studies that are improving prosthetics, which include hips, teeth, structural support of bones, and others. A brief description of structural materials that are made by certain insects and birds is given in this section, whereas Chapters 12 and 14 cover in greater detail the topics of biological materials and their multifunctional characteristics.

1.6.1 Spider Web — Strong Fibers

One of biology's best "manufacturing engineers" with an incredibly effective material-fabrication capability is the spider. It fabricates the web (Figure 1.14) to make a very strong, insoluble, continuous lightweight fiber, and the web is resistant to rain, wind, and sunlight. It is made of very fine fibers that are barely visible allowing it to serve its function as an insect trap. The web can carry significant amount of water droplets from fog, dew, or rain thus making it visible as shown in Figure 1.14. The web structure in the photograph has quite an interesting geometry. It reveals the

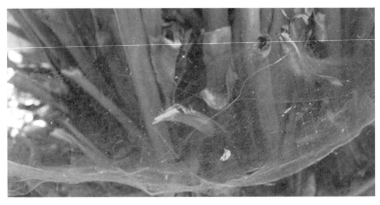

Figure 1.14 (See color insert)   The spider constructs an amazing web made of silk material that for a given weight is five times stronger than steel.

spokes, the length, and density of sticky spiral material for catching bugs. The segments of the photographed web are normally straight, but are seen curved in this figure due to the weight of the accumulated droplets. The net is sufficiently strong to survive this increased load without collapsing.

The spider generates the fiber while at the same time hanging on to it as it emerges cured and flawless from its body. The web is generated at room temperature and at atmospheric pressure. The spider has sufficient supply of raw materials for its silk to span great distances. It is common to see webs spun in various shapes (including flat) between distant trees, and the web is amazingly large compared to the size of the spider. Another interesting aspect of the spider web is the fact that it is a sticky material intended to catch prey, but the spider itself is able to move freely on it without being trapped.

The silk that is produced by a spider is far superior in toughness and elasticity to Kevlar®, which is widely used as one of the leading materials in bullet proof vests, aerospace structures, and other applications where there is a need for strong lightweight fibers. Though produced in water, at room temperature and pressure, spider's silk is much stronger than steel. The tensile strength of the radial threads of spider silk is 1154 MPa, while steel is 400 MPa (Vogel, 2003). Spiders eat flies and digest them to produce the silk that comes out from their back ends, and spool the silk as it is produced while preparing a web for trapping insects. This web is designed to catch insects that cross the net and get stuck due to its stickiness and complexity. While the net is effective in catching insects, the spider is able to maneuver on it without the risk of being caught in its own trap. Recent progress in nano-technology reveals a promise for making fibers that are fine, continuous, and enormously strong. For this purpose, an electrospinning technique was developed (Dzenis, 2004) that allows producing 2-μm diameter fibers from polymer solutions and melts in high electric fields. The resulting nano-fibers were found to be relatively uniform without requiring extensive purification.

1.6.2 Honeybee as a Multiple Materials Producer

Another "material manufacturing engineer" found in nature is the honeybee. This insect can make materials in volumes that far exceed the individual bee's size. Bees are well known for making honey from the nectar that they collect from flowers. They also produce honeycomb from wax. Historically, candles were made using this beeswax, but with the advent of the petroleum industry, candles are now mostly made from paraffin wax. Another aspect of honeybee is that their bodies produce a poison that causes great pain, which is injected, through a stinger, into the body of any intruder who is perceived as a threat to the bee's colony.

1.6.3 Swallow as a Clay and Composite Materials Producer

The swallow makes its nest from mud and its own spit forming a composite structure that is strong. The nest is shaped to fit the area onto which it is built. The swallow builds its nest under roofs and other shelters that provide both protection and concealment. Figure 1.15 is a photograph of two nests of swallow. A flock of swallows have gathered next to the nests. While the two nests are different in shape they have similar characteristics and they both provided sufficient room for the chicks to hatch and reach maturity. It is interesting to note that the birds in the photograph attach themselves to the wall carrying their body weight on their claws, which secure them comfortably to the stucco paint on the wall.

1.6.4 Fluorescence Materials in Fireflies and Road Signs

Fluorescence materials can be found in quite a few living species and these visible light-emitting materials can be divided into two types:

Figure 1.15 A group of swallows gathering next to two nests that are made of a composite mix of mud and straws. These nests were built under the author's roof (July 2004).

(a)   bioluminescence — a voluntary or involuntary light emission, which results from a chemical reaction;

(b)   fluorescence — emission of light under ultraviolet illumination.

Bioluminescence can be found in various beetles (e.g., firefly), marine creatures (e.g., *Pyrocystis lunula*, *Gonyaulax polyedra*, and squids), as well as certain bacteria, and mushrooms. Biolumin-escence materials are used to attract females as in the case of the American firefly (Lloyd, 1984). The male firefly flashes its light in order to "declare" its presence and identity, and to attract females of its own species (Lloyd, 1966). Another example for bioluminescence is the glow-worm, a type of beetle (*Noctiluca*), whose wingless female glows in the dark. Bioluminescence is also used as a deception method, where the *Photuris* females mimic the flashing rate of hetero-specific males and eat them (Lloyd, 1980).

1.6.5 Impact Sensitive Paint Mimicking Bruised Skin

Our skin is sensitive to impact leading to purple color marks in areas of the skin that is hit. This bruise mark indicates the fact that the specific area has suffered an impact. This idea inspired researchers at the South West Research Institute in the mid-1980s (Light et al., 1988) to develop a surface coating as a nondestructive indicator of impact damage in composite materials. The need for such an indicator rose as the use of composite material increased to a level where structures that are critical to the safety of aircraft started to be introduced into military and commercial aircraft.

U.S. Air Force studies showed that these materials are sensitive to impact; a loss of about 80% in the compression static strength was measured when an impact causes an easy-to-see damage to the surface, whereas a loss of 65% when the damage is barely visible (Bar-Cohen, 2000). In order to develop an impact damage indicator, paint was mixed with an encapsulated dye and developer, and was applied to the surface of composite panels. The micro-capsules that were used had a diameter of 1 to 10 μm and in this size they were easy to apply using conventional methods like spraying to increase the practicality of this paint. Tests have shown the feasibility of this concept and the paint was effective in indicating the location and intensity of the impact, where the larger the impacted area, the larger the indication that appeared.

1.6.6 Mimicking Sea Creatures with Controlled Stiffness Capability

Certain sea creatures, such as the sea cucumbers, are capable of controlling the tensile properties of their connective tissues by regulating the stress transfer between collagen fibrils (Trotter et al., 2000; http://www.biochemsoctrans.org/bst/028/0357/0280357.pdf). Trotter et al. (2000) sought to design a synthetic analog with similarly reversible properties, and have been able to demonstrate a pair of synthetic molecules that selectively and reversibly associate with one another under controlled physiological conditions.

1.6.7 Biology as a Source for Unique Properties and Intelligent Characteristics

Materials that are made by animals offer capabilities and properties that are often far superior to any human-made imitations. These material properties include hardness, fracture resistance, and light-weight — as can be found in pearls and shells of various marine species, including the abalone. There are also many body parts (e.g., teeth and eye cornea) that are organized as layered assemblies, which are now emulated by methods such as self-assembly and ink-jet printing. Smart materials are increasingly evolving in various forms, with self-sensing and reaction capabilities that cause them to stretch and contract in response to heat, light, and chemical changes. Another aspect of biomaterials is self-healing, which is increasingly being adapted to polymer and composite materials.

1.6.8 Multifunctional Materials

Nature has made great efforts to use its resources effectively, and besides the use of power in efficient ways including its recycling, nature also assigned multifunctions to its materials and structures. For example, our skin encases blood and other parts of our body, supports the regulation of body temperature, has self-healing capability as well as many other functions. Also, our bones provide the required body stiffness to support it allowing us not only to stand, walk, and conduct various critical mobility functions, but it also produces our blood in the bone marrow. The use of materials that perform multiple tasks allowed nature to make its creatures with a lower body weight. The concepts of multifunctional materials and structures are being studied by many researchers and engineers (see more details in Chapter 12) and has been the subject of a DAPRA program at the end of the 1990s. Increasingly efforts are made to emulate this characteristic, where multiple disciplines are used, for example, applied mechanics (elasticity or plasticity, fracture mechanics, aerodynamics), materials sciences (metallurgy, composites, polymers), electronics (sensors, actuators, controls), photonics (fiber optics), and manufacturing (micro- or macro-structure processing).

1.6.9 Biomimetic Processes

There are many biomimetic processes that were learned from studying the activity of the body of living creatures. The imitation of biological processes ranges from operations at the level of cells to the scale of the full body. Imitated processes, including artificial synthesis of certain vitamins and antibiotics, have been in use for many years. More recently, biomimetics have been used to design navigational systems, data converters, mathematical algorithms (Chapters 4 and 5), and diffusion processes. The neural network (part of the field of AI that was covered earlier) is a hypothetical biomimetic computer that works by making associations and guesses, and that can learn from its own mistakes. Examples of biomimetic processes are described throughout this book.

1.7 BIO-SENSORS

Living creatures are equipped with a sensory system, which provides input to the central nervous system about the environment around and within their body and the muscles are commanded to action after analysis of the received information (Hughes, 1999). Biological sensory systems are extremely sensitive and limited only by quantum effects (Chapter 11; Bill Bialek, 1987). This sensory network is increasingly imitated, where we find our surroundings filled with sensors. Such sensors are monitoring our property to protect it from intruders; releasing soap and water when washing our hands; releasing hot air or paper towels to dry our hands; tracking our driving speed; observing our driving through intersections that are monitored by traffic lights; as well as performing many other tasks that we accept as part of our day-to-day lives. Our cars sense when we close the doors, whether there is sufficient air in the tires, charge in the battery and oil in the engine, and if all the key functions are operating properly. Sensors also control the flow of gasoline to the ignition system in our cars to optimize gas consumption. Similar to the ability of our body to monitor the temperature and keep it within healthy acceptable limits, our habitats, working, and shopping areas have environment control to provide us with comfortable temperatures. These examples are only a small number of the types of sensors that are used in our surroundings and in the instruments that we use today. Pressure, temperature, optical, and acoustical sensors are widely in use and efforts are continuously made to improve their sensing capability and reduce their size and required power while mimicking ideas from biology. These include adapting principles from the eyes to camera, from the whiskers of rodents to sensors for collision avoidance, and from bats to acoustic detectors that imitate their sonar. Specific examples of biomimicked sensors are described below.

1.7.1 Miniature Sensors in Biomimetic Robots

The integration of sensors into mobile systems is critical for their operation, as it is necessary to provide closed-loop feedback to accomplish mobility tasks and other dynamic functions. Emulating the dimensions, density, integration, and distribution of sensors in the human finger will require significant advancements in such fields as MEMS and nano-electro-mechanical systems (NEMS). While currently the packing density of sensors per unit surface using MEMS technology is about 1 to 10 sensors/mm^2 there is still a long way to go before reaching the density level of hundreds of sensors/mm^2 of the skin area of the fingertips. Combining the equivalence of soft skin and integrated sensors is a desired biomimetic development goal. An array of multiple types of sensors will need to be used to provide critical, detailed data about the environment and the performance of the various elements of mobile system. It is also highly desirable to see the development of miniature vision and sound receivers with real-time image and voice recognition allowing rapid response to the environment in a manner akin to living creatures. Moreover, there is increasing need for soft sensors that can support the development of electroactive polymers (EAP) as artificial muscles (Chapter 10). These materials have functional similarities to biological muscles and the use of such sensors as strain gauges is not effective because of the constraining effect that results from the rigidity of the widely used gauges.

1.7.2 MEMS-Based Flow Detector Mimicking Hair Cells with Cilium

On the micron-scale level the monitoring of air (Friedel and Barth, 1997) and water flow (Bond, 1996) in insects and in fish is by clusters of hair cells. These hair cells consist of cilia that are attached to nerve cells, and they sense the bending action that results from the flow. The displacement induces an output response from the attached nerve cell. These hair cell sensors were biomimicked to produce two types of artificial hair cell sensors (Ozaki et al., 2000; Chen et al., 2003b). The first type has a cantilever or paddle that is parallel to the substrate, and is sensitive to flow and forces that act normal to the substrate (Ozaki et al., 2000). The second type has a cantilever that is perpendicular to the substrate, where the early types were made of silicon which is brittle. Improvement has been developed at the University of Illinois at Urbana-Champaign, where robust polymer-based sensor was demonstrated (Chen et al., 2003a). A schematic and graphic view of the developed hair cell is shown in Figure 1.16.

1.7.3 Collision Avoidance Using Whiskers

Another biologically inspired sensor that was adapted is the use of the whisker in various rodents. The whiskers of rats are extremely sensitive helping it avoid collision with obstacles and finding food. Emulating whiskers offers significant advantages to biologically inspired robots and such sensors have already been used in various commercially available robots (Gravagne et al., 2001), such as the BIOBbug toys (Hrynkiw and Tilden, 2002). The BIOBbug is an insect-like toy that operates as a swarm and avoids collision between each other as well as other objects.

1.7.4 Emulating Bats' Acoustic Sensor

The bat can move its ears in all directions, localize sound sources, and avoid obstacles, all while flying at relatively high speed. Ear shapes are different in different bats, indicating that there is no optimal shape, and that each bat species evolved its own biological solution. It is believed that the ear creates interference that is processed by the brain. The bat ear has been the subject of numerous studies including recent efforts to use it to navigate robots (Peremans and Muller, 2003; Muller and Hallam, 2004). The directivity patterns for frequencies from 25 to 75 kHz were studied and the ears of various bats were tested using x-ray to study the internal structure and how sound interacts with the ear. A rapid prototyping method was made to produce pinna-shapes, assuming that the make-up material is not a critical issue because of the large mismatch with air. To convert sound to electric

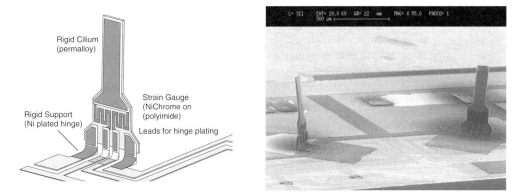

Figure 1.16 A schematic (left) and photo-micrographic view (right) of the cilium that biomimic the hair cells in fish and insects. (Courtesy of Jack Chen and Chang Liu, Mirco Actuators and Sensors Group, University of Illinois at Urbana-Champaign.)

U.S. signals piezoelectric foils were used with an electro-mechanical conversion factor, d_{33}, ranging from 250 pC/N to 400 pC/N (Neugschwandtner et al., 2001). To demonstrate the simulation of the bat capability the developed sensors were used to navigate robots. Current efforts are focused on classifying landmarks, navigation in natural environments, making use of body movement, and echo interpretation (Muller and Hallam, 2004).

1.7.5   Acoustic and Elastic Wave Sensors

Certain animal species are equipped with the ability to sense acoustic or elastic waves at great distances. The elephant can rock its foot and emit vibrations that travel through the ground and are felt and recognized by other elephants at a distance of several kilometers. The whale emits hyper-low frequency sound that travels over great distances in the ocean and can be detected and identified by other whales. Equivalent detectors made by humans include accelerometers used to detect earthquakes and the sonar used in submarines. However, biological capability is still far superior in terms of sensitivity, spectral response, and evaluation capability than any man-made detection instrument.

1.7.6   Fire Monitoring

The jewel beetle (Melanophila) lays its eggs in the bark of freshly burned trees using its ability to detect forest fires from a distance of about 80 km (http://www.uni-bonn.de). The sensory organ that is used for this is located on the underside of the beetle, and consists of a pit that contains a large number of receptors that are extremely responsive to the infrared (IR) radiation created by a forest fire. Recently, zoologists at the University of Bonn have been trying to imitate this sensory capability. The study by Schmitz and Trenner (2003) led to the development of a sensor that is sensitive to infrared and automatically monitors large forest areas to trigger an early warning in the event of fire. As biological imitation of the beetle's cuticulas sensory organ, a polyethylene platelet was developed, which absorbs thermal radiation with wavelength of 3 μm, which is the typical radiation emitted by fierce force fires. This sensor was found to be two orders of magnitude more sensitive than commercially available IR sensors. This new sensor is expected to be produced at lower cost than commercial detectors and efforts to further improve its sensitivity are currently underway.

1.7.7   Sense of Smell and Artificial Nose

The topic of smell sensing has reached a level of interest and progress that led in 2004, to the Nobel Prize Award given to the researchers Buck and Axel (1991). The sense of smell is our analyzer of chemicals of airborne molecules allowing us to determine presence of danger, hazardous chemicals as well as gives us the enjoyment of good food and other pleasant odors. Using receptors in our nose we continuously examine the content of the air we breathe, where the signals are sent through stations called glomeruli that are located in the brain's olfactory bulb. From there, the signals are sent to the brain where patterns of smell memories are formed and compared with previous "records." The sense of smell alerts us of such danger as smoke from fire, leakage of dangerous gases, as well as informs us of other relevant information, such as the presence of food or even perfume from other individuals. The detectable chemicals need to be sufficiently small to be volatile so that they can be vaporized, reach the nose, and then dissolve in the mucus. It is estimated that our nose can distinguish between as many as 10,000 different smells.

Imitating the nose's sensing capability offers important potential applications, and efforts to make such sensors have been explored since the mid-1980s. There are several devices that have been built and tested emulating the nose including some that use chemical sensor array (Bartlett and Gardner,

1999; Dickinson et al., 1998; Nagle, 1998; www.cyranosciences.com). The technology is now at a level where electronic noses are commercially available, and they have been applied to environmental monitoring and quality control in such fields as food processing. Generally, an electronic nose is an array of weakly specific chemical sensors, controlled and analyzed electronically, mimicking the action of the mammalian nose by recognizing patterns of response to vapors. Unlike most existing chemical sensors, which are designed to detect specific chemical compounds, the sensors in an electronic nose are not specific to any one compound, but have overlapping responses. Gases and gas mixtures can be identified by the pattern of the responses of the sensors in the array.

Chemical sensors are made from several different materials that act by several different mechanisms. Conducting polymers such as polyanilines or polypyrroles can be used as the basis for a conductometric sensor, where change at the sensor is read as change in resistance. The ability of conducting polymers to detect a wide variety of compounds can be extended by mixing other polymers with the conductor (Freund and Lewis, 1995). An electronic nose that uses polymers as the basis of the chemical sensors is under development at JPL for such applications as event monitoring on the International Space Station. The polymer-based sensors used in the JPL ENose were developed at Caltech (Lonergan et al., 1996). They are insulating polymers, which have been loaded with a conductive material such as carbon black. A thin film of the polymer or conductor composite absorbs vapor molecules into the matrix, and the matrix changes shape and the relative orientation of the conductive particles. This change results in a change in resistance, which is used to form the pattern of response. The magnitude of the response can be related to the concentration of vapor, and mixtures of a few compounds can be deconvoluted. The library of compound patterns that the ENose contains depends on the particular space in which it is used and the hazards of that space. New compounds can be added to the library as the device is exposed to them. ENoses in different spaces can be equipped with different polymers in the array, and therefore, a different library. The polymers for an array are selected by molecular structure of the polymer and the target compounds for that array.

1.7.8 Sense of Taste and Artificial Tongue

The sense of taste is another chemical analyzer in biology; it examines dissolved molecules and ions and it uses clusters of receptor cells in the taste buds (Craven and Gardner, 1996). Each taste bud has a pore that opens out onto the surface of the tongue enabling molecules and ions taken into the mouth to reach into the receptor cells. Generally, there are five primary taste sensations including: salty, sour, sweet, bitter, and umami. A single taste bud contains 50 to 100 taste cells representing all five taste sensations. Each taste cell has receptors on its apical surface and these are transmembrane proteins that bind to the molecules and ions that give rise to the five taste sensations. Several receptor cells are connected through a synapse to a sensory neuron and from there to the back of the brain, where each sensory neuron responds best to one of the taste sensations (http://users.rcn.com/jkimball.ma.ultranet/BiologyPages/T/Taste.html).

Similar to the electronic nose, researchers explored the development of an electronic tongue that mimics the biological sensory capability (Vlasov and Legin, 1998; Krantz-Ruckler et al., 2001; http://csrg.ch.pw.edu.pl/prepapers/pciosek/etong.html). Generally, the electronic tongue is an automatic system for analysis and recognition (classification) of liquids using nonspecific sensors arrays, data acquisition elements, and analytical tools. The result of E-tongue tests can be the identification of the sample, an estimation of its concentration or its characteristic properties. Using this technology allows overcoming the limitations of human sensing including individual variability, inability to conduct online monitoring, subjectivity, adaptation, infections, harmful exposure to hazardous compounds, and effect of mental state. The artificial taste sensors that mimic the olfactory system consist of various types of sensors including potentiometric sensors, conductivity measurements, voltamperommetry, and optical sensors. Various techniques and methods can be used separately or together to perform the recognition of the samples, where after completing a measurement, procedure signals are transformed by a preprocessing block. The results are then analyzed by various pattern recognition blocks consisting of AI processes (e.g., Cluster Analysis or Artificial Neural Network). The performance of electronic tongues depends on the quality of functioning of its pattern recognition block.

E-tongues are increasingly being used in such applications as monitoring food taste and quality, noninvasive diagnostics (patient's breath, analysis of urine, sweat, and skin odor), searching for chemical or biological weapon, drugs, and explosives, as well as environmental pollution monitoring.

1.8   ROBOTICS EMULATING BIOLOGY

The introduction of the wheel has been one of the most important human inventions — allowing humans to traverse great distances and perform tasks that would have been otherwise impossible within the lifetime of a single human being. While wheel-locomotion mechanisms allow reaching great distances and speeds, wheeled vehicles are subject to great limitations with regard to traversing complex terrain with obstacles. Obviously, legged creatures can perform numerous functions that are far beyond the capability of an automobile. Producing legged-robots is increasingly becoming an objective for robotic developers and using such robots for space applications is currently under consideration. Also, operating robots as colonies or flocks is a growing area of robotic research.

Bio-inspired mechanisms are not only based on legs — since wind is blown throughout Mars, producing a spacecraft that imitates the tumbleweed offers an attractive option. The tumbleweed inspired the design of a mobility system that uses wind rather than a power-consuming mechanism. As shown in Figure 1.17, the tumbleweed has inspired a futuristic lander that is being investigated as a potential vehicle for mobility on Mars.

Industry is increasingly benefited from advancement in robotics and automation that are biologically inspired (Bar-Cohen, 2000; Bar-Cohen and Breazeal, 2003). Crawlers with complex-shaped legs and various manipulation devices are commonly used to perform a variety of non-destructive evaluation (NDE) tasks. At JPL, a multifunctional automated crawling system (MACS) was developed to simplify scanning of aircraft structures in field conditions (Figure 1.18). MACS employs two sets of legs to support mobility and one set that allows for rotation of its platform. MACS was designed to perform scanning by effectively "walking" on the aircraft fuselage while

Figure 1.17   The tumbleweed (left) offered an inspiration for a futuristic design of a Mars lander.

Figure 1.18 (See color insert)   MACS crawling on a wall using suction cups.

adhering to it via suction cups. This locomotive method mimics the mechanism that flies use to walk upside down on ceilings. Other forms of mobility on objects using methods of adhesion include magnetic wheels. The author and his coinvestigator (Bar-Cohen and Joffe, 1997) conceived a rover that can operate on ships and submarines using magnetic wheels. Another legged robot is the JPL's STAR that has four legs and can perform multiple functions, including grabbing objects (Figure 1.19) as well as climbing rocks with the aid of the USDC (Bar-Cohen et al., 1999; Bar-Cohen and Sherrit, 2003; Badescu et al., 2005) on each leg. The USDC is used for this purpose since it requires a relatively low axial force to drill into hard objects. The JPL's legged robots are developed for potential operation in future planetary mission, where a Lemur class robot will be able to autonomously negotiate its way through unknown terrain that is filled with obstacles.

Creating robots that mimic the shape and performance of biological creatures has always been a highly desirable engineering objective. Searching the Internet using the keyword *robot* would point

Figure 1.19 (See color insert)   JPL's Lemur, six-legged robots, in a staged operation. (Courtesy of Brett Kennedy, JPL.)

out many links to research and development projects that are involved with such robots. The entertainment and toy industries have greatly benefited from advancement in this technology. Increasingly, robots are used in movies where creatures are shown to exhibit realistic behavior. The capabilities demonstrated even include creatures that are no longer in existence like the dinosaurs in the movie *Jurassic Park*.

As mentioned earlier, visits to toy stores, show how far technology has progressed in making inexpensive toys that imitate biology. Such store displays include frogs swimming in a fish bowl and dogs walking back and forth. Operating robots that emulate the functions and performance of humans or animals use capabilities of actuators and mechanisms that depend on state-of-the-art technology. Upper-end robots and toys are becoming increasingly sophisticated, allowing them to walk and talk and some robots can be operated autonomously or can be remotely reprogrammed to change its characteristic behavior. Some of the toys and robots can even display expressions and exhibit behaviors similar to humans and animals. An example of such a robot is Kismet, which can express and react to human expressions facially and verbally (Bar-Cohen and Breazeal, 2003). This expression is made as a function of the level of expression that is being emulated or programmed to perform. As this technology evolves, it is likely that in the future, human-like robots may be developed to perform tasks without the possibility of human errors, needing a break, being distracted, or getting tired. Moreover, these robots may be programmed to display happiness, sorrow, etc.

One may even see a day when such robots could become human companions and advisors. At such point, the population of biomimetic robots will increase to possibly become a household "tool" just like personal computers. With the increase in availability of robots as property there will arise a need to protect them as valuables possibly even requiring to equip them with self-defense. Such a capability will raise major concerns related to the limits that will be allowed with regard to their interaction with humans. Another potential issue that may arise with the evolution of such robots is the potential copying of humans forming the equivalence of cloning. In contrast to the genetic procedure which can take many years to grow, a duplicate of the human without ability to control the outcome, the robotic version can potentially be rapidly produced and be programmed to emulate the behavior and response of the original person. Aspects of this mimicking will need to be addressed by future generations as the potential unlawful possibilities that may become possible could pose major concerns to law-enforcement agencies.

1.8.1 Artificial Muscles

Polymers that can be stimulated to change shape and size have been known for years. The functional similarity of such polymers led to their being named artificial muscles. The activation mechanism for such polymers includes electric, chemical, pneumatic, optical, and magnetic. Electrical excitation is one of the most attractive stimulators that can produce elastic deformation in polymers. The convenience and the practicality of electrical stimulation, as well as the improved capabilities, make the EAP one of the most attractive among the activatable polymers (Bar-Cohen, 2001, 2004; Chapter 10).

Generally, EAP materials can be divided into two major categories based on their activation mechanism: electronic and ionic. Most electronic polymers (electrostrictive, electrostatic, piezo-electric, and ferroelectric) require high activation fields (>150 V/μm) close to the breakdown level. However, they can be made to hold the induced displacement under activation of a DC voltage, allowing them to be considered for robotic applications. These materials have a faster response, a greater mechanical energy density, and they can be operated in air. In contrast, ionic EAP materials (gels, IPMC, conductive polymers, and carbon nanotubes) require drive voltages as low as 1 to 5 V, and produce significant bending. However, bending actuators have relatively limited applications for mechanically demanding tasks due to the low force or torque that can be induced. Also, with some exceptions, these materials require maintaining their wetness and when containing water they suffer electrolysis with irreversible effects when they are subjected to voltages above 1.23 V. Except for conductive polymers, it is difficult to sustain DC-induced displacements.

Unfortunately, EAP-based actuators are still exhibiting low force below their efficiency limits, are not robust, and are not available as commercial materials for practical application considerations. Each of the known materials requires adequate attention to the associated unique properties and constraints. In order to be able to take these materials from the development phase to use as effective actuators, there is a need to have an established EAP infrastructure. Effectively addressing the requirements of the EAP infrastructure involves developing its science and engineering basis, namely, having an adequate understanding of EAP materials' behavior, as well as processing and characterization techniques. Enhancement of the actuation force requires understanding the basic principles, computational chemistry models, comprehensive material science, electro-mechanical analysis, and improved material processing techniques. Efforts are on for a better understanding of the parameters that control the EAP electroactivation force and deformation. The processes of synthesizing, fabricating, electroding, shaping, and handling are being established and refined to maximize the EAP materials actuation capability and robustness. In addition, methods of reliably characterizing the response of these materials are being developed. This effort also includes the establishment of a database with documented material properties in order to support design engineers who are considering the use of these materials. Various configurations of EAP actuators and sensors are being modeled to produce an arsenal of effective, smart EAP-driven systems. The development of the infrastructure is multidisciplinary, and requires international collaboration and these efforts are currently underway worldwide.

In 1999, the author challenged the world's research and engineering community to develop a robotic arm that is actuated by artificial muscles (moniker for EAP) to win a wrestling match against a human opponent. The objectives of the match are to promote advances in making EAP actuators that are superior in performance to the performance of human muscles. Also, it is sought to increase the worldwide visibility and recognition of EAP materials, attract interest among potential sponsors and users, and lead to general public awareness since it is hoped that they will be the end users and beneficiaries in many areas including medical, commercial, and military. The first arm-wrestling competition with humans was held against a 17-year-old girl on March 7, 2005 and the girl won against three robotic arms that participated. Even though the arms did not beat the challenge, one of the arms was able to hold against the girl for 26 sec, and this is an important milestone.

1.8.2 Aerodynamic and Hydrodynamic Mobility

Judging from the number of flying insects, birds, and marine creatures, nature has "experimented" extensively with aerodynamics and hydrodynamics. There are several aspects that deserve attention. For instance, birds can catch fish underwater with their eyes closed. They are able to catch fish by taking into account the refraction-effect, which creates an illusion as to the location of the fish. Birds and various mammal predators take into account the vector trajectory of the escaping prey, as in the case of hunting a running rabbit or deer. These trajectories are increasingly the capability of military weapons allowing tanks to destroy a moving target while they are moving too. Sophisticated capabilities are used to track the moving target and either adjust the direction in flight or aim upon launch using high speed missiles or bullets.

The ability of the dragonfly to maneuver at high speed is another aspect of flying that considerably inspired humans. Using a liquid-filled sac that surrounds its cardiac system, the dragonfly adjusts the effects of high G on its body during its flight and incredible maneuvers. G represents unit of gravitational force on Earth where high G is many multiples of one G. This technique inspired a mechanism that allows pilots to fly at high mach speed with significantly lower effects on the ability of the pilot to stay coherent. A liquid-filled, anti-G suit was developed by Life Support Systems, a Swiss company, and the suit is called "Libelle," which means in German "dragonfly" (http://www.airpower.at/news01/0625_libelle/libelle3.htm). The Libelle suit promises advantages over the pneumatic (compressed air) anti-G suits that are currently in use at various air forces including the US Air Force. Instead of air, Libelle uses water to provide counter pressure proportional to the gravitation force. The fluid is contained in expandable, snake-like tubes that run from the neck to the ankles and over the shoulders to the wrists (http://www.txkell.ang.af.mil/news_events/suit.htm).

Like biology, botany also takes aerodynamics into account. The seeds of many plants are designed with features that allow them to disperse away from their origin. The need to disperse can be attributed to the possibility of overcrowding of the specific type of plant in the same local area. Seeds use various aerodynamic techniques to be propelled by the aid of winds, for example, see Figure 1.20 the winged seed of the *Tipuana tipu* (about 6.5-cm long). Such seeds have inspired designs of futuristic missions with spacecraft that would soft-land on planets with atmospheres such as Mars. Adapting this design may allow for designing a parachute with better

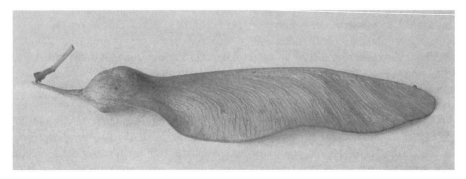

Figure 1.20   Seeds of the *Tipuana tipu*, which has an aerodynamic shape for dispersion by wind.

capability to steer itself to land at selected sites. Some of the issues being studied include the appropriate vehicle size, acceptable descent speed in the Martian atmosphere, mass distribution and platform shape to assure stable autorotation and scalability from operation on Earth to performance on Mars.

1.8.3 Social and Other Biological Behaviors

One of the many characteristics of humans and animals is their sociability, i.e., having the ability to express feelings, respond to stimulation, and make independent decisions, actions, and reactions. Efforts are increasingly being made to imitate such characteristics with robots. These robots are being equipped with autonomous operation, the ability to communicate feelings in the form of facial expressions and voices, to react to feelings expressed by humans, to have defense and attack capabilities, as well as many other characteristics that are considered *biological* (Bar-Cohen and Breazeal, 2003).

Generally, many of our behavioral characteristics are learned throughout our life, with some abilities that are genetically coded and improved through our life experience. Some animals have some abilities that are far superior to those of humans, and these are coded into their genetics. For example, babies of migratory animals begin walking without assistance or guidance minutes after birth. While the baby depends on its mother for milk for survival, it is "equipped" with extensive other abilities that are critical to its survival including seeing, hearing, running, recognizing danger, and even the capability of passive self-defense. Inspired by these characteristics, robots are increasingly being developed with autonomous operations and programmed with social abilities to interact with humans (Breazeal, 2004). Learning to make realistic robots with social skills can have many important benefits including understanding behavior in humans and providing a cure to certain phobias. Recent advances in virtual reality and AI allow studying and treating patients with phobias such as the fear of heights, or closed areas.

Social creatures, including insects and birds, have various approaches for solving difficult computational problems such as discrete optimization. Examples of modeling for optimization include activities in ant colony and the seed-picking process of pigeons. The latter, which is also known as the Particle Swarm optimization algorithm, is an analytical tool that is based on the statistical process of seeds picking. This algorithm is considered very effective in evolving hardware and particularly in designing combinational electric circuits (Amaral et al., 2004). These biologically inspired optimization algorithms are used to solve problems such as vehicle routing and routing in telecommunication networks.

Operating in a group generally gives social creatures advantages in defense that are unavailable to animals that operate individually. Migrant animals move as a herd, warn each other, and even jointly defend themselves. Birds fly in formation to help with long distance travel by taking advantage of vortices that are formed by flapping their wings. Lionesses gather in packs to protect their cubs not only from potential enemies but also from other male lions, as lions will kill cubs that are not their own to have their own genes entered into the species gene pool. Ants and bees live in a social structure and are able to accomplish extensive tasks because of this behavior. Wolves traveling in packs to hunt have the head of the pack "instruct" the other wolfs who obey the head's orders while chasing herds or other prey (Towery, 1996). In some ways, military operations are similar, as the commander in a military operation directs the soldiers, and they also "obey the orders" while executing the war strategy that is dictated by the commander. Generally, wolfs have a highly complex social order where every pack has a male leader, and all wolves in the pack are aware of their positions through communication with various body postures.

Roboticists are now considering the use of multiple small robots that can operate in colonies like the ants. Such robots are made capable to operate both as individuals in cooperative systems, and as inter-connectable parts of a large system. Their capabilities would be far greater than those that can be obtained with an individual robot.

1.9   INTERFACING BIOLOGY AND MACHINES

Interfacing between humans or animals and machines to complement or substitute our biological senses enables important means for medical applications. Of notable significance is the interfacing of machines and the human brain. A development by scientists at Duke University (Wessberg et al., 2000; Mussa-Ivaldi, 2000) enabled this possibility, where electrodes were connected to the brain of a monkey, and using brain waves, the monkey operated a robotic arm, both locally and remotely via the Internet. This research is also in progress at Caltech, MIT, Brown University, and other research institutes. Progress in the past couple of years led to the development of chips that can recognize brain signals for movement and convert them into action (Musallam et al., 2004). Monkeys fitted with such chips were trained to move cursors on computer monitors, where such devices translate signals from the brain's motor cortex, the region that directs physical movement.

Advances in this field have reached the level that recently, the US Food and Drug Administration (FDA) approved, on a limited basis, the conduction of such experiments on humans. For this purpose, Cyberkinetics, in Foxborough, Massachusetts (Serruya et al., 2002) is developing this capability using microchips that are implanted in the motor cortex region of five quadriplegic patients to allow them mouse control and computer access. The near term objective of this study is to develop neural-controlled prosthetics. The current chips last up to a year and efforts are made to develop a longer lasting wireless capability. Using such a capability to control prosthetics would require feedback in order to provide the human operator a "feel" of the environment around artificial limbs. The feedback can be provided with the aid of tactile sensors, haptic devices, and other interfaces. Besides feedback, sensors will be needed to allow users to protect the prosthetics from potential damage (heat, pressure, impact, etc.), just as it is with our biological limbs. Also, it is hoped to provide disabled people with the ability to communicate through speech or sign to control their artificial organs.

Interfacing of visualization and hearing devices and the human brain have already emerged where hearing devices are increasingly implanted and imaging devices are currently at advanced research stages (Chapters 11 and 17). The eye's focusing mechanism as well as the iris and the eyelid have already been mimicked in today's cameras. While significant advances have been already made, the human eyes combined with the brain have far superior capabilities including image interpretation and recognition, ability to rapidly focus without moving the lens location in the eye, 3-D capability, high sensitivity, and operability in a wide range of light intensities from very dark to quite bright light. Such a capability has grown significantly with the emergence of small digital cameras that are now part of many cellular phones and webcams for telecommunication via computers. It is highly desirable to see via such cameras real-time images with performances that approach the capability of the human eye. Also, researchers are working to create implants that can help the vision-impaired regain the ability to see (Chapter 17). Increasingly, sophisticated visualization and image recognition are emerging in security systems. However, while lab demonstrations have been very successful, these systems still have recognition errors at unacceptable levels. One of the benefits of this capability, once the reliability issues are overcome, would be a standard operation as part of homeland security in airports, public areas, or even in our homes.

1.9.1   Telepresence and Teleoperation

Simulators, which involve virtual reality and the ability to "feel" remote or virtual environments are highly attractive and offer unmatched capabilities. To address the need for remote feeling of mechanical forces, the engineering community is developing haptic (tactile and force) feedback systems. Users of such simulators of procedures may immerse themselves in the display medium while being connected through haptic and tactile interfaces to allow them to "feel the action" at the level of their fingers and toes. Thus, an expert can perform various procedures from the convenience of the office

Figure 1.21   Performing virtual reality tasks using the ERF-based MEMICA haptic interface offers the potential of a highly attractive interactive simulation system.

without having to be present at the operation site. Telepresence requires the capability to intuitively project to the user as much sensation of the remote site as possible, including the distribution of reaction forces, temperatures, textures, and other sensations that we associate with our feeling of touching objects. The potential of making such a capability was enabled with a high resolution and large workspace using the novel remote MEchanical MIrroring using Controlled stiffness and Actuators system (MEMICA) (Bar-Cohen, 1999; Fisch et al., 2003). For this purpose, scientists at JPL and Rutgers University used an electro-rheological fluid (ERF), which becomes viscous under electroactivation. Taking advantage of this property, they designed miniature electrically controlled stiffness (ECS) elements and electrically controlled force and stiffness (ECFS) actuators. Using this system, the feeling of the stiffness and forces applied at remote or virtual environments are potentially reflected to the users via proportional changes in ERF viscosity. Figure 1.21 shows a graphic representation of the concept of MEMICA for the simulation of various control procedures — either through virtual reality or as a telepresence. Using such a system, surgeons may be able to conduct a virtual surgery via a virtual reality display while "feeling" the stiffness and forces that are involved with the procedure.

Potential beneficiary of the simulation of medical therapy includes astronauts who operate at a great distance from Earth. The probability that an urgent medical procedure will need to be performed in space is expected to increase with the growth in duration and distance of manned missions. A major obstacle may arise as a result of the unavailability of on-board medical staff capable of handling every possible medical emergency. To conduct emergency treatments and deal with unpredictable health problems, the medical crews will need adequate tools, and the capabilities to practice the necessary procedure in order to minimize risk to the astronauts. With the aid of all-in-one type surgical tools and a simulator, astronaut(s) with medical background would be able to practice the required procedures, and later physically perform the specific procedures. Medical staff in space may be able to sharpen their professional skills by practicing new procedures. Generally, such a capability can also serve people who live in rural and other remote areas with no readily available full medical care capability. As an education tool employing virtual reality, training paradigms can be changed while supporting the trend in medical schools towards replacing cadavers with computerized models of human anatomy. Another potential benefit that MEMICA offers is the ability to provide intuitive control of remote robots. These types of robots include the NASA Johnson Space Center robotic astronaut, which is known as Robonaut.

1.10 CONCLUSIONS

After billions of years of evolution, nature developed inventions that work, which are appropriate for the intended tasks and that last. The evolution of nature led to the introduction of highly effective and power efficient biological mechanisms. Failed solutions often led to the extinction of the specific species that became a fossil. In its evolution, nature archived its solutions in genes of creatures that make up the terrestrial life around us. Imitating nature's mechanisms offers enormous potentials for the improvement of our life and the tools we use. Humans have always made efforts to imitate nature and we are increasingly reaching levels of advancement where it becomes significantly easier to mimic biological methods, processes, and systems. Advances in science and technology are leading to knowledge and capabilities that are multiplying every year. These improvements lead to capabilities that help understand better and implement nature's principles in more complex ways. Effectively, we have now significantly better appreciation of nature's capabilities allowing us to employ, extract, copy, and adapt its inventions.

Benefits from the study of biomimetics can be seen in many applications, including stronger fiber, multifunctional materials, improved drugs, superior robots, and many others. Another aspect of biomimetics is to recognize the importance of protecting species from extinction, lest we lose nature's solutions that have managed to survive, but which we have not yet studied or still do not understand. Nature offers a model for us as humans in our efforts to address our needs. We can learn manufacturing techniques from animals and plants such as the use of sunlight and simple compounds to produce with no prolusion, biodegradable fibers, ceramics, plastics, and various chemicals. Nature has already provided a model for many human-made devices, processes, and mechanisms. One can envision the emergence of extremely strong fibers that are woven as the spider does, and ceramics that are shatterproof emulating the pearl. Besides providing models, nature can serve as a guide to determine the appropriateness of our innovations in terms of durability, performance, and compatibility. Biomimetics has many challenges, including the author's arm-wrestling challenge announced in 1999, which has taken the human muscle as a baseline for the development of artificial muscles. The challenge is still open even after the competition held in 2005; however, advances towards making such arms are helping the field of biomimetic greatly.

Inspirations from nature are expected to continue leading to technology improvements, and the impact is expected to be felt in every aspect of our lives. Some of the solutions may be considered science-fiction in today's capability, but as we improve our understanding of nature and develop better capabilities this may become a reality sooner than we expect.

ACKNOWLEDGMENTS

Research reported in this manuscript was conducted at the Jet Propulsion Laboratory (JPL), California Institute of Technology, under a contract with National Aeronautics and Space Administration (NASA). The author would like to thank Babak Amir Parviz, University of Washington, Seattle, WA, for providing information about his research related to Guided Device-to-Substrate Self-Assembly. The author appreciates the many helpful suggestions that he received from Julian FV Vincent, University of Bath, England. The author also appreciates his daughter's (Limor Bar-Cohen, UCLA) help with extensive editing. Moreover, the author acknowledges the assistance of Greg Gmurczyk from NASA Headquarters, Washington DC, in identifying experts in the field of biomimetics.

REFERENCES

Adami C., *Introduction to Artificial Life*, ISBN 0-387-94646-2, Springer-Verlag, Berlin (1998), pp. 1–374.

Amaral J.F.M., J.L.M. Amaral, C. Santini, R. Tanscheiot, M. Vellasco, and M. Pacheco, "Towards evolvable analog artificial neural networks controllers," *Proceedings of the 2004 NASA/DoD Conference on Evolvable Hardware*, Seattle, Washington, June 24–26, 2004, pp. 46–52.

Autumn K. and A.M. Peattie, "Mechanisms of adhesion in geckos," *Journal of Integrative and Comparative Biology*, Vol. 42, No. 6 (2003), pp. 1081–1090.

Autumn K., M. Sitti, Y.A. Liang, A.M. Peattie, W.R. Hanen, S. Sponberg, T.W. Kenny, R. Fearing, J.N. Israelachvili, and R.J. Full, "Evidence for van der Waals attachment by gecko foot-hairs inspires design of synthetic adhesive," *PNAS*, Vol. 99, No. 19 (2002), pp. 12252–12256.

Badescu M., Y. Bar-Cohen, X.Q. Bao, Z. Chang, B.E. Dabiri, B.A. Kennedy, and S. Sherrit, "Adapting the Ultrasonic/Sonic Driller/Corer (USDC) for walking/climbing robotic applications," *Proceedings of the Industrial and Commercial Applications of Smart Structures Technologies, SPIE Smart Structures and Materials Symposium*, pp. 5762–22, San Diego, CA, March 7–10, 2005.

Bar-Cohen Y. (Ed.), "Automation, miniature robotics and sensors for nondestructive evaluation and testing," *Topics on NDE (TONE) Series*, ISBN 1-57117-043, American Society for Nondestructive Testing, Columbus, Ohio, Vol. 4 (2000), pp. 1–481.

Bar-Cohen Y. (Ed.), *Electroactive Polymer (EAP) Actuators as Artificial Muscles — Reality, Potential and Challenges*, ISBN 0-8194-4054-X, SPIE Press, Bellingham, Washington, Vol. PM98 (March 2001), pp. 1–671.

Bar-Cohen Y., "Emerging NDE technologies and challenges at the beginning of the 3rd millennium — part II," *Material Evaluation*, Vol. 58, No. 2 (February 2000), pp. 141–150.

Bar-Cohen Y., *Electroactive Polymer (EAP) Actuators as Artificial Muscles — Reality, Potential and Challenges*, 2nd Edition, ISBN 0-8194-5297-1, SPIE Press, Bellingham, Washington, Vol. PM136 (March 2004), pp. 1–765.

Bar-Cohen Y. and B. Joffe, "Magnetically Attached Multifunction Maintenance Rover (MAGMER)," NTR, Docket 20229, Item No. 9854, February 6, 1997.

Bar-Cohen Y. and C. Breazeal (Eds), *Biologically-Inspired Intelligent Robots*, ISBN 0-8194-4872-9, SPIE Press, Bellingham, Washington, Vol. PM122 (May 2003), pp. 1–393.

Bar-Cohen Y. and S. Sherrit, "Self-Mountable and Extractable Ultrasonic/Sonic Anchor (U/S-Anchor)," NASA New Technology Report, Docket No. 40827, December 9, 2003 (patent disclosure in preparation).

Bar-Cohen Y., C. Mavroidis, M. Bouzit, C. Pfeiffer, and B. Dolgin, "Remote MEchanical MIrroring using controlled stiffness and actuators (MEMICA)," NASA New Technology Report, Item No. 0237b, Docket 20642 (January 27, 1999). *NASA Tech Briefs*, Vol. 24, No. 2 (February 2000), pp. 7a–7b.

Bar-Cohen Y., S. Sherrit, B. Dolgin, S. Askin, T.M. Peterson, W. Bell, J. Kroh, D. Pal, R. Krahe, and S. Du, "Ultrasonic/Sonic Mechanism of Deep Drilling (USMOD)," JPL New Technology Report, Docket No. 30291 (July 17, 2001), U.S. Patent application No. 10/304,192, filed on November 27, 2003.

Bar-Cohen Y., S. Sherrit, B. Dolgin, T. Peterson, D. Pal, and J. Kroh, "Ultrasonic/sonic driller/corer (USDC) with integrated sensors," NTR, August 30, 1999, Item No. 0448b, Docket No. 20856, November 17 (1999). *NASA Tech Briefs*, Vol. 25, No. 1 (January 2001), pp. 38–39. Patent registration numbers US 10/258007, submitted on May 1, 2001.

Bartlett P.N. and J.W. Gardner, *Electronic Noses: Principles and Applications*, Oxford University Press, Oxford (1999).

Benyus J.M., *Biomimicry: Innovation Inspired by Nature*, ISBN 0688160999, Perennial (HarperCollins) Press, New York, NY, USA (1998), pp. 1–302.

Bill Bialek W., "Physical limits to sensation and perception," *Annual Review of Biophysics, Biophysics Chemistry*, Vol. 16 (1987), pp. 455–478.

Bond C.E., *Biology of Fishes*, 2nd Edition, Saunders, Philadelphia, Pennsylvania (1996).

Breazeal C.L., *Designing Sociable Robots*, ISBN 0262524317, MIT Press, Cambridge, Massachusetts (2004), pp. 1–281.

Buck L. and R. Axel, "A novel multigene family may encode odorant receptors: a molecular basis for odor recognition," *Cell*, Vol. 65 (5 April 1991), pp. 175–187.

Chen J., J. Engel, and C. Liu, "Development of polymer-based artificial haircell using surface micromachining and 3D assembly," *The 12th International Conference on Solid-State Sensors, Actuators and Micro-systems*, Vol. 2, Boston, Massachusetts, June 8–12, 2003a.

Chen J., Z. Fan, J. Engel, and C. Liu, "Towards modular integrated sensors: the development of artificial haircell sensors using efficient fabrication methods," *IEEE/RSJ International Conference on Intelligent Robots and Systems (IROS 2003)*, Las Vegas, Nevada, Vol. 3, October 27–31, 2003b, pp. 2341–2346.

Cott H.B., "Camouflage in nature and war," *Royal Engineers Journal* (December 1938), pp. 501–517.

Craven M.A. and J.W. Gardner, "Electronic noses — development and future prospects," *Trends in Analytical Chemistry*, Vol. 15 (1996), p. 486.

Dickinson T.A., J. White, J.S. Kauer, and D.R. Walt, "Current trends in artificial-nose technology," *Trends in Biotechnology*, Vol. 16 (1998), pp. 250–258.

Dzenis Y., "Spinning continuous fibers for nanotechnology," *Science*, Vol. 304 (25 June 2004), pp. 1917–1919.

Eismer T. and D.J. Aneshansly, *PNAS*, Vol. 97, No. 12 (2000), pp. 6568–6573.

Fisch A., C. Mavroidis, Y. Bar-Cohen, and J. Melli-Huber, "Haptic and telepresence robotics," Chapter 4, in Bar-Cohen, Y. and C. Breazeal (Eds), *Biologically-Inspired Intelligent Robots*, ISBN 0-8194-4872-9, SPIE Press, Bellingham, Washington, Vol. PM122 (May 2003), pp. 73–101.

Freund M.S. and N.S. Lewis, "A chemically diverse conducting polymer-based electronic nose," *Proceedings of National Academic Science U.S.A.*, Vol. 92 (1995), p. 2652.

Friedel T. and F.G. Barth, "Wind-sensitive interneurones in the spider CNS," *Journal of Comparative Physiology A*, Vol. 180 (1997), pp. 223–233.

Gardner M., "Mathematical games — the fantastic combinations of John Conway's new solitaire game life," *Scientific American* (October 1970), pp. 120–123.

Gardner M., *Wheels, Life, and Other Mathematical Amusements*, ISBN 0-7167-1589-9, W.H. Freeman and Company, New York, New York (1983).

Gordon J.E., *The New Science of Strong Materials, or Why You Don't Fall Through the Floor*, 2nd Edition, ISBN 0140209204, Pelican-Penguin, London (1976), pp. 1–287.

Gravagne, I., C. Rahn, and I. Walker, "Good vibrations: a vibration damping setpoint controller for continuum robots," *IEEE International Conference of Robotics and Automation*, Seoul, Korea (May 2001), pp. 3877–3884.

Hanlon R., J.W. Forsythe, and D.E. Joneschild, "Crypsis, conspicuousness, mimicry and polyphenism as antipredator defenses of foraging octopuses on Indo-Pacific coral reefs, with a method of quantifying crypsis from video tapes," *Biological Journal of the Linnean Society*, Vol. 66 (1999), pp. 1–22.

Hrynkiw D. and M.W. Tilden, *JunkBots, Bugbots, and Bots on Wheels: Building Simple Robots with BEAM Technology*, ISBN 0072226013, McGraw-Hill, Osborne (2002), http://www.natur.cuni.cz/~vpetr/Develop.htm

Hughes H.C., *Sensory Exotica A World Beyond Human Experience*, ISBN 0-262-08279-9, MIT Press, Cambridge, Massachusetts (1999), pp. 1–359.

Koza J.R., *Genetic Programming: On the Programming of Computers by Means of Natural Selection*, ISBN 0-262-11170-5, MIT Press, Cambridge, Massachusetts (1992), pp. 1–40.

Krantz-Ruckler C., M. Stenberg, F. Winquist, and I. Lundstrom., "Electronic tongues for environmental monitoring based on sensor arrays and pattern recognition: a review," *Analytica Chimica Acta*, Vol. 426 (2001), p. 217.

Langton C.G., (Ed.), *Artificial Life: an Overview*, ISBN 0262621126, MIT Press, Cambridge, Massachusetts (1995), pp. 1–336.

Levy S., *Hackers: Heroes of the Computer Revolution*, Chapter 7, ISBN 0-385-19195-2, Anchor Press, Doubleday, Garden City, New York (1984).

Light G., W. Schlemeus, and C. Parr, "Development of encapsulated dye for a surface impact-damage indicator system," in Thompson D.O. and D.E. Chimenti (Eds), *Review of Progress in Quantitative Nondestructive Evaluation*, ISBN 0-306-42837-7, Plenum Press, New York, New York, Vol. 7A (1988), pp. 681–689.

Lloyd J.E., "Male photuris fireflies mimic sexual signals of their females' prey," *Science*, Vol. 210 (1980), pp. 669–671.

Lloyd J.E., *On Deception, a Way of All Flesh, and Firefly Signaling and Systematics*, Oxford University Press, Oxford (1984).

Lonergan M.C., E.J. Severin, B.J. Doleman, R.H. Grubbs, and N.S. Lewis, "Array-based sensing using chemically sensitive, carbon black-polymer resistors," *Chemical Materials*, Vol. 8 (1996), p. 2298.

Lowman P., *Long Way East of Eden: Could God Explain the Mess We're In?* ISBN: 1842271083, Paternoster Press, Milton Keynes, UK (2002), pp. 1–390.

Luger G.F., *Artificial Intelligence: Structures and Strategies for Complex Problem Solving*, ISBN 0201648660, Pearson Education Publishers, Upper Saddle River, New Jersey (2001), pp. 1–856.

Mann S., (Ed.), *Biomimetic Materials Chemistry*, ISBN 0-471-18597-3, Wiley Publishers, New York, New York, (1995), pp. 1–400.

Morris C.J., S. Stauth, and B.A. Parviz, "Guided and unconstrained self-assembled silicon circuits," *Proceedings of the NASA 6th Evolvable Hardware Conference*, Washington, District of Columbia, June 24–26, 2004.

Muller R. and J.C.T. Hallam, "From bat pinnae to sonar antennae: augmented obliquely truncated horns as a novel parametric shape model," *Proceeding of the 8th International Conference on the Simulation of Adaptive Behavior*, SAB'04 (2004).

Musallam S., B.D. Corneil, B. Greger, H. Scherberger, and R.A. Andersen, "Cognitive control signals for neural prosthetics," *Science*, Vol. 305 (9 July 2004), pp. 258–262.

Mussa-Ivaldi S., "Real brains for real robots," *Nature*, Vol. 408 (16 November 2000), pp. 305–306.

Nagle H.T., R. Gutierrez-Osuna, and S.S. Schiffman, "The how and why of electronic noses," *IEEE Spectrum* (September 1998), pp. 22–38.

Neugschwandtner G.S., R. Schwödiauer, S. Bauer-Gogonea, S. Bauer, M. Paajanen, and J. Lekkala, "Piezo- and pyroelectricity of a polymer-foam space-charge electret," *Journal of Applied Physics*, Vol. 89, No. 8 (15 April 2001), pp. 4503–4511.

Ozaki Y., T. Ohyama, T. Yasuda, and I. Shimoyama, "An air flow sensor modeled on wind receptor hairs of insects," *Proceedings of MEMS 2000, Miyazaki, Japan* (2000), pp. 531–536.

Peremans, H. and R. Muller, "A comprehensive robotic model for neural and acoustic signal processing in bats," *Proceedings of the 1st International IEEE EMBS Conference on Neural Engineering, IEEE EMBS* (2003), pp. 458–461.

Petr V., "Animal extinctions in the fossil record: a developmental paradigm," *Bulletin of the Czech Geological Survey*, Vol. 71, No. 4, Praha (1996), pp. 351–365.

Ray T., "An approach to the synthesis of life," in C.G. Langton et al. (Eds), *Artificial Life II*, Addison-Wesley, Redwood City, California (1992).

Research Report, "*Global Biodiversity Strategy: Guidelines for action to save, study and use Earth's biotic wealth sustainably and equitably*," ISBN 0-915825-74-0, World Resources Institute, Scientific and Cultural Organization (UNESCO), Washington, District of Columbia (1992), pp. 1–200.

Russell S.J., and P. Norvig, *Artificial Intelligence: A Modern Approach*, ISBN 0137903952, Pearson Education Publishers, Upper Saddle River, New Jersey (2003), pp. 1–1132.

Schmitt O.H., "Some interesting and useful biomimetic transforms," *Proceedings of Third International Biophysics Congress*, Boston, Massachusetts, August 29–September 3, 1969, p. 297.

Schmitz H. and S. Trenner, "Electrophysiological characterization of the multipolar thermoreceptors in the fire-beetle, "*Merimna atrata* and comparison with the infrared sensilla of *Melanophila acuminata* (both Coleoptera, Buprestidae). *Journal of Comparative Physiology A*, Vol. 189 (2003), pp. 715–722.

Schopf J.W., "Microfossils of the early archean apex chert: new evidence of the antiquity of life," *Science*, Vol. 260 (1993), pp. 640–646.

Serruya M.D., N.G. Hatsopoulos, L. Paninski, M.R. Fellows, and J.P. Donoghue, "Instant neural control of a movement signal," *Nature*, Vol. 416, No. 6877 (14 March 2002), pp. 141–142.

Towery T.L., *The Wisdom of Wolves: Natures Way to Organizational Success*, ISBN 0964687208, Wessex House Publishing, Hampshire, UK (February 1996), pp. 1–131.

Trotter J.A., J. Tipper, G. Lyons-Levy, K. Chino, A.H. Heuer., Z. Liu, M. Mrksich, C. Hodneland, W.S. Dillmore, T.J. Koob, M.M. Koob-Emunds, K. Kadler, and D. Holmes, "Towards a fibrous composite with dynamically controlled stiffness: lessons from echinoderms," *Biochemical Society Transactions* Vol. 28, No. 4 (2000), pp. 357–362.

Vincent J.F.V., "Stealing ideas from nature," Pellegrino S. (Ed.), Chapter 3 in *Deployable Structures*, Springer-Verlag, Vienna (2001), pp. 51–58.

Vlasov Y. and A. Legin, "Non-selective chemical sensors in analytical chemistry: from 'electronic nose' to 'electronic tongue'," *Journal of Analytical Chemistry*, Vol. 361 (1998), p. 255.

Vogel S., *Comparative Biomechanics: Life's Physical World*, Princeton University Press, Princeton, New Jersey (2003).

Wessberg J., C.R. Stambaugh, J.D. Kralik, P.D. Beck, M. Lauback, J.C. Chapin, J. Kim, S.J. Biggs, M.A. Srinivasan, and M.A. Nicolelis, "Real-time prediction of hard trajectory by ensembles of cortical neurons in primates," *Nature*, Vol. 408 (16 November 2000), pp. 361–365.

WEBSITES

http://ice.chem.wisc.edu/materials/light/lightandcolor6.html (luminescence and firefly)

http://jasss.soc.surrey.ac.uk/4/1/reviews/hales.html

http://www.alcyone.com/max/links/alife.html

http://link.abpi.net/l.php?20040802A5 (fiber beetle)

http://mcb.harvard.edu/hastings/Images/bioluminescence.html (luminescence and firefly)

http://www.aaai.org/AITopics/html/overview.html (AI)

http://www.airpower.at/news01/0625_libelle/libelle3.htm (in German about the Libelle g-suite)

http://www.bath.ac.uk/mech-eng//biomimetics/ (The Centre for Biomimetic and Natural Technologies)

http://www.bath.ac.uk/mech-eng/biomimetics/Biomimetics.pdf Biomimetics

http://www.biomimicry.org/case_studies_materials.html

http://www.biomimicry.org/case_studies_processes.html

http://www.biomimicry.org/intro.html

http://www.cs.cmu.edu/~listen/ (AI)

http://www.cs.indiana.edu/%7Eleake/papers/p-01-07/p-01-07.html (AI)

http://www.jwst.nasa.gov/project/Groups/Technology/bin/Henderson.pdf (inchworm)

http://www.nature.com/news/2004/040705/full/040705-7.html#B1 (monkey control robotics)

http://www.newscientist.com/hottopics/ai/

http://www.otto-schmitt.org/ (biomimetics)

http://www.txkell.ang.af.mil/news_events/suit.htm (Libelle g-suite)

http://www-gap.dcs.st-and.ac.uk/~history/Mathematicians/Conway.html, http://delta.cs.cinvestav.mx/~mcintosh/newweb/what/node7.html

http://xxx.infidels.org/~meta/getalife/coretierra.html

http://xxx.infidels.org/~meta/getalife/life.html

2

Biological Mechanisms as Models for Mimicking: Sarcomere Design, Arrangement and Muscle Function

Kenneth Meijer, Juan C. Moreno, and Hans H.C.M. Savelberg

CONTENTS

2.1   INTRODUCTION

We marvel at the extraordinary performances of animals. The animal kingdom provides us with inspiring examples of species that move seemingly effortless in unstructured, unpredictable, and ever-changing environments. To appreciate the beauty and the complexity of the motion of animals, one just has to think of a gazelle trying to escape a cheetah on the Serengeti plains. Both animals display a remarkable maneuverability, while running at maximum speed over a bumpy terrain covered by obstacles. Their performance is shaped by their evolution driven by the need to outperform the other in the struggle for survival. Their abilities outdo that of any human-made device and it is not surprising that humans have tried to develop devices that mimic animal locomotion.

As far back as the 15th century people have designed and built machines in an effort to copy aspects of animal locomotion (Breazeal and Bar-Cohen, 2003). The resulting designs were great illustrations of the technological "state of the art" of those times. None of these designs, however, yielded performances that came close to mimicking the agility and performance of biological locomotion. The most likely explanation for this is that those devices were engineered using materials with very different properties compared to biological materials. Only recently, with the development of soft, flexible polymer materials with actuation properties, biological-like locomotion has been made possible (Breazeal and Bar-Cohen, 2003). The benefits are many, provided that we can identify the principles that constitute the basis for biological-like locomotion. Nature can serve as a template for future designs, given that the proper questions are asked and the potential pitfalls are identified. The most important pitfall to consider is the fact that nature does not strive for optimality. Natural designs are built upon their evolutionary history, which may impose considerable constraints. Nature's design process works on a "good enough" basis (Vogel, 1998). Direct copying from Nature is likely to result in suboptimal performances; rather we should strive for understanding the enabling principles and develop them further to achieve optimal performance (Full and Meijer, 2001; Meijer et al., 2003).

Biomimetic design requires that engineers and biologists work closely together. To make this collaboration work, one should understand that both fields have very different approaches as Vincent (2004) concluded: "Engineers look at the problem and try to find an answer, biologists look at the answers and try to find out what the problem was." The starting point of any biomimetic design should be the function to be emulated. For example, for a legged biomimetic robot, one would like to emulate the spring-mass and pendulum characteristics that are exploited by animals (Full and Koditschek, 1999). The technological aim here is to build mobile platforms that are robust, agile, flexible, energy-efficient, self-sustaining, self-repairing, independent movers (no cables), as well as adaptable to requirements set by the task and the environment. To this aim, it is insightful to study the solutions that animals have found to meet these requirements (Full and Meijer, 2001; Meijer et al., 2003). Moving animals exploit various energy-saving mechanisms; they have a redundant set of actuators, they are soft and flexible, and most important they can adapt and repair their tissues in response to injury and changing requirements. The key to successful animal locomotion is the multi-functionality of their muscles.

Primordial biological qualities like adaptation, modularity, robustness are important principles for R&D of new artificial muscles. They represent the basis for new developments in bionics, mechatronics, orthotics, and prosthetics that explore the simplicity of a mechanism or material with the complexity or sophistication of a control system mimicking the biological parts with state-of-the-art actuators. Biomimetic control, in which adaptation of state-of-the-art actuators and design of control systems provide new functionalities to current aids for disabled, is an important new field. Understanding the behavior of the musculoskeletal system will lead to active or semiactive systems for interaction with the human limbs: spring-based actuator system for a knee–ankle–foot orthosis (KAFO) mimicking the lacking functionalities of a certain group of muscles during walking, upper limb orthotics for active treatment of pathological tremor by means of dampers, and ultrasonic motors compensating a certain disorder.

In recent years, material scientists have developed soft and compliant electroactive polymers (EAP) that have actuating abilities (Bar-Cohen, 2001a,b; Kornbluh et al., 2001). It has been argued that these novel technologies will enable the development of artificial muscles and eventually lead to legged robots that outperform their biological counterparts (Bar-Cohen, 2001a,b; Kornbluh et al., 2001). Preliminary comparisons between rudimentary EAP actuators and biological muscles have revealed that their mechanical performance is comparable (Full and Meijer, 2000, 2001; Meijer et al., 2003; Wax and Sands, 1999). Specifically, it has been found that stress, strain, and power capabilities of the EAP actuators are within or even exceed that of natural muscle (Meijer et al., 2001, 2003). Despite the resemblance in these performance metrics, none of these actuators could be called truly "muscle-like" for two reasons. First the working principle of EAP actuators is very different from biological muscle; it will be argued in this chapter that the uniqueness of muscle as an actuator is partly due to its contractile mechanism. Second, muscles are complex and dynamic actuators that are capable of tailoring to specific functional demands by modification of their structure, thus far no human-made actuator possesses this capacity for remodeling.

This chapter focuses on the principles that underlie muscle function and plasticity while considering their potential for new design in actuators. The emphasis will be on the organization of the contractile proteins and how this is related to functional demands. To this aim, a description of the principal contractile unit, the sarcomere, will be given. The various sarcomere designs present in the animal kingdom will be discussed in relation to their functional consequences. Subsequently, the principles of muscle remodeling and repair in response to use and disuse will be discussed. The chapter will end with a discussion of the principles that could prove to be relevant for the design of "muscle-like actuators."

2.2 MUSCLE FUNCTION

Muscle force production is characterized by three contraction modes: concentric, isometric, and eccentric. During concentric contractions muscles generate force while shortening. Force production during concentric contractions is described by the force–velocity relationship in which force production declines with increasing speed. In isometric contractions the muscle generates force without changing its length, for example, when the task requires holding a certain position. In eccentric contractions the muscle generates force while being lengthened, for example, when an animal needs to decelerate a limb.

One of the primary functions of skeletal muscles is to generate force while shortening in order to power the movement of the attached appendages. Comparative studies have revealed the broad range in force generating and shortening abilities of skeletal muscle (Full, 1997; Josephson, 1993; Medler, 2002). Maximal strain ranges from 2 to 200% (Full, 1997). The maximal isometric stress of muscles (Po) varies by three orders of magnitude from 8 to 2200 kN/m^2. The maximal rate of shortening (V_{max}) varies by two orders of magnitude from 0.35 to 38 muscle lengths per second (Josephson, 1993; Medler, 2002). Body size has an important influence on muscle function, with muscles from smaller animals having larger contractile speed (Medler, 2002). It has been suggested that this is a consequence of the higher movement frequencies utilized by small animals (Medler, 2002). Operating frequency varies by three orders of magnitude and ranges from less than one to over a 1000 Hz (Full, 1997).

Recent sophisticated experiments have revealed that during animal locomotion muscles do more than just generating power. In fact, the multi-functionality of muscle is the key explanation for the success of animal locomotion (Dickinson et al., 2000; Full and Meijer, 2001). Driven by technological advances, researchers are now capable of determining muscle function during animal locomotion. One of the approaches involves direct measurement of muscle function using small force and length sensors implanted in the muscle of choice (Biewener et al., 1998a,b; Griffiths, 1991; Roberts et al., 1997). Others have determined *in vivo* 3-D kinematics of animal locomotion and muscle activity patterns, and used this data to replicate the *in vivo* muscle length changes and stimulation patterns in workloop experiments (Ahn and Full, 2002; Josephson, 1985). The emerging picture from these experiments is that muscles are well equipped to meet the basic requirements for successful locomotion, that is power generation, stability, maneuverability, and energy conservation. For example, insect flight muscles operate as tunable springs that keep the thorax at which the wings attach in resonance. The muscles themselves undergo very small strains and the design is very effective for operation at high frequencies (100 Hz and above) that are needed to keep insects airborne. To sustain the high frequencies, these muscles make use of specialized contractile mechanisms (Josephson et al., 2000). In these muscles there is no direct correspondence between muscle contraction and muscle action potential; hence they are called asynchronous muscles (Machin and Pringle, 1959). Some muscles do not even shorten during their daily tasks. For example, during level running, the calf muscle fibers of turkeys generate force without shortening (Roberts et al., 1997). Functionally, they work like struts, transmitting energy between body segments. They use their force to load the elastic structures within the muscle, like the aponeurosis, that takes up most of the length changes while storing and returning elastic energy during the locomotion cycle. Due to the high resilience of these series of elastic structures, this mechanism allows the muscle to operate more efficiently. Several other studies have revealed that muscles are also used as brakes (Ahn and Full, 2002), shock absorbers (Wilson et al., 2001), and even (to push the analogy with motor parts further) as gearboxes (Rome and Lindstedt, 1997; Rome, 1998). In addition to this, recent modeling studies have pointed out the importance of viscoelastic muscle properties for the stability of locomotion (van Soest and Bobbert, 1993; Wagner and Blickhan, 1999). The idea postulated in these latter studies is that, due to their inherent stiffness and damping properties, muscles will act as a first line of defense in response to external perturbations (Loeb et al., 1999). Understanding muscle function requires a systems approach in which the influences of the neural control signals, the muscles biochemistry, and its morphology are studied in relation to the required performance (Dickinson et al., 2000; Full and Meijer, 2001).

2.3 THE FUNCTIONAL UNITS

Muscle function is determined by specific adaptations at all levels of the muscle hierarchy. Muscles are comprised of distinct functional modules called "motor units" which are controlled individually by the central nervous system (CNS) via a network of peripheral nerves. A motor unit consists of motor neuron, which via its axon innervates a distinct set of muscle fibers. From a control perspective, motor units are the building blocks of muscle function. Force production and modulation occur through discrete and sequential recruitment of individual motor units. An important property of motor units is that all muscle fibers belonging to a single unit have an identical biochemical make up. Individual motor units are classified based on their size, speed of contraction, and fatigue resistance. A typical muscle contains a mix of different motor units, which gives the CNS the freedom to tailor function to demand. For example, during slow incremental loading tasks, motor units are recruited according to Henneman's size principle (Henneman et al., 1965). This means that the slow, small, fatigue-resistant motor units are recruited first, followed by faster, larger, and less fatigue-resistant motor units when the load increases. During fast ballistic tasks like jumping, however, recruitment according to the size principle is not sufficient to accelerate the limbs fast enough. It has been shown that under these circumstances motor units are recruited according to a reversed size principle (Wakeling, 2004). Furthermore, motor unit plasticity in response to use or disuse can alter the motor unit profile of a muscle and thereby its function. Muscle function is not just influenced by the amplitude of the neural control signal, but also by the phase of the control signal in relation to the movement kinematics. For example, it has been shown that neuromuscular system of jumping frogs has evolved phase relationships between the control signals and the movement kinematics that yield optimal power output (Lutz and Rome, 1994). Motor unit activity is under control of the CNS, and regulated by reflex activity of several sensory systems. Therefore, it enables a rich pattern of voluntary and autonomous muscle functions.

Besides neural control, muscle morphology at the macroscopic and microscopic level has a major impact on muscle function. Muscle fibers are attached to the skeleton via elastic tendons. Macroscopically, the ratio of muscle fiber length to tendon length is a major determinant of muscle function (Biewener et al., 1998a). For example, the calf muscles of wallabies have very short muscle fibers in series with a long tendon. This design appears to be an adaptation to enhance the storage and return of elastic energy to allow for more efficient locomotion (Biewener et al., 1998a). At the microscopic level, muscle tissue is highly ordered, typically comprising thousands of muscle cells embedded in a matrix of basal lamina (Trotter and Purslow, 1992). The muscle cells, or muscle fibers, are long and slender multinucleated cells in which the contractile proteins are arranged in highly organized structures called "sarcomeres." The sarcomeres are the working units of the muscle fiber. A typical fiber comprises several thousand sarcomeres in series and in parallel. Microscopically, sarcomere design and the arrangement of sarcomeres within a muscle fiber are major determinants of muscle function. Other important structures within muscle cells are the mitochondria that are responsible for the aerobe energy metabolism and the sarcoplasmic reticulum (SR), which plays a crucial role in the activation and relaxation kinetics of muscle. It is known that changes in the volume fraction of mitochondria, SR, and myofibrillar proteins can be utilized to modify muscle function (Conley and Lindstedt, 2002). For example, in high-frequency muscles involved in sound production, the SR fraction is enlarged at the expense of the myofibrillar protein fraction to attain superfast muscle contraction (Conley and Lindstedt, 2002). This kind of specialization will not be dealt with in this chapter. Instead the remainder of this chapter will focus on the design and organization of the sarcomeres, and it will be discussed how the natural design might provide inspiration for artificial muscles.

2.3.1 The Sarcomere

Sarcomeres are anisotropic, hierarchic, liquid crystalline structures comprised of contractile and structural proteins (Figure 2.1). The constituting proteins are responsible for muscle elasticity and its ability to perform work. Under the microscope, sarcomeres are visible as repetitive units of dark and light bands. The light band or I-band contains the thin, actin filaments and the dark or A-band contains the thick, myosin filaments. The sarcomeres are separated by Z-disks, comprised of α-actinin, which segment the myofibrils (Figure 2.1). The actin filaments project from the Z-disks towards myosin filaments in the center of the sarcomere. In the center of the A-band there is a lighter zone, the M-line which is a disk of delicate filaments, and its main function is to keep the myosin filaments aligned. The myosin filaments are also connected to the Z-disks via a protein called titin. Titin is responsible for keeping the myosin filaments aligned, and is the main determinant of passive elasticity in muscle (Tskhovrebova and Trinnick, 2002; Lindstedt et al., 2001, 2002). It also plays an important role in the sarcomerogenisis (Russell et al., 2000). There are several other important proteins present in the sarcomere. Nebulin, for example, is located in the I-band, and is thought to be responsible for determining the length of the actin filament.

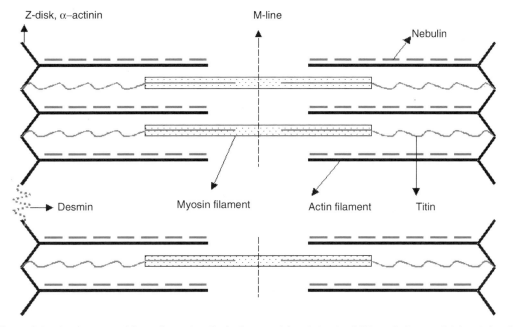

Figure 2.1   Arrangement of the major contractile (actin, myosin) and structural (titin, nebulin, α-actinin) proteins of the vertebrate sarcomere. Adjacent sarcomeres are interconnected via desmin.

Furthermore, there are several structural proteins present in the M-line (i.e., M-protein, myomesin) and the A-band (C-protein) that presumably keep the myosin filaments in register during contraction.

Sarcomere force production stems from the interaction between the actin and myosin filaments. In vertebrate sarcomeres, six actin filaments surround each myosin filament. Myosin filaments of vertebrates consist of approximately 100 myosin molecules, each shaped like a golf club with a double head. The myosin heads protrude from the core of the filament towards the surrounding actin filaments. Actin filaments consist of two helical strands of F-actin twined together like a bead necklace. On each of the beads is a site where myosin can bind. Binding is regulated by the configuration of the proteins troponin and tropomyosin, which is controlled by Ca^{2+}. When a myosin head attaches to an actin-binding site, it undergoes a conformational change resulting in the development of force and sliding of the actin and myosin filaments along each other. Under the influence of adenosine triphosphate (ATP), the crossbridge detaches again. Pumping back calcium ions into the SR via ATP-consuming calcium pumps triggers the relaxation. The formation of connections between myosin and actin is a stochastic process and it is known as the crossbridge theory (Huxley, 1957, 2000).

Force production of the sarcomere unit depends on the length of the sarcomere and the velocity at which the sarcomeres shorten or lengthen. According to the sliding filament theory (Huxley and Niedergerke, 1954; Huxley and Hanson, 1954), the length dependence of force production is determined by the amount of overlap between the actin and the myosin filaments. Sarcomeres have an optimal length for force production ($\pm$ 2.3 μm in vertebrates) at which the filament overlap allows the maximum number of crossbridges to be attached. At lengths over the optimal one, the overlap decreases and thus the amount of force. At lengths less than optimal, internal forces and reduced overlap due to interference of actin filaments of neighboring sarcomeres also result in less force. As a consequence, each sarcomere has a typical length–force relationship (Figure 2.2) whose shape depends on the length and ratio of the actin and myosin filaments. The velocity dependence of sarcomere force production is determined by the probabilities for crossbridge attachment and detachment. For shortening sarcomeres, the relationship is characterized by a hyperbolic function (Figure 2.2). Together the force–length and force–velocity functions determine the maximal work and power that a sarcomere of given dimensions can generate. Theoretical studies have indicated that in many cases sarcomere design is optimized for power production (van Leeuwen, 1991).

Sarcomeres do not operate independently. They are connected to adjacent sarcomeres in series via the Z-disk and until recently it was thought that the series connection was the main pathway to get the force of individual sarcomeres to the outside world. More recently (Patel and Lieber, 1997), it has been found that sarcomeres also make connections with adjacent sarcomeres in parallel and

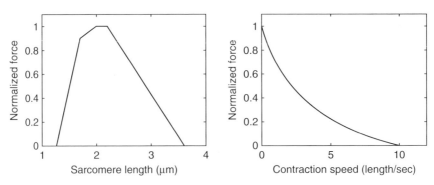

Figure 2.2 Normalized length–force and force–velocity relationships for a vertebrate sarcomere (myosin filament 1.6 μm, actin filament 0.95 μm, Z-line width 0.1 μm, M-line width 0.2 μm, and maximal contraction speed 10 lengths per second).

with the cell membrane via specialized structural proteins like desmin (Figure 2.1). Based on the evidence from animal experiments (Huijing, 1999), it is now thought that the force of individual sarcomeres finds its way to the outside via both serial and parallel pathways.

2.4 MUSCLE DESIGN

Within the animal kingdom, the variety in muscle designs is stunning. There are bulky muscles (m. gluteus maximus), long slender muscles (sartorius), muscles with short fibers attached to long tendons (m. gastrocnemius), pennate muscles, etc. Muscle design is highly variable within an animal and also between species. It appears as if there is a specialized muscle design for each possible function (Otten, 1988). It is beyond the scope of this chapter to review all possible designs and functions, and therefore a few basic design principles of muscle will be discussed. Muscles are built from sarcomeres and as a consequence it has two basic design options to tune into functional demands. It can modify either the design or the arrangement of the sarcomeres. Both options appear to have been explored by Nature.

2.4.1 Not all Sarcomeres Are Alike

Invertebrates appear to have explored the possibilities of sarcomere design to its full potential. Invertebrate sarcomeres range from very short (0.9 μm) as in squid tentacles (Kier, 1985) to very long (20 μm) as in crab claw muscles (Taylor, 2000). This broad range is achieved by the diversity in the length of both the myosin (0.86–10 μm) and actin filaments. In addition, the ratio of actin to myosin filaments is also variable ranging from as low as 2:1 to as much as 7:1 (Figure 2.3 and Figure 2.4).

The diversity of the invertebrate sarcomere design illustrates how nature makes use of slight modifications to a basic design to meet functional demands. From a theoretical point of view, it

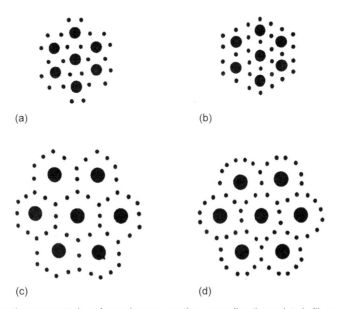

(a)   (b)

(c)   (d)

Figure 2.3 Schematic representation of muscle cross sections revealing the variety in filament lattice and ratio of actin:myosin filaments: (a) vertebrate skeletal muscle, ratio 2:1, (b) insect flight muscle, ratio 3:1, (c) and (d) arthropod leg and trunk muscles, ratio 5–6:1. (From Pringle, J.W.S. (1980) A review of arthropod muscle. In: *Development and Specialization of Skeletal Muscle*, Goldspink, D.F. (Ed.), Cambridge University Press, Cambridge, Massachusetts. With permission.)

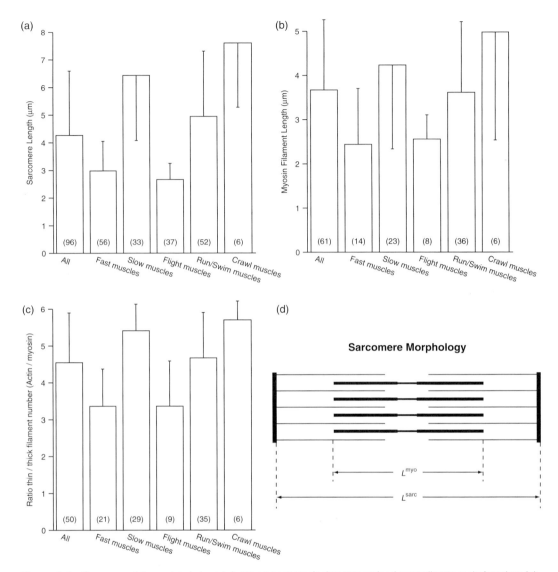

Figure 2.4 Summary of the variety in invertebrate sarcomere design categorized according to main function: (a) range of sarcomere lengths, (b) range of myosin filament length, (c) ratio of actin:myosin filaments, (d) schematic representation of the sarcomere morphology, L^{sarc} represents sarcomere length and L^{myo} represents the length of the myosin filament. (From Full, R.J. (1997) Invertebrate locomotor systems. In: *The Handbook of Comparative Physiology*, Dantzler, W. (Ed.), Oxford University Press, Oxford. With permission.)

could be argued that long sarcomeres with long myosin filaments mean that more crossbridges will be available for force generation (Vogel, 2001; Alexander, 2003). Thus long sarcomeres should be capable of generating large forces. This view is supported by experimental evidence on crustacean claw muscles, where it is shown that muscle stress increases with sarcomere resting length (Taylor, 2000). At the other end of the spectrum, it could also be argued that short sarcomeres are good for fast contractions needed in power-demanding tasks like flying or ballistic movements like jumping or catching a prey. After all, for a given crossbridge stroke, a short sarcomere would shorten relatively more than a long sarcomere, and thus its intrinsic speed would be higher. This is in fact what happens in squid tentacles. The sarcomeres responsible for the fast elongation of squid tentacles are ultra short and can contract very rapidly (Kier, 1985). In an excellent review on invertebrate musculoskeletal design, Full (1997) showed that there are specific sarcomere designs for specific functions or modes of locomotion (Figure 2.4). For example, arthropod limbs have slow and fast muscles. The slow muscles are mainly used during posture, burrowing, and slow locomotion, while the fast muscles are involved in rapid locomotion and escape. Not surprisingly, the slow muscles are the ones that have the longest sarcomeres (Full, 1997).

With respect to sarcomere design, vertebrates are pretty conservative. Their sarcomeres typically have a length between 2 and 3 μm. With myosin filaments having a more or less constant length of 1.6 μm, much of the variability is due to differences in the length of actin filaments. Their length ranges from 0.95 μm in chicken to 1.27 μm in humans (Ashmore et al., 1988; Burkholder and Lieber, 2001; Lieber and Burkholder, 2000; Walker and Schrodt, 1973). Furthermore, in vertebrate sarcomeres, the ratio of actin to myosin filaments is virtually constant at 2:1. As a consequence, vertebrates have only a limited capacity to tailor their sarcomeres to meet functional demands and will have to resort to different mechanisms to achieve this.

2.4.2 Rearranging the Sarcomeres, Muscle Morphology

The function of vertebrate and invertebrate muscle is intimately related to their morphology. To meet functional demands while at the same time accounting for volume and length constraints set by (exo)skeletal dimensions, sarcomeres are arranged in specific ways. The basic design options are the parallel and serial arrangement of the sarcomeres. Figure 2.5 illustrates the functional consequences of these mechanisms. Adding sarcomeres in parallel increases the force of the muscle, whereas serial addition of sarcomeres increases the operating range of the muscle as well as the maximal shortening velocity.

Some muscles, like the human hamstrings, are long and slender. They have long parallelly arranged muscle fibers that contain many sarcomeres in series. They are capable of considerable shortening while maintaining the ability to generate sufficient force. Interestingly, there appears to be a limit to the length of individual muscle fibers; one rarely comes across muscle fibers longer than 10 cm. Muscles whose fleshy belly exceeds this length, like the human and feline sartorius muscle (Loeb et al., 1987), have tendinous plates that interconnect muscle fibers in series. The exact reason for this design is thus far unclear. It has been suggested that it has to do with control problems involved in synchronizing the activation of sarcomeres in very long fibers, but it might also be a solution to ensure structural integrity of the muscle.

Pennate muscles have relatively short muscle fibers that are orientated at an angle with the line of work of the muscle. The advantage of this design is that the number of sarcomeres arranged in

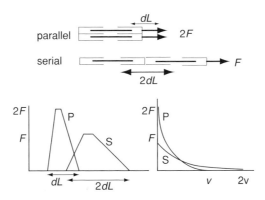

Figure 2.5 (See color insert following page 302)   Functional effects of parallel (P) and serial (S) arrangement of sarcomeres. *F* represents force, *v* represents velocity, and *dL* represents the length ranges over which the muscle can generate force.

parallel for a given muscle length and volume is much larger than what could be obtained with a parallel fibered muscle. Clearly, pennate muscles are built for force. Examples of pennate muscles are the calf muscles of humans (whose main function is to provide enough force to allow storage of elastic energy in the Achilles tendon) and the claw closer muscles of crabs. Interestingly, the latter uses both sarcomere (long sarcomeres) and muscle (pennation) design to generate as much grip force as possible. This may not come as a surprise when one considers the tough shells a crab has to crack. For the invertebrates with their exoskeletons, the pennate muscle design gives one additional advantage. Jan Swammerdam discovered in 1737 that muscles remain constant in their volume during contraction, a fact that falsified the then prevailing hypothesis that contraction came about by a change in muscle volume. For a parallel-fibered muscle, the requirement of constant volume means that the muscle must become thicker when contracting. This can be disadvantageous when you are trapped in an exoskeleton. Pennate muscles offer the solution to this problem. Their fibers rotate when they shorten, thereby making volume available for the thickening fibers without changing the width of the muscle (Vogel, 2002).

2.5 MUSCLE ADAPTATION

Once a muscle has formed and its basic morphological design is set, there still is room for remodeling. The ability to adapt in response to changes in functional demands sets living tissues apart from their engineered counterparts. Muscles grow during development, they remodel in response to use and disuse, and they are able to repair themselves after an injury. Fully grown muscles still posses the ability to more than double their size by increasing either their physiological cross-sectional area (PCSA) or their length. This is achieved by increasing muscle fiber size by adding sarcomeres in parallel or in series, but not by increasing the number of muscle fibers. The first signs of muscle adaptation occur within hours and adaptation can be completed within days (Shah et al., 2001). It is not known whether adaptation involves alterations in sarcomere design.

Whether a muscle adapts by parallel or serial addition of sarcomeres is determined by the functional demands. In strength training where the muscle is subjected to high loads, the adaptation will involve addition of parallel sarcomeres to reduce the load on the individual contractile units (Russell et al., 2000). This mechanism may be responsible for a more than twofold strength gain of the muscle. Alternatively, when an animal grows or when it starts using its limbs in new body configurations, the muscle will start adding sarcomeres in series. This mechanism can be responsible for length changes of the muscle of up to 27% (Shah et al., 2001). There are a number of theories on the mechanism for length adaptation of the muscle. Some studies have provided evidence that a muscle strives to have its optimal muscle length at the most prevalent joint position (Williams and Goldspink, 1973; Burkholder and Lieber, 1998), while others have argued that maintenance of adequate joint excursion is the most important trigger (Koh and Herzog, 1998). Another theory is that muscles adapt their length to prevent injury. In severely injured muscles, entire muscle fibers are replaced, however, in mild injury involving local lesions to sarcomeres just the damaged sarcomere are replaced. Muscle responds to injury with overcompensation probably as a safety precaution to future incidents. Lynn et al. (1998) have shown that injury induced by eccentric contractions results in addition of serial sarcomeres. The consequence of this adaptation is that the recovered muscle will operate at the ascending limb of its length–tension relationship, where it is less prone to lengthening induced injury. It is conceivable that all three mechanisms co-exist, but the length at which the muscle operates determines their action. It has been observed that the operating range of different muscles is scattered over the entire functional length range, some muscles work on the ascending limb and others on the descending limb (Burkholder and Lieber, 2001; Lieber and Burkholder, 2000). This is also reflected in the observation that muscles within a single anatomical group display different adaptations that are triggered by functional demands (Savelberg and Meijer, 2003).

The rules governing muscle adaptation are complex and far from being resolved (Russell et al., 2000). Regulatory pathways are triggered by growth signals (mechanical, hormonal), resulting in gene transcription followed by translation and assembly of the proteins into the contractile architecture (Russell et al., 2000). Several myogenic regulatory factors are involved in the remodeling of muscle, they are triggered by multiple signals and they can activate or inhibit each other's action (Brooks and Faulkner, 2000). Teasing out the exact relationships is experimentally difficult and time consuming. As a consequence, our understanding of the adaptation laws at the molecular level is still fragmentary. Modeling approaches might be helpful in understanding the intricate relationships (Jacobs and Meijer, 1999)

2.6   BIOMIMETICS OF MUSCLE DESIGN

It is unlikely and probably undesirable that future polymer actuators will use the exact working principles as the contractile mechanism of biological muscle. Consequently, current research focuses on the design of polymer actuators that mimic the functionality of muscle based on alternative working principles (Bar-Cohen, 2001b; Kornbluh et al., 2001; Meijer et al., 2003). It is argued in this chapter that it might be useful to look at the design principles that enable the variety in muscle function. Unlike current EAP actuators, muscle design is modular. Muscle function is achieved by concerted action of thousands of functional units called sarcomeres. It has been shown that muscle function is shaped by sarcomere design and arrangement. Hence, an evaluation of the benefits of sarcomeric design in relation to synthetic muscle design may be useful.

Robustness is an important requirement for an actuator. It is crucial that an actuator does not breakdown while functioning, in other words it needs to avoid mechanical failure. Biological materials are remarkably tough, meaning that it requires a lot of energy to break them. They achieve this by using energy release mechanisms that help to avoid crack propagation. As a consequence, small failures do not become catastrophic (Gordon, 1976). Although there is little data on the fracture mechanics of muscle, it can be argued that the sarcomere design of muscle helps to avoid small injuries that may make the muscle nonfunctional. It is well known that muscle injury in response to tensile stresses results in local disruptions of sarcomeres. These lesions are local and do not seem to propagate through the muscle. Morgan (1990) provided an explanation for these lesions and their functional consequences in what is now known as the 'popping sarcomere' theory. He proposed that sarcomeres that are subjected to high tensile stress undergo rapid lengthening that is stopped by the structures responsible for the passive tension of muscles (titin, external membranes). The popping has three functional consequences: (1) the rapid lengthening releases some of the energy, (2) the lengthened sarcomere will act as a spring in series with the remaining sarcomeres and will be able to withstand higher tensile stresses, and (3) the remaining sarcomeres will shorten somewhat and increase their strength as a consequence they will be able to withstand higher tensile stresses as well. In other words, under high tensile stresses individual sarcomeres will be sacrificed to maintain the structural integrity of the muscle. From experience it is known that some EAP actuators break very easily under tensile stresses, it could be argued that a modular design might help to increase the robustness of these actuators.

The modular design of muscle also facilitates the remodeling and repair of the muscle. The self-healing properties of muscle emerge from the integration of muscles into a system that allows wound healing and continuous turnover via transport of nutrients and removal of waste products. It is arguably much simpler to grow and repair individual units than having to adapt the entire structure. Furthermore, it may be argued that the variety in designs is facilitated by the modular design — just like Lego enables designs only limited by one's imagination. Until recently, remodeling and repair was only feasible within the domain of biological materials and systems. However, recent innovations in material science have resulted in self-repairing polymers (Wool,

2001), smart materials that can remodel (Anderson et al., 2004) and be fabricated using molecular self-assembly (Zhang, 2003). If these concepts can be integrated in a system that allows for transport of the necessary components and removal of the waste products then remodeling polymer actuators may become available in the future.

The use of sarcomeres as the basic functional unit also imposes limitations on the functionality of muscles. It is likely that millions of years of evolution have resulted in a full exploration of the sarcomere design. Consequently, it seems unlikely that the design has the potential to generate tensions far above the 2200 kN/m^2, strain larger than 200% or shortening rates above 40 lengths per second. These performance metrics are eventually limited by space requirements and the speed of enzyme actions. For example, to accommodate the large forces generated by the claw closer muscles of the crab, the thickness of the myosin filament has to increase. As a consequence, there is less space for the actin filaments. This will limit the maximum amount of crossbridges in a certain volume and thus the specific tension. Understanding these limitations may be useful for the design of future actuators.

Mimicking the sarcomeric design of muscle in a synthetic muscle may prove to be a first step towards a novel class of robust and functionally diverse actuators, and initial attempts look promising (Frank and Schilling, 1998). The next step will require an integrative systems approach to understand and mimic the functions of biological musculoskeletal systems during natural movements (Full and Koditschek, 1999). This approach will identify the biomechanical principles to be introduced in artificial models. An integrated approach to artificial muscle design has a strong research potential. As an example, realistic biomechanical models of human limbs for analysis of locomotion, with emphasis on understanding the underlying geometries and control problems, provide an interesting basis to conceive a systems-based approach: large groups of muscle tendon complexes have been successfully modeled as simple contractile elements in a functional model (Roberts and Marsh 2003); redundancy problems associated with large muscle numbers are solved with the proper control criteria (Rehbinder and Martin, 2001). Most importantly, the qualitative insight obtained from models of biomechanical and control mechanisms are to be included in the design of novel biomimetic muscular systems.

A biomimetic muscle must be provided with versatility and adaptability; with current state-of-the-art actuator technology and its known limitations, this can be obtained if conceived in an integrated approach. Examples of this can be found in novel applications in the field of biorobotics and prosthetic devices. For example, a force-controllable ankle joint actuator for an ankle–foot orthosis (Blaya and Herr, 2004) conceived as combination of controllable devices (DC motor, springs) with an adaptive control strategy defined upon the biomechanical model of the anatomical joint, can result in an actuator system that can adapt dynamically partially recovering a specific gait disorder (drop foot) suffered by a group of patients. The joint impedance control introduced through the series elastic actuator reduces significantly the foot slap and improves swing phase dynamics in patients, as reported by the authors. Crucial constructive needs expected for such a system — and any biomimetic wearable device — are low volume and size, low energy consumption, quiet operation, low heat dissipation, and high torque (i.e., 3.3 W per body kilogram are required at the beginning of the leg swing). These challenges are to be overcome by new actuators and materials, providing lifelike characteristics. The weak musculoskeletal system in this case not only requires assistance to control the impedance but also power generation (peak demand during gait, 3.3 W per body kilogram) and other compensations to avoid other disorders found under the same muscular disabilities, like dragging of the toe during swing phase, incomplete forefoot rocker and difficulty to raise the foot. Such a biomimetic system can increase its level of functionality by increasing the level of system integration. Following this example of biomimetic actuation and control for orthopedics, a novel system for the impaired lower leg is being developed (Moreno et al., 2004). It includes elements imitating the roles of anatomical parts, like tendons to assist powered accelerations or the roles of biarticular muscles in a limb (Hof, 2001) to include the coordination mechanisms.

2.7 SUMMARY

In recent years, material scientists have developed polymer materials that can be used to develop artificial muscles. To facilitate robotic and prosthetic design, such artificial muscles should be multi-functional, robust, modular, and have the capacity to repair themselves in response to damage. It has been argued that studying the working principles of biological muscle may inspire the design of artificial muscles. This chapter gives an overview of the relationship between muscle form and function, with an emphasis on the sarcomeric design of muscle. The following issues were addressed: (1) muscles are multi-functional actuators; (2) contractile proteins are organized in functional units called sarcomeres; (3) muscle function is modified in two basic ways — (a) modifying the sarcomere design and (b) rearranging the sarcomeres; and (4) muscle adaptation in response to functional demands. The chapter ends with a discussion on how the sarcomeric design of muscle can provide inspiration for the design of artificial muscles.

REFERENCES

Ahn, A. and Full, R.J. (2002) A motor and a brake: two leg extensor muscles acting at the same joint manage energy differently in a running insect. *J. Exp. Biol.* 205(Pt 3):379–389.

Alexander, R.McN. (2003) *Principles of Animal Locomotion*, Princeton University Press, Princeton and Oxford, Ch. 2.

Anderson, D.G., Burdick, J.A. and Langer, R. (2004) Smart biomaterials. *Science* 305:1923–1924.

Ashmore, C.R., Mechling, K. and Lee, Y.B. (1988) Sarcomere length in normal and dystrophic chick muscles. *Exp. Neurol.* 101:221–227.

Bar-Cohen, Y. (2001a) EAP history, current status, and infrastructure. In: *Electroactive Polymers (EAP) as Artificial Muscles, Reality Potential and Challenges*, Y. Bar-Cohen (Ed.), SPIE Press, Bellingham, Washington, Ch. 1, pp. 3–38.

Bar-Cohen, Y. (2001b) EAP applications, potential and challenges. In: *Electroactive Polymers (EAP) as Artificial Muscles, Reality Potential and Challenges*, Y. Bar-Cohen (Ed.), SPIE Press, Bellingham, Washington, Ch. 21, pp. 615–659.

Biewener, A.A., Konieczynski, D.D. and Baudinette, R.V. (1998a) *In vivo* muscle force–length behavior during steady-speed hopping in tammar wallabies. *J. Exp. Biol.* 201(11):1681–1694.

Biewener, A., Corning, W.R. and Tobalske, B.W. (1998b) *In vivo* pectoralis muscle force–length behavior during level flight in pigeons (*Columba livia*). *J. Exp. Biol.* 201:3293–3307.

Blaya, J. and Herr, H. (2004) Adaptive control of a variable-impedance ankle–foot orthosis to assist drop-foot gait. *IEEE Transact. Neural Syst. Rehabil. Eng.* 12:24–31.

Breazeal, C. and Bar-Cohen, Y. (2003) Introduction to biomimetic intelligent robots. In: *Biologically Inspired Intelligent Robots*, Bar-Cohen, Y. and Breazeal, C. (Eds), SPIE Press Monographs Vol. 122, Ch. 1, pp. 1–25.

Brooks, S.V. and Faulkner, J.A. (2000) Tissue engineering of skeletal muscle. In: *The Biomedical Engineering Handbook*, Second Edition, Bronzino, J.D. (Ed.) Vol. II, pp. 123.1–123.14.

Burkholder, T.J. and Lieber, R.L. (1998) Sarcomere number adaptation after retinaculum release in adult mice. *J. Exp. Biol.* 201:309–316.

Burkholder, T.J. and Lieber, R.L. (2001) Sarcomere length operating range of vertebrate muscle during movement. *J. Exp. Biol.* 2004:1529–1536.

Conley, K.E. and Lindstedt, S.L. (2002) Energy-saving mechanisms in muscle: the minimization strategy. *J. Exp. Biol.* 205:2175–2181.

Dickinson, M.H., Farley, C.T., Full, R.J., Koehl, M.A.R., Kram, R. and Lehman, S. (2000) How animals move: an integrative view. *Science* 288:100–106.

Frank, T. and Schilling, C. (1998) The development of cascadable microdrives with muscle-like operating behavior. *J. Micromech. Microeng.* 8: 222–229.

Full, R.J. (1997) Invertebrate locomotor systems. In: *The Handbook of Comparative Physiology*, Dantzler, W. (Ed.), Oxford University Press, Oxford, pp. 853–930.

Full, R.J. and Koditschek, D.E. (1999) Templates and anchors: neuromechanical hypotheses of legged locomotion on land. *J. Exp. Biol.* 202, 3325–3332.

Full, R.J. and Meijer, K. (2000) Artificial muscles versus natural actuators from frogs to flies. *Proceedings of the SPIE — The International Society for Optical Engineering, Vol. 3987 (Smart Structures and Materials 2000: Electroactive Polymer Actuators and Devices (EAPAD), Newport Beach, CA, USA, 6–8 March 2000.) SPIE-International Society of Optical Engineering*, pp. 2–9.

Full, R.J. and Meijer, K. (2001) Metrics of natural muscle. In: *Electroactive Polymers (EAP) as Artificial Muscles, Reality Potential and Challenges*, Y. Bar-Cohen (Ed.), SPIE Press, Bellingham, Washington, Ch. 3, pp. 67–81.

Gordon, J.E. (1976) *The New Science of Strong Materials*, Second Edition, Penguin Books, London.

Griffiths, R.I. (1991) Shortening of muscle fibres during stretch of the active cat medial gastrocnemius muscle: the role of tendon compliance. *J. Physiol.* 436:219–236.

Henneman, E., Somjen, G. and Carpenter, D.O. (1965) Functional significance of cell size in spinal motoneurones. *J. Neurophysiol.* 28:560–580.

Hof, A.L. (2001) The force resulting from the action of mono- and biarticular muscles in a limb. *J. Biomech.* 34(8):1085–1089.

Huijing, P.A. (1999) Muscle as a collagen fiber reinforced composite: a review of force transmission in muscle and whole limb. *J. Biomech.* 32:329–345.

Huxley, A.F. (1957) Muscle structure and theories of contraction. *Prog. Biophys. Biophys. Chem.* 7:255–318.

Huxley, A.F. (2000) Cross-bridge action: present views, prospects and unknowns. *J. Biomech.* 33:1189–1195.

Huxley, A.F. and Niedergerke, R. (1954) Interference microscopy of living muscle fibers. *Nature* 173: 971–973.

Huxley, H.E. and Hanson, J. (1954) Changes in cross-striations of muscle during contraction and stretch and their structural interpretation. *Nature* 173:973–976.

Jacobs, R. and Meijer, K. (1999) A fuzzy model of skeletal muscle adaptation: a tool to study effects of surgical, therapeutic and rehabilitation procedures. *AutoMedica* 18:85–106.

Josephson, R.K. (1985) Mechanical power output from striated muscle during cyclic contraction. *J. Exp. Biol.* 114:493–512.

Josephson, R.K. (1993) Contraction dynamics and power output of skeletal muscle. *Ann. Rev. Physiol* 55:527–546.

Josephson, R.K., Malamud, J.G. and Stokes, D.R. (2000) Asynchronous muscle: a primer. *J. Exp. Biol.* 203:2713–2722.

Kier, W.M. (1985) The musculature of squid arms and tentacles: ultrastructural evidence for functional differences. *J. Morph.* 185:223–239.

Koh, T.J. and Herzog, W. (1998) Excursion is important in regulating sarcomere number in the growing rabbit tibialis anterior. *J. Physiol.* 1:508 (Pt 1):267–280.

Kornbluh, R., Full, R.J., Meijer, K., Pelrine, R. and Shastri, S.V. (2001) Engineering a muscle: an approach to artificial muscle based on field-activated electroactive polymers. In: *Neurotechnology for Biomimetic Robots*. MIT press, Cambridge, Massachusetts, pp. 137–172.

Lieber, R.L. and Burkholder, T.J. (2000) Musculoskeletal soft tissue mechanics. In: *The Biomedical Engineering Handbook*, Second Edition, Bronzino, J.D. (Ed.), Vol. 1, pp. 22.1–22.8.

Lindstedt, S.L., LAStayo, P.C. and Reich, T.E. (2001) When active muscles lengthen: properties and consequences of eccentric contractions. *News Physiol. Sci.* 16:256–261.

Lindstedt, S.L., Reich, T.E., Keim, P. and LaStayo, P.C. (2002) Do muscles function as adaptable locomotor springs? *J. Exp. Biol.* 205:2211–2216.

Loeb, G.E., Praat, C.A., Chanaud, C.M. and Richmond, F.J. (1987) Distribution and innervation of short, interdigitated muscle fibers in parallel-fibered muscles of the cat hindlimb. *J. Morphol.* 191(1):1–15.

Loeb, G.E., Brown, I.E. and Cheng, EJ. (1999) A hierarchical foundation for models of sensorimotor control. *Exp. Brain Res.* 126:1–18.

Lynn, R., Talbot, J.A. and Morgan, D.L. (1998) Differences in rat skeletal muscle after incline and decline running. *J. Appl. Physiol.* 85:98–104.

Lutz, G.J. and Rome, L.C. (1994) Built for jumping: the design of the frog muscular system. *Science* 263(5145):370–372.

Machin, K.E. and Pringle, J.W.S. (1959) The physiology of insect fibrillar muscle II. Mechanical properties of a beetle flight muscle. *Proc. R. Soc. Lond. B* 151:204–225.

Medler, S. (2002) Comparative trends in shortening velocity and force production in skeletal muscles. *Am. J. Physiol.* 283:R368–R378.

Meijer, K., Rosenthal, M. and Full, R.J. (2001) Muscle-like actuators. A comparison between three electro-active polymers. In: *Smart structures and Materials 2001: Electroactive Polymer Actuators and Devices, Proceedings of SPIE, Vol. 4329*, pp. 2–12.

Meijer, K., Bar-Cohen, Y. and Full, R.J. (2003) Biological inspiration for muscle like actuators for robotics. In: *Biologically Inspired Intelligent Robots*, Bar-Cohen, Y. and Breazeal, C. (Eds.) SPIE Press Monographs. Vol. 122, Ch. 2, pp. 26–41.

Moreno, J.C., Meijer, K., Savelberg, H.H.C.M. and Pons, J.L. (2004) Characterization of an actuator system for a controllable knee ankle foot orthosis. *Proceedings Actuator 2004*, Bremen, Germany, 14–16 June, Schneider, H.P. and Borgman, H. (Eds.).

Morgan, D.L. (1990) New insights into the behavior of muscle during active lengthening. *Biophys. J.* 57(2):209–221.

Otten, E. (1988) Concepts and models of functional architecture in skeletal muscle. *Exerc. Sports Sci. Rev.* 16:89–137.

Patel, T.J. and Lieber, R.L. (1997) Force transmission in skeletal muscle: from actomyosin to external tendons. *Exerc. Sport Sci. Rev.*, 25:321–363.

Pringle, J.W.S. (1980) A review of arthropod muscle. In: *Development and Specialization of Skeletal Muscle*, Goldspink, D.F. (Ed.), Cambridge University Press, Cambridge, Massachusetts, pp. 91–105.

Rehbinder, H. and Martin, C. (2001) A control theoretic model of the forearm. *J. Biomech.* 34(6):741–748.

Roberts, T. and Marsh, L. (2003) Probing the limits to muscle-powered accelerations: lessons from jumping bullfrogs. *J. Exp. Biol.* 206:2567–2580.

Roberts, T.J., Marsh, R.L., Weyand, P.G. and Taylor, C.R. (1997) Muscular force in running Turkeys: the economy of minimizing work. *Science* 275:1113–1115.

Rome, L.C. (1998) Some advances in integrative muscle physiology. *Comp. Biochem. Physiol. B*, 120: 51–72.

Rome, L.C. and Lindstedt, S.L. (1997) Mechanical and metabolic design of the muscular system in vertebrates. In: *The Handbook of Comparative Physiology*, Dantzler, W. (Ed.), Oxford University Press, Oxford, pp. 1587–1652.

Russell, B., Motlagh, D. and Ashley, W.W. (2000) Form follows function: how muscle shape is regulated by work. *J. Appl. Physiol.* 88:1127–1132.

Savelberg, H.H.C.M. and Meijer, K. (2003) Contribution of mono- and biarticular muscles to extending knee joint moments in runners and cyclists. *J. Appl. Physiol* 94:2241–2248.

Shah, S.B., Peters, D., Jordan, K.A., Milner, D.J., Friden, J., Capetanaki, Y. and Lieber, R.L. (2001) Sarcomere number regulation maintained after immobilization in desmin-null mouse skeletal muscle. *J. Exp. Biol.* 204:1703–1710.

Taylor, G.M. (2000) Maximum force production: why are crabs so strong? *Proc. R. Soc. Lond. B* 267: 1475–1480.

Trotter, J.A. and Purslow, P.P. (1992) Functional morphology of the endomysium in series-fibered muscles. *J. Morphol.* 212:109–122.

Tskhovrebova, L. and Trinnick, J. (2002) Role of titine in vertebrate striated muscle. *Phil. Trans. R. Soc. Lond. B.* 357:199–206.

van Leeuwen, J.L. (1991) Optimum power output and structural design of sarcomeres. *J. Theor. Biol.* 149: 229–256.

van Soest, A.J. and Bobbert, M.F. (1993) The contribution of muscle properties in the control of explosive movements. *Biol. Cyber.* 69(3):195–204.

Vincent, J. (2004) Life among the formulae of physics. *Science* 304:520.

Vogel, S. (1998) *Cats' Paws and Catapults*, Norton, New York, New York, p. 382.

Vogel, S. (2001) *Prime Mover; A Natural History of Muscle*, Norton, New York, New York.

Vogel, S. (2002) *Comparative Biomechanics. Life's Physical World*, Princeton University Press, Princeton and Oxford. Ch. 23.

Wagner, H. and Blickhan, R. (1999) Stabilizing function of skeletal muscles: an analytical investigation. *J. Theor. Biol.* 199:163–179.

Wakeling, J.M. (2004) Motor units are recruited in a task-dependent fashion during locomotion. *J. Exp. Biol.* 207(Pt 22):3883–3890.

Walker, S.M. and Schrodt, G.R. (1973) Segment lengths and thin filament periods in skeletal muscle fibers of the rhesus monkey and humans. *Anat. Rec.* 178:63–82.

Wax, S.G. and Sands, R.R. (1999) Electroactive polymers and devices. In: *Smart Structures and Materials: Electroactive Polymer Actuators and devices*, Y. Bar-Cohen (Ed.), *Proceedings of SPIE, Vol. 3669*, pp. 2–10.

Williams, P.E. and Goldspink, G. (1973) The effect of immobilization on the longitudinal growth of striated muscle fibers. *J. Anat.* 116:45–55.

Wilson, A.M., McGuigan, M.P., Su, A. and van den Bogert, A.J. (2001) Horses damp the spring in their step. *Nature* 414:895–899.

Wool, R.P. (2001) Polymer science: a material fix. *Nature* 409:773–774.

Zhang, S. (2003) Fabrication of novel biomaterials through molecular self-assembly. *Nat. Biotechnol.* 21:1171–1178.

3

Mechanization of Cognition

Robert Hecht-Nielsen

CONTENTS

3.1   INTRODUCTION

This chapter describes the state of the art in creating animal cognition in machines. It begins with a discussion of the two fundamental processes of cognitive knowledge acquisition — *training* and *education*. The subsequent sections then present some ideas for building key components of cognition (language, sound, and vision). The main point of this chapter is to illustrate how we can now proceed towards the mechanization of key elements of cognition. This chapter assumes that the reader is familiar with the concepts, terminology, and mathematics of elementary *confabulation* (as described in Hecht-Nielsen, 2005) and its hypothesized biological implementation in the human cerebral cortex and thalamus (as described in the Appendix of this chapter).

3.1.1   Mechanized Cognition: The Most Important Piece of AI

As discussed in Section 3.A.1 of the Appendix, human (and higher mammal) intelligence involves a number of strongly interacting, but functionally distinct brain structures. Of these, significant progress has now been made on three: cerebral cortex and thalamus (the *engine* of cognition — and the focus of this chapter), basal ganglia (the *behavioral manager* of the brain — which manages action evaluation, action selection, and skill learning), and cerebellum (the *autopilot* of the brain — which implements detailed control of routine movement and thought processes with little or no need for ongoing cognitive involvement once a process has been launched and until it needs to be terminated). There are a number of other, smaller-scale, brain functions that are also critical for intelligence (e.g., ongoing drive and goal state determination by the limbic system), but these will not be discussed here.

Of all of the components of intelligence, cognition is, by far, the most important. It is also the one that has, until now, completely resisted explanation. This chapter provides the first sketch of how cognition can be mechanized. The approach is based upon the author's theory of vertebrate cognition, which is described in the chapter's Appendix. This chapter is not a historical description of "how cognition was mechanized"; but is instead an "initial plan for mechanizing cognition." Initial progress in implementing this plan in areas such as language and hearing (the subjects of Sections 3.3 and 3.4) has been encouraging.

3.1.2   Lexicon Capabilities

This chapter considers some more sophisticated variants of confabulation that go beyond elementary confabulation. Each lexicon used in our (technological) *cognitive architectures* (collections of lexicons and knowledge bases) will be assumed to possess the machinery for carrying out each of these confabulation variants (or information processing *effects* — the term that will be used for them here), as described below. Thus, from now on, the term *lexicon* implies a capability for implementing a finite set of symbols, maintaining a list of the excitation states of those symbols, and for executing the effects defined below. For the moment, lexicon dynamics will be ignored. (However, in later sections, concepts such as *consensus building* and *symbol interpolation*, which intrinsically require lexicon dynamical behavior, will be briefly mentioned.)

One very important detail that was not discussed in Hecht-Nielsen (2005), and only briefly discussed in the Appendix (because it is not relevant to the biological implementation of elementary confabulation), is that a lexicon can have multiple highly excited symbols at the end of a confabulation (as opposed to just one active symbol or the null symbol). When this occurs, the neurons representing these symbols will be excited at various (high) levels for different symbols.

When such multiple highly excited but not active symbols are used as "assumed facts" transmitting through a knowledge base, their effects on symbols to which they link via this knowledge base will essentially be the product of their excitation times, the link strength. In practice, such multi-symbol "assumed facts" are very important, as they are the key ingredients in *consensus building* (dynamically interacting confabulations taking place contemporaneously in multiple lexicons), which is the dominant mode of use of confabulation in human cognition. However, to keep this chapter at an elementary and introductory level, the mathematics of multiple-symbol "assumed fact sets" will not be discussed in detail. As needed, the qualitative properties of this mode of confabulation will be discussed, which will be sufficient for this introduction.

For the technological purposes of this chapter, confabulation will be taken to be dependent upon the *input excitation* sum $I(\lambda)$ of symbol λ, which is redefined (from the Appendix) to be:

$$\begin{aligned} I(\lambda) &\equiv [\ln{(p(\alpha|\lambda)/p_0)} + B] + [\ln{(p(\beta|\lambda)/p_0)} + B] + [\ln{(p(\gamma|\lambda)/p_0)} + B] \\ &\quad + [\ln{(p(\delta|\lambda)/p_0)} + B] \\ &= \ln{[p(\alpha|\lambda) \cdot p(\beta|\lambda) \cdot p(\gamma|\lambda) \cdot p(\delta|\lambda)]} - 4\ln{(p_0)} + 4B, \end{aligned} \tag{3.1}$$

where ln is the natural logarithm function, B is a positive global constant called the *bandgap* (a term coined by my colleague Robert W. Means), and p_0 is the smallest meaningful $p(\psi|\lambda)$ value. Clearly, for a symbol λ receiving N knowledge links, the value of $I(\lambda)$ ranges over the numerical *interval* from NB to $N[\ln{(1/p_0)} + B]$. It will be assumed that the constant B is selected such that for $N = 1, 2, \ldots, N_{max}$ none of these intervals ever overlap. For example, if we take $p_0 = 0.0005$ and $N_{max} = 10$, then we can select $B = 100$. The intervals upon which $I(\lambda)$ can lie are then given by $[100,107.6], [200,215.2], \ldots, [1000,1076.0]$ for $N = 1, 2, \ldots, N_{max}$, respectively. The utility of this definition will be seen immediately below. Given these preliminaries, we can now discuss variants of confabulation.

The first effect considered is *erasing*, denoted by **E**. Erasing clears the current record of excitation states of the lexicon and prepares the lexicon for a new use. For example, before a lexicon is used as the answer lexicon of a confabulation operation, it must be erased.

Elementary confabulation (as described in Hecht-Nielsen, 2005), denoted by **W**, is carried out by *activating* a single symbol ε with the highest value of $I(\lambda)$ (ties are broken randomly). By activation it is meant that the *final excitation level* $I(\varepsilon)$ of that symbol is set to 1 and the final excitation levels of all other symbols are set to zero. There is also the effect **WK**, which is the same as **W** with the added requirement that the single winning symbol, if there is to be one, must have had at least **K** knowledge link inputs (i.e., the winning symbol must have its input intensity in the **K**th, or higher, $I(\lambda)$ interval). The primary form of confabulation discussed in Hecht-Nielsen (2005) was **W4**.

The effect **CK** (*confabulation conclusions having* **K** *or more knowledge link inputs*), which will be needed for discussions, first zeros the excitation sum $I(\theta)$ of each symbol θ whose $I(\theta)$ is not in the **K**th (or higher) $I(\lambda)$ interval(s) occupied by symbols of the lexicon. The excitation levels of all the symbols are then summed. Finally, each remaining nonzero symbol excitation is then divided by this sum to yield its *final excitation level*. For example, in the above example with $p_0 = 0.0005$ and $N_{max} = 10$, and $B = 100$, if the four highest symbol input excitation sums are 346.8, 304.9, 225.0, and 146.8, then a **C2** will yield only the top three symbols, with final excitation levels of 0.395, 0.3478, and 0.2566, respectively. Clearly, this effect yields the set of confabulation conclusion symbols that had **K** or more knowledge link inputs. Normalizing the sum of the "significant symbol's" excitation levels to 1.0 corresponds to the notions of "activation" and "high excitation" in cortex. It also induces what might be thought of as a probability distribution on the expectation symbols. Cognition must very often conduct a multi-stage process of gradually promoting hypotheses (expectation symbols) which gain significant support from incoming knowledge links and demoting those which fail to gain as much support. Thus, the effect **CK** is very important in cognition. In the brain, "**CK** processing" is continuous in time and happens very rapidly.

The set of symbols of a lexicon having nonzero excitation levels I(λ) following a **W, W**K, or **CK** effect is termed an *expectation*. Expectations are considered to have a short life (i.e., after a "short" time has elapsed after a confabulation the *state* of the lexicon, its collection of I(λ) values, becomes indeterminate). Note that this is a generalization of the term *expectation* used in Hecht-Nielsen (2005). The term *active* is still reserved for the case of a single confabulation conclusion; and *highly excited* will still mean that the expectation has multiple elements.

Another effect is *freezing*, denoted by the letter **F**. Freezing a lexicon causes each symbol with positive final excitation (i.e., after a **W, W**K or **CK**) to have its final excitation I(λ) value preserved for a longer time. During this (still rather brief) period of time that follows **F, only** those symbols which are members of this expectation can receive further knowledge link inputs. In other words, the input excitations of symbols not in the expectation stay at zero during the frozen period. So, for example, if further new link inputs arrive shortly after an **F** has been invoked, and then a **W** is commanded, an expectation symbol (if there are any) which obtains the highest positive I(λ) value will be made active.

As we will see later, building and using expectations is one of the most important elements of cognitive information processing. By using sequences of confabulations to "whittle down" expectations, *constraint knowledge* of various kinds can be applied to rapidly home in on a final conclusion. In effect, each expectation represents the set of all "reasonable conclusions" that are worth considering further. When the expectation is finally reduced to one conclusion, via successive freezes and confabulations, the final, *decisive* conclusion is found (or if the final expectation is empty, then the answer is "I don't know"). Almost every aspect of cognition is implemented by such sequences of such "deductive" confabulation steps (although this is **not** deduction in any formal sense, because it is based on the undecidable (but usually reasonable) assumption of exhaustive knowledge).

Finally, consider a lexicon which, when last used for confabulation (within the past few hours), yielded a decisive conclusion and which, subsequently, has not been erased. If this lexicon now receives a **W, W**K, or **CK** but no knowledge link inputs, that symbol which was its last conclusion will, in isolation, become active. This is a sort of temporary symbol storage mechanism that the theory terms *working memory*. If the lexicon has been erased, an expectation containing some of the symbols which resulted from recent past uses will be expressed (with the chances of appearing depending upon how many times that symbol was frozen in succession when it originally appeared).

3.1.3 Discussion

Technological cognition will be inherently limited without the other functionalities that brains provide (see Appendix Section A.1). Further limitations arise because of the lack of on-line memory formation mechanisms (short-term, medium-term, and long-term memory processes) and the lack of a capability for goal-driven delayed reinforcement learning of thought and movement procedures (see Section 3.6). Yet, despite these limitations, there are probably many high-value early applications of *pure cognition* that will be possible. Pure cognition is the focus of this chapter.

Language is almost surely the faculty which accounts for the dramatic increase in human mental capability in comparison with all other animals. It is in the language faculty, and in the language faculty's interfaces with the other cognitive faculties, that almost all distinctly human knowledge is centered. Thus, language is where the mechanization of cognition must start (see Section 3.3).

However, before discussing language cognition, the next section discusses the currently available general methods of antecedent support knowledge acquisition: *training* and *education*.

3.2 TRAINING AND EDUCATION

As discussed above, current confabulation technology is limited to development of knowledge using some externally guided process; not via dynamic, autonomous goal and drive satisfaction-driven memory formation, as in brains. This section discusses the two main processes currently used in knowledge development: training and education. When dynamic memory formation eventually arrives, training and education will still be important learning processes (but no longer the only ones).

3.2.1 Training

Training is a knowledge acquisition process that is carried out in a batch mode without any significant active supervision or conditional intervention. It is a learning mode that can only be applied when the data set to be used has been carefully prepared. For example, in learning proper English language structure it is possible to take a huge (multi-gigaword) proper text corpus and train knowledge bases between lexicons representing the words in English (e.g., Hecht-Nielsen, 2005 presented an example of this). The corpus used must be near-perfect. It must be purged of words, punctuation, and characters that are not within the selected word list and must not have any strange annotation text, embedded tables, or markup headers that will be inadvertently used for learning. Achieving this level of *cleanliness* in a huge training corpus which, necessarily — for diversity, is drawn from many sources, is expensive and time consuming.

Once a suitably clean text corpus has been created, each sentence is considered as a whole item (up to a chosen maximum allowed number of words — e.g., 20 — after which the sentence is simply truncated). The confabulation architecture to be trained has as many word lexicons (in a linear sequence) as the maximum number of allowed words in a sentence. The words of the sentence are represented by active symbols on the corresponding lexicons of the architecture (see Section 3.3 for more details). Co-occurrence counts are then recorded for each *causal* pair of symbols (i.e., between each symbol and each of the symbols on lexicons further down the temporal sequence of lexicons). Once these counts are recorded, the process moves on to the next sentence of the training corpus.

A beautiful thing about training is that the result is knowledge that presumably has the same origin and legal standing as knowledge obtained from material that a person has read; but which they do not remember in detail. Namely, this knowledge is presumably not subject to source copyright restrictions or other source intellectual property restrictions. Use of raw data for training probably falls under the category of "fair use," which eliminates any need to pay royalties. Confabulation-based systems may thus be able to absorb whole libraries of knowledge without cost. This is fair use because the content of the work is not stored and cannot be recalled. (How much does your library charge you in royalties for reading a book? Answer: Absolutely nothing, because reading a library book is fair use.) This fortuitous loophole may allow cognitive machines to rapidly and efficiently accumulate almost all human knowledge; without having to pay any royalties and without the delays associated with working through legal and bureaucratic objections. Mechanizers of cognition may want to expose their systems to the available libraries of written knowledge at the first possible opportunity; before legal innovators find ways of closing this loophole. It may not be long before intelligent machines are as unwelcome at libraries as blackjack card counters are at casinos.

In the near term, early confabulation entrepreneurs will probably use libraries, web scrapers, or informally obtained e-mail message examples (for text knowledge), informal public volunteer web portals for conversational data (for sound knowledge), public location video (for vision knowledge), and multi-camera video of moving humans with colored dots pasted to their bodies (for motor knowledge). Paying for training data will probably not be feasible for most confabulation startup companies.

The above comments also raise the technical legal question of whether the knowledge in confabulation-based systems can itself be copyrighted (this would seem reasonable); or must it be protected as a trade secret? Methods of training and education can probably be patented. The legal implications and ramifications of confabulation are clearly going to be complicated and probably contentious. An overriding consideration should be the irreplaceable value of the work output that intelligent machines (which can potentially produce prodigiously, but not consume significantly) will quickly add to the world economic product. It will be fun to watch this saga unfold in the courts and in diplomacy over the coming decades.

Knowledge created by training is limited to situations such as that considered above; namely, where extensive, highly conditioned and prepared, data sets exist. In more general situations, online, active, expert human supervision must be employed to carefully select meaningful symbol co-occurrences for use in learning. Such a carefully sequenced program of sophisticated and controlled exposure of the machine to meaningful examples is termed *education*; which is the subject of the next subsection.

3.2.2 Education

A critical aspect of development, particularly in higher mammals, is the limited, deliberately controlled exposure to progressively more complicated stimuli, and intelligent responses thereto, that characterizes the early phases of an animal's life (which in cats, might occupy a few weeks; whereas in humans it occupies tens of years — which is often not enough!). During this development period, the sequence of exposure of the animal to information is in some manner controlled (often by confining the animal to a particular limited range, such as a nest, home, or school and its immediate surround).

For example, a human baby learning to see has eyes that are physically incapable of focusing much beyond its reach. Thus, most visual stimuli are the baby's own limbs or individual objects that the baby itself is holding and manipulating. During this period, the visual system develops its ability to segment individual objects in single views and also develops higher-level visual lexicons containing symbols that are pose-insensitive (see Section 3.5). Knowledge related to the integration of form, color, texture, and internal object motion is also developed during this initial phase. In order for this phase to properly complete, the baby must have spent a large amount of time holding and viewing a reasonably rich collection of objects.

Once the initial phase of human visual development is completed, the baby begins to acquire distant vision and begins to learn about a much richer visual environment. Again, parental provision of appropriate stimuli and response examples during this period is critical. Persons who are deprived of visual input during these early phases (e.g., due to disease that temporarily impairs visual function) are never able to complete their visual development, even if their visual input is restored at some later point. Such persons can respond to light in some limited ways, but can never see. Some persons with restored sight actually voluntarily limit their exposure to visual input (Gregory, 2004).

As with the initial stage of visual development, the most important source of educational input in the later stages of visual development is the children themselves. By holding an object and examining it (e.g., in an exploration of its function or component parts), knowledge in the visual domain, as well as in the linkage of vision to the language (and other) faculties, is expanded. Unlike intellectual knowledge (which is subject to various distortions such as philosophical or ideological brainwashing), visual knowledge is "safe" to rapidly gather and store because it is essentially never erroneous (except in cases where optical distortions exist — which when corrected too late in the development process, often cause a permanent reduction in visual capability). Parents often endlessly admonish their children not to handle everything they fancy in stores; yet, this is probably enriching. Perhaps the admonishment should be to take care not to soil or damage what they handle. If you are punctilious in this regard, have your children wear disposable latex gloves and force them to pay for any damage or breakage out of their allowance. But give them these valuable experiences.

Even more than basic sensory processing, learning to carry out important behavioral tasks requires deliberate provision of examples and supervised rehearsal practice. This often includes feedback on performance; something that will be ignored here since using such feedback requires noncognitive functions, which as yet, we do not understand sufficiently to build. Because of this current lack of a reinforcement learning adjunct to cognition (work is proceeding in this area — see Miyamoto et al., 2004, for example), for the moment, education of confabulation-based cognitive systems will probably be confined to strictly positive examples. In other words, examples, where learning should definitely take place.

For example, consider a confabulation-based vision system viewing cars passing by on a busy road. The visual portion of the system segments each car it fixates on (see Section 3.5) and then rerepresents its visual form, color, and internal motion using high-level symbols that have invariance properties (e.g., pose insensitivity). Thus, the final product of processing one such *look* is activation of a set of high-level symbols, each describing one visual *attribute* of the object.

Imagine that a human educator sitting at a computer screen where each *look* (eyeball snapshot image — see Section 3.5) to be processed by the confabulation-based vision system is being displayed (each subsequent look is processed only after the previous look's use for education has been completed). The human examines the visual object upon which the center (fixation point) of the eyeball image rests and describes it in terms of English phrases (spoken into a noise-canceling microphone connected to an accurate speech transcriber — see Section 3.4). For example, if the object is a green Toyota Tundra truck with a double cab; the educator might speak: "Toyota Tundra truck," "dark green," "two rows of seats; in other words, a full-sized back seat," "driving in the left lane of traffic." After accurate transcription, this text is represented by a set of active symbols in the language module (see Section 3.3). Knowledge links are then established between the active visual symbols representing the visual content of the look and the active language symbols representing the education-supplied language content of the look.

Note that the language description is not exhaustive; it is just a sample of descriptive terms for the visual object. For example, if a similar look of the same truck were presented on another occasion, the educator might add: "oh, and there are four dogs in the bed of the truck." This would add further links.

3.2.3 Discussion

One of the most exasperating things about this theory is that it seems impossible that just forming links between symbols and then using these links to approximately maximize cogency could ever yield anything resembling human cognition. The theory appears to be nothing but a giant mountain of wishful thinking!!! That such a simple construction can do all of cognition is indeed astounding. Yet, that is precisely my claim. Some reasons why confabulation may well be able to completely explain cognition are now discussed.

First is the fact that the number of links that get established (i.e., the number of individual items of knowledge that are employed) is enormous. Even in the narrow domain of single proper English text sentences (see Section 3.3), over a billion individual knowledge items are often employed (contrast this with the world's largest rule bases; which have about 2 million items of knowledge). Slightly more elaborate proper English text confabulation systems (able to deal with two successive sentences — see Section 3.3) often possess multiple billions of items of knowledge.

The value of having such huge quantities of such a simple form of knowledge is best seen in terms of how this knowledge is used in confabulation. The first use of knowledge is to excite those conclusion symbols which strongly support the truth of the set of assumed facts being considered (this is the *duck test*, as described in Hecht-Nielsen, 2005).

But an even more important aspect of cognition is the underlying assumption that the knowledge we possess is *exhaustive* (see Hecht-Nielsen, 2005). In other words, once the available knowledge has been used, we can be reasonably sure that no other possible conclusions exist. This effectively causes the known knowledge to act as an implied constraint. In particular, those possible conclusions identified as known to be supportive of the assumed facts are not just viable alternatives; they are probably the **only** viable alternatives. Thus, in non-Aristotelian information environments, the exhaustive knowledge assumption leads to answers which are "almost logical deductions." Thus, beyond just being an implementation of the duck test, confabulation might be termed a "strong" (although not logically rigorous) form of inductive reasoning.

Another factor that makes confabulation so powerful is its ability to support the construction and use of lexicon hierarchies. In the simplest case, the symbols of a higher-level lexicon each represent an ordered set of symbols that meaningfully co-occur on lower-level lexicons. But much more is possible. For example, in vision (see Section 3.5), higher-level symbols each represent several groups of lower-level symbols. These are symbol groups that are seen in successive "eyeball snapshots" of the same object at the same fixation point (but at slightly varied object poses). In this way, these higher-level symbols respond to the appearance of a localized portion of an object at a number of different poses. They are *pose-insensitive localized visual appearance descriptors*.

In language hierarchies, knowledge can be used to discern symbols which are highly similar in meaning and usage (in a particular given context) to a particular symbol. These are termed *semantically replaceable elements* (SREs). Knowledge possessed about an SRE of a symbol can sometimes be used to augment knowledge possessed about that base symbol. This can significantly extend the "conceptual reach" of a system; without requiring training material covering all possible combinations of all symbols. For example, what if a friend tells you about the food "guyap" that they had for breakfast. They poured the flakes of guyap from its cardboard box into a bowl; added milk and sweetener, and then ate it with a spoon. It was good. By now, you are fairly sure that "guyap" is a breakfast cereal of some kind, and at least in the "breakfast food" context, you can apply your knowledge about breakfast cereal to "guyap."

Hierarchies can work backwards too. For example, if you say you are looking for a ruler on your desk; then links from the word **ruler** (and perhaps some of its SREs) go to the visual system and provide input to high-level visual attribute ("holistic") representation symbols which, in the past, have meaningfully co-occurred with visual sightings of rulers. This is accomplished via a **CKF** effect; which leaves an expectation of all such symbols that have previously been significantly linked to the word **ruler**. During *perception*, which takes place immediately after this expectation symbol set has been generated, knowledge links from primary, and then secondary visual lexicons arrive at this high-level visual lexicon and a **W, CK**, or **WN** is issued at the same time. Only elements of the expectation can be activated by the visual input; and the net result is a set of symbols that are consistent with both the word **ruler** and with the current visual input. As discussed further in Section 3.5, the final step is a rapid bidirectional knowledge link interaction of the higher-level expected symbols with those of the lower-level lexicons to shut off any symbols that are not participating in "feeding" (i.e., are not consistent with) symbols of the high-level expectation. This, in effect, causes low-level symbols representing portions of the visual input that are not part of the ruler to be shut off. It is by this *visual object segmentation* mechanism that sensory objects are almost instantly isolated so that they can be analyzed without interference from surrounding objects (segmentation is also a key part of sound and somatosensory processing). Without an expectation, sensory processing cannot proceed (this point, which seems to be widely unappreciated in the biologically oriented neuroscience disciplines, is the subject of a wonderful book describing clever experiments that well illustrate this point [Mack and Rock, 1998]).

Above all else, cognition works because of the huge hierarchical repertoire of learned and stored *action sequences* (programs of thought and/or movement — see Section 3.6). Appropriate actions or action sequences are triggered instantly each time a lexicon confabulation operation yields a single active symbol (i.e., a *decisive conclusion*). As discussed in Appendix, this is the *conclusion–action principle* of the theory. The action(s) automatically triggered by the winning symbol can be of many characters. They can be immediate *postural goal* outputs that are sent down the spinal cord to motor nuclei and the cerebellum, they can be immediate *lexicon operation* commands, they can be immediate *knowledge base operation* commands, or they can be *candidate actions* (cortically proposed thoughts and movements) which must first be sent to the basal ganglia for evaluation and approval before they are executed.

The main advantage of confabulation-based cognition over traditional *programmed computing* (formal computer programs, rule-based systems, etc.) is a much greater capacity for handling novel arrangements of individually familiar objects. Programmed computing must essentially have a predefined plan for dealing with every situation that is to be handled. For example, a plan for breaking up a complicated ensemble of problems into isolated, disconnected sub-problems, so that each can be handled in a predefined way. Unfortunately, in most real-world situations, this approach fails badly because complicated real-world situations inevitably have unanticipatable interrelations between their elements that disallow pre-defined decompositions. By virtue of their huge stores of general-purpose and low-level knowledge, confabulation-based systems are inherently able to take novel external context into account as each individual conclusion (or ensemble of conclusions — if mutual solution constraints are to be honored — see the discussion of *consensus building* in Section 3.3) is addressed. Confabulation-based systems can also adapt existing action plans (e.g., by replacing specific elements of a stored plan with similar substitutions which are relevant to the current situation) to fit novel circumstances. They do not typically run out of things to try and, instead, tend to press on and do the best they can, given what they know. If a particular approach yields no conclusion, other approaches are typically immediately launched. Yet, because actions are triggered each time a conclusion is reached, almost all behavioral sequences are dramatically novel. Also, each new experience can (with occasional help from a human educator) be added to the knowledge base to further enlarge the system's future repertoire.

More could be said regarding the benefits of the confabulation approach. However, the remaining sections of this chapter present more concrete examples of this. The nature of cognition is very different from that of computing. So much depends upon designing clever architectures of lexicons and knowledge bases and upon using clever, highly threaded, but very simple, thought processes to control these architectures. Since the information processing control which must be exerted at each stage of an action process is triggered by the current *cognitive world state* (the collection of all decisive confabulation conclusions that are active, or accessible from working memory, at that moment), cognition has no need for "computer programs" or "software." In effect, the conclusion of each "cognitive microprogram" (lowest level action sequence) is a GOTO statement. There is no overall program flow defined. Just action sequences completing and then triggering subsequent action sequences (in a pattern that almost never exactly repeats). Things happen as they happen, with no master program controller involved or needed (although a number of subcortical brain structures can execute "interrupts" when certain conditions occur). This brain operating system (or lack thereof, depending on your point of view) seems like an invitation to disaster. However, beyond possible conflicting commands to the same action (movement or thought) resource (which are impossible by design! — see Appendix Section 3.A.2), very little can go wrong.

3.3 LANGUAGE COGNITION

This section discusses the use of confabulation for representing and generating language. This application arena is the most developed, and yet is transparently crude and primitive. An enormous amount of work needs to be done in language. The hope of this section is to illustrate how promising this research direction is.

3.3.1 Phrase Completion and Sentence Continuation

This discussion of language cognition begins with consideration of a class of *confabulation architectures* for dealing with single English sentences. These architectures address the problems of *phrase completion* and *sentence continuation*; simple subcases of *language generation*. This subsection expands upon on the brief introduction to phrase completion provided in Hecht-Nielsen (2005). These architectures provide a good introduction to the "look and feel" of cognitive information processing — which is completely different than the familiar computer paradigm.

Figure 3.1 illustrates a confabulation architecture for phrase completion and sentence continuation in a single sentence of up to 20 words. Each lexicon has about 63,000 symbols; including symbols for the 63,000 most common words in English (as reflected in the training corpus) and eight punctuations (period, comma, semicolon, etc.), which are treated as separate words. Capital letters are used when they appear in words in the training corpus selected for representation within the word lexicons (i.e., mark and Mark are different words with different symbols). Thus, many of the words in the lexicon are represented twice — once capitalized and once not; some have even more than two representations, e.g., EXIT, Exit, and exit; and some, such as e.g., and the punctuations are never capitalized and only have one representation.

Once a suitably "clean" huge proper English text training corpus (typically containing billions of words) has been created, each successive sentence in the corpus is *entered*, in sequence, into the architecture of Figure 3.1. The first word of the sentence is entered into the leftmost lexicon (i.e., the symbol representing this word is made active) and the remaining words of the sentence (or punctuations — which, again, are treated as separate words) are entered successively until the ending period. If the sentence has more than 20 words, those words beyond the first 20 are discarded. Because of the positioning of the words of each sentence in order, this architecture is termed *position-dependent*.

It is also possible to use hierarchical *ring* architectures for representing strings of words; which I believe is probably how the human cortical language architecture is organized. As the words are loaded into the ring of lexicons, they are quickly removed in groups (phrases) and re-represented in lexicons at a higher conceptual level — leaving the lower-level lexicons free for capturing additional words. I believe that this is why humans can only instantly remember "about 7 things ± 2" (Miller, 1956) — we physically only have about seven lexicons at the word level. When required to remember a sequence of things, we repeatedly rehearse the sequence (to firmly store it in short-term memory) by traversing the ring from the beginning lexicon (which is always the same one for each sentence or word sequence) to the last item and then back to the beginning. However, given the lack of limitations of computer implementations of confabulation architectures (at least conceptually), there is no need for us to use these more complicated ring architectures for this chapter's introductory discussion.

The knowledge bases of the architecture of Figure 3.1 are all *causal*; meaning that the symbols of each lexicon are only linked to symbols of later lexicons (i.e., those that lie to the right of it);

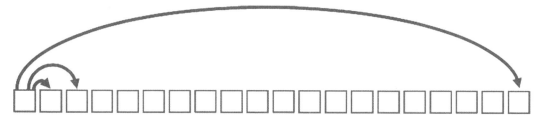

Figure 3.1 Naïve single-sentence confabulation architecture for proper English phrase completion or sentence continuation. Knowledge bases link each of the first 19 of the 20 lexicons to all of the lexicons to their right. Sentences are represented with the first word in the first lexicon on the left; and so on in sequence. This architecture has a total of $19 + 18 + \cdots + 1 = 190$ knowledge bases.

which represent words which occur later in the temporal sequence of the word string. The first (i.e., leftmost) lexicon is connected to all of the 19 lexicons which follow it by 19 individual knowledge bases. The second lexicon to the 18 lexicons to its right, and so forth. Thus, this architecture has a total of 20 lexicons and 190 knowledge bases.

The training process starts with the first sentence of the training corpus and marches one sentence at a time to the last sentence. As each sentence is encountered, it is entered into the architecture of Figure 3.1 (unless its first 20 words include a word not among the 63,000; in which case, for this introduction, the sentence is assumed to be skipped) and used for *training*. The details of training are now discussed.

At the beginning of training, one hundred and ninety $63,000 \times 63,000$ single precision float matrices are created (one for each knowledge base) and all of their entries are set to zero. In each knowledge base's matrix, each row corresponds to a unique source lexicon symbol and each column corresponds to a unique target lexicon symbol. The indices of the symbols of each lexicon are arbitrary, but once set, they are frozen forever. These matrices are used initially, during training on the text corpus, to store the (integer) co-occurrence counts for the (causally) ordered symbol pairs of each knowledge base. Then, once these counts are accumulated, the matrices are used to calculate and store the (floating point) $p(\psi|\lambda)$ antecedent support probabilities. In practice, various computer science storage schemes for sparse matrices are used (in both RAM and on hard disk) to keep the total memory cost low.

Given a training sentence, it is entered into the lexicons of the architecture by activating the symbol representing each word or punctuation of the sentence, in order. Unused trailing lexicons are left blank (*null*). Then, each causal symbol pair is recorded in the matrix of the corresponding knowledge base by incrementing the numeric entry for that particular source symbol (the index of which determines the row of the entry) and target symbol (the index of which determines the column of the entry) pair by one.

After, all of the many tens of millions of sentences of the training corpus have been used ("read") for training (i.e., the entire training corpus has been traversed from the first sentence to the last); the entries (ordered symbol pair *co-occurrence counts*) in each knowledge base's matrix are then used to create the knowledge links of that knowledge base.

Given a knowledge base matrix, what we have traditionally done is to first set to zero any counts which are below some fixed threshold (e.g., in some experiments three, and in others 25 or even 50). In effect, such low counts are thereby deemed random and *not meaningful*. Then, after these low-frequency co-occurrences have been set to zero, we use the "column sum" of each count matrix to determine the *appearance count* $c(\lambda)$ of each target symbol λ for a particular knowledge base. Specifically, if the count of co-occurrences of source symbol ψ with target symbol λ is $c(\psi,\lambda)$ (i.e., the matrix entry in row ψ and column λ), then we set $c(\lambda)$ equal to the *column sum* of the quantities $c(\phi,\lambda)$ over all source lexicon symbols ϕ. Finally, the knowledge link probability $p(\psi|\lambda)$ is set equal to $c(\psi,\lambda)/c(\lambda)$, which approximates the ratio $p(\psi\lambda)/p(\lambda)$, which by Bayes' law is equal to $p(\psi|\lambda)$.

Note that the values of $c(\psi,\lambda)$, $c(\lambda)$ and $p(\psi|\lambda)$ for the same two symbols can differ significantly for different pairs of source and target lexicons within the sentence. This is because the appearances of particular words at various positions within a sentence differ greatly. For example, essentially no sentences begin with the uncapitalized word **and**. Thus, the value of $c(\psi,\lambda)$ will be zero for every knowledge base matrix with the first lexicon as its source region and the symbol $\psi = $ **and** as the source symbol. However, for many other pairs of lexicons and target symbols, this value will be large. (A technical point: these disparities are greatest at the early words of a sentence. At later positions in a sentence, the $p(\psi|\lambda)$ values tend to be very much the same for the same displacement between the lexicons — probably the underlying reason why language can be handled well by a ring architecture.)

After the $p(\psi|\lambda)$ knowledge, link probabilities have been created for all 190 knowledge bases using the above procedure, we have then traditionally set any of these quantities which are below some small value (e.g., in some experiments 0.0001, in others 0.0002, or even 0.0005) to zero; on the basis that such weak links reflect random and meaningless symbol co-occurrences. It is important to state that this policy (and the policy of zeroing out co-occurrence counts below some set number) is arbitrary and definitely subject to refinement (e.g., in the case of high-frequency target symbols, we sometimes accept values below 0.0001 because these low-probability links can still be quite meaningful). The final result of this *training process* is the formation of 190 knowledge bases, each containing an average of a million or so individual items of knowledge.

Given this architecture, with its 20 lexicons and 190 knowledge bases, we can now consider some thought processes using it. The simplest is *phrase completion*. First, we take a coherent, meaningful, contiguous string of fewer than 20 words, and represent them on the lexicons of the architecture; beginning with the first lexicon. The goal is to use these words as context for selecting the next word in the string (which might be a punctuation; since these are represented in each of the lexicons). To be concrete, consider a situation where the first three words of a sentence are provided.

The three words are considered to be *assumed facts* (Hecht-Nielsen, 2005). They must be coherent and "make sense," else the confabulation process will yield no answers. To find the phrase completion, we use the knowledge bases from the first, second, and third lexicons to the fourth. The completion is obtained by carrying out confabulation on the fourth lexicon using a **W**3. The answer, if there is one, is then the symbol expressed on the fourth lexicon after confabulation.

With only three words of context (e.g., **The only acceptable**), the answer that is obtained will often be one of a huge number of viable possibilities (**alternative, person, solution, flight, car, seasoning**, etc., etc., etc. — which can be obtained as an expectation by simply performing a **C**3). Language generation usually involves invoking longer-range or abstract *context* (expressed in some manner as a set, or multiple sets of assumed facts that act as constraints on the completions or continuations) to more precisely focus the *meaning content* of the language construction (which by the inherent nature of confabulation, is generally automatically grammatical and syntactically consistent). This context can arise from the same sentence (e.g., by supplying more or more specific words as assumed facts) or from external bodies of language (e.g., from previous sentences; as considered in Section 3.3.3).

If we supply more assumed facts or more narrowly specific assumed facts, confabulation can then supply the best answer from a much more restricted expectation. For example, **Mickey and Minnie** will yield only one answer: **mouse**.

However, using more words in phrase completion (or in *sentence continuation*; where multiple successive words are added onto a starting string) introduces some new dilemmas. In particular, beyond a range of two or three words, the string of words that emerges is likely to be novel in the sense that some of the early assumed facts may not have knowledge links to distant, newly selected words in the word string. The design of confabulation architectures and thought processes to handle this common situation is a key problem that my research group has solved; at least in a preliminary way. As always, there is no software involved; just proper sequences of thought actions (lexicon confabulations and knowledge base enablements) that are invoked by the conclusions of previous confabulations.

For example, consider the assumed facts **The canoe trip was going smoothly when all of a sudden**. Such partial sentences will almost certainly not have a next-word symbol that receives knowledge links from all of the preceding assumed fact symbols. So what procedure shall we use to select the next word? One answer is to simply go on the preponderance of evidence: select that 12th lexicon symbol that has the highest input intensity among those symbols which have the maximum available number of knowledge links. This is accomplished by **W**. This approach can yield acceptable answers some of the time; but it does not work as well as one would like. If we were to attempt sentence continuation with this approach (i.e., adding multiple words), the results are awful.

The solution is to invoke two new confabulation architecture elements: a *language hierarchy* and *consensus building*. These are sketched next.

3.3.2 Language Hierarchies

There are many reasons why the architecture of Figure 3.1 does not solve the phrase completion and sentence continuation problems. First of all, this architecture disallows the learning and application of *standard language constructions* such as multi-word *conceptual units* (e.g., New York Stock Exchange, which we will refer to as *phrases*), variable element constructions (*VECs*, e.g., ___ went to the ___), and pendent clauses (e.g., **the success of her daughter was**, except for ordinary daily distractions, **foremost on her mind**). Standard constructions are important elements of all human languages (although, they differ and can take on different forms, in different languages) and a comprehensive architecture must include provisions for learning and representing them.

For the problems of phrase completion and sentence continuation, the architecture of Figure 3.2 is much more capable than that of Figure 3.1. For example, consider again the problem of finding the next word for the assumed fact phrase The canoe trip was going smoothly when all of a sudden. Now, the first thing that happens is that this phrase is *parsed*, meaning that the words are re-represented at the phrase level. This happens almost instantly by using the knowledge bases which proceed from the word level to the phrase level. The parsing process, which is described next, proceeds in a rapid "rippling wave" of thought processes running from the beginning of the assumed fact word string to the end.

To start, the first word, The, of the string goes up first. These links (which in accordance with the knowledge base design described in the caption of Figure 3.2) only go to the first phrase lexicon. A **C1F** on this first phrase lexicon yields an expectation consisting of those symbols which represent phrases that begin with the word The. The second word lexicon then sends links upward to the first and second phrase regions from the symbol for canoe. **C1F**s on phrase regions, one and

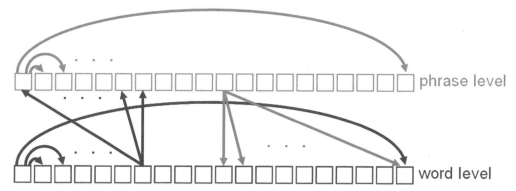

Figure 3.2 Single-sentence hierarchical confabulation architecture for proper English phrase or sentence completion. The lower row of lexicons is used to represent words, as in the architecture of Figure 3.1. Again, knowledge bases link each of the first 19 of these 20 *word-level* lexicons to all of the lexicons to their right. Positioned exactly above the word-level lexicon row is a row of 20 phrase-level lexicons. These phrase *lexicons* represent word groups (and other standard language constructions, although these will not be discussed much in this introductory chapter). Each phrase lexicon has at least 126,000 symbols (63,000 single words and punctuations and the 63,000 most common multiple word groups). Knowledge bases connect each phrase lexicon to each of the phrase lexicons which follow it. Knowledge bases also connect each phrase lexicon to all of the word lexicons except those that lie to its left. Finally, knowledge bases connect each word region with all of the phrase regions except those that lie to its right. This architecture has a total of 800 knowledge bases. On average, each knowledge base contains roughly a million individual items of knowledge. The capability of this architecture is a practical demonstration of the main premise of the author's theory of vertebrate cognition; namely, that lots of simple knowledge, along with a single, simple, information processing operation can implement all of cognition.

two then do two things: on phrase region one, only the symbol for The remains (since The canoe, nor any further extension of it, is not in the phrase lexicon — for brevity, the manner in which the phrase lexicon itself, and the additional knowledge bases of the Figure 3.2 architecture, are derived using word-level knowledge — this process too is totally confabulation-implemented and does not use any linguistic knowledge — is not described here). This parsing process continues down the phrase lexicons (each possessing 126,008 symbols), quickly yielding the parse (with the phrase symbol numbers in parentheses): The(8) canoe(25085) trip(1509) {was going}(63957) smoothly(9723) when(64) all(56) {of a sudden}(69902). Thus, phrase lexicons 1, 2, 3, 4, 6, 7, 8, and 9 have symbols active on them (each phrase is represented on the lexicon immediately above its first word). All the other phrase regions have no symbols active on them. Note that if the last word (sudden) of the assumed fact phrase were not present, that phrase lexicon 9 would not have a single phrase active on it; but would have several (representing all of the phrases that begin with of a: e.g., of a, of a kind, of a sudden, of a sort, etc.). I thank my colleague Robert W. Means for implementing and providing the details of this example.

The above processing sounds like it would take a long time. But remember that thinking is just like moving. When you throw a baseball many tens of muscles are being commanded in parallel in a precisely timed and coordinated way. The above thought process ("parse sentence") is stored, recalled, and executed just like a motor action such as throwing a baseball. The initiation of each involved knowledge base activation and confabulation happens in close succession in a "ripple" of processing that rapidly moves from the left end of the architecture to the right; terminating at the end of the assumed fact phrase. The entire parsing process is completed in just a small multiple of one knowledge base transmission time. Like some movement actions (e.g., dribbling a basketball); thought actions are often divided up into small "macro" segments which, depending upon their outcome (i.e., which symbol wins the confabulation competition), trigger alternative next-segments.

As discussed in the Appendix, a key concept of hierarchical architecture design is the *precedence principle*. There, it was discussed in the context of the constitution of individual symbols within a single lexicon. However, the same principle holds between lower and higher abstraction level lexicons within a hierarchy (such as that of Figure 3.2). In this expanded form, what the precedence principle says is that as soon as content that is represented at a lower level of a hierarchy is re-represented at a higher level, the involved active lower-level lexicon symbols must be shut off. This is implemented in human cerebral cortex by use of the *conclusion–action principle* (see Section 3.A.6).

In the case of the precedence principle, the action which is triggered by the expression of a phrase representation symbol is to shut off the lexicons which supplied words to the phrase that the symbol represents. For example, if the phrase that emerges from the parse has three words, the word lexicon directly underneath the phrase lexicon, as well as the next two lexicons to its right, are shut off (they stop expressing their word symbols). If the phrase has only one word, a different action is triggered: namely, only shutting off the lexicon directly beneath. And so on. Note that these action commands are not issued until the choices have narrowed to a single symbol; since it is only then that the conclusion–action principle operates. This is a concrete example of how thought is not software. It is a series of sets of action commands; each set being immediately *originated* (issued to action nuclei) when a firm confabulation conclusion is reached (i.e., each conclusion has its own set of action commands that are permanently associated with it, and which are originated every time that conclusion is expressed as the lone final result of a confabulation operation by its lexicon).

This example illustrates a thought process that can be launched immediately with no further evaluation (e.g., by basal ganglia). It also illustrates how we will need to implement the action command output portion of cognition from the very outset of research. A great deal more could be said about action command generation and action symbol sequence learning and recall using confabulation architectures. But this topic would take us beyond the introductory sketch being attempted in this chapter. Suffice it to say that quite a lot is known about how action sequences can be learned (by rehearsal), stored, and recalled using confabulation architectures (e.g., the UCSD graduate students in my course built a confabulation-based checker-playing system that learned to play — trigger appropriate actions — by mimicking a skilled human). Confabulation architectures for appropriately modifying action sequences, in real time, in response to changes in the world state that occur during execution (a crucial capability if we are to perform in a complicated, real-world environment) have also been developed.

Obviously, when the conclusion–action principle "branching" capability is combined with an ability to store and retrieve data (e.g., using short-term, medium-term, long-term memory or working memory), the cognitive brain passes the test of being, at least conceptually, capable of *universal computation* in the Turing sense. However, the very limited "RAM memory" or "tape memory" available for immediate reading and writing probably limits the value of this capability. Certainly, as demonstrated in Hecht-Nielsen (2005), logical reasoning in Aristotelian information environments is carried out directly by confabulation (cogency maximization); without need for any recourse to computer principles. Nonetheless, a human with a paper and pencil (to supplement the extremely limited "RAM memory" available in the brain) can easily learn to carry out thought processes that will accurately simulate operation of a computer. However, such a "human-imple-mented computer" is to a modern desktop electronic computer as a unicycle is to racecar.

Given the parse of an assumed fact phrase (say, the first few words of a sentence), we can then use the architecture of Figure 3.2 to carry out "phrase completion" (as in Hecht-Nielsen, 2005). The first step is to build an expectation in the phrase lexicon above the next-word's lexicon by activating the knowledge bases between the last active phrase lexicon of the parse and that "target" phrase lexicon and doing a **C1F**. This first step exploits the fact that adjacent phrases are usually highly coherent and it would be rare indeed for the next phrase to not receive knowledge links from the last known phrase of the parse. The result of this first step is an expectation on the target phrase lexicon containing all of the reasonable next phrases. Note that the last-phrase lexicon of the parse may itself have an expectation containing multiple symbols which themselves could contain the next word.

For example, as above, if sudden were not present in the starting word string phrase: The canoe trip was going smoothly when all of a sudden, the last-phrase lexicon would have an expectation with multiple phrases, including: all of a, all of a sort, all of a sudden, all of a kind, etc. If, for example, all of these symbols represent multi-word phrases then the target phrase lexicon expectation will automatically be empty (since none of the phrases in the last-phrase lexicon's expectation will have any knowledge links to symbols of that lexicon). If this is not clear, using Figure 3.2, work out some examples using a diagram on a piece of paper. This is a perfect example of how all thought processes are conclusion-driven.

The expectations established by the above process then send output through their knowledge links to the first unfilled word lexicon; where an expectation is formed by a **C1F**. Since this word lexicon is the next one after the last assumed fact word lexicon, we can again assume that the symbols in this expectation represent all reasonable possibilities for the next word of the continuation. Then knowledge linking the rest of the parsed phrase symbols to the word lexicon is used with a **W** to select that word symbol in the expectation which is most consistent with this additional context. Here again, there are many possible things that could go on (e.g., knowledge links may or may not exist from various phrase symbols to words of the expectation); yet, whatever the situation, this process works better than that using the architecture of Figure 3.1. A bit of time spent thinking about this phrase completion process with some concrete examples will be most illuminating and compelling. Try to build some meaningful examples where this process will not work. You won't be able to.

Why would this phrase completion process (using the architecture of Figure 3.2) be better than just using the word-level knowledge; as described earlier in connection with Figure 3.1? The answer is that knowledge links from phrases to words generally have two superior characteristics over links at the word level. First, the parse often removes a significant amount of ambiguity that can exist in word-level knowledge. For example, the word lexicon symbol for the word New will have strong links to the symbol for Stock two word lexicons later (independent of what word follows it). However, if the parse has activated the phrase New Orleans no such erroneous knowledge will be invoked. The other advantage of using the parsed representation is that the knowledge links tend to have a longer range of utility; since they represent originally extended conceptual collections that have been unitized.

If, as often occurs, we need to restore the words of a sentence to the word lexicons after a parse has occurred (and the involved word lexicons have been automatically shut off by the resulting action commands), all we need to do is to activate all the relevant downward knowledge bases and simultaneously carry out confabulation on all of the word regions. This restores the word-level representation. If it is not clear why this will work, it may be useful to consider the details of Figure 3.2 and the above description. The fact that "canned" thought processes (issued action commands), triggered by particular confabulation outcomes, can actually do the above information processing, generally without mistakes, is rather impressive.

3.3.3 Consensus Building

For sentence continuation (adding more than just one word), we must introduce yet another new concept: *consensus building*. Consensus building is simply a set of brief, but not instantaneous, temporally overlapping, mutually interacting, confabulation operations that are conducted in such a way that the outcomes of each of the involved operations are consistent with one another in terms of the knowledge possessed by the system. Consensus building is an example of *constraint satisfaction*; a classic topic introduced into neurocomputing in the early 1980s by studies of Boltzmann machines (Ackley et al., 1985).

For example, consider the problem of adding two more sensible words onto the following sentence-starting word string (or simply *starter*): The hyperactive puppy. One approach would be to simply do a **W** simultaneously on the fourth and fifth word lexicons. This might yield: The hyperactive puppy was water; because was is the strongest fourth word choice, and based upon the first three words alone, water (as in drank water) is the strongest fifth word choice. The final result does not make sense.

But what if the given three-word starter was first used to create expectations on both the fourth and fifth lexicons (e.g., using **C3F**s). These would contain all the words consistent with this set of assumed facts. Then, what if **W**'s on word lexicons four and five were carried out simultaneously with a requirement that the only symbols on five that will be considered are those which receive inputs from four. Further, the knowledge links back to phrase lexicons having unresolved expectations from word lexicons four and five, and those in the opposite directions, are used as well to incrementally enhance the excitation of symbols that are consistent. Expectation symbols which do not receive incremental enhancement have their excitation levels incrementally decreased (to keep the total excitation of each expectation constant at 1.0). This multiple, mutually interacting, confabulation process is called *consensus building*. The details of consensus building, which would take us far beyond the introductory scope of this chapter, are not discussed here.

Applying consensus building yields sensible continuations of starters. For example, the starter I was very, continues to: I was very pleased with my team's, and the starter There was little continues to: There was little disagreement about what importance. Thanks to my colleague Robert W. Means for these examples.

3.3.4 Multi-Sentence Language Units

The ability to exploit long-range context using accumulated knowledge is one of the hallmarks of human cognition (and one of the glaring missing capabilities in today's computer and AI systems). This section presents a simple example of how confabulation architectures can use long-range context and accumulated knowledge. The particular example considered is an extension of the architecture of Figure 3.2.

The confabulation architecture illustrated in Figure 3.3 allows the meaning content of a previous sentence to be brought to bear on the continuation, by consensus building following a starter (shown in green in Figure 3.3) for the second sentence. The use of this architecture, following knowledge acquisition, is illustrated in Figure 3.4 (where for simplicity, the architecture of Figure 3.3 is represented as a "purple box"). This architecture, its education, and its use are now briefly explained.

The sentence continuation architecture shown in Figure 3.3 contains two of the sentence modules of Figure 3.2; along with two new sentence meaning content *summary* lexicons (one above each sentence module). The left-hand sentence module is used to represent the context sentence, when it is present. The right-hand sentence module represents the sentence to be continued.

To prepare this architecture for use, it is educated by selecting pairs of topically coherent successive sentences, belonging to the same paragraph from a general coverage, multi-billion-word proper English text corpus. This sentence pair selection process can be done by hand by a human or using a simple computational linguistics algorithm. Before beginning education, each individual sentence module was trained in isolation on the sentences of the corpus.

During education of the architecture of Figure 3.3, each selected sentence pair (of which roughly 50 million were used in the experiment described here) is loaded into the architecture, completely parsed (including the summary lexicon), and then counts were accumulated for all ordered pairs of symbols on the summary lexicons. The long-term context knowledge base linking the first sentence

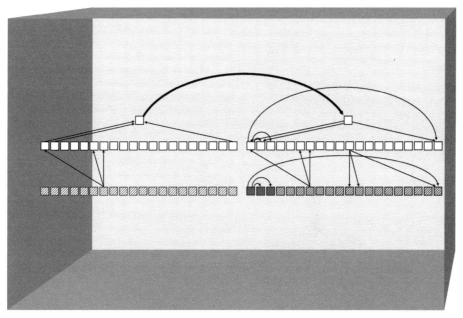

Figure 3.3   Two-sentence hierarchical confabulation architecture for English text analysis or generation, illustrated as the functional machinery of a 'purple box.' The sub-architectures for representing the first sentence (illustrated on the left) and that for the second sentence — the one to be continued — illustrated on the right) are each essentially the same as the architecture of Figure 3.2, along with one new lexicon and 20 new knowledge bases. The one additional lexicon is shown above the *phrase layer* of lexicons of each sub-architecture. This *sentence meaning content summary lexicon* contains symbols representing all of the 126,000 words and word groups of the phrase-level lexicons (and can also have additional symbols representing various other standard language constructions). Once the first sentence has been parsed; its summary lexicon has an expectation containing each phrase-level lexicon symbol (or construction subsuming a combination of phrase symbols) that is active. The (causal) *long-range context* knowledge base connects the summary lexicon of the first sentence to the summary lexicon of the second sentence.

to the second was then constructed in the usual way, using these counts. This education process takes about 2 weeks on a PC-type computer.

Figure 3.4 illustrates the architecture evaluation process. During each testing episode, two *evaluation trials* are conducted: one with no previous sentence (to establish baseline continuation) and one with a previous sentence (to illustrate the changes in the continuation that the availability of context elicited). For example, if no previous sentence was provided, and the first three words of the sentence to be continued were The New York, then the architecture constructed: The New York Times' computer model collapses . . . (where the words added by this sentence continuation process without context are shown in green). However, if the previous context sentence Stocks proved to be a wise investment, was provided, then, again beginning the next sentence with The New York, the architecture constructed The New York markets traded lower yesterday . . . (where, as in Figure 3.4, the words added by the sentence continuation process are shown in red). Changing the context sentence to Downtown events were interfering with local traffic., the architecture then constructs The New York City Center area where. . . . Changing the context sentence to Coastal homes were damaged by tropical storms. yields The New York City Emergency Service System. . . . And so on. Below are some other examples (first line — continuation without context, second line — previous sentence supplied to the architecture, third line — continuation with the previous sentence context):

A) Trial 1 – No previous sentence context supplied

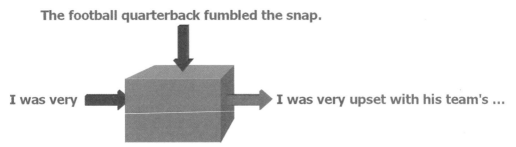

B) Trial 2 – Previous sentence context supplied

Figure 3.4   Use of the "purple box" confabulation architecture of Figure 3.3 for sentence continuation. Following knowledge acquisition (see text), the architecture's capabilities are evaluated by a series of testing events (each consisting of two *trials*). In Trial 1 (part A of the figure), three words, termed a sentence *starter* (shown in blue entering the architecture from the left) are entered into the architecture; without a previous sentence being provided. The architecture then uses its acquired knowledge and a simple, fixed, thought process to add some words; which are shown on the right in green appended to the starting words. In Trial 2 (part B of the figure), a previous context sentence (shown in brown being entered into the top of the architecture) is also provided. This alters the architecture's continuation output (shown in red). The context sentence (if one is being used on this trial) is entered into the left-hand sentence representation module of Figure 3.3 and the starter is entered into the first three words of the right-hand module. A simple, fixed, "swirling" consensus building thought process then proceeds to generate the continuation.

The New York Times' computer model collapses . . .
Medical patients tried to see their doctors.
The New York University Medical Association reported . . .

But the other semifinal match between fourth-seeded . . .
Chile has a beautiful capital city.
But the other cities have their size . . .

But the other semifinal match between fourth-seeded . . .
Japan manufactures many consumer products.
But the other executives included well-known companies . . .

When the United Center Party leader urged . . .
The car assembly lines halted due to labor strikes.
When the United Auto Workers union representation . . .

When the United Center Party leader urged . . .
The price of oil in the Middle East escalated yesterday.
When the United Arab Emirates bought the shares . . .

But the Roman Empire disintegrated during the fifth . . .
She learned the history of the saints.
But the Roman Catholic population aged 44 . . .

But the Roman Empire disintegrated during the fifth . . .
She studied art history and classical architecture.
But the Roman Catholic church buildings dating . . .

The San Francisco Redevelopment Authority officials announced . . .
Their star player caught the football and ran!
The San Francisco quarterback Joe Brown took . . .

The San Francisco Redevelopment Authority officials announced . . .
The pitcher threw a strike and won the game.
The San Francisco fans hurled the first . . .

The San Francisco Redevelopment Authority officials announced . . .
I listen to blues and classical music.
The San Francisco band draws praise from . . .

The San Francisco Redevelopment Authority officials announced . . .
Many survivors of the catastrophe were injured.
The San Francisco Police officials announced Tuesday . . .

The San Francisco Redevelopment Authority officials announced . . .
The wheat crops were genetically modified.
The San Francisco food sales rose 7.3 . . .

I was very nervous about my ability . . .
The football quarterback fumbled the snap.
I was very upset with his team's . . .

I was very nervous about my ability . . .
Democratic citizens voted for their party's candidate.
I was very concerned that they chose . . .

I was very nervous about my ability . . .
Restaurant diners ate meals that were served.
I was very hungry while knowing he had . . .

In spite of yesterday's agreement among analysts . . .
The Mets were not expected to win.
In spite of the pitching performance of some . . .

In spite of yesterday's agreement among analysts . . .
The President was certain to be reelected.
In spite of his statements toward the government . . .

In spite of yesterday's agreement among analysts . . .
She had no clue about the answer.
In spite of her experience and her . . .

In the middle of the 5th century BC . . .
Mike Piazza caught the foul ball.
In the middle of the season came . . .

In the middle of the 5th century BC . . .
The frozen lake was still very dangerous.
In the middle of the lake is a . . .

It meant that customers could do away . . .
The stock market had fallen consistently.
It meant that stocks could rebound later . . .

It meant that customers could do away . . .
I was not able to solve the problem.
It meant that we couldn't do much better . . .

It meant that customers could do away . . .
The company laid off half its staff.
It meant that if employees were through . . .

It meant that customers could do away . . .
The salesman sold men's and women's shoes.
It meant that sales costs for increases . . .

It must not be confused about what . . .
The effects of alcohol can be dangerous.
It must not be used without supervision . . .

It must not be confused about what . . .
The subject was put to a vote.
It must not be required legislation to allow . . .

It was a gutsy performance by John . . .
The tennis player served for the match.
It was a match played on grass . . .

It was a gutsy performance by John . . .
Coastal homes were damaged by tropical storms.
It was a huge relief effort since . . .

It was a gutsy performance by John . . .
The ship's sails swayed slowly in the breeze.
It was a long ride from the storm . . .

She thought that would throw us away...
The tennis player served for the match.
She thought that she played a good...

Shortly thereafter, she began singing lessons...
The baseball pitcher threw at the batter.
Shortly thereafter, the Mets in Game...

Shortly thereafter, she began singing lessons...
Democratic citizens voted for their party's candidate.
Shortly thereafter, Gore was elected vice president...

The president said he personally met French...
The flat tax is an interesting proposal.
The president said he promised Congress to let...

The president said he personally met French...
The commission has reported its findings.
The president said he appointed former Secretary...

The president said he personally met French...
The court ruled yesterday on conflict of interest.
The president said he rejected the allegations...

This resulted in a substantial performance increase...
The state governor vetoed the bill.
This resulted in both the state tax...

This resulted in a substantial performance increase...
Oil prices rose on news of increased hostilities.
This resulted in cash payments of $...

This resulted in a substantial performance increase...
The United States veto blocked the security council resolution.
This resulted in both Britain and France...

Three or four persons who have killed...
The tennis player served for the match.
Three or four times in a row...

We could see them again if we...
The president addressed congress about taxes.
We could see additional spending money bills...

We could see them again if we...
The view in Zion National Park was breathtaking.
We could see snow conditions for further...

We could see them again if we...
We read the children's books out loud.
We could see the children who think...

We could see them again if we...
The U.N. Security Council argued about sanctions.
We could see a decision must soon...

What will occur during the darkest days . . .
Research scientists have made astounding breakthroughs.
What will occur within the industry itself . . .

What will occur during the darkest days . . .
The vacation should be very exciting.
What will occur during Christmas season when . . .

What will occur during the darkest days . . .
I would like to go skiing.
What will occur during my winter vacation . . .

What will occur during the darkest days . . .
There's no way to be certain.
What will occur if we do nothing . . .

When the Union Bank launched another 100 . . .
She loved her brother's Southern hospitality.
When the Union flag was raised again . . .

When the Union Bank launched another 100 . . .
New York City theater is on Broadway.
When the Union Square Theater in Manhattan . . .

A good analogy for this system is a child learning a human language. Young children need not have any formal knowledge of language or its structure in order to generate it effectively. Consider what this architecture must "know" about the objects of the world (e.g., their attributes and relationships) in order to generate these continuations; and what it must "know" about English grammar and composition. Is this the world's first AI system? You decide.

Note that in the above examples the continuation of the second sentence in context was conducted using an (inter-sentence, long-range context) knowledge base educated via exposure to meaning-coherent sentence pairs selected by an external agent. **When tested with context, using completely novel examples, it then produced continuations that are meaning-coherent with the previous sentence** (i.e., the continuations are rarely unrelated in meaning to the context sentence). Think about this for a moment. This is a valuable general principle with endless implications. For example, we might ask: how can a system learn to carry on a conversation? Answer: simply educate it on the conversations of a master human conversationalist! There is no need or use for a "conversation algorithm." Confabulation architectures work on this *monkey-see/ monkey-do* principle.

This sentence continuation example reveals the true nature of cognition: it is based on ensembles of properly phased confabulation processes mutually interacting via knowledge links. Completed confabulations provide assumed facts for confabulations newly underway. Contemporaneous confabulations achieve mutual "consensus" via rapid interaction through knowledge links as they progress (thus the term *consensus building*). There are no algorithms anywhere in cognition. Only such ensembles of confabulations. This illustrates the truly **alien nature** of cognition in comparison with existing neuroscience, computer science, and AI concepts.

In speech cognition (see Section 3.4), elaborations of the architecture of Figure 3.3 can be used to define expectations for the next word that might be received (which can be used by the acoustic components of a speech understanding system); based upon the context established by the previous sentence and previous words of the current sentence which have been previously transcribed. For *text generation* (a generalization of sentence continuation, in which the entire sentence is completed with no starter), the choices of words in the second sentence can now be influenced by the context established by the previous sentence. The architecture of Figure 3.3 generalizes to using larger bodies of context for a variety of cognition processes.

Even more abstract levels of representation of language meaning are possible. For example, after years of exposure to language and co-occurring sensory and action representations, lexicons can form that represent sets of commonly encountered lower-abstraction-level symbols. Via the SRE mechanism (a type of thought process), such symbols take on a high level of abstraction, as they become linked (directly, or via equivalent symbols) to a wide variety of similar-meaning symbol sets. Such symbol sets need not be complete to be able to (via confabulation) trigger activation of such high-abstraction representations. In language, these highest-abstraction-level symbols often represent **words**! For example, when you activate the symbol for the word joy, this can mean joy as a word, or joy as a highly abstract concept. This is why in human thought the most exalted abstract concepts are made specific by identifying them with words or phrases. It is also common for these most abstract symbols to belong to a foreign language. For example, in English speaking lands, the most sublime abstract concepts in language are often assigned to French, or sometimes German, words or phrases. In Japanese, English or French words or phrases typically serve in this capacity.

High-abstraction lexicons are used to represent the meaning content of *objects* of the mental world of many types (language, sound, vision, tactile, etc.). However, outside of the language faculty, such symbols do not typically have names (although they are often strongly linked with language symbols). For example, there is probably a lexicon in your head with a symbol that abstractly encodes the combined taste, smell, surface texture, and masticational feel of a macaroon cookie. This symbol has no name, but you will surely know when it is being expressed!

3.3.5 Discussion

A key observation is that confabulation architectures automatically learn and apply grammar, and honor syntax; without any in-built linguistic structures, rules, or algorithms. This strongly suggests that grammar and syntax are fictions dreamed up by linguists to explain an orderly structure that is actually a requirement of the mechanism of cognition. Otherwise put, for cognition to be able, given the limitations of its native machinery, to efficiently deal with language, that language must have a structure which is compatible with the mathematics of confabulation and consensus building. In this view, every functionally usable human language must be structured this way. Ergo, universal appearance of some sort of grammar and syntactic structure in all human languages.

Thus, Chomsky's (1980) famous long search for a universal grammar (which must now be declared over) was both correct and incorrect. Correct, because if you are going to have a language that cognition can deal with at a speed suitable for survival, grammar and syntactic structure are absolute requirements (i.e., languages that don't meet these requirements will either adapt to do so, or will be extincted with their speakers). Thus, grammar is indeed universal. Incorrect, because grammar itself is a fiction. It does not exist. It is merely the visible spoor of the hidden native machinery of cognition: confabulation, antecedent support knowledge, and the conclusion–action principle.

3.4 SOUND COGNITION

Unlike language, which is the centerpiece and masterpiece of human cognition, all the other functions of cognition (e.g., sensation and action) must interact directly with the outside world. Sensation requires conversion of externally supplied sensory representations into symbolic representations and vice versa for actions. This section, and the next (discussing vision), must therefore discuss not only the confabulation architectures used, but also cover the implementation of this *transduction* process; which is necessarily different for each of these cognitive modalities. Readers are expected to have a solid understanding of traditional speech signal processing and speech recognition.

3.4.1 Representation of Multi-Source Soundstreams

Figure 3.5 illustrates an "audio front end" for transduction of a soundstream into a string of "multi-symbols;" with a goal of carrying out ultra-high-accuracy speech transcription for a single speaker embedded in multiple interfering sound sources (often including other speakers). The description of this design does not concern itself with computational efficiency. Given a concrete design for such a system, there are many well-known signal processing techniques for implementing approximately the same function, often orders of magnitude more efficiently. For the purpose of this introductory treatment (which, again, is aimed at illustrating the universality of confabulation as the mechanization of cognition), this audio front-end design does not incorporate embellishments such as binaural audio imaging.

Referring to Figure 3.5, the first step in processing is analog speech lowpass filtering (say, with a flat, zero-phase-distortion response from DC to 4 kHz, with a steep rolloff thereafter) of the high-quality (say, over 110 dB dynamic range) analog microphone input. Following bandpass filtering, the microphone signal is sampled with an (e.g., 24-bit) analog to digital converter operating at a 16 kHz sample rate. The combination of high-quality analog filtering, sufficient sample rate (well above the Nyquist rate of 8 kHz) and high dynamic range, yield a digital output stream with almost no artifacts (and low information loss). Note that digitizing to 24 bits supports exploitation of the wide dynamic ranges of modern high-quality microphones. In other words, this dynamic range will make it possible to accurately understand the speech of the attended speaker, even if there are much higher amplitude interferers present in the soundstream.

The 16 kHz stream of 24-bit signed integer samples generated by the above preprocessing (see Figure 3.5) is next converted to floating point numbers and blocked up in time sequence into 8000-sample windows (8000-dimensional floating point vectors), at a rate of one window for every 10 ms. Each such *sound sample vector* X thus overlaps the previous such vector by 98% of its length (7840 samples). In other words, each X vector contains 160 new samples that were not in the previous X vector (and the "oldest" 160 samples in that previous vector have "dropped off the left end").

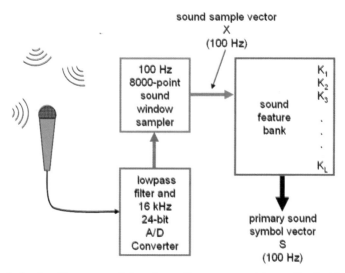

Figure 3.5   An audio front-end for representation of a multi-source soundstream. See text for details.

As shown in Figure 3.5, the 100 Hz stream of sound sample vectors then proceeds to a *sound feature bank*. This device is based upon a collection of L fixed, 8000-dimensional floating point *feature vectors*: $K_1, K_2, \ldots, K_L$ (where L is typically a few tens of thousands). These feature vectors represent a variety of *sound detection correlation kernels*. For example: gammatone wavelets with a wide variety of frequencies, phases, and gamma envelope lengths, broadband impulse detectors; fricative detectors; etc. When a sound sample vector X arrives at the feature bank the first step is to take the inner product of X with each of the L feature vectors; yielding L real numbers: $(X \cdot K_1), (X \cdot K_2), \ldots, (X \cdot K_L)$. These L values form the *raw feature response vector*. The individual components of the raw feature response vector are then each subjected to further processing (e.g., discrete time linear or quasi-linear filtering), which is customized for each of the L components. Finally, the logarithm of the square of each component of this vector is taken. The net output of the sound feature bank is an L-component non-negative *primary sound symbol excitation vector* S (see Figure 3.5). A new S vector is issued in every 10 ms.

The criteria used in selection of the feature vectors are low information loss, sparse representation (a relatively small percentage of S components meaningfully above zero at any time due to any single sound source), and low rate of individual feature response to multiple sources. By this latter it is meant that, given a typical application mix of sources, the probability of any feature which is meaningfully responding to the incoming soundstream at a particular time being stimulated (at that moment) by sounds from more than one source in the auditory scene is low. The net result of these properties is that S vectors tend to have few meaningfully nonzero components per source, and each sound symbol with a significant excitation is responding to only one sound source (see Sagi et al., 2001 for a concrete example of a sound feature bank).

Figure 3.6 illustrates a typical primary sound symbol excitation vector S. This is the mechanism of analog sound input transduction into the world of symbols. A new S vector is created 100 times per second. S describes the content of the sound scene being monitored by the microphone at that moment. Each of the L components of S (again, L is typically tens of thousands) represents the response of one sound feature detector (as described above) to this current sonic scene.

S is composed of small, mostly disjoint (but usually not contiguous), subsets of excited sound symbol components — one subset for each sound source in the current auditory scene. Again, each excited symbol is typically responding to the sound emanating from only one of the sound sources in the audio scene being monitored by the microphone. While this single-source-per-excited-symbol rule is not strictly true all the time, it is almost always true (which, as we will see, is all that matters). Thus, if at each moment, we could somehow decide which subset of excited symbols of the symbol excitation vector to *pay attention* to, we could ignore the other symbols and thereby focus our attention on one source. That is the essence of **all** initial cortical sensory processing (auditory, visual, gustatory, olfactory, and somatosensory): figuring out, in real-time, which primary sensor input representation symbols to pay attention to, and ignoring the rest. This ubiquitous cognitive process is termed *attended object segmentation*.

Figure 3.6 Illustration of the properties of a primary sound symbol excitation vector S (only a few of the L components of S are shown). Excited symbols have thicker circles. Each of the four sound sources present (at the moment illustrated) in the auditory scene being monitored is causing a relatively small subset of feature symbols to be excited. Note that the symbols excited by sources 1 and 3 are not contiguous. That is typical. Keep in mind that the number of symbols, L (which is equal to the number of feature vectors) is typically tens of thousands; of which only a small fraction are meaningfully excited. This is because each sound source only excites a relatively small number of sound features at each moment and typical audio scenes contain only a relatively small number of sound sources (typically fewer than 20 monaurally distinguishable sources).

3.4.2 Segmenting the Attended Speaker and Recognizing Words

Figure 3.7 shows a confabulation architecture for directing attention to a particular speaker in a soundstream containing multiple sound sources and also recognizing the next word they speak. For a concrete example of a simplified version of this architecture (which nonetheless can competently carry out these kinds of functions; see Sagi et al., 2001). This architecture will suffice for the purposes of this introduction; but would need to be further augmented (and streamlined for computational efficiency) for practical use.

Each 10 ms a new S vector is supplied to the architecture of Figure 3.7. This S vector is directed to one of the primary sound lexicons; namely, the next one (moving from left to right) in sequence after the one which received the last S vector. It is assumed that there are a sufficient number of lexicons so that all of the S vectors of an individual word have their own lexicon. Of course, this requires 100 lexicons for each second of word sound input, so a word like antidisestablishmentarianism will require hundreds of lexicons. For illustrative purposes, only 20 primary sound lexicons are shown in Figure 3.7. Here again, in an operational system, one would simply use a ring of lexicons (which is probably what the cortical "auditory strip" common to many mammals, including humans [Paxinos and Mai, 2004], probably is — a linear sequence of lexicons which functionally "wraps around" from its physical end to its beginning to form a ring).

The architecture of Figure 3.7 presumes that we know approximately when the last word ended. At that time, a thought process is executed to erase all of the lexicons of the architecture, feed in expectation-forming links from external lexicons to the next-word acoustic lexicon (and form the next-word expectation), and redirect S vector input to the first primary sound lexicon (the one on the far left). (Note: As is clearly seen in mammalian auditory neuroanatomy, the S vector is wired to all portions (lexicons) of the strip in parallel. The process of "connecting" this input to one selected lexicon (and no other) is carried out by manipulating the operating command of that one lexicon. Without this operate command input manipulation, which only one lexicon receives at each moment, the external sound input is ignored.)

The primary sound lexicons have symbols representing a statistically complete coverage of the space of momentary sound vectors S that occur in connection with auditory sources of interest, when they are presented in isolation. So, if there are, say 12 sound sources contributing to S, then we would nominally expect that there would be 12 sets of primary sound lexicon symbols responding to S (this follows because of the "quasiorthogonalized" nature of S, for example, as depicted in

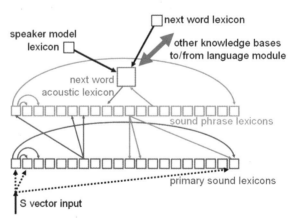

Figure 3.7 Speech transcription architecture. The key components are the *primary sound lexicons*, the *sound phrase lexicons*, and the *next-word acoustic lexicon*. See text for explanation.

Figure 3.6). Mathematically, the symbols of each primary sound lexicon are a vector quantizer (Zador, 1963) for the set of S vectors that arise, from all sound sources that are likely to occur, when each source is presented in isolation (i.e., no mixtures). Among the symbol sets that are responding to S are some that represent the sounds coming from the attended speaker. This illustrates the critically important need to design the acoustic front-end so as to achieve this sort of *quasiorthogonalization of sources*. By confining each sound feature to a properly selected time interval (a subinterval of the 8000 samples available at each moment, ending at the most recent 16 kHz sample), and by using the proper postfiltering (after the dot product with the feature vector has been computed) this quasiorthogonalization can be accomplished. (Note: This scheme answers the question of how brains carry out "independent component analysis" [Hyvärinen et al., 2001]. They don't need to. Properly designed quasiorthogonalizing features, adapted to the pure sound sources that the critter encounters in the real world, map each source of an arbitrary mixture of sources into its own separate components of the S vector. In effect, this is essentially a sort of "one-time ICA" feature development process carried out during development and then essentially frozen (or perhaps adaptively maintained). Given the stream of S vectors, the confabulation processing which follows (as described below) can then, at each moment, ignore all but the attended source-related subset of components, independent of how many, or few, interfering sources are present. Of course, this is exactly what is observed in mammalian audition — effortless segmentation of the attended source at the very first stage of auditory (or visual or somatosensory, etc.) perception.

The expectation formed on the next-word acoustic lexicon of Figure 3.7 (which is a huge structure, almost surely implemented in the human brain by a number of physically separate lexicons) is created by successive **C1Fs**. The first is based on input from the speaker model lexicon. The only symbols (each representing a stored acoustic model for a single word — see below) that then remain available for further use are those connected with the speaker currently being attended to.

The second **C1F** is executed in connection with input from the language module word lexicon that has an expectation on it representing possible predictions of the next word that the speaker will produce (this next-word lexicon expectation is produced using essentially the same process as was described in Section 3.3 in connection with sentence continuation with context). (Note: This is an example of the situation mentioned above and in the Appendix, where an expectation is allowed to transmit through a knowledge base.) After this operation, the only symbols left available for use on the next-word acoustic lexicon are those representing expected words spoken by the attended speaker. This expectation is then used for the processing involved in recognizing the attended speaker's next word.

As shown in Figure 3.7, knowledge bases have previously been established (using pure source, or well-segmented source, examples) to and from the primary sound symbol lexicons with the sound phrase lexicons and to and from these with the next-word acoustic lexicon. Using these knowledge bases, the expectation on the next-word acoustic lexicon is *transferred* (as described immediately above) via the appropriate knowledge bases, to the sound phrase lexicons, where expectations are formed; and from these to the primary sound lexicons, where additional expectations are formed. It is easy to imagine that, since each of these transferred expectations is typically much larger than the one from which it came, that by the time this process gets to the primary sound lexicons, the expectations will encompass almost every symbol. THIS IS NOT SO! While these primary lexicon expectations are indeed large (they may encompass many hundreds of symbols), they are still only a small fraction of the total set of tens of thousands of symbols. Given these transfers, which actually occur as soon as the recognition of the previous word is completed — which is often long before its acoustic content ceases arriving, the architecture is prepared for detecting the next word spoken by the attended speaker.

As each S vector arrives at the architecture of Figure 3.7, it is sent to the proper lexicon in sequence. For simplicity, let us assume that the first S vector associated with the initial sound content of the next word is sent to the first primary sound lexicon (if it goes to the "wrong" lexicon or is missed altogether, it does not matter much — as will be explained below). Given that the first primary sound lexicon has an expectation, and that the only symbols in this expectation are those that represent sounds that a speaker of this type would issue (we each have hundreds of "canonical models" of speakers having different accents and vocal apparati, and most of us add to this store throughout life) when speaking early parts of one of the words we are expecting. Again note that, because of the orthogonalized nature of the S vector and the pure-signal nature of the primary feature symbols, each of the symbols in this expectation will typically represent sounds having only a tiny number of S vector components that are nonzero. Each symbol in a primary sound lexicon is expressed as a unit vector having these small number of components with coefficients near 1, and all other components at zero. The lexicon takes the inner product of each symbol's vector expression with S and this is then used as that symbol's *initial input excitation* (**this** is how symbols get excited by sensory input signals; in contrast to how symbols get excited by knowledge links from other symbols, which was discussed in Section 3.1). We have now completed the transition from acoustic space to symbol space.

Notice that the issue of signal level of the attended source has not been discussed. As described in Section 3.3.1, each S vector component has its amplitude expressed on a logarithmic scale (based on "sound power amplitudes" ranging across many orders of magnitude). Thus, on this scale, the inner product of S with a particular symbol's unit vector will still (because of the linear nature of the inner product) be substantial, even if the attended source sounds are tens of dB below those of some individual interferers. Thus, with this design, attending to weak, but distinct, sources is generally possible. These are, of course, the characteristics we as humans experience in our own hearing. Further, in auditory neuroscience, such logarithmic coding of sound feature response signals (in particular, those from the brainstem auditory nuclei to the medial geniculate nucleus, which are the auditory signals analogous to the components of S) is well established (Oertel et al., 2002).

During the entire time of the word detection processes, all of the lexicons of the Figure 3.7 architecture are operated in a consensus building mode. Thus, as soon as the S-input excitations are established on the expectation element symbols of the first primary sound lexicon, only those symbols which received these expectations remain in the expectation (the consensus building is run faster on the primary sound lexicons, somewhat slower on the sound phrase lexicons, and even slower on the next-word acoustic lexicon). This process of expectation refinement that occurs during consensus building is termed *honing*.

After acoustic input has arrived at each subsequent primary sound lexicon (the pace of the switching is set by a separate part of the auditory system, which will not be discussed further here, which synchronizes the pace of S vector formation — no it is not always exactly every 10 ms — to the recent pace of speech production of the attended speaker), that lexicon's expectation is thereby honed and this revised expectation is then automatically transferred to all of the sound phrase regions that are not on its right (during consensus building, all of the involved knowledge bases remain operational). This has the effect of honing some of the sound phrase lexicon expectations, which then are transferred to the next-word acoustic lexicon; honing its expectation.

This process works in reverse also. As higher-level lexicon expectations are honed, these are transferred to lower levels, thereby refining those lower-level expectations. Note that if occasional erroneous symbols are transferred up to the sound phrase lexicons, or even from the phrase lexicons to the next-word acoustic lexicon, this will not have much effect. That is because the process of consensus building effectively "integrates" the impact of all of the incoming transfers on the symbols of the original expectation. Only when a phrase region has honed its symbol list down to one symbol (which then becomes active) is a final decision made at that level. Similarly, only at the point where the expected word duration has been reached does the next-word acoustic lexicon make a decision (or, it can even transfer a small expectation back to the language module – which is one way that robust operation can be aided).

Figure 3.8 illustrates the consensus building process on the next-word acoustic lexicon as the S vector is directed to each subsequent primary sound lexicon in turn. As honed expectations are transferred upward, the expectation of the next-word acoustic lexicon is itself honed. This honed expectation is then transferred downward to refine the expectations of the as-yet-unresolved sound phrase and primary sound layer lexicons. These consensus building interactions happen dynamically in continuous time as the involved operation commands are slowly tightened. This again illustrates the almost exact analogy between thought and movement. As with a movement, these smoothly changing, precisely controlled, consensus building lexicon operate commands are

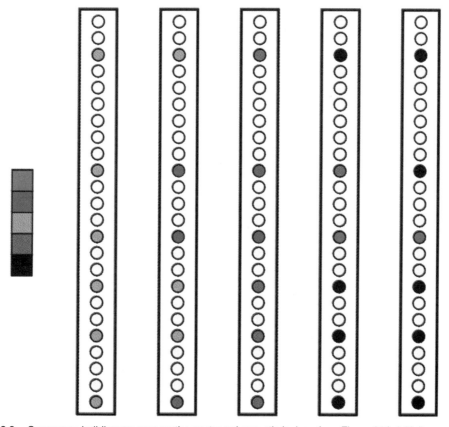

Figure 3.8 Consensus building process on the next-word acoustic lexicon (see Figure 3.7). Initially, symbols in the next-word expectation (green dots in the left-most representation of the lexicon state) are established by knowledge link inputs from the speaker model lexicon and from the language module. As consensus building progresses, transfers of honed expectations from sound phrase lexicons (which themselves are receiving transfers from primary sound lexicons) hone this initial expectation, as illustrated here moving from left to right. Yellow-filled circles represent symbols that were not part of the initial expectation. These are locked at zero excitation. The color chart on the left shows the positive excitation scale from lowest on the bottom to highest on top. Some of the initial expectation symbols become progressively promoted to higher levels of excitation (the sum of all symbol excitations is roughly constant during consensus building). Others go down in excitation (it is possible for a symbol to change nonmonotonically, but that is not illustrated here). In the end state of the lexicon (far right) one symbol (red) has become active — this symbol represents the word that has been detected. Keep in mind that in a real architecture there would typically be tens of thousands of symbols and that only a few percent, at most, would be part of the initial expectation.

generated by a set of lexicons (in frontal cortex) that specialize in storing and recalling action symbol sequences.

A common objection about this kind of system is that as long as the expectations keep being met, the process will keep working. However, if even one glitch occurs, it looks like the whole process will fall apart and stop working. Then, it will somehow have to be restarted (which is not easy — for example, it may require the listener to somehow get enough signal to noise ratio to allow a much cruder trick to work). Well, this objection is quite wrong. Even if the next word and the next-after-that word are not one of the expected ones, this architecture will often recover and ongoing speechstream word recognition will continue; as we already proved with our crude initial version (Sagi et al., 2001). A problem that can reliably make this architecture fail is a sudden major change in the pace of delivery, or a significant brief interruption in delivery. For example, if the speaker suddenly starts speaking much faster or much slower the mentioned subsystem that monitors and sets the pace of the architecture's operation will cause the timing of the consensus building and word-boundary segmentation to be too far off. Another problem is if the speaker gets momentarily tongue-tied and inserts a small unexpected sequence of sounds in a word (try this yourself by smoothly inserting the brief meaningless sound "BRYKA" in the middle of a word at a cocktail party — the listener's Figure 3.7 architecture will fail and they will be forced to move closer to get clean recognitions to get it going again).

A strong tradition in speech recognition technology is an insistence that speech recognizers be "time-warp insensitive" (i.e., insensitive to changes in the pace of word delivery). Well Figure 3.7 certainly is not strongly "time-warp insensitive," and as pointed out immediately above, neither are humans! However, modest levels of time warp have no impact, since this just changes the location of the phrase region (moves it slightly left or right of its nominal position) where a particular phrase gets detected. Also note that since honed phrase expectations are transferred, it is not necessary for all of the primary sound symbols of a phrase to be present in order for that phrase to contribute significantly to the "promotion" of the next-word acoustic lexicon symbols that receive links from it. Thus, many primary symbols can be missed with no effect on correct word recognition. This is one of the things which happens when we speak more quickly: some intermediate sounds are left out. For example, say Worcestershire sauce at different speeds from slow to fast and consider the changes in the sounds you issue.

3.4.3 Discussion

This section has outlined how sound input can be transduced into a symbol stream (actually, an expectation stream) and how that stream can, through a consensus building process, be interpreted as a sequence of words being emitted by an attended speaker.

One of the many Achilles' heels of past speech transcription systems has been the use of a vector quantizer in the sound-processing front end. This is a device that is roughly the same as the sound feature bank described in this section, except that its output is one and only one symbol at each time step (10 ms). This makes it impossible for such systems to deal with multi-source audio scenes.

The sound processing design described in this section also overcomes the inability of past speech recognition systems to exploit long-range context. Even the best of today's speech recognizers, operating in a totally noise-free environment with a highly cooperative speaker, cannot achieve much better than 96% sustained accuracy with vocabularies over 60,000 words. This is primarily because of the lack of a way to exploit long-range context from previous words in the current sentence and from previous sentences. In contrast, the system described here has full access to the context-exploitation methods discussed in Section 3.3; which can be extended to arbitrarily large bodies of context.

Building a speech recognizer for colloquial speech is much more difficult than for proper language. As is well known, children essentially cannot learn to understand speech unless they can also produce it (in some way). Undoubtedly, this will hold for systems of the type considered in this section. Thus, to solve the whole speech language understanding problem, we must also solve the speech language production problem.

In summary, the confabulation theory of vertebrate cognition seems to provide the basis for mechanizing sound cognition in a manner that has the familiar characteristics of human sound cognition.

3.5   VISUAL COGNITION

As with sound, the key challenge of vision is to usefully transduce incoming image information into symbolic form. Another key part of vision is to build symbolic representations of individual visual objects that are invariant to useful combinations of selected visual attributes such as pose, lighting, color, and form. These are the main topics of this section. Ancillary subjects, such as the highly specialized visual human face recognition system and binocular vision are not discussed. Readers are expected to have a solid understanding of traditional machine vision.

3.5.1   Building an Eyeball Vision Sensor and its Gaze Controller

Vertebrate vision is characterized by the use of eyeballs. A *gaze controller* is used to direct the eye(s) to (roughly repeatable) key points on objects of interest. In this section, we will consider only monocular, panchromatic, visual cognition in detail.

Figure 3.9 illustrates the basic elements of the confabulation-based vision architecture that will be discussed in this section. For simplicity, the subject of how pointing of the video camera sensor will be controlled is ignored. It is assumed that the wide-angle large image camera is fixed and that everything we want to see and visually analyze is within this sensor's fixed visual field of view and is of sufficient size (number of pixels) to make its attributes visible at the sensor's resolution. For example, imagine a wide-angle, high-resolution video camera positioned about 8 ft above the pavement at a busy downtown street intersection, pointed diagonally across the inter-section, viewing the people on the sidewalks and the vehicles on the streets.

Assume that the visual sensor (i.e., video camera) gathers digital image frames, each with many millions of pixels, at a rate of 30 frames per second. For simplicity, each pixel will be assumed to have its panchromatic (grayscale) brightness measured on a 16-bit linear digital scale.

The *gaze controller* of this visual system (see Figure 3.9) is provided with all of the pixels of each individual frame of imagery. Using this input, it decides whether to select a *fixation point* (a particular pixel of the frame) for that frame (it can select at most one). The manner in which a gaze controller can be built (my laboratory has built one [Hecht-Nielsen and Zhou, 1995] and so have a number of others) is described next. To make the discussion which follows concrete, consider a situation where our video camera sensor is monitoring a street scene in a busy downtown area. Each still frame of video contains tens of people and a number of cars driving by.

The basic idea of designing a gaze controller is to mimic human performance. Let an attentive human visual observer watch the output of the video sensor on a computer screen. Attach an eye tracker to the screen to monitor the human's eye movements. These movements will typically be *saccades* — jumps of the eye position between one *fixation point* and the next. At each fixation point, the human eye gathers image data from a region surrounding the fixation point. This can be viewed as taking a "snapshot" or "eyeball" image centered at that fixation point. The human visual system then processes that eyeball image and jumps to the next fixation point selected by its gaze director (a function which is implemented, in part, by the *superior colliculus* of the brainstem). Visual processing is not carried out during these eyeball jumps. While the human

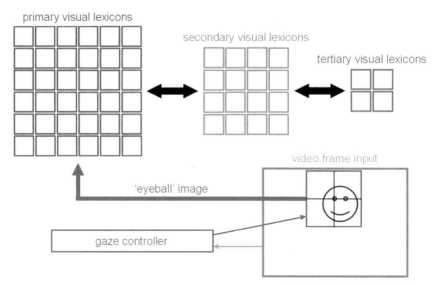

Figure 3.9 Vision cognition architecture. The raw input to the visual system is a wide-angle high-resolution video camera (large frame shown in the lower right of the figure). A subimage, of a permanently fixed size (say 1024 × 1024 pixels) of a single video frame (shown as a square within the large frame), termed the *eyeball image*, is determined by the location of its center (depicted by the intersection of crosshairs), known as the *fixation point*. The gaze controller uses the entire large frame to select a single fixation point, if it deems that such a selection is warranted for this large frame (it only attempts to select a fixation point when processing of the last eyeball image has been completed). For simplicity, it is assumed that the video camera is fixed and is able to see the entire visual scene of interest (e.g., a camera viewing a busy downtown intersection). The confabulation architecture used for visual processing is described in the text.

observer is viewing the video it is important that they be carrying out whatever specific task or tasks that the automated vision system will be asked to carry out (e.g., spotting people, pets, bicycles, and cars).

After many tens of hours of video have been viewed by the human observer **carrying out the function that the machine visual cognition system will later perform**, and their eye movements have been recorded, this provides a record of their fixation point choices for each still frame of specific scene content when that choice was made. This record is then used to train a multi-layer perceptron (Hecht-Nielsen, 2004) to carry out the gaze control function. The basic idea is simple. Each frame of high-resolution video is described by an *image feature vector* V. This feature vector is produced by first taking the inner product of each of a collection of Gabor logons with the image frame (both considered as vectors of the same dimension). The specific Gabor logons used in forming V (each logon is defined by the constants E, F, and G, and by its position and angle of plane rotation in the image — see Figure 3.10) are now described.

First, we create a fixed rectangular set of gridpoints located at equal pixel spacings across the entire high-resolution video camera frame (Caid and Hecht-Nielsen, 2001, 2004; Daugman, 1985, 1987, 1988a,b; Daugman and Kammen, 1987; Hecht-Nielsen, 1990; Hecht-Nielsen and Zhou, 1995). For example, if each video camera image frame were a 8,192 × 8,192 pixel digital image, with a 16-bit panchromatic grayscale, or equivalently, a 67,108,864-dimensional floating point real vector with integer components between 0 and 65,535, then we might have gridpoints spaced every 16 pixels vertically and horizontally, with gridpoints on the image edges, for a total of 513 × 513 = 263,169 gridpoints.

At each gridpoint we create a set of Gabor logons centered at that position, each having a specified rotation angle and E, F, and G values. The set of logons at each gridpoint is exactly the same, save for their translated position. This set, which is now described, is termed a *jet* (von der

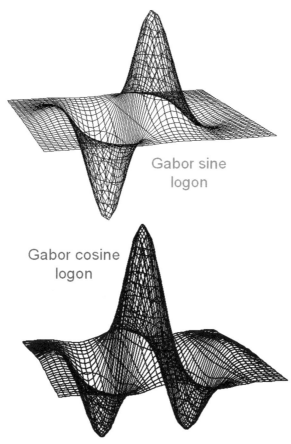

Figure 3.10 *Gabor logon* local image features. Logons are defined as images with real-valued pixel brightnesses (i.e., both positive and negative values are allowed) defined by geometrical plane rotations and translations (in the image plane) of the canonical two-dimensional functions $\sin(Gx)/\exp(Ex^2 + Fy^2)$ (called a *sine logon*) and $\cos(Gx)/\exp(Ex^2 + Fy^2)$ (termed a *cosine logon*); where E, F, and G are positive constants and x and y are image plane coordinates in the translated and rotated coordinates of the logon frame. Note that E and F define the principal axis lengths of a two-dimensional Gaussian-type ellipsoid and G defines the spatial frequency of a plane grating (with oscillations along the x-axis). The ratio E/F is fixed for all logons used. Each individual logon is considered as a (real valued) digital image, that is as image vectors of the same dimension as the wide-angle video camera frames.

Malsburg, 1990). In the vision work done in my lab we have typically set the ratio E/F to 5/8 and the G/E ratio to $3\pi/2$ for every logon in every jet (these are the values that seem to be used by domestic cats; Hecht-Nielsen, 1990).

Each jet consists, for example, of pairs of a sine logon and a cosine logon at each of seven scales (E = 2, 3, 5, 9, 15, 20, and 35 pixel units) and 16 regularly spaced angular orientations, including having the major ellipse axis of one logon pair vertical. Thus, each jet at each gridpoint has 224 logons. Again, each individual logon is viewed as a 67,108,864-dimensional floating point real vector, with each component value given by the evaluation of its formula (Figure 3.10, properly translated and rotated) evaluated at the pixel location corresponding to that component (obviously, with most of its values at pixels distant from the gridpoint very close to zero). Thus, there are a total of $224 \times 263,169 = 58,949,856$ logons in all of the gridpoint jets; almost as many as there are pixels in the high-resolution camera image.

The image feature vector V of a single camera (assumed to employ a progressive scan) frame is defined to be the 29,474,928-dimensional vector obtained by first calculating the inner product of each logon of each jet with the image vector, and then, to get each component of V, adding the squares of the sine and cosine inner products of the logons of the same scale and rotational orientation in each jet (which reduces the total dimensionality of V to half that of the total number of logons). (Note: Other mathematical transformations are then applied to each of these sums to make their values insensitive to lighting gradient slopes and other lighting-dependent effects — but these details go beyond the scope of this sketch and so are left out — see Hecht-Nielsen and Zhou, 1995 for examples of such transformations.)

Each component of V essentially represents an estimate of the localized spatial frequency content of the camera image (at the position of the associated gridpoint) at the spatial frequency of the involved logon pair, in the direction of oscillation of that pair. It is on the basis of local spatial frequency structure (which V accurately defines) that fixation points are chosen by the gaze controller.

The job of the gaze controller is to learn to mimic the performance of a skilled human observer performing the visual task that is to be mechanized. The manner in which the gaze controller works and the method used to train it are now described.

The gaze controller (a perceptron; Hecht-Nielsen, 2004) has 224 inputs and two outputs. The inputs represent the components of V corresponding to the jet at a particular image gridpoint (the current *position of regard* of the gaze controller). The outputs of the gaze controller are estimates of the *a posteriori* probability of this gridpoint being chosen by the skilled human as a fixation point along with the *a posteriori* probability of this gridpoint not being chosen by the skilled human as a fixation point. Training of the gaze controller is discussed below; but, to set the stage, the manner in which the gaze controller is used operationally is described first.

Once trained, the gaze controller is used to select a fixation point in a newly acquired video frame by evaluating each of the V component sets from each of the 263,169 gridpoints of the frame. If the first output of the controller is above a fixed threshold (say, 0.8), and the second output is below a fixed threshold (say, 0.2), then that gridpoint is selected as a *candidate fixation point*. If there are no candidate fixation points for the frame, then that frame is skipped. If there are one or more, the one with the highest first output value is selected as the fixation point. The gaze controller also has provisions for creating multiple successive "looks" at the same object during visual training to facilitate learning of pose insensitivity (see below). In operational use, when a visual object of interest has been fixated on and described, the gaze controller tracks that object's fixation points and prevents return to it until the other visual objects of interest in the scene have been described.

To train the gaze controller, each fixation point example (for which a *reference frame* is selected as the definitive "image input" that the human used — by taking a frame a fixed time increment right before the beginning of their saccade) has its pixel coordinates (supplied by the frequently-recalibrated eye tracker) stored with its reference frame. Eventually, many thousands of such fixation point and reference frame pairs are produced, randomly scrambled to remove possible content correlations between them, and stored. The V vector for each reference frame is also calculated and stored with it.

The gaze controller perceptron is trained by marching through the fixation point or reference frame examples, in sequence, many times. At each training episode, the next fixation point and reference frame example in sequence is selected and the gridpoint nearest to the fixation point is located. The jet components of the reference frame V vector for that gridpoint are then extracted and provided to the perceptron, along with desired outputs 1 and 0, and one backpropagation training episode using these specified inputs and outputs is carried out. Another gridpoint, distant from the fixation point, is then selected and its jet V components are provided to the perceptron, along with desired outputs 0 and 1, and a second perceptron training episode is carried out using these inputs and outputs. The training process then moves on to the next fixation point or reference image example. Thus, this training procedure beneficially utilizes *oversampling* of the examples of the class of human-supplied fixation points (Hecht-Nielsen, 2004).

Training is continued until the perceptron learning curve (as calculated by considering the performance of the perceptron when tested on, say, the last 1000 training examples) reaches a sufficiently low value (say, 80% of the training example pairs would be declared as fixation points and not fixation points, respectively). Final testing is carried out on hundreds of fresh examples not used in testing. If, say, 70% of the final testing examples are classified correctly, then the gaze controller is frozen and ready for service. If not, then additional training is called for. After training, the outputs are scaled to reflect operational class *a priori* probabilities (Hecht-Nielsen, 2004).

It is natural to doubt that the above procedure would produce a functional gaze controller that would mimic the performance of a skilled human. But it can! The reason is probably that the human superior colliculus is essentially a fixed neuronal machine (at least in autonomous operational mode where no external control is exerted — there are several brain nuclei that can send "commands" to the superior colliculus which override its indigenous decisions) that is not all that "smart" (it operates very fast, in what looks like a "flow through" processing mode). Thus, its natural internal function is capable of being fairly accurately mimicked by a perceptron.

3.5.2 Building the Primary Visual Lexicons and Knowledge Bases

After the gaze controller has finished its training, it is time to build the rest of the visual system (and link it up with the language module). The first step is to set up the camera and start feeding frames to the gaze controller. Every time it chooses a fixation point (which is of necessity, a grid point), the V components of the gridpoints lying within the eyeball image centered at that fixation gridpoint (Figure 3.9) are gathered to form the eyeball image description vector (or just *eyeball vector*) U.

Just as in the design of mammalian primary visual cortex, each of the primary visual lexicons is responsible for monitoring a small local neighborhood of the eyeball image (these neighborhoods are all regularly spaced, they overlap somewhat, and they completely cover the eyeball image). For example, using the example numbers provided above, each primary visual lexicon (of which, for illustration purposes, Figure 3.9 shows 36, but there might actually be say, 81) would monitor the U components from say, 4900 gridpoints within and adjacent to its neighborhood of the eyeball image. The vector formed by these selected U components constitutes the *input vector* to that lexicon.

Now comes the tricky part! In order to train the primary visual lexicons, it is essential that, while this training is underway, the images being gathered by the high-resolution video camera have only **ONE VISUAL OBJECT** (an object of operational interest) in them; and **nothing else**. Further, all visual objects that will ever be of interest to the system must be presented in this manner during this training phase. As mentioned in the Appendix, in humans, this requirement is met by physically altering the characteristics of the baby's eyes after it passes through this stage (during which its vision is limited in range out to about arm length). Similarly in other mammals. For artificial visual cognition, a way must be found to meet this critical requirement. For many applications, motion segmentation, and rejecting eyeball images with more than one object fragment in them (as determined by a human educator supervising visual knowledge acquisition), will work.

The symbols of each lexicon are built by collecting input vectors from a huge collection of eyeball images (selected by the gaze controller from images gathered in the operational visual environment), but where each eyeball image contains only one object (as described in the previous paragraph). These input vectors for each lexicon are then used to build a VQ codebook for that lexicon (Zador, 1963) which is sufficiently large so that, as training progresses, very few input vectors are relatively far (more than the local intra-codebook vector distance) from a codebook vector. Once this criterion is met, the codebook is frozen and one symbol is created for, and uniquely associated with, each codebook vector. This is how the primary visual lexicon symbol sets are developed. As discussed in the Appendix, it can also be useful (but it is not essential) to develop "complex feature detector" symbols and invoke the precedence principle, as in mammalian primary visual cortex. However, this possibility will be largely ignored here.

Once the primary lexicon symbol sets are developed, the next step is to develop the knowledge bases between these lexicons. For simplicity, we can assume that every primary visual lexicon is connected to every other by a knowledge base.

The *primary visual layer* (i.e., the primary visual lexicons and the knowledge bases linking them) knowledge bases are trained using large quantities of new video gathered from the operational source, with the gaze controller selecting fixation points. Again, it is somehow arranged that each eyeball image contains only an object of operational interest at the fixation point and no visual elements of other objects (i.e., the rest of the eyeball image is blank).

As each eyeball image vector U is created and its selected subsidiary components (making up the 81 primary visual lexicon input vectors) are sent to the primary visual lexicons, each lexicon expresses an expectation with the, say, 10 symbols whose associated codebook vectors lie closest to its input vector. Count accumulation then takes place for all (unidirectional) links between pairs of these expectation symbols lying on different lexicons.

The idea of using the ten closest symbols is based upon the discovery (Caid and Hecht-Nielsen, 2001, 2004) that jet correlation vectors which are near to one another in the Euclidean metric (i.e., in the VQ space of a lexicon) represent local visual appearances that are (to a human observer) visually similar to each other; **AND VICE VERSA**. This valuable fact was pointed out in the 1980s by John Daugman (Daugman, 1985, 1987, 1988a,b; Daugman and Kammen, 1987) (Daugman also invented the iris scan biometric signature). This way, symbols which could reasonably occur together meaningfully within the same object become linked. This is much more efficient and effective than if each lexicon simply expressed the one closest symbol; and yet, because of Daugman's important principle, no harm can come of this expansion to multiple symbols. The key point is that counts are kept between each of the combinatorially many ordered excited symbol pairs (of symbols on different lexicons) involved. The process of deriving the $p(\psi|\lambda)$ knowledge link strengths ensures that only the meaningful links are retained in the end.

As training progresses, the $p(\psi|\lambda)$ knowledge link strengths are periodically calculated from the symbol co-occurrence count matrices (of which there is one for each knowledge base). When the meaningful $p(\psi|\lambda)$ values stop changing much, training is ended. The primary visual layer is now complete.

3.5.3 Building the Secondary and Tertiary Visual Layers

After completion of the primary visual layer, it is time to build the secondary and tertiary visual layers. However, this process again requires that the primary visual layer representation of each eyeball image pertain to only one object — which can now be accomplished using the primary layer's knowledge bases, as described next.

Figure 3.11 shows a portion of a frame from the wide-angle high-resolution panchromatic video camera containing an eyeball image that has been selected by the gaze controller. Each of the 81 primary visual lexicons shown is receiving its input vector from this eyeball image. The first thing that happens is that each lexicon expresses an expectation consisting of those (again, say, 10) symbols which were closest to that lexicon's input vector. (Note: This is similar to a **C1F** effect, except that the inputs are not coming from knowledge links, but from "extra-cortical sensory afferents." This illustrates, as does the handling of the S vector by primary sound lexicons discussed in Section 3.4, how the handling of these special external sensory inputs is very similar to the handling of knowledge link inputs.)

Once the primary visual lexicon expectations are established, knowledge links proceeding from the central lexicon of the primary layer, and its immediate neighboring lexicons, outward are enabled (allowing all symbols of all expectations of those lexicons to transmit) and the *distal lexicons* that these links target receive **C1F** commands. Those distal lexicons that do not receive links to symbols of their (previously established and frozen) expectations describing their portion of

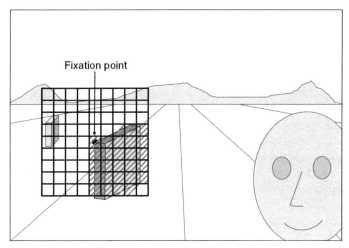

Figure 3.11   Object segmentation example. A portion of a wide-angle camera frame is shown. The gaze controller has fixated upon the upper left back corner of a rectangular solid (shown here in color for clarity). The eyeball image is shown surrounding the fixation point with the 81 (overlapping — they actually overlap a bit more than is shown here) fields of view of the 81 primary visual lexicons shown. As explained in the text, the primary visual layer knowledge bases can be used to eliminate the lexicon responses to all visual objects except the one upon which the fixation point lies.

the eyeball image have all of their symbols shut down and thereby become null (this follows from the fact, discussed in Section 3.1, that the only symbols of a lexicon with a frozen expectation which can receive input excitation from a knowledge link are those which belong to the expectation). In general, the only way that the expectation of an outlying lexicon can have any symbols retained is if one or more of its expectation symbols codes a local appearance that has been meaningfully seen before in conjunction with one or more of those expectations of the lexicons proximal to the fixation point.

A more elaborate version of this process can also be used, in which a "wave" of confabulations moves outward from the middle of the primary lexicon array to the periphery; with only knowledge bases spanning one or two inter-lexicon distances being enabled as the wave progresses. This improves performance because closer-distance related appearances are more likely to have appeared enough during training to be considered meaningful and be retained.

The astounding thing about this process (which is very fast because all of the distal lexicon confabulations happen in parallel) is that it effectively *SEGMENTS* **THE OBJECT UPON WHICH THE FIXATION POINT LIES** from all the other image content of the eyeball image. In other words, ideally, after this segmentation procedure, which is virtually instantaneous, only symbols describing local appearance of the *attended object* (the one selected by the gaze controller having the fixation point sitting on it) remain in the expectations of the primary visual layer lexicons. In Figure 3.11, these nonnull lexicons (representing the rectangular solid shown) are illustrated as diagonally-hatched in magenta. In other words, the only visual appearance data left on the primary visual layer is that describing the attended object, which has thereby effectively been *segmented* and *isolated* from the surrounding objects (as if cut out by scissors).

Note that, given the reasonably long reach of the knowledge bases projecting radially outward from the center of the primary visual layer, even objects which are interrupted by an occluding foreground object will, in principle, have all of their visible components represented by primary lexicons (and those coding the interrupting object(s) will be nulled). Also note that the smaller each primary visual lexicon is (in terms of the fraction of the eyeball image it covers), the better this process will work. Thus, segmentation might work even better if we had 625 primary visual lexicons (a **25** × **25** array). The use of more "complex" features, based upon more localized "simple" features (see Appendix), and the precedence principle is one design approach to achieving some of the benefits of more, smaller, lexicons without actually building them.

Multiple natural questions arise at this point. First, how well does this design actually work in practice? In other words, how thoroughly does this segmentation process null lexicons coding other objects and how reliably are the lexicons that code the attended object retained? The short answer is that I don't know. The only evidence I have is based upon experiments done in my lab with a very simple image environment (images of capital Latin alphabetical characters moving about, with slowly randomly changing spatial and angular velocities, on a plane). In this case, a segmentation scheme of this basic type worked very well.

In reality, probably not all fixation point objects will segment cleanly. Sometimes irrelevant lexicons will not be nulled, and relevant lexicons will be. However, because of the nature of development process for the secondary and tertiary visual lexicon symbol sets, which is described next, such errors will not matter; as long as these lapses occur randomly and as long as the general quality of the segmentation is fairly good. We will proceed with the assumption that these conditions are satisfied.

The goal for the secondary lexicon symbols is twofold. First, each such symbol should be somewhat *pose insensitive* (i.e., if it responds strongly to an object at one pose it will respond strongly to the same object at nearby poses). Also, each secondary lexicon symbol should represent a larger "chunk" of an object than any primary symbol. Such symbols are said to be more *holistic* than primary lexicon symbols. Tertiary layer lexicon symbols are to be even more holistic than secondary layer symbols.

For secondary and tertiary layer development, sequences of camera images containing the same (operationally relevant) visual object are used. At the beginning of each sequence, we assume that the gaze director has selected a fixation point on the object. In the subsequent frames of the sequence, we check to see that in each, one point near the initial fixation point is also given a high score by the gaze director. If this is true for a significant sequence of frames (say, 10 to 20 or more), then these nearby points on the subsequent frames are designated as the fixation points for those frames and this sequence of eyeball images is added to our training set for layers two and three. It is assumed that this set of sequences provides good statistical coverage of the set of all operationally relevant objects, and that each object is seen in many different, operationally characteristic poses in the sequences. It is also assumed that the poses of the fixated object in each sequence are dynamically changing. (Note: This dynamic variation in pose is needed for training, but is not a requirement for operational use; where objects can be stationary, and yet can still usually be described with a single look.)

As shown in Figure 3.9, the secondary layer lexicons receive knowledge links from primary layer lexicons. The arrangement of these links is that a secondary lexicon symbol can only receive a link from symbols lying on primary lexicons surrounding the position of the secondary lexicon in the second layer lexicon array (i.e., like the primary lexicon array, the secondary array is envisioned as also representing, with a regular "tiling," the eyeball image content of the attended object, but with each secondary lexicon representing a larger "chunk" of this object than a primary layer lexicon — since the secondary layer has fewer lexicons than the primary layer). These knowledge links connect every symbol belonging to each primary lexicon within the "field of view" of a secondary lexicon to every symbol of that secondary lexicon. For each such forward knowledge link, a link between the same two symbols in the reverse (secondary to primary) direction is also created. All of these links start out with zero strength.

As mentioned in the Appendix, not all knowledge bases need to have graded $p(\psi|\lambda)$ strengths. For many purposes in cognition, it is sufficient for knowledge links to simply be present (essentially with strength 1) or absent (with effective "strength" 0). These inter-visual-level knowledge links are of this *binary* character.

During each secondary layer training episode, sequences of say, four to six consecutive eyeball images of the same fixation point on a dynamically pose-changing object of interest (extracted at random from one of the training set sequences) are used in order. As described above, as each eyeball image in the sequence is entered, the above segmenting process is applied to it — yielding expectations on a subset of the primary layer lexicons. After the first eyeball image of the first training episode sequence has been so represented, a symbol is formed in each secondary lexicon and that symbol is bi-directionally connected from and to each of the primary layer symbols to which it is connected (by setting the relevant knowledge link strengths to 1). The second eyeball image of the sequence is then entered and segmented. The same secondary lexicon symbols created using the first eyeball image of the sequence are then connected from and to all of the primary symbols of this processed second eyeball image for which connections exist. And so forth for the remaining eyeball images of the sequence used on this first training episode. On subsequent training episodes we proceed in exactly the same manner.

Clearly, one symbol is typically going to be added to each secondary lexicon on each training trial. We stop training when the vast majority of new secondary symbols turn out to be equivalent to existing symbols — as measured by noting that, of those secondary lexicons which are receiving knowledge link inputs from primary symbols, each such secondary lexicon already has a symbol that simultaneously receives links from at least one expectation symbol of each nonnull primary lexicon from which that secondary lexicon receives a knowledge base. In other words, training is stopped when the vast majority of segmented eyeball images can be well represented by secondary symbols which have already been created.

Once training has been stopped, we then use the same training set again for *consolidating* the symbols. This involves using the primary symbols representing each eyeball image as inputs to the secondary lexicons to which they connect by strengthened connections. Whenever multiple symbols of a secondary lexicon receive links from one primary symbol of each primary lexicon from which that secondary lexicon receives links, then those secondary lexicon symbols are *merged*. Merging simply means that all of the primary to secondary links that went to each of the symbols being merged now go to the merged symbol (and vice versa for the secondary to primary links). What merging does is combine symbols which represent intersecting pose–space trajectories for the same object; thus, increasing the pose-insensitivity of the merged symbol.

Once the secondary layer lexicons are built and merged (and the knowledge bases between the primary and secondary layers frozen), the last step is to train the knowledge bases between the secondary layer lexicons. This is done by entering single eyeball images from the training set, segmenting and representing each image using the primary layer (as during training), carrying out a **W** on each secondary lexicon, and recording the symbol co-occurrence counts for each secondary layer knowledge base.

When all of this is done, the secondary to tertiary knowledge bases (and their inverses) are built using the same method as described above for the primary to secondary knowledge bases. Except this time, each training episode uses the entire set of eyeball images of each training set sequence. The resulting tertiary lexicon symbols are then merged and the tertiary layer inter-lexicon knowledge bases are built. This completes development of the vision module.

3.5.4   How Is the Visual Module Used?

After all of the lexicons and knowledge bases of the visual module of Figure 3.9 are built, the module is ready for use. This subsection briefly sketches how it can be used.

Given a new frame of imagery in which the gaze controller has found a fixation point, the primary layer of the visual module segments and represents the attended object with expectations; just as during the later phases of training and education. The symbols of the non-null expectations of primary lexicons then transmit to other primary lexicons and to secondary layer lexicons via the established knowledge bases. The other primary layer and the secondary layer lexicons then create expectations in response to **C1Fs**. The secondary visual layer expectation symbols then transmit to other secondary lexicons without expectations (if any there be) and to tertiary lexicons, again using the knowledge links established during training, and **C1Fs** establish expectations on all relevant lexicons. Finally, the knowledge links of the third layer are used to transmit from the tertiary expectations to any lexicons without expectations, followed by a final round of **C1Fs**.

The expectations formed by this initial "feedforward" interaction represent all of the symbols that are known (i.e., established by the knowledge) to be compatible with the combinations of the symbols in the primary lexicon expectations. At this point, a consensus building process is launched involving all nonnulled lexicons on all layers and all knowledge bases linking those lexicons. This consensus building process hones all the expectations until each of the involved lexicons has at most one symbol left (which is, of necessity, active). This collection of symbols is the vision module's representation of the attended visual object.

This tertiary visual object representation has three important properties. First, it has significant pose insensitivity. With high probability, if you changed the pose of the object somewhat, almost the same set of symbols would be obtained as the object's representation.

Second, the object has been *completed*; meaning that the representation has removed the effects of occluding objects that blocked the view of some portions of the object (of course, the visible portions of the object must be sufficient for completion by this method).

Third, the representation of the object at the lower levels contains details. For example, if the object is a truck being viewed from the front, the front grille and headlamps will typically be visible and will be represented at the primary level. Whereas, the representation of the object at the tertiary level will not have these details. It will be more abstract (many more specific truck images would invoke this same, or a very similar, representation).

3.5.5 Linking the Visual Module with the Language Module

Once the visual module is built, what good is it? By itself, not much. It only becomes useful when it is linked by knowledge with other cognitive modules. This subsection presents a brief sketch of an example of how, via instruction by a human educator, a vision module could be usefully linked with a language module.

A problem that has been widely considered is the automated text annotation of video describing objects within video scenes and some of those object's attributes. For example, such annotations might be useful for blind people if the images being annotated were taken by a camera mounted on a pair of glasses (and the annotations were synthesized into speech provided by the glasses to the wearer's ears via small tubes issuing from the temples of the glasses near the ears).

Figure 3.12 illustrates a simple concept for such a text annotation system. Video input from the eyeglasses-mounted camera is operated upon by the gaze controller and objects that it selects are segmented and represented by the already-developed visual module, as described in the previous subsection. The objects that were used in the visual module development process were those that a blind person would want to be informed of (curbs, roads, cars, people, etc.). Thus, by virtue of its development, the visual module will search each new frame of video for an object of operational interest (because these were the objects sought out by the human educator who's examples were used to train the gaze controller perceptron) and then that object will be segmented, and after consensus building, represented by the module on all of its three layers.

To build the knowledge links from the visual module to the text module, another human educator is used. This educator looks at each fixation point object selected by the vision module (while it is being used out on the street in an operationally realistic manner), and if this is indeed an object that would be of interest to a blind person, types in one to five sentences describing that object. These sentences are designed to convey to the blind person useful information about the nature of the object and its visual attributes (information that can be extracted by the human educator just by looking at the visual representation of the object).

To train the links from the vision module to the language module (every visual lexicon is afforded a knowledge base to every phrase lexicon), the educator's sentences are entered, in order, into the word lexicons of the sentence modules (each of which represents one sentence — see Figure 3.12); each sentence is parsed into phrases (see Section 3.4); and these phrases are represented on the sentence summary lexicon of each sentence. Counts are accumulated between the symbols active on the visual module's tertiary lexicons and those active on the summary lexicons. If the educator wishes to describe specific visual subcomponents of the object, they may designate a local window in the eyeball image for each subcomponent and supply the sentence(s) describing each such subcomponent. The secondary and tertiary lexicon symbols representing the subcomponents within each image are then linked to the summary lexicons of the associated sentences. Before being used in this application, all of the internal knowledge bases of the language module have already been trained using a huge text training corpus.

After a sufficient number of education examples have been accumulated (as determined by final performance — described below), the link use counts are converted into $p(\psi|\lambda)$ probabilities and frozen. The knowledge bases from the visual module's lexicons to all of the sentence summary lexicons are then combined (so that the available long-range context can be exploited by a sentence in any position in the sequence of sentences to be generated). The annotation system is now ready for testing.

The testing phase is carried out by having a sighted evaluator walk down the street wearing the system (yes, the idea is that the entire system is in the form of a pair of glasses!). As the visual module selects and describes each object, knowledge link inputs are sent to the language module. These inputs are used, much as in the example of Section 3.3: as context that drives formation of a sentence (only now there is no starter). Using consensus building (and separate sentence starter generator and sentence terminator subsystems — not shown in Figure 3.12 and not discussed here

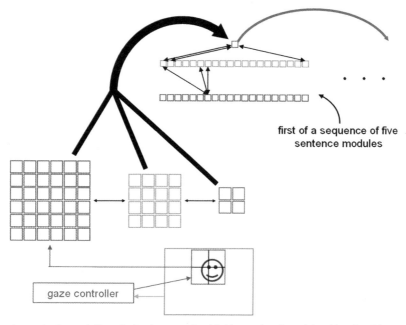

first of a sequence of five
sentence modules

gaze controller

Figure 3.12   Image text annotation. A simple example of linking a visual module with a (text) language module. See text for description.

— for starting and ending the sentence), the language module composes one or more grammatical sentences that describe the object and its attributes.

The number of sentences is determined by a meaning content critic subsystem (not shown in Figure 3.12) which stops sentence generation when all of the distinctive, excited, sentence summary lexicon symbols have been "used" in one or more of the generated sentences.

This sketch illustrates the monkey-see/monkey-do principle of cognition: there is never any complicated algorithm or software; no deeply principled system of rules or mathematical constraints; just confabulation and consensus building. It is a lot like that famous cartoon where scientists are working at a blackboard, attempting, unsuccessfully, to connect up a set of facts on the left with a desired conclusion on the right via a complicated scientific argument spanning the gap between them. In frustration, one of the scientists erases a band in the middle of the argument and puts in a box (equipped with input and output arrows) labeled "And Then a Miracle Occurs." THAT is the nature of cognition.

3.6 DISCUSSION

This chapter has reviewed a "unified theory of cognition" which purports to explain all aspects of this vast subject with one type of knowledge and one information processing operation. The hope is that this discussion has convinced you that this approach to cognition is worthy of more extensive investigation. Only after language, sound, and vision systems such as those described here have been built, and widely evaluated and criticized, will a sense begin to emerge that the mechanization of cognition is truly possible. I am hopeful that the arguments and discussion presented here are sufficiently compelling to make such a research program sensible.

REFERENCES

Ackley D.H., G.E. Hinton and T.J. Sejnowski, A learning algorithm for Boltzmann machines, *Cognitive Science*, 9 (1985), pp. 147–169.

Caid W.R. and R. Hecht-Nielsen (Inventors), Representation and retrieval of images using context vectors derived from image information elements, U.S. Patent No. 6,173,275 (January 9, 2001).

Caid W.R. and R. Hecht-Nielsen (Inventors), Representation and retrieval of images using context vectors derived from image information elements, U.S. Patent No. 6,760,714 (July 6, 2004).

Chomsky N., *Rules and Representations*, Columbia University Press, New York, New York (1980).

Daugman J.G., Uncertainty relation for resolution in space, spatial frequency, and orientation optimized by two-dimensional visual cortical filters, *Journal of Optical Society of America A*, 2(7) (1985), pp. 1160–1169.

Daugman J.G., Image analysis and compact coding by oriented 2-D Gabor primitives, in: Eamon B. Barrett and James J. Pearson (eds), *Image Understanding and the Man–Machine Interface, SPIE Proceedings*, vol. 758, Society for Photo-Optical Instrumentation Engineers, Bellingham, Washington (1987), pp. 19–30.

Daugman J.G., Complete discrete 2-D Gabor transforms by neural networks for image analysis and compression, *IEEE Transactions of Acoustics, Speech Signal Processing*, 36(7) (1988a), pp. 1169–1179.

Daugman J.G., Relaxation neural network for non-orthogonal image transforms, *Proceedings of the International Conference on Neural Networks*, vol. 1, IEEE Press, New York, New York (1988b), pp. 547–560.

Daugman J.G. and D.M. Kammen, Image statistics, gases, and visual neural primitives, *Proceedings of the International Conference on Neural Networks*, vol. 4, San Diego, California (1987), pp. 163–175.

Gregory R., The blind leading the sighted, *Nature*, 430 (2004), p. 836.

Hecht-Nielsen R., *Neurocomputing*, Addison-Wesley, Reading, Massachusetts (1990); Corrected Printing, 1991; Japanese Edition, Addison-Wesley, Toppan, Tokyo (1993).

Hecht-Nielsen R., Perceptrons, UCSD Institute for Neural Computation, Report No. 0403 (2004) (available at http://inc2.ucsd.edu/addedpages/techreports.html).

Hecht-Nielsen R., Cogent confabulation, *Neural Networks*, 18 (2005), pp. 111–115.

Hecht-Nielsen R. and Y.-T. Zhou, VARTAC: A foveal active vision ATR system, *Neural Networks*, 8 (1995), pp. 1309–1321.

Hyvärinen A., J. Karhunen and E. Oja, *Independent Component Analysis*, Wiley Interscience, New York, New York (2001).

Mack A. and I. Rock, *Inattentional Blindness*, MIT Press, Cambridge, Massachusetts (1998).

Miller G.A. The magical number seven, plus or minus two: some limits on our capacity for processing information, *Psychological Reviews*, 63 (1956), pp. 81–97.

Miyamoto H., J. Morimoto, K. Doya and M. Kawato, Reinforcement learning with via-point representation, *Neural Networks*, 17 (2004), pp. 299–305.

Oertel D., R.R. Fay and A.N. Popper, *Integrative Functions in the Mammalian Auditory Pathway, Handbook in Auditory Research*, Springer-Verlag, New York, New York, (2002).

Paxinos G. and J.K. Mai (eds), *The Human Nervous System*, Second Edition, Elsevier, San Diego, California (2004).

Sagi B., S.C. Nemat-Nasser, R. Kerr, R. Hayek, C. Downing and R. Hecht-Nielsen, A biologically motivated solution to the cocktail party problem, *Neural Computation*, 13(7) (2001), pp. 1575–1602.

von der Malsburg C., Considerations for a visual architecture, in: Eckmiller R. (ed.), *Advanced Neural Computers*, Elsevier, North Holland, Amsterdam (1990), pp. 303–312.

Zador P.L. Development and evaluation of procedures for quantizing multi-variate distributions, PhD Dissertation, Stanford University, Palo Alto, California (1963).

APPENDIX: BIOLOGICAL COGNITION

3.A.1 Introduction

This Appendix sketches the author's *confabulation theory* of animal cognition. The discussion is focused on the biological implementation of cognition in human cerebral cortex and thalamus (hereinafter, often referred to jointly as *thalamocortex*).

The enormous diversity of animal life, currently ranging in size from single cells (the smallest animals which have ever lived) to blue whales (the largest), and ranging in adaptation across a huge range of biomes, obfuscates its unity. All animal cells function using very similar basic biochemical mechanisms. These mechanisms were developed once and have been genetically conserved across essentially all species. Mentation is similar. The basic mechanism of cognition is, in the view of this theory, the same across all vertebrates (and possibly invertebrates, such as octopi and bees, as well).

The term *cognition*, as used in this Appendix, is not meant to encompass all aspects of mentation. It is restricted to (roughly) those functions carried out by the human cerebral cortex and thalamus. Cognition is a big part of mentation for certain vertebrate species (primates, cats, dogs, parrots, ravens, etc.), but only a minor part for others (fish, reptiles, etc.). Frog cognition exists, but is a minor part of frog mentation. In humans, cognition is the part of mentation of which we are, generally, most proud; and most want to imitate in machines.

An important concept in defining cognition is to consider function; not detailed physiology. In humans, the enormous expansion of cerebral cortex and thalamus has allowed a marked segregation of cognitive function to those organs. Birds can exhibit impressive cognitive functions (Pepperberg, 1999; Weir et al., 2002). However, unlike the case in humans, these cognitive functions are probably not entirely confined to a single, neatly delimited, laminar brain nucleus. Even so, the theory hypothesizes that the underlying mathematics of cognition is exactly the same in all vertebrate species (and probably in invertebrates); even though the neuronal implementation varies considerably. (This is much as in electronics: the same logic circuit can be implemented with electromechanical relays, in silicon CMOS circuits, using vacuum tubes, or even using fluidic devices. While these implementations are physically dissimilar; they are mathematically identical.)

Although not enough is known to create a definitive list of specific human cognitive functions, the following items would certainly be on such a list:

- Language
- Hearing
- Seeing
- Somatosensation
- Action (movement process and thought process) origination

This Appendix focuses upon the implementation of cognitive knowledge links, confabulation, and action command origination by the human cerebral cortex and thalamus. It is assumed that the reader is familiar with the concepts, terminology, and mathematics of elementary confabulation (e.g., as discussed in Hecht-Nielsen, 2005) and with elementary human neuroanatomy and neurophysiology (e.g., as presented in Mai et al., 2004; Mountcastle, 1998; Nicholls et al., 2001; Nolte, 1999; Paxinos and Mai, 2004; Steward, 2000). The theory hypothesizes that all human cognitive functions, including those listed above, are implemented using the basic confabulation machinery sketched in this Appendix. To keep this Appendix focused, the manner in which confabulation can be used to carry out specific cognitive functions (such as those listed above) will not be discussed here, as this is essentially the material covered in the main body of the chapter.

To keep the size of this Appendix reasonable, and to avoid speculations about fine details, the treatment will avoid extensive discussion at the level of individual neurons, synapses, and axonal signals. For example, only simplified gross or summary aspects of interneuronal signaling processes and neurodynamic processes will be discussed. Yet, the theory contends that the fine details jibe with these slightly larger-scale functional descriptions. *Consensus building* (dynamically interacting confabulations taking place contemporaneously in multiple lexicons), which the theory hypothesizes is the dominant mode of use of confabulation in human cognition, will only be briefly mentioned, as a detailed treatment would go beyond the introductory scope of this sketch of the theory's biological implementation. At the current time, the theory presented in this Appendix is the only existing detailed explanation of the operation of cerebral cortex or thalamus, and of human cognition.

3.A.2 Summary of the Theory

The fundamental hypotheses of the theory are summarized in this section. Subsequent sections elaborate.

All information processing involved in human cognition is hypothesized to be carried out by thousands of separate thalamocortical *modules*; each consisting of a particular small localized patch of cortex (possibly consisting of disjoint, nonadjacent, subpatches), and a particular, uniquely paired, small localized zone of thalamus, which are reciprocally connected axonally. These *feature attractor modules* (of which human thalamocortex has many thousands) are hypothesized to each implement a list of (typically thousands of) discrete symbols (which is stable over time, and can be added to) and to carry out a single symbolic information processing operation called confabulation. Each symbol is represented by a specific collection of neurons within the module. These collections are all about the same size within any single module; but this size varies considerably, from tens to hundreds of neurons per symbol — a genetically determined value, between modules located in different parts of the cortex. Any pair of such neuron collections of the same module, representing two different symbols, typically have a few neurons in common. Each neuron which participates in such a collection typically participates in many others as well. When considering their function (rather than their biological implementation), each such thalamocortical module will be referred to as a *lexicon*.

Each feature attractor module is hypothesized to be controlled by a single graded (i.e., analog-valued) excitatory control input; exactly in analogy with an individual muscle (each muscle has a single-graded excitatory control input that, by its value across time, specifies the muscle's contraction force history). The theory hypothesizes that properly phased and timed sequences of such thought control inputs to each member of an ensemble of cortical modules cause them to carry out a thought process. These thought processes are "data-independent," much like computer operations such as numerical addition and Boolean XOR. It is hypothesized that vast numbers of such thought processes (and movement processes) are learned by rehearsal training and stored in a hierarchical organization within knowledge bases linking yet other lexicons of cerebral cortex.

Confabulation is implemented in parallel by the neurons of a feature attractor module and is often completed in a few tens of milliseconds. This is a "winners take all" style of dynamical competitive interaction between symbols that does not require a "referee" or "controller" to be in charge. The states of the involved neurons evolve dynamically and autonomously during confabulation via the massively parallel mutual interactions of the involved neurons. The final state of each involved neuron is either *excited* or *active* (a small minority of neurons), or almost completely *inactive* (the vast majority). The term *active* (implying a momentary, maximally communicating state) is deliberately undefined as it involves neuronal signaling details which are not yet known; as does the term *excited* (implying a highly, but not maximally, communicating state).

If the outcome of a confabulation is a single symbol, the neurons representing that symbol will automatically be made active and all other neurons of the module are *inactive* (not communicating). However, if multiple symbols result from a confabulation (the outcome is dependent upon multiple factors, including the time profile of the module control signal — see below), these will be at different levels of excitation (but not active) and all other symbols will be inactive. Those few neurons which end up in the excited or active state represent the symbol(s) which "won" the confabulation competition. These symbols are termed the *conclusions* of that confabulation operation. Confabulations frequently end with no excited or active neurons — a conclusion termed the *null symbol* — which signifies that no viable conclusion was reached. This ability to decide that "I don't know" is one of the great strengths of cognition.

The theory hypothesizes that the only knowledge stored and used for cognition within thalamocortex takes the form of (indirect, parallel) unidirectional axonal connections between the population of neurons within one feature attractor module used to represent one symbol and neurons used to represent a symbol in another feature attractor. Each such *link* between a pair of symbols is termed an *item of knowledge*. The average human is hypothesized to possess billions of such items of knowledge.

The theory hypothesizes that items of knowledge are immediately established on a temporary basis when a novel, meaningful, co-occurrence of symbol pair activity occurs during a period of wakefulness (assuming that those symbols are equipped with the necessary axonal paths with which a link can be established). If this *short-term memory* link is selected for deliberate rehearsal during the next period of sleep, it gets promoted to the status of a *medium-term memory*. If this medium-term memory link is revisited on near-term subsequent sleep periods it then gets promoted to a *long-term memory*, which will typically last years; as long as the involved tissue remains patent and not redeployed.

It is hypothesized that each time a thalamocortical module carries out a confabulation which concludes with the expression of a single active symbol (as opposed to no symbols or multiple excited symbols), an *action command* associated with that symbol is immediately issued by a specialized cortical component of that module (this is the theory's *conclusion–action principle*). Action commands cause muscle and thalamocortical thought module control signals to be sent. In other words, every time a thought process successfully reaches a single conclusion, a new movement process and/or thought process is launched (some of which may undergo additional evaluation before being finally *executed*). This is the theory's explanation for the origin of all nonautonomic animal behavior.

As with almost all cognitive functions, actions are organized into a hierarchy, where individual symbols belonging to higher-level lexicons typically each represent a time-ordered sequence of multiple lower-level symbols.

Evolution has seen to it that symbols, which when expressed alone launch action commands that could conflict with one another (e.g., carrying out a throwing motion at the same time as trying to answer the telephone), are grouped together and collected into the same lexicon (usually at a high level in the action hierarchy). That way, when one such *action symbol* wins a confabulation (and has its associated lower-level action commands launched), the others are silent — thereby automatically *deconflicting* all actions. This is why all aspects of animal behavior are so remarkably *focused* in character. Each complement of our moving and thinking "hardware" is, by this mechanism, automatically restricted to doing one thing at a time. *Dithering* (rapidly switching from one decisive action (behavioral program) to another, and then back again) illustrates this perfectly.

The thought processes at the lowest level of the action hierarchy are typically carried out unconditionally at high speed. If single symbol states result from confabulations which take place as part of a thought process, these symbols then decide which actions will be carried out next (this happens both by the action commands the expression of these symbols launch, and by the influence of these symbols — acting through knowledge links — on the outcomes of subsequent confabulations; for which these symbols act as assumed facts). Similarly for movements, as ongoing movements bring about changes in the winning symbols in confabulations in somatosensory cortex — which then alter the selections of the next action symbols in modules in motor and premotor cortex. This ongoing, high-speed, dynamic contingent control of movement and thought helps account for the astounding reliability and comprehensive, moment-by-moment adaptability of animal action.

All of cognition is built from the above discussed elements: lexicons, knowledge bases, and the action commands associated with the individual symbols of each lexicon. The following sections of this Appendix discuss more details of how these elements are implemented in the human brain. See Hecht-Nielsen and McKenna (2003) for some citations of past research that influenced this theory's development.

3.A.3 Implementation of Lexicons

Figure 3.A.1 illustrates the physiology of thalamocortical feature attractor modules. In reality, these modules are not entirely disjoint, nor entirely functionally independent, from their physically neighboring modules. However, as a first approximation, they can be treated as such; which is the view which will be adopted here.

Figure 3.A.2 shows more details of the functional character of an individual lexicon. The cortical patch of the module uses certain neurons in Layers II, III, and IV to represent the symbols of the module. Each symbol (of which there are typically thousands) is represented by a roughly equal number of neurons; ranging in size from tens to hundreds (this number deliberately varies, by genetic command, with the position of the cortical patch of the module on the surface of cortex). The union of the cortical patches of all modules is the entire cortex, whereas the union of the thalamic zones of all modules constitutes only a portion of thalamus.

Symbol-representing neurons of the module's cortical patch can send signals to the glomeruli of the paired thalamic zone via neurons of Layer VI of the patch (as illustrated on the left side of Figure 3.A.2). These downward connections each synapse with a few neurons of the thalamic reticular nucleus (NRT) and with a few glomeruli. The NRT neurons themselves (which are inhibitory) send axons to a few glomeruli. The right side of Figure 3.A.2 illustrates the connections back to the cortical patch from the thalamic zone glomeruli (each of which also synapses with a few neurons of the NRT). These axons synapse primarily with neurons in Layer IV of the patch, which

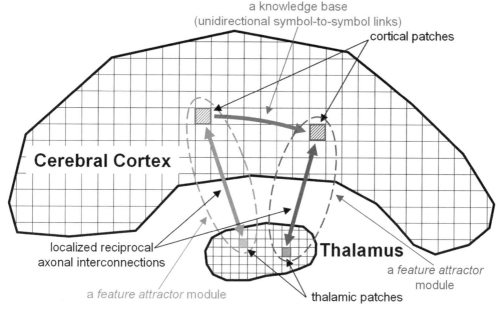

Figure 3.A.1   Thalamocortical modules. All cognitive information processing is carried out by distinct, modular, thalamocortical circuits termed *feature attractors*; of which two are shown here. Each feature attractor module (of which human cortex has many thousands) consists of a small localized patch of cortex (which may be comprised of disjoint, physically separated, sub-patches), a small localized zone of thalamus, and the reciprocal axonal connections linking the two. When referring to its function (rather than its implementation, a feature attractor is termed a *lexicon*). Each feature attractor module implements a large stable set of attractive states called *symbols*, each represented by a specific collection of neurons (all such collections within a module are of approximately the same size). Neuron overlap between each pair of symbols is small, and each neuron involved in representing one symbol typically participates in representing many symbols. One *item of knowledge* is a (parallel, two-stage synfire) set of unidirectional axonal connections collectively forming a *link* between the neurons representing one symbol within one feature attractor (e.g., the green one shown here) and neurons representing one symbol on a second feature attractor (e.g., the blue one shown here). The collection of all such links between the symbols of one module (here the green one), termed the *source lexicon*, and that of a second (here the blue one), termed the *target lexicon*, are termed a *knowledge base* (here represented by a red arrow spanning the cortical portions of the green and blue modules).

subsequently excite other neurons of Layers II, III, and IV. As mentioned above, no attempt to discuss the details of this module design will be made, as these details are not yet adequately established and, anyway, are irrelevant for this introductory sketch. Instead, a discussion is now presented of a simple mathematical model of an attractor network to illustrate the hypothesized dynamical behavior of a thalamocortical model in response to proper knowledge link and operation command inputs.

The theory hypothesizes that each thalamocortical module carries out a single information processing operation — confabulation. This occurs whenever appropriate knowledge link inputs and the operation command input arrive at the module at the same time. The total time required for the module to carry out one confabulation operation is roughly 100 msec. Ensembles of mutually interacting confabulations (instances of *consensus building* — see the main Chapter) can often be highly overlapped in time. By this means, the "total processing time" exhibited by such a consensus building ensemble of confabulations can be astoundingly short — often a small multiple of the involved axonal and synaptic delays involved; and not much longer than a small number of individual confabulations. This accounts for the almost impossibly short "reaction times" often seen in various psychological tests.

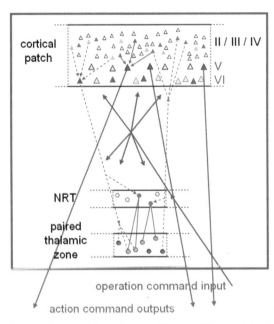

Figure 3.A.2 A single thalamocortical module; side view. The module consists of a full-depth patch of cortex (possibly comprised of multiple separate full-depth disjoint sub-patches — not illustrated here); as well as a paired zone of thalamus. The green and red neurons in cortical layer II, III or IV illustrate the two collections of neurons representing two symbols of the module (common neurons shared by the two collections are not shown; nor are the axons involved in the feature attractor neuronal network function used to implement confabulation). The complete pool of neurons within the module used to represent symbols contains many tens, or even hundreds, of thousands of neurons. Each symbol-representing neuron collection has tens to hundreds of neurons in it. Axons from cortical layer VI to NRT (NRT) and thalamus are shown in dashed blue. Axons from thalamic glomeruli to NRT and cortical layer IV are shown in dashed red. Axons from NRT neurons to glomeruli are shown in pink. An axon of the operation command input, which affects a large subset of the neurons of the module, and which arrives from an external subcortical nucleus, is shown in green. The theory only specifies the overall information processing function of each cortical module (implementation of the list of symbols, confabulation, and origination or termination of knowledge links). Details of module operation at the cellular level are not known.

The mathematical model discussed below illustrates the dynamical process involved in carrying out one confabulation. Keep in mind that this model might represent strictly cortical neuron dynamics, module neurodynamics between the cortical and thalamic portions of the module, or even the overall dynamics of a group of smaller attractor networks (e.g., a localized version of the "network of networks" hypothesis of Sutton and Anderson in Hecht-Nielsen and McKenna, 2003; Sutton and Anderson, 1995).

In 1969, Willshaw and his colleagues (Willshaw et al., 1969) introduced the "nonholographic" associative memory. This "one-way" device ("retrieval key" represented on one "field" of neurons and "retrieved pattern" on a second), based on Hebbian learning, is a major departure in concept from the previous (linear algebra-based) associative memory concepts (Anderson, 1968, 1972; Gabor, 1969; Kohonen, 1972). The brilliant Willshaw design (an absolutely essential step towards the theory presented in this Appendix) is a generalization of the pioneering Steinbuch *learnmatrix* (Steinbuch, 1961a,b, 1963, 1965; Steinbuch and Piske, 1963; Steinbuch and Widrow, 1965); although Willshaw and his colleagues were not aware of this earlier development. For efficiency, it is assumed that the reader is familiar with the Willshaw network and its theory (Amari, 1989; Kosko, 1988; Palm, 1980; Sommer and Palm, 1999). A related important idea is the "Brain State in a Box" architecture of Anderson et al. (1977).

In 1987, I conceived a hybrid of the Willshaw network and the Amari or Hopfield "energy function" attractor network (Amari, 1974; Amit, 1989; Hopfield, 1982, 1984). In effect, this hybrid

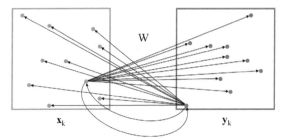

Figure 3.A.3 Simple attractor network example. The left, x, neural field has N neurons; as does the right, y, neural field. One Willshaw stable state pair, x_k and y_k is shown here (actually, each x_k and y_k typically has many tens of neurons — e.g., Np = 60 for the parameter set described in the text — of which only 10 are shown here). Each neuron of each state sends connections to all of the neurons of the other (only the connections from one neuron in x_k and one neuron in y_k are shown here). Together, the set of all such connections for all L stable pairs is recorded in the connection matrix W. Notice that these connections are not knowledge links — they are internal connections between x_k and y_k — the two parts of the neuron population of symbol k within a single module. Also, unlike knowledge link connections (which, as discussed in the next section, are unidirectional and for which the second stage is typically very sparse), these interpopulation connections must be reciprocal and dense (although they need not be 100% dense — a fact that you can easily establish experimentally with your model).

network was two reciprocally connected Willshaw networks; however, it also had an energy function. Karen Haines and I theoretically investigated the dynamics of this network (Haines and Hecht-Nielsen, 1988) [in 1988 computer exploration of the dynamics of such networks, at scales sufficiently large to explore their utility for information processing, was not feasible]. We were able to show theoretically that this hybrid had four important (and unique) characteristics. First, it would, with very high probability, converge to one of the Willshaw stable states. Second, it would converge in a finite number of steps. Third, there were no "spurious" stable states. Fourth, it could carry out a "winner take all" kind of information processing. This hybrid network might thus serve as the functional implementation of (in the parlance of this Appendix) a symbolic lexicon. This was the first result on the trail to the theory presented here. It took another 16 years to discover that, by having antecedent support knowledge links deliver excitation to symbols (i.e., stable states) of such a lexicon, this simple one-winner-take-all information processing operation (*confabulation*) is sufficient to carry out all of cognition.

By 1992 it had become possible to carry out computer simulations of reciprocal Willshaw networks of interesting size. This immediately led to the rather startling discovery that, even without an energy function (i.e., carrying out neuron updating on a completely local basis, as in Willshaw's original work), even significantly "damaged" (the parlance at that stage of discovery) starting states (Willshaw stable states with a significant fraction of added and deleted neurons) would almost always converge in one "round-trip" or "out-and-back cycle." This made it likely that this is the functional design of cortical lexicon circuits.

As this work progressed, it became clear that large networks of this type were even more robust and would converge in one cycle even from a small incomplete fragment of a Willshaw stable state. It was also at this point that the issue of "threshold control" (Willshaw's original neurons all had the same fixed "firing" threshold — equal to the number of neurons in each stable state) came to the fore. If such networks were operated by a threshold control signal that rose monotonically from a minimum level, it could automatically carry out a global "most excited neurons win" competition without need for communication between the neurons. The subset of neurons which become active first then inhibit others from becoming so (at least in modules in the brain; but not in these simple mathematical models, which typically lack inhibition). From this came the idea that each module must be actively controlled by a graded command signal, much like an individual muscle. This eventually led to the realization that the control of movement and the control of thought are implemented in essentially the same manner; using the same cortical and subcortical structures (indeed, the theory postulates that there are many combined movement and thought processes which are represented as unitized symbols at higher levels in the action hierarchy — e.g., a back dive action routine in which visual perception must feed corrections to the movement control in order to enter the water vertically).

To see what attractor networks of this unusual type are all about, the reader is invited to pause in their reading and build (e.g., using C, LabVIEW, MATLAB, etc.) a simple working example using the following prescription. If you accept this invitation, you will see first-hand the amazing capabilities of these networks (which will help you appreciate and accept the theory). While simple, this network possesses many of the important behavioral characteristics of the hypothesized design of biological feature attractor modules.

We will use two N-dimensional real column vectors, $\mathbf{x}$ and $\mathbf{y}$, to represent the states of N neurons in each of two "neural fields." For good results, N should be at least 10,000 (even better results are obtained for N above 30,000). Using a good random number generator, create L pairs of $\mathbf{x}$ and $\mathbf{y}$ vectors $\{(\mathbf{x}_1,\mathbf{y}_1), (\mathbf{x}_2,\mathbf{y}_2),\ldots, (\mathbf{x}_L,\mathbf{y}_L)\}$ with each $\mathbf{x}_i$ vector and each $\mathbf{y}_i$ vector having binary (0 and 1) entries selected independently at random; where the probability of each component being 1 is p. Use, for example, p = 0.003 and L = 5,000 for N = 20,000. As you will see, these $\mathbf{x}_i$ and $\mathbf{y}_i$ pairs turn out to be *stable states* of the network. Each $\mathbf{x}_k$ and $\mathbf{y}_k$ vector pair, k = 1, 2,..., L *represents* one of the L *symbols* of the network. For simplicity, we will concentrate on the $\mathbf{x}_k$ vector as the representation of symbol k. Thus, each symbol is represented by a collection of about Np "active" neurons. The random selection of the symbol neuron sets and the deliberate processes of neuronal interconnection between the sets correspond to the development and refinement processes in each thalamocortical module that are described later in this section.

During development of the bipartite stable states $\{(\mathbf{x}_1,\mathbf{y}_1), (\mathbf{x}_2,\mathbf{y}_2),\ldots, (\mathbf{x}_L,\mathbf{y}_L)\}$ (which happens gradually over time in biology, but all at once in this simple model), connections between the neurons of the $\mathbf{x}$ and $\mathbf{y}$ fields are also established. These connections are very simple: each neuron of $\mathbf{x}_k$ (i.e., the neurons of the $\mathbf{x}$ field whose indices within $\mathbf{x}_k$ have a 1 assigned to them) sends a connection to each neuron of $\mathbf{y}_k$ and vice versa. This yields a connection matrix W given by

$$W = U\left(\sum_{i=1}^{N} \mathbf{y}_k\mathbf{x}_k^T\right) \tag{3A.1}$$

where the matrix function U sets every positive component of a matrix to 1 and every other component to zero. Given these simple constructions, you are now ready to experiment with your network.

First, choose one of the $\mathbf{x}_k$ vectors and modify it. For example, eliminate a few neurons (by converting entries that are 1 to 0s) or add a few neurons (by converting 0s to 1s). Let this modified $\mathbf{x}_k$ vector be called $\mathbf{u}$. Now, "run" the network using $\mathbf{u}$ as the initial $\mathbf{x}$ field state. To do this, first calculate the *input excitation* I_j of each $\mathbf{y}$ field neuron j using the formula $\mathbf{I} = W\mathbf{u}$; where $\mathbf{I}$ is the column vector containing the input excitation values I_j, j = 1, 2,..., N. In effect, each active neuron of the $\mathbf{x}$ field (i.e., those neurons whose indices have a 1 entry in $\mathbf{u}$) sends output to neurons of the $\mathbf{y}$ field to which it has connections (as determined by W). Each neuron j of the $\mathbf{y}$ field sums up the number of connections it has received from active $\mathbf{x}$ field neurons (the ones designated by the 1 entries in $\mathbf{u}$) and this is I_j.

After the I_j values have been calculated, those neurons of the $\mathbf{y}$ field which have the largest I_j values (or very close to the largest — say within 3 or 4 — this is a parameter you can experiment with) are made *active*. As mentioned above, this procedure is a simple, but roughly equivalent, surrogate for active global graded control of the network. Code the set of active $\mathbf{y}$ field neurons using the vector **v** (which has a 1 in the index of each active **y** field neuron and zeros everywhere else). Then calculate the input intensity vector $W^T\mathbf{v}$ for the **x** field (this is the "reverse transmission" phase of the operation of the network) and again make active those neurons with largest or near-largest values of input intensity. This completes one cycle of operation of the network. Astoundingly, the state of the **x** field of the network will be very close to $\mathbf{x}_k$, the vector used as the dominant base for the construction of **u** (as long as the number of modifications made to $\mathbf{x}_k$ when forming **u** was not too large).

Now expand your experiments by letting each **u** be equal to one of the **x** field stable states $\mathbf{x}_k$ with many (say half) of its neurons made inactive plus the union of many (say, 1 to 10) small fragments (say, 3 to 8 neurons each) of other stable **x** field vectors, along with a small number (say, 5 to 10) of active "noise" (randomly selected) neurons (see Figure 3.A.4). Now, when operated, the network will converge rapidly (again, often in one cycle) to the $\mathbf{x}_k$ symbol whose fragment was the largest. When you do your experiments, you will see that this works even if that largest fragment contains only a third of the neurons in the original $\mathbf{x}_k$. If **u** contains multiple stable **x** field vector fragments of roughly the same maximum size, the final state is the union of the complete **x** field vectors (this is an important aspect of confabulation not mentioned in Hecht-Nielsen, 2005). As we will see below, this network behavior is essentially all we need for carrying out confabulation.

Again, notice that to achieve the "neurons with the largest or near-largest, input excitation win" information processing effect, all that is needed is to have an excitatory operation control input to the network which uniformly raises all of the involved neurons' excitation levels (towards a constant fixed "firing" threshold that each neuron uses) at the same time. By ramping up this input, eventually a group of neurons will "fire"; and these will be exactly those with the largest or

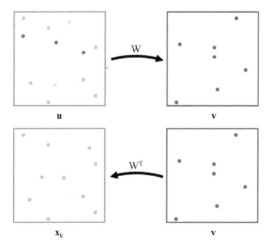

Figure 3.A.4   Feature attractor function of the simple attractor network example. The initial state (top portion) of the **x** neural field is a vector **u** consisting of a large portion (say, half of its neurons) of one particular $\mathbf{x}_k$ (the neurons of this $\mathbf{x}_k$ are shown in green), along with small subsets of neurons of many other **x** field stable states. The network is then operated in the **x** to **y** direction (top diagram). Each neuron of **u** sends output to those neurons of the **y** field to which it is connected (as determined by the connection matrix W). The **y** field neurons which receive the most, or close to the most, connections from active neurons of **u** are then made active. These active neurons are represented by the vector **v**. The network is then operated in the **y** to **x** direction (bottom diagram), where the **x** field neurons receiving the most, or close to the most, connections from active neurons of **v** are made active. The astounding thing is that this set of active **x** field neurons is typically very close to $\mathbf{x}_k$, the dominant component of the initial **u** input. Yet, all of the processing is completely local and parallel. As will be seen below, this is all that is needed to carry out confabulation. In thalamocortical modules this entire cycle of operation (which is controlled by a rising operation command input supplied to all of the involved neurons of the module) is probably often completed in roughly 100 msec. The hypothesis of the theory is that this *feature attractor* behavior implements *confabulation* — the universal information processing operation of cognition.

near-largest input intensity. Localized mutual inhibition between cortical neurons (which is known to exist, but is not included in the above simplified model) then sees to it that there are no additional winners; even if the control input keeps rising. Note also that the rate of rise of the control signal can control the width of the band of input excitations (below maximum) for which neurons are allowed to win the competition: a fast rate allows more neurons (with slightly less input intensity than the first winners) to become active before inhibition has time to kick in. A slow rate of rise restricts the winners to just one symbol. Finally, the operation control input to the network can be limited to be less than some deliberately chosen maximum value: which will leave no symbols active if the sum of the all neuron's input excitation, plus the control signal, are below the fixed "threshold" level. Thus, an attractor network confabulation can yield a null conclusion when there are no sufficiently strong answers. Section 3.1 of the main chapter discusses some of these information processing *effects*; which can be achieved by judicious control of a lexicon's operation command input signal.

An important difference between the behavior of this simple attractor network model and that of thalamocortical modules is that, by involving inhibition (and some other design improvements such as unifying the two neural fields into one), the biological attractor network can successfully deal with situations where even hundreds of stable $\mathbf{x}$ field vector fragments (as opposed to only a few in the simple attractor network) can be suppressed to yield a fully expressed dominant fragment $\mathbf{x}_k$. This remains an interesting area of research.

The development process of feature attractors is hypothesized by the theory to take place in steps (which are usually completed in childhood; although under some conditions adults can develop new feature attractor modules).

Each feature attractor module's set of symbols is used to describe one *attribute* of objects in the mental universe. Symbol development starts as soon as meaningful (i.e., not random) inputs to the feature attractor start arriving. For "lower-level" attributes, this *self-organization* process sometimes starts before birth. For "higher-level" attributes (modules), the necessary inputs do not arrive (and lexicon organization does not start) until after the requisite lower-level modules have organized and started producing assumed fact outputs.

The hypothesized process by which a feature attractor module is developed is now sketched. At the beginning of development, a sizable subset of the neurons of cortical layers II, III, and IV of the module happen by chance to preferentially receive extra-modular inputs and are stimulated repeatedly by these inputs. These neurons develop, through various mutually competitive and cooperative interactions, responses which collectively cover the range of signal ensembles the region's input channels are providing. In effect, each such feature detector neuron is simultaneously driven to respond strongly to one of the input signal ensembles it happens to repeatedly receive; while at the same time, through competition between feature detector neurons within the module, it is discouraged from becoming tuned to the same ensemble of inputs as other feature detector neurons of that module. This is the classic insight that arose originally in connection with the mathematical concepts of *vector quantization* (*VQ*) and *k-means*. These competitive and cooperative VQ feature set development ideas have been extensively studied in various forms by many researchers from the 1960s through today (e.g., see Carpenter and Grossberg, 1991; Grossberg, 1976; Kohonen, 1984, 1995; Nilsson, 1965, 1998; Tsypkin, 1973; Zador, 1963). The net result of this first stage of feature attractor circuit development is a large set of feature detector neurons (which, after this brief initial plastic period, become largely frozen in their responses — unless severe trauma later in life causes recapitulation of this early development phase) that have responses with moderate local redundancy and high input range coverage (i.e., low information loss). These might be called the *simple* feature detector neurons.

Once the simple feature detector neurons of a module have been formed and frozen, additional *secondary* (or "complex") feature detector neurons within the region then organize. These are neurons which just happen (the wiring of cortex is locally random and is essentially formed first, during early organization and learning, and then is soon frozen for life) to receive most of their input from simple feature detector neurons (as opposed to primarily from extra-modular inputs, as with the simple feature detector neurons themselves).

In certain areas of cortex (e.g., primary visual cortex) secondary feature detector neurons can receive inputs from primary feature detector neurons "belonging" to other nearby modules. This is an example of why it is not correct to say that modules are disjoint and noninteracting (which nonetheless is exactly how we will treat them here).

Just as with the primary neurons, the secondary feature detector neurons also self-organize along the lines of a VQ codebook — except that this codebook sits to some degree "on top" of the simple cell codebook. The net result is that secondary feature neurons tend to learn statistically common **combinations** of multiple coexcited simple feature detector neurons, again, with only modest redundancy and with little information loss.

A new key principle postulated by the theory relative to these populations of feature detector neurons is that secondary (and tertiary — see below) feature detector neurons also develop inhibitory connections (via growth of axons of properly interposed inhibitory interneurons that receive input from the secondary feature detector neurons) that target the simple feature detector neurons which feed them. Thus, when a secondary feature detector neuron becomes highly excited (partly) by simple feature detector neuron inputs, it then immediately shuts off these simple neurons. This is the theory's *precedence principle*. In effect, it causes groups of inputs that are statistically "coherent" to be re-represented as a whole ensemble; rather than as a collection of "unassembled" pieces. For example, in a visual input, an ensemble of simple feature detector neurons together representing a straight line segment might be re-represented by some secondary feature detector neurons which together represent the whole segment. Once activated by these primary neurons, these secondary neurons then, by the precedence principle, immediately shut off (via learned connections to local inhibitory interneurons) the primary neurons that caused their activation.

Once the secondary feature detectors of a module have stabilized they too are then frozen and (at least in certain areas of cortex) tertiary feature detectors (often coding even larger complexes of statistically meaningful inputs) form their codebook. They too obey the precedence principle. For example, in primary visual cortical regions, there are probably tertiary feature detectors which code long line segments (probably both curved and straight) spanning multiple modules. Again, this is one example of how nearby modules might interact — such tertiary feature detectors might well inhibit and shut off lower-level feature detector neurons in other nearby modules. Of course, other inhibitory interactions also develop — such as the line "end stopping" that inhibits reactions of line continuation feature detectors beyond its end. In essence, the interactions within cortex during the short time span of its reaction to external input (20 to 40 msec) are envisioned by this theory as similar to the "competitive and cooperative neural field interactions" postulated by Stephen Grossberg and Gail Carpenter and their colleagues in their visual processing theories (Carpenter and Grossberg, 1991; Grossberg, 1976, 1987, 1997; Grossberg et al., 1997). When external input (along with an operate command) is provided to a developed module, the above brief interactions ensue and then a single symbol (or a small set of symbols, depending upon the manner in which the operate command to the module is manipulated) representing that input is expressed. The process by which the symbols are developed from the feature detector neuron responses is now briefly discussed.

Once the feature detector neurons (of all orders) have had their responses frozen, the next step is to consider the sets of feature detector neurons which become highly excited together across the cortical region due to external inputs. Because the input wiring of the feature detector neurons is random and sparse; the feature detector neurons function somewhat like VQ codebook vectors with many of their components randomly zeroed out (i.e., like ordinary VQ codebook vectors projected into randomly selected low-dimensional subspaces defined by the relatively sparse random axonal wiring feeding the feature detector neurons of the module). In general, under these circumstances, it can be established that any input to the region (again, whether from thalamus, from other cortical regions, or from other extracortical sources) will cause a roughly equal number of feature detector neurons to become highly excited. This is easy to see for an ordinary VQ codebook. Imagine a probability density function in a high-dimensional input space (the raw input to the region). The feature detector responses can be represented as points spread out in a roughly equiprobable manner within this data cloud (at least before projection into their low-dimensional subspaces) (Kohonen, 1995). Thus, given any specific input, we can choose to highly excite a roughly uniform number of highest appropriate precedence feature detector vectors that are closest in angle to that input vector.

In effect, if we imagine a rising externally supplied operation control signal (effectively supplied to all of the feature detector neurons that have not been shut down by the precedence principle), as the sum of the control signal and each neuron's excitation level (due to the external inputs) climbs, the most highly excited neurons will cross their fixed "thresholds" first and "fire" (there are many more details than this, but this general idea is hypothesized to be correct). If the rate of rise of the operate signal is constant, a roughly fixed number of not-inhibited feature detector neurons will begin "firing" before local inhibition from these "early winners" prevents any more winners from arising. This leaves a fixed set of active neurons of roughly a fixed size. The theory presumes that such fixed sets will, by means of their coactivity and the mutually excitatory connections that develop between them, tend to become established and stabilized as the internal feature attractor circuit connections gradually form and stabilize. Each such neuron group, as adjusted and stabilized as an attractor state of the module over many such trials, becomes one of the symbols in the lexicon.

Each final symbol can be viewed as being a localized "cloud" in the VQ external input representation space composed of a uniform number of close-by coactive feature detector responses (imagine a VQ where there is not one winning vector, but many). Together, these clouds cover the entire portion of the space in which the external inputs are seen. Portions of the VQ space with higher input vector probability density values automatically have denser clouds. Portions with lower density have more diffuse clouds. Yet, each cloud is represented by roughly the same number of vectors (neurons). These clouds are the symbols. In effect, the symbols form a Voronoi-like partitioning of the occupied portion of the external input representation space (Kohonen, 1984, 1995); except that the symbol cloud partitions are not disjoint, but overlap somewhat.

Information theorists have not spent much time considering the notion of having a cloud of "winning vectors" (i.e., what this theory would term a *symbol*) as the outcome of the operation of a vector quantizer. The idea has always been to only allow the single VQ codebook vector that is closest to the "input" win. From a theoretical perspective, the reason clouds of points are needed in the brain is that the connections which define the "input" to the module (whether they be sensory inputs arriving via thalamus, knowledge links arriving from other portions of cortex, or yet other inputs) only connect (randomly) to a sparse sampling of the feature vectors. As mentioned above, this causes the feature detector neurons' vectors to essentially lie in relatively low-dimensional random subspaces of the VQ codebook space. Thus, to comprehensively characterize the input (i.e., to avoid significant information loss) a number of such "individually incomplete," but mutually complementary, feature representations are needed. So, only a cloud will do. Of course, the beauty of a cloud is that this is exactly what the stable states of a feature attractor neuronal module must be in order to achieve the necessary confabulation "winner-take-all" dynamics.

A subtle point the theory makes is that the organization of a feature attractor module is dependent upon which input data source is available first. This first-available source (whether from sensory inputs supplied through thalamus or active symbol inputs from other modules) drives development of the symbols. Once development has finished, the symbols are largely frozen (although they sometimes can change later due to symbol disuse and new symbols can be added in response to persistent changes in the input information environment). Since almost all aspects of cognition are hierarchical, once a module is frozen, other modules begin using its assumed fact outputs to drive their development. So, in general, development is a one-shot process (which illustrates the importance of getting it right the first time in childhood). Once the symbols have been frozen, the only synaptic modifications which occur are those connected with knowledge acquisition, which is the topic discussed next.

3.A.4 Implementation of Knowledge

As discussed in Hecht-Nielsen (2005), all of the knowledge used in cognition (e.g., for vision, hearing, somatosensation, language, thinking, and moving) takes the form of unidirectional weighted links between pairs of symbols (typically, but not necessarily, symbols residing within different modules). This section sketches how these links are implemented in human cortex (all knowledge links used in human cognition reside entirely within the white matter of cortex).

Figure 3.A.5 considers a single knowledge link from symbol ψ in a particular cortical *source* module (lexicon) to symbol λ in a particular *target* or *answer* lexicon. The set of all knowledge links from symbols of one particular source lexicon to symbols of one particular target lexicon are called a *knowledge base*. The single knowledge link considered in Figure 3.A.5 belongs to the knowledge base linking the particular source lexicon shown to the particular target lexicon shown.

When the neurons of Figure 3.A.5 representing symbol ψ are active (or highly excited if multiple symbols are being expressed, but this case will be ignored here), these *ψ neurons* send their action potential outputs to millions of neurons residing in cortical regions to which the neurons of this source region send axons (the gross statistics of this axon distribution pattern are determined genetically, but the local details are random). Each such active symbol-representing neuron sends action potential signals via its axon collaterals to tens of thousands of neurons. Of the millions of neurons which receive these signals from the ψ neurons, a few thousand receive not just one such axon collateral, but many. These are termed *transponder* neurons. They are strongly excited by this simultaneous input from the ψ neurons; causing them to send strong output to all of the neurons to which they in turn send axons. In effect, the first step of the link transmission starts with the tens to hundreds of active neurons representing symbol ψ and ends with many thousands of excited transponder neurons, which also (collectively) uniquely represent the symbol ψ. In effect, transponder neurons momentarily *amplify* the size of the ψ symbol representation. It is hypothesized by the theory that this *synfire chain* (Abeles, 1991) of activation does not propagate further because

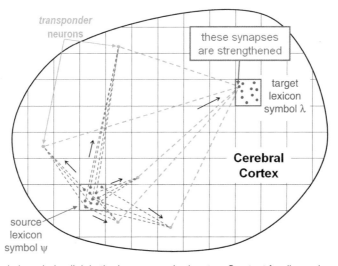

Figure 3.A.5   A single knowledge link in the human cerebral cortex. See text for discussion.

only active (or highly excited) neurons can launch such a process and while the transponder neurons are *excited*, they are not active or highly excited (i.e., active, or highly excited, neurons — a rare state that can only exist following a confabulation information processing operation — are the only ones that can unconditionally excite other neurons) However, as with transponder neurons, if a neuron receives a high-enough number of simultaneous inputs from active neurons — even through unstrengthened synapses, and in the absence of any operation command input — it will become excited. Finally, excited neurons *can* excite other neurons if those other neurons reside in a lexicon which is simultaneously also receiving operation command signal input (this is what happens when knowledge is used and when short-term memory learning takes place, as will be discussed below).

The wiring of the symbol and transponder neuron axons is (largely) completed in childhood and then remains (at least for our purposes here) essentially fixed for life. Again, the gross statistics of this wiring are genetically determined; but the local details are random.

A relatively small number (say, 1 to 25% — a genetically controlled percentage that deliberately varies across cortex) of the target region neurons representing symbol λ will just happen to each receive many synaptic inputs from a subset of the transponder neurons (Figure 3.A.5 illustrates the axonal connections from ψ transponder neurons for only one of these few λ neurons). These particular λ neurons *complete* the knowledge link. If all of the neurons representing symbol λ are already active at the moment these synaptic inputs arrive, then (in the event that they have not been previously permanently strengthened) the transponder neuron synapses that land on this subset of them will be temporarily strengthened (this is called *short-term memory*). During the next sleep period, if this causal pairing of symbols ψ and λ is again deliberately rehearsed, these temporarily strengthened synapses may be more lastingly strengthened (this is *medium-term memory*). If this link is subsequently rehearsed more over the next few days, these synapses may be permanently strengthened (this is *long-term memory*). It is important to note that the synapses from the ψ neurons to the ψ transponder neurons are generally not strengthened. This is because the transponder neurons are not meaningfully active at the time when these inputs arrive. **Only deliberate usage of a link with immediately prior co-occurrence of both source symbol and target symbol activity causes learning**. This was, roughly, the learning hypothesis that Donald Hebb advanced 56 years ago (Hebb, 1949).

Note again that the transponder neurons that represent a symbol ψ will always be the same; independent of which target lexicon(s) are to be linked to. Thus, ψ transponder neurons must send a sufficiently large number of axons to all of the lexicons containing symbols to which symbol ψ might need to connect. The theory posits that genetic control of the distribution of axons (nominally) ensures that all of the potentially necessary knowledge links can be formed. Obviously, this postulated design could be analyzed, since the rough anatomy and statistics of cortical axon fascicles are known. Such an analysis might well be able to support, or raise doubts, that this hypothesis is capable of explaining cortical knowledge.

Cognitive functions where confabulations always yield zero or one winners, because at most one symbol has anything close to enough knowledge links from the assumed facts, do not need precisely weighted knowledge links. In cortical modules which only require such confabulations, knowledge links terminating within that module are hypothesized by the theory to be essentially binary in strength: either completely *unstrengthened* (i.e., as yet unused) or *strong* (strengthened to near maximum). Such modules together probably encompass a majority of cortex.

However, other cognitive functions (e.g., language) **do** require each knowledge link to have a strength that is directly related by some fixed function to $p(\psi|\lambda)$. The theory's hypothesis as to how these weightings arise is now sketched.

Although the mechanisms of synaptic modification are not yet well understood (particularly those connected with medium-term and long-term memory), research has established that "Hebbian" synaptic strengthening does occur (Cowan et al., 2001). This presumably can yield a transponder neuron to target symbol neuron synapse strength directly related to the joint probability

$p(\psi\lambda)$ (i.e., roughly, the probability of the two involved symbols being coactive). In addition, studies of postsynaptic neurotransmitter depolarization transduction response (i.e., within the neuron receiving the synaptic neurotransmitter output; separate from the transmitting synapse itself) by Marder and her colleagues (Marder and Prinz, 2002, 2003) and by Turrigiano and her colleagues (Desai et al., 2002; Turrigiano and Nelson, 2000, 2004; Turrigiano et al., 1998) suggest that the postsynaptic apparatus of an excitatory cortical synapse (e.g., one landing on a target symbol neuron) is independently modifiable in efficacy, in multiplicative series with this Hebbian $p(\psi\lambda)$ efficacy. This "post-synaptic signaling efficacy" is expressed as a neurotransmitter *receptivity* proportional to a direct function of the reciprocal of that target neuron's average firing rate, which is essentially $1/p(\lambda)$. The net result is implementation by this *Hebb/Marder/Turrigiano* learning process (as I call it) of an overall link strength directly related to $p(\psi\lambda)/p(\lambda)$, which by Bayes law is $p(\psi|\lambda)$. Thus, it is plausible that biological learning processes at the neuron level can accumulate the knowledge needed for confabulation.

3.A.5 Implementation of Confabulation

Since only a small subset of the neurons representing target lexicon symbol λ are excited by a knowledge link from source lexicon symbol ψ, how can confabulation be implemented? This section, which presents the theory's hypothesized implementation of confabulation, answers this question and shows that these "internally sparse" knowledge links are an **essential element** of cortical design. Counterintuitively, if these links were "fully connected," cortex could not function.

Figure 3.A.6 schematically illustrates how confabulation is implemented in a thalamocortical (answer lexicon) module. The four boxes on the left are four cortical lexicons, each having exactly one assumed fact symbol active (symbols α, β, γ, and δ respectively). Each of these active symbols is represented by the full complement of the neurons which represent it, which are all active (illustrated as a complete row of filled circles within that assumed fact symbol's lexicon module, depicted in the figure in colors green, red, blue, and brown for α, β, γ, and δ, respectively). As will be seen below, this is how the symbol(s), which are the conclusions of a confabulation operation are biologically expressed (namely, all of their representing neurons are active and all other symbol representing neurons are inactive).

In Figure 3.A.6 the neurons representing each symbol of a module are shown as separated into their own rows. Of course, in the actual tissue, the neurons of each symbol are scattered randomly within the relevant layers of the cortical portion of the module implementing the lexicon. But for clarity, in Figure 3.A.6 each symbol's neurons are shown collected together into one row. The fact that the same neuron appears in multiple rows (each symbol-representing neuron typically participates in representing many different symbols) is ignored here, as this small pairwise *overlap* between symbol representations causes no significant interference between symbols.

(Note: This is easy to see: consider the simplified attractor you built and experimented with above. It always converged to a single pure state x_k (at least when the initial state u was dominated by x_k); meaning that all of the neurons which represent x_k are active and all other neurons are inactive. However, each of the neurons of x_k also belongs to many other stable states x_i, but this does not cause any problems or interference. You may not have seen this aspect of the system at the time you did your experiments — go check! You will find that even though the overlap between each pair of x field stable states is relatively small, each individual neuron participates in many such stable states. The properties of this kind of attractor network are quite astounding; and they do not even have many of the additional design features that thalamocortical modules possess.)

The answer lexicon for the elementary confabulation we are going to carry out (based upon assumed facts α, β, γ, and δ, just as described in Hecht-Nielsen, 2005) is shown as the box on

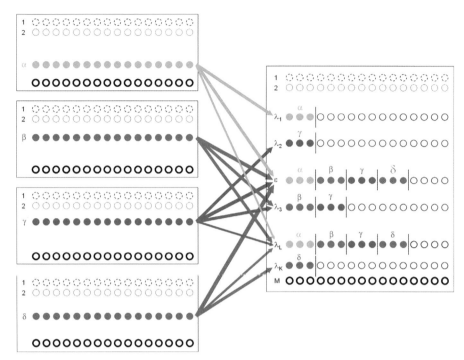

Figure 3.A.6 The implementation of confabulation in human cerebral cortex. See text for explanation.

the right in Figure 3.A.6. Each assumed fact symbol has knowledge links to multiple symbols of the answer lexicon; as illustrated by the colored arrows proceeding from each source lexicon to the answer lexicon. The width of each such knowledge link arrow corresponds to the link *strength*; i.e., the value of its $p(\psi|\lambda)$ probability. Each assumed fact symbol in this example is assumed to be the sole conclusion of a previous confabulation on its lexicon. Thus, symbols α, β, γ, and δ are all active (maximally transmissive).

The symbols of the answer lexicon which receive one or more links from the assumed facts are denoted by ε, λ_1, λ_2, λ_3, and so forth, and for clarity, are grouped in Figure 3.A.6. As discussed in the previous section, the actual percentage of neurons of each target symbol which receive synaptic inputs from the assumed fact's transponder neurons is approximately the same for all symbols (this is a function of the roughly uniform — at least for each individual answer lexicon — binomial connection statistics of the locally random cortico-cortical axons implementing each knowledge link). And as mentioned earlier, this percentage is low (from 1 to 25%, depending on where the module is located in cortex).

As shown in Figure 3.A.6, symbol λ_1 receives only one link (it is a medium-strength link from assumed fact symbol α). In accordance with Figure 3.A.5, only a fraction of the neurons of the answer lexicon which represent symbol λ_1, are actually being excited by this input link. These are shown as green filled circles with α above them (again, for clarity, the target symbol neurons which happen to receive input excitation from a particular assumed fact, which are actually randomly located, are grouped together in the figure, and labeled above with the symbol of that assumed fact). Note that, in the case of this group of green neurons of symbol λ_1 receiving input from assumed fact symbol α, that a medium-sized font α is shown above the group; reflecting the fact that the knowledge link delivering this assumed fact excitation only has medium strength $p(\lambda_1|\alpha)$. Similarly, the neurons representing symbol λ_2 are also receiving only one medium-strength link; namely from assumed fact symbol γ.

Only two of the answer lexicon symbols shown in Figure 3.A.6, namely ε and λ_L are receiving links from all four assumed facts. However, note that the links impinging on the neurons of symbol ε are stronger than those impinging on symbol λ_L. Now this discussion of the biological implementation of confabulation will pause momentarily for a discussion of synapses.

Despite over a century of study, synapse function is still only poorly understood. What is now clear is that synapses have dynamic behavior, both in terms of their responses to incoming action potentials, and in terms of modifications to their transmission efficacy (over a wide range of time scales). For example, some synapses seem to have transmission efficacy which "droops" or "fades" on successive action potentials in a rapid sequence (such are sometimes termed *depressing* synapses — which has nothing to do with the clinical condition of depression). Other synapses (termed *facilitating*) increase their efficacies over such a sequence; and yet, others exhibit no change. However, it has been learned that even these categorizations are too simplistic and do not convey a true picture of what is going on. That clear picture awaits a day when the actual modulations used for information transmission, and the "zoo" of functionally distinct neurons and synapses, are better understood. Perhaps this theory can speed the advent of that day by providing a comprehensive vision of overall cortical function, which can serve as a framework for formulating scientific questions.

Even though little is known about synapses, it is clear that many synapses are weak (unstrengthened), quite likely unreliable, and marginally capable of signaling (this theory claims that roughly 99% of synapses must be in this category, see Section 3.A.7). This is why it takes a pool of highly excited or active neurons representing a symbol (such neurons possess the ultimate in neural signaling power) to excite transponder neurons (each of which receives many inputs from the pool). No lesser neural collection is capable of doing this through unstrengthened synapses (which is why cortical synfire chains have only two stages). However, it is also known that some synapses (this theory claims that these represent fewer than 1% of the total of cortical excitatory synapses, see Section 3.A.7) are much stronger. These stronger synapses (which the theory claims are the seat of storage of all cortical knowledge) are physically larger than unstrengthened synapses and are often chained together into multiple-synapse groups that operate together (see Figure 3.A.7). One estimate (Henry Markram, personal communication) is that such a strengthened synapse group can be perhaps 60 times stronger than the common unstrengthened synapse (in terms of the total depolarizing effect of the multi-synapse on the target cell at which they squirt glutamate neurotransmitter). These strong synapses are probably also much more reliable. Figure 3.A.7 illustrates these two hypothesized types of cortical excitatory synapses.

The theory hypothesizes that synapses which implement knowledge links (as in Figure 3.A.5) are always strengthened greatly in comparison with unstrengthened synapses. When the knowledge link requires that a transponder–neuron-to-target–symbol–neuron synapse code the graded probability $p(\psi|\lambda)$ (as opposed to just a binary "unstrengthened" or "strong"), the dynamic range of such a strengthened synapse is probably no more than a factor of, say, 6. In other words, if the weakest strengthened synapse has an "efficacy" 10 times that of an unstrengthened synapse, the strongest possible synapse will have an efficacy of 60. Thus, we must code the smallest meaningful $p(\psi|\lambda)$ value as 10 and the strongest as 60.

In our computer confabulation experiments (e.g., those reported in Hecht-Nielsen, 2005 and many others), the smallest meaningful $p(\psi|\lambda)$ value (define this to be a new constant p_0) turns out to be about $p_0 = 0.0001$ and the largest $p(\psi|\lambda)$ value seen is almost 1.0. As it turns out, the smaller $p(\psi|\lambda)$ values need the most representational precision; whereas little error is introduced if the larger $p(\psi|\lambda)$'s are more coarsely represented. Clearly, this is a situation that seems ripe for using logarithms! The theory indeed proposes that nonbinary strengthened synapses in human cortex have their $p(\psi|\lambda)$ probabilities coded using a logarithmic scale (i.e., $y = \log_b(cx) = a + \log_b(x)$, where $a = \log_b(c)$). This not only solves the limited synaptic dynamic

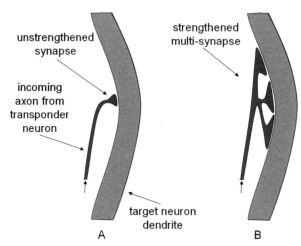

Figure 3.A.7 Synapse strengthening — the fundamental storage mechanism of cortical knowledge links. Subfigure A illustrates a weak, unreliable, unstrengthened synapse making a connection from a transponder neuron axon to a target neuron dendrite. The theory hypothesizes that roughly 99% of human cortical synapses with this connectivity are unstrengthened. Subfigure B illustrates the same synapse after learning (i.e., the progression from short-term memory to medium-term memory to long-term memory has been completed). Now, the synapse has blossomed into three parallel synapses, each physically much larger than the original one. This multi-synapse (perhaps what has been recently termed a *ribbon synapse*) is more reliable and has an efficacy ranging from perhaps 10 to 60 times that of the original unstrengthened synapse (learning always yields a great increase in efficacy — the theory posits that there are no such knowledge storage synapses which are only slightly strengthened).

range problem mentioned above, but it is also a key part of making confabulation work (as we will see below)!

So, given the above estimates and hypothesis, let us determine the base b of the logarithms used for synaptic knowledge coding in the human cerebral cortex, as well as the constant c (actually, we will instead estimate $a = \log_b(c)$). We want $p(\psi|\lambda) = 0.0001$ to be represented by a synaptic strength of 10; and we want $p(\psi|\lambda) = 1.0$ to be represented by a synaptic strength of 60. In other words, we need to find positive constants a and b such that:

$$a + \log_b (0.0001) = 10 \qquad\qquad (3A.2)$$

and

$$a + \log_b (1.0) = 60 \qquad\qquad (3A.3)$$

Clearly, from the second equation, $a = 60$ (since the log of 1 is zero for every b). Then the first equation yields $b = 1.2023$. Thus, when a highly excited transponder neuron representing source symbol ψ delivers its signal to a neuron of answer lexicon symbol λ, the signal delivered to that neuron will be proportional to $a + \log_b(p(\psi|\lambda))$ (where the constant of proportionality is postulated to be the same for all target neurons of a single module; and where nearby modules typically have very similar proportionality constants).

You might wonder why the signal delivered is not the "product" of the transponder neuron output signal and the synaptic efficacy (as was common in classical "neural network" models such as the Perceptron [Hecht-Nielsen, 2004]). Well, it is! However, exploring this aspect of the theory would quickly take us beyond the scope of this introductory sketch. Since transponder neurons coding a single active symbol (assumed fact) on a lexicon, essentially anywhere in cortex, always fire at about the same signal level (namely, the maximum possible) when they are implementing a link, we can consider this link input signal as constant. Thus, for the purposes of discussing elementary confabulation (the process of reaching conclusions based upon sets of assumed facts), we need not worry about this issue here. Another issue that can be ignored is the influence of the many nonstrengthened synapses impinging on target lexicon symbol neurons. This effect can be ignored because the inputs due to this prolific, but unreliable, source is very uniformly distributed across all neurons of all symbols and so it affects them all equally. In other words, this input acts as a low-variance, roughly constant, uniform "background noise."

The main conclusion of the above argument is that those neurons which represent answer lexicon symbol λ, which happen to receive a sufficient number of ψ transponder neuron inputs to allow them to respond, will all have about the same response to that input; namely a response proportional to $a + \log_b(p(\psi|\lambda))$.

Recall from the discussion of Figure 3.A.6 that the number of neurons of each answer lexicon symbol which receive sufficient synaptic inputs from the transponder neurons of a source symbol ψ are about the same for each knowledge link and each symbol. You may wonder why only λ neurons having this maximum number of synapses from ψ transponders will respond. It has to do with the events of the confabulation process. As the operate command input rises, these "sufficient" neurons will become active first. In the operation of the feature attractor (which is very fast) only those neurons with a sufficient number of inputs from an assumed fact will be able to participate in the dynamical convergence process. Another good question is why the variance in this number of synapses turns out to be small. This is because the binomial statistics of random transponder neuron axons make it such that neurons with unusually large numbers of synapses are extremely unlikely. Otherwise put, binomial (or Poisson) probability distributions have "thin tails." Thus, the set of all λ neurons which have strengthened synapses — the ones which participate in the (strength-weighted) excitation of λ — are those that lie in a narrow range at the top end of the Poisson density right before it plummets.

The binomial statistics of the locally random cortical connections also keep the number of target symbol neurons with near-maximum complements of input synapses very close to being constant for all symbols. Let this number of neurons be K. Then the total excitation of the K neurons which represent answer lexicon symbol λ that are receiving input from ψ symbol transponders (where ψ is one of the assumed facts) is proportional to $K[a + \log_b(p(\psi|\lambda)]$ (again, with a universal constant of proportionality that is the same for all the symbols of one module).

Finally, since the subsets of λ-representing neurons which receive inputs from different links typically do not overlap, the total excitation of the entire set of neurons representing answer lexicon symbol λ (assuming that λ is receiving knowledge link inputs from assumed facts α, β, γ, and δ) is approximately proportional to (again, with a universal constant of proportionality) the *total input excitation* sum $I(\lambda)$:

$$I(\lambda) \equiv K \cdot [a + \log_b (p(\alpha|\lambda))] + K \cdot [a + \log_b (p(\beta|\lambda))]$$
$$+ K \cdot [a + \log_b (p(\gamma|\lambda))] + K \cdot [a + \log_b (p(\delta|\lambda))]$$
$$= 4K \cdot a + K \cdot \log_b [p(\alpha|\lambda) \cdot p(\beta|\lambda) \cdot p(\gamma|\lambda) \cdot p(\delta|\lambda)] \qquad (3A.4)$$

Recall from the discussion of Section 3.A.2, that when the answer lexicon feature attractor is operated (and yields only one winning symbol), all of the neurons representing the winning symbol (which will be the one with the highest total input excitation) are left in the active state and all other symbol neurons are left inactive. By virtue of the above formula, we see that this winning symbol will be the symbol λ with the highest confabulation product $p(\alpha|\lambda) \cdot p(\beta|\lambda) \cdot p(\gamma|\lambda) \cdot p(\delta|\lambda)$ value (e.g., in the specific case of Figure 3.A.6, this will be symbol ε). This is the theory's explanation for how thalamocortical modules can carry out confabulation.

Since not all symbols of the answer lexicon of Figure 3.A.6 receive knowledge links from all four assumed facts α, β, γ, and δ, what will be the input excitation sums on symbols that receive fewer than four link inputs (total excitation level of the entire ensemble of neurons representing that symbol in the answer module)? For example, consider an answer lexicon symbol θ which only receives links from assumed facts β and δ. The total input excitation sum I(θ) of the set of neurons which represent θ will be:

$$I(\theta) \equiv K \cdot [a + \log_b (p(\beta|\theta))] + K \cdot [a + \log_b (p(d|\theta))]$$
$$= 2K \cdot a + K \cdot \log_b [p(\beta|\theta) \cdot p(d|\theta)] \tag{3A.5}$$

Thus, given that each individual term in the first lines of Equations (3A.4) and (3A.5) lies between $K \cdot 10$ and $K \cdot 60$, the value of I(θ) (Equation (3A.5)) could, in extreme cases, be larger than that of I(λ) of Equation (3A.4) (although in most cases I(θ) will be smaller and θ will not be the winning symbol). In any event, the symbol with the highest I value will win the confabulation.

Note that in cognitive functions which employ binary knowledge (every knowledge link transponder neuron synapse is either unstrengthened or is "strong"), I(λ) is roughly proportional to the number of links that symbol λ receives. Thus, in these cortical areas, confabulation devolves into simply choosing the symbol with the most knowledge link inputs. Although it is not discussed in this chapter, this is exactly what such cognitive functions demand.

The seeming problem identified above of having symbols which are missing one or more knowledge links win the confabulation competition is not actually a problem at all. Sometimes (e.g., in early visual processing) this is exactly what we want, and at other times, when we want to absolutely avoid this possibility, we can simply carry out multiple confabulations in succession to form a sequence of expectations. Also, some portions of cortex probably have smaller dynamic ranges (e.g., 40 to 60 instead of 10 to 60) for strengthened synapses, which also helps solve this potential problem.

As discussed in Section 3.1 of the main chapter, in mechanizing cognition we explicitly address this issue by appropriately defining a constant called the *bandgap* (related to quantity above).

In summary, the theory claims that the above-sketched biological implementation of confabulation meets all information processing requirements of all aspects of cognition; yet, it is blazingly fast and can be accurately and reliably carried out with relatively simple components (neurons and synapses) which operate independently in parallel. Confabulation is my candidate for the **greatest evolutionary discovery of all time** (with strong runners-up being DNA and photosynthesis).

3.A.6   Action Commands

At the end of a confabulation operation, there is often a single symbol active. For example, the triangular red cortical neurons (belonging to Layers II, III, and IV) shown in Figure 3.A.2 represent one particular symbol of the module which is now active following a confabulation. Of course, in a real human thalamocortical module, such an active symbol would be represented by tens to hundreds (depending on the location of the module in cortex) of "red" neurons, not the few shown in the figure.

A key principle of the theory is that at the moment a single symbol of a module achieves the active state at the end of a confabulation operation, a specific set of neurons in Layer V of the cortical portion of that module (or of a nearby module — this possibility will be ignored here) become highly excited. The outputs of these cortical Layer V neurons (shown in brown in Figure 3.A.2) leave cortex and proceed immediately to subcortical *action nuclei* (of which there are many, with many different functions). This is the theory's *conclusion–action* principle. In effect, every time cognition reaches a definitive single conclusion, a behavior is launched. This is what keeps us moving, thinking, and doing, every moment we are awake.

The Layer V neurons which become highly excited when a symbol wins a confabulation cause a very specific set of actions to be *executed* (or at least to be considered for execution; depending on the function of the action nucleus that receives the Layer V efferents). This is the origin of all behavior — each successful (one winning symbol) confabulation causes the launch of a set of *associated* action commands. These actions can be part of a movement process, part of a thought process, or both.

During development, the genetically determined program for creating the brain is, barring problems, executed. This program causes the development of axons from neurons in Layer V of each cortical module portion which proceed to genetically directed subcortical action nuclei (of which there are tens). In other words, genetics can ensure that a module has the Layer V neurons it needs to launch those actions, which that particular module should be empowered to execute. Thus, each of us has a range of behavioral potentialities which are in this sense predetermined. Undoubtedly, this is how various talents and personality traits are transferred from parents to children. This is part of the "nature" portion of the human equation.

Given the behavioral potentialities established by the genetically directed wiring of the axons of the Layer V neurons of a module to action nuclei, the big question is how exactly the correct ones of these Layer V neurons end up getting "wired" **from** the population of neurons representing each symbol. Given the exact specificity of effect each Layer V neuron produces, there is no room for error in this wiring from each symbol to the action commands it should launch. Since the local geometrical arrangement of the symbol-representing neurons and action command neurons within their respective layers is random, and their local axonal wiring is largely random, this wiring from symbol representing neurons to Layer V action-command-generating neurons cannot be genetically determined. **These associations must be learned and they must be perfect**. Figure 3.A.8 illustrates the theory's hypothesized mechanism for implementing these precise *symbol to action associations*. This figure will be referred to extensively below.

The learning of symbol to action command associations is almost certainly a totally different learning process from that used in development of module symbol sets or in the establishment of knowledge links. This symbol to action association learning process is hypothesized to take place extensively during childhood; but also very frequently during adulthood. Cognitive lexicon development, cognitive knowledge acquisition, and symbol to action command association learning together make up the most "glamorous" parts of the "nurture" portion of the human equation (there are a number of other, quite different, learning processes that go on in other parts of the brain; e.g., learning to sense when we should use the toilet).

Notice that in Figure 3.A.2, every cortical layer of a module is mentioned except Layer I (the most superficial). Layers II, III, and IV are primarily involved in symbol representation, precedence principle interactions among feature detector neurons, and the receipt of afferents from thalamus. Layer V is where the action command output neurons reside. And Layer VI is where the cortical efferents to thalamus arise. The theory hypothesizes that Layer I is where the wiring between the symbol representation neuron sets and the Layer V action command output neurons takes place (and quite possibly some of the wiring for the feature attractor module function as well). It is well known (Paxinos and Mai, 2004) that the neurons of Layer V (typically these are of the pyramidal category) have *apical* dendrites that ascend to Layer I and then branch profusely. Further, neurons of Layers II, III, and IV typically send large numbers of axon collaterals to Layer I (and also frequently have apical dendrites too — but these will not be discussed here). Further, the *basal ganglia* (*BG* — a complicated set of brain nuclei known to be involved in multiple types of action learning (Paxinos and Mai, 2004)), specifically, the BG substructure known as the *striatum* sends signals in great profusion to Layer I of cortex via the thalamus (see Figure 3.A.8). This radiation is principally concentrated in frontal cortex (where most behaviors seem to originate), but other cortical areas also receive some of these inputs.

Given the random nature of cortical wiring, the only way to establish correct symbol to action associations is via experimentation. This experimentation is carried out (starting with the simplest

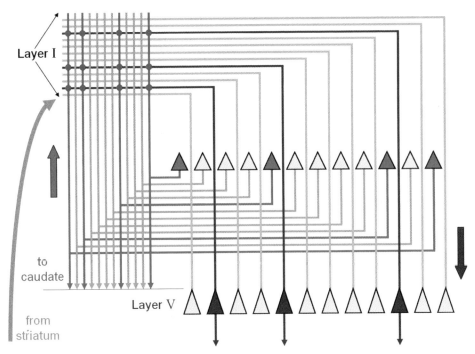

Figure 3.A.8 Learning and using the precise associations from symbols to action commands within a single cortical module. Keep in mind that the neuron populations involved in these associations, illustrated here as small sets, are, in the brain, extremely large sets (tens of thousands of neurons in every case). See text for explanation of the figure.

actions and then constructing an *action hierarchy*). At beginning of development of each module, the first item on the agenda is development of the module's symbols (which was discussed in Section 3.A.3). As this lexicon development process begins to produce stable symbols, the problem of associating these with actions is addressed.

At first, action command neurons are randomly triggered when a particular single symbol is being *expressed* by the lexicon (i.e., that symbol was the lone outcome of a confabulation operation by the module). As this occurs, the BG monitors the activity of this lexicon (via efferents from Layers III and V — see Figure 3.A.8). When a randomly activated action command happens to cause an action that the basal ganglia judge to be particularly "good" (meaning that a reduction in a drive or goal level was observed — which the basal ganglia know about because of their massive input from the limbic system), that action is then associated with the currently expressed symbol via the mechanism of Figure 3.A.8.

(Note: Reductions in drive and goal states are almost never immediate following an action. They are usually delayed by seconds or minutes; sometimes by hours. One of the hypothesized functions of the BG [Miyamoto et al., 2004] is that it develops a large number of predictive models, called *critics* [Barto et al., 1983], that learn [via delayed *reinforcement learning* methods; Sutton and Barto, 1998] to accurately predict the eventual goal-or-drive-state-reduction "value" or "worth" of an action at the time the action is suggested or executed. It is by using such critic models that the BG is hypothesized by the theory to immediately assess the worth of action commands produced by Layer V outputs.)

When an action command that is randomly launched is indeed judged worthy of association from the currently expressed symbol of a module, a special signal (the green arrow in Figure 3.A.8) is sent (via thalamus) from the striatum of the BG to cortical Layer I of the module. This green signal causes the synapses (blue circles) connecting axon collaterals of the neurons representing the currently expressed symbol (these neurons are shown in red in Figure 3.A.8, and reside in Layer II, III, or IV [I am not sure]) with the apical dendrites of the now-validated action command neurons of Layer V (shown in brown in Figure 3.A.8) to be incrementally strengthened. Essentially every neuron representing the expressed symbol gets its direct synaptic connections with the action command neurons incrementally strengthened.

Notice how different the situation of Figure 3.A.8 is from that of knowledge links. In a knowledge link, the source symbol must first amplify its signal by briefly recruiting thousands of transponder neurons to retransmit it. Even then, when the knowledge link signals arrive at the target lexicon module, only a relatively small fraction of each target symbol's neurons receive a sufficient number of inputs to complete the link. In Figure 3.A.8, we presume that almost **all** of the expressed symbol's representing neurons synapse directly with the apical dendrites of **each** Layer V action command neuron. The reason that this is a sensible hypothesis is that Layer I is well known to be fed extensively with axons from the neurons below it (i.e., neurons of the module that represent symbols), and to be profusely supplied with dense apical dendrites from Layer V neurons.

The synapses from symbol representing neurons to action command neurons are hypothesized to be quite different from those used in knowledge links. In particular, these synapses can slowly and gradually get stronger (if repeatedly strengthened over many trials over time), and can slowly and gradually get weaker (if not strengthened very often, or not at all, over time). This is why "skill knowledge" decays so fast (in comparison with cognitive knowledge, which lasts for very long periods of time, even if not used). A major benefit of this dynamic synapse characteristic is that occasionally erroneous strengthening of synapses (e.g., when a random action command set includes some irrelevant commands along with some effective ones) will, in general, not cause problems (as long as the vast majority of strengthenings are warranted). This is very different from cognition, where correction of erroneous knowledge is often impossible (and then the only solution is to specifically learn not to use the erroneous knowledge).

The universal truism that "practice makes perfect" is thus exactly correct when it comes to behavior. And for a difficult skill (e.g., landing a jet fighter on an aircraft carrier at night) to be usable; that practice must have been recent. The associations from symbols to action command sets are constantly being reshaped during life. If we live in a highly stable information environment we might not notice much change in our behavioral repertoire over many years. If we are subjected to a frequently and radically changing information environment, our behavior patterns are constantly changing. In some respects, people who undergo such changes are being constantly "behaviorally remade." The workings of the neuronal network of Figure 3.A.8 are now briefly discussed.

Clearly, the size of the set of specific Layer V action command neurons which need to be triggered by the expression of a particular symbol is arbitrary. One symbol's association might involve activating a set of five specific Layer V neurons, another might involve activating 79, and yet another might activate no Layer V neurons. Keep in mind that each individual neuron in the population of tens to hundreds of neurons which together represent one particular symbol in a lexicon also participates in many other such representations for other symbols. So, this association must be between the **population** representing a symbol and a specific set of Layer V neurons.

This requirement suggests a unidirectional Willshaw-type associative network structure wherein the "retrieval keys" all have almost exactly the same number of neurons (which is exactly what the symbol representation neuron sets are like); but where the "output" neurons activated by each key have an arbitrary number of neurons. This is exactly what a Willshaw structure can do — the retrieval keys ("stable states") x_k MUST be random and MUST each have almost the same number of neurons; but there can be as many or as few "output neurons" in the associated y_k as desired, with no restriction; and the individual neurons making up each x_k population can appear in many other such populations.

(Note: If you don't see this, consider again the computer experiments you performed in Section 3.A.3. You will see that it does not matter how many y_k neurons there are for each x_k, as long as we are not implementing the second, **y** field to **x** field, part of the cycle (this is not well known, because for analytical simplicity, the original Willshaw model used the same number of neurons in both the $\mathbf{x}_k$ and $\mathbf{x}_k$ vectors). Further, as long as the $\mathbf{x}_k$ keys are random and have almost exactly the same numbers of active neurons, the reliability of the $\mathbf{y}_k$ neuron responses is extremely high.)

However, as mentioned earlier, unlike the situation in knowledge links (where only a few of the target symbol neurons receive connections from the transponder neurons of a source symbol), in this case, **all** of the neurons of each active symbol must connect to each of the desired action command neurons. Partial connectivity will not work here, since there is no feedback to implement a "convergence" process. But the enormous local connectivity within Layer I is hypothesized to make, achieving a sufficient level of this connectivity no problem.

By incorporating inhibitory neurons into its intrinsic design, such a one-way Willshaw network (with inhibition added) will only respond with a $\mathbf{y}_k$ when its input is a newly active SINGLE symbol (multiple symbols will fail to yield any association output because they induce excessive inhibition; which shuts down all of the Layer V neurons). This is hypothesized to be why action commands are only issued when a confabulation produces a single winning symbol. Also, when considering what action to take for a given $\mathbf{x}_k$ input, only those Layer V neurons having a sufficient input excitation will respond (much like in confabulation competitions). In other words, even near the beginning of learning, when behavioral symbol to action associations are all weak, the Layer V response will be based upon this "competitive" criterion, not a fixed threshold.

That a vast majority of cortex would be involved in issuing thought action commands, as opposed to movement action commands, makes sense because there are many more feature attractor modules (and knowledge bases) than muscles. (It is not discussed here, but each knowledge base may also need to receive an "operate" command in order to function — if this is true, this function probably involves the large "higher order" [Sherman and Guillery, 2001] portion of the thalamus that is not included in the thalamocortical modules.) So it probably requires a much larger portion of cortex to producing such thought process control action commands (muscle action commands come mostly from Layer V of lexicons located within the relatively small *primary motor area* of cortex).

Most action commands represent "low-level housekeeping functions" that are executed reflexively whenever a single symbol (often one of a large set of symbols that will elicit the same action command set) is expressed on a lexicon. For example, if a confabulation in a lexicon that is recalling a stored action sequence (such lexicons are typically located in frontal cortex) ends in the expression of a single action symbol, then that lexicon must be immediately erased and prepared for generating the next sequence symbol. This is an action command that is issued along with the expression of the current action sequence symbol. Overriding such reflexive thought progressions is possible; but generally involves shutting off tonic cortical arousal (one of multiple adjuncts to the lexicon operation command input) in a general cortical area via action commands issued to brainstem thought nuclei. The result is a momentary freezing of the halted function as a new thought process stream is inaugurated. This is what happens when we see that we are about to step on dog poop. It takes a only fraction of a second for us to recover from the suspension of the ongoing action and activate an alternative. Further, since muscle tone and rhythmic actions such as walking are nominally controlled by other brain nuclei (not cortex and thalamus), all the cortex typically needs to do (once the prior action sequence has been suspended) in such instances is issue a momentary set of *corrective alteration* action commands which are instantly executed as a momentary perturbation to the ongoing (subcortically automated) process, which then typically resumes.

It is important to note that the details of how sequences of "action" symbols — each representing (via its symbol to action command association) a particular specific set of action commands that will be launched every time that symbol is the sole conclusion of a confabulation — are learned, stored, and recalled are the same as with all other cognitive knowledge. However, unlike many other types of knowledge (e.g., sensory, event, or factual knowledge), only the action symbol replay knowledge is rehearsed and solidified at night. The action symbol to action command associations can only be learned and refined via awake rehearsal. This accounts for the fact that anyone learning a new skill will frequently find themselves (either through vague memories of dreams upon waking, or via reports from their sleep partner) carrying out "silent practice" of those skills in their sleep. These do not involve launching the involved actions (a function that is normally suppressed during sleep), but simply running through the involved action symbol sequences. Such activities can help solidify the symbol sequences and this often yields improved skill performance the next day.

Quite a bit of experience has been gained with learning and recalling action symbol sequences in one of my UCSD graduate courses. For example, a checker-playing system that learns by expert-guided rehearsal has been demonstrated. However, issues surrounding the replay of action sequence hierarchies are complicated and not within the scope of an introductory chapter (e.g., provisions for automatic real-time, moment-by-moment modification of an ongoing lower-level action sequence replay in response to the exact current state of the world; with no modification at the higher level — a process called *instantiation* — must be introduced). So action symbol sequence learning and recall are not discussed in this chapter.

In summary, the theory proposes that the unidirectional symbol pair links used in confabulation are the only knowledge learned and stored in cortex that is used in cognition. However, as described in this section, there is a second kind of knowledge learned and stored in cortex: the associations between each symbol and the action commands that its expression should launch. This knowledge is not really part of cognition. It is the mapping from *decisive cognitive conclusions* (single active symbols resulting from confabulations) to *behaviors*. Thus, the ultimate end product of cognition is the origination of action commands; some of which are unconditionally *executed* immediately and others, termed *suggested actions*, must be approved (*vetted*) by the basal ganglia before they can be executed.

3.A.7 Discussion

The theory's hypothesized cortical implementation of knowledge links has some important universal properties. First, the locally random wiring of the cortical axons can be established during development and then frozen, essentially for life (although there may be a very slow *replenishment* of some types of neurons throughout life that helps keep the brain functional as neurons slowly die; but this has not been established — the vast majority of neurons probably live a very long time, perhaps for the full life span of the individual). Knowledge links, by means of a parallel, two-step synfire chain communication process through the random cortical signaling network, can be immediately formed, as appropriate, between almost any two symbols in any two lexicons that genetics have provided connection possibilities for. A link can be temporarily established instantly (via the short-term memory mechanism) and then, if it is warranted, the link can be progressively transformed into permanent knowledge during the subsequent few sleep periods.

The price of this ability to instantly learn almost anything without need for rewiring (to carry out such wiring by growing new axons would take days and would require the involved axons to have unbelievable navigation abilities) is probably a **vast over-wiring of cortex**. A prediction of the theory is that only roughly 1% of cortical synapses are actually used to store knowledge (i.e., have been strengthened). The rest are there to provide the capacity for *instant arbitrary learning*. Thus, the old saw that "we only use 10% of our brain" is probably wrong on the high side. Ninety nine percent of unstrengthened synapses are hypothesized to simply be sitting around waiting to be needed. This may seem wasteful; but unstrengthened cortical knowledge synapses and axon collaterals are small, and humans have about 10^{14} to 10^{15} of them (Mountcastle, 1998; Nicholls et al., 2001; Nolte, 1999; Steward, 2000). Clearly, the survival value of instant arbitrary learning vastly outweighs whatever inefficiency is incurred. This hypothesis helps explain one of the most puzzling findings of neuroscience: the vast majority of synapses that have ever been individually evaluated (e.g., by manipulating them, and monitoring their effects on the target cell, using multiple patch clamps [Cowan et al., 2001]) have turned out to be very unreliable and only marginally functional. This is exactly what you would expect to find if 99% of synapses are in a state of minimal existence, awaiting the possible moment that they will be needed.

Humans live for roughly 3×10^9 sec. So, for example, if we acquire an average of one item of knowledge during every second of life (86,400 knowledge items per day), and if an average of 300 transponder neuron synapses are used to implement each knowledge item, far less than 1% of all synapses will ever be used (of course, not all cortical synapses are available for knowledge storage, but most probably are, so this conclusion is still probably correct). So, the theory proposes that the potential amount of cognitive knowledge that can be stored is huge.

In my laboratory's computer implementations of confabulation, a startling fact (which is consistent with the above numbers) has emerged: a staggeringly large number of knowledge items is needed to do even simple cognitive functions. The theory postulates that the average human must possess billions of items of knowledge. This has many startling and profound implications, and assuming that the theory gains acceptance, many philosophical and educational views of humans (and other animals) will likely be completely altered. For example, the theory implies that children (and adults too!) probably accumulate tens of thousands, or more, new individual items of knowledge **every day**. Thus, the process of reconsidering each day's short- and medium-term memories and converting selected ones into a more permanent form is a huge job. It is no wonder that we must sleep a third of the time.

To appreciate the vast storage capacity of your cerebral cortex, imagine for a moment that you are being asked a long series of detailed questions about the kitchen in your home. Describe all of the spoons and where they are kept; then the forks, the drinking glasses, and so on. Describe how you select and employ each item. Where and when you obtained it, and some memorable occasions when it was used. Obviously, such a process could go on for tens of hours and still turn up lots of new kitchen information. Now consider that you could probably answer such detailed questions for thousands of mental arenas. Humans are phenomenally smart.

Another cortical property, which the theory's hypothesized design of cortex imparts, is an insensitivity to occasional random neuronal death. If a few of the transponder neurons which represent a particular symbol randomly die, the remaining knowledge links from this symbol continue to function. Newly created *replenishment neurons* (which the theory proposes arise throughout life) which turn out to have the appropriate connectivity (once they have spread out and connected up and reached maturation), can be incorporated into such a weakened link to replace lost neurons; assuming the link is used from time to time.

If a link is not used for a long time, then as the transponder neurons of its source symbol slowly get *redeployed* (see below) or die, the axons to the target symbol neurons of the link will not be replenished and the link will become gradually weaker (other links having the same source symbol, which are used, will not suffer this fate because they will be replenished). Eventually, the unused link will become so weak that it cannot function by itself. Sometimes, when a link has become weak, but is not completely gone, it can be used if accompanied by additional assumed fact inputs to the same target symbol — a faded-memory recall trick known popularly as *mnemonics*. This is the theory's explanation for why we forget long-disused knowledge.

Another aspect of the hardware failure tolerance of cortex is the primary representation of each symbol within its own lexicon. With tens or hundreds of neurons representing each symbol, the lexicons symbols too have some redundancy and failure tolerance.

When new inputs to a cortical lexicon arise which do not fit any of the existing symbols well, and continue to appear repeatedly, new symbols can be formed, even in adulthood. Depending on how close to capacity the involved lexicon is, these new symbols may or may not displace existing symbols. This *lexicon rebuilding* process is often used to add new symbols to lexicons when we learn a subject in more depth (e.g., when we take Calculus III after having already taken Calculus

I and Calculus II). Total rebuilding of a lexicon typically only occurs in the event of trauma (e.g., stroke), where the entire information input environment to the lexicon has dramatically changed. Total rebuilding takes weeks and requires lots of practice with the new symbols. This is why recovery of function after a stroke takes so long and why intensive physical and mental therapy based upon practice and use is so important. Aspects of childhood development are being recapitulated on an abbreviated schedule.

Lexicons also slowly incorporate replenishment neurons into existing symbol representations that are used. As with forgetting of knowledge; long-disused symbols eventually have their sets of representing neurons redeployed (see below) or eroded beyond functionality. A person who spoke French when he was a child, but who has not used French at all for 40 years, will likely have many of the French word representation symbols eroded beyond recovery.

The only instance of deliberate fast cognitive knowledge erasure in human cortex is *redeployment*, where a source symbol in a lexicon, which used to be linked to a particular set of target symbols in other lexicons, suddenly has an entirely new ensemble of links to new target symbols arise for it, and these new links persist (and the old ones are disused). For example, when we move to a new home, it may be necessary to learn that the alarm clock is now on the left side of the bed, not the right. What happens in this instance is that the sets of transponder neurons representing the involved source symbol have a finite limit to the number of highly strengthened synapses that they can have at any time (this probably has to do with a total individual cellular limit on synthesis of certain consumable biochemicals — the critical ones of which are produced only in the neuron's soma and dendrites, where the ribosomes reside). (Note: The ultimate limit to knowledge storage capacity is not synapses; it is the number of strengthened synapses that each transponder neuron can support at one time. There are probably people [e.g., perhaps the author] who have spent their entire lives studying and who reached this capacity limit long ago.) As the transponder neuron synapses implementing the many new links are learned and strengthened, many of the old, now unused, links must be immediately sacrificed (their synapses shrivel to the unstrengthened state). Within a few weeks, we instinctively reach left. The old knowledge has been effectively erased. The synapses of many of the old knowledge links have shriveled (but not all of them; some remnant knowledge links often remain — which you can experience by revisiting one of your old haunts and trying to carry out formerly familiar patterns; like skipping down stairs at a childhood residence). Fragments of your former knowledge will still be there.

Redeployment is a critical cognitive capability that allows us to adapt to environmental change quickly. It is also hypothesized to be the only mechanism of deliberate forgetting in cognition.

Finally, it is important to note that any global theory of human cerebral cortex and thalamus is bound to be vastly oversimplified. For example, it is well known (Paxinos and Mai, 2004) that different areas of cortex have some Layers dramatically attenuated (e.g., Layer IV in certain areas of frontal cortex). Others have Layers that are dramatically elaborated (e.g., in primary visual cortex, Layer IV becomes tripartite). These local modifications almost certainly must have significant meaning for the nuances of function. However, the theory proposes that these are all relatively small variations of the same overall grand theme.

The central notion of the theory: that cognition, that greatest engine of animal ennoblement, is universally mechanized by one information processing operation (confabulation) employing a single form of knowledge (antecedent support), with each singular conclusion reached launching an associated set of action commands; seems to me to now be secure. The concreteness and specificity of this theory guarantees that it is testable.

ACKNOWLEDGEMENTS

Thanks to Fair Isaac Corporation for long-term research support and to Kate Mark for help with the manuscript. Thanks to Robert F. Means, Syrus Nemat-Nasser, and Luke Barrington of my

Figure 3.A.9   Zeus Hecht-Nielsen.

laboratory for help with computer confabulation experiments. Domestic cat Zeus Hecht-Nielsen (Figure 3.A.9) and I have spent part of most mornings since May 1990 exploring our family compound and eating breakfast together. These thousands of hours of personal interaction and behavioral observation have yielded many valuable insights into the workings of cognition; some of which were indispensable in the development of this theory.

APPENDIX REFERENCES

Abeles M., *Corticonics*, Cambridge University Press, Cambridge, UK (1991).

Amari S., A mathematical theory of nerve nets, *Advances in Biophysics*, 6 (1974), pp. 75–120.

Amari S., Characteristics of sparsely encoded associative memory, *Neural Networks*, 2 (1989), pp. 451–457.

Amit D., *Modeling Brain Function: The World of Attractor Networks*, Cambridge University Press, Cambridge, UK (1989).

Anderson J.A., A memory storage model utilizing spatial correlation functions, *Kybernetik*, 5 (1968), pp. 113–119.

Anderson J.A., A simple neural network generating an interactive memory, *Mathematical Biosciences*, 14 (1972), pp. 197–220.

Anderson J.A., J.W. Silverstein, S.A. Ritz and R.S. Jones, Distinctive features, categorical perception, and probability learning: some applications of a neural model, *Psychological Reviews*, 84 (1977), pp. 413–451.

Barto A.G., R.S. Sutton and C.W. Anderson, Neuronlike adaptive elements that can solve difficult learning problems, *IEEE Transactions on Systems, Man, and Cybernetics*, SMC–13 (1983), pp. 834–846.

Carpenter G.A. and S. Grossberg (eds), *Pattern Recognition by Self-Organizing Neural Networks*, MIT Press, Cambridge, Massachusetts (1991).

Cowan W.M., T.C. Sudhof and C.F. Stevens (eds), *Synapses*, Johns Hopkins University Press, Baltimore, Maryland (2001).

Desai N.S., R.H. Cudmore, S.B. Nelson and G.G. Turrigiano, Critical periods for experience-dependent synaptic scaling in visual cortex, *Nature Neuroscience*, 8 (2002), pp. 783–789.

Gabor D., Associative holographic memories, *IBM Journal of Research and Development*, March (1969), pp. 156–159.

Grossberg S., Adaptive pattern classification and universal recoding, *Biological Cybernetics*, 23 (1976), pp. 121–134.

Grossberg S. (ed.), *The Adaptive Brain*, Volumes I and II, Elsevier, Amsterdam (1987).

Grossberg S., Cortical dynamics of three-dimensional figure-ground perception of two-dimensional patterns, *Psychological Reviews*, 104 (1997), pp. 618–658.

Grossberg S., E. Mingolla and W.D. Ross, Visual brain and visual perception: how does the cortex do perceptual grouping? *Trends in Neurosciences*, 20 (1997), pp. 106–111.

Haines K. and R. Hecht-Nielsen, A BAM with increase information storage capacity, *Proceedings 1988 International Conference on Neural Networks*, IEEE Press, Piscataway, New Jersey, vol. I (1988), pp. 181–190.

Hebb D., *The Organization of Behavior*, Wiley, New York, New York (1949).

Hecht-Nielsen R., Perceptrons, UCSD Institute for Neural Computation, Report #0403 (2004) (available at http://inc2.ucsd.edu/addedpages/techreports.html).

Hecht-Nielsen R., Cogent confabulation, *Neural Networks*, 18 (2005), pp. 111–115.

Hecht-Nielsen R. and T. McKenna (eds), *Computational Models for Neuroscience*, Springer-Verlag, London (2003).

Hopfield J.J., Neural networks and physical systems with emergent collective computational abilities, *Proceedings of the National Academic Sciences*, 79 (1982), pp. 2554–2558.

Hopfield J.J., Neurons with graded response have collective computational properties like those of two-state neurons, *Proceedings of the National Academic Sciences*, 81 (1984), pp. 3088–3092.

Kohonen T., Correlation matrix memories, *IEEE Transactions on Computers*, C21 (1972), pp. 353–359.

Kohonen T., *Self-Organization and Associative Memory*, Springer-Verlag, Berlin (1984).

Kohonen T., *Self-Organizing Maps*, Springer-Verlag, Berlin (1995).

Kosko B., Bidirectional associative memories, *IEEE Transactions on Systems, Man, and Cybernetics*, SMC–18 (1988), pp. 49–60.

Mai J.K., J. Assuher and G. Paxinos, *Atlas of the Human Brain*, Second Edition, Elsevier, San Diego, California (2004).

Marder E. and A.A. Prinz, Modeling stability in neuron and network function: the role of activity in homeostasis, *BioEssays*, 24 (2002), pp. 1145–1154.

Marder E. and A.A. Prinz, Current compensation in neuronal homeostasis, *Neuron*, 37 (2003), pp. 2–4.

Miyamoto H., J. Morimoto, K. Doya and M. Kawato, Reinforcement learning with via-point representation, *Neural Networks*, 17 (2004), pp. 299–305.

Mountcastle V.B., *Perceptual Neuroscience: The Cerebral Cortex*, Harvard University Press, Cambridge, Massachusetts (1998).

Nilsson N.J., *Learning Machines*, McGraw-Hill, New York, New York (1965).

Nilsson N.J., *Artificial Intelligence: A New Synthesis*, Morgan Kaufman Publishers, San Francisco, California (1998).

Nicholls J.G., A.R. Martin, B.G. Wallace and P.A. Fuchs, *From Neuron to Brain*, Fourth Edition, Sinauer, Sunderland, Massachusetts (2001).

Nolte J., *The Human Brain*, Fourth Edition, Mosby, St Louis, Missouri (1999).

Palm G., On associative memory, *Biological Cybernetics*, 36 (1980), pp. 19–31.

Paxinos G.J. and K. Mai (eds), *The Human Nervous System*, Second Edition, Elsevier, San Diego, California (2004).

Pepperberg I.M., *The Alex Studies: Cognitive and Communicative Abilities of Grey Parrots*, Harvard University Press, Cambridge, Massachusetts (1999).

Sherman S.M. and R.W. Guillery, *Exploring the Thalamus*, Academic Press, San Diego, California (2001).

Sommer F.T. and G. Palm, Improved bidirectional retrieval of sparse patterns stored by Hebbian learning, *Neural Networks*, 12 (1999), pp. 281–297.

Steinbuch K., *Automat und Mensch*, Springer-Verlag, Heidelberg (1961a).

Steinbuch K., Die Lernmatrix, *Kybernetik*, 1 (1961b), pp. 36–45.

Steinbuch K., *Automat und Mensch*, Second Edition, Springer-Verlag, Heidelberg (1963).

Steinbuch K., *Automat und Mensch*, Third Edition, Springer-Verlag, Heidelberg (1965).

Steinbuch K. and U.A.W. Piske, Learning matrices and their applications, *IEEE Transactions on Electronic Computers*, December (1963), pp. 846–862.

Steinbuch K. and B. Widrow, A critical comparison of two kinds of adaptive classification networks, *IEEE Transactions on Electronic Computers*, October (1965), pp. 737–740.

Steward O., *Functional Neuroscience*, Springer-Verlag, New York, New York (2000).

Sutton J.P. and J.A. Anderson, Computational and neurobiological features of a network of networks, in: Bower, J.M. (ed.), *Neurobiology of Computation*, Kluwer Academic, Boston, Massachusetts (1995), pp. 317–322.

Sutton R.S. and A.G. Barto, *Reinforcement Learning: An Introduction*, MIT Press, Cambridge, Massachusetts (1998).

Tsypkin Y.Z., *Foundations of the Theory of Learning Systems*, Academic Press, New York, New York (1973).

Turrigiano G.G. and S.B. Nelson, Hebb and homeostasis in neuronal plasticity, *Current Opinons of Neurobiology*, 10 (2000), pp. 358–364.

Turrigiano G.G. and S.B. Nelson, Homeostatic plasticity in the developing nervous system, *Nature Reviews of Neuroscience*, 5 (2004), pp. 97–107.

Turrigiano G.G., K.R. Leslie, N.S. Desai, L.C. Rutherford and S.B. Nelson, Activity-dependent scaling of quantal amplitude in neocortical pyramidal neurons, *Nature*, 391 (1998), pp. 892–895.

Weir A.A.S., J. Chappell and A. Kacelnik, Shaping of hooks in New Caledonian crows, *Science*, 297 (2002), p. 981.

Willshaw D.J., O.P. Buneman and H.C. Longuet-Higgins, Non-holographic associative memory, *Nature*, 222 (1969), pp. 960–962.

Zador P., Development and evaluation of procedures for quantizing multivariate distributions, PhD Dissertation, Stanford University, Palo Alto, California (1963).

4

Evolutionary Robotics and Open-Ended Design Automation

Hod Lipson

CONTENTS

4.1 INTRODUCTION

Can a computer ultimately augment or replace human invention?

Imagine a Lego set at your disposal: Bricks, rods, wheels, motors, sensors, and logic are your "atomic" building blocks, and you must find a way to put them together to achieve a given high-level functionality: A machine that can move itself, say. You know the physics of the individual components' behaviors; you know the repertoire of pieces available, and you know how they are allowed to connect. But how do you determine the combination that gives you the desired functionality? This is the problem of Synthesis. Although engineers practice it and teach it all the time, we do not have a formal model of how open-ended synthesis can be done automatically. Applications are numerous. This is the meta-problem of engineering: Design a machine that can design other machines.

The example above is confined to electromechanics, but similar synthesis challenges occur in almost all engineering disciplines: circuits, software, structures, robotics, control, and MEMS to name a few. Are there fundamental properties of design synthesis that cut across engineering fields? Can a computer ultimately augment or replace human invention?

While we may not know how to synthesize things automatically, nature may give us some clues; after all, the fascinating products of nature were designed and fabricated autonomously.

In the last two centuries, engineering sciences have made remarkable progress in the ability to analyze and predict physical phenomena. We understand the governing equations of thermo-dynamics, elastics, fluid flow, and electromagnetics to name a few domains. Numerical methods such as finite elements allow us to solve these differential equations, with good approxi-mation, for many practical situations. We can use these methods to investigate and explain observations, as well as to predict the behavior of products and systems long before they are physically realized.

But progress in systematic synthesis has been much slower. For example, the systematic synthesis of a kinematic machine for a given purpose is a long-standing problem, and perhaps one of the earliest general synthesis problems to be posed. Robert Willis, a professor of natural and experimental philosophy at Cambridge, wrote in 1841:

> [A rational approach to synthesis is needed] to obtain, by direct and certain methods, all the forms and arrangements that are applicable to the desired purpose. At present, questions of this kind can only be solved by that species of intuition that which long familiarity with the subject usually confers upon experienced persons, but which they are totally unable to communicate to others. When the mind of a mechanician is occupied with the contrivance of a machine, he must wait until, in the midst of his meditations, some happy combination presents itself to his mind which may answer his purpose.
>
> Robert Willis, *Principles of Mechanism* (Willis, 1841)

Almost two centuries later, a rational method for synthesis in many domains is still not clear. Though many best-practice design methodologies exist, at the end of the day they rely on elusive human creativity. Product design is still taught today largely through apprenticeship: engineering students learn about existing solutions and techniques for well-defined, relatively simple problems, and then — through practice — are expected to improve and combine these to create larger, more complex systems. How is this synthesis process achieved? We do not know, but we cloak it with the term "creativity."

The question of how synthesis of complex systems occurs has been divided in a dichotomy of two views. One view is that complex systems emerge through successive adaptations coupled with natural selection. This Darwinian process is well accepted in biology, but is more controversial in engineering (Basalla, 1989; Ziman, 2003). The alternative explanation is intelligent design, mostly rejected in biology, but still dominant in engineering — as the celebrated revolutionary inventor.

The process of successive adaptation by improvement and recombination of basic building blocks is evolutionary in its nature. Unlike classical genetic algorithms (e.g., Goldberg, 1989, Chapter 5), however, it is *open-ended*: We do not know *á priori* what components we will need and how many of them. The permutation space is exponential, and complexity is unbounded. This is perhaps a subtle but key difference between optimization (e.g., Papadimitriou and Steiglitz, 1998) and synthesis. In optimization problems, we tune the values of a set of parameters in order to maximize a target function. The set of parameters, their meaning, and their ranges are predetermined. Synthesis, on the other hand, is an open-ended process, where we can add more and more components, possibly each with their own set of parameters. Consider, for example, a case where we need to design a new electronic circuit that performs some target function. One approach would be to manually provide a basic layout of resistors, capacitors, and coils, and then try to automatically tweak their values so as to maximize performance. Alternatively, we could start with a bucket of components, and use an algorithm to automatically compose them into a circuit that performs the target function. The former would be a case of optimization, and the latter an example of synthesis. There are numerous examples and many books dedicated to the application of evolutionary optimization in almost any engineering domain (e.g., Gen and Cheng, 1999; Zalzala and Fleming, 1999; Jamshidi et al., 2002; Karr and Freeman, 1998; Mazumder and Rudnick, 1998), but the use of evolution for open-ended design remains relatively unexplored, yet has the highest potential impact in its ability to "think outside the box."

4.1.1 Structure of This Chapter

As we shall see in the next few sections, we can use many of the ideas of biological evolutionary adaptation to inspire computational synthesis methods. To keep things intuitive, we shall describe some of these methods in the context of designing electromechanical machines, such as robots, and in particular legged robots. But these methods can be (and indeed have been) applied to numerous engineering application areas. This chapter is not intended to be a comprehensive review of evolutionary robotics or of evolutionary design research. Instead I have chosen a small set of results that portray an interesting perspective of the field and where it is going. These results are not necessarily in chronological order — scientific discoveries are not always made in an order most conducive for learning. Interested readers are encouraged to see the "further reading" section for more in-depth and broader reviews.

4.2 A SIMPLE MODEL OF EVOLUTIONARY ADAPTATION

There are a variety of computational models of open-ended synthesis loosely inspired by natural evolutionary adaptation. Perhaps the simplest approach uses a direct representation. We start off with a large set of initial candidate designs — this is the initial *population*. These designs may be random, blank, or may be seeded with some prior knowledge in the form of solutions we think are good starting points. We then begin evolving this population through repeated *selection* and *variation*. To perform selection, we first measure the performance of each solution in the population. The performance, *fitness* in evolutionary terminology, captures the merit of the design with respect to some target performance we are seeking as designers. The fitness metric needs to be solution-neutral, i.e., measure *the extent* to which the target task has been achieved, regardless of *how* it was achieved. We select better solutions (*parents*) and use them to create a new generation of solutions (*offspring*). The offspring are variations of the parents, created through variation operators like mutation and recombination. The process is repeated generation after generation until good solutions are found.

In practice, there are many modifications to the simple process described above. We use special representations, clever selection methods, sophisticated variation, evaluation methods, as well as multiple co-evolving populations. Most interestingly, we let the representations and the evaluation methods evolve too, to allow for a more open-ended search. Mitchell (1996) provides a review of many of these processes. Let us look at some simple examples applied to robotics.

4.3   MACHINE BODIES AND BRAINS

Many systems, including robotic systems in particular, are often viewed as comprising two major parts: the morphology and the controller. The morphology is the physical structure of the system, and the controller is a separate unit that governs the behavior of the morphology by setting the states of actuators and reading sensory data. In nature, we often refer to these as the body and brain, respectively. In control theory, we refer to these as the plant and the control (the term *plant,* as in "manufacturing plant," is used because of the original industrial applications). In computer engineering terms, this often translates into hardware and software. This distinction is semantic; we simply tend to refer to the part which is more easily adaptable as control and the part that is fixed as the morphology. In practice, both the morphology and control contribute to the overall behavior of the system and the distinction between them is blurred. Very often a particular morphology accounts for some of the control and the control is embedded in the morphology. Nevertheless, in describing the application of evolutionary design to systems, we find this distinction pedagogically useful.

In the following sections, we will see a series of examples of the application of evolutionary processes to open-ended synthesis. These examples were chosen to illustrate the design of robotic systems for their intuitiveness, starting at control and moving on to both control and morphology. Following these examples, we will take a look at the common principles, and future challenges.

4.3.1   Evolving Controllers

It is perhaps easier, both conceptually and technically, to explore application of evolutionary techniques to the design of robot controllers before using it to evolve their morphologies too. Robot controllers can be represented in any one of a number of ways: as logic functions ("if–then–else" rules), as finite state machines, as programs, as sets of differential equations, or as neural networks to name a few. Many of the experiments that follow represent the controller as a neural network that maps sensory input to actuator outputs. These networks can have many architectures, such as feed-forward or recurrent. Sometimes the choice of architecture is left to the synthesis algorithm.

Some of the early experiments in this area performed by Beer and Gallagher (1992). Nolfi and Floreano (2004), Harvey et al. (1997), and Meyer (1998) review many interesting experiments evolving controllers for wheeled and gantry robots, but let us look at some examples with legged robots. Consider a case where we have a legged robot morphology fitted with actuators and sensors, and we would like to use evolutionary methods to evolve a controller that would make this machine move (locomote) towards an area of high chemical concentration. Bongard (2002) explored this concept on a legged robot in a physically realistic simulator. The robot has four legs and eight rotary actuators as shown in Figure 4.1a. It has four touch sensors at the feet, which output a binary signal depending on weather or not they are touching the ground. The machine also has four angle sensors at the knees, outputting a graded signal depending on the actual angle of the knee. There are two chemical sensors at the top, which output a value corresponding to the chemical level they sense locally.

The behavior of the machine is determined by a neural controller that maps sensors to actuators, as shown in Figure 4.1b. Inputs of candidate neural controllers were connected to the sensors, and their output connected directly to the eight motors. Machines were rewarded for their ability to reach the area with high concentration. The fitness was evaluated by trying out a candidate controller in four different concentration fields, and summing up the distance between the final position of the robot and the highest concentration point. The shorter the distance the better — and in this sense the total distance is a performance error. In this experiment, 200 candidate controllers were evolved for 50 generations. The variation operators could decide if and how to connect the neurons. Figure 4.1c shows the progress of this error over generational time. The performance of

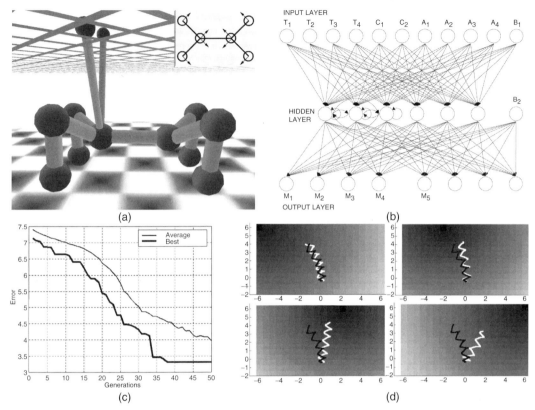

Figure 4.1 Evolving a controller for a fixed morphology. (a) The morphology of the machine contains four legs actuated with eight motors, four ground touch sensors, four angle sensors, and two chemical sensors. (b) The machine is controlled by a recurrent neural net whose inputs are connected to the sensors and whose outputs are connected to the motors. (c) Evolutionary progress shows how the target misalignment error reduces over generations. (d) White trails show the motion of the machine towards high concentration (darker area). Black trail shows strack when the chemical sensors are turned off. (From Bongard, J. C. (2002) Evolved Sensor Fusion and Dissociation in an Embodied Agent, *Proceedings of the EPSRC/BBSRC International Workshop Biologically-Inspired Robotics: The Legacy of W. Grey Walter*. With permission.)

one successful controller in four different chemical concentration fields is shown in Figure 4.1d. The white trails, which mark the progress of the center of mass of the robot over time, show clearly how the robot moves towards high concentration.

But what is more striking about this experiment is that the robot learned to perform essentially two tasks: to locomote and to change orientation towards the high concentration. When the chemical sensors are disabled, the robot moves forward but not towards the chemical concentration (see black trail in Figure 4.1d). This shows that the network evolved two *independent* functions: locomotion and gradient tracking.

Can this process also work for a real (not simulated) legged robot? We recently tried evolving controllers for a dynamical, legged robot (Zykov et al., 2004). The nine-legged machine is composed of two Stewart platforms back to back. The platforms are powered by 12 pneumatic linear actuators, with power coming from an onboard 4500 psi paintball canister. While most robotic systems use position-controlled actuators whose exact extension can be set, pneumatic actuators of the kind used here are force-controlled. Like biological muscle, the controller can specify the force and duration of the actuation, but not the position. It is therefore a challenging control problem. The controller architecture for this machine was an open-loop pattern generator that determines when to open and close pneumatic valves. The on–off pattern was evolved;

candidate controllers were evaluated by trying them out on the robot in a cage, and measuring fitness using a camera that tracks the red ball on the foot of one of the legs of the machine (see inset in Figure 4.2b for a view from the camera). Snapshots from one of the best evolved gates are shown in Figure 4.2c. Walker et al. (2004) provide a review of controller evolution on both simulated and physical machines.

Figure 4.2 (See color insert following page 302)   Evolving a controller for physical dynamic legged machine. (a) The nine-legged machine is powered by 12 pneumatic linear actuators arranged in two Stewart platforms. The controller for this machine is an open-loop pattern generator that determines when to open and close pneumatic valves. (b) Candidate controllers are evaluated by trying them out on the robot in a cage, and measuring fitness using a camera that tracks the red foot (see inset). (c) Snapshots from one of the best evolved gates. (From Zykov, V., Bongard, J., Lipson, H., (2004) Evolving dynamic gaits on a physical robot, *Proceedings of Genetic and Evolutionary Computation Conference*, Late Breaking Paper, GECCO'04. With permission.)

Manual design of a neural controller for a legged machine of this sort is possible, but not easy. The advantage of design automation here is that a design was found with minimal prior information on how it should be done. We could now reverse engineer the evolved controller to find out exactly how it works — like biologists. Should the morphology or the task change, we can have the process redesign new controllers. The evolutionary architecture described here was rather simple; many more sophisticated neural controller architectures and evolutionary processes are being explored, such as the use of plasticity (controllers that can learn after they have been evolved), controllers that grow, and other types of neurons such as spiking neurons (Nolfi et al., 1994; Floreano and Urzelai, 2001; Floreano et al., 2001, 2005).

4.3.2 Evolving Controllers and Some Aspects of the Morphology

Design of a robot involves not only the design of controller, but the morphology as well. What happens if some aspects of the morphological design are also allowed to evolve? For example, Lund et al. (1997) explored the effect of evolutionary adaptation of physical placement of sensors in a wheeled robot and showed improved performance. Let us examine this process in context of a legged machine.

Paul and Bongard (2001) used evolutionary adaptation to evolve designs for a bipedal robot in simulation, as shown in Figure 4.3a. The machine comprises the bottom half of a walker with six motors (two at each hip and one in each knee), a touch sensor at each foot and an angle sensor at each joint. The fitness of a controller was the net distance it could make a machine travel. The controllers had architecture similar to that shown in Figure 4.1b, with the appropriate number of inputs and outputs.

Evolving 300 controllers over 300 generations created various controllers that could make the machine move while keeping it upright. Figure 4.3b shows the maximum fitness per generation for a number of independent runs. While many did not make much progress, some runs were able to find good controllers, as evident by the curves with high fitness. More importantly, however, was that this time the evolutionary process was also allowed to vary the mass distribution of the robot morphology and that this new freedom allowed it to find good solutions. This may suggest that evolving a controller for a fixed morphology may be too restrictive, and that better machines might be found if both the controller and the morphology are allowed to coevolve, as they do in nature. This lends some credibility to the notion of concurrent engineering, where several aspects of a

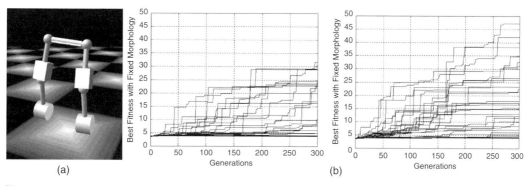

(a)   (b)

Figure 4.3 Evolving a controller and some morphology parameters for bipedal locomotion: the morphology of the machine consists of six motors (four at the hip and two at the knees), six angle sensors, and two touch sensors. The controller is a recurrent network similar to Figure 4.1b. (a) One of the evolved machines, (b) a comparison of fitness over generations for the fixed morphology (left) and a variable morphology (right). (From Paul, C., Bongard, J. C. (2001) The road less traveled: morphology in the optimization of biped robot locomotion, *Proceedings of the IEEE/RSJ International Conference on Intelligent Robots and Systems (IROS2001)*, Hawaii, U.S.A. With permission.)

product are engineered in concert rather than sequentially. Some small changes to the morphology may make the controller design task much simpler and vice versa.

4.3.3 Evolving Bodies and Brains

One may wonder what happens if the evolutionary process is given even more freedom in the design of both the morphology and control. Sims (1994) explored this idea in simulation using 3D cubes and oscillators as building blocks. Inspired by that work, we were interested in exploring physically realizable machines and start with lower-level building blocks, such as simple neurons and 1D elements (Lipson and Pollack, 2000). We used a design space consisting of bars and linear actuators for the morphology and neurons for the control (Figure 4.4a). The design space we used comprised bars and actuators as building blocks of structure and artificial neurons as building blocks of control. Bars connected with free joints can potentially form trusses that represent arbitrary rigid, flexible, and articulated structures, as well as multiple detached structures, and emulate revolute, linear, and planar joints at various levels of hierarchy. Similarly, sigmoidal neurons can connect to create arbitrary control architectures such as feed-forward and recurrent nets, state machines and multiple independent controllers. The bars can connect to each other through ball-and-socket joints, neurons can connect to other neurons through synaptic connections, and neurons can connect to bars. In the latter case, the length of the bar is governed by the output of the neuron by means of a linear actuator. No sensors were used. Variation operators used in the evolutionary process were allowed to connect, disconnect, add, remove, or modify any of the components.

Starting with a population of 200 blank machines that were comprised initially of zero bars and zero neurons, we conducted evolution in simulation. The fitness of a machine was determined by its locomotion ability: the net distance its center of mass moved on an infinite plane in a fixed duration. The process iteratively selected fitter machines, created offspring by adding, modifying, and removing building blocks and replaced them into the population. This process typically continued for 300 to 600 generations. Both body (morphology) and brain (control) were thus coevolved simultaneously. The simulator we used for evaluating fitness supported quasi-static motion in which each frame is statically stable. This kind of motion is simpler to transfer reliably into reality, yet is rich enough to support low-momentum locomotion.

Typically, several tens of generations passed before the first movement occurred. For example, at a minimum, a neural network generating varying output must assemble and connect to an actuator for any motion at all (see sequence in Figure 4.4a, for an example). A sample instance of an entire generation, thinned down to unique individuals is shown in Figure 4.4b. Various patterns of evolutionary dynamics emerged, some of which are reminiscent of natural phylogenic trees. Figure 4.4c presents examples of extreme cases of convergence, speciation, and massive extinction, and Figure 4.4d shows progress over time of one evolutionary run. Figure 4.4e shows some of the fitter machines that emerged from this process; these machines were "copied" from simulation into reality using rapid-prototyping technology (Figure 4.4f). The machines performed in reality, showing the first instance of a *physical* robot whose entire design — both morphology and control — were evolved.

In spite of the relatively simple task and environment (locomotion over an infinite horizontal plane), surprisingly different and elaborate solutions were evolved. Machines typically contained around 20 building blocks, sometimes with significant redundancy (perhaps to make mutation less likely to be catastrophic). Not less surprising was the fact that some exhibited symmetry, which was neither specified nor rewarded for anywhere in the code; a possible explanation is that symmetric machines are more likely to move in a straight line, consequently covering a greater net distance and acquiring more fitness. Similarly, successful designs appear to be robust in the sense that changes to bar lengths would not significantly hamper their mobility. The three samples shown in Figure 4.4d exploit principles of ratcheting, anti-phase synchronization, and dragging. Others (not

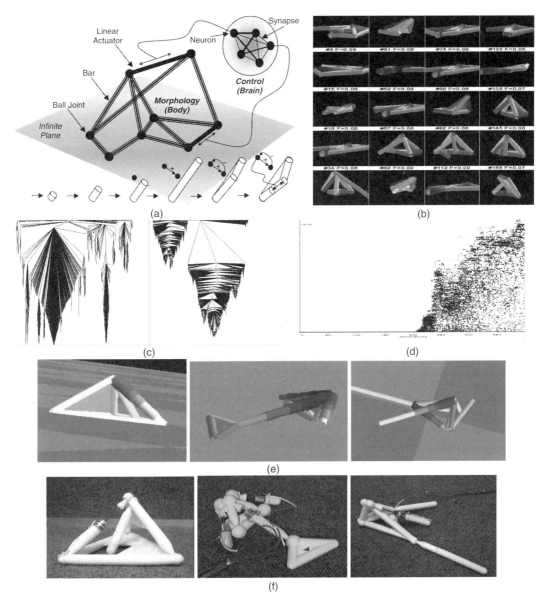

Figure 4.4 (See color insert following page 302) Evolving bodies and brains: (a) schematic illustration of an evolvable robot, (b) an arbitrarily sampled instance of an entire generation, thinned down to show only significantly different individuals, (c) phylogenetic trees of two different evolutionary runs, showing instances of speciation and massive extinctions from generation 0 (top) to approximately 500 (bottom), (d) progress of fitness versus generation for one of the runs. Each dot represents a robot (morphology and control), (e) three evolved robots, in simulation (f) the three robots from (e) reproduced in physical reality using rapid prototyping. (From Lipson, H., Pollack, J. B. (2000) *Nature*, 406, 974–978. With permission.)

shown here) used a sort of a crawling bi-pedalism, where a body resting on the floor is advanced using alternating thrusts of left and right "limbs." Some mechanisms used sliding articulated components to produce crab-like sideways motion. Other machines used a balancing mechanism to shift friction point from side to side and advance by oscillatory motion. Taylor and Massey (2001) provide a review of several works on evolution of morphologies.

4.4   MORPHOLOGY REPRESENTATIONS

The examples above used mostly a direct encoding — a representation of the morphology and control that evolution uses to explicitly modify each aspect of the design, adding, removing, and modifying components and parameters directly. Clearly, however, such an approach would not work in nature, for an average animal body contains billions of cells. Nature typically uses a more compact representation — a *genotype* — to encode for a much more complex machine — the *phenotype*. The genotype does not directly encode the phenotype, but instead it encodes information for growing or *developing*, a phenotype. This is one form of an *indirect* representation that maps between a genotype and a phenotype. In nature, these maps are evolving themselves and several hierarchical layers of mappings are used before a real DNA yields a working phenotype.

The use of a genotype–phenotype mapping allows for many advantages, primarily the compactness of a description and the ability to reuse components (more on that later). How can we use these representations computationally?

Mechanisms and neural networks can both be described as graphs. Luke and Spector (1996) survey a number of different representations used to describe or "grow" graphs, such as neural networks. Some methods use context-free grammars, L-systems, and parse trees operating on nodes and edges. Most of the existing representations for encoding networks generate highly connected architectures that are suitable for computational networks, but which are less suitable for kinematic machines because they over-constrain the motion and create deadlocked mechanisms. Using these representations, the likelihood of generating a mechanism with a specific number of degrees of freedom (DoF) is vanishingly small. In order to allow an evolutionary algorithm to explore the space of one DoF mechanisms more efficiently, a more suitable representation is required.

A second consideration in the choice of representation is *evolvability*. Many of the representations cited above result in context-sensitive and order-sensitive description of a network. For example, the structure generated by a branch in Gruau's cellular encoding depends on whether it is parsed before or after its sibling branch. If that branch is transplanted by cross-over into another tree, it may produce an entirely different structure. Such behavior hampers the effectiveness of recombinative operators by precluding the formation of modular components that are discovered by the search in one place and then reused elsewhere. A representation where the structure produced by a branch of the tree is minimally affected by its context may thus be more evolvable.

4.4.1   Tree Representations

Tree-based representations can describe a set of operations to construct a phenotype in a top-down or bottom-up manner. A top-down representation starts with an initial structure (an embryo) and specifies a sequence of operations that progressively modify it into its final form. Figure 4.5a shows a top-down tree that specifies the construction of an electric circuit, starting with an initial circuit and recursively replacing circuit segments with serial and parallel arrangements of electrical components (Koza, 1992). Each node of the tree is either an operator that modified the circuit and passes segments to its child nodes, or a terminal electrical component. The specific parallel and serial operators cannot be used for construction of mechanisms as they will immediately create over- and under-constrained kinematic chains. Because of the physics of electric circuits, ordering of children under a parent does not matter. This tree is thus both order independent and context independent. In a top-down tree, parent nodes must be constructed before their children. Figure 4.5a also shows a bottom-up construction of a symbolic expression. Here terminal nodes represent constants or variables, and parent nodes represent mathematical operators. Because of the nature of mathematical expressions, parsing order is important, and swapping order of some child nodes would result in a mathematically different expression. The terms are unchanged, however, by the content of their siblings. This tree is thus order dependent but context independent. In a bottom-up tree, child nodes must be constructed before their parents.

How could a tree representation be used to describe robot morphologies? Top-down construction of a mechanism starts with an embryonic kinematic basis with the desired number of DoFs, such as the four-bar mechanism shown in Figure 4.5c. A tree of operators then recursively modifies that mechanism by replacing single links (DoF = −1, i.e., over-constrained) with assemblies of links with an equivalent DoF, so that the total number of DoF remains unchanged. Two such transformations are shown in Figure 4.5c: the D and T operators. The D operator creates a new node and connects it to both the endpoints of a given link, essentially creating a rigid triangular

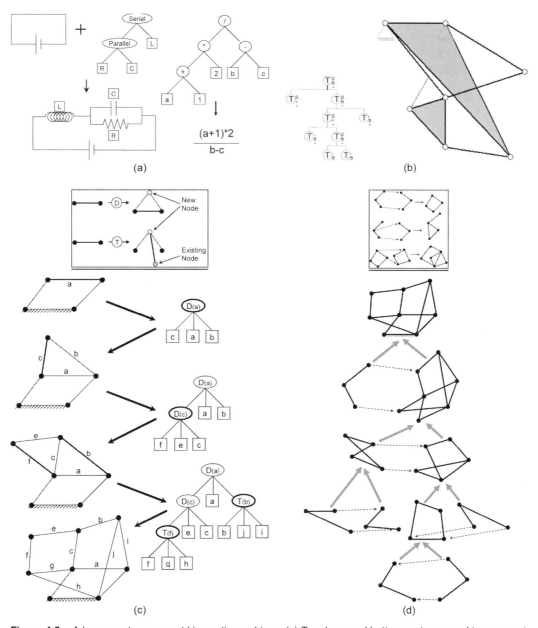

Figure 4.5 A language to represent kinematic machines: (a) Top-down and bottom-up trees used to represent structure, (b) a tree used to represent a kinematic machine; this machine traces a nearly-exact straight line. These mechanisms can be represented as top-down trees (c), or as bottom-up trees (d). (From Lipson, 2006. With permission.)

component. The T operator replaces a given link with two links that pass through a newly created node. The new node is also connected to some other existing node. In both operators, the position of the new nodes is specified in coordinates local to link being modified.

The T operator specifies the external connecting node by providing coordinates relative to link being modified; the closest available node from the parent structure is used. This form of specification helps assure the operators remain as context and order independent as possible. Figure 4.5c shows how a certain sequence of operators will transform a dyad into a triad. Figure 4.5c also shows how application of a tree of operators to the embryonic mechanism will transform it into an arbitrary compound mechanism with exactly one DoF. Terminals of the tree are the actual links of the mechanism.

Alternatively, bottom-up construction of a one-DoF mechanism begins at the leaves of the tree with atomic building blocks and hierarchically assembles them into components. The atomic building block is a dyad as shown in Figure 4.4a, and has exactly one DoF when grounded. The composition operator ensures that the total number of DoF is not changed when two subcomponents are combined, and thus the total product of the tree will also be a mechanism with exactly one DoF. When combining two components, each of one DoF, the resulting assembly will have five DoF (one DoF from each, plus three DoF released by ungrounding one of the components). The total DoF is restored to one by eliminating four DoF through the merging of two point pairs. An example of this process is shown in Figure 4.5d. Note that points must be merged in a way that avoids overlapping constraints, such as causing two links to merge. The components may need to be scaled and oriented for the merger to work. The ground link of the entire structure is specified at the root of the tree.

Figure 4.5b shows an application of this representation to the design of a single DoF mechanism that when actuated traces a nearly exact straight line, without reference to an existing straight line. This problem may seem somewhat arbitrary, but it was of major practical importance in the 19th century and many notable inventors, including James Watt, spent a considerable amount of time developing mechanisms to meet this requirement as the bootstrap of precision manufacturing. It therefore serves as a nice benchmark for the "inventiveness" of the algorithm. Using evolutionary computation based on tree representations, we were able to evolve machines, from scratch, that infringe and outperform previous established designs (Lipson, 2006).

4.4.2 Developmental Representations

Other types of representations allow the robot's morphology to develop from a basic "seed" and a set of context-free development rules. Consider, for example, the two rules "A→B" and "B→AB." If we start with the seed "A," and apply these two rules wherever they are applicable, the seed will develop as follows: A→B→AB→BAB→ABBAB→BABABBAB ... , and so forth. A seed and two simple rules can thus create very complex and elaborate structures. This type of representation, similar to an L-system or cellular automaton, can be applied to evolving morphologies and controllers of robots.

We start with a constructor that can build a machine from a sequence of build commands. The language of build commands is based on instructions to a LOGO-style turtle, which direct it to move forward, backward or rotate about a coordinate axis. Robots are constructed from rods and joints that are placed along the turtle's path (Figure 4.6a). Actuated joints are created by commands that direct the turtle to move forward and place an actuated joint at its new location with oscillatory motion and a given offset. The operators "[" and "]" push and pop the current state — consisting of the current rod, current orientation, and current joint oscillation offset — to and from a stack. Forward moves the turtle forward in the current direction, creating a rod if none exists or traversing to the end of the existing rod. Backward goes back up the parent of the current rod. The rotation commands turn the turtle about the Z-axis in steps of $60°$, for 2D robots, and about the X, Y or Z-axes, in steps of $90°$, for 3D robots. Joint commands move the turtle forward, creating a rod, and end with an actuated joint. The parameter to these commands specifies the speed at which the joint

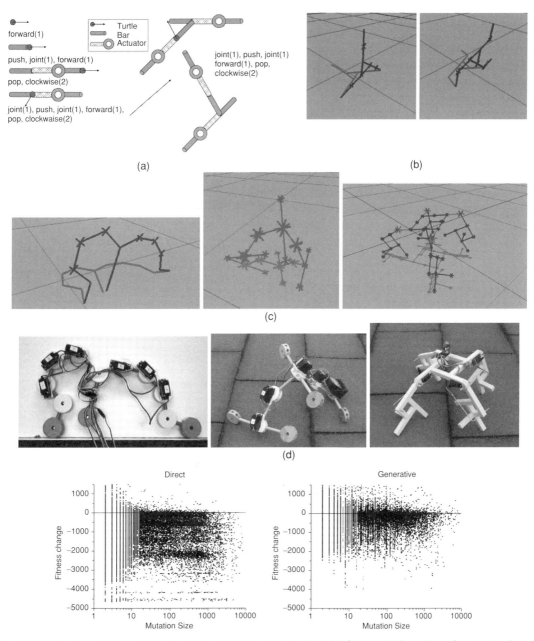

Figure 4.6 Evolving bodies and brains using generative encodings: (a) Schematic illustration of a construction sequence and (b) the resulting robot with actuated joints. (c) Three examples of robots produced by evolving L-systems that produce construction sequences, and (d) their physical instantiations. (e) A comparison of effects of mutation in the direct encoding versus the generative encoding shows that the generative encoding has transformed the space in a way that makes mutation more effective. (From Hornby, G. S., Lipson, H., Pollack, J. B. (2003) Generative encodings for the automated design of modular physical robots, *IEEE Transactions on Robotics and Automation*, 19(4). With permission.)

oscillates, using integer values from 0 to 5, and the relative phase-offset of the oscillation cycle is taken from the turtle's state. The commands "increase-offset" and "decrease-offset" change the offset value n the turtle's state by $\pm 25\%$ of a total cycle. Command sequences enclosed by "{ }" are repeated a number of times specified by the brackets' argument.

For example, the string *{joint(1) [joint(1) forward(1)] clockwise(2)}(3)* produces the robot in Figure 4.6b, through the development process shown in Figure 6a. Constructed robots do not have a central controller; rather each joint oscillates independent of the others. In Figure 4.6 large crosses are used to show the location of actuated joints and small crosses show unactuated joints. The left image shows the robot with all actuated joints in their starting orientation and the image on the right shows the same robot with all actuated joints at the other extreme of their actuation cycle. In this example, all actuated joints are moving in phase.

These strings were generated using an L-system. The L-systems are a set of rules like the "A→B" and "B→AB" rules discussed above. However, this time these "rewrite" rules are parametric (i.e., may pass parameters), and have conditions (are executed only when the parameters meet some conditions).

For example, the L-system to produce the robot in Figure 4.6b consists of two rules with each rule containing two condition-successor pairs:

$$P0(n): n > 2 \rightarrow \quad \{P0(n-1)\}(n)$$
$$n > 0 \rightarrow \quad joint(1) \, P1(n \times 2) \, clockwise(2)$$

$$P1(n): n > 2 \rightarrow \quad [P1(n/4)]$$
$$n > 0 \rightarrow \quad joint(1) \, forward(1)$$

If the L-system is started with P0(3), the resulting sequence of strings is produced:

P0(3)
{P0(2)}(3)
{joint(1) P1(4) clockwise(2)}(3)
{joint(1) [P1(1)] clockwise(2)}(3)
{joint(1) [joint(1) forward(1)] clockwise(2)}(3)

which produces the robot in Figure 4.6b.

An evolutionary algorithm was used to evolve individual L-systems, that when executed produced a build sequence which produced the machine. Approximately half the runs produced "interesting" viable results. The two main forms of locomotion found used one or more oscillating appendages to push along, or had two main body parts connected by a sequence of rods that twisted in such a way that first one half of the robot would rotate forward, then the other. Some examples of successful machines are shown in Figure 4.6c and their physical instantiations are shown in Figure 4.6d.

A comparison of robots evolved using the developmental encoding to robots whose construction sequence was evolved directly revealed that robots evolved with the generative representation not only had higher average fitness, but also tended to move in a more continuous manner. In general, robots evolved using the generative representation increased their speed by repeating rolling segments to smoothen out their gaits, and increasing the size of these segments or appendages to increase the distance moved in each oscillation.

One of the fundamental questions is whether the actual grammar evolved in the successful L-systems has captured some of the intrinsic properties of the design space. A way to quantify this is to measure the correlation between fitness change and a random mutation of various sizes, and compare this with the correlation observed in random mutations on the nongenerative representation as a control experiment. If the observed correlation is distinguishable and better for the generative system than it is for the blind system, then the generative system must have captured some useful properties.

The plot in Figure 4.6e is a comparison of the fitness-mutation correlation between a generative representation and a random control experiment on the same substrate and on the same set of randomly selected individuals. For this analysis, 80,000 individuals were selected uniformly from 16 runs and over 100 generations using a generative representation. Each point represents a particular fitness change (positive or negative) associated with a particular mutation size. The points on the left plot of Figure 4.6e were carried out on the nongenerative representation generated by the generative representation and serve as the control set. For these points, 1 to 6 mutations were applied so as to approximate mutations of similar phenotypic-size as those on the generative representation. Each mutation could modify or swap a sequence of characters. The points on the right of Figure 4.6e were also carried out randomly but on the generative representations of the same randomly selected individuals. Only a single mutation was applied to the generative representation, and consisted of modifying or swapping a single keyword or parameter. Mutation size was measured in both cases as the number of modified commands in the final construction sequences.

The two distributions in Figure 4.6e have distinct features. The data points separate into two distinguishable clusters, with some overlap. Mutations generated on the generative representations clearly correlate with both positive fitness and negative fitness changes, whereas most mutations on the nongenerative representation result in fitness decrease. Statistics of both systems, averaged over 8 runs each, reveal that the two means are different with at least 95% confidence. Cross-correlation showed that in 40% of the instances where a nongenerative mutation was successful, a generative mutation was also successful, whereas in only 20% of the instances where a generative mutation was successful, was a nongenerative mutation successful too. In both cases smaller mutations are significantly more successful than larger mutations. However, large mutations (>100) were an order of magnitude more likely to be successful in the generative case than in the nongenerative case. All these measures indicate that the generative representation is more efficient in exploiting useful search paths in the design space.

4.4.3 Regulatory Network Representations

The way that morphologies of organisms develop in biology is not only dependent on their genotype; many other environmental effects play an important role. The ontology of an organism depends on chains of productions that trigger other genes in a complex regulatory network. Some of these triggers are intracellular, such as one set of gene products resulting in expression of another group of genes, while other products may inhibit certain expressions creating feedback loops and several tiers of regulation. Some signaling pathways transduce extracellular signals that allow the morphology to develop in response to particular properties of its extracellular environment. This is in contrast to the representations discussed earlier, where the phenotype was completely defined by the genotype. Through these regulatory pathways, a genotype may encode a phenotype with variations that can compensate, exploit, and be more adaptive to its target environment.

Bongard and Pfeifer (2003) explored a regulatory network representation for evolving both a body and a brain of a robot. The machines were composed of spherical cells, which could each contain several angular actuators, touch sensors, and angular sensor, as seen in Figure 4.7a. The actuators and sensors were connected through a neural network as in Figure 4.1b, but the specific connectivity of the network was determined by an evolved regulatory network. The regulatory network contained genes which could sprout new connections and create new spherical cells, as well as express or inhibit "chemical" signals that would propagate through the structure. These chemical signals could also trigger the expression of other genes, giving rise to complex signaling and feedback pathways. Some machines evolved in response to a fitness rewarding the ability to push a block forward are shown in Figure 4.7b. These machines grow until they reach the block and have a firm grasp of the ground; their regulatory nature would allow them to attain a slightly different morphology if they would be growing in the presence of a slightly differently shaped block. It is interesting to note that an analysis of the regulation pattern (who regulates who, Figure 4.7c) shows that genes that regulate growth of neurons (colored red) and genes that regulate growth if new

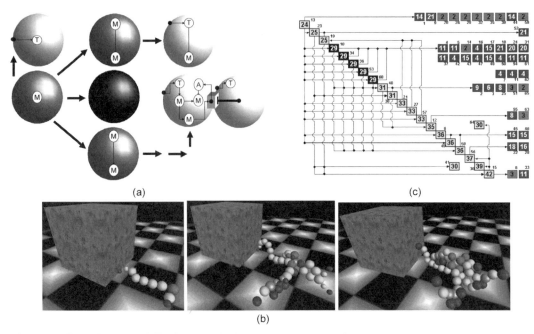

Figure 4.7 (See color insert following page 302)   Artificial ontogeny: Growing machines using gene regulatory networks. (a) An example of cells that can differentiate into structural, passive cells (dark), or active cells (bright) which contains neurons responsible for sensing (T = touch, A = angle) and motor actuation (M). The connectivity of the neurons is determined by propagation of "chemicals" expressed by genes and sensors, who are themselves expressed in response to chemicals in a regulatory network. (b) Three machines evolved to be able to push a block. (c) The distribution of genes responsible for neurogenesis (red) and morphogenesis (blue) shows a clear separation that suggests an emergence of a "body" and a "brain." (From Bongard, J. C., Pfeifer, R. (2003) Evolving complete agents using artificial ontogeny. In: Hara, F., Pfeifer, R. (eds), *Morpho-Functional Machines: the New Species (Designing Embodied Intelligence)*, Springer-Verlag, New York, New York. With permission.)

cells (colored blue) are relatively separated, suggesting an initial emergence of what we call "body" and "brain."

4.5 EVOLVING MACHINES IN PHYSICAL REALITY

Though many robotic experiments are carried out in simulation, a robot must ultimately reside in physical reality. Applying evolutionary processes to physical machines is difficult for two reasons. First, even if we are only evolving controllers for a fixed machine, each evaluation of a candidate controller involves trying it out in reality. This is a slow and costly process that also wears out the target system. Performing thousand of evaluations is usually impractical. Second, if we are evolving morphology as well, then how would these morphological changes take place in reality? Changes to the controller can be done simply by reprogramming, but changes to the morphology require more sophisticated processes. Nature has some interesting solutions to this problem, such as growing materials, or self-assembling and self-replicating basic building blocks like cells. Let us examine these two approaches.

4.5.1 Evolving Controllers for Physical Morphologies

One approach to evolving controllers for fixed morphologies is to make a simulator that is so perfect, that whatever works in simulation will work in reality equally well. Unfortunately, such a simulator has not yet been constructed. It is unlikely that one could be constructed, given the chaotic nature of machine dynamics and their sensitivity to initial conditions and many small parameter variations. Even if such simulators existed, creating accurate models would be pains-takingly difficult, or may be impossible if the target environment is not perfectly known.

An alternative approach to "crossing the reality gap" is to use a crude simulator that captures the salient features of the search space. Techniques have been developed for creating such simulators and using noise to cover uncertainties so that the evolved controllers do not exploit these uncer-tainties (Jakobi, 1997). Yet another approach is to use plasticity in the controller: allow the robot to learn and adapt in reality. In nature, animals are born with mostly predetermined bodies and brains, but these have some ability to learn and make final adaptations to whatever actual conditions may arise.

A third approach is to coevolve simulators so that they are increasingly predictive. Just as we use evolution to design a controller, we can use evolution to design the simulator so that it captures the important properties of the target environment. Assume we have a rough simulator of the target morphology, and we use it to evolve controllers in simulation. We then take the best controller and try it — once — on the target system. If successful, we are done; but if the controller did not produce the anticipated result (as is likely to happen since the initial simulator was crude), then we observed some unexpected sensory data. We then evolve a new set of simulators, whose fitness is their ability to reproduce the actual observed behavior when the original controller is tested on them. Simulators that correctly reproduce the observed data are more likely to be predictive in the future. We then take the best simulator, and use it to evolve a new controller, and the cycle repeats. If the controller works in reality, we are done. If it does not work as expected, we now have more data to evolve better simulators, and so forth. The coevolution of controllers and simulators is not necessarily computationally efficient, but it dramatically reduces the number of trials necessary on the target system.

The coevolutionary process consists of two phases: evolving the controller (or whatever we are trying to modify on the target system) — we call this the exploration phase. The second phase tries to create a simulator, or model of the system — we call this the estimation phase. To illustrate the estimation–exploration process, consider a target robot with some unknown, but critical, morpho-logical parameters, such as mass distribution and sensory lag times. Fifty independent runs of the algorithm were conducted against the target robot. Figure 4.8a shows the 50 series of 20 best simulator modifications output after each pass through the estimation phase. Figure 4.8a makes clear that for all 50 runs, the algorithm was better able to infer the time lags of the eight sensors than the mass increases of the nine body parts. This is not surprising in that the sensors themselves provide feedback about the robot. In other words, the algorithm automatically, and after only a few target trials, deduces the correct time lags of the target robot's sensors, but is less successful at indirectly inferring the masses of the body parts using the sensor data. Convergence towards the correct mass distribution can also be observed, but even with an approximate description of the robot's mass distribution, the simulator is improved enough to allow smooth transfer of controllers from simulation to the target robot. Using the default, approximate simulation, there is a complete failure of transferal: the target robot simply moves randomly, and achieves no appreciable forward locomotion. It is interesting to note that the evolved simulators are not perfect; they capture well only those aspects of the world that are important for accomplishing the task.

The exploration–estimation approach can be used for much more than transferring controllers to robots — it could be used by the robot itself to estimate its own structure. This would be particularly useful if the robot may undergo some damage that changes some of its morphology in unexpected ways, or some aspect in its environment changes. As each controller action is taken, the actual sensory data is compared to that predicted by the simulator, and new internal simulators are evolved to be more predictive. These new simulators are then used to try out new, adapted controllers for the new and unexpected circumstances. Figure 4.8b shows some results applying this process to design controllers for a robot which undergoes various types of drastic morphological damage, like losing

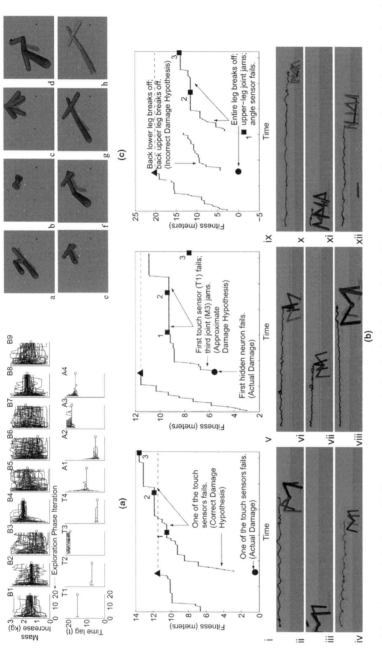

Figure 4.8 Co-evolving robots and simulators: (a) Convergence towards the physical characteristics of the target robot. Each pass through the estimation phase, produces a set of mass changes for each of the nine body parts of the robot (top row), and a set of time lags for each of the eight sensors (bottom row). The open circles indicate the actual differences between the target robot and the starting default simulated robot. (From Bongard, J. C., Lipson, H. (2004a) Once more unto the breach: automated tuning of robot simulation using an inverse evolutionary algorithm, *Proceedings of the Ninth International Conference on Artificial Life (ALIFE IX)*. With permission.) (b) Three typical damage recoveries. (i) The evolutionary progress of the four sequential runs of the estimation EA (all of which are correct) are shown. The dots indicate the fitness of the best controller from each generation of the exploration EA. The triangle shows the fitness of the first evolved controller on the target robot (the behavior of the 'physical' robot with this controller is shown in (ii); the filled circle shows the fitness of the robot after the damage occurs (the behavior is shown in (iii); the squares indicate the fitness of the "physical" robot for each of the three subsequent hardware trials (the behavior of the physical robot during the third trial is shown in (iv)). (v–viii) The recovery of the quadrupedal robot when it experiences unanticipated damage. (ix–xii) The recovery of the hexapedal robot when it experiences severe, compound damage. The trajectories in (ii–iv), (vi–viii), and (x–xii) show the change in the robot's center of mass over time (the trajectories are displaced upwards for clarity). (From Bongard, J. C., Lipson, H. (2004b) Automated damage diagnosis and recovery for remote robotics, *IEEE International Conference on Robotics and Automation (ICRA04)*. With permission.) (c) "he simulator progressively learns the entire robot morphology from scratch. Panels (i–vii) are progressive intermediate self-inference stages, panel (viii) is the true target system. (From Bongard, J. C., Lipson, H. (2004c) Integrated design, deployment, and inference for robot ecologies, *Proceedings of Robosphere 2004*, November 2004, NASA Ames Research Center, Moffett Field, California. With permission.)

a leg, motor, or sensor, or combinations of these. In most cases, the estimation–exploration process is able to reconstruct a new simulator that captures the actual damage using only 4 to 5 trials on the target robot, and then use the adapted simulator to evolve compensatory controllers that recover most of the original functionality. There are numerous applications to this identification and control process in other fields.

4.5.2 Making Morphological Changes in Hardware

An evolutionary process may require a change of morphology or production of a new physical morphology altogether. One approach for generating new morphology is to use reconfigurable robots (Yim et al., 2002). Reconfigurable robots are composed of many modules that can be connected, disconnected, and rearranged in various topologies to create machines with variable body plans. Self-reconfigurable robots are able to rearrange their own morphology, and thus adapt in physical reality. Figure 4.9a shows one example of a self-reconfiguring robot composed of eight identical cubes (Zykov et al., 2005). Each cube can swivel around its (1,1,1) axis, and connect and disconnect to other cubes using electromagnets on its faces. Though this robot contains only eight units, it is conceivable that a future machine will be composed of hundreds and thousands of modules of smaller scales, allowing much greater control and flexibility in morphological change. As scales decrease, one may need to switch from classical deterministic reconfiguration processes to stochastic processes that exploit Brownian motion, not mechanisms in the biological cell. Figure 4.9b shows some large scale robot prototypes which operate on these stochastic principles (White et al., 2004).

An alternative approach to varying morphology is to produce the entire robot morphology automatically. For example, the robots shown in Figure 4.4f were produced using rapid prototyping equipment: These are 3D printers, that deposit material layer by layer to gradually build up a solid object of arbitrary geometry, as shown in Figure 4.9c. This "printer," when coupled to an evolutionary design process, can produce complex geometries that are difficult to produce any other way, and thus allow the evolutionary search much greater design flexibility. Nevertheless, even when using such automated fabrication equipment, we needed to manually insert the wires, logic, batteries, and actuators. What if the printer could print these components too? Future rapid prototyping systems may allow deposition of multiple integrated materials, such as elastomers, conductive wires, batteries, and actuators, offering evolution an even larger design space of integrated structures, actuators and sensors, not unlike biological tissue. Figure 4.9d shows some of these printed components (Malone and Lipson, 2004).

4.6   THE ECONOMY OF DESIGN AUTOMATION

The examples shown so far are all related to design and fabrication of robotic systems, but the principles described here are applicable in many other domains. Is there a way to know *á priori* where these methods will be successful? Several decades of experience have shown that there are a number of conditions that suggest such problem domains.

- *Known physics.* Most evolutionary systems use some form of simulation to determine the consequence of various design choices. Evolutionary algorithms are fruitful when the physics are understood well enough that simulations are predictive, there is no question about the underlying physical phenomena, and that simulation can be carried out in reasonable time.
- *Well-defined search space.* The basic "atomic" building blocks comprising potential solutions are known, and it is clear how they are allowed to fit together. These two aspects define a search space in which the evolutionary algorithm can operate. Knowing the building blocks and interfaces does not imply knowing the solution. It is important to realize that "building blocks" are not necessarily discrete components — they can be features of a solution or partial solutions.

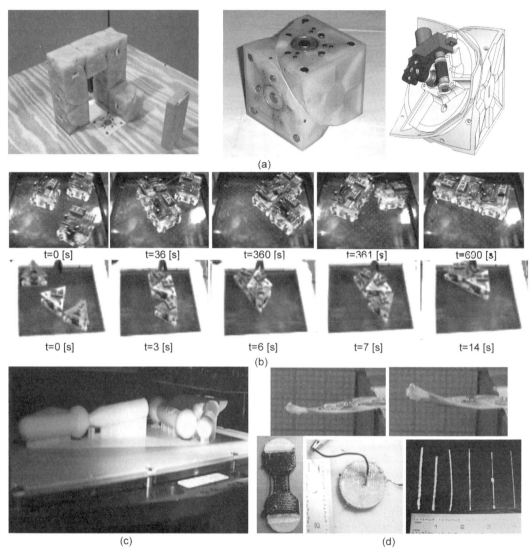

Figure 4.9 (See color insert following page 302) (a) Reconfigurable *molecube* robots. (From Zykov, V., Mytilinaios, E., Lipson, H., (2005) *Nature*, 435 (7038), 163–164. With permission.) (b) Stochastic modular robots reconfigure by exploiting Brownian motion, and may allow reconfiguration at a micro-scale in the future. (From White, P. J., Kopanski, K., Lipson, H. (2004) Stochastic self-reconfigurable cellular robotics, *IEEE International Conference on Robotics and Automation (ICRA04)*. With permission.) (c) Rapid prototyping. (d) Future rapid prototyping systems will allow deposition of multiple integrated materials, such as elastomers, conductive wires, batteries, and actuators, offering evolution of a larger design space of integrated structures, actuators, and sensors, not unlike biological tissue. (From Malone, E., Lipson, H. (2004) Functional freeform fabrication for physical artificial life, Ninth International Conference on Artificial Life (ALIFE IX), *Proceedings of the Ninth International Conference on Artificial Life (ALIFE IX)*. With permission.)

•   *Little formal design knowledge.* Evolutionary algorithms are "knowledge sparse"; they essentially generate knowledge through search. They are thus able to work in the absence of formal knowledge in the problem domain. Given enough time and resources, one may be able to design a specialized algorithm that takes advantage of specific domain knowledge and outperforms an evolutionary algorithm, but often this is time consuming, costly, and too difficult.

- *An approximate solution will suffice*. Evolutionary algorithms do not provide guarantees on the solution optimality, and do not necessarily find the optimal solution. In many practical problem domains, we do not require strict optimality; a solution is good enough if it is better than the competition.

Computer scientists often compare algorithms on the basis of their computational efficiency — how fast can the algorithm solve a problem. A second (often neglected) factor affecting the usefulness of algorithms is their economy: How many resources do you need to invest in designing and implementing an algorithm before it can produce useful results? Evaluating algorithms based on performance alone is equivalent to pricing products without amortizing their development costs. As the cost of development labor increases and the cost of computing power decreases, we begin to favor algorithms that are easy to implement and require little formal knowledge in the problem domain, even if they are computationally less efficient. These trends are becoming more pronounced as we venture into new scientific and engineering domains where human intuition is poor.

Robotics is one area where these criteria are met: the physics are well understood, the building blocks are well known, there is little formal design knowledge, and approximate solutions will suffice. There are many more examples, especially in emerging fields. A typical domain where these criteria are met is micro-scale design. For example, microphotonics devices manipulate light (like microelectronic devices manipulate electrons). Their function depends on manipulating photons at the quantum level. The physics is well understood, as we can predict the behavior of photons by solving Maxwell's equations; the building blocks are well defined, as we know the capabilities of microfabrication tools; little design knowledge exists as few people have the intuition to design structures that manipulate light at subwavelength scales; and approximate solutions will suffice as current solutions are suboptimal anyway. One notable challenge is the design of regular (periodic) structures that confine light, known as *photonic crystals*. Figure 4.10a shows a way of representing the geometry of photonic cells as a hierarchy of partitions. Evolving cell representations and checking their ability to confine light using a simulator produced an interesting pattern shown in Figure 4.10b. This pattern outperforms human-designed patterns by

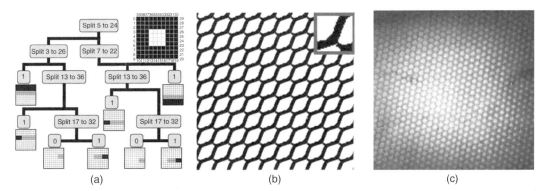

(a)   (b)   (c)

Figure 4.10 Evolving photonic crystal geometries. (a) A tree representation is used to encode geometry of a photonic cell by specifying a hierarchy of partition lines. The tree shown encodes the square ring shown at the top right. (b) Evolving structures with large photonic bandgaps produced this structure (unit cell shown in inset). This structure has a bandgap that is 10% larger than any human-designed pattern. (From Preble, S. F., Lipson, H., Lipson, M. (2005) *Applied Physics Letters*, 86 (c) A transmission electron micrograph of the Sea Mouse spine (notoseta). The dark areas are chitin and the light areas are voids, with a spacing of 510 nm. (From Parker, A. R., McPhedran, R. C., McKenzie, D. R., Botten, L. C., Nicorovici, N.-A. P. (2001) *Nature*, 409, 36–37. With permission.)

10% (bandgap size). It is interesting to note that similar skewed-hexagonal pattern also appears in nature for the same purpose (Figure 4.10c). The number of domains where open-ended synthesis algorithms are producing human-competitive designs is growing rapidly (Koza, 2003).

4.7 FUTURE CHALLENGES

Parametric evolutionary optimization has been successfully applied in almost every engineering domain (e.g., Gen and Cheng, 1999; Zalzala and Fleming, 1999; Jamshidi et al., 2002; Karr and Freeman, 1998; Mazumder and Rudnick, 1998), but the use of evolution for open-ended design will likely have an even higher impact. It is also one of the most poorly understood areas of evolutionary computation. We are seeking to understand what underlies the complexity limits of what can be designed automatically, and what allows natural systems to evolve systems so much more advanced than what we can evolve artificially. Is it simply a matter of computational power — that nature is performing an immeasurable number of evaluations every second? Or is there something more fundamental about the evolutionary process that we have failed to capture? What are the implications of physical embodiment and self-replication that we often bypass in our simulations? What are the implications of external fitness measures that we impose on the system, and of arbitrary inductive biases we introduce thorough our choices of atomic building blocks and representations? Does complexity require complex ecosystem with coevolution, symbiosis, competition, and co-operation? Can we outperform natural evolution by using analytical shortcuts through its weak statistical processes?

These are long standing problems that are not unique to evolutionary computation. The question of how complex systems are synthesized is fundamental from three perspectives: AI research interested in automating discovery processes, engineering research in understanding the design process, and biology research interested in the origin of complexity. These perspectives are captured well in the following statments:

> One may wonder, [...] how complex organisms evolve at all. They seem to have so many genes, so many multiple or pleiotropic effects of any one gene, so many possibilities for lethal mutations in early development, and all sorts of problems due to their long development.
>
> (Bonner, J. T., (1988) *The Evolution of Complexity*, p. 173.)

> Today more and more design problems are reaching insoluble levels of complexity ... these problems have a background of needs and activities which is becoming too complex to grasp intuitively ... The intuitive resolution of contemporary design problems simply lies beyond a single individual's integrative grasp.
>
> (Alexander, C. A., *Notes on the Synthesis of Form*, 1964, pp. 3–5.)

I believe that scalability of open-ended evolutionary processes depends on their ability to exploit functional modularity, structural regularity, and hierarchy (Lipson, 2004). Functional modularity creates a separation of function into structural units, thereby reducing the amount of coupling between internal and external behavior on those units and allowing evolution to reuse them as higher-level building blocks. Structural regularity is the correlation of patterns within an individual. Examples of regularity are repetition of units, symmetries, self-similarities, smoothness, and any other form of reduced information content. Regularity allows evolution to specify increasingly extensive structures while maintaining short description lengths. Hierarchy is the recursive com-position of function and structure into increasingly larger and adapted units, allowing evolution to search efficiently increasingly complex spaces.

The existence of modular, regular, and hierarchical architectures in naturally evolved systems is well established (Wagner and Altenberg, 1996; Hartwell et al., 1999). Though evolutionary processes have been studied predominantly in biological contexts, they exist in many other domains, such as language, culture, social organization, and technology (Basalla, 1989; Ziman, 2003), among many others. Principles of modularity, regularity, and hierarchy are, however, nowhere as dominant as they are in engineering design. Tracing the evolution of technology over generations of products, one can observe numerous instances of designs being encapsulated into modules, and those modules being used as standard higher-level building blocks elsewhere. Similarly, there is a pressure to reduce the information content in designs, by repeating or reusing the same modules where possible, using symmetrical and regular structures, and stand-ardizing on components and dimensions. These and other forms of regularity translate into reduced design, fabrication, and operation costs. The organization of engineering designs, especially as complexity increases, is typically hierarchical. The hierarchy is often organized such that the amount of information is distributed uniformly across levels, maintaining a "manageable" extent of information at each stage. These principles of modularity, regularity, and hierarchy are cornerstones of engineering design theory and practice (e.g., Suh, 1990). Though these principles are well established, there is — like biological evolution — still a lack of a formal understanding of how and why modular, regular, and hierarchical structures emerge and persist, and how can we computationally emulate these successful principles in the design automation processes.

- *Functional modularity* is the structural localization of function.
  In order to measure functional modularity, one must have a quantitative definition of function and structure. It is then possible to take an arbitrary chunk of a system and measure the dependency of the system function on elements of that chunk. The more that the dependency *itself* depends on elements outside the chunk, the less the function of that chunk is localized, and hence the less modular it is. If we represent dependencies as second derivatives of function with respect to pairs of parameters (i.e., the Hessian matrix of the fitness), then modules will be collections of parameters that can be arranged with lighter off-diagonal elements (Wyatt and Lipson, 2003).
- *Structural regularity* is the compressibility of the description of the structure.
  The more the structure contains repetitions, near-repetitions, symmetries, smoothness, self-similarities, etc., the shorter its description length will be. The amount of regularity can thus be quantified as the inverse of the description length or of its Kolmogorov complexity.
- *Hierarchy* of a system is the recursive composition of structure and/or function.
  The amount of hierarchy can be quantified given the connectivity of functional or structural elements (e.g., as a connectivity graph). The more the distribution of connectivities path lengths among pairs of elements approximates a power law distribution, the more hierarchical the system is.

4.7.1 Principles of Design

Modularity and regularity are independent principles. Principles of modularity and regularity are often confused in the literature through the notion of *reuse*. Indeed, modularity has several advantages, one of which is that modules can be used as building blocks at higher levels, and therefore can be *repeated*. Nonetheless, it is easy to imagine a system that is composed of modules, where each module appears only once. For example, opening the hood of a car reveals a system composed of a single engine, a single carburetor, and a single transmission. Each of these units appears only once (i.e., is not reused anywhere else in the system), but can be considered a module as its function is localized. Its evolutionary advantage is that it can be adapted more independently, with less impact of the adaptation on the context. A carburetor may be swapped to a newer technology without affecting the rest of the engine system.

Similarly, there are instances of regularity without modularity. The smoothness of the hood of the car, for example, reduces the information content of the structure but does not involve the reuse of a particular module.

Though these principles are independent, they often appear in tandem and hence the confusion: we tend to speak of useful modules being reused as building blocks, and indeed recurrence of a pattern may be an indication of its functional modularity, though not a proof of it.

An inherent tradeoff exists between modularity and regularity through the notion of *coupling*. Modularity by definition reduces coupling, as it involves the localization of function. Regularity, however, increases coupling as it reduces information content. For example, if a module is reused in two different contexts, then the information content of the system has reduced (the module needs to be described only once and then repeated), but any change to the module will have an effect on both places. Software engineers are well aware of this tradeoff. As a function is encapsulated and called from an increasing number of different contexts in a program, so does modifying it become increasingly difficult because it is entangled in so many different functions.

The tradeoff between modularity and reuse is also observed in engineering as the tradeoff between modularity and optimality. Modularity often comes at the expense of optimal performance. Systems that are less modular, that is more integrated, can be more efficient in their performance as information, energy, and materials can be passed directly within the system, at the expense of increased coupling. Software engineers are familiar with "long jumps" and "global variables" that have this effect, similarly, mechanical products will often achieve optimality of performance or cost through integration of parts into monolithic components wherever possible. The increased performance gained by reduction of modularity is often justified in the short term, whereas increased modularity is often justified in longer time scales where adaptation becomes a dominant consideration.

It is not clear whether modularity, regularity, and hierarchy are properties of the system being evolved (i.e., the "solution"), or of the target fitness specification (i.e., the "problem"). It may well be that there is a duality between these viewpoints. The evolutionary computation literature contains several instances of test functions that are themselves modular (separable, e.g., Royal Roads [Mitchell, 1996]), hierarchical (e.g., Hierarchical-IFF [Watson and Pollack, 1999]), and regular (e.g., one-max). It is not surprising then to see corresponding algorithms that are able to exploit these properties and find the solutions to these problems.

Engineers often go to great lengths to describe design goals in a way that is solution-neutral, that is it describes target functionality while placing the least constraints on the solution. Indeed engineering design is notorious for having multiple — even many — solutions to any given problem, without any solution being clearly superior. The fact that modular, regular, and hierarchical solutions are more attractive is because — we conjecture — the design process itself tends to prefer those for reasons of scalability. It is therefore plausible that in search of scalable algorithms for synthesizing solutions bottom up, we should avoid test functions that have an inherent modular or hierarchical reward, and have these solution properties emerge from the search process itself.

4.7.2 Research Methodology

Though robotic systems provide an intuitive and appealing substrate to explore many of these questions, they also pose many difficulties. They are computationally expensive to simulate and difficult to construct physically. More importantly, like biology, they contain many beautifully complex but arbitrary details that obscure the universal principles that we are looking for. There is always a temptation to increase the fidelity of the simulators, adding more biologically realistic details, in hopes that this would lead to more life-like behaviors. However, it is sometimes more fruitful to investigate these questions in a simpler, more transparent substrate. In fact we look for the *minimal substrate* that still exhibits the effects we are investigating. Many insights can be gained by looking at these simplified systems, and the lessons learned brought to bear on the complex problems of practical importance. Thus much of the research in evolutionary design and evolutionary robotics is disguised as experiments in much more abstract systems.

4.8. CONCLUSIONS

We have followed through a number of cases where principles of biological evolution have been used to automate the design of machines — from relatively simple examples in controller design to design and fabrication of complete functional machines in physical reality, sometimes outperforming the human designs. Unlike other forms of biomimicry, however, we are not seeking to *imitate the solutions* that present themselves in nature — like the gecko's feet, a bird's wing, or a human's muscle — because these solutions were optimized for very specific needs and circumstances that may not reflect our requirements and unique capabilities. Instead, we chose to *imitate the process* that led to these solutions, as biology's design process has shown time and again its ability to discover new opportunities.

It is clear that the complexity of engineering products is increasing to the point where traditional design processes are reaching their limits. More manpower is being invested in managing and maintaining large systems than designing them, and this ratio is likely to increase because no single person can fathom the complexities involved. Alexander's quote (above) is truer today than it was in the 1960s. Engineering and science are moving into scales and dimensions where people have little or no intuitions and the complexities involved are overwhelming. One way out of this conundrum is to design machines that can design for us — this is the future of engineering.

4.9 FURTHER READING

Digital Biology, Peter Bentley, Simon and Schuster, 2004.
Evolutionary Robotics: The Biology, Intelligence, and Technology of Self-Organizing Machines, Stefano Nolfi and Dario Floreano, Bradford Books, 2004.
Out of Control: The New Biology of Machines, Social Systems and the Economic World, Kevin Kelly, Perseus Books Group, 1995.

REFERENCES

Alexander, C. A. (1964) *Notes on the Synthesis of Form*, Harvard University Press, Cambridge, Massachusetts.
Basalla, G. (1989) *The Evolution of Technology*, Cambridge University Press, Cambridge, Massachusetts.
Beer, R. D., Gallagher, J. C. (1992) Evolving dynamical neural networks for adaptive behavior, *Adaptive Behavior*, 1(1), 91–122.
Bongard, J. C. (2002) Evolved Sensor Fusion and Dissociation in an Embodied Agent, *Proceedings of the EPSRC/BBSRC International Workshop Biologically-Inspired Robotics: The Legacy of W. Grey Walter*, pp. 102–109.
Bongard, J. C., Lipson, H. (2004a) Once more unto the breach: automated tuning of robot simulation using an inverse evolutionary algorithm, *Proceedings of the Ninth International Conference on Artificial Life (ALIFE IX)*, pp. 57–62.
Bongard, J. C., Lipson, H. (2004b) Automated damage diagnosis and recovery for remote robotics, *IEEE International Conference on Robotics and Automation (ICRA04)*, pp. 3545–3550.
Bongard, J. C., Lipson, H. (2004c) Integrated design, deployment and inference for robot ecologies, *Proceedings of Robosphere 2004*, November 2004, NASA Ames Research Center, Moffett Field, California.
Bongard, J. C., Pfeifer, R. (2003) Evolving complete agents using artificial ontogeny. In: Hara, F., Pfeifer, R., (eds), *Morpho-Functional Machines: the New Species (Designing Embodied Intelligence)*, Springer-Verlag, New York, New York, pp. 237–258.
Bonner, J. T. (1988) *The Evolution of Complexity by Means of Natural Selection*, Princeton University Press, Princeton, New Jersey.
Floreano, D., Urzelai, J. (2001) Evolution of plastic control networks, *Autonomous Robots*, 11, 311–317.

Floreano, D., Nolfi, S., Mondada, F. (2001) Co-evolution and ontogenetic change in competing robots. In: Patel, M., Honavar, V., Balakrishnan, K. (eds), *Advances in the Evolutionary Synthesis of Intelligent Agents*, MIT Press, Cambridge, Massachusetts.

Floreano, D., Zufferey, J. C., Nicoud, J. D. (2005) From wheels to wings with evolutionary spiking neurons, *Artificial Life*, 11(1–2), 121–138.

Gen, M., Cheng, R. (1999) *Genetic Algorithms and Engineering Optimization (Engineering Design and Automation)*, Wiley-Interscience, Hoboken, New Jersey.

Goldberg, D. E. (1989) *Genetic Algorithms in Search, Optimization, and Machine Learning*, Addison-Wesley, Madison, Massachusetts.

Hartwell, L. H., Hopfield, J. H., Leibler, S., Murray, A. W. (1999) From molecular to modular cell biology, *Nature*, 402, C47–C52.

Harvey, I., Husbands, P., Cliff, D., Thompson, A., Jakobi, N. (1997) Evolutionary robotics: the Sussex approach, *Robotics and Autonomous Systems*, 20, 205–224.

Hornby, G. S., Lipson, H., Pollack, J. B. (2003) Generative encodings for the automated design of modular physical robots, *IEEE Transactions on Robotics and Automation*, 19(4), 703–719.

Jakobi, N. (1997) Evolutionary robotics and the radical envelope of noise hypothesis, *Adaptive Behavior*, 6(1), 131–174.

Iamshidi, M., Krohling, R. A., dos Santos Coelho, L. (2002) *Robust Control Systems with Genetic Algorithms*, CRC Press, Boca Raton, Florida.

Karr, C. L. Freeman, L. M. (1998) *Industrial Applications of Genetic Algorithms*, CRC Press, Boca Raton, Florida.

Koza, J. (2003) Human-competitive applications of genetic programming. In: Ghosh, A., Tsutsui, S. (eds), *Advances in Evolutionary Computing: Theory and Applications*, Springer-Verlag, New York, New York, pp. 663–682.

Koza, J. (1992) *Genetic Programming: On the Programming of Computers by Means of Natural Selection*, MIT Press, Cambridge, Massachusetts.

Lipson, H. (2006) Evolutionary synthesis of kinematic mechanisms, *Journal of Computer Aided Design*.

Lipson, H. (2004) Principles of modularity, regularity, and hierarchy for scalable systems, *Genetic and Evolutionary Computation Conference (GECCO'04) Workshop on Modularity, Regularity and Hierarchy*.

Lipson, H., Pollack, J. B. (2000) Automatic design and manufacture of artificial lifeforms, *Nature*, 406, 974–978.

Luke, S., Spector, L. (1996) Evolving graphs and networks with edge encoding: preliminary report. In Late Breaking Papers at the Genetic Programming 1996 Conference (GP96). Koza, J. (ed.), Stanford Bookstore, Stanford, pp. 117–124.

Lund, H., Hallam, J., Lee, W. (1997) Evolving robot morphology, *Proceedings of IEEE Fourth International Conference on Evolutionary Computation*, IEEE Press, New York, New York, pp. 197–202.

Malone, E., Lipson, H. (2004) Functional freeform fabrication for physical artificial life, Ninth International Conference on Artificial Life (ALIFE IX), *Proceedings of the Ninth International Conference on Artificial Life (ALIFE IX)*, pp. 100–105.

Mazumder, P., Rudnick, E. (1998) *Genetic Algorithms for VLSI Design, Layout and Test Automation*, Prentice Hall, Englewood cliffs, New Jersey.

Meyer, J. -A. (1998) Evolutionary approaches to neural control in mobile robots, *Proceedings of the IEEE International Conference on Systems, Man and Cybernetics*.

Mitchell, M. (1996) *An Introduction to Genetic Algorithms*, MIT Press, Cambridge, Massachusetts.

Nolfi, S., Floreano, D. (2004) *Evolutionary Robotics: The Biology, Intelligence, and Technology of Self-Organizing Machines*, Bradford Books, Cambridge, Massachusetts.

Nolfi, S., Elman, J. L., Parisi, D. (1994) Learning and evolution in neural networks, *Adaptive Behavior*, 3(1), 5–28.

Papadimitriou, C. H., Steiglitz, K. (1998) *Combinatorial Optimization: Algorithms and Complexity*, Dover Publications, New York, New York.

Parker, A. R., McPhedran, R. C., McKenzie, D. R., Botten, L. C., Nicorovici, N. -A. P. (2001) Aphrodite's iridescence, *Nature*, 409, 36–37.

Paul, C., Bongard, J. C. (2001) The road less traveled: morphology in the optimization of biped robot locomotion, *Proceedings of The IEEE/RSJ International Conference on Intelligent Robots and Systems (IROS2001), Hawaii, USA*.

Preble, S. F., Lipson, H., Lipson, M. (2005) Two-dimensional photonic crystals designed by evolutionary algorithms, *Applied Physics Letters*, 86.

Sims, K. (1994) Evolving 3D morphology and behaviour by competition, *Artificial Life*, IV, 28–39.

Suh, N. P. (1990) *The Principles of Design*, Oxford University Press, Oxford, UK.

Taylor, T., Massey, C. (2001) Recent developments in the evolution of morphologies and controllers for physically simulated creatures, *Artificial Life*, 7(1), 77–87.

Wagner, G. P. Altenberg, L. (1996) Complex adaptations and the evolution of evolvability, *Evolution*, 50, 967–976.

Walker, J., Garrett, S., Wilson, M. (2004) Evolving controllers for real robots: a survey of the literature, *Adaptive Behavior*, 11(3), 179–203.

Watson, R. A., Pollack, J. B. (1999) Hierarchically-consistent test problems for genetic algorithms, Angeline, P. J., Michalewicz, Z., Schoenauer, M., Yao, X., Zalzala, A. (eds), *Proceedings of 1999 Congress on Evolutionary Computation (CEC 99)*, IEEE Press, New York, New York, pp. 1406–1413.

White, P. J., Kopanski, K., Lipson, H. (2004) Stochastic self-reconfigurable cellular robotics, *IEEE International Conference on Robotics and Automation (ICRA04)*, pp. 2888–2893.

Willis, R. (1841) *Principles of Mechanism*, London (available online at Syalor et al., ibid).

Wyatt, D., Lipson, H. (2003) Finding building blocks through eigenstructure adaptation, *Genetic and Evolutionary Computation Conference (GECCO'03)*.

Yim, M., Zhang, Y., Duff, D. (2002) Modular reconfigurable robots, machines that shift their shape to suit the task at hand, IEEE Spectrum Magazine cover article, February 2002.

Zalzala, A. M. S., Fleming, P. J. (1999) *Genetic Algorithms in Engineering Systems (Control Series)*, Institution of Electrical Engineers.

Ziman, J. (2003) *Technological Innovation as an Evolutionary Process*, Cambridge University Press, Cambridge, Massachusetts.

Zykov, V., Bongard, J., Lipson, H. (2004) Evolving dynamic gaits on a physical robot, *Proceedings of Genetic and Evolutionary Computation Conference*, Late Breaking Paper, GECCO'04.

Zykov, V., Mytilinaios, E., Adams, B., Lipson, H. (2005) Self-reproducing machines, *Nature*, 435 (7038), 163–164.

Genetic Algorithms: Mimicking Evolution and Natural Selection in Optimization Models

Tammy Drezner and Zvi Drezner

CONTENTS

5.1 INTRODUCTION

Optimization problems are defined by an objective function. The objective function is a dependent variable, i.e., it is a function of several independent variables. The objective is to find the best combination of the independent variables such that the objective function is either minimized or maximized (for clarity of the discussion we assume in this chapter that the problem is minimization). Typically, the objective function is subject to a set of constraints that must be satisfied.

Some optimization models are based on continuous variables (linear or nonlinear programming), that is the variables are allowed to obtain any value (whether integer or fractional). Other optimization models require integer variables, that is variables that must be whole numbers 0, 1, 2, . . . (integer programming, Nemhauser and Wolsey, 1988). Other models require 0 or 1 variables, that is variables that must be binary, which can assume either a value of 0 or a 1 (such as "yes" or "no," build or not build, select or not select). For such 0 or 1 (sometimes called Boolean) programming problems, there is a finite number of combinations. For example, 10 binary variables can have $2^{10} = 1,024$ combinations that can usually be evaluated within a reasonable computer time. Theoretically, one can calculate the value of the objective function for all possible combinations of the independent variables and select the best combination (such a process is called total enumeration). However, the number of possible combinations is often prohibitively large, as it will take thousands of years to evaluate all of them making it nontractable. For 100 binary variables, the number of possible combinations is $2^{100} = 1,267,650,600,228,229,401,496,703,205,376$ which will take over 40,000 trillion years to evaluate even if the evaluation of each combination takes one millionth of a second.

For instance, consider the location of instruments such as cellular telephone transmitters (towers) to cover 1000 sites. There are 100 candidate locations (a preselected set) for placing the instruments. For each candidate location, there is a subset of the 1000 sites that are covered by it. The objective is to find the minimum number of instruments required to cover all 1000 sites. This is the set covering problem (Daskin, 1995; Current et al., 2002). Alternatively, one may wish to maximize the number of covered sites with a given number of instruments (the maximum cover problem, Daskin, 1995; Current et al., 2002). The latter objective is appropriate when there is a limited budget (a constraint). The variables are assigned a value of "0" if the candidate location is not selected and "1" if it is selected. Each possible combination is described by a vector of 100 zeroes and ones. If one needs to check all possible subsets of the candidate locations, one has to check 2^{100} combinations, which is prohibitive (see above).

In cases where total enumeration is impractical, various optimization techniques have been developed. One approach for finding the optimal solution is the branch and bound approach (Land and Doig, 1960; Nemhauser and Wolsey, 1988). The branch and bound approach implicitly evaluates all possibilities, but a bound is used to eliminate the need to evaluate all of them. The number of remaining combinations is significantly reduced thereby making it possible to evaluate them in a reasonable computer time.

When robust approaches (that guarantee finding the optimal solution) are not efficient enough to be performed in a reasonable computer time, *heuristic* algorithms are used. Heuristic algorithms are designed to find a "good" solution but do not guarantee that the optimal solution is found. In the last three decades, researchers have developed "metaheuristic" methods (Salhi, 1998) that are problem-independent general heuristic approaches to be applied in any optimization problem.

Following a brief overview of major metaheuristic approaches, we will focus on genetic algorithms, also referred to as evolutionary algorithms.

It is interesting to note that many of these metaheuristic methods mimic biological and other natural processes, most notably genetic algorithms mimic evolutionary processes, natural selection, and survival of the fittest.

5.1.1 Common Metaheuristic Methods

1.   *Steepest descent* is the "classic" local search for a minimum. A "neighborhood" of nearby combinations is defined for each particular combination. The number of combinations in the neighborhood is usually quite small. The algorithm starts at one (usually constructed randomly) combination and proceeds iteratively. At each iteration, (i) all combinations in the neighborhood are evaluated, and (ii) if a better combination is found, the search moves to the best combination in the neighborhood and the iteration is completed. The next iteration applies the same process to the

2.   *Tabu search* (Glover, 1986; Glover and Laguna, 1997) is based on artificial intelligence. The search starts as a steepest descent algorithm but continues after the steepest descent algorithm has been terminated. Unlike the steepest descent, tabu search may take upward moves in the hope that a sequence of upward moves will lead to a better downward move and eventually to a better solution. The direction of the search is determined by the recent history of moves that are "memorized." Once a move is performed, the reverse move (i.e., moving back to the previous combination) is forbidden for some iterations (hence the name tabu which can also be spelled taboo), thus pushing the search away from previous combinations. Imagine a search on a plane with many craters. One of these craters is the deepest one, and that one is the desired solution (the global optimum). The steepest descent performs only downward moves and may land at a shallow crater (a local optimum) and not at the global one. Tabu search attempts to get out of a shallow crater in the hope of getting into a better one. Therefore, when the steepest descent algorithm terminates at a bottom of a crater, upward moves are taken in tabu search while sliding back into the same crater is disallowed with the hope of sliding into deeper craters and eventually reaching the global optimum.

3.   *Simulated annealing* (Kirkpatrick et al., 1983) simulates the annealing of metals from a very hot liquid phase to a cool solid phase. Borrowing the metaphor of the plane full of craters, simulated annealing is like a "bouncing" rubber ball which we hope will settle at the deepest crater because it is more difficult to get out of it. The cooling of the temperature means that the "height" of the bounce diminishes as the process continues.

4.   The *ant colonies* metaheuristic (Colorni et al., 1992; Dorigo and Gambardella, 1997; Gambardella et al., 1999) is based on the behavior of ants when they find a food source (the optimal solution). Ants return to their nest discharging pheromone. Other ants follow the pheromone and eventually form a line leading to the food source. The premise of the algorithm is that more pheromone will be discharged on the way to the food source than on the way to other points in the plane.

5.   *Genetic algorithms* (Holland, 1975; Goldberg, 1989; Reeves, 1993) simulate evolution and survival of the fittest (attributed to Charles Darwin even though Alfred Russell Wallace discovered it first). A population (made of individual combinations) evolves over time (generations). Pairs of population members (combinations) mate and produce an offspring (two combinations are merged to produce a new combination — details below). Good offspring are kept in the population whereas unfit population members are discarded (the survival of the fittest). The population evolves, and at the end of the process, the population usually consists of fairly good solutions (without a guarantee that the optimal solution is found). Details of genetic algorithms follow.

5.2   THE FRAMEWORK OF GENETIC ALGORITHMS

In nature, while two dandelions or two squirrels may look the same to us, no two individuals of any species are identical (including human identical twins, Segal, 2000); some are larger, some healthier, some faster, some more aggressive in behavior, and the number of traits that distinguish individuals is seemingly limitless. In any environment, some traits are naturally selected, making an individual more fit, thus able to live longer and produce viable offspring. The fittest individuals will thus produce more offspring than the less fit, and thus certain traits are naturally selected in any population over time. A well-known example is that of the peppered moth in England during the industrial revolution. This moth is light colored with black spots (hence its name), and rests on lichen-covered trees of the same coloring. Birds prey upon the moth, therefore good camouflage is critical for survival as those individuals that stand out are quickly spotted and eaten. With the industrial revolution and the burning of coal, the trees became dark in color from the release of pollutants into the air. Those moths that were slightly darker in color, survived more successfully than the slightly lighter colored individuals. Thus in this case, "fitness" was related to the ability of an individual to blend with the environment, that is coloring (among other traits). The darker colored moths had higher survival rates and lived longer, and as a result produced more offspring. Thus the next generation was, on average, relatively darker in color. After several generations, most of the peppered moths were dark in the cities while remaining lighter in rural England. The darker color was "naturally selected" in urban populations. Since the passage of the Clean Air Act and the resulting reduction of coal burning, numbers of light colored moths are now increasing in England's industrial towns.

There are many differences between natural processes and genetic algorithms. In nature, fitness varies across species, across space, and over time. The fittest individuals of a particular species 18,000 years ago during the last ice age changed as climate warmed and the environment changed, and as different species' ranges shifted with the warming climate. In a sense, "fitness" is a moving target in nature. However, in genetic algorithms the fit function is well defined and does not change over time.

Another major difference between natural evolution and genetic algorithms is that in nature there is interaction between different species while none exists in genetic algorithms. In the moth case, if there is an increase in the bird population, more moths are eaten. In nature, an important measure of fitness relates to survival from predators. In genetic algorithms, there is only one defined "species" and the evolution occurs in a vacuum.

In genetic algorithms, when two combinations are compared, a mechanism for determining which one is "better" is needed. This criterion is called the "fit function" named after the concept of the survival of the fittest in nature. Typically, this fit function is the objective function. The general framework of genetic algorithms is an evolving population of selected combinations. A starting population (typically randomly generated) of combinations is established. Each generation follows a sequence of steps:

1.   Two population members are selected as parents. In most algorithms the two parents are randomly selected but the selection may be governed by some other rule.
2.   The two parents are "merged" (mate) and produce an offspring. A successful merging process for producing an offspring is probably the most important feature of a genetic algorithm.
3.   The population is updated. Some offspring are added to the population as some members are removed. The rules by which offspring are added to the population and population members "die" also affect the effectiveness of the algorithm.

The process continues for a prespecified number of generations and the best member of the final population is the result of the algorithm.

5.3 MODIFICATIONS OF THE GENETIC ALGORITHM FRAMEWORK

Over the years a variety of modifications have been proposed to the basic genetic algorithm described above. It remains essential to design a good merging process for a successful genetic algorithm. Once a satisfactory merging process is designed, these modifications may improve even a good genetic algorithm.

Typically, as the population evolves, the genetic diversity among members declines and the population becomes more homogeneous. This is because a "good" trait tends to remain in the population and is transferred to new offspring who are more likely to join the population. "Bad" traits tend to disappear from the population because offspring with bad traits are less likely to survive and join the population, and if they do join the population, they are more likely to be discarded from it later in the evolutionary process. Most proposed modifications tend to increase genetic diversity among population members thereby slow down the convergence of the population to similar members. Increasing genetic diversity slows down the convergence as it increases the chance to obtain the global optimum. If one of the necessary traits required for the global optimum is missing from the population, the global optimum will be missed. Higher diversity increases the probability that all necessary traits do exist in the population gene pool (possibly in different members, though). If all traits exist in the gene pool, the combination that is needed for creating the global optimum is more likely to be obtained by mating the "right" parents. On the other hand, early convergence to a homogeneous population increases the probability that a necessary trait (for the global optimum) will disappear from the population and the algorithm will terminate in a local optimum rather than in the global one.

One simple way to retain greater genetic diversity is not to add to the population an offspring that is identical to an existing population member. Even though it is rare to get two identical combinations, if it happens and the two identical population members are selected as parents, they will generate an identical offspring and before long all population members will be identical to this member. If it is the best population member, it may be at the bottom of a "shallow" crater. If it is not the best population member, the population may evolve into one with many identical members and only a few better ones. The probability of generating better members under such circumstances is significantly reduced, and the population may become stagnant.

The modifications listed below attempt to improve some aspects of the genetic algorithm. All are rooted in biological or evolutionary processes:

1. The existence of different populations with movements between them (Parallel Genetic Algorithms [PGA], Cantu-Paz, 1998).
2. Improving the creation of the starting population (compounded genetic, Drezner, 2005a).
3. Improving the newly generated offspring by using a local search such as a descent algorithm or tabu search. The combination of a local search with a genetic algorithm is called a hybrid genetic algorithm or a memetic algorithm (Moscato, 2002).
4. The introduction of mutations that affect the newly generated offspring (Spears, 2000).
5. Invasions of "foreign" combinations that affect the population (Goldberg, 1989).
6. Improving parent selection procedures by assigning two genders to population members (Drezner and Drezner, 2005) or distance-based selection (Drezner and Marcoulides, 2003).
7. Improving the selection criteria for the removal of population members (Drezner, 2005b).

These modifications are discussed in detail below.

5.3.1 Parallel Genetic Algorithms

PGA allow for parallel populations of the same species with movements between them. They are especially suited for parallel computing. Parallel GAs are easy to implement and they have great potential for substantial improvement in search performance. For a review of PGA, see Cantu-Paz (1998) and Cantu-Paz and Goldberg (2000).

Multiple-population GAs are more sophisticated, as they consist of several subpopulations which occasionally are allowed to exchange individuals (movement). This exchange of individuals is controlled by several parameters (Cantu-Paz, 1999):

•   Movement rate determines how many individuals leave a population.
•   Movement frequency (movement interval) determines how often a move occurs.
•   Movement topology determines the destination of the move.
•   Movement policy determines which individuals move (and are added) and which are replaced at the receiving population.

Multiple-population PGA are known by different names. Sometimes they are known as "distributed" GAs, because they are usually implemented in distributed-memory Multiple Instruction or Multiple Data (MIMD) computers. Since the computation to communication ratio can be relatively high, they are occasionally called coarse-grained GAs.

5.3.2 Compounded Genetic Algorithms

The compounded genetic algorithm (Drezner, 2005a) is similar to the PGA. The difference is that in the compounded genetic algorithm there is no movement between the isolated populations. Genetic algorithms are applied in two phases, generating the starting population for phase 2 by repeating genetic algorithms in phase 1. This mimics evolving parallel populations at several isolated locations. The best species in each location are moved to a common location thus creating a "high quality" starting population. Suppose that a population of P members is applied in phase 2. Genetic algorithms are run K times in phase 1 using K randomly generated starting populations (it is convenient but not necessary to have integer P/K). The population size of the phase 1 genetic algorithm should be at least P/K. The best P/K population members of each run are compiled to construct the starting population for phase 2. Phase 2 genetic algorithm is run once. It is recommended that a "quick" genetic algorithm is used for phase 1 and an "effective" and possibly "slow" genetic algorithm is used for phase 2.

For example, if a population of $P = 100$ members is required for phase 2, phase 1 is run $K = 20$ times (each with a population of at least five members), the best $P/K = 5$ population members are selected from each run and compiled to create a starting population for phase 2.

Note that the best solution found in phase 1 by any of the runs can only be improved by the compounded genetic algorithm because the best solution found during phase 1 is a member of the starting population of phase 2 and can only be removed from the population by better solutions.

5.3.3 Hybrid Genetic Algorithms

Hybrid genetic algorithms (also referred to as memetic algorithms, Moscato, 2002) employ an improvement algorithm on the newly established offspring before considering it for inclusion in the population. This is analogous to training the offspring to improve its fitness just like training dogs to follow orders or teaching pupils in order to enhance their knowledge. One may apply a steepest descent or a tabu search procedure on every offspring before considering it for inclusion in the population. Even a simple approach may "correct" a few traits in the offspring and provide an improved solution. The hybrid modification tends to accelerate the convergence of the population because offspring tend to be more similar to one another.

Since such an improvement algorithm can be time consuming, one has to balance the benefit of employing an improvement algorithm against the time required to do it.

5.3.4 Mutations

Mutations in the formation of offspring occur quite frequently in nature. Most mutations are not beneficial to the species and can, in fact, be rather harmful. However, on rare occasions, a mutation is beneficial to the species and improves the offspring. It can be argued that evolution could not have succeeded without mutations. Mutations may create new beneficial traits that did not exist before. If a trait is beneficial to the species, the offspring will be "more fit," will be more likely to produce better offspring, and will stay in the population longer.

The common way to apply mutations (Spears, 2000) is to occasionally (e.g., 10% of the time, a parameter of the approach) introduce, by randomly selecting, a new gene in the newly created offspring (by changing its value).

5.3.5 Invasions

Throughout the history of mankind, invasion of foreign tribes has been quite common, that is "the Barbarian invasions." Such invasions increase the gene pool of the local tribe and may lead to a more advanced population. Following this phenomenon some researchers (Goldberg, 1989) suggest that occasionally, new randomly generated combinations are added to the population, replacing below average population members, mimicking an invasion. This "new blood" may generate better and improved offspring and may introduce new traits that are not found in the population (the local tribe). Such invasions tend to enhance the genetic diversity of the population with the positive effect genetic diversity has on evolution. Some researchers refer to this modification as injection. Salhi and Gamal (2003) propose to inject into the population once in a while a few chromosomes, which are either randomly generated or obtained from another heuristic. The injection rate is not necessarily constant but can slightly decrease with the number of generations.

Invasion is also common in nature in the distribution of plants and animals. In such cases, competition for resources ensues. This competition with the invading species may signal the extinction of the native species if they are less fit.

5.3.6 Gender

In nature, most advanced species require two genders in order to mate and reproduce. The gender modification attempts to mimic this natural process. One can argue that the division into two genders was selected over time as the preferred way for producing offspring and is therefore superior to other possible mating schemes. In gender-specific genetic algorithms, the diversity of the population is better maintained with no detrimental effects on run time.

It is easy to "convert" a given genetic algorithm to a gender-specific one. Three minor modifications are suggested (Drezner and Drezner, 2005):

1. When the starting population is generated, half the population members are designated as males and half are designated as females. The assignment of gender is done at random and no characteristic of the population member is used for such determination.
2. When selecting two parents, the first parent is randomly selected while the second is randomly selected from the pool of the opposite gender.
3. When an offspring is generated, it is randomly assigned a gender with a 50% probability of being assigned a male gender and 50% probability a female. Again, no characteristic of the offspring should be used to determine its gender.

No extra effort is required for the implementation of the gender-specific modification. A vector of genders for population members needs to be maintained, along with the gender determined for each offspring. In Drezner and Drezner (2005) it has been statistically shown that the gender-specific algorithm significantly improves the solutions on four sets of problems.

Note that it is important that an offspring's gender is randomly determined. An early attempt (Allenson, 1992) for the gender line a modification failed because it was suggested that the offspring is assigned the gender of the discarded population member. The rationale for his rule is to keep the population half males and half females. But this rule is inconsistent with nature. The concern is that if the population becomes all males or all females no further evolution is possible. The evolutionary process must be terminated prematurely if such a population structure evolves. In Drezner and Drezner (2005), it is shown that for a sufficiently large population (50 or more members), the probability that all population members will have the same gender is extremely low and can be ignored.

5.3.7 Distance-Based Parent Selection

All human cultures prohibit marriage between siblings or between parents and children (genetically similar pairs). In societies where marriages are arranged, similarity in socio-economic standing, but not genetic make-up, is prevalent. Some plants avoid pollination from genetically similar or identical individuals because self-pollination or pollination by "siblings" is typically unsuccessful, a phenomenon referred to in biology as "inbreeding depression." Mating between close relatives often results in less fit offspring. Another, less well known biological fact, is that mating between genetically distant members of the same species can lead to a decline in offspring fitness, a condition known as "outbreeding depression" or "hybrid breakdown." Some species avoid pollination from individuals that are geographically distant or genetically dissimilar, as offspring may be less suited to the local conditions and may be poorer competitors locally. Edmands (1999, 2002) observed that parental divergence (parents who are genetically distant) leads to less fit offspring.

In genetic algorithms, if dissimilar individuals mate, the offspring is more genetically diverse which is critical in maintaining a population's genetic diversity. However, parents who are too dissimilar produce less fit offspring. Using the distance criterion for parent selection, Drezner and Marcoulides (2003) crafted a rule attempting to find dissimilar but not too distant parents. A parameter $K = 1, 2, 3, \ldots$ is used. The first parent is randomly selected and K candidates for mating are then randomly selected. The distance (number of variables with different values, the Hamming distance metric) between the first parent and all candidate mates is calculated. The farthest mate among these K candidates is selected as the second parent. Note that $K = 1$ is the "standard" parent selection. Drezner and Marcoulides (2003) found that the efficiency of the modification for a set of test problems peaks for $K = 2, 3$. Selecting the farthest population member as a mate does not work well. It was also found that run time increases with K which further reduces the appeal of larger Ks.

5.3.8 Removal of Population Members

In most standard genetic algorithms, when an offspring is generated, it is compared with the worst population member, and if the offspring is better than the worst population member, it replaces it. Some genetic algorithms employ a rule according to which if the offspring is identical to an existing population member, it is not considered for inclusion in the population. This precludes the possibility of having two identical population members.

Drezner (2005c) suggested a different rule for the removal of population members. Two rules for the removal of a population member, once a better offspring (who is not identical to an existing population member) is found, are used. Rule 1 is the standard approach and Rule 2 is a new one.

Rule 1: Remove the worst population member.
Rule 2: Distances between all pairs of population members are calculated. Suppose that the shortest distance among all pairs of population members is d. All existing population members who are at distance d from another population member form a subset. This subset must have at least two members (at least one pair of population members are at distance d from one another). Remove the worst population member in this subset.

In the experiments performed in Drezner (2005c), it was found that Rule 2 is not necessarily better than Rule 1. The suggested rule is to select Rule 2 with probability p, and to select Rule 1 otherwise. Note that $p = 0$ is the standard rule (Rule 1), and $p = 1$ is Rule 2. The mix between the two rules by selecting $0 < p < 1$ seems to work well.

5.4   AN ILLUSTRATION

Three facilities (such as post office branches) are to be constructed to serve 20 communities. For simplicity, we assume that the communities have the same number of residents. Of the 20 communities, seven communities can house a branch and 13 communities cannot (for various reasons). See Figure 5.1, where communities that can house a branch are denoted by black circles and the rest by empty circles. Table 5.1 gives the x–y coordinates of the seven potential facilities. Distances are straight-line (Euclidean) distances. Each community is served by the closest branch. The objective is to locate the three branches so as to minimize the sum of the service distances to all 20 communities thus providing the best overall service to all residents. This problem is a p-median problem. For a discussion of p-median problems, see Current et al. (2002).

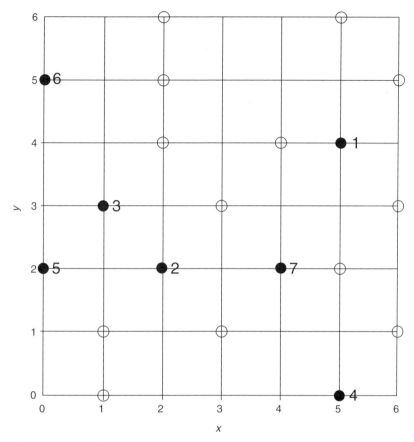

Figure 5.1 The *p*-median problem.

This particular problem is very simple. There are only 35 possible combinations for selecting three communities out of seven and thus total enumeration is straightforward. We list all possible combinations in Table 5.2. A "1" indicates a community selected for a branch and "0" indicates that it is not. To simplify the illustration, we calculate the objective function values using an Excel spreadsheet. These values are depicted in the last column of Table 5.2. Since the objective is the minimization of the sum of service distances to all communities, total enumeration will find that the optimal solution is combination #27 (communities #1, #3, and #7, see Figure 5.1) with a total service distance of 31.01. To illustrate the genetic algorithm, we "pretend" that the optimal solution is not known and whenever the value of the fit function (objective function) is needed, it is calculated for that combination.

Table 5.1   Location of Candidate Communities

#	x	Y
1	5	4
2	2	2
3	1	3
4	5	0
5	0	2
6	0	5
7	4	2

Table 5.2   Total Enumeration of Branch Locations Combinations

#	x_1	x_2	x_3	x_4	x_5	x_6	x_7	$f(x)$
1	0	0	0	0	1	1	1	36.04
2	0	0	0	1	0	1	1	39.92
3	0	0	0	1	1	0	1	40.41
4	0	0	0	1	1	1	0	46.07
5	0	0	1	0	0	1	1	35.98
6	0	0	1	0	1	0	1	36.61
7	0	0	1	0	1	1	0	55.88
8	0	0	1	1	0	0	1	36.32
9	0	0	1	1	0	1	0	43.79
10	0	0	1	1	1	0	0	44.42
11	0	1	0	0	0	1	1	35.80
12	0	1	0	0	1	0	1	39.57
13	0	1	0	0	1	1	0	47.41
14	0	1	0	1	0	0	1	39.11
15	0	1	0	1	0	1	0	41.14
16	0	1	0	1	1	0	0	44.90
17	0	1	1	0	0	0	1	36.01
18	0	1	1	0	0	1	0	46.83
19	0	1	1	0	1	0	0	48.81
20	0	1	1	1	0	0	0	41.95
21	1	0	0	0	0	1	1	34.61
22	1	0	0	0	1	0	1	33.79
23	1	0	0	0	1	1	0	36.16
24	1	0	0	1	0	0	1	41.60
25	1	0	0	1	0	1	0	39.39
26	1	0	0	1	1	0	0	35.61
27	1	0	1	0	0	0	1	**31.01**
28	1	0	1	0	0	1	0	35.53
29	1	0	1	0	1	0	0	36.17
30	1	0	1	1	0	0	0	32.59
31	1	1	0	0	0	0	1	33.40
32	1	1	0	0	0	1	0	32.73
33	1	1	0	0	1	0	0	36.09
34	1	1	0	1	0	0	0	33.35
35	1	1	1	0	0	0	0	33.54

The p-median problem is well researched in the literature. Various optimization techniques have been developed for its solution (see Daskin, 1995; Current et al., 2002). Typically, it is formulated as integer programming (Daskin, 1995; Current et al., 2002). We present here a different, intuitive formulation. Suppose that $x = \{x_1, x_2, \ldots, x_7\}$ are seven binary variables. A variable is assigned a value of "1," if the corresponding community is selected for a branch and "0," if it is not. The set I is defined to include all variables with a value of "1." For a feasible (acceptable) solution, the set I must be of cardinality 3. The service distance for community j is defined as $d_j = \min_{i \in I} \{d_{ij}\}$, where d_{ij} is the distance between communities i and j. By these definitions, the formulation is:

$$\text{Minimize}\left\{ f(x) = \sum_{j=1}^{20} d_j \right\} \tag{5.1}$$

$$\text{Subject to:} \sum_{i=1}^{7} x_i = 3 \tag{5.2}$$

$$x_i \in \{0, 1\} \tag{5.3}$$

5.4.1 The Genetic Algorithm Process

For the illustrative genetic algorithm problem (post office branches), we selected the following parameters. We maintain a population of five members. The parents are randomly selected to produce an offspring. An offspring replaces the worst population member if (a) it is better than the worst population member and (b) it is not identical to an existing population member. The merging rule is as follows: the two parents are compared and all variables (genes) common to both retain this common value. For the variables with different values (one parent has a "0" and the other has a "1"), we randomly select the necessary number of "1"s to bring the total number of "1"s to 3. Thus, the offspring satisfies Equations (5.2) and (5.3). For example, the two parents are 1001001 and 0111000. We first determine the common genes creating $\frac{0111001}{1001000} \Rightarrow {***}100{*}$, where an asterisk denotes an undetermined value which may be either "0" or "1." Two "1"s out of the four "*"s are randomly selected, while the rest get a "0." The first two stars were randomly selected as "1"s, and the offspring is therefore 1101000. Its value of the fit function is calculated by Equation (5.1). This process closely simulates nature with one exception. In the algorithm, if both parents have the same trait for a specific gene, it is passed on to the offspring. If they have different traits, one of them is randomly selected for the offspring. In nature, half of each chromosome is passed on to the offspring. Also, there are dominant and recessive genes that determine the trait of the offspring.

The step-by-step description of the illustrative genetic algorithm is as follows.

1. Randomly construct five different combinations to form the starting population (we employ a population of five members). Next to each member is its value of the objective function calculated by Equation (5.1). Since there are only 35 possible combinations we can construct them all and take the fit function values from Table 5.2. Typically, such a table cannot be constructed and the fit function is calculated for each combination separately. The random starting population is:

1	0001011	39.92
2	0010110	55.88
3	0101001	39.11
4	0101100	44.90
5	1010010	35.53

2. *Generation 1*: the second and fourth combinations are randomly selected. The merge is: $\frac{0010110}{0101100} \Rightarrow 0{***}1{*}0 \Rightarrow 0011100$. The objective value is 44.42. It replaces the worst population member, which is combination #2 above with a fit function of 55.88. The evolved population is:

1	0001011	39.92
2	0011100	44.42
3	0101001	39.11
4	0101100	44.90
5	1010010	35.53

3. *Generation 2*: the first and third population members are randomly selected. The merge is: $\frac{0001011}{0101001} \Rightarrow 0{*}010{*}1 \Rightarrow 0001011$. The objective value is 39.92. It is better than the worst population member, but it is identical to the first one, so the population is not changed and the offspring is discarded.

4. *Generation 3*: the third and fifth members are randomly selected. The merge is: $\frac{0101001}{1010010} \Rightarrow {****}0{**} \Rightarrow 1010001$. The objective value is 31.01. It replaces the worst population member, the fourth one. The evolved population is:

1	0001011	39.92
2	0011100	44.42
3	0101001	39.11
4	1010001 ·	31.01
5	1010010	35.53

5.   The process continues until the number of generations reaches a prespecified limit. Note that once the optimal solution is a member of the population, it can never be removed from it, because no offspring can be "superior." So, the optimal combination (now #4 in the population) will stay in the population until we stop the evolution process and will be the best population member and thus the solution.

6.   *Observe the evolutionary process.* The fittest survives and the quality of population members improves throughout the generations. There is no guarantee that the optimal solution will be detected, but experience shows that the evolutionary process is very efficient and results in very good solutions.

5.4.2 Illustrating the Steepest Descent Algorithm

A steepest descent approach is a search procedure that checks all "nearby" solutions in the neighborhood. We define the neighborhood as the set of all combinations for which one "1" is moved to another variable with a "0" value (thus retaining three "1"s in the combination). Every combination has 12 neighbors. Suppose that the offspring is 0101010 with an objective function of 41.14 (see Table 5.2). The neighbors of 0101010 along with their objective function values are listed in Table 5.3. For example, combination 1001010 was obtained by moving the "1" from second place to first place.

   The best neighbor is 1101000 with an objective function of 33.35. Since 33.35 is lower than 41.14 (the original value of the objective function), the change is accepted. The steepest descent continues until there is no improved combination in the neighborhood. Such a steepest descent algorithm may or may not find the optimal solution.

5.4.3 Illustrating a Mutation

Suppose that the offspring 0101010 is selected for a mutation. A mutation is moving one "1" to another place to replace a "0." It is equivalent to selecting one of the 12 neighboring members (see Table 5.3) as the mutated offspring. The mutation may or may not improve the value of the objective function. On occasion, the value of the objective function deteriorates. Reviewing Table 5.3, one finds that 5 out of 12 possible mutations improve the offspring (i.e., have an objective function value lower than 41.14) while 7 of the 12 possible mutations are harmful. Therefore, randomly performing a mutation on 0101010, there is a $5/12 = 41.7\%$ probability that the offspring is improved. Some researchers suggest that only improving mutations should be accepted, while the majority has the opinion that every mutation should be accepted, as is the case in nature.

Table 5.3   The Neighbors of 0101010

Combination	Objective
1001010	39.39
0011010	43.79
0001110	46.07
0001011	46.07
1100010	32.73
0110010	46.83
0100110	47.41
0100011	35.80
1101000	**33.35**
0111000	41.95
0101100	44.90
0101001	39.11

5.4.4 Calculating Diversity

Genetic diversity can be measured by calculating the distances between population members. The distance between any two members is defined as the number of genes with different values. For example, the distances between the members of the initial population in the post office, for example

1	0001011
2	0010110
3	0101001
4	0101100
5	1010010

are:

	#1	#2	#3	#4	#5
#1	0	4	2	4	4
#2	4	0	6	4	2
#3	2	6	0	2	6
#4	4	4	2	0	6
#5	4	2	6	6	0

The total distance is 80. The expected distance between any two different combinations is $60/17 \approx 3.53$, therefore the expected sum of (20) distances between members of a population of 5 (all different from one another) is $1200/17 \approx 70.59$. This means that the randomly generated starting population (with a total distance of 80) is more genetically diverse than expected. When the genetic algorithm progresses, distances tend to decrease, therefore genetic diversity (total distance) decreases as well. The five best combinations in Table 5.2 have a total distance of 56, that means less diversity than average. Note that the lowest possible diversity is 40. A population consisting of the last five combinations in Table 5.2 has a distance of "2" between any two different members, and thus a sum of 40. Since all population members are different from each other, the minimum possible distance between any two different population members is "2" and thus 40 is the lowest possible sum of distances for a population of 5.

When higher genetic diversity is maintained, convergence of the population is slowed down and more generations are required. Therefore, the run time of the algorithm increases. It is therefore important to balance both the advantages of achieving greater diversity and the disadvantages of increased run time.

5.5 APPLICATION: BALANCING A TURBINE ENGINE

In Chapter 4 of this book, Lipson describes various applications of genetic algorithms mainly in the field of robotics. We describe here an engineering application and demonstrate the effectiveness of genetic algorithms.

The turbine balancing problem, suggested by Laporte and Mercure (1988), is a good example of the effectiveness of genetic algorithms. Consider the manufacturing of a turbine engine, such as a hydro turbine or a jet engine, with n blades. The blades are inserted into equally spaced slots. To function properly, the turbine must be balanced. If all blades are identical, the turbine engine is balanced. In reality, there are slight variations in the weights of different blades, therefore, the turbine is not perfectly balanced. Suppose that the weights are designed to be 5 kg each and the variations across blades are in the order of magnitude of milligrams. The problem is to find the "correct" assignment of blades into slots so that the turbine will be as balanced as possible.

Let the blades' deviation from the ideal weight be: δ_i for $i = 1, \ldots, n$. The angles of the slots are: $\frac{2\pi i}{n}$ for $i = 1, \ldots, n$. Let blade $p(i)$ (p denotes a permutation) be installed in slot i. A perfectly balanced turbine will satisfy:

$$\sum_{i=1}^{n} \delta_{p(i)} \cos \frac{2\pi i}{n} = 0$$

$$\sum_{i=1}^{n} \delta_{p(i)} \sin \frac{2\pi i}{n} = 0$$

Since it may be impossible to achieve a perfect balance, the objective is to minimize the length of the resultant vector, whose square is:

$$\left(\sum_{i=1}^{n} \delta_{p(i)} \cos \frac{2\pi i}{n} \right)^2 + \left(\sum_{i=1}^{n} \delta_{p(i)} \sin \frac{2\pi i}{n} \right)^2 \tag{5.4}$$

by selecting the best permutation $p(i)$.

Expression (5.4) can be simplified as follows:

$$\left(\sum_{i=1}^{n} \delta_{p(i)} \cos \frac{2\pi i}{n} \right)^2 + \left(\sum_{i=1}^{n} \delta_{p(i)} \sin \frac{2\pi i}{n} \right)^2 = \sum_{i=1}^{n} \sum_{j=1}^{n} \delta_{p(i)} \delta_{p(j)} \left\{ \cos \frac{2\pi i}{n} \cos \frac{2\pi j}{n} + \sin \frac{2\pi i}{n} \sin \frac{2\pi j}{n} \right\}$$

$$= \sum_{i=1}^{n} \sum_{j=1}^{n} \delta_{p(i)} \delta_{p(j)} \cos \frac{2\pi(i-j)}{n}$$

$$= \sum_{i=1}^{n} \delta_i^2 + 2 \sum_{i=1}^{n-1} \sum_{j=i+1}^{n} \delta_{p(i)} \delta_{p(j)} \cos \frac{2\pi(i-j)}{n}$$

Since $\sum_{i=1}^{n} \delta_i^2$ is constant, the objective is to minimize

$$\sum_{i=1}^{n} \sum_{j \neq i=1}^{n} \delta_{p(i)} \delta_{p(j)} \cos \frac{2\pi(i-j)}{n} \tag{5.5}$$

The turbine engine example is a symmetric quadratic assignment problem (QAP). The QAP was first proposed by Koopmans and Beckmann (1957) and is well researched (Burkard, 1990; Taillard, 1991, 1995; Burkard et al., 1997; Cela, 1998; Rendl, 2002). The general QAP formulation is as follows. Given two $n \times n$ matrices (a_{ij}) and (b_{ij}), find a permutation $p(i)$ which minimizes:

$$\sum_{i=1}^{n} \sum_{j=1}^{n} a_{p(i)p(j)} b_{ij}$$

A symmetric problem fulfills $a_{ij} = a_{ji}$ and $b_{ij} = b_{ji}$. In our turbine balancing problem $a_{ij} = \delta_i \delta_j$ except $a_{ii} = 0$; and $b_{ij} = \cos \frac{2\pi(i-j)}{n}$ except $b_{ii} = 0$.

There are many applications for the QAP. For example, planning the layout of an office complex. There are n planned offices and n employees to be assigned to them. The interaction between any two employees (a_{ij}) and the physical distance between any two offices (b_{ij}) are known. The problem is to assign employees to offices such that those who interact extensively are as close as possible to each other. The objective is to find the best assignment of employees to offices so that the sum of the products of the interactions and the distances is minimized. A variant of this setting is incorporating the probability that a customer of this complex needs to visit two or more offices to complete the service. The objective in this case is to minimize the total distance to the customer.

Many pairs of offices will have zero interaction whereas others may have a negative interaction (such as in planning a military base, top secret intelligence offices should be as far as possible from the cafeteria or other frequently visited offices). Planning a keyboard of 26 letters (Burkard and Offermann, 1977) is a QAP. The interactions are the probabilities of typing a letter following another letter, and the distances are the distances between the letter-keys on the keyboard. Different languages require different key configurations (even for the same letters). Planning an airplane dashboard (Drezner, 1980) requires consideration of the frequency that a pilot uses an instrument in conjunction with another instrument as the interaction factor. The wiring problem of an electronic board or the construction of a computer chip are additional examples of QAPs, where the wiring distance between components that send signals to one another has to be minimized (Steinberg, 1961). For a discussion of QAP applications, see Cela (1998).

The QAP is considered one of the most complicated combinatorial optimization problems. Algorithms that guarantee finding the optimal solution are very inefficient. Only recently have problems of up to $n = 36$ facilities (blades in our example) been optimally solved. Anstreicher et al. (2002) solved a problem with $n = 30$ facilities using multiple computers requiring a total computer time of over 6 years. The number of permutations (ignoring identical solutions due to symmetry) is $n!$. For $n = 20$, evaluating all of the permutations will take over 77,000 years if every permutation is evaluated in a millionth of a second. Therefore, a large body of research focuses on heuristic algorithms. There are several proposed genetic algorithms for the solution of the QAP (Fleurent and Ferland, 1994; Tate and Smith, 1995; Ahuja et al., 2000; Drezner, 2003, 2005b).

The best available hybrid genetic algorithms for the solution of QAP problems are Drezner (2003, 2005c). The merging procedure of the parents is the same in both algorithms. It is referred to as the cohesive merging procedure (Drezner, 2003). The algorithms differ in the improvement algorithm applied to the offspring before it is considered for inclusion in the population. For complete details, see Drezner (2005b).

5.5.1 A Turbine Balancing Example

We randomly generated a turbine-balancing problem with 20 blades. For simplicity, we multiply the cosine values by 10,000 and round them off to the nearest integer. The data are given in Table 5.4. We add the sum of the squares of the δ_i to the objective function. Because of rounding-off, errors introduced by rounding the cosine values, the best found solution is negative ($-1,550$).

Table 5.4   Parameters of the Turbine Blade Problem

i	d_i	$\cos \frac{2\pi i}{n} \times 10{,}000$
1	−4	9,511
2	3	8,090
3	0	5,878
4	2	3,090
5	5	0
6	2	−3,090
7	5	−5,878
8	−3	−8,090
9	6	−9,511
10	6	−10,000
11	−8	−9,511
12	6	−8,090
13	−5	−5,878
14	−1	−3,090
15	2	0
16	−8	3,090
17	−10	5,878
18	1	8,090
19	−8	9,511
20	9	10,000

Table 5.5   Results for Non-Genetic Heuristics

Heuristic	Minimum	Average	Maximum	Time per 10,000 runs (sec)
Steepest descent	−1362	+588.59	+25,500	0.57
0 Levels	−1462	+148.86	+25,358	3.71
5 Levels	−1460	−24.36	+13,200	18.92
10 Levels	−1456	−58.39	+10,532	33.75
15 Levels	−1452	−80.51	+11,910	48.54
20 Levels	−1464	−93.11	+40,182	62.66
25 Levels	−1468	−104.07	+13,162	77.19

We solved the QAP problem using a Fortran code compiled by Microsoft PowerStation 4.0 for Windows. The code uses integer values for the parameters. The experiments were performed on a desktop PC with 2.8 GHz CPU and 256 MB RAM.

To demonstrate the effectiveness of the genetic algorithm, we first report in Table 5.5 the results of the steepest descent and the Ring Moves (RM) (Drezner, 2005b) for various levels without using any evolution. Each algorithm was run 10,000,000 (ten million) times. The best solution, average solution, and time in seconds for 10,000 runs are reported. The best solution, found with hybrid genetic algorithms reported below of −1,550, has never been found by any of these 10,000,000 experiments. The quality of the solution increases as the number of levels increases, but it does not seem to merit the extra computer time.

For the genetic algorithms we employ the hybrid genetic algorithm described in Drezner (2005b) using the steepest descent, and RM with 10 and 25 levels. The population size is 100, with 10,000 generations for each solution. Each replication is run 1,000 times thus a total of 10,000,000 offspring are generated. In Table 5.6 we report the number of times the best known solution (−1550) was found, the average solution, the maximum solution, and the run time required for one run. To demonstrate some of the modifications, we repeated the runs with the gender-specific modification (Drezner and Drezner, 2005) and the distance modification using $K = 1, 2, \ldots, 5$, for a total of 30 experiments. The results of the experiments are depicted in Table 5.6.

A comparison of Tables 5.5 and 5.6 clearly shows the benefit derived from employing the evolutionary process. The average result obtained with the evolutionary process is about the same as the best results obtained in 10,000,000 runs without evolution and in most cases in a shorter computer time. For example, the ten-levels case required about 9.4 h to evaluate all 10,000,000 replications without evolution. The best solution of −1,550 was never found in 10,000,000 replications. However, using a genetic algorithm with a $K = 3$ distance modification and no gender modification required a total of only 7.9 h for 1,000 replications and the best known solution was found in 63.1% of those replications. Even the steepest descent version with $K = 2$ found the best-known solution in 12.5% of the replications in a total run time of 9.5 min.

5.6   DISCUSSION

In this chapter we describe genetic algorithms for solving optimization problems. These algorithms mimic natural selection and the survival of the fittest principles. Modifications of the "standard" genetic algorithm are also presented.

Genetic algorithms are based on evolutionary premises, and attempt to simulate natural selection and survival of the fittest. To survive, species must reproduce and regenerate. This requires new members of the population to be fit and adaptable to changing environmental conditions. Only the fittest individuals survive while the weak members perish or are killed by their natural enemies. Inherent to this theory is, therefore, the definition of what constitutes the "fittest." In nature, the

Table 5.6   Results for Hybrid Genetic Algorithms

Distance	Gender	*	Average	Maximum	Time per run (sec)
Steepest descent					
$K = 1$	No	86	−1452	−1264	0.57
$K = 2$	No	125	−1471	−1254	0.62
$K = 3$	No	114	−1469	−1214	0.64
$K = 4$	No	79	−1458	−1170	0.65
$K = 5$	No	50	−1449	−1180	0.66
$K = 1$	Yes	78	−1452	−1236	0.57
$K = 2$	Yes	115	−1469	−1334	0.62
$K = 3$	Yes	74	−1460	−1262	0.65
$K = 4$	Yes	61	−1448	−1164	0.66
$K = 5$	Yes	56	−1440	−1120	0.67
10 Levels					
$K = 1$	No	430	−1510	−1420	25.54
$K = 2$	No	584	−1524	−1432	27.15
$K = 3$	No	631	−1528	−1428	28.29
$K = 4$	No	607	−1527	−1400	28.82
$K = 5$	No	626	−1528	−1428	29.32
$K = 1$	Yes	448	−1511	−1412	25.51
$K = 2$	Yes	612	−1525	−1422	27.14
$K = 3$	Yes	591	−1525	−1398	28.17
$K = 4$	Yes	594	−1526	−1420	28.91
$K = 5$	Yes	568	−1524	−1402	29.36
25 Levels					
$K = 1$	No	538	−1518	−1426	58.46
$K = 2$	No	616	−1526	−1438	62.01
$K = 3$	No	691	−1532	−1448	64.06
$K = 4$	No	688	−1532	−1450	65.49
$K = 5$	No	664	−1531	−1420	66.56
$K = 1$	Yes	516	−1517	−1402	58.58
$K = 2$	Yes	632	−1527	−1448	61.86
$K = 3$	Yes	660	−1530	−1420	64.27
$K = 4$	Yes	673	−1531	−1428	65.60
$K = 5$	Yes	639	−1529	−1384	66.46

*Number of times out of 1000 that the best known solution of −1550 was obtained.

definition keeps evolving with changing environmental conditions and across species. For example, the male bird of paradise in New Guinea is the fittest when his feathers and tail are very colorful and attractive to the female bird of paradise. The same colorful and beautiful male would not be the fittest in a different environment (off the island), one that is predator rich. Similarly, the peppered moth, in England, during the Industrial Revolution would not have survived without a color adaptation. In urban areas, the fittest was the darker peppered moth that adapted to the new gray, ash-covered trees on which it rests. By blending into the tree, it protected itself from predators, while at the same time, in rural areas, the peppered moth continued to thrive and survive on lichen-covered tree branches. Unlike nature, in genetic algorithms the definition of the "fittest" is stable. The more stable definition of "fittest" in genetic algorithms, in turn, allows for the ultimate achievement of an "ideal" population, a situation not paralleled in nature.

In nature, species have to cope with invasion of other species and competition for resources. Species diversity is rampant as genetic diversity is instrumental to adaptation. The survival of the fittest individual leads to survival of the species. In genetic algorithms, by comparison, there is one species only. Occasionally generating offspring who are "fitter" than existing members in order to "enrich" the population "gene pool" incorporates invasion in genetic algorithms. PGA allow population movements, but those are of the same species. In compounded genetic algorithms, there is no population movement between the isolated populations.

Offspring mutation is another natural selection tenet incorporated in genetic algorithms. Mutations occur quite frequently in nature. Most mutations are not beneficial to the species, while some, on rare occasions, are beneficial, are instrumental in adaptation, and, in fact, produce the "fittest" offspring. Some genetic algorithms incorporate mutations, some of which are beneficial.

Other similarities consist of gender-based traits and procedures for replacing and removing population members. In addition, nature allows for population or species growth and decline. These are occasionally practiced in genetic algorithms as well.

We demonstrate the effectiveness of evolutionary processes on a turbine balancing problem. Solving the problem without employing the evolutionary process does not result in good solutions. Seventy million attempts at its solution did not find the best known solution even once. Applying the hybrid genetic algorithm 1,000 times (which is faster than 10,000,000 replications of a non-evolutionary algorithm), found the best known solution (in one variant) 691 times out of 1,000 replications. The superiority of the evolutionary process is clearly demonstrated.

REFERENCES

Ahuja, R.K., J.B. Orlin, and A. Tiwari (2000). A descent genetic algorithm for the quadratic assignment problem. *Computers and Operations Research, 27*, 917–934.

Allenson, R. (1992). Genetic algorithms with gender for multi-function optimisation, Technical Report EPCC-SS92-01, Edinburgh Parallel Computing Centre, Edinburgh, Scotland.

Anstreicher, K., N. Brixius, J.-P. Goux, and J. Linderoth (2002). Solving large quadratic assignment problems on computational grids. *Mathematical Programming, 91*, 563–588.

Burkard, R.E. (1990). Locations with spatial interactions: the quadratic assignment Problem. In: Mirchandani, P.B., R.L. Francis (Eds), *Discrete Location Theory*, Wiley, Berlin.

Burkard, R., S.E. Karisch, and F. Rendl (1997). QAPLIB — a quadratic assignment problem library. *Journal of Global Optimization, 10*, 391–403, electronic update: http://www.seas.upenn.edu/qaplib/(revised 02.04.2003).

Burkard, R.E. and J. Offermann (1977). Entwurf von Schreibmaschinentastaturen mittels quadratischer Zuordnungsprobleme. *Zeitschrift für Operations Research, 21*, B121–B132.

Cantu-Paz, E. (1998). A survey of parallel genetic algorithms. *Calculateurs Paralleles, Reseaux et Systems Repartis, 10*, 141–171.

Cantu-Paz, E. (1999). Migration policies and takeover times in genetic algorithms. In: Benzhaf, W., J. Daida, A.E. Eiben, M.H. Garzon, V. Honavar, M. Jakeila, R.E. Smith (Eds), *Proceedings of the Genetic and Evolutionary Computation Conference*, Morgan-Kaufman, San Francisco, California, p. 775.

Cantu-Paz, E. and D.E. Goldberg (2000). Parallel genetic algorithms: theory and practice. *Computer Methods in Applied Mechanics and Engineering*, Elsevier, New York, New York.

Cela, E. (1998). *The Quadratic Assignment Problem: Theory and Algorithms*, Kluwer Academic Publishers, Dordrecht, The Netherlands.

Colorni, A., M. Dorigo, and V. Maniezzo (1992). Distributed optimization by ant colonies. In: Varela, F.J., P. Bourgine (Eds), *Proceedings of ECAL'91 — European Conference on Artificial Life*, MIT Press, Cambridge, Massachusetts, pp. 134–142.

Current, J., M. Daskin, and D. Schilling (2002). Discrete network location models. In: Drezner, Z., H.W. Hamacher (Eds), *Location Analysis: Applications and Theory*, pp. 81–118.

Daskin, M. (1995). *Network and Discrete Location: Models, Algorithms and Applications*, John Wiley, New York, New York.

Dorigo, M., and L.M. Gambardella (1997). Ant colony system: a cooperative learning approach to the traveling salesman problem. *IEEE Transanctions on Evolutionary Computation, 1*, 53–66.

Drezner, Z. (1980). DISCON — A new method for the layout problem. *Operations Research, 28*, 1375–1384.

Drezner, Z. (2003). A new genetic algorithm for the quadratic assignment problem. *INFORMS Journal on Computing, 15*, 320–330.

Drezner, Z. (2005a). Compounded genetic algorithms. *Operations Research Letters, 33*, 475–480.

Drezner, Z. (2005b). Extended concentric tabu for the quadratic assignment problem. *European Journal of the Operational Research, 160*, 416–422.

Drezner, Z. (2005c). A distance based rule for removing population members in genetic algorithms. *4oR*, In press.

Drezner, T. and Z. Drezner. (2005). Gender-specific genetic algorithms. *Information Systems and Operational Research*, accepted for publication.

Drezner, Z. and G.A. Marcoulides (2003). A distance-based selection of parents in genetic algorithms. In: Mauricio, G., C. Resende, Jorge P. de Sousa (Eds), *Metaheuristics: Computer Decision-Making, Combinatorial Optimization Book Series*, Kluwer Academic Publishers, Dordrecht, The Netherlands, pp. 257–278.

Edmands, S. (1999). Heterosis and outbreeding depression in interpopulation crosses spanning a wide range of divergence, *Evolution*, *53*, 1757–1768.

Edmands, S. (2002). Does parental divergence predict reproductive compatibility? *Trends in Ecology and Evolution*, *17*, 520–527.

Fleurent, C. and J. Ferland (1994). Genetic hybrids for the quadratic assignment problem. *DIMACS Series in Mathematical Theoretical Computer Science*, 16, *190–206*.

Gambardella, L., E. Taillard, and M. Dorigo (1999). Ant colonies for the quadratic assignment problem. *Journal of the Operational Research Society*, *50*, 167–176.

Glover, F. (1986). Future paths for integer programming and links to artificial intelligence. *Computers and Operations Research*, *13*, 533–549.

Glover, F. and M. Laguna (1997). *Tabu Search*, Kluwer Academic Publishers, Boston, Massachusetts.

Goldberg, D.E. (1989). *Genetic Algorithms in Search, Optimization and Machine Learning*, Addison-Wesley, Wokingham, England.

Holland, J.H. (1975). *Adaptation in Natural and Artificial Systems*, University of Michigan Press, Ann Arbor, Michigan.

Kirkpatrick, S., C.D. Gelat and M.P. Vecchi (1983). Optimization by simulated annealing. *Science*, *220*, 671–680.

Koopmans, T. and M.J. Beckmann (1957). Assignment problems and the location of economics activities. *Econometrica*, *25*, 53–76.

Land, A.H. and A.G. Doig (1960). An automatic method of solving discrete programming problems. *Econometrica*, *28*, 497–520.

Laporte, G. and H. Mercure (1988). Balancing hydraulic turbine runners: a quadratic assignment problem. *European Journal of Operational Research*, *35*, 378–381.

Moscato, P. (2002). Memetic algorithms. In: Pardalos, P.M., M.G.C. Resende (Eds), *Handbook of Applied Optimization*, Oxford University Press, Oxford, U.K.

Nemhauser, G.L. and L.A. Wolsey (1988). *Integer and Combinatorial Optimization*, Wiley, New York, New York.

Reeves, C.R. (1993). Genetic algorithms. In: Reeves, C.R. (Ed.), *Modern Heuristic Techniques for Combinatorial Problems*, pp. 151–196.

Rendl, F. (2002). The quadratic assignment problem. In: Drezner, Z., H. Hamacher (Eds), *Facility Location: Applications and Theory*, Springer, Berlin.

Salhi, S. (1998). Heuristic search methods. In: Marcoulides, G.A. (Ed.), *Modern Methods for Business Research*, Lawrence Erlbaum Associates, Mahwah, New Jersey.

Salhi, S. and M.D.H. Gamal (2003). A GA-based heuristic for the multi-Weber problem. *Annals of Operations Research*, *123*, 203–222.

Segal, N.L. (2000). News, views and comments, *Twin Research*, *3*, 338–343.

Spears, W.M. (2000). *The Role of Mutation and Recombination*, Natural Computing Series, Springer-Verlag, Berlin.

Steinberg, L. (1961). The backboard wiring problem: a placement algorithm. *SIAM Review*, *3*, 37–50.

Taillard, E.D. (1991). Robust tabu search for the quadratic assignment problem, *Parallel Computing*, *17*, 443–455.

Taillard, E.D. (1995). Comparison of iterative searches for the quadratic assignment problem, *Location Science*, *3*, 87–105.

Tate, D.E. and A.E. Smith (1995). A genetic approach to the quadratic assignment problem. *Computers and Operations Research*, *22*, 73–83.

Robotic Biomimesis of Intelligent Mobility, Manipulation, and Expression

David Hanson

CONTENTS

6.1 INTRODUCTION

A revolution is quietly brewing. The numerous systems and subsystems of intelligent, agile synthetic organisms are rapidly improving in efficacy. And yet, the tremendous scope of this action may be less than obvious to the casual observer, because the various points of evidence are rarely examined in tandem. This chapter seeks to highlight trends that indicate that biorobotics stands poised to strongly affect our new century.

Our biology and our technology are converging on numerous fronts: genetic engineering, augmented cognition, artificial life, and bio-inspired robotics are some examples. While this chapter focuses on robotics, the boundaries between disciplines grow increasingly porous. This chapter does not try to limit its purview to technologies that are relevant to robotics exclusively; much of biomimetic robotics overlaps with other disciplines. Yet this chapter does consider robots in particular, especially robotics' ambition to physically interact with the world usually guided by bio-inspired intelligent control systems.

Though many robots bear little semblance to animals, in some regards, robotics' relationship to biology is essential. In their more complete, integrated incarnations, robots can be considered to be synthetic animals containing the perceptual systems, intelligence, motility, and mobility needed to operate in the world in ways that are derivative of animals or inspired by them. To wit, the accelerating functionality of robots can be largely traced to the widening intersection of biology and engineering.

The biosciences and robotics are each exhibiting unprecedented, explosive rates of discovery and innovation (Kurzweil, 2002), and their junction increasingly operates as a hybrid discipline known by several alternate names: biomimetics, biorobotics, and bio-inspired robotics being among them. As bio-discoveries are translated into robotics, the resulting robots look, act, and function ever more like animals — synthetic organisms which then offer biologists opportunities to test theories of animal locomotion, intelligence, and materials. This synergistic interplay represents a feedback loop that propels the pace of discovery and innovation in both biology and technology and this interplay between biology and robotics is largely the subject of this chapter.

The concept is not new. For thousands of years, people have sought to emulate animals and people in various media; and as far back as ancient Egypt and Greece, the media included mechanical automata (Cassell, 2001). In the 19th and 20th centuries, advances in electromotors, batteries, materials, and manufacturing enabled these artificial creatures to move with increasingly lifelike grace and autonomy. While bodies of robots do not yet possess the full capabilities of humans or animals, the accomplishments and the pace of progress in these areas exceed those of any other time in history. Already, robots walk and run bipedally in the fashion of humans (Doi, 2004), affect realistic facial expressions (Hanson et al., 2003), fly like hummingbirds (Dickinson, 2001), and perform many other animal-like feats.

In parallel with the progress of synthetic bodies, artificial brains have evolved swiftly as well. With the 20th century information sciences, technology began to emulate the nervous system, beginning with the work of McCulloch and Pitts, Turing, Walter and others (McCorduck, 1979). From these pioneers, the work has continued with new generations of robotics and artificial intelligence (AI) researchers. In AI research, engineers coordinate with biosciences under the rubric of cognitive science to the profound benefit of both robotics and neuroscience. Since the foundations of AI research in the 1940s, AI systems have increased enormously in functionality, and have become widely deployed in commercial applications, such that AI now constitutes a nearly $9 billion market (BCC, 2003a,b).

While most pioneering AI research focused on a "top-down" symbolic approach that largely disregarded the importance of a body or embodiment, the work of Brooks and others pushed the paradigm of intelligence as forming from "bottom-up" — highly distributed and physically-embodied architectures (Brooks, 1991). This perspective reinvigorated the emulation of animal and human bodies in robotics, and is validated by much bio and neuroscience, including the work of Damasio (1994) , which shows that the mind and body are integrally connected, that the abstract mind does not float mysteriously above the organism as Descartes postulated. In addition to validating research in locomotion, sensing and grasping, this paradigm shift frees researchers and companies to consider biomimetic displays of anthropomorphic emotion, animated by software models of "emotional intelligence"; and together these tools are characteristic of the booming field of social robotics (Breazeal, 2002).

Although rhetoric generally discusses terms of "top-down" or "bottom-up", most progress is clearly being rendered in patches of the middle actually, somewhere between top and bottom. The spreading complexity of this patchwork progress mirrors the "systems paradigm" of biosciences. In this paradigm, an organism, even an intelligent human organism, is a highly integrated web of systems. For example, it is well demonstrated that the intelligence of human brain suffers terribly should any of several "lower" brain systems (such as the amygdala) be damaged (Damasio, 2005).

Largely a reapplication of information processing theory to biology (arising from the hybrid discipline of bioinformatics), this systems paradigm of life as an integrated web of interaction at many scales, has largely moved to center of the biological sciences, complementing the reductionist view of the organism (Dickinson et al., 2000). This chapter adopts this paradigm to frame a biorobot as an artificial organism — a mélange of composites, subsystems, and integrated supersystems, which are assembled to achieve a desired function, or multiple functions as the case may be.

To be definitively a robot though, an artificial organism must be capable of motility or mobility or both. Otherwise, the phrase "artificial organism" can apply to biorobot's close cousins: artificial life, computer animated agents, and the like. Physical embodiment, the ability to move, the ability to grasp and handle objects, to locomote: these distinguish a robot. And yet clearly, the overlapping of technologies makes it difficult, and almost arbitrary, to define "robotics" at this stage. Indeed, many technologies are relevant to robotics function and are useful without mobile physical bodies, with examples including machine vision, automatic speech recognition (ASR), and natural language processing (NLP). Alternately, the robot-like body may exist perfectly well without intelligence, as is the case with animatronics — entertainment robots used in themeparks and films. Truthfully, many technologies relevant to robotics have yet to be integrated into robots or into whole synthetic organisms. Such integration represents one of the grand challenges in biomimetic robotics.

As such integration inches forward, robots will reap benefits from the numerous bio-inspired technologies that are flowering presently. Bio-inspired tunable optics (Buckley et al., 2004), bio-inspired sensors, actuators (Bar-Cohen, 2002), and materials promise to boost the performance of robots. Make no mistake, robots also benefit from technology advances that are not bio-inspired, such as advanced manufacturing technologies like MicroElectricalMechanical systems (MEMS) and shape deposition manufacturing (SDM).

Many of these technologies can be well used to emulate biological systems. For example, advanced manufacturing technologies result in faster, iterative robot design (Bailey et al., 2000), more complex yet affordable robots, and designs at meso, micro, and even nanometer scales. Advanced digital visualization tools are enabling better design and finite element testing before the expense and trouble of manufacturing. Numerous shifts in design paradigms can be associated with the new tools of manufacturing, design, and the biological sciences.

Biological systems are more capable than technology in countless ways. And yet, technology is capable in some ways that biology is not. Moreover, when performing the same task as an organism, humans can often use very different strategies, for example, aerofoil wings versus flapping wings to achieve flying. So, the most effective robotic creatures can be an amalgam of pure technology and translated or emulated biology.

When emulating bio-systems, the design approach may be either directly imitative (biomimetics) or inspired by physical principles derived from the study of organisms (bio-inspired). In the first approach, technology approximates the end result or function of an organ or organism, for example: when rotary motors turn leg joints, the mechanics are very different from those of biological leg muscles, though the effected end motion of the leg is similar. In the second approach, the principles extracted from bio-systems may be applied in ways very much unlike those exhibited in the originating organism; for example, the crystalline structures of starfish eyes are being used to create far more efficient fiber-optic routing hubs (Aizenberg et al., 2001). The two sets are not mutually exclusive — there are many derivations and combinations of these techniques, and the end objective is similar to create machines that move and function usefully by transferring bio-systems into techno-systems.

Truly, as physical nature manifested through biological evolution, it solved a staggering number of design and engineering problems. The job of the bioroboticist, then, is to discover and to abstract these solutions, and to implement them in the limited media of human technology.

Computers grow more powerful and cheaper. Materials are growing more diverse and effective in many ways. Manufacturing technologies are growing more automated, efficient, effective, and flexible.

Additionally, the markets for robotics and AI are increasing; they are up tenfold from 1992 to 2002 (UNECE-A) and are projected to expand tenfold again by 2010. The market for personal and mobile robots, which tend to be more bio-inspired, appears poised to surpass industrial robots this year (Kara, 2004). According to the U.N., the market for robotics was up 34% each year from 2002 to 2004, and is projected to expand sevenfold from 2004 to 2007, with the largest gains in the consumer and entertainment robotics (UNECE, 2004).

It does appear that the age of biorobotics is neatly unfolding.

Please note: the specific examples of robots and technology cited in this chapter are far from comprehensive. The chapter's chosen examples are not necessarily more highly meritorious than others not mentioned, but rather represent an arbitrary cross-section of research that embodies bio-inspired principles. The field is rich with quickly evolving research and researchers. The author extends apologies to all those researchers whose work is not represented in this chapter, and encourages readers to research independently to obtain a more effective image of work in this rich field.

6.2   MOBILITY AND MOTILITY: FLYING, WALKING, CRAWLING, MANIPULATION

For many kinds of interaction with the real physical world, technology must have the ability to move. Such mechanically effected, controlled motion is a distinguishing aspect of robotics. This motion can be merely motile — anchored like the motion of an industrial robotic arm (biologically analogous to the anchored motile arms of an anemone), or the motion may be locomotive (like the flying of a bumblebee).

In all motive and locomotive modes of action, certain issues must be considered — sensing, mechanical forces and resonances at play, and controls. Beyond these, the end application will guide the modes of motion required, and together these will determine the specific, associated engineering challenges (Bar-Cohen, 1999, 2002).

Undeniably, creating machines that move like animals can involve significant challenges. For one thing, most available motors do not behave like animal muscles, neither mechanically nor in their controls. Unlike conventional electromotors, biological muscles are inherently linear in actuation, and are compliant. The linear action of muscles more efficiently effects animal-like motion than do rotary electromotors, and the compliant properties appear key to biological motor control schemes (Full and Meijer, 1999).

In another extreme difference from biological motor systems, today's robot technology cannot maintain or repair itself. This means that many of the materials and motors used in robots must be even more robust than similar-function materials in animals for the devices to have useful lifespans. While someday robots may be endowed with self-repair capabilities (perhaps thanks to improving genetic or protein engineering, MEMS automation, or nano-machines), for now, robots must be overengineered for durability, and they must be regularly serviced or replaced.

The end application will determine the readiness of today's technology. Long-term constant duty applications, like robots used in theme parks or outer space exploration, require very robust materials and construction (Hanson and Pioggia, 2001). However, low-duty applications, such as short promotional appearances or film special effects, may operate in considerably shorter duty cycles, in less difficult environment conditions, and with lower requirements on lifespan.

Accordingly, various applications have varying motive requirements. For exploring Mars, stability over irregular terrain is a critical priority, and therefore a hexapod like McGill's Rhex might be more apropos than a biped. For entertainment applications on the other hand, aesthetic impact on an audience is the critical priority, and so biped humanoids like Honda's Asimo would likely serve better.

Fortunately for robotics applications engineers, an unprecedented bounty of robotic motion technologies is available to choose from, and still others are emerging encouragingly.

6.2.1 Locomotion Principles

Robots presently display a wide variety of legged locomotion strategies (Hanson et al., 2003). Some engineers explore sprawled-leg hexapod configurations akin to those of cockroaches, while others build robots that hop on a singular leg. Others yet build robots that walk or run on two like a turkey, a dinosaur, or even in the fashion of a human. And still others create robots that self-reconfigure to adopt multiple locomotion styles (Fitch et al., 2000; Butler et al., 2002).

These robots are sometimes built just to resemble animals, while others utilize the physics of animal locomotion and controls.

Unlike animals, most current legged robots do not employ mechanical compliance, instead using rigid mechanical systems and, sometimes, computationally intensive digital PID loop controls (although PID loops may be rendered much less expensive and intensive by using analog elements). Animals, on the other hand, use a number of techniques involving mechanical compliance to simplify controls and absorb shock, regardless of the number of legs. Full and Meijer (1999) show that during animal-legged locomotion energy is stored in the spring-like compliant materials as the animal lands in a stride, and then the energy is reemitted as the animal springs forward in the stride, a dynamic that is akin to an inverted pendulum.

More than 20 years ago, Marc Raibert developed springy, dynamically stable-legged robots, utilizing the principles described in the preceding paragraph (Raibert, 1983); numerous versions of these robots bounced, ran, and turned flips to the world's delight (see Figure 6.1). Since then, many robots have used this principle, as MIT's spring series elastic actuators, in Stanford's urethane–rubber Sprawl series, and in many others.

In addition to absorbing energy, though, the viscous properties of the compliant material in leg biomaterial also dampen the vibrations associated with locomotion (Full, 2000), thus stabilizing the

Figure 6.1 Early dynamic stability in action. (Image provided courtesy of Marc Raibert and The MIT Leg Laboratory copyright 1990. With permission.)

locomotion. And so the visco-elastic biomaterials serve several functions during locomotion: absorbing shock, stabilizing, and improving locomotion energy efficiency (Full and Meijer, 1999).

Moreover, the resonance of the compliant biomaterials appears to perform as the oscillating elements of analog computers, dynamically stabilizing animal locomotion in response to the terrain — a low-level intelligence function labeled "preflex" by Brown et al. (1996). Note that such preflex would function separately from nerve activity, involving only well-tuned mechanical compliance.

In a dramatic display of this principle transferred into a robot, Stanford's Sprawlita robots traverse highly complex terrain with no sensor feedback at all, stabilized only by the feedforward-responsive, rubbery compliance of its body and legs (Kim et al., 2004). The latest robot in the Sprawl series, the small (0.3 kg) autonomous iSprawl runs at 15 body lengths per second — several times faster than earlier, slower bio-inspired robots (see Figure 6.2). And yet examination and analysis of iSprawl shows the same bouncing, stable locomotion patterns in as those seen in insects (Kim et al., 2004).

Comparable to iSprawl, the Mini-Whegs™ of Case Western Reserve University is smaller and lighter than iSprawl, predates iSprawl and runs at 10 body lengths per second. Mini-Whegs also run rapidly over obstacles that are taller than their legs. The Whegs robots are discussed further in the next section.

Full and Meijer (1999) show that the principles of compliant locomotion hold regardless of leg number, but with some variations. Full and Meijer also find that in locomotion with four or more legs, the leading two legs absorb shock and stabilize an animal, while the rear two legs provide propulsion. In animals with six or more legs, the middle legs provide combined propulsion and stabilization. In animals with two legs or mono-style hoppers (like a kangaroo), the legs serve for both shock absorption and propulsion.

Implementing these biological strategies will be more challenging in biped robots, so, many researchers are initially implementing them in hexapod robots. Such an insect-like hexapod architecture offers inherent stability, especially when employed with a ground-hugging, sprawl-legged posture. One sees such a posture in lizards, crabs, cockroaches, and spiders. Both the

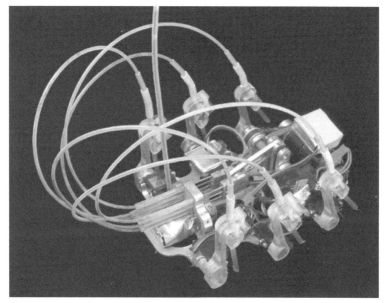

Figure 6.2 Stanford's dynamically stable iSprawl. (Image courtesy of Stanford's Biomimetic Robotics Laboratory.)

stability of a sprawled posture and of an increased number of legs reduce the complexity of controls and decrease likelihood of catastrophic fall or inversion. Full hypothesizes that six legs offer an optimal accord between stability and complexity (1999), extra legs being expensive to build, creating potential failure points, and requiring extra energy to operate.

On the other hand, multipurpose extra limbs could serve as alternate legs for the six in case of emergency, and in the interim they could operate as sensor antennae, or as arms tipped with manipulators or instruments.

Similar to Sprawl, in the Rhex robot of McGill the compliance of the six legs improves the stability and controls of the robot; but Rhex can alternately also run on two legs, in the manner of a toddler. Such reconfigurable design promises to enable robots to be less domain-specific, and more capable of meeting the demands of real-world operations. Reconfigurable robots are further discussed in Section 6.2.2.

6.2.2 Flexibility, Hybrids, and Reconfiguration

Real-world performance can require flexibility and multipurpose applicability in mechanical architecture. To achieve this flexibility, roboticists have introduced novel hybrid systems such as wheel–leg combinations (such as the Whegs of Case Western), or reconfigurable architectures that can morph between walking, crawling, and slithering morphologies (such as the crystal robots of Dartmouth and MIT).

Slithering morphologies are inherently flexible, enabling robots to crawl through tight spaces, such as arteries, caves, or sewers, in a snakelike or centipede-like manner. The snake robot S5 (1998–1999) built by Gavin Miller, demonstrates untethered legless snakelike slithering locomotion (see Figure 6.3). Miller's S1–S5 build on earlier work by Shigeo Hirose of Japan, in which oscillatory deflections travel down the length of the snake causing the snake to move forward, while offset to oscillations steer the device.

MAKRO, a snake robot the Fraunhofer Institute for Autonomous Intelligent Systems in St Augustin, Germany, boasts an articulated-wheeled, segmented design that enables maneuverability

Figure 6.3   S5 by Gavin Miller © 1999 and MAKRO of Fraunhofer, DE.

Figure 6.4   Whegs II of Case Western Reserve University.

around sharp bends, over refuse, and up stairs. Most other sewer robots are tethered by power and control cables and navigate uni-directionally. The wireless, waterproof, and two-headed MAKRO, however, can autonomously navigate bi-directionally, monitoring its own position, detecting sewer landmarks, and enacting mission tasks, such as taking sewage samples.

Case Western Reserve University robot Whegs II combines wheels and legs, words that contract to spell "whegs," now a trademarked word for wheel–leg combinations (see Figure 6.4). In Whegs II, a bio-inspired design of the leg is made more functional by the addition of the useful, nonbiological structure of the wheel. Whegs II boasts a segmented body that enables complex, insect-like flexures that allow the climbing of large obstacles with improved stability. What makes Whegs really interesting is that they use only one propulsion motor, yet passively adapt their gait to the terrain using preflexive components. Because of these advantages, Whegs II can climb obstacles taller than twice their leg lengths, accelerate so rapidly as to jump, and also run faster than 3 body lengths per second.

Other compound wheel–leg combinations show promise as well. McGill's ANT hybrid architecture, with springy, biologically inspired legs that are tipped with actuated wheels, results in vehicles that can substantially outperform vehicles which use wheels or tank-treads alone (Steeves et al., 2002).

To contend with the mutability of the real world, a robot may need to change shape and function on the fly. Many bio-systems self-reconfigure to meet environmental circumstances, and this ability is the inspiration for reconfigurable robotics (Fukuda and Kawakuchi, 1990; Rus and Vona, 2001). For example, a robot could assume a flat or snakelike form to squeeze though a slim passage, but then morph into a hexapod to traverse rough terrain upon exiting the passage. Reconfigurable robotics researchers do not just focus on macroanimal forms though. Some researchers draw reconfiguration analogies to molecules and in particular folding proteins, in robots labeled "molecule robots" (Kotay and Rus, 1999) (see Figure 6.5).

6.2.3 Walking and Running

Other roboticists choose a four-legged or quadruped configuration. While not as stable as a hexapod, especially over rugged terrain, four legs do offer better inherent static stability than two- or one-leg architectures.

In 1999, Sony introduced the Aibo robot dog (Figure 6.6) a quadruped robot with an RISC processor, camera, and quadruped locomotion, and in this regard Aibo is distinguished from low-

Figure 6.5   A molecule robot developed in the Dartmouth Robotics Laboratory consists of four modules that can locomote by using a "tumbling" gait.

Figure 6.6   Sony AIBO. (Image courtesy of the Sony Corporation.)

cost toys by being a truly sophisticated biomimetic robot, with a battery of sensors and nimble legs. Since the robot's introduction, over 140,000 units have sold mostly to tech enthusiasts, but also to numerous robotics and AI labs, such as Luc Steels, who use the Aibo's sophisticated feature set to investigate navigation, and the vision plus speech sensor fusion in learning (Steels, 2001).

Though emulating humanlike locomotion is a much more challenging task than hexapod locomotion, numerous Japanese firms have long poured resources into biped locomotion research. In recent years, these efforts have borne fruit, with many firms revealing stable humanlike bipeds, including Honda, Sony, Toyota, Kawada, Fujitsu, ERATO, VStone, and Wow Wee Toys (see Figure 6.7). While such robots may not be ready for applications such as rescue operations, etc., biped robots are great for entertainment and publicity.

These robots dynamically stabilize in response to live data from six-axis gyro-accelerometer sensors, taking advantage of fast computation and small footprint of today's processors, and wireless uplink to more computation.

When walking, these robots employ a "controlled fall," where the walking speed maintains a zero moment point (zmp) such that with each footstep, the figure catches itself from the fall. While this controlled fall is also exhibited in human walking, in other ways, these above bipeds resemble humans in form and not in physics. As discussed earlier in this chapter, animals utilize springy gaits, whereas most biped robots do not, instead relying on rigid mechanical systems. Thus, most

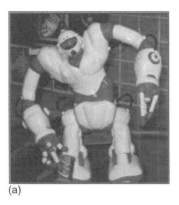

(a)   (b)   (c)

Figure 6.7   Contemporary bipedal robots: (a) Wow Wee's RoboSapien, (b) Sony's Qrio © 2003, and (c) AIST's HRP-2P.

bipeds do not receive the shock absorption, "preflex" stabilizing advantages, or inverse-pendulum energy-recycling advantages of mechanically compliant animal locomotion systems.

At least one humanlike biped does take advantage of mechanical compliance: the low cost Wow Wee "RoboSapien" biped (Figure 6.7a) engineered by physicist Mark Tilden. As a result of this, Robosapien is more energy efficient than other biped humanoids, running for several hours on a charge (versus Sony's and Honda's bipeds, which run for ~30 min on a charge).

Of the publicly-shown biped robots, the 18 in. tall Sony Qrio (Figure 6.7b) shows the greatest range of biped functionality and cognitive abilities. As a biped, the robot cannot just walk, but run, hop, and right itself from a fall. Its movements are extremely swift, graceful, and humanlike, and its grasping hands enable Qrio to toss a ball.

Of human-scale bipeds, the HRP-2P of Kawada, Japan is remarkable for its ability to climb to its feet from a lying posture (Figure 6.7c).

The rate of progress in the ability of legged robots and bipeds, in particular, is extremely encouraging. It is easy to imagine that such devices will be commonplace in entertainment and service applications in coming years.

6.2.4   Flying, etc.

Many biomimetic or bio-inspired flying robots have shown swift progress of late. Many of these imitate the small scale flight of insects, while others fly at larger scales, after the fashion of birds.

The Micromechanical Flying Insect (MFI) Project at University of California, Berkeley has demonstrated a 30-mm MFI prototype (see Figure 6.8), using piezoelectric actuators and a flexible thorax structure to incur notable thrust force (Dickinson, 2001) and demonstrate flight (MFI Website, 2002). This performance results from extensive studies of the aerodynamics of fly-wings in motion, which explain how flies generate three times as much lift as was previously understood (Dickinson, 2001). This work demonstrates that insect flight results from three phenomena: the leading-edge vortex (or delayed stall), the rotational lift, and the wake capture. Funded by DARPA, ONR, and MURI, the ultimate goal of the MFI project is to produce autonomous military robots.

The Mentor robot (1998 to 2002) of SRI International and University of Toronto was the first flapping wing microaviation vehicle (MAV) capable of hovering in place. Weighing 435 g, and 30 cm in height, the double-hummingbird X-wing model flew in a tethered mode, demonstrating both hovering and forward flight (DeLaurier, 2003). University of Toronto designed and built the complete aircraft system and SRI investigated actuating the device with dielectric-type electro-active polymers (EAP) artificial muscles.

Figure 6.8   MFI at University of California, Berkeley. (Image courtesy of R. Fearing, University of California, Berkeley.)

Many other bio-inspired morphologies of locomotion exist in robots, from fish robots that can swim like manta rays, to lobster robots, and to robots that tumble like tumbleweeds. The reader is encouraged to investigate further the astonishing diversity of robotic locomotion that is currently flowering in the world.

6.2.5 Grasping and Manipulation

Many robots need the ability to grasp and handle objects. On a large scale, this may be done in the manner of hands, claws, or teeth, while at a smaller scale the task may be accomplished in the manner of proteins or the sticky feet of flies or geckos.

Numerous issues play into grasping and manipulation. First, the manipulator element must be guided to the object to be handled, a task that can require machine vision and a high resolution control system. Second, the mechanical system must be able to reach objects with sufficient flexibility and dexterity. And when the manipulator handles the object, the manipulator must contain enough sensors and the right control software to apply just the right amount of force in the object.

Many robots around the world have accomplished grasping and manipulation tasks, and some are even developing the ability to learn manipulation tasks by learning and imitation such as Ripley at MIT, and the juggling robots of USC Georgia Tech and MIT (Atkeson and Schaal, 1997).

Shadow Robotics of the U.K. has developed a robotic hand with 24 degrees of freedom (DOF), driven by McKibben air muscles. Most of the joints are driven by an opposing pair of muscles, permitting variable compliance at the joint; however, some of the finger joints are driven by a single muscle with return spring. The Shadow hand operates at about half speed of a human.

The NASA Robonaut, a humanoid torso for outer space applications, boasts a pair of dexterous arms enabling dual-arm operations, and 14 DOF hands that interface directly with a wide range of tools. Using a humanlike model for autonomous grasping, the Robonaut relies heavily on feedback from its robotic fingers and palm. Numerous sensor technologies are employed including piezo-films, capacitive pressure sensors, and Force Sensing Resistor (FSR) technology. To achieve the fine resolution of feedback, tactile sensors for grasping, and the smart algorithms for interpretation of the signals, are systemically dubbed tactile perception. Robonaut's manipulation capabilities are being automated, and can also be run by teleoperation to enable delicate space operations via distant human presence.

6.2.6 Motion Conclusion Summary

Physically actuated embodiment is certainly essential for robots to interface with the real world. Additionally, many roboticists debate that interface with the complex, nonlinear real world is important for the formation of intelligence (Brooks, 1991). Whether that is true or not, intelligence and perceptive systems are of great importance for robots to be effective in the real world.

6.3 BEHAVIOR, EXPRESSIVITY

6.3.1 Intelligence and Perception

The emulation of human and animal central nervous systems (CNS) stands as the most challenging domain of bio-inspired robotics. While neuroscience is deciphering the mysteries of mind at unprecedented rates, thanks largely to novel imaging techniques such as fMRI, many components of machine perception and intelligence are coming into functional maturity. Though not nearly as capable as humans, many "human-emulation" technologies have sprouted substantially in the last decade, showing remarkable surges in functionality including face tracking, feature tracking, visual biometric identification, bipedal locomotion, and semantically rich NLP (Kurzweil, 1999; Menzel, 2001; Bar-Cohen and Breazeal, 2003). With these tools, we can sketch crude models of simulated mind in technological media. The emphasis, however, is on the word "crude": it must be acknowledged that most of the mysteries of the CNS are well beyond science at this time.

Accordingly, machine intelligence is decidedly below that of most animals and certainly humans. But our machines must be judged on their own standards. After all, a machine can understand speech better than a dog can, and what's more, the machine can speak back. Many of the intelligent and perceptive systems available today have yet to be integrated into functional whole. This section first considers intelligent systems as parts, and then discusses their integration into a systematically emulated animal intelligence, with a focus on social intelligence.

6.3.1.1 Language, Ontologies, Top-Down

At the foundation of human–machine language interaction lie ASR, automated speech synthesis (ASS), and various approaches to NLP. Although only capable of rudimentary language interactions, machine language has recently shown a remarkable rate of progress, both in successful academic research and in deployed speech solutions.

For many years, basic speech recognition and synthesis were major obstacles even to the most elementary human–computer language interactions. However, progress in the late 1980s and 1990s led to a large number of deployed speech applications, ranging from dictation software such as IBM's Viavoice to natural language ticketing and customer service agents, such as those offered by ATT. Companies now marketing commercial speech applications include SpeechWorks, Sensory, Nuance, and Dragon Naturally Speaking. Another highly effective system is open-source to researchers: Carnegie Mellon's Sphinx is highly functional, robust, user-independent ASR software (Carnegie Mellon, website, 2002).

Several common features operate rather naturally; the "barge-in" capability allows users to interrupt the system and still have speech recognized. "Rejection and keyword spotting" recognizes a speaker's keywords without prompts. Using Bayesian analysis, "N-best" sorts through possibilities of what a speaker might have said to locate a correct word, while the statistical language modeling of "N-gram" creates a sizable vocabulary and natural language recognition.

Word recognition and synthesis is only the first step toward endowing machines with humanlike language intelligence. Text-to-speech (TTS) software outputs increasingly natural-sounding speech, with off-the-shelf solutions including Rhetorical, Elan, Nuance, and the open-source Festival.

Between recognition and synthesis, an intelligent system needs to process language, cross-relate language to vision and other senses (a task known as multimodal sensor fusion), and make decisions about how to act in this world. Many labs tackle this problem with natural language as the nexus of the above, an approach known as NLP.

Some NLP researchers ambitiously attempt to completely model human grammars, while others such as the Cyc project of Austin, Texas model ontological relationships into expert systems — an approach that has proven successful for some limited applications. Many functional natural language applications, such as electronic ticketing agents or IBM's Natural Language Assistant (NLA) search engine (Chai et al., 2002), compensate for their inability to understand full, general language by relying on the constraints specific to the application's situations. Other ambitious language-engine projects attempt to model the emergence of language — the paths by which one (a human or a machine) can acquire language from a social environment.

Under the hypothesis that language is inherently an emergent phenomenon, Luc Steels and other researchers at the Sony Computer Science Lab in Paris are teaching Sony AIBO robots to recognize objects via natural language games (Steels and Kaplan, 2002; Boyd, 2002). The results are promising. While these robots are learning only the simplest of grammars and words, they are doing so under highly variable conditions, and can recognize learned objects independent of lighting or viewing angle. In fact, this method has considerably outperformed other language acquisition systems that used neural networks or symbolic learning labeling theory (Steels and Kaplan, 2002). Here emphasis is made that such a natural language system is an integration of many cognitive components: vision, gesturing, pattern recognition, speech analysis and synthesis, conceptualization, interpretation, behavioral recognition, action, etc.

6.3.1.2 Vision, Other Sensing, Sensor Fusion

The work of computational neuroscientist Cristoph von der Malsberg's theories of complex, nonlinear behavior in neurons has driven the development of numerous successful vision algorithms (Von der Malsberg and Schneider, 1986). One descendant of Von der Malsberg's work developed by Malsberg's student Hartmut Nevin, stands out as the most successful tracker of human facial expressions from live streaming video is sold as NevenVision FFT. NevenVision modules use these theories to accomplish numerous other vision tasks as well, including biometric face recognition, object, and gesture recognition as well. The author of this chapter is currently investigating the use of this software to endow social robots with emotional-expression recognition in context-driven conversation.

The automated face analysis (AFA) software system developed in the Carnegie Mellon University Face Lab determines the emotional state of a subject by automatically analyzing images against Ekman's facial action coding system (FACS) (Xiao et al., 2002). While this AFA FACS analysis is not in real time, if optimized and integrated with quick and robust expression recognition software, this software will greatly advance progress toward complete and effective sociable robot systems.

Using the work of Steels and Kaplan (2002) described in section 6.3.1.1 and others, Sony has demonstrated the integration of many visual and perceptual systems and speech in its Qrio biped. The Qrio can biometrically identify a face, recognize, and respond to a person's facial expressions, and recognize objects and environmental attributes. The visual ontologies are fused with the semantic language ontologies, allowing Qrio to converse in a simple but lifelike way about a number of subjects. This work is a forerunner of integrated machine intelligence systems with nimble humanlike embodiment.

6.3.2 Social Intelligence, Social Robots, and Robot Visual Identity

Social robots particularly require the fusion of many perceptual, language, and physical embodiment systems — requirements that drive the systematic integration of these components into a whole that is greater than the sum of parts (Breazeal, 2002).

Sparked largely by the mid-1990s MIT graduate work of Cynthia Breazeal, sociable robots integrate many of these human-emulation technologies into singular synthetic organisms, designed to communicate more effectively with people (Breazeal, 2002). While these robots only crudely simulate social cognition, they are being actively used as modeling tools in cognitive science (Fong et al., 2003). Since Breazeal's seminal work, a sizable number of sociable robots have sprung into existence. Although a comprehensive list is beyond the scope of this paper, a few sociable robots include: Ridley at MIT (lead by Deb Roy), Nursebot Pearl of CMU, Kismet and Leonardo of MIT, and Mabel at the University of Rochester (built by a team of undergraduates). Additionally, companies including Panasonic, Sony, and Honda have lately pursued sociable humanoid robots. Although these robots all seek to achieve bio-inspired communicative interaction with humans, none has a realistic humanlike face.

In social robots that do have faces, the hardware mimics the expressive action of the human face — humanity's primary mode of expressing affective states (Ekman, 1989). Whether depicting a realistic human or an abstract character (like a cartoon or animal), the expressive animated motions of the character should be humanlike in order to be sensible to a human, because the human nervous system is innately and finely attuned to understand the human face's visual language (Levenson et al., 1990; Bruce et al., 2002).

As discussed later in the chapter, this can be a challenging hardware task, and even harder can be the socially interactive use of facial expressions. Better mechanization and automation of this social expression could unlock many useful service and entertainment applications from toys to comforting companions for the elderly. Even in a military scenario, wherein a robot must communicate swiftly with human soldiers, the power of emotive communications cannot be over-estimated.

The mechanics of the biological human face are well studied; meanwhile the semiotics of human-facial communication have been preliminarily defined by anthropologist P. Ekman and others in the aforementioned FACS (Ekman and Friesen, 1971). Body language, also well studied (Birdwhistle, 1970), can further enable robotics' sociable applications. While further work remains to decipher the cognitive systems that underlie dynamic facial effect (including their complex relation to language), these robots can still be interesting as entertainment, training devices, and as quantitative tools for the better study of social cognition.

Most hardware technology for simulating facial expression springs from the entertainment special effects industry, where the technology is used to animate characters in movies, theme parks, etc. Stan Winston Studios, Walt Disney Imagineering, and Jim Henson Creature Shop, and many other "animatronics" (themed animation robots) shops, utilize the power of nonverbal communication, by simulating human and animal faces and figures in story-telling context. In these applications, the complexity of social cognition is theatrically designed by animator and writers, and is not interactive or intelligent. Nevertheless, these approaches that emphasize commercially presentable results have achieved the highest degree of mechanical aesthetic biomimetics in history (see Figure 6.9), as shown in feature films such as *AI* and *Jurrassic Park* among others.

Animatronics seems like a natural match for sociable robotics. Indeed, the merger of animatronics and sociable robotics has begun; in 2002, one of the leading shops in animatronics built the mechanical and aesthetic systems of Cynthia Breazeal's Leonardo robot (Bar-Cohen and Breazeal, 2003; Landon, 2003) (see Figure 6.10).

As with the work of Luc Steels and Qrio, the Leonardo project uses learning algorithms with vision–tactile–language knowledge fusion to accomplish learning by imitation. This work is collinear to that of other MIT groups such as the Cognitive Machines Group (CGM) led by Deb Roy, in which the robot Ripley uses a grasping mouth to manipulate objects. Ripley clearly has a machine identity, and no facial expressions. Leonardo, by contrast, boasts 32 DOF in the face, achieving very agile facial effect. While Leonardo is anthropomorphic, Leonardo is conscientiously not human in form (Breazeal, 2002). Leonardo team leader Cynthia Breazeal expresses that realistic animatronic technology is not quite human enough to be convincing, and just human enough to push people's expectations of the intelligence of the machine and too high to be met with today's

Figure 6.9 (See color insert following page 302) "Mask," a 5 in. self-portrait by Ron Mueck, a graduate of Jim Henson Creature Shop, a leading animatronics studio. (Photo by Mark Feldman [Feldman, 2002 website]. With permission.)

Figure 6.10 Sociable robot, Leonardo of MIT; on the right shown in a learning game or task (From Breazeal, C., Buchsbaum, D., Gray, J., Gatenby, D. and Blumberg, B. Rocha, L. and Almedia e Costa, F. (Eds), *Artificial Life*, Forthcoming 2005. With permission.)

technology. This hesitation is a point of some debate in the AI community, with most roboticists believing that cartoon identities are OK, but humanlike ones are not.

One of the most cited manifestations of this argument, the theory of the Uncanny Valley, appeared in a 1975 essay by Japanese robotics researcher Masahiro Mori (Reichardt, 1978). Here, Mori speculated that as an anthropomorphic object looks and acts more realistically human, it will receive increasingly favorable reaction, but only up to a limited point of realism (see Figure 6.11). After this node, however, the viewer starts to become more distracted by flaws in humanoid demeanor, such that the object will soon become highly disturbing to a person. This graphic depression in favorable opinion is the valley of Mori's Uncanny Valley.

Mori further speculated that should the object increase sufficiently in realism, viewer opinion will eventually rise back out of the valley, cross the neutral threshold of viewer opinion, and ultimately,

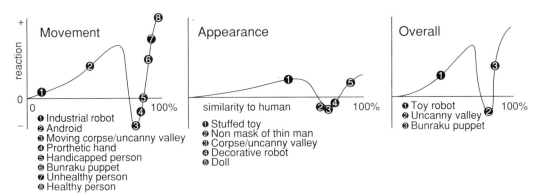

Figure 6.11   Reichardt Jasia's Uncanny Valley of anthropomorphic rejection. (Illustration by Bryant, 2003. With permission.)

(once appearance rivals realism) will turn into complete acceptance. Based upon this theory, though, Mori concluded that anthropomorphic robot designs should always stop short of the Uncanny Valley to avoid public fear and loathing or worse yet: the complete rejection of robots by the public.

Although no data has been collected to substantiate the Uncanny Valley theory, it is the closest thing to an engineering principle that exists for guiding the design of anthropomorphic robot identity. It is important to emphasize that the chart provided in Figure 6.3 is conceptual, and not based on data. Other aspects of the Uncanny Valley theory are specious as well. For example, in the theory, verisimilitude is not well defined. Many dimensions of aesthetic percept fluctuate widely in the examples given with the theory. Examples of rising realism leap from a "stuffed toy" to a "walking corpse" to a "decorative robot," without quantifying any of the characteristics of the above. Clearly, disturbing and unhealthy percepts are associated with corpses, which would represent *retrograde* in verisimilitude.

Unfortunately, in the intervening years, hard data on the purported phenomenon has yet to be gathered, in spite of regular debate of the dogma of the Uncanny Valley. In his original theory, Mori posits that the valley works in both static representation and dynamic. Yet, clearly the fine arts represent humans in every possible shade of realism, in paintings and illustrations. People are not horrified by Van Gogh paintings because they are not perfectly realistic. And the realistic motion of Sony's Qrio shows that a robot can move just like a real person without being horrific. In short, although the Uncanny Valley theory may hint at real phenomena, it is not real science.

The author proposes that the Uncanny Valley effect arises from a distributed network of brain-systems that, in concert, function as an "emergency alarm." This alarm system becomes acutely enabled by the detection of high-verisimilitude anthropomorphic stimuli, and rings with alarm if patterns that signal crisis are detected. But the alarm also will ring (provided it is enabled) if certain patterns that signal a healthy social presence are *not* detected. This revised theory is renamed the theory of Bridge of Engagement (BOE).

Recent brain imaging (LaBar et al., 2003) has found that visual stimulus of moving fear expressions shown to test subjects activates a distributed pattern involving the pSTS, right lateral fusiform gyrus (FFA), and the amygdala. Kesler-West et al. (2001) found similar results, but found that "happy" expressions activated a very different distributed pattern in test subjects than did visual percepts of negative affect, and this pattern did not include elevated activity in the amygdala, but instead involved elevations in activity of the medial frontal or cingulate sulcus, an area that has been found to be critical to the initiation of language (Crosson et al., 1999). These findings support the notion that crisis stimuli such as a fearful expression will trigger a neural alarm of fear, whereas facial stimuli that do not cause fear lead to preparations for social engagement. Additionally, LaBar et al. (2003) also show that sliding "identity morphs," which animate identity change from one individual to another, activate similar distributed neural patterns as do expressions of fear, notably with elevated activity in the amygdala. This also supports the concept that if high-resolution identity cues fall outside expected patterns, the brain will signal alarm. While these studies support the hypothesis of a social emergency alarm that is more sensitive to high-verisimilitude cues, future experiments are required.

Intuitively, it seems probable that humans' visual expertise to human faces helps us to receive the full bandwidth of paralinguistic semantics. If this is true, then realistic faces in robots will simply be more communicative.

There are other reasons to consider making robots more realistically humanlike. Clearly, realistic human depictions have been highly successful in arts, film, video games, and toys; so the question naturally arises: why wouldn't the appeal of realism extend to robots too, if the robots look good enough? Shades of realism would certainly be critical for many applications, for example: training models that enact face-to-face exchange, such as medical, police, or psychology simulation. Moreover, accurately modeling the human face allows scientific investigation of human communication.

Several groups use computer-simulations of realistic faces for their robots and autonomous agents, for instance: Cassell's Rea at MIT and Vikea of CMU's Sociable Robots Group. Perhaps this is because video games have made humanlike simulations more acceptable. Two exceptions that pursue realistic faced robot identities include: Fumio Hara's lab at the Science University of Tokyo (Hara, 1998) and the work of the author of this paper at the University of Texas at Dallas (Hanson, 2003; Ferber, 2003) (see Figure 6.12).

Historically, the most successful mechanically emulated faces have been entertainment animatronics, but these robots actually reinforce the bias against realism by being inadequate in their expressivity, consuming a great deal of power, and being bulky, weighty, and costly. Each of these faults prohibits widespread deployment beyond high-end niche markets of theme parks and film. Yet, these drawbacks all spring from one cause: the great force required by animatronics' solid-elastomer skins to deform into facial expressions.

The solid elastomer materials that are currently in use to emulate facial soft tissues require relatively large amounts of force to move, which leads to high costs, high power requirements, and aesthetic movement unlike human facial tissue. In spite of the pliable, elastic qualities of solid elastomers, the molecules of such elastomer tangle with their near neighbors, fundamentally constraining the material when deformed. The molecules of facial flesh are not so constrained, being in effect an array of liquid-filled tissue-sacs, which allows molecules of liquid to flow into any

Figure 6.12 High-verisimilitude anthropomorphic robots, left: the Science University of Tokyo; right: the University of Texas at Dallas.

topological form that cells and fascia will tolerate. Squeeze a water balloon in one hand and a sphere of solid rubber in the other, and the differences become quickly apparent (Hanson and White, 2004). The use of currently available materials causes the resulting mechanism to be power hungry, massive, and costing more than $80,000 to manufacture (Menzel and D'Aluisio, 2000); yet, animatronic-face hardware nevertheless sells vigorously in niche-markets of movie and theme park animatronics. However, to impact wider markets and to realize the potential of social robots, new facial materials are necessary.

To overcome this essential hurdle, the author's work advances novel materials that affect human facial expressions more realistically, with approximately 1/20th the force required by solid elastomers (see Figure 6.13). These materials are low cost, nontoxic (made of medical grade materials), and altogether practicable for mass production.

The first pass at the material was a urethane-based foamed elastomer which elongated approximately 500%, yet compressed like a conventional sponge or foam rubber. Dubbed "F'rubber" (a contraction of "face" and "rubber"), the material exhibits physical characteristics much closer to human skin versus a solid elastomer. The cells of the material are filled with air rather than liquid, which cause the volume of the simulated tissue to be variable, unlike the practically invariable volume of liquid-filled facial tissues. Nevertheless, these new materials fold, wrinkle, and bunch in ways that are highly naturalistic, much more so than can be achieved with solid elastomer. In addition to improved verisimilitude in simulated facial expressions, this material decreases the force requirements by an order of magnitude, enabling lower power, lower cost expressive robots, rendering them applicable to a wider range of art and science.

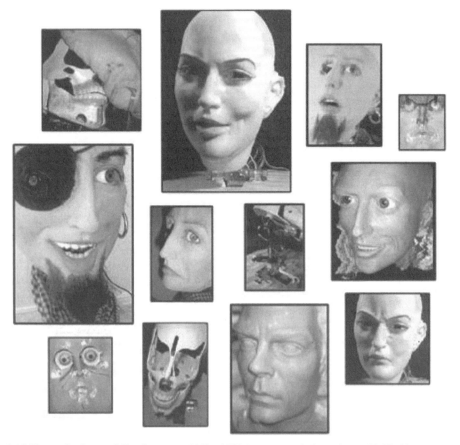

Figure 6.13 (See color insert following page 302)   UTD human emulation robots with F'rubber.

More recently, new F'rubber material has been created out of silicone, which is softer, can elongate 1050%, and is tolerant to a wider range of environmental conditions. Also, computer-numeric-controlled (CNC) deposition of thermoplastic elastomer (TPE) into a designed matrix has shown promising preliminary results, exhibiting elongation up to 1250%. The pores in such a material need not be spherical; they can be shaped into complex manifolds for improved mechanical and expressive behavior. They may even contain closed cells filled with liquid for still more advanced expressive emulation of facial tissues. These processes are described further in Section 6.4.

These new material approaches may help to satisfy people's discriminating taste for verisimilitude. They may also enable to convey more relevant sociable perceptual patterns when deviating from verisimilitude. Additionally, however, the novel materials may be used for nonrealistic cartoon and animal faces, which will benefit from the likeness to animal soft tissues, and the decreased weight and energy requirements.

6.4   ROBOTIC MATERIALS, STRUCTURES, AND MANUFACTURABILITY

Robots are inherently integrated systems, and will be beneficiaries of the numerous bio-inspired technologies that can be integrated into robots. Make no mistake, robots also benefit from technology advances that are not bio-inspired. Advancing manufacturing technologies, such as MEMS and SDM, though not themselves bio-inspired, are being used to produce more lifelike robots.

Multifunctional materials promise to make robots more effective. Single multifunctional materials, such as carbon nanotubes, offer mechanical strength and flexibility, provide computation, emit light, capture sensory data, provide electrically actuated motion (Baughman et al., 1999). If utilized in a robot's skin, such electroactuative polymers (EAPs) may greatly streamline a robot's manufacturing and cost (Bar-Cohen, 2002). Likewise, multiplicity of functions in the locomotive systems, as may be provided by biomimetic viscoelastic mechanical compliance (Full, 1999) and reconfigurable designs, can allow a robot to transit through more diverse, complex terrain.

Advanced manufacturing is leading to faster robot design, more complex yet affordable robots, and design control at meso, micro, and nanoscales. Rapid progress is occurring in both materials sciences (including recent advances in EAP) and manufacturing technologies on multiple scales. The overlaps with biomimetic engineering are increasing in number impressively.

Rapid prototyping and digital design tools are enabling complex concepts to be turned into physical objects in very short cycles. Advanced silicon manufacturing techniques, largely innovated for manufacturing microprocessors, have resulted in burgeoning techniques of MEMS in several interesting robot projects (Dickinson et al., 2000). A comprehensive list of techniques that may be pertinent to manufacturing and prototyping entertainment robots with EAP actuators would be prohibitively long to include here, but are covered in more detail in Chapter 18 of Bar-Cohen (2002/2003) and Hanson et al. (2003).

Fusing several rapid prototyping technologies with mold-making and advanced materials, SDM has been described in Full (2000) as particularly interesting for use with biorobotics. SDM uses various computer-aided manufacturing (CAM) technologies to layer and refine materials into arbitrarily complex configurations, bonded without the use of mechanical fasteners. Actuators and sensors can be imbedded directly into the "flesh" of a device, composed of materials of varied elasticity and rigidity, layered and bonded *in situ*. This process can achieve fully functional rapid prototypes, as well as highly efficient manufacturing procedures (Amon et al., 1996). Images of the SDM process are shown in Figure 6.14.

SDM may be effective for MEMS microscale devices, and may be extensible to nano-devices. The SDM process is proving decidedly useful in rubbery macroscale robots, which are simplified by the absence of mechanical fasteners. As an example, the 15-cm (6-in.) legged robot "Sprawlita" at

Figure 6.14 Stages in shape deposition manufacture of a small robot at Stanford's Rapid Prototyping and Biomimetic Robotics Labs.

Stanford's SDM Rapid Prototyping Lab was shown in Figure 6.2. This SDM robot is extremely robust, capable of receiving quite a pounding (Bailey et al., 2000). This approach promises benefits over traditional, rigid mechanical engineering — wherein systems fail from shock, friction among parts, and fatigue — factors reduced or absent in such elastomeric devices.

In order to improve robotic facial materials, the author and White (Hanson and White, 2004) developed a series of fabrication techniques akin to SDM, but utilizing a sacrificial matrix perfused with elastomer, to create controlled cellular pores in the elastomer. The collection of techniques is dubbed structured porosity elastomer manufacturing (SPEM). In preliminary experiments using these materials in robots, the materials have shown dynamic aesthetics vastly more akin to those of the human face, while requiring only 1/20th the force to actuate relative to solid elastomers.

Emulation of human skin in appearance and properties is one of the most challenging aspects of creating lifelike robotic faces. HER's patent-pending "structured porosity elastomer manufacturing" (SPEM) process results in materials that very closely approximate the properties of human skin, in a way amenable to mass production. SPEM produces a composite elastomer material with a controlled 3D chamber geometry. The process can produce chambers that are analogous to those in open-cell foam. Alternately, the pores need not be spheroids and can instead be rectangular, star-shaped, or any topology that is useful to achieve novel material properties. By implementing hierarchical chamber sizes, SPEM can relieve localized nodes of stress accumulation in a foamlike material, thereby increasing the overall elastic strain of the material. With this technique, the author increased strain in a silicone SPEM foam from 280 to greater than 800% — a value that is 85% of the solid constituent elastomer, one that enables facial expressions (Hanson and White, 2004) (Figure 6.15). Realistic facial expressions require up to 400% strain.

Because the enhanced compression of SPEM materials more closely matches that of facial soft tissues than do solid elastomers (Hanson and White, 2004), SPEM-based faces wrinkle and bunch more naturalistically, as can be seen at www.human-robot.org. SPEM techniques extend the power of SDM (Amon et al., 1996; Bailey et al., 2000; Hanson and White, 2004) and rapid prototyping (Figure 6.16).

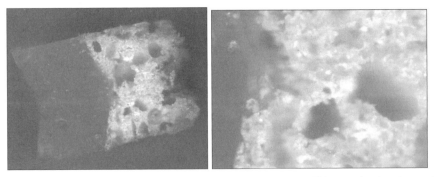

Figure 6.15 A cross-section of silicone SPEM, with 0.5 and 3 mm pores. This sample also demonstrates the composite possibilities of the material, as the material transitions into nonporous solid silicone toward the left.

We seek to address remaining challenges that were identified with the use of SPEM. These challenges include the need to find ways to remove the sacrificial matrix more effectively and quickly. There is a need to optimize the pore structure for expressive robot faces. Also, it is necessary to design the SPEM to interoperate with other mechanical systems, actuators and anchor placement, and mechanical attachments such that the face moves truthfully. Once we have developed the required material and procedures, we will have to evaluate the robustness, stability, and mechanical performance of SPEM materials.

Even more dramatically, the material requires less than 1/30th the force to compress relative to nonporous casts of the same silicone material. In facial expression robotics, this is particularly advantageous as all expressions both compress facial soft tissues and elongate facial soft tissues.

Figure 6.17 shows HER's latest robot EVA, which has been rendered using several of the silicone SPEM techniques described above. Because of the silicone SPEM, EVA requires a tiny fraction of the force to move into facial expressions, relative to other animatronic materials. For this reason, this robot's 36 DOF will run for hours on four AA batteries, consuming less than 10 W average and 40 W peak.

As of the time of this chapter's writing, the author's robots can be seen in action at the following urls: http://ndeaa.jpl.nasa.gov/nasa-nde/biomimetics/Biomimetic-robot-Hanson.mov and http://androidworld.com/HansonHead.wmv.

For widespread applicability, bio-inspired robotic materials, structures, and systems need to be manufacturable inexpensively and in bulk quantities. The toolset for creating bio-inspired robots is clearly maturing, though much work remains.

6.5   CONCLUSORY REMARKS

Clearly the future glows for bio-inspired robotics. Many trends are showing high degrees of functionality, and yet are increasing rapidly in function: computational hardware, materials, software, and mobility are examples. Yet, daunting quantities of work remain to create robots that are as capable as animals or humans. In fact, the increasing functionality of biomimetic robots and AI results in humbling insights regarding the complexity of the bio-systems, which let us know how much we yet know about life.

Future work includes the improved software integration that would accelerate the functionality of social robots and the advancement of automated design and prototyping systems for robotic systems from macro scale facial expressions and locomotion systems, to micro scale actuation and electronics. Additionally, the biosciences need to further discover what makes animals so effective, and engineers need to replicate these discoveries in technology. As the economy of biorobotics continues to expand, largely bolstered by the ongoing trends of increasing functionality, there should be ample resources for future research in this exciting field.

 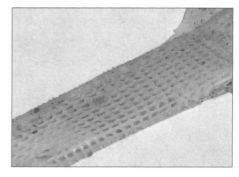

Figure 6.16   Rectangular-celled SPEM made via rapid prototyping. Elongated state shown on the right. The sample is 2 cm in width, 10 cm length, and 0.6 cm depth.

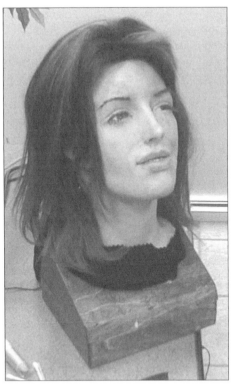

Figure 6.17 (See color insert following page 302)   Author's latest robot EVA. Because SPEM silicone requires little force to move, this robot's 36 DOF run for hours on four AA batteries.

At some point if the trends continue to extremes, our biology may be difficult to distinguish from our technology, and in many ways our human identity will be challenged. As the technology advances, it will be worthwhile to ask questions regarding the ethical employment of the technology. Will robots, like animals and humans, be afforded rights? What happens if they become conscious? What happens if they get smarter than us? Accelerating progress in robotics clearly implies that these questions are migrating out of speculative fiction and philosophy, and into reality, faster than many may think.

ACKNOWLEDGMENTS

The author would like to especially acknowledge Yoseph Bar-Cohen for continued dedication, support, and patience. He also would like to acknowledge the dedicated support of Victor White, Thomas Linehan, Alice O'Toole of the University of Texas at Dallas; Kristen Nelson, Elaine Hanson, and Dennis Kratz.

REFERENCES

Aizenberg, J., Tkachenko, A., Weiner, S., Addadi, L. and Hendler, G. Calcitic microlenses as part of the photoreceptor system in brittlestars. *Nature*, 412, 819–822, 2001.
Amon, C. H., Beuth, J.L., Merz, R., Prinz, F.B. and Weiss, L.E. Shape deposition manufacturing with microcasting: processing, thermal and mechanical issues. *Journal of Manufacturing Science and Engineering*, 1996.

Atkeson, C. G. and Schaal, S. Robot learning from demonstration. In: *Machine Learning: Proceedings of the Fourteenth International Conference (ICML '97)*, Douglas H. Fisher, Jr. (Ed.), Morgan Kaufmann, San Francisco, California, pp. 12–20, 1997.

Bailey, S. A., Cham, J. G., Cutkosky, M. R. and Full, R. J. Biomimetic robotic mechanisms via shape deposition manufacturing, *Robotics Research: the Ninth International Symposium*, Hollerbach, J. and Koditschek, D. (Eds), Springer-Verlag, London, pp. 403–410, 2000.

Bar-Cohen, Y. (Ed.), *Proceedings of the SPIE's Electroactive Polymer Actuators and Devices Conferenece*, 6th Smart Structures and Materials Symposium, vol. 3669, ISBN 0-8194-3143-5, pp. 1–414, 1999.

Bar-Cohen, Y. (Ed.), EAPAD Worldwide Web Hub, Newsletter, Spring 2002.

Bar-Cohen, Y. and Breazeal, C. *Biologically Inspired Intelligent Robotics*, SPIE Press, London, 2003.

Baughman, R. H., Cui, C., Zakhidov, A. A., Iqbal, Z., Barisci, J. N., Spinks, G. M., Wallace, G. G., Mazzoldi, A., De Rossi, D., Rinzler, A. G., Jaschinski, O., Roth, S. and Kertesz, M. Carbon nanotube actuators. *Science*, 284, 1340–1344, 1999.

BCC-A. *Global Robotics to Nearly Double to a $16 billion Industry by 2007*, RG-270 Robots/Automation Devices, BCC Business Communication Company, Inc, April 2003a.

BCC-B. *Artificial Intelligence: Burgeoning Applications in Industry*, RG-275, BCC Business Communication Company, Inc, June 2003b.

Birdwhistle, R. L. *Kinesics and Context: Essays on Body Motion and Communication*, University of Pennsylvania Press, Philadelphia, Pennsylvania, 1970.

Boyd, R. S. Scientists give language lessons to robots, Mercury News Washington Bureau, Sun, June 23, 2002.

Breazeal, C. *Designing Sociable Robots*, MIT Press, Cambridge, Massachusetts, 2002.

Breazeal (Ferrell), C. and Scassellati, B., Infant-like social interactions between a robot and a human caretaker. In: *Special issue of Adaptive Behavior on Simulation Models of Social Agents*, Kerstin Dautenhahn (guest editor), 2000.

Breazeal, C., Buchsbaum, D., Gray, J., Gatenby, D. and Blumberg, B. Learning from and about others: towards using imitation to bootstrap the social understanding of others by robots. Rocha, L. and Almedia e Costa, F. (Eds), *Artificial Life*, 11(1–2), 2005.

Brooks, R. A. Intelligence without reason. MIT AI Lab Internal Report, 1991.

Brown, I. E., Scott, S. H. and Loeb, G. E. Mechanics of feline soleus: II. Design and validation of a mechanical model. *Journal of Muscle Research and Cell Motility*, 17, 221–233, 1996.

Bruce, A., Nourbakhsh, I. and Simmons, R. The role of expressiveness and attention in human–robot interaction. In: *Proceedings of the IEEE International Conference on Robotics and Automation*, Washington, District of Columbia, May 2002.

Buckley, L., Sands, R., Scribner, D., Zucarello, G. and Hamilton, B. Technology imitates life, *OE Magazine*, SPIE Press, Amherst, New Hampshire, September 2004.

Butler, Z., Murata, S. and Rus, D. Distributed replication algorithms for self-reconfiguring modular robots. In: DARS 5, 2002.

Carnegie, Mellon Website: http://fife.speech.cs.cmu.edu/sphinx, 2002.

Cassell, J. Embodied conversational agents, *AI Magazine*, 22(4), Winter 2001.

Chai, J., Pan, S. and Zhou, M.X. MIND: a semantics-based multimodal interpretation framework for conversational systems. *Proceedings of the International CLASS Workshop on Natural, Intelligent and Effective Interaction in Multimodal Dialogue Systems*, Copenhagen, pp. 37–46, June 2002.

Damasio, A. R. *Descartes' Error*. The Grosset Putnam, New York, New York, 1994.

Damasio, A. R. *The Feeling of What Happens: Body and Emotion in the Making of Consciousness*, Harcourt Brace, New York, New York, 2005.

Darwin, C. and Ekman, P. (Eds), *The Expression of the Emotions in Man and Animals*, Oxford University Press, New York, New York (1998/1872).

DeLaurier, J., *Proceedings of Interdisciplinary Workshop on Micro-Aerial Vehicles at Schloss Elmau/ Germany*, September 22 to 24, 2003.

Dickinson, M. H., Solving the mystery of insect flight, *Scientific American*, June 2001.

Dickinson, M. H., Farley, C. T., Full, R. J., Koehl, M. A. R., Kram, R. and Lehman, S. How animals move: an integrative view. *Science*, 288, 100–106, 2000.

Ekman and Friesen, 1971. *Basic Emotions*.

Ekman, P., The argument and evidence about universals in facial expressions of emotion. In: *Handbook of Psychophysiology*, Wagner, H. and Manstead, A. (Eds), John Wiley, London, 1989.

Feldman, M. Website: 8am.com, 2002.

Ferber, D. *The Man Who Mistook his Girlfriend for a Robot*, Popular Science, September 2003.

Fitch. R., Rus, D. and Vona, M. A basis for self-repair robots using self-reconfiguring crystal modules. *Intelligent Autonomous Systems*, 6, 2000.

Fong, T., Nourbakhsh, I. and Dautenhahn, K. A survey of socially interactive robots. *Robotics and Autonomous Systems*, 42, 143–166, 2003.

Fukuda, T. and Kawakuchi, Y. Cellular robotic system (CEBOT) as one of the realization of self-organizing intelligent universal manipulator. In: *Proceedings of IEEE ICRA*, pp. 662–667, 1990.

Full, R.J., Biological inspiration: lessons from many-legged locomotors. In: *Robotics Research 9th International Symposium*, Hollerbach, J. and Koditschek, D. (Eds), Springer-Verlag, London, pp. 337–341, 2000.

Full, R.J and Meijer, K. Artificial muscles and natural actuators from frogs to flies. *SPIE*, 1999.

Hanson, D. and White, V, Converging the requirements of facial robotics and the capabilities of EAP actuators. *Proceedings of the SPIE Smart Materials Conference*, March 2004.

Hanson, D. Rus, D., Canvin, S. and Schmierer, G. Ch. 18: Biologically inspired robotic applications. In *Biologically Inspired Intelligent Robotics*, Yoseph Bar-Cohen and Cynthia Breazeal (Eds), SPIE Press, Bellingham, Washington, 2003.

Hanson, D. and Pioggia, G. Ch. 18: Entertainment applications for electrically actuated polymer actuators. In: *Electrically Actuated Polymer Actuators as Artificial Muscles*, SPIE Press, International Society of Optical Engineers, Bellingham, Washington, vol. PM98, March 2001.

Kotay, K. and Rus, D. Locomotion versatility through self-reconfiguration. *Robotics and Autonomous Systems*, 26, 217–232, 1999.

Kurzweil, R. *The Age of the Spiritual Machines*, Viking Press, Tonbridge, Kent, 1999.

Kurzweil, R. Website: KurzweilAI.net, 2002.

London, R. *Biologically Inspired Intelligent Robotics*, SPIE Press, 2003.

Levenson, R., Ekman, P. and Friesen, W. Voluntary facial action generates emotion-specific autonomic nervous system activity. *Psychophysiology*, 27(4), 363–383, 1990.

McCorduck, P. *Machines Who Think*, 1979.

Menzel, P. and D' Aluisio, F. *Robo Sapiens: Evolution of a New Species*, MIT Press, Boston, 2000.

MFI website, http://robotics.eecs.berkeley.edu/~ronf/mfi.html, 2002.

Raibert, M. H. Dynamic stability and resonance in a one-legged hopping machine. In: *Theory and Practice of Robots and Manipulators, Proceedings of RoManSy'81*, Morecki, A., Bianchi, G. and Kedzior, K. (Eds), Polish Scientific Publishers, Warsaw, pp. 352–367, 1983.

Reichardt, J., *Robots: Fact, Fiction, and Prediction*, Penguin Books Ltd., Harmondsworth, Middlesex, England, 1978.

Steels, L. Social learning and verbal communication with humanoid robots. *Proceedings of the IEEE-RAS International Conference on Humanoid Robots*, pp. 335–342, 2001.

Steels, L. and Kaplan, F. *AIBO's First Words. The Social Learning of Language and Meaning*. SONY Computer Science Laboratory, Paris (2), VUB Artificial Intelligence Laboratory, Brussels, steels@arti.vub.ac.be, 2002.

Steeves, C., Buehler, M. and Penzes, S. G. Dynamic behaviors for a hybrid leg-wheel mobile platform. *Proceedings SPIE Conference*, 2002.

UNECE, United Nations Economic Commission for Europe, Press Release. ECE/STAT/03/P01, Geneva, 17 October 2003.

Von der Malsberg, C. and Schneider, W. A neural cocktail-party processor. *Biological Cybernetics*, 54, 29–40, 1986.

Xiao, J., Kanade, T. and Cohn, J. Robust full motion recovery of head by dynamic templates and re-registration techniques. *Proceedings of the Fifth IEEE International Conference on Automatic Face and Gesture*, 2002.

Bio-Nanorobotics: A Field Inspired by Nature

Ajay Ummat, Atul Dubey, and Constantinos Mavroidis

CONTENTS

7.1   INTRODUCTION

The underlying principle of biomimetics deals with the understanding, conceptualization, and mimicking nature's way of handling various problems and situations. Nature has inspired mankind for ages and has been a key source from which we can learn and adapt. Natural processes are extremely efficient in terms of energy and material usage and provide us with many inspiring and thought provoking designs and principles. This chapter discusses biomimetics at the nano-scale, where we talk about nanorobotics and its design principles, which are inspired by nature's way of doing things at that scale.

Figure 7.1 describes the biomimetics principle and details the various aspects of mimetics. It explains the mimetics at two levels when nano-scale is considered. One is the "*machine nanomimetics*" principle meaning the creation of nanomachine components inspired by the equivalent machine components at the macro-scale and the other is the "*bionanomimetics*" principle where biological entities such as proteins and DNA are used to create the nanomachine components. The field of nanorobotics hence encapsulates these two mimetic principles and inherits their various characteristics, design logic, and advantages.

Nanotechnology can best be defined as a description of activities at the level of atoms and molecules that have applications in the real world. A nanometer is a billionth of a meter, that is, about 1/80,000 of the diameter of a human hair, or ten times the diameter of a hydrogen atom. The size-related challenge is the ability to measure, manipulate, and assemble matter with features on the scale of 1 to 100 nm. In order to achieve cost-effectiveness in nanotechnology, it will be necessary to automate molecular manufacturing. The engineering of molecular products needs to be carried out by robotic devices, which have been termed *nanorobots* (Freitas, 1999, 2003). A nanorobot is essentially a controllable machine at the nanometer or molecular scale that is composed of nano-scale components and algorithmically responds to input forces and information.

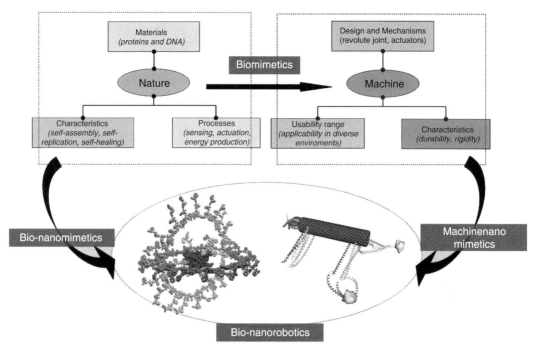

Figure 7.1 (See color insert following page 302)   Biomimetics — bio-nanorobotics, inspired by nature and machine.

The field of nanorobotics studies the design, manufacturing, programming, and control of the nano-scale robots.

This review chapter focuses on the state of the art in the emerging field of nanorobotics and its applications and discusses in brief some of the essential properties and dynamical laws which make this field more challenging and unique than its macro-scale counterpart. This chapter is only reviewing nano-scale robotic devices and does not include studies related to nano-precision tasks with macro-robotic devices that usually are also included in the field of nanorobotics (e.g., Atomic Force Microscope (AFM) and other forms of proximal probe microscopy).

Nanorobots would constitute any active structure (nano-scale) capable of actuation, sensing, signaling, information processing, intelligence, and swarm behavior at nano-scale. These function-alities could be illustrated individually or in combinations by a nanorobot (swarm intelligence and cooperative behavior). So, there could be a whole genre of actuation and sensing or information processing nanorobots having ability to interact and influence matter at the nano-scale. Some of the characteristic abilities that are desirable for a nanorobot to function may include:

(i)   *swarm intelligence* — decentralization and distributive intelligence;
(ii)   self-assembly and replication — assemblage at nano-scale and "nano-maintenance";
(iii)   *nano-information processing and programmability* — for programming and controlling nano-robots (autonomous nanorobots);
(iv)   nano- to macro-world *interface architecture* — an architecture enabling instant access to the nanorobots and its control and maintenance.

There are many differences between macro- and nano-scale robots. However, they occur mainly in the basic laws that govern their dynamics. Macro-scaled robots are essentially in the Newtonian mechanics domain whereas the laws governing nanorobots are in the molecular quantum mechanics domain. Furthermore, uncertainty plays a crucial role in nanorobotic systems. The fundamental barrier for dealing with uncertainty at the nano-scale is imposed by the quantum and the statistical mechanics and thermal excitations. For a certain nanosystem at some particular temperature, there are positional uncertainties that cannot be modified or further reduced (Drexler, 1992).

The nanorobots are invisible to the naked eye, which makes them hard to manipulate and work with. Techniques like scanning electron microscopy (SEM) and atomic force microscopy (AFM) are being employed to establish a visual and haptic interface to enable us to sense the molecular structure of these nano-scaled devices. Virtual reality (VR) techniques are currently being explored in nano-science and biotechnology research as a way to enhance the operator's perception (vision and haptics) by approaching more or less a state of "full immersion" or "telepresence." The development of nanorobots or nanomachine components presents difficult fabrication and control challenges. Such devices will operate in microenvironments whose physical properties differ from those encountered by conventional parts. Since these nano-scale devices have not yet been fabricated, evaluating possible designs and control algorithms requires using theoretical estimates and virtual interfaces or environments. Such interfaces or simulations can operate at various levels of detail to trade-off physical accuracy, computational cost, number of components, and the time over which the simulation follows the nano-object behaviors. They can enable nano-scientists to extend their eyes and hands into the nano-world, and they also enable new types of exploration and whole new classes of experiments in the biological and physical sciences. VR simulations can also be used to develop virtual assemblies of nano and bio-nano components into mobile linkages and to predict their performance.

Nanorobots with completely artificial components have not been realized yet. The active area of research in this field is focused more on molecular machines, which are thoroughly inspired by nature's way of doing things at nano-scale. Mother Nature has her own set of molecular machines that have been working for millions of years, and have been optimized for performance and design over the ages. As our knowledge and understanding of these numerous machines continues to increase, we now see a possibility of using the natural machines or creating synthetic ones from scratch using nature's components. This chapter focuses more on molecular machines and explores various designs and research prevalent in this field. The main goal in the field of molecular machines is to use various biological elements — whose function at the cellular level creates motion, force, or a signal — as machine components. These components perform their preprogrammed biological function in response to the specific physiochemical stimuli but in an artificial setting. In this way proteins and DNA could act as motors, mechanical joints, transmission elements, or sensors. If all these different components were assembled together in the proper proportion and orientation, they would form nanodevices with multiple degrees of freedom, able to apply forces and manipulate objects in the nano-scale world. The advantage of using nature's machine components is that they are highly efficient and reliable.

Nanorobotics is a field which calls for collaborative efforts between physicists, chemists, biologists, computer scientists, engineers, and other specialists to work towards this common objective. Figure 7.2 details the various fields which come under the field of bio-nanorobotics (this is just a representative figure and not exhaustive in nature). Currently this field is still developing, but several substantial steps have been taken by great researchers all over the world who are contributing to this ever challenging and exciting field.

The ability to manipulate matter at the nano-scale is one core application for which nanorobots could be the technological solution. A lot has been written in the literature about the significance and motivation behind constructing a nanorobot. The applications range from medical to environmental sensing to space and military applications. Molecular construction of complex devices could be possible by nanorobots of the future. From precise drug delivery to repairing cells and fighting tumor cells, nanorobots are expected to revolutionize the medical industry in the future. These applications come under the field of nanomedicine (Freitas, 1999, 2003), which is a very active area of research in nanotechnology. These molecular machines hence form the basic enablers of future applications.

In the next section, we shall try to understand the principles, theory, and utility of the known molecular machines and look into the design and control issues for their creation and modification. A majority of natural molecular machines are protein-based which involve using the exact replica

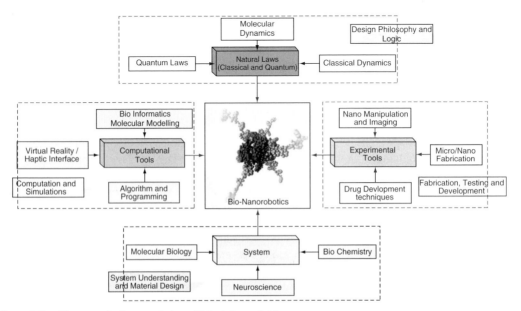

Figure 7.2   Bio-nanorobotics — a truly multidisciplinary field.

of nature's mechanism, while the DNA-based molecular machines use the basic properties of DNA to design various synthetic mechanisms (which might not be present in the nature). Nature deploys proteins to perform various cellular tasks — from moving cargo to catalyzing reactions, while it has kept DNA as an information carrier. It is hence understandable that most of the natural machinery is built from proteins. With the powerful crystallographic techniques available in the modern world, the protein structures are clearer than ever. The ever increasing computing power makes it possible to dynamically model protein folding processes and predict the conformations and structure of lesser known proteins (Rohl et al., 2004). All this helps unravel the mysteries associated with the molecular machinery and paves the way for the production and application of these miniature machines in various fields including medicine, space exploration, electronics, and military.

7.2 BIOMOLECULAR MACHINES: BACKGROUND AND SIGNIFICANCE

7.2.1 Significance

The recent explosion of research in nanotechnology, combined with important discoveries in molecular biology, has created a new interest in biomolecular machines and robots. The main goal in the field of biomolecular machines is to use various biological elements — whose function at the cellular level creates motion, force, or a signal, or stores information — as machine components. These components perform their preprogrammed biological function in response to the specific physiochemical stimuli but in an artificial setting. In this way proteins and DNA could act as motors, mechanical joints, transmission elements, or sensors. If all these different components were assembled together in the proper proportion and orientation, they would form nanodevices with multiple degrees of freedom, able to apply forces, and manipulate objects in the nano-scale world. The advantage of using nature's machine components is that they are highly efficient (Kinosita et al., 2000) and reliable. Just as conventional macro-machines are used to generate forces and motions to accomplish specific tasks, bionanomachines can be used to manipulate nano-objects to assemble and fabricate other machines or products and to perform maintenance, repair, and inspection operations.

Such bio-nanorobotic devices will hopefully be part of the arsenal of future medical devices and instruments that will: (1) perform operations, inspections, and treatments of diseases inside the body, and (2) achieve ultra-high accuracy and localization in drug delivery, thus minimizing side effects. Figure 7.3 shows an idealized rendition of a biomolecular nanorobot repairing an infected cell in a blood vessel. The bio-nanorobot will be able to attach to the infected cell alone and deliver a therapeutic drug that can treat or destroy only the infected cell, sparing the surrounding healthy cells.

Development of robotic components composed of simple biological molecules is the first step in the development of future biomedical nanodevices. Since the planned complex systems and devices will be driven by these components, we must first develop a detailed understanding of their operation. From the simple elements such as structural links to more advanced concepts as motors, each part must be carefully studied and manipulated to understand its functions and limits.

Figure 7.4 lists the most important components of a typical robotic system or machine assembly and the equivalence between macro and potential bio-nano components. Beyond the initial component characterization is the assembly of the components into robotic systems. Figure 7.5 shows one such concept of a nano-organism, with its "feet" made of helical peptides and its body using carbon nanotubes while the power unit is a biomolecular motor. For this phase to be successful, a library of biological elements of every category must be available. At that point, conventional robotics can be used as a guide for fabrication of bio-nanorobots that function in the same manner. There will be systems that have mobile characteristics to transport themselves, as well as other objects, to desired locations.

Figure 7.3 A "nanorobot" flowing inside a blood vessel finds an infected cell. The nanorobot attaches to the cell and projects a drug to repair or destroy the infected cell.

Some bio-nanorobots can be conceived as able to manufacture additional elements and various structures. There may also be robots that not only perform physical labor, but also sense the environment and react accordingly. There is no doubt that biomedical applications will be both a driving force and a beneficiary of these developments.

7.2.2 Brief Review of Biomolecular Machines

While the majority of the prior research in this field has largely focused on biomolecular motors, several other nano components such as sensors and even assemblies of components in the form of mechanisms have been studied. In the macroscopic world, what we understand by a "motor" is a machine capable of imparting motion associated by the conversion of energy. Biomolecular motors have attracted a lot of attention recently because: (1) they operate at high efficiency, (2) some could be self-replicating and hence cheaper in mass usage, and (3) they are readily available in nature (Boyer, 1998). A number of enzymes function as nano-scale biological motors, such as kinesin (Block, 1998; Schnitzer and Block, 1997), RNA polymerase (Wang et al., 1998), myosin (Kitamura et al., 1999), and adenosine triphosphate (ATP) synthase, function as nano-scale biological motors (Montemagno and Bachand, 1999; Bachand and Montemagno, 2000; Soong et al., 2000; Noji et al., 1997; Yasuda et al., 1998; Walker, 1998).

Component	MacroRobots	Bio-Nano Robots
Structural Elements- Links	Metal, Plastic Polymer	DNA 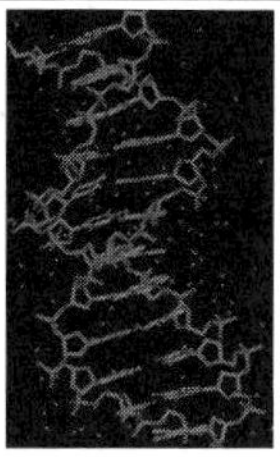 [PDB file:119D] Nanotubes
Joints	Metal, Plastic Polymer material *Revolute joints* *Prismatic joints* *Spherical joints* *Cylindrical joints*	*DNA* hinge *Molecular* bonds, Synthetic joints
Actuators	Electric motors, Pneumatic motors, Hydraulic motors, Smart material-based actuators	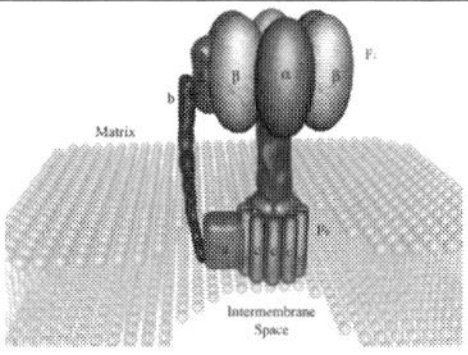 ATPase protein flagella motors, DNA actuators, Viral protein motors etc.
Transmission Elements	Springs (Metal, Polyvinyl) Bearings Gears	β Sheets Molecular camshaft design Smith SS (2001). United States Patent No. 6,200,782 13 March 2001.
Sensors	Light sensors, force sensors, position sensors, temperature sensors	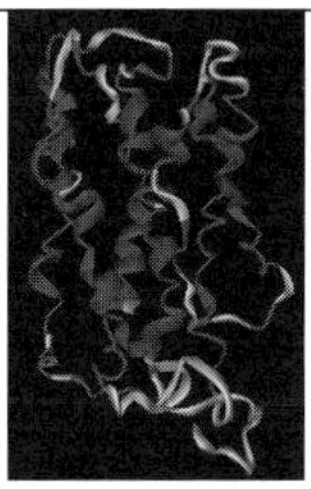 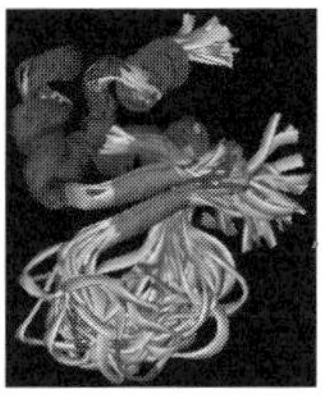 Rhodopsin [PDB file-1JFP] Heat Shock Factor [PDB file–3HSF]

Figure 7.4   Macro- and bio-nano-equivalence of robot components.

7.2.2.1 The ATPase Motor

One of the most abundant rotary motors found in life forms is F_0F_1 ATP synthase, commonly known as the "ATPase motor." Oxidative phosphorylation was demonstrated over 50 years ago as an important process by which our bodies capture energy from the food we eat. The mechanism of this process was not known until 1997, when Boyer and Walker described the key role that ATP plays in the process (Boyer, 1998; Walker, 1998). Noji et al. published the structural and performance data of the ATPase motor in 1997 (Noji et al., 1997; Yasuda et al., 1998). According to this study, the γ-subunit, which is about 1 nm in diameter, rotates inside the F_1 subunit, which is about 5 nm in diameter, to produce approximately 40 pN-nm of rotary torque. Montemagno and his group were the first to indicate that the rotation of the γ-subunit of the ATPase motor could be mechanically useful based on fabricated nanomechanical inorganic devices, which could be compatible with the force

Figure 7.5 The biological elements will be used to fabricate robotic systems. A vision of a nano-organism: carbon nanotubes (CNT) form the main body; peptide limbs can be used for locomotion and object manipulation, a biomolecular motor located at the head can propel the device in various environments.

production and dimensions of the molecular motors (Montemagno and Bachand, 1999; Bachand and Montemagno, 2000; Soong et al., 2000). Frasch's group is currently studying the binding of metals to amino acids of the motor protein. These experiments are providing new insights into the means by which the energy obtained from the hydrolysis of ATP can be converted into the physical action of pumping a proton in a unilateral direction (Frasch, 2000).

7.2.2.2 Kinesin and Myosin

Motor proteins are tiny vehicles that transport molecular cargoes within cells. These minute cellular machines exist in three families: the kinesins, the myosins, and the dyneins (Farrell et al., 2002). Conventional kinesin was found to be a highly processive motor that could take several hundred steps on a microtubule without detaching (Block et al., 1990; Howard et al., 1989), whereas muscle myosin has been shown to execute a single "stroke" and then dissociate (Finer et al., 1994). A detailed analysis and modeling of these motors has been done (Vale and Milligan, 2000). Hackney's group has concentrated upon the usage of ATP energy by motors like kinesin, myosin, dynenin, and related motor families (Hackney, 1996). Unger's group is currently working towards developing a microtubule–kinesin system as a biological linear-motoric actuator. Their work is aimed at producing force multiplication by parallel action of numerous single driving units as well as a more efficient means for system control (Bohm et al., 1997). Other researchers have discovered a new member of the myosin-V family (Myo5c) and have implicated this myosin in the transport of a specific membrane compartment (Mehta et al., 1999). The role of ATP hydrolysis in kinesin motility has also been recently described (Farrell et al., 2002).

7.2.2.3 The Flagella Motors

Escherichia coli and similar organisms are equipped with a set of rotary motors only 45 nm in diameter. Each motor drives a long, thin, helical filament that extends several cell body lengths out into the external medium. In addition to rotary engines and propellers, *E. coli*'s standard accessories include particle counters, rate meters, and gearboxes and thus have been described as a nanotechnologist's dream (Berg, 2000). Berg developed one of the earliest models for the rotary motor (Berg, 1974). Improved models came in 1992 (Ueno et al., 1992, 1994). Flagella motor analysis coupled to real-time computer assisted analysis of motion has also been performed (Khan et al., 1998). Researchers in Japan have applied crystallographic studies in order to understand the molecular structure of flagella motors as well as that of kinesin (Namba and Vonderviszt, 1997). Finally, Hess' group is attempting to build a nano-scale train system, complete with tracks, loading docks and a control system. Since motor proteins are a thousand times smaller than any man-made motor, they aim to utilize them in a synthetic environment as engines powering the nanotrains (Hess and Vogel, 2001).

7.2.2.4 *Other Motors and Mechanisms*

In addition to work on naturally existing motors, considerable effort is also being applied to develop synthetic molecular motors. The structure of the ATP synthase, a rod rotating inside a static wheel, suggests the use of rotaxanes as potential artificial models for natural motors (Harada, 2001). Rotaxanes are organic compounds consisting of a dumbbell-shaped component that incorporates one or more recognition sites in its rod section and is terminated by bulky "stoppers," encircled by one or more ring components. The possibility of manufacturing specific forms of rotaxane and creating molecular motors capable of guided rotary motion and the possibility of fueling such a motor by light, electrons, and chemical energy have been proposed (Schalley et al., 2001).

Schemes for using pseudorotaxanes, rotaxanes, and catenanes as molecular switches to perform chemical, electrochemical, and photochemical switching and controllable molecular shuttles have also been proposed recently (Balzani et al., 1998). Molecular shuttles have been reported using α-cyclodextrin — a parent of rotaxanes and catenanes (Harada, 2001). A light-driven monodirectional rotor made of helical alkene, with rotation around a central carbon–carbon covalent bond due to chirality has been reported (Koumura et al., 1999). Another simple way to convert chemical energy into mechanical motion in a controlled fashion is by using a metal ion which can be translocated reversibly between two organic compartments with the change of its ionization state, controllable by redox reaction or pH change (Amendola et al., 2001). Motility of unicellular organisms like vortecellids reminds us of energy storage and release by mechanical springs on a macromolecular scale. Spring-like action has been observed in sperm cells of certain marine invertebrates during fertilization. Springs and supramolecular ratchets by actin polymerization have yet to be built *in vitro*, but they theoretically can be generalized, as recently demonstrated (Mahadevan and Matsudaira, 2000).

7.2.2.5 *DNA-Based Molecular Nanomachines, Joints, and Actuators*

Several researchers are exploring the use of DNA in nano-scale mechanisms. DNA is small, relatively simple, and homogeneous and its structure and function is well understood. The predictable self-assembling nature of the double helix makes it an attractive candidate for engineered nanostructures. This property has been exploited to build several complex geometric structures, including knots, cubes, and various polyhedra (Seeman, 1998). Mathematical analyses of the elastic structure of DNA using energy minimization methods have been performed to examine its molecular stability, wherein short DNA strands were treated as an elastic rod (Tobias et al., 2000). Initial experiments on DNA visualization and manipulation using mechanical, electrical, and chemical means have been underway for a decade (Yuqiu et al., 1992; Hu et al., 2002). A dynamic device providing atomic displacements of 2–6 nm was proposed in Mao et al. (1999), wherein the chemically induced transition between the B and Z DNA morphologies acts as a moving nano-scale device. A method for localized element-specific motion control was seen in the reversible transition between four stranded topoisomeric DNA motifs (PX and JX2) thereby producing rotary motion (Yan et al., 2002). A very important, though simple DNA machine that resembles a pair of tweezers has been successfully created, whose actuation (opening and closing) is also fueled by adding additional DNA fuel strands (Yurke et al., 2000).

7.2.3 Nanosensors

The technology of nanosensing is also under development. For example, silicon probes with single walled carbon nanotube (CNT) tips are being developed (MIT Media Laboratory Nanoscale Sensing, http://www.media.mit.edu/nanoscale/). For sensing certain analytes, genetically engineered versions of pore-forming proteins like *Staphylococcus aureus* α-hemolysin are also being studied Braha et al., 1997. Light sensors could be made using certain photoreceptive polypeptides containing azobenzene or spyropyran units as they respond to light or dark environmental conditions by undergoing conformational change, for example, transition from random coil to a α-helix (Pieroni et al., 2001). An optical DNA biosensor platform has been reported using etched optical fiber bundles filled with oligonucleotide-functionalized microsphere probes (Ferguson et al., 1996). Finally, work is in progress to develop sensors for brain implantation, which would foretell the development of a stroke and be useful for perioperative online monitoring during coronary by-pass surgery (Manning and McNeil, 2001).

In addition to many of the examples mentioned above which generally correspond to one degree of freedom (DOF) rotary actuators, there are many other machine elements, the functional capabilities of which have not yet been represented by biomolecular elements. In addition, the assembly of different molecules in a multi-degree of freedom machine or the formation of hybrid systems composed of biomolecules and synthetic nonorganic elements has not yet been explored. In this context, our long term goal is to identify novel biomolecules that can be used as different types of machine components and to assemble them into controlled multi-degree of freedom systems using organic and synthetic nonorganic parts.

7.3 DESIGN AND CONTROL PHILOSOPHIES FOR NANOROBOTIC SYSTEMS

The design of nanorobotic systems requires the use of information from a vast variety of sciences ranging from quantum molecular dynamics to kinematic analysis. In this chapter we assume that the components of a nanorobot are made of biological components, such as proteins and DNA strings. So far, no particular guideline or a prescribed manner that details the methodology of designing a bio-nanorobot exists. There are many complexities that are associated with using biocomponents (such as protein folding and presence of aqueous medium), but the advantages of using these are also quite considerable. These biocomponents offer immense variety and functionality at a scale where creating a man-made material with such capabilities would be extremely difficult. These biocomponents have been perfected by nature through millions of years of evolution and hence these are very accurate and efficient. As noted in the review section on Molecular Machines, F_1-ATPase is known to work at efficiencies which are close to 100%. Such efficiencies, variety, and form are not existent in any form of material found today. Another significant advantage in protein-based bio-nano components is the development and refinement over the last 30 years of tools and techniques enabling researchers to mutate proteins in almost any way imaginable. These mutations can consist of anything from simple amino acid side-chain swapping, amino acid insertions or deletions, incorporation of nonnatural amino acids, and even the combination of unrelated peptide domains into whole new structures. An excellent example of this approach is the use of zinc to control F_1-ATPase, which is able to rotate a nanopropeller in the presence of ATP. A computational algorithm (Hellinga and Richards, 1991) was used to determine the mutations necessary to engineer an allosteric zinc-binding site into the F_1-ATPase using site-directed mutagenesis. The mutant F_1-ATPase would rotate an actin filament in the presence of ATP with average torque of 34 pNm. This rotation could be stopped with the addition of zinc, and restored with the addition of a chelator to remove the zinc from the allosteric binding site (Liu et al., 2002). This type of approach can be used for the improvement of other protein-based nano components.

These biocomponents seem to be a very logical choice for designing nanorobots. In addition, since some of the core applications of nanorobots are in the medical field, using biocomponents for these applications seems to be a good choice as they both offer efficiency and variety of functionality. This idea is clearly inspired by nature's construction of complex organisms such as bacteria and viruses which are capable of movement, sensing, and organized control. Hence, our scope would be limited to the usage of these biocomponents in the construction of bio-nanorobotics. A roadmap is proposed which details the main steps towards the design and development of bio-nanorobots.

7.3.1 The Roadmap

The roadmap for the development of bio-nanorobotic systems for future applications (medical, space, and military) is shown in Figure 7.6. The roadmap progresses through the following main steps:

Step 1: Bio-Nano Components

Development of bio-nano components from biological systems is the first step towards the design and development of an advanced bio-nanorobot, which could be used for future applications (see

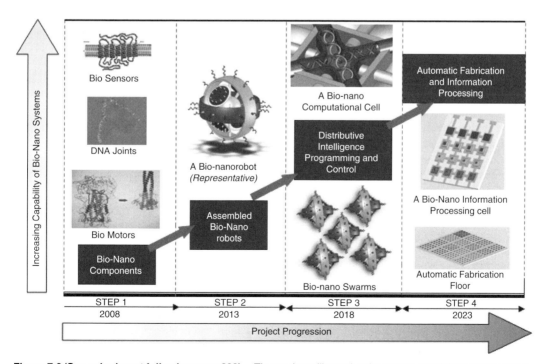

Figure 7.6 (See color insert following page 302)   The roadmap illustrating the system capability targeted as the project progresses.

(a)   (b)   (c)

Figure 7.7 (Step 1) Understanding of basic biological components and controlling their functions as robotic components. Examples are: (a) DNA which may be used in a variety of ways such as a structural element and a power source; (b) hemagglutinin virus may be used as a motor; (c) bacteriorhodopsin could be used as a sensor or a power source.

Figure 7.7). Since the planned systems and devices will be composed of these components, we must have a sound understanding of how these behave and how they could be controlled. From the simple elements such as structural links to more advanced concepts such as motors, each component must be carefully studied and possibly manipulated to understand the functional limits of each one of them. DNA and carbon nanotubes are being fabricated into various shapes, enabling possibilities of constructing newer and complex devices. These nanostructures are potential candidates for integrating and housing the bio-nano components within them. Proteins such as *rhodopsin* and *bacteriorhodopsin* are a few examples of such bio-nano components. Both of these proteins are naturally found in biological systems as light sensors. They can essentially be used as solar collectors to gather abundant energy from the sun. This energy could either be harvested (in terms of proton motive force) for later use or could be consumed immediately by other components, such as the ATP synthase nano rotary motor. The initial work is intended to be on the biosensors, such as heat shock factor. These sensors will form an integral part of the proposed bionano assemblies, where these will be integrated within a nanostructure and will get activated as programmed, for gathering the required information at the nano-scale. Tools and techniques from *molecular modeling* and *protein engineering* will be used to design these modular components.

Step 2: Assembled Bio-Nanorobots

The next step involves the assembly of functionally stable bio-nano components into complex assemblies. Some examples of such complex assemblies are shown in Figure 7.8. Figure 7.8A shows a bio-nanorobot with its "*feet*" made of helical peptides and its body of CNT, while the power unit is a biomolecular motor. Figure 7.8B shows a conceptual representation of *modular organization* of a bio-nanorobot. The modular organization defines the hierarchy rules and spatial arrangements of various modules of the bio-nanorobots, such as the inner core (the brain or energy source for the robot), the actuation unit, the sensory unit, and the signaling and information processing unit. By the beginning of this phase, a "*library of bio-nano components*" will be developed, which will include various categories, such as actuation, energy source, sensory, signaling, etc. Thereafter, one will be able to design and develop such bionanosystems that will have enhanced mobile characteristics and will be able to transport themselves as well as other objects to desired locations at nano-scale. Furthermore, some bio-nanorobots need to assemble various biocomponents and nanostructures, including *in situ* fabrication sites and storage areas; others will manipulate existing structures and maintain them. There will also be robots that not only perform physical labor, but also sense the environment and react accordingly. There will be systems that will sense an oxygen deprivation and stimulate other components to generate oxygen creating an environment with stable homoeostasis.

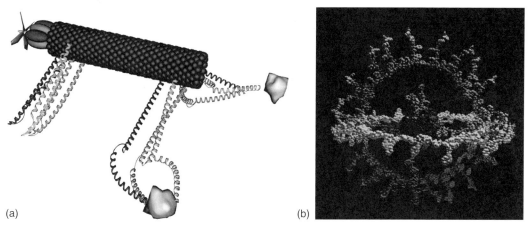

Figure 7.8 (Step 2)   (a) The bio-nano components will be used to fabricate complex bio-robotic systems. A vision of a nanorobot: carbon nanotubes (CNT) form the main body; peptide limbs can be used for locomotion and object manipulation and the biomolecular motor located at the head can propel the device in various environments. (b) Modular organization concept for the bio-nanorobots. Spatial arrangements of the various modules of the robots are shown. A single bio-nanorobot will have actuation, sensory, and information processing capabilities.

Step 3: Distributive Intelligence, Programming and Control

With the individual bio-nanorobots capable of basic functions, we would now need to develop concepts that would enable them to collaborate with one another to develop "colonies" of similar nanorobots. This design step could lay the foundation towards the concept of *bionanoswarms* (distributive bio-nanorobots) (see Figure 7.9A). Here work has to be done towards the control and programming of such swarms. This will evolve concepts like distributive intelligence in the context of bio-nanorobots. Designing swarms of bio-nanorobots capable of carrying out complex tasks and capable of computing and collaborating amongst them will be the focus of this step. Therefore, the basic computational architectures need to be developed and rules need to be evolved for the bio-nanorobots to make intended decisions at the nano-scale.

To establish an interface with the macro-world, the computers and electronic hardware have to be designed as well. Figure 7.10 shows the overall electronic communication architecture. Humans should be able to control and monitor the behavior and action of these swarms. This means that basic computational capabilities of the swarms will need to be developed. A representative computational bionanocell, which will be deployed within a bio-nanorobot, is shown in Figure 7.9B. This basic computational cell will initially be designed for data retrieval and storage at the nano-scale. This capability will enable us to program (within certain degrees of freedom) the swarm behavior in the bio-nanorobots. We will further be able to get their sensory data (from nano-world) back to the macro-world through these storage devices. This programming capability would control the bio-nanorobotics system and hence is very important.

Step 4: Automatic Fabrication and Information Processing Machines

Specialized bio-nano robotic swarms would need to be designed to carry out complex missions, such as sensing, signaling, and data storage. The next step in nanorobotic designing would see the emergence of automatic fabrication methodologies (see Figure 7.11, *which only shows the floor concept of assembling bio-nanorobots*) of such bio-nanorobots *in vivo* and *in vitro*. Capability of information processing will be a key consideration of this step. This would enable bio-swarms to have capability of *adjusting* based on their interacting environment they will be subjected to. These

(a)

(b)

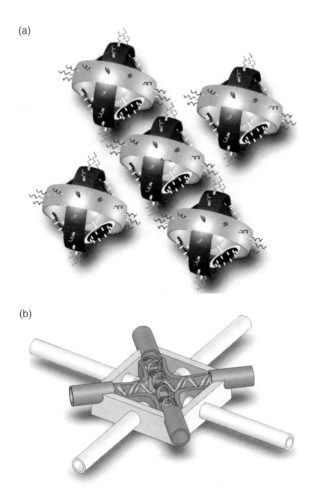

Figure 7.9 (Step 3)   (a) Basic bio-nanorobot forming a small swarm of five robots. The spatial arrangement of the individual bio-nanorobot will define the arrangement of the swarm. These swarms could be re-programmed to form bindings with various other types of robots. The number of robots making a swarm will be determined by the mission. Such swarms will attach additional bio-nanorobots at run time and replace any non-functional ones. (b) A basic bio-nano computational cell. This will be based on one of the properties of the biomolecules, which is "reversibility."

swarms could be programmed for more than one energy source and hence would have an ability to perform in an alternate environment. Energy management, self-repairing, and evolving will be some of the characteristics of these swarms.

7.3.2   Design Architecture for the Bio-Nanorobotic Systems

(a) Modular Organization: Modular organization defines the fundamental rule and hierarchy for constructing a bio-nano robotic system. Such construction is performed through *stable integration* (energetically in the most stable state) of the individual "*bio-modules or components*", which constitute the bio-nanorobot. For example, if the entity **ABCD** defines a bio-nanorobot having some *functional specificity* (as per the Capability Matrix defined in Table 7.1) then A, B, C, and D are said to be the basic bio-modules defining it. The basic construction will be based on the techniques of molecular modeling with emphasis on principles such as *energy minimization* on the hypersurfaces

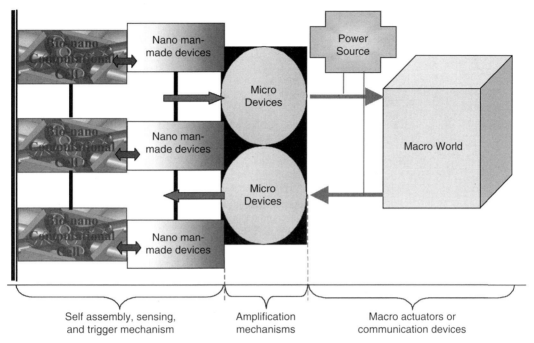

Figure 7.10 (See color insert following page 302) Feedback path from nano- to macro-world route.

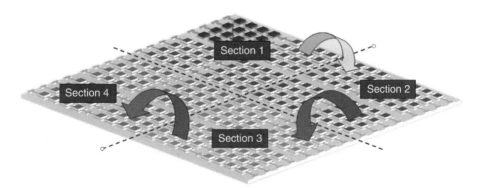

Figure 7.11 (Step 4) An automatic fabrication floor layout. Different colors represent different functions in automatic fabrication mechanisms. The arrows indicate the flow of components on the floor layout. Section 1 → Basic stimuli storage — control expression; Section 2 → Biomolecular component manufacturing (actuator or sensor); Section 3 → Linking of bio-nano components; Section 4 → Fabrication of bio-nanorobots (assemblage of linked bio-nano components).

of the bio-modules, *hybrid quantum-mechanical and molecular mechanical* methods, *empirical force field* methods, and *maximum entropy production* in least time.

Modular organization also enables the bio-nanorobots with capabilities, such as organizing into *swarms*, a feature, which is extremely desirable for various applications. Figure 7.12A and Figure 7.12B show the conceptual representation of modular organization. Figure 7.12C shows a more realistic scenario in which all the modules are defined in some particular spatial arrangements based on their functionality and structure. A particular module could consist of other group of modules, just like a fractal structure (defined as *fractal modularity*). The concept of *bionanocode* has been devised, which basically describes the unique functionality of a bio-nano component in

Table 7.1   Defining the Capability Matrix for the Bio-Modules

Functionality	Bionanocode	Capabilities Targeted	General Applications
Energy storage and carrier	E	Ability to store energy from various sources such as solar and chemical for future use and for its own working	Supplies the energy required for the working of all the bio-chemical mechanisms of the proposed bio-nanorobotic systems
Mechanical	M	Ability to precisely move and orient other molecules or modules at nano-scale — includes the ability to mechanically bind to various target objects and carry them to desired locations	1. Carry moities and deliver them to the precise locations in correct orientations 2. Move micro-world objects with nano-precision
Input sensing	S	Sensing capabilities in various domains such as chemical, mechanical, visual, auditory, electrical, and magnetic	Evaluation and discovery of target locations based on either chemical properties, temperature, or other characteristics
Signaling	G	Ability to amplify the sensory data and communicate with bio-systems or with the micro-controllers and ability to identify their locations through various trigger mechanisms such as fluorescence	Imaging for medical applications or for imaging changes in nanostructures
Information storage	F	Ability to store information collected by the sensory element — behave similar to a read–write mechanism in computer field	1. Store the sensory data for future signaling or usage 2. Read the stored data to carry out programmed functions 3. Back bone for the sensory bio-module 4. Store nano-world phenomenon currently not observed with ease
Swarm behavior	W	Exhibit binding capabilities with "similar" bio-nanorobots so as to perform distributive sensing, intelligence, and action (energy storage) functions	All the tasks to be performed by the bio-nanorobots will be planned and programmed keeping in mind the swarm behavior and capabilities
Information processing	I	Capability of following algorithms (Turing equivalent)	Programmable
Replication	R	Replicate themselves depending on the situation and requirement	Replicate by assembling raw components into nanorobots, and programming newly-made robot to form swarms that form automated fabricators consistent with the Foresight Guidelines for safe replicator development (Foresight Institute, 2000)

terms of alphabetic codes. Each bionanocode represents a particular module defining the structure of the bio-nanorobot. For instance, a code like **E-M-S** will describe a bio-nanorobot having capabilities of energy storage, mechanical actuation, and signaling at the nano-scale. Such representations will help in general classifications and representative mathematics of bio-nanorobots and their swarms. Table 7.1 summarizes the proposed capabilities of the bio-modules along with their targeted general applications. The bio-nano code **EIWR ‖ M ‖ S ‖ FG** representing the bio-nano system shown in Figure 7.12B which could be decoded as shown in Figure 7.13.

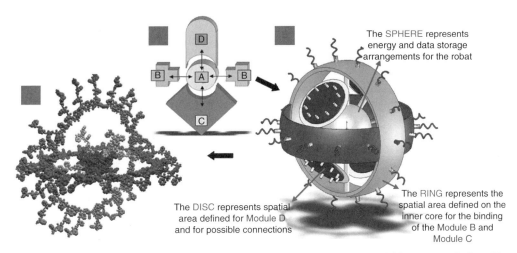

Figure 7.12 (See color insert following page 302)   (A) A bio-nano robotic entity "ABCD", where A, B, C, and D are the various bio-modules constituting the bio-nanorobot. In our case these bio-modules will be set of stable configurations of various proteins and DNAs. (B) A bio-nanorobot (representative), as a result of the concept of modular organization. All the modules will be integrated in such a way so as to preserve the basic behavior (of self-assembly and self-organization) of the biocomponents at all the hierarchies. The number of modules employed is not limited to four or any number. It is a function of the various capabilities required for a particular mission. (C) A molecular representation of the figure in part B. It shows the red core and green and blue sensory and actuation bio-modules.

(b) The Universal Template — BioNano STEM System: The modular construction concept involves designing a universal template for bionanosystems, which could be "programmed and grown" into any possible bionanocoded system. This concept mimics the embryonic stem cells found in the human beings, that are a kind of primitive human cells which give rise to all other specialized tissues found in a human foetus and ultimately to all the three trillion cells in an adult human body. Our BioNano STEM system will act in a similar way. This universal growth template will be constituted of some basic bionanocodes, which will define the BioNano STEM system. This STEM system will be designed in a manner that could enable it to be programmed at runtime to any other required bio-module. Figure 7.14 shows one such variant of

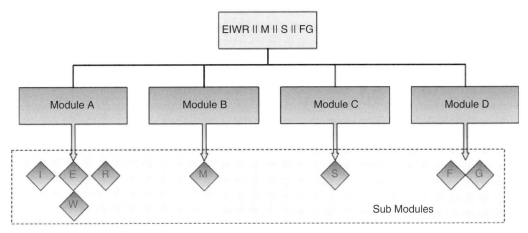

Figure 7.13   Showing the bio-nano code and the fractal modularity principle. The letter symbols have the values specified in Table 7.1. The "‖" symbol integrates the various bio-modules and collectively represents a higher order module or a bio-nanorobot.

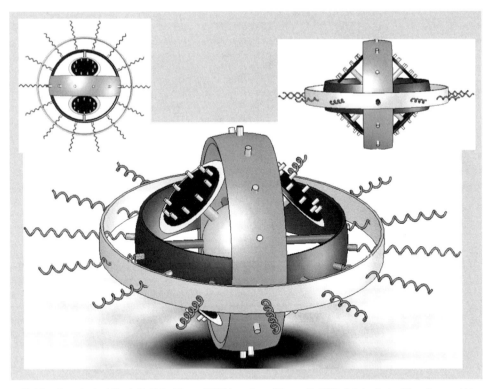

Figure 7.14   A variant of the initial Bio-Nano STEM system (Figure 7.12B), fabricated with enhanced bio-nano code S, which defines it as a bio-nanorobot having enhanced sensory capabilities. The other features could be either suppressed or enhanced depending upon the requirement at hand. The main advantage of using Bio-Nano STEM system is that we could at runtime decide which particular type of bio-nanorobots we require for a given situation. The suppression ability of the Bio-Nano STEM systems is due to the property of "Reversibility" of the biocomponents found in living systems.

the BioNano STEM system, having the bionanocode: EIWR ‖ M ‖ S ‖ FG and having enhanced sensory abilities.

7.4   SELF-REPLICATION — MIMETICS: A NOVEL PROPERTY OF LIVING SYSTEMS

7.4.1   Significance

Mimetics of self-replication, as exhibited by nature, would influence any nano-level application. We would need an army of nanorobots (or living systems), mass-produced via techniques of self-replication (or life), to carry out many meaningful tasks at the nano-scale. These applications involving nanorobots demand that these machines are manufactured in millions or billions and in a timeframe reasonable for a particular application. One of nature's noblest properties is that of life. It is how nature progresses through its environment ever adapting and evolving. Although philosophically what life means and what constitutes it is not very clear, what is clear is how nature propagates itself with time and survives, every day and every moment! This is one attribute of nature which is of prime importance to us as researchers of science and engineering, and which, if understood, would bring a unique revolution that in a way would change the course of our lives.

The concept of self-replicating mechanisms (SRM) or mimetics of life is not new (Freitas and Merkle, 2004). We are perfect examples of these kinds of systems. We are wounded and our internal mechanisms heal it with some differences in some cases. Taking the example of the wound and its healing process, we move ahead and try to analyze how we can achieve such behaviors in the mechanisms that we design. At the core of the concept of self-replication lies the basic material (DNA/RNA) which undergoes such activity. Though we hardly know why these materials behave in this fashion, what we do know is how they behave and this provides the stepping stone for us to move ahead.

Before looking at some of the possible designs, a brief discussion on the application of such mechanisms is necessary. Why do at all we need living systems or self-replications? Where would they be best suited?

7.4.2 Applications

(a) Consider that our application depends upon a particular part — mechanical or electrical, or any other physical, biological, or chemical element — which fails or starts developing problems. We need self-rectifying mechanisms within our application to detect the problem and rectify it. It is similar to our example of the wound. We can think of many applications where we would desire such behavior. Given some initial material feedstock, it would be desired that the self-replicating mechanisms would rectify the problems. Having said that, we can classify the self-replicating mechanisms in the order we classify our main mechanisms or machines:

- mechanical self-replicating mechanisms;
- electrical self-replicating mechanisms;
- chemical or biological self-replicating mechanisms.

There could be other classifications as well and numerous other examples following the lines of our wound example, can be thought of. It just depends on our imagination. For example, suppose that we build some SRM which mimics the living system. Its function is to detect the crucial defect in a mechanical element and then mend that defect. If we are able to devise such an application, it could significantly enhance the life and performance of the system. The system in this example would be designed and constructed to work at nano-scale, and therefore it would have an ability to detect the slightest of defects and start working towards rectifying it.

(b) Remote Applications would also benefit from SRM systems. Maintaining remote applications requires constant human interaction. If these systems mimic the coded logic and goals of the living system, then it would be able to perform optimally with minimal human interventions. For example, deep space explorations would require circuits, machines, and equipments to adjust and adapt with time and as per the conditions they would be subjected to.

(c) Applications at the nano-scale. This category of applications would be most influenced by our biomimetic systems because they could lay the foundations of nanodevices that have the capability of manipulating molecular matter.

In the following section we try to define some of the guidelines and working philosophies for designing and fabricating such replicating systems. The details are the thoughts and ideas of the authors and are not verified or supported by experimental facts.

7.4.3 The Design of Life-Mimetic Systems

The design of life-mimetic systems requires new innovative materials to be designed that behave in the same fashion as that designed by nature. These new materials are termed "intrinsic materials" from here on.

7.4.3.1 Intrinsic Material

The unique arrangement of the constituent atoms of intrinsic materials would give rise to:

- unique potential field surface around them;
- unique charge distribution;
- unique internal energy gradients.

It is through these internal energy gradients that two particular intrinsic materials would interact with each other. Hence the behavior of the intrinsic material would be a direct function of its internal energy gradients.

7.4.3.2 Interaction Laws

The final objective of any interacting intrinsic materials would be to achieve the intrinsic balance of the resulting system (termed *self-balancing*). This would translate to achieving minimum energy gradients in all directions for all interacting materials. The final system would then be defined by its new achieved internal energy gradients. These intrinsic energy gradients would also be influenced by the external fields.

7.4.3.3 Self-Balancing

It implies that the materials considered would tend to align with its intrinsic energy gradients and would try to minimize the formed unbalance. The classical instance of self-replication via energy-minimized self-assembly was first demonstrated in the late 1950s. The canonical example of this approach is called the Penrose Blocks (Penrose and Penrose, 1957; Penrose, 1958). The unbalance and the property of self-balancing are similar in essence to what is postulated by the law of entropy. This concept of self-balancing is motivated from the law of maximum entropy production according to which a system follows a path which minimizes the potential or maximizes entropy at the maximum rate (Archives of Science, 2001).

7.4.3.4 Growth and the Reproductive Limit

An intrinsic material would have a property of growth (an important variable for replication). This property of growth only occurs when the system is provided with some energy maybe in the form of additional intrinsic material or external gradients. Growth cannot happen in isolation. This implies that in the process of self-balancing, it is possible that the particular configuration of the intrinsic material is stable up to a particular level. This level would be governed by the strength of the potential gradients for that intrinsic material and the extrinsic gradients. Therefore, the growth implies that the intrinsic material can achieve higher state by not disturbing its self-balance or increasing it further. But this growth can only be achieved to a particular extent; beyond it, it tends to disintegrate by following the paths defined by the laws of maximum entropy production. And this particular limit of growth is termed Reproductive Limit.

7.4.3.5 Self-Filtering and Self-Healing

The concept of replication further demands that the materials thus designed should exhibit the property of self-filtering and self-healing. Self-filtering implies that the material involved in the systems exhibiting self-replication will not allow any kind of growth pattern but only a particular one. This particular growth pattern (which inherently depends on the interactions of the components) will determine which material it binds to and to which it does not. Hence such systems will only interact with certain intrinsic materials.

In the process of this growth, a realignment of the intrinsic material occurs. As the addition of further intrinsic material takes place, the whole system tries to realign itself towards the most stable state and in the fastest possible time. This realignment goes on with the self-filtering process. If at any incremental stage system does not find the kind and type of material it is looking for, it does not realign and rejects that particular material and does not grow. By rejection, it might imply that either the system does not align with the material or marks it as an unstable configuration and seeks for the opportunity to replace it immediately. Hence it acts as a self-healer.

7.4.4 Self-Replication — A Thought Experiment

Let us try to replicate a system, based on the above properties. Consider a system A, consisting of some intrinsic materials (Figure 7.15). The system A is completely defined by the way these intrinsic materials are associated and aligned within it. According to the above properties of self-replicating mechanisms, it is observed that the intrinsic materials of the system could be broken down into further fundamental intrinsic materials.

Any stable configuration of the individual intrinsic material is in sync with the property of self-balancing. Now when a particular intrinsic material (say 1) gets in an interactive distance of another intrinsic material (say 2), then these two intrinsic materials try to form another subsystem A1 within the super system A, following the property of self-balancing. These two intrinsic materials combined will have some other function of intrinsic energy gradient and could be the sum of the individual intrinsic energy gradients of the intrinsic materials and the applied external fields.

Now this argument could be extended to the situation when the third intrinsic material (say 3) comes into the picture. This intrinsic material 3 would not only interact with intrinsic material 1 but also with intrinsic material 2. Finally, a system A comes into generation, because of self-balancing acts of these three intrinsic materials. The configuration they achieve becomes highly stable for that particular situation. Now let us introduce more *energy* to the system A. It would be in the form of introducing intrinsic materials or applying external gradients to the system A or both. Figure 7.16 explains the concept.

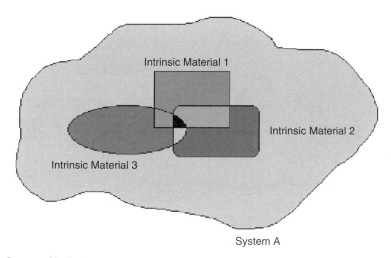

Figure 7.15   System of intrinsic materials — a self-replicative system A.

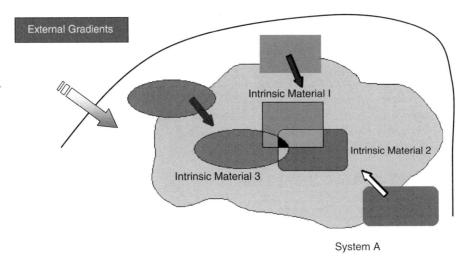

System A

Figuro 7.16 Energy being added to the self-replicative system A in the form of newer intrinsic material (1, 2, and 3) and external gradient (this external gradient is applied either to aid the interaction between the intrinsic materials or to impart a particular dynamics to the system for favorable environment for the interaction).

Here because of the process of self-filtering, copies of intrinsic material 1, 2, and 3 are introduced. The property of self-balancing comes into dominance and the systems try to adjust itself into the most stable state. As defined earlier, the initial state is the most stable state; following is what happens tozthe system A. Two subsystems within the main systems are made as shown in Figure 7.17. The alignment of *subsystems A1* and *A2* is similar to the one of the initial system, that is, *A*. Please note that such system is possible, because we can control the external parameters, namely,

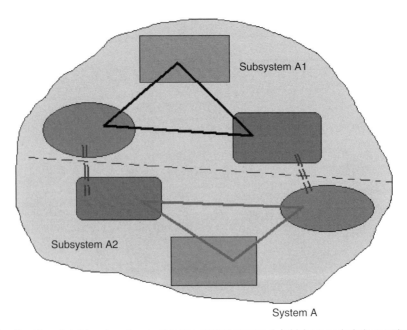

System A

Figure 7.17 Creation of stable subsystems within the original system A (which as a whole is marginally unstable under the external gradients and two independently stable subsystems). This step is the most crucial in the process of attaining a self-replicative super system. This demands a unique selection of such replicative intrinsic materials in the initial place, namely, 1, 2, and 3.

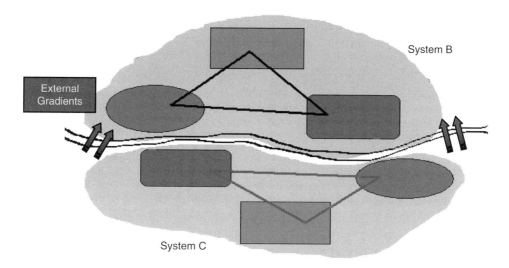

Figure 7.18   Replicating stage of the system A into system B and system C. Systems B and C which could be called the child systems are similar to system A in function and its configurations.

extrinsic gradients and the intrinsic material introduced. The triangles drawn in the figure above show the configuration of the intrinsic materials of *subsystems A1* and *A2*. The dotted lines depict the interaction between the old intrinsic materials and the new ones and the possible configuration that could be achieved. Now because the external gradients are still applicable a unique instability in the system occurs. The system tries to self-balance and in the process leads to its most stable configurations, which was its initial one (the initial configuration, system A). Figure 7.18 explains the concept.

In the end, the original *system A* replicates into *systems B* and *C*. Both these new systems have the same functionalities as defined by the original system A, because they have received the same configuration and the same intrinsic materials.

7.4.5 Design Parameters for Self-Replicating Systems

Following are the various design parameters that need to be considered while designing a self-replicating system.

7.4.5.1 Selection of Intrinsic Materials

This is, of course, the most important parameter in designing the desired system. The obvious choices would be biomaterials and chemicals found in the human body, which have exhibited self-replication. Their choice is mainly because of their availability and the fact that they themselves are the materials resulting from a replicative process. This does not limit the selection of other replicative materials. Also a lot of data on nature's biomaterials is available from the experiments performed in the field of biology and genetic engineering. The field of nanotechnology is the biggest area where the concept of self-replication system would be a success and the biomaterials could be managed at that scale.

7.4.5.2 Defining the External Gradient Parameters

It is extremely important to define the external gradient parameters within which the system needs to perform. Our choice of the intrinsic materials would be greatly impacted by their behavior. Also their sensitivities to these external gradients need to be calculated so as to fine tune the system.

7.4.5.3 Generating Stable Alignment and Internal Gradients

Selection of the appropriate intrinsic materials for our system implies that we need to also select the appropriate internal gradient functions and the alignment generated by these intrinsic materials. We need to calculate the most stable configuration for our system at no external gradient level and then fine tune our alignment as it is applied. Application of these external gradients could generate a situation where no stable configuration is possible within our operating conditions. This calls for adding some further intrinsic materials to the system, which would help us to get to the stable configuration (closure engineering) (Freitas and Merkle, 2004). This variation of the intrinsic gradient in this manner is termed as *intrinsic variational gradients* to distinguish it from the inherent intrinsic gradients generated due to the intrinsic materials.

The parameters mentioned above create the foundation for the development of mathematics for this field. To create any system with self-replicating mechanism we need to first find out its most stable state, then we need to calculate its behavior in the extrinsic gradients and then we need to excite it with energy and supply of intrinsic materials so that it replicates. Though these methodologies are not verified, further research in this area is being carried on by the authors and their collaborators.

7.5 CONCLUSIONS

Biomimetics and its principles would greatly influence the field of nanorobotics and nanotechnology. The way nature is designed and the way nature solves its problems is of great interest to us because they allow us to understand basic principles that would pave the way to practical nanotechnology.

The recent explosion of research in nanotechnology, combined with important discoveries in molecular biology, has created a new interest in bio-nanorobotic systems. The preliminary goal in this field is to use various biological elements — whose function at the cellular level results in a motion, force or signal — as nanorobotic components that perform the same function in response to the same stimuli — but in an artificial setting. This way proteins and DNA could act as motors, mechanical joints, transmission elements, or sensors. Assembled together, these components would form nanorobots with multiple degrees of freedom, with the ability to apply forces and manipulate objects at the nano-scale, and transfer information from the nano- to the macro-scale world.

The first research area is in determining the structure, behavior, and properties of basic bio-nano components such as proteins. Specific problems include the precise mechanisms involved in molecular motors like ATP Synthase, and of protein folding. The next step is combining these components into complex assemblies. Next concepts in control and communication in swarms need to be worked out. Again, we plan to follow nature's path, mimicking the various colonies of insects and animals, and transforming principles learned to our domain. Since it would require specialized colonies of nanorobots to accomplish particular tasks, the concepts of cooperative behavior and distributed intelligence need to be developed, possibly by using known hierarchical and other techniques.

Principles like self-replication are the ones of greatest importance for the field of nanorobotics. It is this life mimetics which will enable us to design and fabricate the future nanorobots having immense capabilities and potential. These would require innovative materials (intrinsic materials) and fabrication methodologies, with due regard to well-known manufacturing- and applications-related safety concerns. The safety issue is of paramount importance in this field for researchers and scientists. The proposed bio-nanorobots would be completely controlled molecular devices and are far from being dangerous to society. Though these devices would have many unique capabilities, which are not seen currently, they are harmful as projected in science fiction movies and books. There is an increasing need for educating the community about the exact nature of this research and its essential differences with the projections of the science fiction community.

REFERENCES

Archives of Science. (2001). All about entropy, the laws of thermodynamics, and order from disorder. http://www.entropylaw.com, (c) Copyright 2001.

Amendola V, Fabbrizzi L, Mangano C, Pallavicini P. (2001). Molecular machines based on metal ion translocation. *Acc. Chem. Res.* 34: 488–93.

Bachand GD, Montemagno CD. (2000). Constructing organic/inorganic NEMS devices powered by biomolecular motors. *Biomed. Microdev.* 2: 179–84.

Balzani V, Lopez MG, Stoddart JF. (1998). Molecular machines. *Acc. Chem. Res.* 31: 405–14.

Berg HC. (1974). Dynamic properties of bacterial flagellar motors. *Nature* 249: 77–9.

Berg HC. (2000). Motile behavior of bacteria. *Phys. Today* 53: 24–9.

Block SM. (1998). Kinesin: what gives? *Cell* 93: 5–8.

Block SM, Goldstein LS, Schnapp BJ. (1990). Bead movement by single kinesin molecules studied with optical tweezers. *Nature* 348: 348–52.

Bohm KJ, Steinmetzer P, Daniel A, Baum M, Vater W, Unger E. (1997) Kinesin-driven microtubule motility in the presence of alkaline-earth metal ions: indication for a calcium ion-dependent motility. *Cell Motil. Cytoskeleton* 37: 226–31.

Boyer PD. (1998). Energy, life and ATP (Nobel Lecture). *Angewandte Chemie International Edition* 37: 2296–307.

Braha O, Walker B, Cheley S, Kasianowicz JJ, Song L, Gouaux JE, Bayley H. (1997). Designed pores as components for biosensors. *Chem. Biol.* 4: 497–505.

Drexler EK. (1992). *Nanosystems: Molecular Machinery, Manufacturing and Computation*, John Wiley and Sons.

Farrell CM, Mackey AT, Klumpp LM, Gilbert SP. (2002). The role of ATP hydrolysis for kinesin processivity. *J. Biol. Chem.* 277: 17079–87.

Ferguson JA, Boles TC, Adams CP, Walt DR. (1996). A fiber-optic DNA biosensor microarray for the analysis of gene expression. *Nat. Biotechnol.* 14: 1681–4.

Finer JT, Simmons RM, Spudich JA. (1994). Single myosin molecule mechanics: piconewton forces and nanometre steps. *Nature* 368: 113–9.

Foresight Institute. (2000). Molecular Nanotechnology Guidelines: Draft Version 3.7, 4 June 2000.

Frasch WD. (2000). Vanadyl as a probe of the function of the F1-ATPase-Mg^{2+} cofactor. *J. Bioenergetics Biomembr.* 32: 2000.

Freitas Jr., RA. (1999). *Nanomedicine, Volume I: Basic Capabilities*, Landes Bioscience, Georgetown, Texas, 1999.

Freitas Jr., RA. (2003). *Nanomedicine, Volume IIA: Biocompatibility*, Landes Bioscience, Georgetown, Texas, 2003.

Freitas Jr., RA, Merkle. RC. (2004). *Kinematic Self-Replicating Machines*, Landes Bioscience, Georgetown, Texas; *http://www.MolecularAssembler.com/KSRM.htm*

Hackney DD. (1996). The kinetic cycles of myosin, kinesin, and dynein. *Annu. Rev. Physiol.* 58: 731–50.

Harada A. (2001). Cyclodextrin-based molecular machines. *Acc. Chem. Res.* 34 (16): 456–64.

Hellinga HW, Richards FM. (1991). Construction of new ligand binding sites in proteins of known structure. I. Computer-aided modeling of sites with pre-defined geometry. *J Mol. Biol.* 222: 763–85.

Hess H, Vogel V. (2001). Molecular shuttles based on motor proteins: active transport in synthetic environments. *J. Biotechnol.* 82: 67–85.

Howard J, Hudspeth AJ, Vale RD. (1989). Movement of microtubules by single kinesin molecules. *Nature* 342: 154–8.

Hu J, Zhang Y, Gao H, Li M, Hartman U. (2002). Artificial DNA patterns by mechanical nanomanipulation. *Nanoletters* 2: 55–7.

Khan S, Zhao R, Reese TS. (1998). Architectural features of the Salmonella typhimurium flagellar motor switch revealed by disrupted C-rings. *J. Struct. Biol.* 122: 311–9.

Kinosita K Jr., Yasuda R, Noji H, Adachi K (2000). A rotary molecular motor that can work at near 100% efficiency. *Phil. Trans. R. Soc. Lond. B* 355: 473–489.

Kitamura K, Tokunaga M, Iwane AH, Yanagida T. (1999). A single myosin head moves along an actin filament with regular steps of 5.3 nanometres. *Nature* 397: 129–34.

Koumura N, Zijlstra RW, van Delden RA, Harada N, Feringa BL. (1999). Light-driven monodirectional molecular rotor. *Nature* 401: 152–5.

Liu H, Schmidt JJ, Bachand GD, Rizk SS, Looger LL, Hellinga HW, Montemagno CD. (2002). Control of a biomolecular motor-powered nanodevice with an engineered chemical switch. *Nat. Mater.* 1: 173–7.

Mahadevan L, Matsudaira P. (2000). Motility powered by supramolecular springs and ratchets. *Science* 288: 95–100.

Manning P, McNeil C. (2001). Microfabricated Multi-Analyte Amperometric Sensors. http://nanocentre.ncl.ac.uk/

Mao C, Sun W, Shen Z, Seeman NC. (1999). A nanomechanical device based on the B–Z transition of DNA. *Nature* 397: 144–6.

Mehta AD, Rock RS, Rief M, Spudich JA, Mooseker MS, Cheney RE. (1999). Myosin-V is a processive actin-based motor. *Nature* 400: 590–3.

MIT Media Laboratory Nanoscale Sensing. http://www.media.mit.edu/nanoscale/

Montemagno CD, Bachand GD. (1999). Constructing nanomechanical devices powered by biomolecular motors. *Nanotechnology* 10: 225–331.

Namba K, Vonderviszt F. (1997). Molecular structure of bacterial flagellum. *Quart. Rev. Biophys.* 30(1): 1–65.

Noji H, Yasuda R, Yoshida M, Kinosita K, Jr. (1997). Direct observation of the rotation of F1-ATPase. *Nature* 386: 299–302.

PDB: 1JFP, Yeagle PL, Choi G, Albert AD. (2001). Studies on the structure of the G-protein-coupled receptor rhodopsin including the putative G-protein binding site in unactivated and activated forms. *Biochemistry* 40: 11932.

PDB: 119D, Leonard GA, Hunter WN. (1993). Crystal and molecular structure of d(CGTAGATCTACG) at 2.25 A resolution. *J. Mol. Biol.* 234: 198.

PDB: 3HSF, Damberger FF, Pelton JG, Liu C, Cho H, Harrison CJ, Nelson HCM, Wemmer DE. (1995). Refined solution structure and dynamics of the DNA-binding domain of the heat shock factor from *Kluyveromyces lactis. J. Mol. Biol.* 254: 704.

Penrose LS, Penrose R. (1957). A self-reproducing analogue. *Nature* 179: 1183.

Penrose LS. (1958). Mechanics of self-reproduction. *Ann. Hum. Genet.* 23: 59–72.

Pieroni O, Fissi A, Angelini N, Lenci F. (2001). Photoresponsive polypeptides. *Acc. Chem. Res.* 34: 9–17.

Rohl CA, Strauss CE, Misura KM, Baker D. (2004). Protein structure prediction using Rosetta. *Methods Enzymol.* 383: 66–93.

Schalley CA, Beizai K, Vogtle F. (2001). On the way to rotaxane-based molecular motors: studies in molecular mobility and topological chirality. *Acc. Chem. Res.* 34: 465–76.

Schnitzer MJ, Block SM. (1997). Kinesin hydrolyses one ATP per 8-nm step. *Nature* 388: 386–90.

Seeman NC. (1998). DNA nanotechnology: novel DNA constructions. *Annu. Rev. Biophys. Biomol. Struct.* 27: 225–48.

Soong RK, Bachand GD, Neves HP, Olkhovets AG, Craighead HG, Montemagno CD. (2000). Powering an inorganic nanodevice with a biomolecular motor. *Science* 290: 1555–8.

Smith SS. (2001). United States Patent No. 6,200,782, 13 March 2001.

Tobias I, Swigon D, Coleman BD. (2000). Elastic stability of DNA configurations. I. General theory. *Phys. Rev. E Stat. Phys. Plasmas Fluids Relat. Interdiscip. Topics* 61: 747–58.

Ueno T, Oosawa K, Aizawa S. (1992). M ring, S ring and proximal rod of the flagellar basal body of *Salmonella typhimurium* are composed of subunits of a single protein, FliF. *J. Mol. Biol.* 227: 672–7.

Ueno T, Oosawa K, Aizawa S. (1994). Domain structures of the MS ring component protein (FliF) of the flagellar basal body of *Salmonella typhimurium. J. Mol. Biol.* 236: 546–55.

Vale RD, Milligan RA. (2000). The way things move: looking under the hood of molecular motor proteins. *Science* 288: 88–95.

Wang MD, Schnitzer MJ, Yin H, Landick R, Gelles J, Block SM. (1998). Force and velocity measured for single molecules of RNA polymerase. *Science* 282: 902–7.

Walker JE. (1998). ATP Synthesis by Rotary Catalysis (Nobel Lecture). *Angewandte Chemie Intternational Edition* 37: 2308–19.

Yan H, Zhang X, Shen Z, Seeman NC. (2002). A robust DNA mechanical device controlled by hybridization topology. *Nature* 415: 62–5.

Yasuda R, Noji H, Kinosita K, Jr., Yoshida M. (1998). F1-ATPase is a highly efficient molecular motor that rotates with discrete 120 degree steps. *Cell* 93: 1117–24.

Yuqiu J, Juang CB, Keller D, Bustamante C, Beach D, et al. (1992). Mechanical, electrical, and chemical manipulation of single DNA molecules. *Nanotechnology* 3: 16–20.

Yurke B, Turberfield AJ, Mills AP, Jr., Simmel FC, Neumann JL. (2000). A DNA-fuelled molecular machine made of DNA. *Nature* 406: 605–8.

8

Molecular Design of Biological and Nano-Materials

Shuguang Zhang, Hidenori Yokoi, and Xiaojun Zhao

CONTENTS

8.1   DESIGN, SYNTHESIS, AND FABRICATION OF BIOLOGICAL AND NANO-MATERIALS AT THE MOLECULAR SCALE

Nature is the grandmaster when it comes to building extraordinary materials and molecular machines — one atom and one molecule at a time. Masterworks include such materials as minerals, well-ordered clays, and photonic crystals, and in the biological world, composites of inorganic or organic shells, pearls, corals, bones, teeth, wood, silk, horn, collagen, muscle fibers, and extracellular matrices. Multifunctional macromolecular assemblies in biology, such as hemoglobin, polymerases, ATP synthase, membrane channels, the splicesome, the proteosome, ribosomes, and photosystems are all essentially exquisitely designed molecular machines (Table 8.1).

Through billions of years of prebiotic molecular selection and evolution, Nature has produced a basic set of molecules that includes 20 amino acids, a few nucleotides, a dozen or so lipid molecules, and a few dozens of sugars as well as naturally modified building blocks or metabolic intermediates.

Table 8.1   What do they have in Common? Machines and Molecular Machines

Machines (Made by Humans)	Molecular Machines (Made by Nature)
Car, train, plane, space shuttle	Hemoglobin
Assembly lines	Ribosomes
Motors or generators	ATP synthases or photosystems
Train tracks	Actin filament network or intermediate filaments
Train controlling center	Centrosome
Digital database	Nucleosomes
Copy machines	Polymerases
Chain couplers	Ligases
Bulldozer or destroyer	Proteases or proteosomes
Mail-sorting machines	Protein sorting system
Electric fences	Membranes
Gates, keys, or passes	Ion channels, pumps, or receptors
Internet or World Wide Web	Neuron synapses

With these seemingly simple molecules, natural processes are capable of fashioning an enormously diverse range of fabrication units, which can further self-organize into refined structures, materials and molecular machines that not only have high precision, flexibility, and error correction, but also are self-sustaining and evolving.

Indeed, Nature shows a highly-flavored bottom-up design, building up molecular assemblies, bit by bit, more or less simultaneously on a well-defined scaffold. Take for example egg formation in oviparous animals. The fabrication of an egg involves not only the creation of the ovum, its various protective membranes, and accompanying nutritive materials (e.g., yolk) but also simultaneous synthesis of the eggshell from an extremely low concentration of calcium and other minerals, all in a very limited space. Oviparous animals synthesize eggshell against an enormous ionic and molecular concentration gradient due to the high levels of minerals at the site of eggshell assembly. Dental tissue formations face similar challenges not only when sharks repeatedly form new teeth, but also when humans form teeth during early childhood.

Nature accomplishes these feats effortlessly, yet recreating them in the laboratory presents an enormous challenge to the human engineer. The sophistication and success of natural bottom-up fabrication processes inspire our attempts to mimic these phenomena with the aim of creating new and varied structures, with novel utilities well beyond the gifts of Nature.

8.1.1   Two Distinctive and Complementary Fabrication Technologies

Two distinctive and complementary fabrication technologies are employed in the production of materials and tools. In the "top-down" approach, materials and tools are manufactured by stripping down an entity into its parts, for example, carving a boat from a tree trunk. This contrasts sharply with the "bottom-up" approach, in which materials and tools are assembled part by part to produce supra-structures, for example, building a ship using wooden strips (Figure 8.1) and complex architectures, construction of a building complex. The bottom-up approach is likely to become an integral part of materials manufacture in the coming decades. This approach requires a deep understanding of individual molecular building blocks, their structures, assembling properties, and dynamic behaviors. Two key elements in molecular material manufacture are chemical complementarity and structural compatibility, both of which confer the weak and noncovalent interactions that bind building blocks together during self-assembly. Following nature's leads, significant advances have been made at the interface of materials, chemistry and biology, including the design of helical ribbons, peptide nanofiber scaffolds for three-dimensional cell cultures and tissue engineering, peptide surfactants, peptide detergents for solubilizing, stabilizing, and crystallizing diverse types of membrane proteins and their complexes.

Figure 8.1 Two distinctive and complementary fabrication technologies: Top-down vs. bottom-up. In the top-down approach, the boat is limited by the size of the tree. On the other hand, the bottom-up approach, the boat is built with smaller parts of the tree. There is no size limit to the boat for which parts are used to build it.

8.2 NANOBIOTECHNOLOGY THROUGH MOLECULAR SELF-ASSEMBLY AS A FABRICATION TOOL

Design of molecular biological materials requires detailed structural knowledge to build advanced materials and complex systems. Using basic biological building blocks and a large number of diverse peptide structural motifs (Branden and Tooze, 1999; Petsko and Ringe, 2003), it is possible to build new materials from bottom-up.

One of the approaches is through molecular self-assembly using these construction units (Branden and Tooze, 1999; Petsko and Ringe, 2003). Molecular self-assembly is ubiquitous in nature, from lipids (that form oil droplets in water) and surfactants (that form micelles and other complex structures in water) to sophisticated multiunit ribosome and virus assemblies. Molecular self-assembly has recently emerged as a new approach in chemical synthesis and materials fabrication in polymer science, nanotechnology, nanobiotechnology, and various other engineering pursuits. Molecular self-assembly systems lie at the interface of molecular and structural biology, protein science, chemistry, polymer science, materials science, and engineering. Many self-assembling systems have been developed. These systems range from organic supramolecular systems, bi-, tri-block copolymers (Lehn, 1995), and complex DNA structures (Seeman, 2003, 2004), simple and complex proteins (Petka et al., 1998; Nowak et al., 2002; Schneider et al., 2002) to peptides (Aggeli et al., 2001; Hartgerink et al., 2001; Zhang et al., 1993, 1995, 2002; Zhang, 2003). Molecular self-assembly systems represent a significant advance in the molecular engineering of simple molecular building blocks for a wide range of material and device applications.

8.3 BASIC ENGINEERING PRINCIPLES FOR MICRO- AND NANO-FABRICATION BASED ON MOLECULAR SELF-ASSEMBLY PHENOMENA

Programmed assembly and self-assembly are ubiquitous in nature at both macroscopic and microscopic scales. The ancient Great Wall of China, the Pyramids of Egypt, the schools of fish in the ocean, flocks of birds in the sky, herds of wild animals on land, protein folding and oil droplets on water are all such examples. Programmed assembly describes predetermined planned structures. On the other hand, self-assembly describes the spontaneous association of numerous individual entities into a coherent organization and well-defined structures to maximize the benefit of the individual without external instruction (Figure 8.2).

Just like the construction of a wall, a house, or a building, many other parts of structures can be prefabricated and program assembled according to architectural plans (Figure 8.3). If we shrink

The Great Wall was program-assembled
one brick at a time (~5,600 km!)
Each brick has a dimension ~10 x 20 x 30 cm
The Great Wall used ~3 billion bricks!

Self-assembly is ubiquitous in Nature
Each fish in about 5–50 cm in length

(a)   (b)

Figure 8.2   The programmed assembly and self-assembly. (a) The Great Wall was program-assembled over 2200 years ago, one brick at a time, with a defined plan, thus an ordered structure, using approximately 3 billion bricks (similar in number to the DNA bases in the human genome). (b) On the other hand, numerous individual fish self-assembled into well-ordered structure without external instructions. The power and ubiquitousness of self-assembly is witnessed everywhere in nature (images of the fish are courtesy of the National Geographic Society).

What do they have in common? Stone walls & Proteins

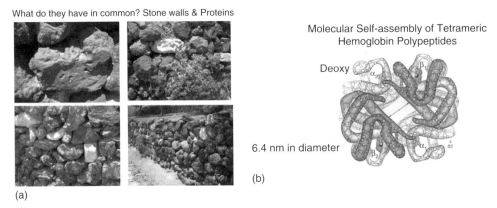

Molecular Self-assembly of Tetrameric
Hemoglobin Polypeptides

Deoxy

6.4 nm in diameter

(b)

(a)

Figure 8.3 (See color insert following page 302)   (a) The stone wall is built one stone at a time with different sizes and colors of stones. It has a defined function. (b) The protein — hemoglobin consisting of four chains — is built one amino acid at a time with 20 amino acids of all shapes and chemical properties. It also has a defined function to carry oxygen.

construction units by many orders of magnitude into nano-scale, such as structurally well-ordered protein fragments, or peptides (Fields, 1999; Yu et al., 1997), we can apply similar principles to construct molecular materials and devices, through molecular self-assembly and programmed molecular assembly. This field is growing at a rapid pace and it is impossible to summarize all aspects of the work being done by others in this limited space, and hence this chapter focuses on a few examples especially from the author's laboratory.

In this chapter, two distinct classes of self-assembling peptide construction motifs are described (Figure 8.4). The first class belongs to amphiphilic peptides that form well-ordered nanofibers (Zhang et al., 1993, 1995). The first member of such self-assembling peptides, EAK16, was discovered in a segment from yeast protein, zuotin (Zhang et al., 1992). These peptides have two distinctive sides, one hydrophobic and the other hydrophilic. The hydrophobic side forms a double sheet within the fiber and hydrophilic side the outside of the nanofibers that interact with water molecules. The hydrophilic side can form extremely high water content hydrogel, containing as high as 99.9% water similar as the water content of a jhellyfish. At least three types of molecules can be made, with −, +, −/+ on the

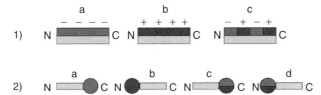

Figure 8.4 (See color insert following page 302)   Two distinct classes of self-assembling peptide construction motifs are shown here. (a) The first class belongs to amphiphilic peptides that form well-ordered nanofibers. These peptides have two distinctive sides, one hydrophobic and the other hydrophilic. The hydrophobic side forms a double sheet inside of the fiber and hydrophilic side forms the outside of the nanofibers that interact with water molecules and they can form extremely high water content hydrogel, containing as high as 99.9% water. At least three types of molecules can be made, with $-$, $+$, $-/+$ on the hydrophilic side. (b) The second class of self-assembling peptide belongs to surfactant-like molecules. These peptides have a hydrophilic head and a hydrophobic tail, much like lipids or detergents. They sequester their hydrophobic tail inside of micelle, vesicles or nanotube structures and expose their hydrophilic heads to water. At least three kinds of molecules can be made, with $-$, $+$, $-/+$ heads.

hydrophilic side (Zhang and Altman, 1999; Zhang, 2002). The second class of self-assembling peptide belongs to a surfactant-like molecule (Vauthey et al., 2002; Santoso et al., 2002; von Maltzahn et al., 2003). These peptides have a hydrophilic head and a hydrophobic tail, much like lipids or detergents. They sequester their hydrophobic tail inside of micelle, vesicles or nanotube structures and expose their hydrophilic heads to water. As in the previous case, at least three kinds of molecules can be made, with $-$, $+$, $-/+$ heads.

The first class includes: "Peptide Lego" that forms well-ordered nanofiber scaffolds and can be used not only for 3-D tissue cell culture but also for regenerative medicine, namely to promote healing and replacing damaged tissues. The second class includes peptide surfactants and detergents that can be used not only for drug, protein and gene deliveries, but also for solubilizing, stabilizing, and crystallizing membrane proteins. Membrane proteins are crucial for biological energy conversations, cell–cell communications, specific ion channels and pumps including our senses, sight, hearing, smell, taste, touch, and temperature sensing.

Like bricks and architectural construction units, these designed peptide construction motifs are structurally simple, versatile for a wide spectrum of applications.

8.4   CHEMICAL COMPLEMENTARITY AND STRUCTURAL COMPATIBILITY THROUGH NONCOVALENT WEAK INTERACTIONS

Molecular self-assembly, by definition, is the spontaneous organization of numerous molecules under thermodynamic and kinetic conditions into structurally well-defined and rather stable arrangements through a number of noncovalent interactions. These molecules undergo self-association forming hierarchical structures. The ribosome is one of the most sophisticated molecular machines nature has ever remarkably self-assembled (Figure 8.5). It has more than 50 different kinds of proteins and 3 different size and functional RNAs, all through weak interactions to form the remarkable assembly line (Stillman, 2002). The other molecular machines include the photosystems I and II that collect photos to convert into electrons in order to produce energy needed for nearly all living systems on Earth (Barber, 1992).

Molecular self-assembly is mediated by weak, noncovalent bonds — notably hydrogen bonds, ionic bonds (electrostatic interactions or salt bridges), hydrophobic interactions, van der Waals interactions, and water-mediated hydrogen bonds. Although these bonds are relatively insignificant in isolation, when combined together as a whole, they govern the structural conformation of all biological macromolecules and influence their interaction with other molecules (Pauling, 1960). The water-mediated hydrogen bond is especially important for living systems as all biological materials interact with water (Ball, 2001).

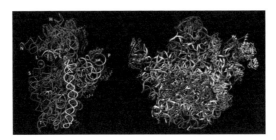

Figure 8.5 (See color insert following page 302)   The bacterial ribosome. 30S ribosome (left panel) and 50S ribosome (right panel). The ribosome is one of the most sophisticated molecular machines nature has ever self-assembled. It has more than 50 different kinds of proteins and 3 different size and functional RNAs, all through weak interactions to form the remarkable assembly line. (*Source*: http://www.molgen.mpg.de/~ag_ribo/ag_franceschi/.)

These weak interactions promote the assembly of molecules into units of well-defined and stable hierarchical macroscopic structures. Although each of the bonds or interaction is rather weak, the collective interactions can result in very stable structures and materials. The key elements in molecular self-assembly are chemical complementarity and structural compatibility. Like hands and gloves, both the size or shape and the correct orientation, that is chirality, are important in order to have a complementary and compatible fitting (Schnur, 1993).

The key engineering principle for molecular self-assembly is to artfully design the molecular building blocks that are able to undergo spontaneously stepwise fine-tuned interactions and assemblies through the formations of numerous noncovalent week chemical bonds.

8.5   SELF-ASSEMBLING SYSTEMS — MODELS TO STUDY MOLECULAR ANTENNA FOR PROGRAMMED ASSEMBLY, SURFACE ENGINEERING, AND FABRICATION OF NANOSCAFFOLD TO NANOBIOTECHNOLOGY

8.5.1   Fabricating Nanowires using Bioscaffolds

In the computing industry, the fabrication of nanowires and nanofeatures using the "top-down" approach increasingly faces tremendous challenges. Thus, the possibility of molecular fabrication of conducting nanowires using DNA (Braun et al., 1998; Keren et al., 2003) peptides and protein scaffolds is of particular interest to electronics industry. One can readily envision that nanotubes, nanofibers, actin filaments, yeast prion nanofibers made from self-assembling peptides and proteins may serve as templates for metallization (Djalali et al., 2002; Scheibel et al., 2003; Reches and Gazit, 2003; Mao et al., 2003). Once the organic scaffold is removed, a pure conducting wire is left-behind and immobilized on a surface. There is great interest in developing various methods for attaching conducting metal nanocrystals to DNA, peptides, and proteins for such a purpose. Furthermore, the coupled DNA, peptides and proteins may not only respond to electronic signals but may also be used as antennae for a wide range of applications including to study detailed molecular interactions and fabricate miniature devices (Hamad-Schifferli et al., 2002; Sung et al., 2004).

8.5.2   Molecular Ink and Nanometer Coatings on Surfaces

Molecular assembly can be targeted to alter the chemical and physical properties of a material's surface (Whitesides et al., 1991; Mrksich and Whitesides, 1996; Whitesides and Grzybowski, 2002). Surface coatings instantly alter a material's texture, color, compatibility with, and responsiveness to the environment. Conventional coating technology is typically accomplished through

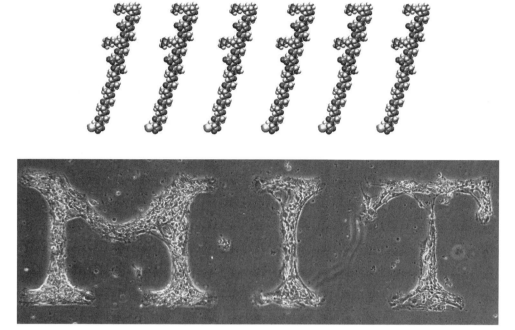

Figure 8.6 Self-assembling peptide molecular ink and the printed MIT. The ink molecules are 4 nm long with a linker that can be directly anchored on surface (Top). The molecular ink was used to print specific patterns for cells and neurons (Bottom). The MIT letters are ~400 μm tall.

painting or electroplating. These coatings are usually in the tens and hundreds micron range and the interface is often not complementary at the molecular level. Thus, erosion is common.

We have developed a class of biologically active molecular ink (Figure 8.6), self-assembling peptides with linkers that anchor on surfaces (Zhang et al., 1999). In conjunction with self-assembled monolayers prepared through microcontact printing, we can place molecules (nanometer scale) and cells (micron scale) into complex patterns. This approach may facilitate research into detail molecular interactions and cell–cell communication. Recently, we have moved one step further: using peptides and proteins as ink, we have directly microprinted specific features onto the nonadhesive surface of polyethylene glycol to fabricate any arbitrary patterns rapidly without preparing a mask or stamps. The process is similar to using an ink pen for writing — here, the microprinting device is the pen and the biological or chemical substances are the inks (Sanjana and Fuller, 2004).

8.5.3 Nanofiber Peptide and Protein Scaffolds

We have also focused on fabricating three-dimensional peptide scaffolds using the self-assembling peptides (Figure 8.7) by exposing them to a salt solution or to physiological media that accelerate the formation of macroscopic structures (Zhang et al., 1993, 1995; Holmes et al., 2000). Scanning electron microscopy (SEM), transmission EM (TEM), and atomic force microscopy (AFM) (Marini et al., 2002) reveal that the matrices formed are made of interwoven nanofibers having a diameter of ~10 nm and pores of ~5 to 200 nm in size. If the alanines are changed to more hydrophobic residues, such as valine, leucine, isoleucine, phenylalanine, or tyrosine, the molecules have a greater tendency to self-assemble and form peptide matrices. These simple, defined and tailor-made self-assembling peptides have provided the first *de novo* designed scaffolds for three-dimensional cell culture, with potential implications for basic studies of cell growth and applied studies in tissue engineering and ultimately regenerative medicine (Kisiday et al., 2002; Zhang, 2004).

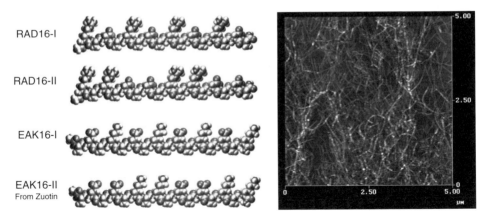

Figure 8.7 The individual self-assembling peptide molecules are 5 nm long (left). The first such peptide, EAK16-II, was discovered from a yeast protein, zuotin (Zhang et al., 1992). This peptide inspired us to design a large class of self-assembling peptide construction motifs. Upon dissolving in water in the presence of salt, they spontaneously assemble into well-ordered nanofibers, further into scaffolds. The AFM Image of peptide RAD16-I nanofibers and PuraMatrix scaffold is shown (right).

We have shown that a variety of tissue cells encapsulated and grown in three-dimensional peptide scaffolds exhibit interesting functional cellular behaviors, including proliferation, functional differentiation, active migration, and extensive production of their own extracellular matrices (Kisiday et al., 2002; Zhang, 2003, 2004). When primary rat hippocampal neuron cells are allowed to attach to the peptide scaffolds, the neuron cells not only project lengthy axons that follow the contours of the scaffold surface, but also form active and functional synaptic connections (Figure 8.8). (Holmes et al., 2000). Furthermore, when the peptide scaffold was injected into brain of animals, it bridged the gap and facilitated the neural cells to migrate across the deep canyon. The animals regained their visual function. Without the peptide scaffold, the gap remains, and the animals did not regain visual function (Ellis-Behnke et al., unpublished results).

8.5.4 Designer Peptide Surfactants or Detergents

We designed another new class of peptide surfactants or detergents with short hydrophobic tail and hydrophilic head (see Figure 8.4, lower panel), taking advantage of the self-assembly properties in water (Vauthey et al., 2002; Santoso et al., 2002; von Maltzahn et al., 2003). Several peptide surfactants have been designed using the nature lipid as a guide. These peptides have a hydrophobic tail with various degrees of hydrophobicity and a hydrophilic head, either negatively charged aspartic and glutamic acids or positively charged lysine or histidine (Figure 8.9). These peptide monomers contain 7 to 8 amino acid residues and have a hydrophilic head composed of aspartic acid and a tail of hydrophobic amino acids such as alanine, valine, or leucine. The length of each peptide is approximately 2 nm, similar to that of biological phospholipids (Vauthey et al., 2002; Santoso et al., 2002; von Maltzahn et al., 2003). The length can also be varied by adding more amino acid, one at a time to a desired length as shown in Figure 8.9.

Although individually these peptide surfactants or detergents have completely different compositions and sequences, these peptides share a common feature: the hydrophilic heads have 1 to 2 charged amino acids and the hydrophobic tails have four or more consecutive hydrophobic amino acids. For example, A_6D (AAAAAAD), V_6D (VVVVVVD) peptide has six hydrophobic alanine or valine residues at the N-terminus followed by a negatively charged aspartic acid residue, thus having two negative charges, one from the side chain and the other from the C terminus; likewise, G_8DD (GGGGGGGGDD), has eight glycines followed by two asparatic acids with three negative charges. In contrast, KV_6 (KVVVVVV) and V_6K (VVVVVVK) have one positively charged lysine as the

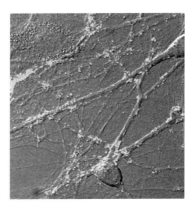

Figure 8.8 (See color insert following page 302) When primary rat hippocampal neuron cells are allowed to attach to the peptide scaffolds, the neuron cells not only project lengthy axons that follow the contours of the scaffold surface, but also form active and functional synaptic connections, each green dot is a functional neuronal connection (upper panel). Furthermore, when the peptide scaffold was injected into brain of animals, it bridged the gap and facilitated the neural cells to migrate across the deep canyon (lower panel). The animals regained their visual function. Without the peptide scaffold, the gap remains, and the animals did not regain visual function.

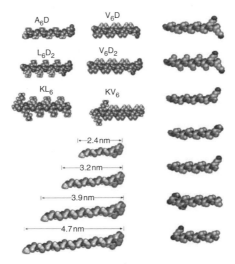

Figure 8.9 A few of the self-assembling peptide surfactant or detergent molecules are modeled here. These peptides have a hydrophilic head and a hydrophobic tail, much like lipids or detergents. They sequester their hydrophobic tail inside of micelle, vesicles or nanotube structures and their hydrophilic heads expose to water. At least three kinds of molecules have been made, with $-$, $+$, $-/+$ heads.

hydrophilic head and six valines as the hydrophobic tail. Leucine and isoleucines are also used as tails. Positively charged lysine and histidine and negatively charged aspartic acid and glutamic acids have also been used as heads. (Vauthey et al., 2002; Santoso et al., 2002; von Maltzahn et al., 2003).

These peptides undergo self-assembly in water to form nanotubes and nanovesicles having an average diameter of 30 to 50 nm (Vauthey et al., 2002; Santoso et al., 2002; von Maltzahn et al., 2003). The tails consisting of alanines and valines produce more homogeneous and stable structures than those of glycines, isoleucine, and leucine. This property may be due to their hydrophobic and hydrophilic ratios. These monomer surfactant peptides were used for molecular modeling. The negatively charged aspartic acid is modeled as red and positively lysine is blue with the green as the hydrophobic tails.

Quick-freeze or deep-etch sample preparation where the sample is instantly flash-frozen below $-190°C$ produced a 3-D structure with minimal structural disturbance. Using transmission electron microscopy, it revealed a network of open-ended nanotubes with three-way junction to connect the nanotubes (Figure 8.10) (Vauthey et al., 2002; Santoso et al., 2002; von Maltzahn et al., 2003). They seem to be dynamic molecular entities overtime. Likewise, A_6K cationic peptide also

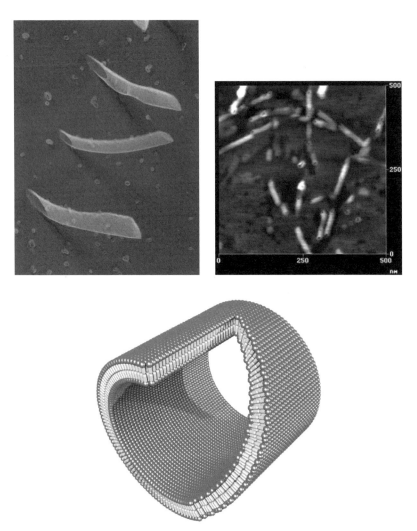

Figure 8.10 (See color insert following page 302)   Self-assembling peptide nanotubes. Peptide detergents: V_6D with the tube diameter ~30 to 50 nm (left panel), A_6K with the tube diameter ~20 to 30 nm (middle panel) and the model for V_6D. The openings of the nanotubes are clearly visible. The wall of the tube has been determined using neutron scattering as ~5 nm, suggestive of a bi-layer structure modeled here.

exhibited similar nanotube structures with the opening ends clearly visible. The wall of the tube has been determined by neutron scattering as ~5 nm, suggestive of a bi-layer structure modeled here.

It is interesting that these simple peptides surfactants can produce remarkable complex and dynamic structures. This is another example to build materials from the bottom-up.

One may ask how could these simple peptide detergents form such well-ordered nanotubes and nanovesicles? The answer may lie in the molecular and chemical similarities between lipids and the peptides since both have a hydrophilic head and a hydrophobic tail. Organic detergents have been well studied over last few decades. The key lies in the molecular packing. However, the packing between lipids and peptides is likely to be quite different. In lipids, the hydrophobic tails pack tightly against each other to completely displace water, without formation of hydrogen bonds at all. On the other hand, in addition to hydrophobic tail packing between the amino acid side chains, peptide detergents also interact through intermolecular hydrogen bonds along the backbone. Some of these peptide detergents displayed typical beta-sheet structures, implying the backbone extended. Thus, the tails are likely packed in the beta-sheet form with certain curvature due to the repulsion charged heads.

8.6 PEPTIDE DETERGENTS STABILIZE MEMBRANE PROTEINS AND COMPLEXES

Many grand challenges remain in the postgenomic era, one of which is the fundamental understanding of membrane biology, namely, the study of the structure and function of membrane proteins, and specifically, the elucidation of high-resolution structures of integral membrane proteins.

Nearly all cellular signal transduction cascades occur through membrane proteins (Brann, 1992; Haga et al., 1999; Wess, 1999). All our senses including sight, smell, hearing, taste, touch, and temperature sensing, use membrane proteins for us to communicate with the external world. Many important drugs used as human therapeutics act through their interaction with membrane proteins. Yet, despite much effort in last few decades, very little is known about the intricacies and function of many membrane proteins. Thus, meticulous and systematic determination of high-resolution membrane protein structure will not only further our understanding of proteins as a whole, but will also enhance our knowledge of signal transduction and accelerate development of ultra-sensitive sensing devices.

Although membrane proteins are composed of approximately one-third of total cellular proteins (Wallin and von Heijne, 1998; Loll, 2003) and carry out some of the most important functions in cells, only ~170 membrane protein structures have been elucidated. This is in sharp contrast to over 30,000 nonmembrane protein structures that have been solved (http://www.rcsb.org/pdb/). The main reason for this delay is difficulty to purify and crystallize membrane proteins because removal of lipids from membrane proteins affects protein solubility and conformation stability. Despite a variety of detergents and lipids as surfactants being used to facilitate, solubilize, stabilize, purify, crystallize, and manipulate the membrane proteins for over the several decades, how detergents interact with membrane protein to impact its structure and functions and how to choose good detergents for the right membrane proteins remain largely unknown. This is partly due to complexity of membrane protein–detergent–lipid interactions and lack of "magic material" detergents. Therefore, the need to develop new material is urgent.

Recent experiments show that these peptide detergents are excellent materials for solubilizing, stabilizing (Kiley et al., 2005), and crystallizing several classes of diverse membrane proteins (Figure 8.11). These simply designed peptide detergents may now open a new avenue to overcome one of the biggest challenges in biology — to obtain large number of high-resolution structures of membrane proteins.

Study of the membrane proteins will not only enrich and deepen our knowledge of how cells communicate with their surroundings since all living systems respond to their environments, but these membrane proteins can also be used to fabricate the most advanced molecular devices, from energy harness devices, extremely sensitive sensors to medical detection devices, we cannot now even imagine. Following nature's lead, as the late legendary Francis Crick best put it: "*You should always ask questions, the bigger the better. If you ask big questions, you get big answers.*"

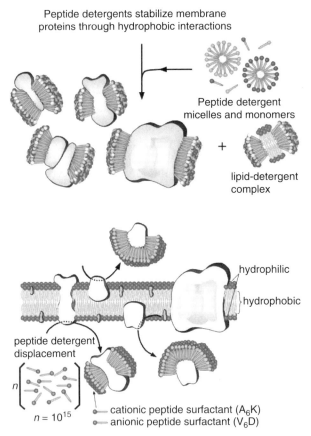

Peptide detergents stabilize membrane
proteins through hydrophobic interactions

Peptide detergent
micelles and monomers

lipid-detergent
complex

hydrophilic

hydrophobic

peptide detergent
displacement

$n = 10^{15}$

cationic peptide surfactant (A_6K)
anionic peptide surfactant (V_6D)

Figure 8.11 (See color insert)   Schematic illustration of designed peptide detergents used to solubilize and stabilize membrane proteins. When mixed with membrane proteins, they solubilize and stabilize them, presumably at the belt domain where the membrane proteins are embedded in lipid membranes.

ACKNOWLEDGMENTS

We also would like to thank members of the lab, past and present, for making discoveries and conducting exciting research. We gratefully acknowledge the supports by grants from ARO, ONR, DARPA (BioComputing), DARPA or Naval Research Labs, DARPA or AFOSR, MURI or AFOSR, NIH, NSF-MIT BPEC and NSF CCR–0122419 to MIT Media Lab's Center for Bits and Atoms, the Whitaker Foundation, DuPont-MIT Alliance, Menicon, Ltd, Japan, Olympus Biomaterials Corp. and Mitsubishi Corp. Research Center. We also acknowledge the Intel Corporation's educational donation of computing cluster to the Center for Biomedical Engineering at MIT.

REFERENCES

Aggeli, A., et al. Hierarchical self-assembly of chiral rod-like molecules as a model for peptide beta-sheet tapes, ribbons, fibrils, and fibers. *Proc. Natl Acad. Sci. USA* 98 (2001) 11857–11862.

Ball, P. *Life's Matrix: A Biography of Water.* (2001) University of California Press, Berkeley, California.

Barber, J. *The Photosystems: Structure, Function and Molecular Biology: Topics in Photosynthesis.* Vol. 11 (1992) Elsevier Science Pub Co.

Branden, C.-I. and Tooze, J. Introduction to Protein Structure. 2nd ed. (1999) Garland Publishing, New York, New York.

Brann, M.R. *Molecular Biology of G-Protein-Coupled Receptors: Applications of Molecular Genetics to Pharmacology.* (1992) Birkhauser, Boston.

Braun, E., Eichen, Y., Sivan, U., and Ben-Yoseph, G. DNA-templated assembly and electrode attachment of a conducting silver wire. *Nature* 391 (1998) 775–778.

Djalali, R., Chen, Y.F., and Matsui, H. Au nanowire fabrication from sequenced histidine-rich peptide. *J. Am. Chem. Soc.* 124 (2002) 13660–13661.

Fields, G.B. Induction of protein-like molecular architecture by self-assembly processes. *Bioorg. Med. Chem.* 7 (1999) 75–81.

Haga, T., Berstein, G., and Bernstein, G. *G Protein-Coupled Receptors.* (1999) CRC Press, Boca Raton, Florida.

Hamad-Schifferli, K. Schwartz, J., Santos, A., Zhang, S., and Jacobson, J. Remote electronic control of DNA hybridization through inductive coupling to an attached metal nanocrystal antenna. *Nature* 415 (2002) 152–155.

Hartgerink, J.D., Beniash, E., and Stupp, S.I. Self-assembly and mineralization of peptide-amphiphile nano-fibers. *Science* 294 (2001) 1684–1688.

Holmes, T.C., et al. Extensive neurite outgrowth and active neuronal synapses on peptide scaffolds. *Proc. Natl Acad. Sci. USA* 97 (2000) 6728–6733.

Keren, K., Berman, R.S., Buchstab, E., Sivan, U., and Braun, E. DNA-templated carbon nanotube field-effect transistor. *Science* 302 (2003) 1380–1382.

Kiley, P., Zhao, X., Vaughn, M., Baldo, M., Bruce, B. and Zhang, S. Self-assembling peptide detergents stabilize isolated photosystem I on a dry surface for an extended time. *PLOS Biol.* 3 (2005) 1180–1186.

Kisiday, J., et al. Self-assembling peptide hydrogel fosters chondrocyte extracellular matrix production and cell division: implications for cartilage tissue repair. *Proc. Natl Acad. Sci. USA* 99 (2002) 9996–10001.

Lehn, J.-M. *Supramolecular Chemistry : Concepts and Perspectives.* (1995) John Wiley and Sons, New York, New York.

Loll, P.J. Membrane protein structural biology: the high throughput challenge. *J. Struct. Biol.* 142 (2003) 144–153.

Mao, C., et al. Viral assembly of oriented quantum dot nanowires. *Proc. Natl Acad. Sci. USA* 100 (2003) 6946–6951.

Marini, D., et al. Left-handed helical ribbon intermediates in the self-assembly of a beta-sheet peptide. *NanoLetters* 2 (2002) 295–299.

Mrksich, M. and Whitesides, G.M. Using self-assembled monolayers to understand the interactions of man-made surfaces with proteins and cells. *Annu. Rev. Biophys. Biomol. Struct.* 25 (1996) 55–78.

Nowak, A.P., et al. Rapidly recovering hydrogel scaffolds from self-assembling diblock copolypeptide amphiphiles. *Nature* 417 (2002) 424–428.

Pauling, L. *The Nature of the Chemical Bond*, 3rd ed. (1960) Cornell University Press, Ithaca, New York.

Petka, W.A., Harden, J.L., McGrath, K.P., Wirtz, D., and Tirrell, D.A. Reversible hydrogels from self-assembling artificial proteins. *Science* 281 (1998) 389–392.

Petsko, G.A. and Ringe, D. *Protein Structure and Function.* (2003) New Science Press Ltd., London, UK.

Reches, M. and Gazit, E. Casting metal nanowires within discrete self-assembled peptide nanotubes. *Science* 300 (2003) 625–627.

Sanjana, N. and Fuller, S.B. A fast flexible ink-jet printing method for patterning dissociated neurons in culture. *J. Neurosci. Methods* 136 (2004) 151–163.

Santoso, S., et al. Self-assembly of surfactant-like peptides with variable glycine tails to form nanotubes and nanovesicles. *NanoLetters* 2 (2002) 687–691.

Scheibel, T., et al. Conducting nanowires built by controlled self-assembly of amyloid fibers and selective metal deposition. *Proc. Natl Acad. Sci. USA* 100 (2003) 4527–4532.

Schneider, J.P., et al. Responsive hydrogels from the intramolecular folding and self-assembly of a designed peptide. *J. Am. Chem. Soc.* 124 (2002) 15030–15037.

Schnur, J.M. Lipid tubules: a paradigm for molecular engineered structures. *Science* 262 (1993) 1669–1676.

Seeman, N.C. DNA in a material world. *Nature* 421 (2003) 427–431.

Seeman, N.C. Nanotechnology and the double helix. *Sci. Am.* 290 (2004) 64–69.

Stillman, B. *The Ribosome: Cold Spring Harbor Symposia on Quantitative Biology.* Vol. 66 (2002) Cold Spring Harbor Laboratory Press, Cold Spring Harbor, New York.

Sung, K.-M., Mosley, D.W., Peelle, B., Zhang, S. and Jacobson, J.M. Synthesis of monofunctionalized gold nanoparticles by F-moc solid-phase reactions. *J. Am. Chem. Soc.* 126 (2004) 5064–5065.

Vauthey, S., et al. Molecular self-assembly of surfactant-like peptides to form nanotubes and nanovesicles. *Proc. Natl Acad. Sci. USA* 99 (2002) 5355–5360.

von Maltzahn, G., et al. Positively charged surfactant-like peptides self-assemble into nanostructures. *Langmuir* 19 (2003) 4332–4337.

Wallin, E. and von Heijne, G. Genome-wide analysis of integral membrane proteins from eubacterial, archaean, and eukaryotic organisms. *Protein Sci.* 7 (1998) 1029–1038.

Wess, J. *Structure–Function Analysis of G Protein-Coupled Receptors.* (1999) Wiley-Liss, New York, New York.

Whitesides, G.M. and Grzybowski, B. Self-assembly at all scales. *Science* 295 (2002) 2418–2421.

Whitesides, G.M., et al. Molecular self-assembly and nanochemistry: a chemical strategy for the synthesis of nanostructures. *Science* 254 (1991) 1312–1319.

Yu, et al. Construction of biologically active protein molecular architecture using self-assembling peptide-amphiphiles. *Methods Enzymol.* 289 (1997) 571–587.

Zhang, S. Emerging biological materials through molecular self-assembly. *Biotechnol. Adv.* 20 (2002) 321–339.

Zhang, S. Fabrication of novel materials through molecular self-assembly. *Nat. Biotechnol.* 21 (2003) 1171–1178.

Zhang, S. Beyond the Petri dish. *Nat. Biotechnol.* 22 (2004) 151–152.

Zhang, S. and Altman, M. Peptide self-assembly in functional polymer science and engineering. *React. Funct. Polym.* 41 (1999) 91–102.

Zhang, S., et al. Zuotin, a putative Z-DNA binding protein in *Saccharomyces cerevisiae. EMBO J.* 11 (1992) 3787–3796.

Zhang, S., et al. Spontaneous assembly of a self-complementary oligopeptide to form a stable macroscopic membrane. *Proc. Natl Acad. Sci. USA* 90 (1993) 3334–3338.

Zhang, S., et al. Self-complementary oligopeptide matrices support mammalian cell attachment. *Biomaterials* 16 (1995) 1385–1393.

Zhang, S., et al. Biological surface engineering: A simple system for cell pattern formation. *Biomaterials* 20 (1999) 1213–1220.

Zhang, S., et al. Design nano biological materials through self-assembly of peptide and proteins. *Curr. Opin. Chem. Biol.* 6 (2002) 865–871.

Engineered Muscle Actuators: Cells and Tissues

Robert G. Dennis and Hugh Herr

CONTENTS

9.1 INTRODUCTION

Muscle tissue mechanical actuators have evolved over millions of years within animals as nature's premier living generators of force, work, and power. The unparalleled efficiency and plasticity of form of living muscles arise from the properties of biomolecular motors. Muscle *cells* serve to self-organize, maintain and repair, and control the mechanical actions of large arrays of biomolecular motors. Muscle *tissues* provide the chemomechanical interface between muscle cells and the environment. It is at the *tissue* level that muscle becomes a practical, responsive, and robust actuator because of the presence of the critical tissue interfaces: the neuromuscular interface, the myotendinous junction (MTJ), and the vascular bed. Tendon tissue is an extension of the muscle extracellular matrix (ECM) and muscle tendon junctions at the end of each muscle fiber. The mechanical structures that make up this transition from muscle to tendon are critical for the transduction of force, work, and power between muscle tissue and the external environment. Systematic derangements of these structures at any level result in severe and sometimes lethal disease resulting from the impairment of the contractility of skeletal muscle and the increased susceptibility to contraction-induced injury.

A detailed description of the biology of muscle development and morphology is beyond the scope of this chapter. The interested reader is referred to the definitive text on this subject: *Myology* (Volume I, Chapters 1, 2, 3, and 4, Engel and Armstrong, eds., 1994, McGraw-Hill).

Embedded within the genetic code of naturally occurring muscles lies the potential to build mechanical actuators that are adaptable, self-healing smart materials (with integrated sensors for position, force, and velocity) from the submillimeter to meter size scale in the form of tubes, rods, sheets, hollow spheres, cones, and many other physical configurations. The key to engineering efficient, robust, and practical muscle actuators lies in understanding the mechanisms by which to control muscle *phenotype*, that is, the size, shape, fiber type, and architecture of the muscle tissue itself. The environmental signals that control muscle phenotype are mediated by the tissue interfaces, and thus it is critical to understand, and to ultimately engineer, adequate tissue interfaces for muscle actuators.

There are four basic approaches (classes) to engineering functional living muscle actuators: the use of whole surgically explanted muscles, recellularized muscle within a muscle-derived ECM, scaffold-based engineered muscle, and self-organized muscle tissue engineered culture. When considering muscle tissue as a functional element in an engineered system it is important to formulate well-defined quantitative Figures of Merit (FoM). It is also important to note that at the time of this writing, practical living muscle actuators are an as-yet unachieved research objective. The remainder of this chapter will discuss many of the reasons why living muscle is being given serious consideration for use as a mechanical actuator in hybrid robotic systems, as well as the many special considerations involved when attempting to employ living actuators in an engineered biohybrid system. The incorporation of functional living elements into otherwise synthetic engineered systems is called *biomechatronics*.

9.2   SYSTEMS ENGINEERING OF LIVING MUSCLE ACTUATORS

Tissue engineering of skeletal muscle could be broadly defined to include any alteration to or enhancement of the musculature of a living organism. This definition, though interesting, would not be specific enough to be useful, as it would include the agricultural use of steroids to rapidly increase the total lean body mass of livestock, the use of resistance training by athletes to induce hypertrophy, and surgical procedures including transplants and flaps in which preexisting skeletal muscle is modified and utilized in clinically relevant procedures (including graciloplasty, cardiomyoplasty, and musculoskeletal reconstructive surgery). Though all of these approaches to the modification and use of skeletal muscle are of interest, this chapter will only address skeletal muscle tissue engineering to generate functional muscle tissues actuators.

Successful tissue engineering must include a focus on the organization of large numbers of cells into higher-order structures that confer emergent properties, which are an important aspect of the tissue-level function. Thus, the engineering of functional tissues is by definition within the domain of "systems engineering." These living structures may be known as tissues or organs depending on the level of anatomical complexity and structural integration. Though all tissue functions arise from fundamental cellular mechanisms, the organization of tissues and organs confers function that is not possible to achieve with individual cells or masses of unorganized cells in a scaffold. By analogy, a pile of bricks does not provide the functionality of a house, nor does a crate full of car parts function as an automobile. Furthermore, when removed from an organism, muscle tissue in general does not persist for long periods. Isolated from its proper environment, muscle tissue tends to degenerate rapidly. The environment that is required to maintain healthy, adult phenotype muscle is highly complex and incompletely understood, involving many chemical, structural, and mechanical signals. In order to understand both natural and tissue-engineered skeletal muscle, we must have a clear working definition of muscle function and understand how the structure of muscle contributes to the emergence of that function. A major challenge facing the use of muscle tissue as a practical living actuator is the identification of suitable tissue interfaces to allow the application of external cues (such as mechanical forces and growth factors) to guide tissue development and to allow the controlled generation of mechanical power.

9.3   MUSCLE: NATURE'S ACTUATOR

Skeletal muscle accounts for nearly half of the total mass of the average adult human and is unique in its ability to actively modify its mechanical properties within tens of milliseconds to allow animals to rapidly react to their environment. Muscle tissues have evolved over the last several billion years as nature's premier living generators of force, work, and power. The success of muscle tissue actuators hinges in part upon the very favorable efficiency of biomolecular motors. Biomolecular motors are the mechanically functional units of muscle cells and tissues, providing motility and mobility for organs and organisms. Muscle *cells* (also known as muscle *fibers*) serve to self-organize, maintain and repair, and control the mechanical actions of large arrays of biomolecular motors. The tremendous plasticity of form of muscle actuators is first realized at the level of cells: biomolecular motors are added in parallel to allow greater force generation, and are added in series to permit more rapid movements over larger displacements. Damaged biomolecular motors are repaired or replaced by intracellular mechanisms. Muscle *tissues* provide the chemomechanical interface between muscle cells and the environment. It is at the tissue level that muscle becomes a practical, responsive, and robust actuator. Tissue level organization provides an ECM for the mechanical support and mechanotransduction of signals to and from the biomolecular motors within each cell. Specialized transmembrane structures transduce force from the arrays of biomolecular motors to the external environment. Failures of these tissue-level structures result in severe pathology of muscle.

Muscle phenotype is known to be a result of a complex interaction between the muscle and its environment. In the absence of the proper signals, muscle will rapidly degenerate. These signals must include chemical signals, mechanical signals, and the activation pattern of the muscle itself. The most important point is that these signals are mediated by the tissue interfaces, and thus it is critical to understand, and to ultimately engineer, adequate tissue interfaces for muscle actuators. The critical interfaces are:

1.   *Vascular:* the primary chemical interface, necessary for sections larger than 0.4 mm in diameter. Perfusion of muscle tissue is important for many reasons, including the removal of metabolically-generated heat, delivery of circulating hormones and metabolic substrates, and removal of metabolic byproducts.
2.   *Myotendinous (MTJ):* the primary mechanical interface, necessary for mechano-transduction in muscle, transmission of force and power to the environment without damage to the muscle cells, and transmission of environmental loads to the muscle cells in such a way that the tissue can respond favorably through functional adaptation. In fact, force and power are transmitted transversely into the ECM surrounding each myofibril as well as directly into the tendon. The ECM extends to meet the tendon and transmit this additional force and power. Derangements of these paths of force transduction at any level in general will lead to pathologies of muscle or tendon or both, often resulting in contraction-induced injury to muscle, as is the case in Duchenne muscular dystrophy.
3.   *Neuromuscular (NMJ):* the primary sensing and control interface, nerve input to muscle plays a dominant role in the control of muscle metabolism and phenotype.

9.3.1 Potential Classes of Living Muscle Actuators

There are four basic approaches to the use of muscle as a mechanical actuator: whole explanted muscles, recellularized muscle ECM, muscle engineered in an artificial ECM, and self-organized muscle tissue engineered *in vitro*. Each class of muscle actuator has technical advantages and presents technical challenges:

9.3.1.1 Whole Explanted Muscles

Whole muscles are frequently explanted to *in vitro* test systems to carry out muscle tissue evaluations. This is common practice in the pharmaceutical industry as well as in muscle research laboratories around the world. These preparations do not qualify as muscle actuators, as they generally have no provision to maintain the muscle explant for longer than a few hours, and they are not configured in such a way that the muscle could perform useful external work. Such preparations are a far cry from any practical actuator embodiments. It is possible, however, to remove whole muscles from a variety of animals and maintain their contractile function for long periods of time (weeks). The use of such explants as practical mechanical actuators was the focus of preliminary work in biomechatronics at MIT in the year 2000.

Advantages: the tissue interfaces are intact and muscle can often be removed with neurovascular pedicles to allow perfusion *ex vivo.*

Disadvantages: architecture is limited to that available in nature. Most natural muscles do not have an architecture suitable for use external to the animal, often due to the tendon geometry or lack of suitable tendons.

Potential applications: drug testing, actuator applications limited by natural architectures.

9.3.1.2 Recellularized Muscle Extracellular Matrix

Under ideal conditions in this process, muscle cells are chemically removed from the tissue, leaving the ECM intact. The matrix would then presumably provide a perfect scaffold for the reintroduction of suitable myogenic cells. In preliminary experiments it has been demonstrated that the acellularized muscle matrix is entirely nonantigenic, so scaffolds can be removed from one animal and implanted in another without fear of tissue rejection.

Advantages: The ECM retains much of the complex physical architecture of the tissue interfaces, so currently it is hypothesized that it will facilitate the reformation of suitable myotendinous and neuromuscular junctions and vasculature for the creation of tissues suitable for surgical repair of lost or damaged muscle tissue. Because the ECM is nonantigenic, it will be possible to remove intact muscle structures from cadavers and acellularize them to form scaffolds for the reengineering of living muscle tissue from the intended recipient, using the recipient's cells (from a biopsy or other method) to preclude subsequent postsurgical tissue rejection. Genetically engineered muscle cells, cells from established cell lines, and primary cells may be reintroduced, as dictated by the actuator application. The existing ECM structure of the acellularized vascular bed allows the acellular muscle to be directly perfused.

Disadvantages: Like whole explanted muscles, the architecture of these actuators is defined by the ECM, and therefore is limited to those forms available in nature. The acellularization process may damage some of the important chemical messages on the matrix, so this method needs to be optimized with this in mind.

Potential applications: Recellularized ECM actuators have the complex architecture of whole muscles *in vitro*, and can be recellularized using cells isolated from any animal, so they would be perfectly suited for engineering complex muscles for surgical transplantation, such as facial muscles. The acellularization process can be readily carried out on cadaveric muscle, so donor tissue availability should present no difficulties whatsoever, thus this class of muscle actuators presents a very promising approach for engineering muscle for surgical transplantation. The acellularized matrix could be repopulated with cells donated (and subsequently amplified in culture) by the recipient of the transplant, thereby totally eliminating the risk of tissue rejection.

9.3.1.3 Muscle Cultured in an Artificial Matrix

A wide range of matrices are available for engineered tissues, but most are unsuitable for engineered muscle due to their limited ability to tolerate repeated macrostrain ($\pm 15\%$ or more the physiologic range for muscle).

Advantages: This is the simplest class of engineered muscle, typically involving the casting of isolated myogenic precursor cells into a gel. It is still the most commonly employed method for engineering muscle in culture, only because it is the easiest method to carry out with the resources available in a typical molecular biology laboratory.

Disadvantages: These constructs in the current state of the art tend to have very weak mechanical interfaces and are thus prone to damage at their points of attachment. In addition, the cellular density in these constructs tends to be well below that of the other three classes, thus they pose significant challenges when their performance is normalized by tissue volume for any functional metric, including protein production, force generation, or sustained power output. To date, these constructs have failed to perform adequately as mechanical actuators. Finally, the most commonly used matrix materials inhibit myocyte fusion into myotubes, arresting the process of muscle development and thereby limiting force and power output. The synthetic matrix materials tend to mechanically fail (tensile failure at the tissue interface) within approximately 2 weeks in culture, whereas self-organized engineered muscle (see below) will persist in culture for approximately 4 months, or longer.

Potential applications: Drug delivery when used as an implanted device, *in vitro* model for basic research in cultured muscle cells.

9.3.1.4 *Self-Organized Muscle Tissue Engineered* In Vitro

Isolated myogenic cells are cultured under conditions to provide cues that promote self-assembly of the cells into functional three-dimensional (3-D) tissues.

Advantages: Self-organizing muscle tissues can take full advantage of genetic engineering combined with the inherent phenotypic potential of all muscle tissues. Thus, a virtually limitless range of tissue architectures are possible. In principle, any myogenic cell type from any species can be employed. The authors (Dennis et al.) have successfully engineered skeletal and cardiac precursor cells into functional 3-D tissues in culture from a range of animal species.

Disadvantages: The cells within the tissues tend to remain at an arrested stage of development (neonatal phenotype), exhibiting low levels of contractility and excitability. The mechanical and chemical environment during development must be emulated in order to promote the formation of adequate tissue interfaces.

Potential applications: With appropriately engineered tissue interfaces and the application of the correct external signals, self-organized muscle actuators can be used in any application for which muscle tissue is needed. This is the most general form of engineered muscle, and has the greatest ultimate potential for many applications. Correspondingly, this class of actuators presents the greatest number of technical challenges.

9.4 BIOMECHATRONICS: WHY USE LIVING MUSCLE IN MACHINES?

The ability to engineer muscle actuators may have significant impact on many areas including: (1) drug testing and screening in *in vitro* bioreactors, (2) drug delivery when implanted as a living "protein factory," (3) the ability to construct practical hybrid mechanical actuators and robotic devices using both motile cells and self-organized tissues, (4) the ability to build biohybrid prosthetic devices, (5) engineered tissue for surgical transplantation, including both skeletal muscle (~45% of adult human body mass) and cardiac, (6) the ability to harvest high-quality animal protein for food from a controlled bioreactor environment. The importance of the last application becomes clear in light of recent concerns about prion disease, a growing social pressure to reduce animal suffering, and the need for closed ecosystems for long-duration space flight and exploration.

The focus of this chapter is the use of living muscle as a mechanical actuator in engineered systems. The main reason that living muscle is seriously considered for such use is simply because the performance of muscle tissue as an actuator is generally quite favorable when quantitatively compared with synthetic actuator technologies. Direct quantitative functional metrics of various mechanical actuator systems, including muscle, have been reviewed in detail by Hollerbach et al. (1991). The benchmark for most of the synthetic muscle actuator systems currently under development is living muscle. Muscle has considerable advantages over many synthetic actuator technologies both in terms of quantifiable FoM, as well as in terms of many qualitative features unique to living muscle. The potential qualitative advantages of muscle are many: muscle operates almost silently, generates biodegradable substances when converting fuel to mechanical work, can functionally adapt to changing demands, and can take many forms and sizes unlike any traditional synthetic actuator technology. The potential quantitative advantages of living muscle as a mechanical actuator are principally the high-chemomechanical efficiency when operating at nearly room temperature, and the high power density with peak values ranging from 50 to 150 W/kg, depending upon the muscle tested and the method of evaluation. There are many synthetic actuator systems with much higher power density, but in these cases often excluded is the external power supply and related hardware that are required to drive the actuator. Examples include hydraulic and pneumatic actuators, as well as some types of electro-magnetic actuator systems.

9.5 QUANTITATIVE ASSESSMENT OF THE FUNCTION OF LIVING MUSCLE ACTUATORS

There are many FoM that have been formulated to quantify the performance of muscle actuators to allow comparisons between each class of muscle actuator and synthetic actuators. These standardized FoM may be employed when evaluating a new engineered muscle construct or any living muscle-based actuator system.

9.5.1 Efficiency (Volumetric, Metabolic, Excitatory)

9.5.1.1 Volumetric

Natural muscle tissue is characterized by an extraordinarily efficient packaging of biomolecular motors. Histologic cross-sections of healthy muscle clearly demonstrate that approximately 95% of the muscle CSA is comprised of tightly packed filaments of biomolecular motors (the contractile proteins *actin* and *myosin*) in a hexagonal lattice. There is little opportunity for improvement upon nature with respect to the volumetric efficiency of the packaging of biomolecular motors into functional macroscopic actuators. Synthetically organized contractile proteins are likely to have an advantage only in single-molecule or molecular monolayer applications, and are likely to be extremely disadvantaged when compared with natural muscle, in terms of volumetric efficiency. Current cultured muscle tissues suffer from low volumetric efficiency in terms of contractile proteins, typically 5 to 10% of the value of adult phenotype healthy control muscle. Also, muscle actuators do not require external support machinery to operate in the same way that many synthetic actuators do. One could reasonably argue that muscle requires many of the other physiologic systems of the body to operate (pulmonary, cardiovascular, neural, gastro-intestinal, etc.), so consider the relative masses of the actuators and the external support system. In an adult human, approximately half of the body mass is muscle tissue. This is supported entirely by the remaining mass of the body, which comprises all other physiologic systems. Compare this with hydraulic or pneumatic systems, for example, for which the power generation system often weighs many times the total mass of all actuators in the system.

9.5.1.2 Metabolic (Chemomechanical Transduction)

The metabolic efficiency is readily mapped into the most commonly defined form of thermodynamic efficiency: work OUTPUT ÷ energy INPUT. In the case of muscle, this would translate simply into the mechanical work done by the muscle actuator divided by the caloric content of the fuel (e.g., glucose) consumed plus the energy required to excite the muscle to contract. Corrections must be made for the glucose stored within the muscle prior to the measurement, and for this and a number of other reasons several indirect measures of metabolism are well advised, such as lactate production. The metabolic efficiency of the muscle actuators will of course be sensitive to many factors, including the mechanical load, muscle phenotype, fuel source, pH, temperature, diffusion distances within the tissues, etc. The sensitivity of the actuator to these factors should be considered, in addition to "peak" or "optimal" efficiency values. For example, certain species of amphibians have muscles that operate relatively efficiently over large temperature ranges.

9.5.1.3 Excitatory (Excitation–Contraction Coupling)

This aspect of efficiency is often overlooked in muscle research, but it can easily be the dominant form of energy loss in engineered muscle actuators. This is because muscle rapidly degenerates when maintained in an inactive state. In addition to loss of mass (atrophy), muscle tissue also experiences a loss of excitability. In order to elicit a contraction, muscle is subjected to electrical pulses characterized by a specified pulse width and pulse amplitude at a specified duty cycle and duration. For any given level of contractile activation, reduced excitability manifests as either increased pulse width, pulse amplitude, or both that are required to elicit the desired force or power output. Based upon our extensive preliminary data with engineered muscle, developing muscle, injured and aging muscle, denervated muscle *in vivo*, and denervated–stimulated muscle both *in vivo* and *in vitro*, we calculate that unless care is taken, muscle tissue can degenerate to the point, where the excitability is reduced by three orders of magnitude, thus requiring approximately 1000 times the electrical energy to elicit any given level of contraction. We have also reported that the excitability of denervated muscle can be maintained at control levels by applying the correct form of electrical stimulation.

9.5.2 Static Contractility

Static measures of contractility are readily made, and allow repeatable quantitative evaluation of living muscle function and normalized comparisons between muscle preparations of vastly differing size and architecture. These metrics include: peak twitch force (P_t), peak tetanic force (P_o), the force–frequency relationship, specific force (sP_o), baseline force (P_b), excitability (rheobase, R_{50} and chronaxie, C_{50}), and the length–tension relationship.

> Principal FoM (physical units follow definition):
> peak normalized twitch force: P_t = maximum force, single pulse input (kPa);
> specific force: sP_o = peak tetanic contractile force or physiologic CSA (kPa);
> specific baseline force: sP_b = baseline tensile force or physiologic CSA (kPa);
> rheobase: R_{50} = pulse field amplitude to elicit $0.5P_t$ at wide pulse width (V/m);
> chronaxie: C_{50} = pulse width to elicit $0.5P_t$ at field amplitudes = $2R_{50}$ (s).

9.5.3 Dynamic Contractility

Measures of dynamic contractility are considerably more experimentally challenging than measures of static contractility; however, they provide considerably more insight into the function of living muscle as a practical actuator. For this purpose, it will in general be necessary to develop bioreactors specifically to monitor these values during the extended *ex vivo* maintenance of each class of living muscle actuator (whole explanted or engineered). Dynamic contractility is generally evaluated using one or more the following metrics: peak power, sustained power, power density (W/kg), maximum velocity (L_f/s, where L_f = muscle fiber length), rate of force development (dp/dt), fatigue resistance (metabolic), and work loop performance (net power output during cyclic loading).

> Principal FoM (physical units follow definition):
> peak normalized power density: i.e., peak power output/tissue mass (W/kg);
> sustained power: i.e., power output at 20% duty cycle continuous (W/kg);
> maximum contractile velocity: V_{max} = maximum contractile velocity, unloaded (L_f/s);
> rate of force development: dp/dt (where p is relative force, P/P_{max}) (s^{-1})

In general it is necessary to assemble the instrumentation systems that are necessary for quantitative evaluation of muscle actuator function. For larger muscles that generate at least 10 mN of force these systems can employ commercially available components that are typically used in muscle physiology research. For smaller muscles, or muscles at early stages of development, it is in general necessary to build many of the components. The necessary instrumentation and methods have been reviewed by Dennis and Kosnik (2002).

In addition to the quantitative performance FoM above, the following additional evaluation tools can be employed to quantitatively evaluate living muscle actuator systems:

- *Functional resilience:* Total work output capacity per unit mass of actuator over the functional lifetime of the actuator $(\text{J/kg})_{\text{lifetime}}$.
- *Cellular function and phenotype*, as determined by quantitative histology and molecular biology. This will include the presence of known adult isoforms of myosin heavy chain, mitochondrial density, prevalence of central nuclei in myotubes and muscle fibers, cross-sectional density of contractile protein lattice, and indications of cellular necrosis or apoptosis.
- *Failure mode analysis* and the demonstrated efficacy of countermeasures that have been engineered into the living actuator system.

9.6 PRACTICAL CONSIDERATIONS FOR THE USE OF LIVING MUSCLE ACTUATORS

When considering the use of living muscle in an engineered system, it is important to take into account a number of factors that are generally not significant challenges in traditional mechatronic system design.

9.6.1 Fuel Sources

Muscle *ex vivo* can operate on a range of fuel sources that are inexpensive and readily available. Ultimately, biomechatronic designers envision the ability of a robot to eat while it travels, much like a fish or a horse. It will be necessary to experimentally evaluate each fuel source for use with muscle powered actuators, since each fuel source has practical limitations and advantages. These include specific energy (kJ/kg), solubility, thermal lability, chemical stability, toxicity, transmembrane transport ability in the absence of systemic metabolic modulatory hormones, and second-order effects such as undesired chemical reactivity. The two principal fuel groups utilized by living muscle are fatty acids and sugars. Evaluation criteria should include quantitative comparisons of muscle actuator efficiency and contractility.

9.6.2 Failure Modes

Based upon our experience with each class of living muscle actuators, the following modes of failure have been identified. For each failure mode, the muscle actuator class(es) that are subject to the failure mode is identified, the theoretical basis for the failure is addressed with supporting experimental verification (if available), and corrective actions are proposed that could be implemented in a biomechatronic system.

9.6.2.1 *Septic Degradation of Tissue Structure*

This mode of failure affects all types of muscle actuators and at room temperature typically results in rapid functional deterioration within 24 h in the absence of countermeasures. Barrier asepsis is probably not practical in the ultimate field applications of living muscle actuators. Chemical countermeasures using broad-spectrum antibiotic or antimycotic formulations in the culture media are effective (Dennis et al., 2000, 2001). These are commercially available for tissue and organ culture and when properly employed are effective for long-term maintenance of living tissue *ex vivo*.

9.6.2.2 Mechanical Failure within the Tissue (Intracellular, ECM)

Also known as contraction-induced injury, this mode of failure is prevalent in muscle tissue subjected to maximal contractions during forced lengthening, and affects all classes of muscle actuators. The effective countermeasure involves employing control algorithms that prevent repeated eccentric contraction of fully activated muscle actuators. Living muscle can functionally adapt to tolerate lengthening contractions if the proper maintenance protocols are employed. An attempt can be made to implement such protocols in the muscle actuator bioreactors using feedback control.

9.6.2.3 Mechanical Failure at the Tissue Interface

Less common for muscle *in vivo*, this is a major failure mode for explanted and engineered tissues in general. For whole explanted muscles, the interface typically involves suture or adhesive applied to the preexisting tendons. Lack of process control in this tissue or synthetic junction leads to unpredictable mechanical failures over time. In engineered tissues the problem is more serious, as tissue failure frequently occurs at the tissue or synthetic interface under relatively mild mechanical conditions. We have extensive experimental data on this failure mode in engineered muscle tissue subjected to external loading. We hypothesize the failure to be due to stress concentration at the tissue or synthetic interface, compounded by inadequate force transduction from the appropriate intracellular force generating machinery to the extracellular synthetic load bearing fixtures, leading to cell membrane damage at the interface with subsequent rapid tissue degradation and necrosis. The best countermeasure requires the engineering of a muscle–tendon interface (MTJ), which is a major objective of current research in muscle tissue engineering. Tendon tissue is 80 to 90% ECM, composed chiefly of parallel arrays of collagen fibers. The tendon-to-synthetic interface, where biology meets machine, is a separate and equally important technical challenge.

9.6.2.4 Metabolic Failure

This failure mode results most frequently from inadequate delivery of metabolic substrates and inadequate clearance of metabolic byproducts, and is exacerbated at elevated temperatures. The best countermeasure for this failure mode is to restrict the muscle actuator cross-section to more than approximately 200 μm diameter, or to provide perfusion through a vascular bed in the case of larger cross-sections. This mode of failure typically initiates at the axial core of cylindrical muscle actuators. For this reason, sustained angiogenesis and perfusion is a major technical objective in current tissue engineering research.

9.6.2.5 Cellular Necrosis and Programmed Cell Death

Several controllable circumstances can lead to this general mode of failure in all classes of muscle actuators. Cellular hypercontraction and hyperextension in muscle results in rapid necrosis. This mechanism will occur more or less uniformly across the muscle cross-section, but will theoretically occur more frequently in areas with reduced physiologic cross-sectional area or inhibited sarcomeric function. This failure mode can be prevented by control of the internal mechanical compliance and stroke of the muscle actuator. Muscle maintained at an inappropriate length, either too short or too long, will deteriorate, even if the muscle is quiescent. In explanted muscles, maintenance at lengths greater than the plateau of the length–tension curve appears to be the most damaging over time.

9.6.2.6 Fatigue (Mechanical and Metabolic)

These failure modes apply to all classes of living muscle actuators. For metabolic fatigue the preferred countermeasures will include genetic engineering of the muscle to promote fatigue-resistant fiber types, the provision of adequate perfusion of the tissue actuator, and the development of protocols for actuator control that optimize total work output, such as the intermittent locomotory behavior of both terrestrial and aquatic animals. It is in terms of mechanical fatigue that living actuators have an enormous advantage over fully synthetic actuators. By monitoring the state of health of the actuator and modifying the mechanical demands accordingly, it is possible to promote functional adaptation of the living component of the actuator as well as the tissue or synthetic interface. It will be necessary to identify biomarkers of mechanical fatigue, such as reduced or altered contractility, to actively detect these markers, and to respond with appropriate modifications of the embedded excitation and control algorithms to allow tissue functional adaptation. In principle a properly monitored and controlled living muscle actuator will exhibit improved dynamic performance and structural resilience with use over a period of decades, unlike any synthetic actuator technology currently available.

9.6.2.7 Toxicity

A serious problem for all classes of living muscle actuators, the best countermeasure is barrier exclusion of exogenous toxic agents, the use of biocompatible materials in the fluid-space of the hybrid actuator assembly, and the clearance of toxic metabolic byproducts via a perfusion and filtration system integrated with the living actuator.

9.6.2.8 Electrochemical Tissue Damage

This failure mode affects all classes of living muscle actuators when exposed to chronic electrical stimulation. The single best countermeasure is to promote and maintain tissue phenotype exhibiting very high excitability. In addition to vastly improving the excitation efficiency of the tissue, adult muscle phenotype excitability can yield as much as a 99.9% reduction in electrical pulse energy requirements for any given level of muscle activation, when compared with chronically denervated or tissue engineered muscle tissue arrested at early developmental stages. For this reason, the development of electro-mechanical muscle bioreactor systems and maintenance stimulation protocols form a core component of all current research on muscle tissue engineering. Additional countermeasures include the selection of appropriate electrode materials, the use of minimally energetic stimulation protocols, the use of pure bipolar stimulation pulses with careful attention to charge balancing, and the use of high-impedance outputs to the electrodes when not stimulating.

9.6.2.9 Damage from Incidental Mechanical Interference

The living actuator will require electrodes to be placed in contact with the tissue, the presence of tubing for perfusion, and other structures required within the hybrid actuator. Lateral mechanical contact between these synthetic objects and the living muscle tissue can result in a range of mechanical failures, including abrasion, incision, and chronic pressure atrophy. The appropriate countermeasure for this is careful mechanical design of the hybrid actuator assembly, with these considerations explicitly included in the system Design Specification.

9.6.2.10 Retrograde or Arrested Phenotype (Failure to Thrive)

Effective countermeasures for this failure mode have been reported for denervated whole muscles *in vivo*, employing a long-term electrical stimulation protocol (Dennis et al., 2003; Dow et al.,

2004). This failure mode is most prevalent in engineered muscle maintained in culture. There are two approaches to dealing with this in engineered muscle: (1) genetic enhancement and (2) development of electromechanical tissue maintenance protocols. In the case of genetic enhancement, the approach is to forcibly express desired genes in an attempt to promote the desired tissue phenotype. The effectiveness of this approach is the core issue in gene therapy for diseases of muscle, but this approach has not yet been demonstrated to be effective for engineered muscle *ex vivo*. Optimal tissue maintenance protocols are a much more natural and subtle approach, based upon the fact that all viable muscle cells contain the necessary genetic machinery to develop any desired muscle phenotype, if the correct signals and growth conditions prevail. In addition to genetic engineering of myocytes to enhance performance of tissue-based actuators, other potential countermeasures include: (1) development of appropriate tissue interfaces to permit signal transduction to the cellular machinery, (2) development of tissue and organ culture bioreactors to allow the experimental determination of optimal control and maintenance protocols for *ex vivo* muscle tissue, (3) use of these protocols to guide tissue development (cell phenotype and tissue architecture), and (4) implementation of this technology into the hybrid actuator system. This topic is currently an area of very active research. Success in terms of counteracting this failure mode in engineered muscle will constitute an extraordinarily significant scientific contribution, as well as providing the key enabling technology to the further development of practical living actuators.

9.7 SELF-ORGANIZING MUSCLE TISSUES

Self-organization within developing animals gives rise to an enormous array of muscle actuator architectures. Each myogenic precursor cell contains the genetic potential to self-organize into muscle tissue with the desired phenotype and tissue interface. The ability to guide the development of self-organizing muscle tissues in culture will provide the systems engineer with the greatest level of design flexibility, since it will in principle be possible to start with a small population of muscle progenitor cells and guide them to self-organize into a muscle actuator of any imaginable geometry. It will also be possible to construct hybrid actuators not found in nature, containing regionally organized tissue structures, perhaps even consisting of fundamentally different types of muscle tissue (skeletal, cardiac, or smooth), depending upon the functional requirements of the actuator system. It is implicit in most muscle tissue engineering research programs that skeletal muscle self-organization and development can be guided by the application of the correct external cues. The general method of guided tissue self-organization in culture (Figure 9.1) briefly is:

- Isolate and coculture the desired cells. The cells may be primary or from cell lines.
- Engineer a cell culture substrate with controlled adhesion properties for the cells.
- Provide permanent anchor points and surfaces to guide tissue architecture formation.
- Culture the cells to permit the formation of a cohesive monolayer.
- Induce monolayer delamination from the substrate at the appropriate point in cell differentiation (the monolayer remains attached to the anchor points).
- Promote tissue self-organization and further development by applying external signals: chemical, electrical, mechanical.

Self-organization of tissues in culture is one effective way to produce small functional tissue constructs from a range of tissues. Examples include:

- *Cardiac myocytes* cocultured at confluence with fibroblasts will self-organize into long cylinders and tapered cones in culture in 340 to 400 h. These constructs are electrically excitable and also spontaneously contract as a syncytium to continuously generate significant mechanical work cycles. Such constructs could be engineered to power cell-scaled implantable pumps, pumps for

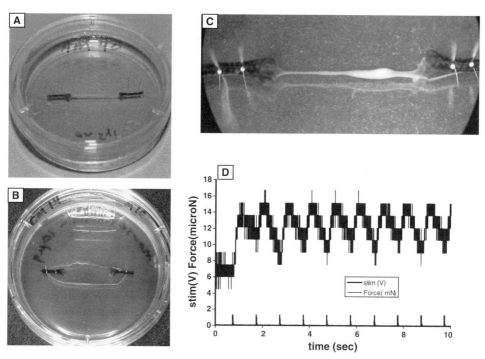

Figure 9.1 (See color insert following page 302) (A) Self-organized skeletal muscle construct after 3 months in culture, length ~12 mm. (B) Rat cardiac myocyte + fibroblast monolayer in the process of delaminating and self-organizing into a functional cardiac muscle construct, 340 h in culture. (C) Self-organized cardiac muscle construct, attached to laminin-coated suture anchors, 380 h in culture. (D) Electrically elicited force trace from the cardiac muscle construct shown in C, stimulation pulses shown below, contractile force trace shown above (raw data, unfiltered).

stand-alone hybrid tissue actuators, or to engineer cardiac tissue for surgical transplantation in cardiac reconstructive surgery.

- *Tendon (Ligament)* tissue will self-organize in culture under the appropriate conditions. The fibro-blasts within the tissue produce a prodigious amount of ECM material, with collagen fibers that are oriented along lines of tensile stress, particularly at locations within the tissue where mechanical interfaces are present (such as suture anchor materials, metal posts, etc.). Self-organization is driven by loss of substrate adhesion and the generation of internal tensile stress by the action of the fibroblasts on the order of 0 to 6 Pa, which can be experimentally controlled by external factors such as the presence of ascorbic acid, serum concentration in the cell culture medium, pH, etc.

- *Muscle Chimeras:* One additional interesting technical possibility is the *in vitro* fusion of myogenic precursor cells from different tissue sources to form chimeric self-organized engineered muscles. Preliminary experiments demonstrate that skeletal muscle satellite cells from differing species will fuse to form multinucleated myotubes with desirable contractile function. In addition, isolated cardiac myocytes will fuse into preexisting myotubes in culture, to produce a skeletal–cardiac muscle hybrid. Such chimeric muscle tissues are not known to exist in nature, but our preliminary data indicate that they are both stable and functional in culture. The contractile function of such chimeric cells and tissues could potentially be engineered to produce tissue-based actuators with combinations of desired characteristics that would be advantageous for use in hybrid bioactuator applications.

9.8   ACELLULARIZED–RECELLULARIZED ECM ENGINEERED MUSCLES

The native ECM of muscle tissue occupies approximately less than 5% of the tissue volume, yet it contains information about the complex architecture of muscle and the corresponding soft tissue interfaces. The cellular components of muscle can be chemically removed while retaining the detailed architecture of the muscle ECM. Preliminary results indicate the success of the reintroduction of myogenic cells into these natural ECM scaffolds. This approach to engineering muscles as actuators has several advantages, among these are that heterogenic cells can be introduced into the preexisting matrix. For example, skeletal–cardiac chimeric muscles could be employed or myogenic precursors from an entirely different species. The main advantage of the use of natural ECM scaffolds is that the fine architecture of the entire muscle organ is retained by the acellularized ECM scaffold. It is possible to perfuse the scaffold using the remnant vascular bed ECM to reintroduce cells and later to provide perfusion to the reengineered muscle organ. The acellularized muscle ECM also has matrix architecture specific to the MTJ and tendon, which may be advantageous in the development of this very critical tissue interface. The principal disadvantage of this approach is that the ECM scaffold architecture is limited to those architectures that are available in nature.

9.9 TISSUE INTERFACES: TENDON, NERVE, AND VASCULAR

For any type of muscle actuator, it will be essential to provide appropriate tissue interfaces. In some cases, the tissue interfaces are already in place and specific measures must be taken to maintain them properly. In other cases, their formation must be guided and facilitated. Based upon our *in vivo* work, we have demonstrated that muscle phenotype can be controlled and maintained in the absence of innervation via electrical stimulation. A considerable volume of published research has been directed toward the promotion of adult phenotype in muscle tissue in culture directly by electrical stimulation, in the absence of nerve-derived trophic factors or depolarization via the neuromuscular junction and related synaptic structures. It remains to be demonstrated, however, that muscle can be guided through the necessary developmental stages in the absence of innervation to achieve adult phenotype. Adequate and functional vascular and tendon interfaces to muscle engineered *in vitro* are also yet to be demonstrated, although they are the topic of intensive research.

9.9.1 Vascular Tissue Interface

Nutrition and oxygen delivery in static culture conditions always limit the cross-sectional area, particularly for tissues with high metabolic demand, such as muscle. Therefore, a 3-D organ culture system with perfusion of a vascular bed within the muscle tissue is a core objective of current research. Cell types associated with angiogenesis, such as endothelial cells, are also crucial players in organ development (Bahary and Zon, 2001). Endothelial progenitor cells from peripheral blood are readily isolated, and have been shown to incorporate into neovessels (Asahara et al. , 1997) and also have potential to expand to more than 10^{19}-fold *in vitro* (Lin et al., 2000). Furthermore, functional small-diameter neovessels can be created in culture by using endothelial progenitor cells (Kaushal et al., 2001).

9.9.2 Strategies for Engineering Functional Vascularized Muscle Tissue

There are three strategies for generating vascularized muscle constructs:

(1) *Recellularization* of an acellular muscle construct.
(2) *Coculture* of myoblasts with endothelial cells and growth factor stimulation for induction of the endothelial cells to form capillary like structures.
(3) *Induction* of sprouting of microvessels into temporarily implanted tissues or from vascularized and perfused tissue explants (such as adipose) cultured adjacent to the engineered muscle.

The strategies for generating functional muscle tissue can be broadly divided into *in vitro* and *in vivo* strategies, the ultimate outcome of which would be a vascularized muscle construct. In any case, once a vascular bed is established, the constructs need to be maintained in a bioreactor to provide further electrical, mechanical, and chemical stimulation, thus guiding both the phenotype and resulting in the development of a fully function muscle construct.

9.9.2.1 Recellularization of an Acellular Muscle Construct

This experimental approach involves harvesting muscle tissue from any natural source and using chemical acellularization to remove myoblasts and fibroblasts leaving behind an intact ECM. The ECM should be evaluated for structural integrity and immunogenic behavior and its ability to support myoblast growth and differentiation. The ECM should then be used as scaffolding material for seeding primary myoblast and the construct will be placed in a perfusion bioreactor allowing formation of functional skeletal muscle tissue (Hall, 1997). Immunohistochemical studies should be performed to determine which ECM components are present in the acellular construct, such as collagen types I and IV, fibronectin, laminin, vitronectin, entactin, heparin sulfate, proteoglycan, and elastin. The acellular muscle can be repopulated by obtaining a purified sample of myogenic precursor cells, which may be injected or perfused into the acellular muscle. Although some initial success has been reported with this general approach, it has not yet been possible to maintain perfusion of the tissue samples in culture for a period sufficiently long to promote and maintain full cellular infiltration into the acellular scaffold.

9.9.2.2 Coculture Systems

Since the early 1990s, there have been reports of the use of various coculture systems to study cell–cell interactions and the formation of tissue interfaces. For vasculogenesis, the cells in question are presumed to be myoblast and endothelial cells. Although promising initial reports have been published, a truly successful demonstration of a vascular bed self-organizing within a tissue construct has yet to be demonstrated. The design of bioreactors for such a technology must stimulate the myoblasts to form functional muscle tissue and simultaneously guide the endothelial cells to form capillary-like structures within the newly forming muscle tissue, while providing perfusion during development. The environment, which the bioreactor provides together with soluble growth factor stimulation, will presumably allow formation of a functional muscle construct (Vernon, 1999).

9.9.2.3 Induced Microvessel Sprouting

This approach can be attempted either *in vivo* or *ex vivo* using small vascularized tissue explants which are cannulated and perfused while adjacent to an avascular tissue such as engineered skeletal muscle. This is an active area of current research. For the *in vivo* approach, it is necessary to mechanically support the muscle tissue while implanted to prevent hypercontraction and subsequent tissue damage. It is also necessary to take measures to prevent tissue rejection to implantation into syngenic animals, or the use of immune-suppressive agents, is required. Otherwise, this method is relatively quite simple and often yields satisfactory results. In addition to vascularization of the implanted muscle tissue, there are collateral effects, as yet not fully understood, that also tend to drive the muscle phenotype toward an adult phenotype, with enhanced contractility. For this reason, it is likely that the future of tissue engineering will see increasingly common application of the approach where the intended recipient is used as a ready-made bioreactor vessel. The engineered tissues would be implanted within the person, presumably along with means to enhance tissue development and to prevent tissue degeneration or resorption while implanted. The tissue need not be implanted at the ultimate site for which it is intended, however, it is essential to consider the morbidity of the site at which the disuse will be initially developed.

9.9.3 Engineered Tissue Interface: Tendon

The MTJ is critical for the ability of muscle tissue to transduce force to and from the external environment, and to produce maximal power without subsequent injury to the muscle cells in the contractile tissue. The MTJ contains specialized structures at the cell membrane which facilitate transmembrane transmission of force from the contractile proteins (biomolecular motors) within the cell to the surrounding collagen fibrils in the ECM (Trotter, 1993). These structures include a large number of infoldings of the muscle cell membrane at the MTJ, increasing the membrane surface area and acting to transfer stress from the cytoskeleton to the ECM in the tendon. These structures have also been demonstrated to occur when myotubes are cocultured with fibroblasts concentrated near the ends of the muscle constructs *in vitro* (Swasdison and Mayne, 1991). In the case of whole explanted muscle actuators, the MTJ already exists, and it is necessary to maintain this structure *in vitro*. In all other classes of muscle actuator it is necessary to generate or regenerate the MTJ and tendon structures. Currently, attempts to engineer tendon-like structures and muscle–tendon junctions in culture follow one of three distinct approaches:

(1)   Scaffold-based tendon, used as an anchor material for engineered muscle.
(2)   Self-organizing tendon and muscle-tendon structures in co-culture.
(3)   Direct laser transfer of muscle and tendon cells into defined 3-D structures.

9.9.4 Nerve–Muscle Interfaces

Skeletal muscle phenotype is defined largely by the motor nerve which innervates each muscle fiber. Adult muscles may be either fast- or slow-twitch, but in general in humans muscles are mixed, containing significant populations of both fast- and slow-twitch fibers. Denervated muscle rapidly loses tissue mass and the adult phenotype, with contractility eventually dropping to essentially zero. Although it is possible to maintain adult phenotype of adult skeletal muscle in the absence of innervation, it is not yet clear whether it is possible to guide skeletal muscle tissue development to an adult phenotype in an entirely aneural culture environment. For that reason, nerve–muscle synaptogenesis in culture is an area of active research in tissue engineering. Putative synaptic structures *in vitro* have been reported for decades (Ecob et al., 1983; Ecob, 1983, 1984; Ecob and Whalen, 1985), in some cases axon sprouting from nerves to muscle tissue in culture is clearly visible (Figure 9.2) and verified upon histologic examination; however, functional nerve–muscle *in vitro* systems that result in advanced tissue development have yet to be demonstrated.

9.9.5 Tissue–Synthetic Interfaces

Another key challenge is to develop means to mechanically interface living muscle cells and tissues to synthetic fixtures in such a way that the tissue development and function will not be inhibited. The technical challenge is to provide a transition of mechanical stiffness and cell density in the region between the contractile tissue and the synthetic fixture, to reduce stress concentrations at the tissue interface and provide mechanical impedance matching. Several approaches are currently under investigation, including the chemical functionalization of synthetic surfaces to bind collagen, and the use of porous scaffolds to promote tissue in-growth at the desired tissue or synthetic interface.

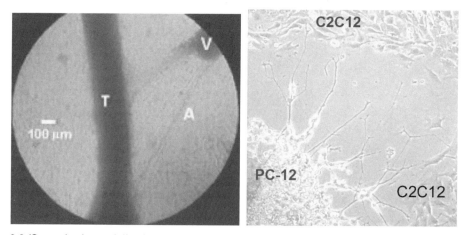

Figure 9.2 (See color insert following page 302) Left: axonal sprouting (A) from an explanted motor neuron cell cluster (V) toward a target tissue (T), in this case, an aneural cultured skeletal muscle "myooid." Right: a simple cell culture system demonstrating axonal sprouting between neural (PC–12) and myogenic (C2C12) cell lines. This co-culture system allows the study of synaptogenesis in culture. (Photographs taken by members of the Functional Tissue Engineering Laboratory at the University of Michigan: Calderon, Dow, Borschel, Dennis.)

9.10 MUSCLE BIOREACTOR DESIGN FOR THE IDENTIFICATION, CONTROL, AND MAINTENANCE OF MUSCLE TISSUE

The engineering of complex functional tissues such as skeletal muscle is by definition a systems engineering problem. Functional muscles are composed of a number of highly integrated tissue systems, none of which is known to function in isolation for any significant period of time without massive deterioration in performance. Any attempt to engineer a functional muscle tissue system *ex vivo*, and to employ that muscle system as a source of motility in robots or prostheses, will by necessity require the development of bioreactor technologies to (1) guide the tissue development to the desired phenotype *ex vivo*, (2) maintain the tissue at the desired phenotype while it is performing its function, and (3) control the mechanical output of the tissue through electrical stimulation. Critical to these three objectives are bioreactor technologies that are capable of monitoring and controlling a muscle's mechanical and electrical environment.

In Figure 9.3, a muscle bioreactor is shown that can implement muscle identification, control, and maintenance protocols under generalized boundary conditions while also providing flexible feedback control of electrical stimulation parameters (Farahat and Herr, 2005). These features are accomplished by having two real-time control loops running in parallel. The first loop, or the mechanical boundary condition (MBC) control loop, ensures that the mechanical response of the servo simulates the dynamics of the associated muscle boundary condition. For example, if the desired boundary condition is a second order, mass–spring–damper system, the MBC control loop controls the motion of the end points of the muscle–tendon unit as if the muscle–tendon were actually pulling against physical mass–spring–damper mechanical elements. The MBC control loop allows for a whole host of boundary conditions, from finite (but nonzero) to infinite impedance conditions. Clearly, to understand muscle tissue performance, muscle dynamics, and the dynamics of the load for which the muscle acts upon must be taken into consideration. Examples of finite-impedance boundary conditions include loads such as springs, dampers, masses, viscous friction, coulomb friction, or a combination thereof. Such loads prescribe boundary conditions that are generally defined in terms of dynamic relationships between force and displacement. Under these loading conditions, it would be expected that the dynamics of the load will interact with the contraction dynamics of the muscle, leading to a behavior that is a resultant of both. This is

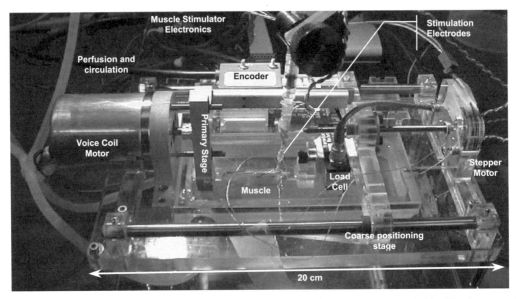

Figure 9.3 (See color insert following page 302) Muscle Bioreactor Technology. Muscle identification, control, and maintenance apparatus is shown with the primary sensors and actuators noted. The coarse positioning stage is adjusted at the beginning of the experiment to accommodate different tissue lengths, but is typically kept at a constant position during a particular contraction. The primary stage provides the motion that simulates the boundary condition force law with which the muscle specimen pulls against. The vertical syringe has a suction electrode at its tip that is connected to the stimulation electronics in the background. The encoder and load cell measure muscle displacement and force, respectively, and are employed as sensory control inputs during FES control experimentation. Silicone tubing recirculates solution via a peristaltic pump, while oxygen is injected in the loop.

primarily because the force generated by muscle is dependent on its mechanical state, namely its length and velocity.

The second control loop for the bioreactor design of Figure 9.3 implements the electrical stimulus (ES) control based on measurements of the muscle's mechanical response. This loop, referred to as the ES control loop, offers simultaneous real-time modulation of pulse width, amplitude, frequency, and the number of pulses per cycle. There is increasing experimental interest in real-time control of muscles, primarily in the context of functional electrical stimulation (FES) (Chizeck et al., 1988; Veltink et al., 1992; Eser et al., 2003; Jezernik et al., 2004). In these investigations, attempts were made to control the response of muscle(s) and associated loads to a desired trajectory by varying electrical stimulation parameters as a function of time. Electrical stimulation patterns are typically square pulses characterized by frequency, amplitude, pulse width, and number of pulses per trigger (considering the cases of doublets, triplets, or more generally N-lets). For testing a variety of FES algorithms, the ES control loop is designed for real-time modulation of these stimulation parameters as a function of a muscle's mechanical response, including tissue length, contraction velocity, and borne muscular force.

9.11 CASE STUDY IN BIOMECHATRONICS: A MUSCLE ACTUATED SWIMMING ROBOT

Biomechatronics is the integration of biological materials with artificial devices, in which the biological component enhances the functional capability of the system, and the artificial component provides specific environmental signals that promote the maintenance and functional adaptation of the biological component. Recent investigations have begun to examine the feasibility of using *animal-derived* muscle as an actuator for artificial devices in the millimeter to centimeter size scale

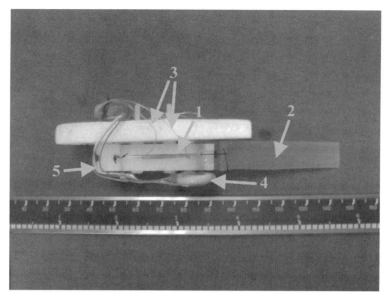

Figure 9.4 (See color insert following page 302)   Biomechatronic swimming robot. To power robotic swimming, two frog semitendinosus muscles (1), attached to either side of elastomeric tail (2), alternately contract to move the tail back and forth through a surrounding fluid medium. Two electrodes per muscle (3), attached near the myotendonous junction, are used to stimulate the tissues and to elicit contractions. To depolarize the muscle actuators, two lithium ion batteries (4) are attached to the robot's frame (5). During performance evaluations, the robot swam through a glucose-filled ringer solution to fuel muscle contractions.

(Herr and Dennis, 2004). Although a great deal of research has been conducted to develop an actuator technology with muscle-like properties, engineering science has not yet produced a motor system that can mimic the contractility, energetics, scalability, and plasticity of muscle tissue (Hollerbach et al., 1991; Meijer et al., 2003). As a demonstratory proof of concept, Herr and Dennis (2004) designed, built, and characterized a swimming robot actuated by two explanted frog semitendinosus muscles and controlled by an embedded microcontroller (Figure 9.4). Using open loop stimulation protocols, their robot performed basic swimming maneuvers such as starting, stopping, turning (turning radius ~ 400 mm), and straight-line swimming (max speed > 1/3 body lengths/sec). A broad-spectrum antibiotic or antimycotic ringer solution surrounded the muscle actuators for long-term maintenance, *ex vivo*. The robot swam for a total of 4 h over a 42-h lifespan (10% duty cycle) before its velocity degraded below 75% of its maximum. The mechanical swimming efficiency of the biomechatronic robot, as determined by a slip value of 0.32, was within the biological efficiency range. Slip values increase with swimming speed in biological swimming, ranging from 0.2 to 0.7 in most fish (Gillis, 1997, 1998).

The development of functional biomechatronic prototypes with integrated musculoskeletal tissues is the first critical step toward the long-term objective of controllable, adaptive, and robust biomechatronic robots and prostheses. The results of the swimming robot study of Herr and Dennis (2004), although preliminary, suggest that some degree of *ex vivo* robustness and controllability is possible for natural muscle actuators if adequate chemical and electromechanical interventions are supplied from a host robotic system. An important area of future research will be to establish processes by which optimal intervention strategies are defined to maximize tissue longevity for a given hybrid-machine task objective. Another important research area is tissue control. It is well established that natural muscle changes in size and strength depending on environmental workload, and when supplied with appropriate signals, changes frequency characteristic or fiber type (Green et al., 1983, 1984; Delp and Pette, 1994). Hence, an important area of future work will be to put forth strategies by which muscle tissue plasticity can be monitored and controlled. Still further, strategies must also be devised to control the force and power output of muscle, in the context of robotic systems, through the modulation of electrical pulses to the muscle cell. Also, for the development of controllable, adaptive and robust biomechatronic systems, feedback control systems that monitor and adapt the mechanical, electrical, and chemical environment of muscle actuators are of critical importance.

9.12 CONCLUDING REMARKS

Muscle tissue as a mechanical actuator has great, though as-yet unrealized potential for use in engineered systems. Synthetic technologies such as electroactive polymers are rapidly emerging as quantitatively functional equivalents to muscle tissue, and it is likely that the technological evolution of EAP muscles will soon out-pace the natural functional evolution of living muscle tissue. This means that the quantitative performance advantages that muscle tissue has over some forms of synthetic actuators· in terms of efficiency, power density, and so forth are not likely to remain the case for very much longer. One then invariably must ask why it is advantageous to even consider the use of living muscle tissue as a mechanical actuator. It is easy to point out that the many disadvantages of muscle outweigh the few performance advantages it may have. The answer lies chiefly in the qualitative differences between muscle and competing synthetic actuator technologies, among these are those qualities that arise from muscle being a living tissue: its ability to functionally adapt and to potentially integrate seamlessly with other living structures. So it is likely that living muscle actuators will only be employed in practical systems where their qualitative advantages as living tissue can be exploited to maximum benefit, such as in hybrid biomechatronic prosthetic systems and implants, and perhaps in bioreactors where their biological products (such as edible proteins) are of primary importance. Certainly though, living muscle tissue serves as the explicit benchmark against which the performance of synthetic actuator technologies will be evaluated for many decades to come.

FURTHER READING

The following list of papers and book chapters comprises a set of useful references for further work in this area. These were not referenced directly in the text, but have been included because the authors have found them to be useful during the course of the development of the technology discussed in this chapter.

Agoram, B. and Barocas, V.H. Coupled macroscopic and microscopic scale modeling of fibrillar tissues and tissue equivalents. *J. Biomech. Eng.* 2001, 123(4): 362–369.

Askew, G.N., Marsh, R.L. et al. The mechanical power output of the flight muscles of blue-breasted quail (*Coturnix chinensis*) during take-off. *J. Exp. Biol.* 2001, 204: 3601–3619.

Barrett, S. Propulsive Efficiency of a Flexible Hull Underwater Vehicle. PhD Thesis, Massachusetts Institute of Technology, Cambridge, Massachusetts, 1996.

Biewener, A.A., Dial, K.P. et al. Pectoralis muscle force and power output during flight in the starling. *J. Exp. Biol.* 1992, 164: 1–18.

Broadie, K.S. Development of electrical properities and synaptic transmission at the embryonic neuromuscular junction. *Neuromuscular Junctions Drosophila* 1999, 43: 45–67.

Brown, K.J. et al. A novel *in vitro* assay for human angiogenesis. *Lab. Invest.* 1996, 75(4): 539–555.

Calve, S., Arruda, E., Dennis, R.G., and Grosh, K. Influence of mechanics on tendon and muscle development. WCCM Abstracts, 2002.

Campbell, P.G., Durham, S.K., Hayes, J.D., Suwanichkul, A., and Powell, D.R.. Insulin-like growth factor-binding protein–3 binds fibrinogen and fibrin. *J. Biol. Chem.* 1999, 274(42): 30215–30221.

Cederna, P.S., Kalliainen, L.K., Urbanchek, M.G., Rovak, J.M., and Kuzon, W.M. "Donor" muscle structure and function following end-to-side neurorrhaphy. *Plast. Reconstr. Surg.* 2001, 107: 789–796.

Close, R. Effects of cross-union of motor nerves to fast and slow skeletal muscles. *Nature* 1965, 206: 831.

Close, R. Dynamic properties of fast and slow skeletal muscles of rat after nerve cross-union. *J. Physiol. Lond.* 1969, 204: 331.

Close, R. and Hoh, J.F.Y. Effects of nerve cross-union on fast-twitch and slow-graded muscle fibres in toad. *J Physiol. Lond.* 1968, 198: 103.

Close, R. and Hoh, J.F.Y. Post-tetanic potentiation of twitch contractions of cross-innervated rat fast and slow muscles. *Nature* 1969, 221: 179.

Condorelli, G., Borello, U., De Angelis, L., Latronico, M., Sirabella, D., Coletta, M., Galli, R., Balconi, G., Follenzi, A., Frati, G., Cusella De Angelis, M.G., Gioglio, L., Amuchastegui, S., Adorini, L., Naldini, L., Vescovi, A., Dejana, E., and Cossu, G. Cardiomyocytes induce endothelial cells to trans-differentiate into cardiac muscle: implications for myocardium regeneration. *Proc. Natl Acad. Sci. USA* 2001, 98(19): 10733–10738.

Conrad, G.W. et al. Differences *in vitro* between fibroblast-like cells from cornea, heart, and skin of embryonic chicks. *J. Cell Sci.* 1977, 26: 119–137.

Dennis, R.G. Bipolar implantable stimulator for long-term denervated muscle experiments. *Med. Biol. Eng. Comput. Med. Biol. Eng. Comput.* 1998 March, 36: 225–228.

Dennis, R.G. Engineered skeletal muscle: nerve and tendon tissue interfaces, contractility, excitability, and architecture. In: *Functional Tissue Engineering*, Guilak, F., Butler, D., Mooney, D., and Goldstein, S. (eds) Springer-Verlag, New York, 2003.

Dennis, R.G. and Kosnik, P. Excitability and isometric contractile properties of mammalian skeletal muscle constructs engineered *in vitro. In Vitro Cell. Dev. Biol. Anim.* 2000, 36(5): 327–335.

Dennis, R.G., Dow, D.E., Hsueh, A., and Faulkner, J.A. Excitability of engineered muscle constructs, denervated and denervated-stimulated muscles of rats, and control skeletal muscles in neonatal, young, adult, and old mice and rats. *Biophys. J.* 2002, 82: 364A.

Dial, K.P. and Biewener, A.A. Pectoralis muscle force and power output during different modes of flight in pigeons (*Columba livia*). *J. Exp. Biol.* 1993, 176: 31–54.

Dickinson, M.H., Farley, C.T., Full, R.J., Koehl, M.A.R., Kram, R., and Lehman, S. How animals move: an integrative view. *Science* 2000, 288: 100–106.

Doane, K.J. and Birk, D.E. Fibroblasts retain their tissue phenotype when grown in three-dimensional collagen gels. *Exp. Cell Res.* 1991, 195(2): 432–442.

Drew, A.F., Liu, H., Davidson, J.M., Daugherty, C.C., and Degen, J.L. Wound-healing defects in mice lacking fibrinogen. *Blood* 2001, 97(12): 3691–3698.

El Oakley, R.M., Ooi, O.C., Bongso, A., and Yacoub, M.H. Myocyte transplantation for myocardial repair: a few good cells can mend a broken heart. *Ann. Thorac. Surg.* 2001, 71(5): 1724–1733. Review.

Evans, C.E. and Trail, I.A. Fibroblast-like cells from tendons differ from skin fibroblasts in their ability to form three-dimensional structures *in vitro. J. Hand Surg. Br. Eur.* 1998, 23B(5): 633–641.

Haase, S., Cederna, P., Dennis, R.G., and Kuzon, W. Peripheral nerve reconstruction using acellular nerve grafts. *Surg. Forum* 2000, 51: 607–609.

Hatta, I., Sugi, H., et al. Stiffness changes in frog skeletal muscle during contraction recorded using ultrasonic waves. *J. Physiol. Lond.* 1988, 403: 193–209.

Hedenstrom, A. and Alerstram, T. Climbing performance of migrating birds as a basis for estimating limits for fuel-carrying capacity and muscle work. *J. Exp. Biol.* 1992, 164: 19–38.

Hoh, J.F.Y. and Close, R. Effects of nerve cross-union on twitch and slow-graded muscle fibres in toad. *Austr. J. Exp. Biol. Med. Sci.* 1967, 45: 51.

Kardon, G. Muscle and tendon morphogenesis in the avian hind limb. *Development* 1998, 125: 4019–4032.

Kosnik, P. and Dennis, R.G., Mesenchymal cell culture: functional mammalian skeletal muscle constructs. In: *Methods in Tissue Engineering*, Chapter 23, Atala, A. and Lanza, R. (eds), Harcourt, Academic Press, San Diego, California, 2002, 299–306.

Kosnik, P. Jr., Dennis, R.G., and Faulkner, J.A. Functional development of engineered skeletal muscle from adult and neonatal rats. *Tissue Eng.* 2001, 7(5): 573–584.

Loeb, G.E. and Gans, C. *Electromyography for experimentalists.* University of Chicago Press, Chicago, Illinois, 1986.

Lutz, G.J. and Rome, L.C. Built for jumping: the design of the frog muscular system. *Science* 1994, 263: 370–372.

Menasche, P., Hagege, A.A., Scorsin, M., Pouzet, B., Desnos, M., Duboc, D., Schwartz, K., Vilquin, J.T., and Marolleau, J.P. Myoblast transplantation for heart failure. *Lancet* 2001, 357(9252): 279–280.

Mensinger, A.F., Anderson, D.J., Buchko, C.J., Johnson, M.A., Martin, D.C., Tresco, P.A., Silver, R.B., and Highstein, S.M. Chronic recording of regenerating VIIIth nerve axons with a sieve electrode. *J. Neurophysiol.* 2000, 83: 611–615.

Mott, P.H. and Roland, C.M. Mechanical and optical behavior of double network rubbers. *Macromolecules* 2000, 33: 4132–4137.

Orlandi, C., Paskinshurlburt, A.J., and Hollenberg, N.K. The microvascular response to growth-factors in the Hamster-Cheek pouch. *Basic Res. Cardiol.* 1986, 81(3): 238–243.

O'Toole, G. et al. A review of therapeutic angiogenesis and consideration of its potential applications to plastic and reconstructive surgery. *Br. J. Plastic Surg.* 2001, 54(1): 1–7.

Peplowski, M.M. and Marsh, R.L. Work and power output in the hindlimb muscles of Cuban tree frogs *Osteopilus septentrionalis* during jumping. *J. Exp. Biol.* 1997, 200: 2861–2870.

Powell, C.A., Smiley, B.L., and Vandenburgh, H.H. Novel techniques for measuring tension development in organized tissue constructs. *FASEB J.* 2000, 14: A444.

Putnam, A.J., Cunningham, J.J., Dennis, R.G., Linderman, J.J., and Mooney, D.J. Microtubule assembly is regulated by externally-applied strain in cultured smooth muscle cells. *J. Cell Sci.* 1998, 111: 3379–3387.

Sahni, A. and Francis, C.W. Vascular endothelial growth factor binds to fibrinogen and fibrin and stimulates endothelial cell proliferation. *Blood* 2000, 96(12): 3772–3778.

Sahni, A., Odrljin, T., and Francis, C.W. Binding of basic fibroblast growth factor to fibrinogen and fibrin. *J. Biol. Chem.* 1998, 273(13): 7554–7559.

Sahni, A., Sporn, L.A., and Francis, C.W. Potentiation of endothelial cell proliferation by fibrin(ogen)-bound fibroblast growth factor–2. *J. Biol. Chem.* 1999, 274(21): 14936–14941.

Saunders, D.K. and Klemm, R.D. Seasonal changes in the metabolic properties of muscle in blue-winged teal, *Anas discors. Comp. Biochem. Physiol.* 1993, 107A: 63–68.

Schneider, B.L., Peduto, G., and Aebischer, P. A self-immunomodulating myoblast cell line for erythropoietin delivery. *Gene Ther.* 2001, 8(1): 58–66.

Suzuki, K., Murtuza, B., Suzuki, N., Smolenski, R.T., and Yacoub, M.H. Intracoronary infusion of skeletal myoblasts improves cardiac function in doxorubicin-induced heart failure. *Circulation* 2001a,104(12 Suppl 1): I213–I217.

Suzuki, K., Brand, N.J., Allen, S., Khan, M.A., Farrell, A.O., Murtuza, B., Oakley, R.E., and Yacoub, M.H. Overexpression of connexin 43 in skeletal myoblasts: relevance to cell transplantation to the heart. *J. Thorac. Cardiovasc. Surg.* 2001b, 122(4): 759–766.

Vandenburgh, H.H., Swasdison, S., and Karlisch, P. Computer-aided mechanogenesis of skeletal muscle organs from single cells *in vitro. FASEB J.* 1991, 5: 2860–2867.

Woledge, R., Curtin, N., and Homsher, E. Energetic aspects of muscle contraction. *Monographs of the Physiological Society, No. 41.* Academic Press, San Diego, California, 1985.

REFERENCES

Asahara, T. et al. Isolation of putative progenitor endothelial cells for angiogenesis. *Science* 1997, 275: 964–967.

Bahary, N. and Zon, L.I. Development: enhanced: endothelium — chicken soup for the endoderm. *Science* 2001, 294(5542): 530–531.

Chizeck, H.J., Crago, P.E., and Kofman, L. Robust closed loop control of isometric muscle force using pulsewidth modulation. *IEEE Trans. Biomed. Eng.* 1988, 35(7): 510–517.

Delp, M.D. and Pette, D. Morphological-changes during fiber-type transitions in low-frequency-stimulated rat fast-twitch muscle. *Cell Tissue Res.* 1994, 277: 363–371.

Dennis, R.G. and Kosnik, P. Mesenchymal cell culture: instrumentation and methods for evaluating engineered muscle. In: *Methods in Tissue Engineering*, Chapter 24, Atala, A. and Lanza, R. (eds), Harcourt, Academic Press, San Diego, California, 2002, pp. 307–316.

Dennis, R.G., Kosnik, P., Gilbert, M.E., and Faulkner, J.A. Excitability and contractility of skeletal muscle engineered from primary cultures and cell lines. *Am. J. Physiol. Cell. Physiol.* 2001, 280: C288–C295.

Dennis, R.G., Dow, D.E., and Faulkner, J.A. An implantable device for stimulation of denervated muscles in rats. *Med. Eng. Phys.* 2003, 25(3), 239–253.

Dow, D.E., Cederna, P.S., Hassett, C.A., Faulkner, J.A., and Dennis, R.G. Numbers of contractions to maintain mass and force of a denervated rat muscle, muscle and nerve. 2004, 30(1): 77–86.

Ecob, M.S. The application of organotypic nerve cultures to problems in neurology with special reference to their potential use in research into neuromuscular diseases. *J. Neurol. Sci.* 1983, 58: 1–15.

Ecob, M. The location of neuromuscular-junctions on regenerating adult-mouse muscle in culture. *J. Neurol. Sci.* 1984, 64: 175–182.

Ecob, M. and Whalen, R.G. The role of the nerve in the expression of myosin heavy-chain isoforms in a nerve muscle-tissue culture system. *J. Muscle Res. Cell Motil.* 1985, 6: 56.

Ecob, M.S., Butler Browne, G.S., and Whalen, R.G. The adult fast isozyme of myosin is present in a nerve–muscle tissue-culture system. *Differentiation* 1983, 25: 84–87.

Eser, P.C., Donaldson, N., Knecht, H., and Stussi, E. Influence of different stimulation frequencies on power output and fatigue during FES-cycling in recently injured SCI people. *IEEE Trans. Neural Syst. Rehabil. Eng.* 2003, 11(3), 236–240.

Farahat, W. and Herr, H. An apparatus for generalized characterization and control of muscle. *IEEE Transact. Neural Syst. Rehabil. Eng.* 2005 (in press).

Gillis, G.B. Anguilliform locomotion in an elongate salamander (*Siren intermedia*): effects of speed on axial undulatory movements. *J. Exp. Biol.* 1997, 200: 767–784.

Gillis, G.B. Environmental effects on undulatory locomotion in the American eel *Anguilla rostrata*: kinematics in water and on land. *J. Exp. Biol.* 1998, 201: 949–961.

Green, H.J., Reichmann, H., and Pette, D. Fiber type specific transformations in the enzyme-activity pattern of rat vastus lateralis muscle by prolonged endurance training. *Pflugers Archiv-Eur. J. Physiol.* 1983, 399: 216–222.

Green, H.J., Klug, G.A., Reichmann, H., Seedorf, U., Wiehrer, W., and Pette, D. Exercise-induced fiber type transitions with regard to myosin, parvalbumin, and sarcoplasmic-reticulum in muscles of the rat. *Pflugers Archiv-Eur. J. Physiol.* 1984, 400: 432–438.

Hall, S. Axonal regeneration through acellular muscle grafts. *J. Anat.* 1997, 190(Pt 1): 57–71.

Herr, H. and Dennis, B.A. Swimming robot actuated by living muscle tissue. *J. NeuroEng. Rehabil.* 2004, 1: 6.

Hollerbach, J.M., Hunter, I.W., and Ballantyne, J. A comparative analysis of actuator technologies for robotics. In: *The Robotics Review*. Khatib, O., Craig, J., and Lozano-Perez, T. (eds), MIT Press, Cambridge, Massachusetts, 1991, pp. 301–342.

Jezernik, S., Wassink, R., and Keller, T. Sliding mode closed loop control of FES: Controlling the shank movement, *IEEE Trans. Biomed. Eng.* 2004, 51(2): 263–272.

Kaushal, S., Amiel, G.E., Guleserian, K.J., Shapira, O.M., Perry, T., Sutherland, F.W., Rabkin, E., Moran, A.M., Schoen, F.J., Atala, A., Soker, S., Bischoff, J., and Mayer, J.E. Jr. Functional small-diameter neovessels createdusing endothelial progenitor cells expanded *ex vivo*. *Nat Med.* 2001, 7(9): 1035–1040.

Lin, Y., Weisdorf, D.J., Solovey, A., and Hebbel, R.P. Origins of circulating endothelial cells and endothelial outgrowth from blood. *J. Clin. Invest.* 2000, 105: 71–77.

Meijer, K., Bar-Cohen, Y., and Full, R. Biological inspiration for musclelike actuators of robots. In: *Biologically Inspired Intelligent Robots*. Bar-Cohen, Y. and Breazeal, C. (eds), SPIE Press, Bellington, Washington, 2003, pp. 25–41.

Murphy, W.L., Dennis, R.G., Kileny, J., and Mooney, D.J. Salt fusion: a method to improve pore interconnectivity within tissue engineering scaffolds. *Tissue Eng*, 2002, 8(1): 43–52.

Swasdison, S. and Mayne, R. *In vitro* attachment of skeletal muscle fibers to a collagen gel duplicates the structure of the myotendinous junction. *Exp. Cell Res.* 1991, 193: 220–231.

Trotter, J.A. Functional morphology of force transmission in skeletal muscle. *Acta Anat.* 1993, 146: 205–222.

Veltink, P.H., Chizeck, H.J., Crago, P.E., and El-Bialy, A. Nonlinear joint angle control of artificially stimulated muscle, *IEEE Trans. Biomed. Eng.* 1992, 39(4): 368–380.

Vernon, R.B. and Sage, E.H. A novel, quantitative model for study of endothelial cell migration and sprout formation within three-dimensional collagen matrices. *Microvasc. Res.* 1999, 57(2): 118–133.

WEBSITES

http://www.bme.unc.edu/~bob/
http://biomech.media.mit.edu/

Artificial Muscles Using Electroactive Polymers

Yoseph Bar-Cohen

CONTENTS

10.1 INTRODUCTION

Natural muscles are one of the most important actuators in biological systems larger than a bacterium. They are highly optimized since they are fundamentally driven by the same mechanism in all animals and the difference between species is relatively small. The drive mechanism of muscles is complex and they are capable of lifting large loads with short time response in the range of milliseconds. Human made actuators that most closely emulate muscles are the recently emerged electroactive polymers (EAP) that exhibit a large strain in response to electrical stimulation. For this response, EAP have earned the moniker "artificial muscles" (Bar-Cohen, 2001, 2004). The impressive advances in improving their actuation strain capability are attracting the attention of engineers and scientists from many different disciplines. These materials are particularly attractive to biomimetics since they can be used to mimic the movements of humans, animals, and insects for making biologically inspired intelligent robots (Bar-Cohen and Breazeal, 2003) and other biomimetic mechanisms. Increasingly, engineers are able to develop EAP-actuated mechanisms that were previously imaginable only in science fiction.

For many decades, it has been known that certain types of polymers can change shape in response to electrical stimulation. Initially, these EAP materials were capable of producing only a relatively small strain. Since the beginning of the 1990s, new EAP materials emerged exhibiting large strains and leading to a great paradigm change with regard to the capability of EAP and their potential. Generally, EAP materials can generate strains that are as high as two orders of magnitude greater than the striction-limited, rigid, and fragile piezoelectric ceramics. Further, EAP materials are superior to shape memory alloys (SMA) in higher response speed, lower density, and greater resilience. They can be used to make mechanical devices without the need for traditional components like gears, and bearings, which are responsible for the current high costs, weight, and premature failures. The current limitations of EAP materials that include low actuation force, mechanical energy density, and robustness constrain the practical application but improvements in the field are expected to overcome these limitations.

In 1999, in recognition of the need for international cooperation among the developers, users, and potential sponsors, the author initiated a related annual SPIE conference as part of the Smart Structures and Materials Symposium (Bar-Cohen, 1999). This conference was held in Newport Beach, California, USA and was the largest ever on this subject, marking an important milestone and turning the spotlight onto these emerging materials and their potential. The SPIE EAP Actuators and Devices (EAPAD) conferences are now organized annually and have been steadily growing in number of presentations and attendees. Also, the author releases the semiannual WW-EAP Newsletter electronically, and mentors a website that archives related information and includes links to homepages of EAP research and development facilities worldwide (http://eap.jpl.nasa.gov). In the past few years, in addition to the SPIE conferences, several other conferences and special sessions within conferences focusing on EAP actuators have also taken place.

10.2 HISTORY AND CURRENTLY AVAILABLE ACTIVE POLYMERS

The beginning of the field of EAP can be traced back to an 1880-experiment that was conducted by Roentgen using a rubber-band with fixed end and a mass attached to the free end that was subjected to electric field across the rubber-band (Roentgen, 1880). Sacerdote (1899) followed this experiment with a formulation of the strain-response to electric field activation. Further milestone progress was recorded in 1925 with the discovery of a piezoelectric polymer called electret when carnauba wax, rosin, and beeswax were solidified by cooling while subjected to a DC-bias field (Eguchi, 1925). Generally, electrets are polymer materials with aligned electrical dipole moments equivalent to magnets, and they are deformed when subjected to voltage across them. However, their strain and work output is generally too low to be applicable as actuators and therefore their use has been limited to sensors.

Generally, electrical excitation is only one of the types of stimulators to cause elastic deformation in polymers. Other activation mechanisms include chemical (Kuhn et al., 1950; Otero et al., 1995), thermal (Li et al., 1999), pneumatic (Chou and Hannaford, 1994), optical (van der Veen and Prins, 1971), and magnetic (Zrinyi et al., 1997). Polymers that are chemically stimulated were discovered over half-a-century ago when collagen filaments were demonstrated to reversibly contract or expand upon dipping into acidic or alkaline aqueous solutions, respectively (Katchalsky, 1949). Even though relatively little has since been done to apply such "chemomechanical" as practical actuators, this early work pioneered the development of synthetic polymers that mimic biological muscles (Steinberg et al., 1966). The convenience and practicality of electrical stimulation, and technology progress led to a growing interest in EAP materials. Following the 1969 observation of a substantial piezoelectric activity in polyvinylidene fluoride (PVDF) (Bar-Cohen et al., 1996; Zhang et al., 1998), investigators started to examine other polymer systems, and a series of effective materials have emerged. The largest progress in EAP materials development has been reported in the last 15 years (Bar-Cohen, 2001, 2004) where materials that can create linear strains that can reach up to 380% have been developed (Pelrine et al., 2000; Kornbluh et al., 2004).

Generally, EAP can be divided into two major groups based on their activation mechanism: ionic (involving mobility or diffusion of ions) and electronic (driven by electric field or Maxwell Forces) (Bar-Cohen, 2001, 2004). A summary of the advantaged and disadvantages of these two group of materials are listed in Table 10.1. The electronic polymers (electrostrictive, electrostatic, piezoelectric, and ferroelectric) are driven by Maxwell Forces and can be made to hold the induced displacement under activation of a DC voltage, allowing them to be considered for robotic applications. Also, these materials have a greater mechanical energy density and they can be operated in air with no major constraints. However, they require a high activation field (>100-V/μm) close to the breakdown level. In contrast, ionic EAP materials (gels, polymer–metal composites, conductive polymers, and carbon nanotubes) are driven by diffusion of ions and they require an electrolyte for the actuation mechanism. Their major advantage is the

TABLE 10.1   A Summary of the Advantages and Disadvantages of the Two Basic EAP Groups

EAP type	Advantages	Disadvantages
Electronic EAP	• Can operate for a long time in room conditions • Exhibits rapid response (milliseconds) • Can hold strain under DC activation • Induces relatively large actuation forces • Exhibits high mechanical energy density	• Requires high voltages (~100 MV/m). Recent development allowed for (~20 MV/m) in the Ferroelectric EAP • Independent of the voltage polarity, it produces mostly monopolar actuation due to associated electrostriction effect
Ionic EAP	• Natural bi-directional actuation that depends on the voltage polarity • Requires low voltage • Some ionic EAP like conducting polymers have a unique capability of bi-stability	• Requires using an electrolyte • Requires encapsulation or protective layer in order to operate in open air conditions • Low electromechanical coupling efficiency • Electrolysis occurs in aqueous systems at more than 1.23 V • Except for CPs and NTs, ionic EAPs do not hold strain under DC voltage • Slow response (fraction of a second) • Bending EAPs induce a relatively low actuation force • Except for CPs, it is difficult to produce a consistent material (particularly IPMC) • High currents require rare earth electrodes such as gold or platinum

requirement for drive voltages as low as 1 to 2 V. However, it is necessary to maintain wetness, and except for conductive polymers and carbon nanotubes it is difficult to sustain DC-induced displacements. The induced displacement of both the electronic and ionic EAP can be geometrically designed to bend, stretch, or contract. Any of the existing EAP materials can be made to bend with a significant curving response, offering actuators with an easy to see reaction and an appealing response. However, bending actuators have relatively limited applications due to the low force or torque that can be induced.

10.3 TYPES OF ELECTROACTIVE POLYMERS

The types of EAP in each of the two material groups are given below:

10.3.1 Electronic EAP

The electronic types of EAP consist of materials that are squeezed by the attraction force between the charged electrodes. Some materials in this group are internally affected to augment the squeezing effect. The dielectric elastomer sustains a deformation under the stress induced by the electric field driven by Maxwell Forces (Zhang et al., 2004). The other type of electronic EAP materials is internally active and sustains realignment of the molecular structure (e.g., pendant group on a backbone chain) or shift of atoms in the crystallographic structure in response to the field. The electronic type of EAP materials include:

10.3.1.1 Dielectric Elastomer EAP

Polymers with low elastic stiffness and high dielectric breakdown strength can be used to produce a large actuation strain by subjecting them to an electrostatic field. Such EAP materials also known as dielectric elastomer EAP can be represented by a parallel plate capacitor as shown in Figure 10.1 (Pelrine et al., 1998; Kornbluh et al., 2004). The electrodes are highly compliant so as not to impede the deformation of the film. The strain that results from applying an electric field is proportional to the square of the electric field square, multiplied by the dielectric constant, and it is inversely proportional to the elastic modulus. To reach the required electric field levels, one needs to use high voltage or employ thin films or both. Under an electric field, the film is squeezed in the thickness direction, causing expansion in the transverse planar direction.

The above characteristic allows for the production of linear actuators using dielectric elastomer films that appear to act similar to biological muscles. For this purpose, SRI International scientists constructed actuators using two elastomer layers with carbon electrodes on both sides of one of the layers, where the layers were rolled to form a cylindrical actuator. The rolled actuator expands longitudinally as a result of the Poisson effect of transverse expansion of the film that is caused by its being squeezed by Maxwell Forces. Thereby, the lateral expansion leads to a lateral stretch of the

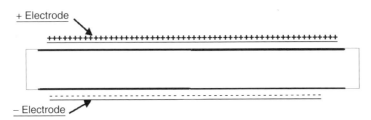

Figure 10.1   Under electric field a dielectric elastomer with electrodes on both surfaces expands laterally.

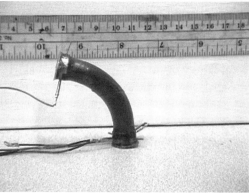

Figure 10.2 (See color insert following page 302) Two-DOF Spring Roll. (Courtesy of SRI International, Menlo Park, CA, U.S.A.)

EAP. A variety of other actuator configurations are also possible (Kornbluh et al., 2004). Dielectric elastomer EAP actuators require large electric fields (~100 V/μm) and can induce significant levels of strain (10 to about 380%). Overall, the associated voltages are close to the breakdown strength of the material, and a safety factor that lowers the potential is used. Moreover, the relatively small breakdown strength of air (2 to 3 V/μm) presents an additional challenge for packaging the actuator. Another concern that is associated with the use of dielectric elastomer EAP is that the actuator requires prestraining the elastomer film and over time the prestrain is released due to creep degrading the actuator performance. Research at Sungkyunkwan University, Korea, shows promise with regard to actuators that do not require prestrain (Jung et al., 2004). Recently, an SRI International research team designed a multi-functional electroelastomer roll (MER) in which highly prestrained dielectric elastomer EAP films were rolled around a compression spring to form an actuator (Pei et al., 2002, 2004). By selectively actuating only certain regions of electrodes around the periphery of the actuator, the actuator can be made to bend as well as elongate. An example of this MER actuator is shown in Figure 10.2 and it represents advancement in making practical EAP-based actuators with a standard configuration.

10.3.1.2 Ferroelectric Polymers

Piezoelectricity was discovered in 1880 by Pierre and Paul-Jacques Curie, who found that when certain types of crystals are compressed (e.g., quartz, tourmaline, and Rochelle salt), along certain axes, a voltage is produced on the surface of the crystal. The year afterward, they observed the reverse effect that upon the application of an electric voltage these crystals sustain an elongation. Piezoelectricity is found only in noncentro-symmetric materials and the phenomenon is called ferroelectricity when a nonconducting crystal or dielectric material exhibits spontaneous electric polarization. There are also polymers with ferroelectric behavior and the most widely exploited one is the poly(vinylidene fluoride), which is known as PVDF or PVF2, and its copolymers (Bar-Cohen et al., 1996). These polymers are partly crystalline, with an inactive amorphous phase, having a Young's modulus near 1 to 10 GPa. This relatively high elastic modulus offers a relatively high mechanical energy density. A large applied electric field (~200 MV/m) can induce electrostrictive (nonlinear) strains of nearly 2% (Zhang et al., 2004). Unfortunately, this level of field is dangerously close to dielectric breakdown, and the dielectric hysteresis (loss, heating) is very large. In 1998, Zhang and his coinvestigators introduced defects into the crystalline structure using electron radiation to increase the dielectric constant of the copolymer P(VDF-TrFE). As a result, electrostrictive strains as large as 5% were demonstrated at low frequency drive fields having amplitudes of about 150 V/μm. Furthermore, the polymer has a high elastic modulus (~1 GPa), and the field-induced strain can operate at frequencies higher than 100 kHz, resulting in a very high elastic power density compared with other EAP materials.

The irradiation of P(VDF-TrFE) into a relaxor ferroelectric with high electrostriction also introduces many undesirable defects to the copolymer including the formation of crosslinkings, radicals, and chain scission (Mabboux and Gleason, 2002; Bharti et al., 2000). By a proper molecular design which enhances the degree of molecular level conformational changes in the polymer, terpolymers are produced that can exhibit a higher electromechanical response than the high energy electron irradiated copolymer (Zhang et al., 2004).

Ferroelectric EAP polymer actuators can be operated in air, vacuum, and water. To reduce the, level of voltage that is needed to activate the ferroelectric EAP, Zhang et al. (2004) used an all-organic composite with organic particulates that have a high dielectric constant ($K > 10,000$). These researchers used a blend of the particulates in a polymer matrix to increase the dielectric constant from single digits to the range from 300 to 1,000 (at 1 Hz). This approach led to an EAP that requires significantly lower voltage as predicted by the author (Section 14.2.2 in Bar-Cohen, 2001). For example, a strain of about 2% is generated by a field of 13 V/μm for a CuPc-PVDF-based terpolymer composite having an elastic modulus of 0.75 GPa. A photographic view of this EAP in passive and activated states is shown in Figure 10.3. One of the challenges to using this material in practical applications is the large dielectric losses involved.

10.3.1.3 Electrostrictive Graft Elastomers

In 1998, a graft-elastomer EAP was developed at NASA Langley Research Center (Su et al., 1999). This EAP material exhibits a large electric-field-induced strain due to electrostriction (Zhang

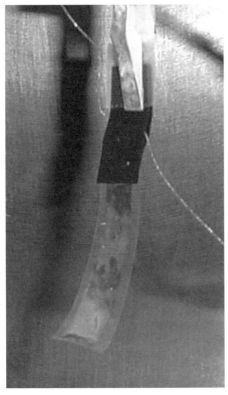
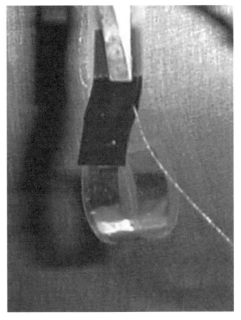

Figure 10.3   A photographic view of ferroelectric EAP in passive (left) and activated (right) states. The material was constructed in a bimorph configuration to turn the contraction to bending.

et al., 2004). This electrostrictive polymer consists of two components, a flexible backbone macromolecule and a grafted polymer that can form a crystalline structure. The grafted crystalline polar phase provides moieties in response to an applied electric field and cross-linking sites for the elastomer system. This material offers high electric-field-induced strain (~4%), relatively high electromechanical power density, and excellent processability. Combination of the electrostrictive-grafted elastomer with a piezoelectric poly(vinylidene fluoride-trifluoroethylene) copolymer yields several compositions of a ferroelectric–electrostrictive molecular composite system. Such a combination can be operated both as a piezoelectric sensor and as an electrostrictive actuator. Careful selection of the composition allows the creation and optimization of the molecular composite system with respect to its electrical, mechanical, and electromechanical properties.

10.3.1.4 Electrostrictive Paper

The use of paper as an electrostrictive EAP actuator was demonstrated at Inha University, Korea (Kim et al., 2000). Paper is composed of a multitude of discrete particles, mainly of a fibrous nature, which form a network structure. Since paper is produced in various mechanical processes with chemical additives, it is possible to prepare a paper that has enhanced electroactive properties. Such an EAP actuator has been prepared by bonding two silver laminated papers with silver electrodes placed on the ouside surface. When an electric voltage is applied to the electrodes the actuator produces bending displacement, and its performance depends on the excitation voltages, frequencies, type of adhesive, and the host paper. Studies indicate that the electrostriction effect that is associated with this actuator is the result of electrostatic forces and an intermolecular interaction of the adhesive. The demonstrated actuator is lightweight and simple to fabricate. Various applications that are currently being considered include active sound absorbing materials, flexible speakers, and smart shape-control devices. The energy density of these materials and electromechanical coupling is relatively small.

10.3.2 Ionic EAP

Ionic EAP materials consist of a polymer with electrolyte and two electrodes where the phenomenon of electroactivation involves diffusion of ions through the thickness of the polymer (Bar-Cohen, 2004). This group includes of the following types of EAP materials:

10.3.2.1 Ionic Polymer Gels

Polymer gels can be synthesized to produce significant actuation forces potentially matching the force and energy density of biological muscles (Calvert, 2004). These materials (e.g., polyacrylonitrile) are generally activated by a chemical reaction, changing from an acidic to an alkaline environment causing the gel to become dense or swollen, respectively. This reaction can be stimulated electrically as was shown by researchers at the University of Arizona, U.S.A. (Liu and Calvert, 2000). When activated, these gels bend as the cathode side becomes more alkaline and the anode side more acidic. However, the response of this multilayered gel structure is relatively slow because of the need to diffuse ions through the gel. A significant amount of research and development efforts and application considerations using ionic gel polymers (IPG) were explored at the Hokkaido University, Japan (Osada and Kishi, 1989). The polymers that were explored include electrically induced bending of gels and electrical induced reversible volume change of gel particles. Further, Schreyer and his coinvestigators at the University of New Mexico, Albuquerque used a combination of ionic gel and conductive polymer electrodes to demonstrate an effective EAP actuator (Schreyer et al., 2000; Section 10.6).

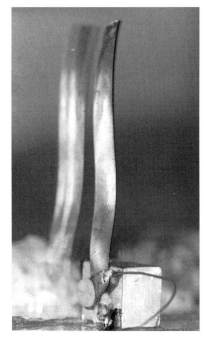

Figure 10.4   IPMC in relaxed (left) and activated states (right).

10.3.2.2 Ionomeric Polymer–Metal Composites

Ionomeric polymer–metal composites (IPMC) is an EAP that bends in response to an electrical activation (Figure 10.4) as a result of mobility of cations in the polymer network (Nemat-Nasser and Thomas, 2004). In 1992, IPMC was realized to have this electroactive characteristic by three groups of researchers: Oguro et al. (1992) in Japan, as well as Shahinpoor (1992) and Sadeghipour et al. (1992) in the U.S. The operation as actuators is the reverse process of the charge storage mechanism associated with fuel cells (Heitner-Wirguin, 1996; Holze and Ahn, 1992). A relatively low voltage is required to stimulate bending in IPMC, where the base polymer provides channels for mobility of positive ions in a fixed network of negative ions on interconnected clusters. Two types of base polymers are widely used to form IPMC: Nafion$^{\circledR}$ (perfluorosulfonate made by DuPont, U.S.A.) and Flemion$^{\circledR}$ (perfluorocarboxylate, made by Asahi Glass, Japan). In order to chemically electrode the polymer films, metal ions (platinum, gold, or others) are dispersed throughout the hydrophilic regions of the polymer surface and are subsequently reduced to the corresponding zero-valence metal atoms.

Generally, the ionic content of the IPMC is an important factor in the electromechanical response of these materials (Bar-Cohen et al., 1999; Nemat-Nasser and Li, 2000). Examining the bending response shows that using low voltage (1 to 5 V) induces a large bending at frequencies below 1 Hz, and the displacement significantly decreases with the increase in frequency. The bending response of IPMC was enhanced using Li+ cations that are small and have higher mobility or large tetra-n-butylammonium cations that transport water in a process that is still being studied. The actuation displacement of IPMC was further increased using gold metallization as a result of the higher electrode conductivity (Abe et al., 1998; Oguro et al., 1999).

10.3.2.3 Conductive Polymers

Conductive polymers (CP) typically function via the reversible counter-ion insertion and expulsion that occurs during redox cycling (Otero et al., 1995; Sansiñena and Olazabal, 2004). Oxidation and reduction occur at the electrodes, inducing a considerable volume change due mainly to the exchange of ions with an electrolyte. A sandwich of two conductive polymer electrodes (e.g., polypyrrole or polyaniline (PAN) doped in HCl) with an electrolyte between them forms an EAP actuator. When a voltage is applied between the electrodes, oxidation occurs at the anode and reduction at the cathode. Ions (e.g., H^+ in PAN) migrate and diffuse between the electrolyte and the electrodes to balance the electric charge. Addition of the ions causes swelling of the polymer and conversely their removal results in shrinkage and as a result the sandwich bends (Figure 10.5). One of the parameters that affects the response is the thickness of the layers; thinner layers are faster (as fast as 40 Hz) (Madden et al., 2000, 2001) but induce lower force. Since strong shear forces act on the electrolyte layer, attention is needed to protect the material from premature failure. In addition, actuators made of this type of EAP materials are sensitive to cyclic operation and they tend to fail after several tens of cycles. Conductive polymer actuators generally require voltages in the range of 1 to 5 V, and the speed increases with the voltage having relatively high mechanical energy densities of over 20 J/cm^3 but with low efficiencies at the level of 1% if no electrical energy is recovered (Madden et al., 2002).

In recent years, several conductive polymers were reported, including polypyrrole, polyethylenedioxythiophene, poly(p-phenylene vinylene)s, polyanilines, and polythiophenes (Anquetil et al., 2002). Complexes between polypyrrole and sulfonated detergents offer relatively good stability in aqueous media, but they are relatively soft compared to other conjugated polymers. Most actuators that use conductive polymers exploit voltage-controlled swelling to induce bending. Conjugated polymer microactuators were first fabricated at Linköpings University, Sweden. Among the devices that were demonstrated include a miniature box that can be opened and closed electrically (Smela et al., 1995). Efforts are also being made to develop linear CP actuators where a number of groups have used polypyrrole for this purpose. The reported actuators were measured to produce a stress of approximately 5 MPa with moderate strains (~ 2%) (Kaneto et al., 1995). Current efforts at Eamax, Japan, seem to suggest significant increase in the actuation capability that can be obtained using conductive polymer EAP materials.

Figure 10.5   Conductive EAP actuator is shown bending under stimulation of 2 V, 50 mA.

10.3.2.4 Carbon Nanotubes

In 1999, carbon nanotubes (CNT) with diamond-like mechanical properties emerged as formal EAP (Baughman et al., 1999; Spinks et al., 2004). The carbon–carbon bond in nanotubes (NT) that are suspended in an electrolyte and the change in bond length are responsible for the actuation mechanism. A network of conjugated bonds connects all carbons and provides a path for the flow of electrons along the bonds. The electrolyte forms an electric double layer with the nanotubes and allows injection of large charges that affect the ionic charge balance between the NT and the electrolyte. The more the charges injected into the bond the larger the dimension changes. Removal of electrons causes the nanotubes to carry a net positive charge, which is spread across all the carbon nuclei causing repulsion between adjacent carbon nuclei and increasing the C–C bond length. Injection of electrons into the bond also causes lengthening of the bond resulting in an increase in nanotube diameter and length. These dimension changes are translated into macroscopic movement in the network element of entangled nanotubes and the net result is extension of the CNT.

Considering the mechanical strength and modulus of the individual CNTs and the achievable actuator displacements, this actuator has the potential of producing higher work per cycle than any previously reported actuator technologies and of generating much higher mechanical stress. However, to date such energy densities and forces have not been realized in macro-scale devices. The material consists of nanometer-size tubes and was shown to induce strains in the range of 1% along the length. A CNT actuator can be constructed by laminating two narrow strips that are cut from a CNT sheet using an intermediate adhesive layer, which is electronically insulated. The resulting "cantilever device" is immersed in an electrolyte such as a sodium chloride solution, and an electrical connection is made in the form of two nanotube strips. Application of about 1.0 V is sufficient to cause bending of the actuator, and the direction depends on the polarity of the field with a response that is approximately quadratic relationship between the strain and charge.

10.4 EAP CHARACTERIZATION

Accurate and detailed information about the properties of EAP materials is critical to designers of related mechanisms or devices. To assess the competitive capability of EAPs, a performance matrix that consists of comparative performance data is necessary. Such a matrix needs to show the properties of EAP materials as compared to other classes of actuators, including piezoelectric ceramic, shape memory alloys, hydraulic actuators, and conventional motors. Studies are currently underway to define a unified matrix and establish effective test capabilities (Sherrit et al., 2004). Test methods are being developed to allow measurements with minimum effect on the EAP material. While the electromechanical properties of electronic-type EAP materials can be addressed with some of the conventional test methods, ionic-type EAPs such as IPMC pose technical challenges. The response of these materials suffers complexities associated with the mobility of the cation on the microscopic level, strong dependence on the moisture content, and hysteretic behavior. Video cameras and image processing softwares allow the study of the deformation of IPMC strips under various mechanical loads. Simultaneously, the electrical properties and the response to electrical activation can be measured. Nonlinear behavior has been clearly identified in both the mechanical and electrical properties and efforts were made to model this behavior (Sherrit et al., 2004).

10.5 APPLICATIONS OF EAP

Compared to existing actuators, EAPs have properties that potentially make them very attractive for a wide variety of biomimetic applications. As polymers, EAP materials can be easily formed in various shapes, their properties can be engineered and they can potentially be integrated with microelectromechanical system (MEMS) sensors to produce smart actuators. Unfortunately, the materials that have been developed so far still exhibit low conversion efficiency and are not robust; besides there are no standard commercial materials available for consideration in practical applications. To be able to take these materials from the development phase to application as effective actuators, it is necessary to adequately establish the field infrastructure (Bar-Cohen, 2004).

In recent years, significant progress has been made in the field of EAP towards making practical actuators, and commercial products are beginning to emerge. The first milestone product, a biomimetic device in the form of a fish-robot, was announced by Eamax, Japan, at the end of 2002. Moreover, a growing number of organizations are exploring potential applications for EAP materials, and cooperation across disciplines is helping overcome related challenges. The mechanisms and devices that are being considered or developed are applicable to aerospace, automotive, medical, robotics, exoskeletons, articulation mechanisms, entertainment, animation, toys, clothing, haptic and tactile interfaces, noise control, transducers, power generators, and smart structures. Some of the biologically inspired applications of EAP are discussed in the following section.

10.5.1 Artificial Organs and Other Medical Applications

Considering the use of EAP for artificial organs requires addressing a number of challenges. These challenges include biological compatibility — avoiding rejection — and ability to meet the stringent functional requirements to operate as organ replacements. Currently, electronic EAP materials seem to be most applicable since they are highly robust and they generate the largest actuation forces. However, the required voltage range — from hundreds to thousands of voltage — presents concerns. Even though the electric current is relatively low, the use of high voltage levels can cause such dangers as producing blood clots or injury due to potential voltage breakdown and short circuits in the body. On the other hand, the ionic group of EAP materials is chemically sensitive requiring careful protection. It is also difficult to maintain their static position, particularly for the IPMC, because these materials involve chemical reaction and even DC voltage causes a reaction.

Interfacing between human and machine to complement or substitute our senses may enable important capabilities for medical applications. A number of such interfaces have been investigated or considered. Of notable significance in this area is the ability to interface machines and the human brain. A development by scientists at Duke University (Wessberg et al., 2000; Mussa-Ivaldi, 2000) enabled this possibility where electrodes were connected to the brain of a monkey, and using brain waves, the monkey operated a robotic arm, both locally and remotely via the Internet. Success in developing EAP-actuated robotic arms with the strength and dexterity to win a wrestling match against a human opponent (Bar-Cohen, 2004) can greatly benefit from this interface development by neurologists. Using such a capability to control prosthetics would require feedback to allow the human operator to "feel" the remote or virtual environment at the artificial limbs. Such feedback can be provided with the aid of tactile sensors, haptic devices, and other interfacing mechanisms. Besides providing feedback, sensors will allow users to protect the prosthetic from potential damage (heat, pressure, impact, etc.) just as we do our biological limbs. The development of EAP materials that can provide tactile sensing is currently under way as described in (Bar-Cohen, 2004).

The growing availability of EAP materials that exhibit high actuation displacement and force is opening new avenues to bioengineering in terms of medical devices for diagnosis, treatment, and assistance to humans in overcoming different forms of disability. Applications that are currently being considered include catheter steering mechanism (Della Santa et al., 1996), vein connectors for repair after surgery (Jager et al., 2000; http://www.micromuscle.com/1024.htm), smart prosthetics (Herr and Kornbluh, 2004), Braille displays (Bar-Cohen, 2004), and others. Recent research at the Sungkyunkwan University, Korea, has led to the development of a series of mechanisms and devices that use dielectric elastomer EAP (Jung et al., 2004). These devices include a smart pill, a tube-like

Figure 10.6   Different views of the smart pill that is a tube-like biomimetic moving mechanism. (Courtesy of Hyoukryeol Choi, Sungkyunkwan University, Suwon, Kyunggi-do, Korea.)

biomimetically moving mechanism with inchworm motion for traveling inside the gastrointestinal track. The flexible skin of the smart pill was fabricated using a 3-D molding technique. A photographic view of this smart pill is shown in Figure 10.6. Using dielectric EAP, another application is currently investigated at the Sungkyunkwan University, Korea, is the development of a Braille display for the visually impaired. It was designed to be compatible with existing Braille devices and its performance is under evaluation. Blind patients are given display patterns of letters and symbols and they are asked to recognize them. A photographic view of the display and its mode of test with a finger placed on the device are shown in Figure 10.7. In Figure 10.8, a photographic view of a blind person testing the new EAP Braille display is seen. The use of dielectric elastomer EAP for Braille display has also been a subject of study at SRI (Heydt and Chhokkar, 2003; Kornbluh et al., 2004) where a simple mechanism was constructed taking advantage of the large strains and high energy density of this EAP material. For this purpose, individually addressable diaphragm actuators were developed at the scale of Braille dots of: 1.5 mm diameter and 2.3 mm center-to-center spacing. The resulting 2-mm-diameter diaphragm actuators that were made of acrylic films exhibited pressures of up to 25 kPa (3.7 psi) resulting in 10–25 g of actuation force on the Braille dot that is needed for easy reading. This approach is scalable to large numbers of cells, and is expected to enable the building of refreshable displays with many lines of characters at an affordable price.

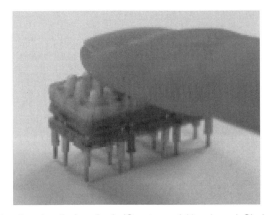

Figure 10.7   Braille display for visually impaired. (Courtesy of Hyoukryeol Choi, Sungkyunkwan University, Suwon, Kyunggi-do, Korea.)

Figure 10.8 Evaluation of the EAP-based braille display device for the visually impaired. (Courtesy of Hyoukryeol Choi, Sungkyunkwan University, Suwon, Kyunggi-do, Korea.)

10.5.2 EAP-Actuated Biomimetic Robots

With current technology, the appearance and behavior of biological creatures are animated in various simulation studies, computer generated imagery (CGI), and commercial movies (Blumberg, 2003). However, in past years, engineering such biomimetic intelligent creatures as realistic robots was a significant challenge due to the physical and technological constraints and the shortcomings of available technology. Making robots that can hop and land safely without risking damage to the mechanism, receive pushes while staying stable or make body and facial expressions of joy and excitement — which are very natural for humans and animals — are extremely complex to engineer. The use of AI, effective artificial muscles, and other biomimetic technologies is expected to make the possibility of realistically looking and behaving robots into practical engineering models (Bar-Cohen and Breazeal, 2003). Mimicking nature would immensely expand the collection and functionality of the robots allowing performance of tasks that are impossible with existing capabilities.

The field of artificial muscles offers many important capabilities for the engineering of robots that are inspired by biological models and systems. The capability to produce EAP in various shapes and configurations can be exploited using such methods as stereolithography and ink-jet processing techniques. Potentially, a polymer can be dissolved in a volatile solvent and ejected drop-by-drop onto various substrates. Such rapid prototyping processing methods may lead to mass-produced robots in full 3D details including the actuators allowing rapid prototyping and quick transition from concept to full production (Bar-Cohen et al., 2004).

Using EAP actuators, biologically inspired robots may be developed with capabilities that are far superior to natural creatures since they are not constrained by evolution and survival needs. Examples may include artificial bugs that walk, swim, hop, crawl, and dig while reconfiguring themselves as needed. An important addition to this capability can be the application of telepresence combined with virtual reality using haptic interfacing. While such capabilities are expected to significantly change future robots, additional effort is needed to develop robust and effective polymer-based actuators. Potential actuators may include the dielectric elastomer or the ferroelectric EAP.

To promote the development of effective EAP actuators in support of future development of such future biomimetic robots, toys, and animatronics, two testbed platforms were developed (see Figure 10.9). These platforms are available at the author's JPL's NDEAA lab and they include an

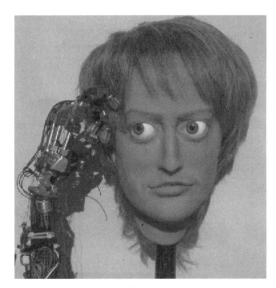

Figure 10.9 (See color insert following page 302)   An android head and a robotic hand that serve as biomimetic platforms for the development of artificial muscles. (*Acknowledgement*: This photograph was taken at JPL where the head was sculptured and instrumented by D. Hanson, University of Texas, Dallas. The hand was made by G. Whiteley, Sheffield Hallam University, U.K.)

android head that can make facial expressions and a robotic hand with activatable joints. Currently, conventional electric motors produce the deformations required for the android to make relevant facial expressions of the Android. Once effective EAP materials are available they will be modeled into the head. There they can receive control instructions for the creation of desired facial expressions. The robotic hand is equipped with sensors for the operation of the various joints and is capable of mimicking the human hand. The index finger of this hand is currently being driven by conventional motors to establish a baseline. The motors would be replaced by EAP when they are developed as effective actuators.

10.5.2.1 Gripper and Robotic Arm Lifter

To mimic a biological hand using simple elements, the author and his coinvestigators constructed a miniature robotic arm that was lifted by a rolled dielectric elastomer EAP (Section 10.3.1.1) as a linear actuator and four IPMC-based fingers as a bending actuator (Chapter 21 in Bar-Cohen, 2004). The linear actuator was used to raise and drop a graphite or epoxy rod which served as a simplistic representation of a robotic arm. Unfortunately, after activating this actuator, the arm sustains a series of oscillations that need to be dampened to allow accurate positioning. This requires sensors and a feedback loop to support the kinematics of the system control. Several alternatives were explored, including establishment of a self-sensing capability, but more work is needed before such an arm can become practical. To produce an end-effector for the arm, a four finger gripper was developed (see Figure 10.10). These bending actuator fingers were made of IPMC (Section 10.3.2.2) strips with hooks at the bottom emulating fingernails. As shown in Figure 10.10, this gripper grabs rocks just like the human hand.

10.5.2.2 Biologically Inspired Robots for Planetary Robotics

The evolution in the capabilities that are inspired by biology has increased to a level where more sophisticated and demanding fields, such as space science, are considering the use of such robots. At JPL, four- and six-legged robots are currently being developed for consideration in future

Figure 10.10 (See color insert following page 302)   Four-finger EAP gripper lifting a rock.

missions to such planets as Mars. Such robots include the Limbed Excursion Mobile Utility Robot (LEMUR) shown in Figure 10.11. These robots are still mobilized by conventional motors. In contrast, the use of EAP can lead to capabilities that are identified as biomimetic including running on complex terrains at the speed of a horse or even a Cheetah.

Such robots would potentially be mobile in complex terrains, acquire and analyze samples, and perform many other functions that are attributed to legged animals including grasping and object manipulation. This evolution may potentially lead to the use of life-like robots in future NASA missions that involve landing on various planets. Equipped with multi-functional tools and multiple cameras, the LEMUR robots are intended to inspect and maintain installations beyond humanity's easy reach in space. This spider-like robot has six legs, each of which has interchangeable end-effectors to perform the required mission. The radially symmetric layout is a lot like a starfish or octopus, and it has a panning camera system that allows omni-directional movement and manipulation operations.

The possibility of making an aircraft that flaps its wing as a bird was recently envisioned by (Colozza et al., 2004) who conceived a flying machine that uses EAP as an actuator supported by polymeric solar cells on the wings. This solid state aircraft does not use any conventional moving parts, where the airfoil, propulsion, energy production, energy storage, and control are integrated into the structure (Figure 10.12). For the purpose of producing the required bending deformation at low voltage, IPMC-based wings were considered for the development of the concept of Solid State Aircraft. Using a flight profile similar to a hawk or eagle, the conceived aircraft can potentially soar for long periods of time and utilize flapping to regain lost altitude. By analyzing the glide and flap durations, wing length, and distance of travel, it has been determined that a number of design configurations can be produced to enable flight over a range of latitudes and times of the year on Earth, Venus, and possibly Mars. The implementation of this capability will require significant advances in the development of IPMC towards producing significantly higher energy efficient actuators.

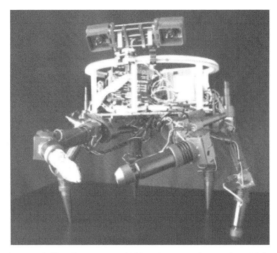

Figure 10.11 (See color insert following page 302) A new class of multi-limbed robots called Limbed Excursion Mobile Utility Robot (LEMUR) is under development at JPL. (Courtesy of Brett Kennedy, JPL.)

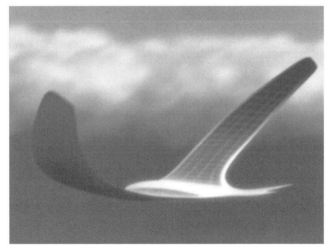

Figure 10.12 Artists' drawing of the solid state IPMC-based aircraft concept. (Courtesy of Anthony Colozza, Ohio Aerospace Institute, Cleveland, Ohio, U.S.A.)

10.6 MILESTONE FOR THE FIELD

Improved collaboration among developers, users, and sponsors as well as increased resources with a growing number of investigators have led to rapid progress in the field. In the last two years several milestones have been made in this field:

(a)   In December 2002, the first commercial product emerged in the form of a Fish-Robot (Eamex, Japan). An example of this Fish-Robot can be seen at http://eap.jpl.nasa.gov where there is also a link to a video showing Fish-Robots swimming in a fish tank. These robots swim without batteries or motor and use EAP materials that simply bend on stimulation. For power they use inductive coils that are energized from the top and bottom of the fish tank.

In 1999, in an effort to promote worldwide development towards the realization of the potential of EAP materials an arm wrestling challenge was posed to the engineers worldwide (Bar-Cohen,

Figure 10.13 An artistic interpretation of the Grand Challenge for the development of EAP-actuated robotics.

2001). Success in developing such an arm will lead to the possible use of EAP to replace damaged human muscles, that is making "bionic human." A remarkable contribution of the EAP field would be to see, one day, a handicapped person jogging to the grocery store using this technology. A graphic rendering of this challenge that was posed by the author is illustrated in Figure 10.13. The intent of posing this challenge was to use the human arm as a baseline for the implementation of the advances in the development of EAP materials. Success in wrestling against humans will enable biomimetic capabilities that are currently considered impossible. It would allow applying EAP materials to improve many aspects of our life where some of the possibilities include smart implants and prosthetics (also known as cyborgs), active clothing (de Rossi et al., 1997), realistic biologically inspired robots as well as fabricating products with unmatched capabilities. Recent advances in understanding the behavior of EAP materials and the improvement of their efficiency led to the historical first competition held in March 2005. In this competition, three robotic arms participated and the human opponent was a 17-year-old female student. The three arms were made by Environmental Robots Incorporated (ERI), New Mexico; Swiss Federal Laboratories for Materials Testing and Research, EMPA, Dubendorf, Switzerland; and three senior students from the Engineering Science and Mechanics Department, Virginia Tech.

1. The arm that was made by Environmental Robots Incorporated (ERI), New Mexico held for 26 sec against the 17-year-old student. This wrestling arm (see Figure 10.14) had the size of an average human arm and it was made of polypropylene and Derlin. This arm was driven by two groups of artificial muscle. One group consisted of dielectric elastomeric resilient type that was used to maintain an equilibrium force and the other was composed of ionic polymer–metal composites (IPMC) type strips that flex to increase or decrease the main resilient force.
2. The Materials Testing and Research, EMPA, Dubendorf, Switzerland, arm (see Figure 10.15) held for 4 sec before losing. This arm was driven by the dielectric elastomer type using multi-layered scrolled actuators that were organized in four groups. A photo of one of the group lifting two 5-gallon water containers (about 20-kg) is shown in Figure 10.16. Using electronic control, these actuators were operated similar to human muscles, where two of these groups acted as protagonists and the other two operated as antagonists. The arm had an outer shell made of fiberglass that was used as a shield for the electric section. The arm structure was made of composite sandwich consisting of fiberglass and carbon fibers.

Figure 10.14   The ERI arm wrestling with the 17-year-old human opponent, Panna Felsen. This arm has the size of an average human arm and it managed to last for 26 sec against Panna.

Figure 10.15   The arm that was made by the Swiss Company, EMPA, is shown wrestling with Panna Felsen. The rubber glove that the Panna is using provided her electrical insulation for protection.

3.   The arm that was made by the three senior students from the Engineering Science and Mechanics Department, Virginia Tech (see Figure 10.17) managed to last 3 sec. As an EAP actuator they constructed batches of polyacrylonitrile (PAN) gel fibers that were designed to operate as artificial muscles. This EAP material was shown experimentally to produce close to 200% linear strain and

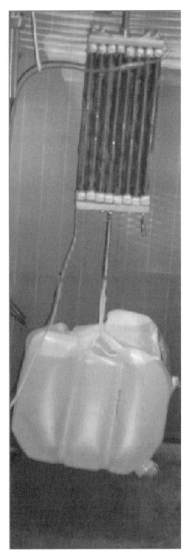

Figure 10.16   One of the groups of EAP actuators made by EMPA lifting two 5-gallon water containers.

pulling strength that is higher than human muscles (Schreyer et al., 2000). To encase the fibers and chemicals that make up their EAP actuator, they designed an electrochemical cell. For the skeleton of the arm they used a structure that is made of composite material and, for support, this structure was connected to an aluminum base.

This competition has been a very important milestone for the field and helped accomplish the goals of this challenge, namely:

1. promote advances towards making EAP actuators that are superior to the performance of human muscles;
2. increase the worldwide visibility and recognition of EAP materials;
3. attract interest among potential sponsors and users;
4. lead to general public awareness since it is hoped that they will be the end users and beneficiaries in many areas including medical, commercial, and other fields.

Figure 10.17 (See color insert following page 302)   The Virginia Tech students' arm being prepared for the match against Panna Felsen, the 17-year-old student from San Diego.

10.7   SUMMARY AND OUTLOOK

For many years, EAP received relatively little attention due to their limited actuation capability and the small number of available materials. In the last 15 years, a series of new EAP materials have emerged that exhibit large displacement in response to electrical stimulation. The capability of these new materials is making them attractive as actuators for their operational similarity to biological muscles, particularly their resilience, damage tolerance, and ability to induce large actuation strains (stretching, contracting, or bending). The application of these materials as actuators to drive various manipulation, mobility, and robotic devices involves multi-disciplines including materials, chemistry, electromechanics, computers, and electronics. Even though the force of actuation of existing EAP materials and their robustness require further improvement, there has already been a series of reported successes in the development of EAP-actuated mechanisms. Successful devices that have been reported include a fish-robot, audio speakers, catheter-steering element, miniature manipulator and gripper, active diaphragm, and dust wiper. The field of EAP has enormous potential in many application areas, and, judging from the range of inquiries that the author has received since his start in this field in 1995, it seems that almost any aspect of our lives can potentially be impacted. Some of the considered applications are still far from being practical, and it is important to tailor the requirements to the level that current materials can address. Using EAP to replace existing actuators may be a difficult challenge and therefore it is highly desirable to identify niche applications where EAP materials would not need to compete with existing technologies.

Space applications are among the most demanding in terms of the harshness of the operating conditions, requiring a high level of robustness and durability. Making biomimetic capability using EAP material will potentially allow NASA to conduct missions in other planets using robots that emulate human operation ahead of a landing of human. For an emerging technology, the requirements and challenges associated with making hardware for space flight are very difficult to overcome. However, since such applications usually involve producing only small batches, they can provide an important avenue for introducing and experimenting with new actuators and devices. This is in contrast to commercial applications, for which issues of mass production, consumer demand, and cost per unit can be critical to the transfer of technology to practical use. Some of the challenges that are facing the users of EAP materials in expanding their potential applications to space include their capability to respond at low or high temperatures. Space applications are of great need for materials that can operate at single digit degrees of Kelvin or at temperatures as high as hundreds of Celsius as on Venus. Another challenge to EAP is the development of large scale EAP in the form of films, fibers, and others. The required dimensions can be as large as several meters or kilometers and in such dimensions they can be used to produce large gossamer structures such as antennas, solar sails, and various large optical components.

In order to exploit the highest benefits from EAP, multidisciplinary international cooperative efforts need to grow further among scientists, engineers, and other experts (e.g., medical doctors, etc.). Experts in chemistry, materials science, electromechanics or robotics, computer science, electronics, etc., need to advance the understanding of the material behavior, as well as develop EAP materials with enhanced performance, processing techniques, and applications. Effective feedback sensors and control algorithms are needed to address the unique and challenging aspects of EAP actuators. If EAP-driven artificial muscles can be implanted into a human body, this technology can make a tremendously positive impact on many human lives.

This field of EAP is far from mature and progress is expected to change the field in future years. Recent technology advances led to the development of three EAP-actuated robotic arms that wrestled with a 17-year-old female student who was the human opponent in the competition held on March 7, 2005. Even though the 17-year-old student won against the three arms the competition helped increase the visibility of the field worldwide and the recognition of its potential. While more work is needed to reach the level of winning against humans it is inevitable that this would happen just like the chess game between the champion and the Big Blue IBM computer (http://www.geocities.com/siliconValley/lab/7378/comphis.htm). Initially, the challenge is to win a wrestling match against a human (any human) using a simple shape arm with minimum functionality. However, the ultimate goal is to win against the strongest human using as close as possible a resemblance of the shape and performance of the human arm. Once such a robotic arm wins against humans, it would become clear that EAP performance has reached a level where devices can be designed and produced with the many physical functions of humans with far superior capability. Such a success is one of the ultimate goals of the field of biomimetics.

ACKNOWLEDGMENT

Some of the research reported in this chapter was conducted at the Jet Propulsion Laboratory (JPL), California Institute of Technology under a contract with National Aeronautics and Space Administration (NASA).

REFERENCES

Abe Y., A. Mochizuki, T. Kawashima, S. Tamashita, K. Asaka, and K. Oguro, Effect on bending behaviour of counter cation species in perfluorinated sulfonate membrane–platinum composite, *Polymers for Advanced Technologies*, Vol. 9 (1988), 520–526.

Anquetil P.A., H.-H. Yu, P.G. Madden, J.D. Madden, T.M. Swager, and I.W. Hunter, Thiophene-based molecular actuators, in Bar-Cohen Y. (ed.), *Proceedings of SPIE 9th Annual Symposium on Smart Structures and Materials: Electroactive Polymer Actuators and Devices*, SPIE Press, San Diego, California (March 2002) pp. 424–434.

Bar-Cohen Y. (ed.), *Proceedings of the First SPIE's Electroactive Polymer Actuators and Devices (EAPAD) Conference, Smart Structures and Materials Symposium*, Vol. 3669, ISBN 0-8194-3143-5 (1999), pp. 1–414.

Bar-Cohen Y. (ed.), *Electroactive Polymer (EAP) Actuators as Artificial Muscles — Reality, Potential and Challenges*, ISBN 0-8194-4054-X, SPIE Press, San Diego, California, Vol. PM98 (March 2001), pp. 1–671.

Bar-Cohen Y., *Electroactive Polymer (EAP) Actuators as Artificial Muscles — Reality, Potential and Challenges*, 2nd Edition, ISBN 0-8194-5297-1, SPIE Press, San Diego, California, Vol. PM136 (March 2004), pp. 1–765.

Bar-Cohen Y. and C. Breazeal (eds), *Biologically-Inspired Intelligent Robots*, SPIE Press, San Diego, California, Vol. PM122, ISBN 0-8194-4872-9 (May 2003), pp. 1–393.

Bar-Cohen Y., T. Xue, and S.-S. Lih, Polymer Piezoelectric Transducers For Ultrasonic NDE, *First International Internet Workshop on Ultrasonic NDE, Subject: Transducers*, organized by R. Diederichs, UTonline Journal, Germany (September 1996).

Bar-Cohen Y., S. Leary, M. Shahinpoor, J.O. Harrison, and J. Smith, Electro-active polymer (EAP) actuators for planetary applications, *Proceedings of SPIE's EAPAD Conference, 6th Annual International Symposium on Smart Structures and Materials*, Vol. 3669 (1999), pp. 57–63.

Bar-Cohen Y., V. Olazábal, J. Sansiñena, and J. Hinkley, Processing and support techniques, Chapter 14, in Bar-Cohen Y. (ed.), *Electroactive Polymer (EAP) Actuators as Artificial Muscles — Reality, Potential and Challenges*, 2nd Edition, ISBN 0-8194-5297-1, SPIE Press, San Diego, California, Vol. PM136 (March 2004), pp. 431–463.

Baughman R.H., C. Cui, A.A. Zakhidov, Z. Iqbal, J.N. Basrisci, G.M. Spinks, G.G. Wallace, A. Mazzoldi, D. de Rossi, A.G. Rinzler, O. Jaschinski, S. Roth, and M. Kertesz, Carbon nanotube actuators, *Science*, Vol. 284 (1999), 1340–1344.

Bharti V., H. Xu, G. Shanthi, Q.M. Zhang, and K. Liang, Polarization and structural properties of high energy electron irradiated P(VDF-TrFE) copolymer films, *Journal of Applied Physics*, Vol. 87 (2000) 452–461.

Blumberg B., Biomimetic animated creatures, Chapter 3, in Bar-Cohen Y. and C. Breazeal (eds), *Biologically-Inspired Intelligent Robots*, SPIE Press, San Diego, California, Vol. PM122, ISBN 0-8194-4872-9 (May 2003), pp. 47–71.

Calvert P., Electroactive polymer gels, Chapter 5, in Bar-Cohen, Y. (ed.), *Electroactive Polymer (EAP) Actuators as Artificial Muscles — Reality, Potential and Challenges*, 2nd Edition, ISBN 0-8194-5297-1, SPIE Press, San Diego, California, Vol. PM136 (March 2004), pp. 95–148.

Chou C.P. and B. Hannaford, Static and dynamic characteristics of McKibben pneumatic artificial muscles, *Proceedings of the IEEE International Conference on Robotics and Automation*, San Diego, California, May 8–13 (1994), Vol. 1, pp. 281–286.

Colozza A., C. Smith, M. Shahinpoor, K. Isaac, P. Jenkins, and T. DalBello, Solid state aircraft concept overview, *Proceedings of the 2004 NASA/DoD Conference on Evolvable Hardware*, Seattle, Washington, June 24–26 (2004), pp. 318–324.

de Rossi D., A. Della Santa, and A. Mazzoldi, Dressware: wearable piezo- and thermo-resistive fabrics for ergonomics and rehabilitation, Proceedings of the 19th Annual International Conference of the IEEE Engineering in Medicine and Biology Society, Chicago, Illinois (1997).

Della Santa A., A. Mazzoldi, and D. De Rossi, Steerable microcatheters actuated by embedded conducting polymer structures, *Journal of Intelligent Material Systems and Structures*, ISSN 1045–389X, Vol. 7, No. 3 (1996), 292–301.

Eguchi M., On the permanent electret, *Philosophical Magazine*, Vol. 49 (1925), 178.

Heitner-Wirguin C., Recent advances in perfluorinated ionomer membranes: structure, properties and applications, *Journal of Membrane Science*, Vol. 120, No. 1 (1996), 1–33.

Herr H.M. and R.D. Kornbluh, New horizons for orthotic and prosthetic technology: artificial muscle for ambulation, in Bar-Cohen Y. (ed.), *Proceedings of SPIE's Smart Structures and Materials 2004: Electroactive Polymer Actuators and Devices (EAPAD)*, Vol. 5385 (2004), 1–9.

Heydt R. and S. Chhokar, Refreshable Braille display based on electroactive polymers, *Record of the 23rd International Display Research Conference*, sponsored by the Society for Information Display, Phoenix, Arizona (16–18 September 2003), pp. 111–114.

Holze R., and J.C. Ahn, Advances in the use of perfluorinated cation-exchange membranes in integrated water electrolysis and hydrogen — oxygen fuel systems, *Journal of Membrane Science*, Vol. 73, No. 1, 87–97 (1992).

Jager E.W.H., O. Inganäs, and I. Lundström, Microrobots for micrometer-size objects in aqueous media: potential tools for single cell manipulation, *Science*, Vol. 288 (2000), 2335–2338.

Jung K., J. Nam, and H. Choi, Micro-inchworm robot actuated by artificial muscle actuator based on dielectric elastomer, *Proceedings of the 2004 SPIE's EAP Actuators and Devices (EAPAD)*, paper number 5385–47, Vol. 5385, San Diego, California, March 14–18 (2004).

Kaneto K., M. Kaneko, Y. Min, and A.G. MacDiarmid, Artifical muscle: electromechanical actuators using polyaniline films, *Synthetic Metals*, Vol. 71 (1995), 2211–2212.

Katchalsky A., Rapid swelling and deswelling of reversible gels of polymeric acids by ionization, *Experientia*, Vol. V (1949), 319–320.

Kim J., J.-Y. Kim, and S.-J. Choe, Electro-active papers: its possibility as actuators, in Bar-Cohen Y. (ed.), *Proceedings of the SPIE's EAPAD Conference, Part of the 7th Annual International Symposium on Smart Structures and Materials*, Vol. 3987, ISBN 0-8194-3605-4 (2000), pp. 203–209.

Kornbluh K., R. Pelrine, Q. Pie, M. Rosenthal, S. Standford, N. NBowit, R. Heydt, H. Prahlad, and S.V. Sharstri, Application of dielectric elastomer EAP actuators, Chapter 16, in Bar-Cohen, Y. (ed.), *Electroactive Polymer (EAP) Actuators as Artificial Muscles — Reality, Potential and Challenges* (2004), pp. 529–581.

Kuhn W., B. Hargitay, A. Katchalsky, and H. Eisenburg, Reversible dilatation and contraction by changing the state of ionization of high-polymer acid networks, *Nature*, Vol. 165 (1950), 514–516.

Li F.K., W. Zhu, X. Zhang, C.T. Zhao, and M. Xu, Shape memory effect of ethylene-vinyl acetate copolymers, *Journal of Applied Polymer Science*, Vol. 71, No. 7 (1999), 1063–1070.

Liu Z. and P. Calvert, Multilayer hydrogels and muscle-like actuators, *Advanced Materials*, Vol. 12, No. 4 (2000), pp. 288–291.

Mabboux P., and K. Gleason, F-19 NMR characterization of electron beam irradiated vinyllidene fluoride-trifluoroethylene copolymers, *Journal of Fluorine Chemistry*, Vol. 113 (2002), 27.

Madden J.D.W., R.A. Cush, T.S. Kanigan, C.J. Brenan and I.W. Hunter, Fast-contracting conducting polymer-based actuators, *Synthetic Metals*, Vol. 113 (2000), 185–192.

Madden J.D.W., P.G.A. Madden, and I.W. Hunter, Characterization of polypyrrole actuators: modeling and performance, in Bar-Cohen Y. (ed.), *Proceedings of SPIE 8th Annual Symposium on Smart Structures and Materials: Electroactive Polymer Actuators and Devices*, SPIE Press, San Diego, California (March 2001), pp. 72–83.

Madden J.D.W., P.G.A. Madden, and I.W. Hunter, Conducting polymer actuators as engineering materials, in Bar-Cohen Y. (ed.), *Proceedings of SPIE 9th Annual Symposium on Smart Structures and Materials: Electroactive Polymer Actuators and Devices*, SPIE Press, San Diego, California (March 2002), pp. 176–190.

Mussa-Ivaldi S., Real brains for real robots, *Nature*, Vol. 408 (16 November 2000), 305–306.

Nemat-Nasser S. and Li, J.Y. Electromechanical response of ionic polymer–metal composites, *Journal of Applied Physics*, Vol. 87, No. 7 (2000), 3321–3331.

Nemat-Nasser S. and C.W. Thomas, Ionomeric polymer–metal composites, Chapter 6, in Bar-Cohen, Y. (ed.), *Electroactive Polymer (EAP) Actuators as Artificial Muscles,* 2nd Edition, ISBN 0-8194-5297-1, SPIE Press, San Diego, California, Vol. PM136 (March 2004), pp. 171–230.

Oguro K., Y. Kawami and H. Takenaka, Bending of an ion-conducting polymer film-electrode composite by an electric stimulus at low voltage, *Transactions Journal of Micromachine Society*, Vol. 5 (1992), 27–30.

Oguru K., N. Fujiwara, K. Asaka, K. Onishi, and S.Sewa, Polymer electrolyte actuator with gold electrodes, *Proceedings of the SPIE's 6th Annual International Sympsium on Smart Structures and Materials*, SPIE Proc. Vol. 3669 (1999), pp. 64–71.

Osada Y. and R. Kishi, Reversible volume change of microparticles in an electric field, *Journal of Chemical Society*, Vol. 85 (1989), 665–662.

Otero T.F., H. Grande, and J. Rodriguez, A new model for electrochemical oxidation of polypyrrole under conformational relaxation control, *Journal of Electroanalytical Chemistry*, Vol. 394 (1995), 211–216.

Pei Q., R.M. Rosenthal, R. Perline, S. Stanford, and R. Kornbluh. 3-D multifunctional electroelastomer roll actuators and thier application for biomimetic walking robots, in McGoWan A. (ed.), *Proceedings of SPIE's Smart Structers and Materials 2002: Industrial and Commerical Applications of Smart Structures Technology*, Vol. 4698 (2002), 246–253.

Pei Q., R. Pelrine, M. Rosenthal, S. Stanford, H. Prahlad, and R. Kornbluh. Recent progress on electroelastomer artificial muscles and their application for biomimetic robots, in Bar-Cohen, Y. (ed.) *Proceedings of SPIE EAPAD Conference*, Vol. 5385 (2004), pp. 41–50.

Pelrine R., R. Kornbluh, and J.P. Joseph, Electrostriction of polymer dielectrics with compliant electrodes as a means of actuation, *Sensor Actuator A*, Vol. 64 (1998), 77–85.

Pelrine R., R. Kornbluh, Q. Pei, and J. Joseph, High speed electrically actuated elastomers with strain greater than 100%, *Science* Vol. 287 (2000), 836–839.

Roentgen W.C., About the changes in shape and volume of dielectrics caused by electricity, Section III in Wiedemann G. (ed.), *Annual Physics and Chemistry Series*, Vol. 11, John Ambrosius Barth Publisher, Leipzig, German (1880), pp. 771–786 (in German).

Sacerdote M.P., On the electrical deformation of isotropic dielectric solids, *Journal of Physics*, 3 Series t, VIII, 31 (1899), 282–285 (in French).

Sadeghipour K., R. Salomon, and S. Neogi, Development of a novel electrochemically active membrane and 'smart' material based vibration sensor/damper, *Journal of Smart Materials and Structures*, Vol. 1, No. 1 (1992), 172–179.

Sansiñena J.M. and V. Olazabal, Condictive polymers, Chapter 7, in Bar-Cohen Y. (ed.), *Electroactive Polymer (EAP) Actuators as Artificial Muscles*, 2nd Edition, ISBN 0-8194-5297-1, SPIE Press, San Diego, California, Vol. PM136 (March 2004), pp. 231–259.

Schreyer H.B., N. Gebhart, K.J. Kim, and M. Shahinpoor, Electric activation of artificial muscles containing polyacrylonitrile gel fibers, *Biomacromolecules Journal*, ACS Publications, El Cajon, California, Vol. 1 (2000), 642–647.

Shahinpoor M., Conceptual design, kinematics and dynamics of swimming robotic structures using ionic polymeric gel muscles, *Journal of Smart Materials and Structures*, Vol. 1, No. 1 (1992), pp. 91–94.

Sherrit S., X. Bao and Y. Bar-Cohen, Methods of testing and characterization, Chapter 15, in Bar-Cohen Y. (ed.), *Electroactive Polymer (EAP) Actuators as Artificial Muscles*, 2nd Edition, ISBN 0-8194-5297-1, SPIE Press, San Diego, California, Vol. PM136 (March 2004), pp. 467–526.

Smela E., O. Inganäs, and I. Lundström, Controlled folding of micrometer-size structures, *Science*, Vol. 268 (1995), 1735–1738.

Spinks G.M., G.G. Wallace, R.H. Baughman, and L. Dai, Carbon nanotube actuators: synthesis, properties and performance, Chapter 8, in Bar-Cohen Y. (ed.), *Electroactive Polymer (EAP) Actuators as Artificial Muscles*, 2nd Edition, ISBN 0-8194-5297-1, SPIE Press, San Diego, California, Vol. PM136 (March 2004), pp. 261–295.

Steinberg I.Z., A. Oplatka, and A. Katchalsky, Mechanochemical engines, *Nature*, Vol. 210 (1966), 568–571.

Su J., J.S. Harrison, T. St. Clair, Y. Bar-Cohen, and S. Leary, Electrostrictive graft elastomers and applications, *MRS Symposium Proceedings*, Vol. 600, Warrendale, Pennsylvania (1999), pp. 131–136.

van der Veen G. and W. Prins, Light-sensitive polymers, *Nature, Physical Science*, Vol. 230 (1971), 70–72.

Wessberg J., C.R. Stambaugh, J.D. Kralik, P.D. Beck Nicolelis, M. Lauback, J.C. Chapin, J. Kim, S.J. Biggs, M.A. Srinivasan, and M.A. Nicolelis, Real-time prediction of hard trajectory by ensembles of cortical neurons in primates, *Nature*, Vol. 408 (2000), pp.361–365

Zhang Q.M., V. Bharti, and X. Zhao, Giant electrostriction and relaxor ferroelectric behavior in electron-irradiated poly(vinylidene fluoride-trifluoroethylene) copolymer, *Science*, Vol. 280 (1998), 2101–2104.

Zhang Q.M., C. Huang, F. Xia and J. Su, Electric EAP, Chapter 4, in Bar-Cohen, Y. (ed.), Electroactive polymer (EAP) actuators as artificial muscles, 2nd Edition, ISBN 0-8194-5297-1, SPIE Press, San Diego, California, Vol. PM136 (March 2004), pp. 95–148.

Zrinyi M., L. Barsi, D. Szabo, and H.G. Kilian, Direct observation of abrupt shape transition in ferrogels induced by nonuniform magnetic field, *Journal of Chemical Physics*, Vol. 106, No. 13 (1997), 5685–5692.

11

Biologically Inspired Optical Systems

Robert Szema and Luke P. Lee

CONTENTS

11.1   INTRODUCTION

Of the five senses, the mechanism of sight is perhaps the most diverse in the animal kingdom. There exist at least eight generalized types of optical systems with numerous variations within each classification. This is to be expected, as each animal-eye is tailored to the specific needs of its owner. From the defense-oriented pinhole clam eye to the night-adapted owl eye, nature has provided a plethora of examples to study and emulate.

The ability to reproduce biological optical systems using man-made materials has applications in navigation systems, specialized detectors, and in surveillance cameras. Of late, there has been particular interest within the military which has provided much of the funding towards research in this field. Advancements in materials science and manufacturing technologies have shown to be invaluable in the construction of biomimetic optics.

Biomimetic optics is a relatively new and expanding field, although it can be argued that older technologies such as photographic cameras already mimic biology by having analogous structures (i.e., glass lens to biological lens, film to retina, etc.). However, this chapter is devoted to those optical devices, which, by their design, seek to imitate living organisms. The examples that follow offer a survey of the ever-increasing attempts to reconstruct biological eyes. They are roughly divided by the general classifications of the eyes they seek to replicate; that is, they are separated into biomimetic camera (single lens) eyes, compound eyes, and others.

11.2   CAMERA EYES

Certainly the most familiar example, the human eye is but one of many forms of camera-type eyes and generally relies on a single lens to focus images onto a retina for image acquisition. The lens material properties, structure, and focusing mechanism vary from organism to organism. For example, some amphibian species have eyes which accommodate by moving a lens closer or farther to the retina. By contrast, the human eye adjusts the curvature of the lens itself to accomplish the same task. Birds have the added benefit of being able to reshape the cornea as well as the lens for accommodation. Some of the various camera eye designs are shown below (Figure 11.1).

These natural eyes have provided the inspiration for a number of optical systems with specific capability requirements. These include different approaches for adaptive optics, efficient image processing, and size-constrained wide-angle views.

11.2.1   A Fluidic Adaptive Lens

Some forms of camera eyes, oftentimes in amphibious animals, use hydraulics to adjust their focal lengths. A chamber behind the lens is filled or emptied with fluid depending on the desired focus (Figure 11.2). One example of this is the whale eye, where this design allows for good vision both in and out of the water. In addition, the fluid also compensates for increased pressure at deeper aquatic environments.

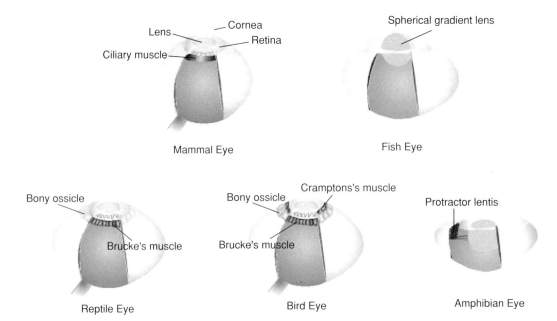

Figure 11.1 (See color insert following page 302)   Various types of camera-type eyes. The arrangement of the mammalian ciliary muscle allows for passive changes in lens thickness. Brucke's muscles attached to bony ossicles in reptiles and birds, on the other hand, actively change the lens thickness. Birds have an additional muscle, Crampton's muscle, which can alter the shape of the cornea. The protractor lentis in some amphibian eyes moves a fixed-shape lens closer to or farther from the retina for accommodation.

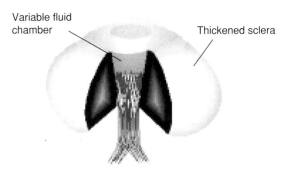

Figure 11.2 Biological fluidic adaptive lens schematic. The thickened sclera allows the eye to withstand pressures at increased diving depths.

By using similar principles, Zhang et al. (2003) at the University of California at San Diego have created an adaptive fluidic lens. The lens itself is made of an inexpensive polymer, polydimethyl-siloxane (PDMS), processed using soft lithography to include a fluid chamber and injection port. The 60-μm thick PDMS membrane is then bonded to a glass substrate using oxygen plasma bonding technology. By filling the chamber, Zhang et al. were able to demonstrate a focal length range from 41 to 172 mm, with corresponding numerical aperture values of 0.24 to 0.058. The highest recorded resolution was 25.39 lp/mm in both horizontal and perpendicular directions. These results are shown in Figure 11.3.

This unique design allows for a variable focal length system in a rather compact and robust arrangement. It should be noted, however, that it may be possible to improve upon their design by studying nature further. It is well known that a homogenous spherical lens will suffer from spherical aberration, when the peripheral light rays are refracted more than the axial ones. In the biological world, this problem has been managed in two ways. The first, and most appropriate to this design, is to have a nonspherical profile such that the periphery of the lens is flatter than the center. By following nature's example again, Zhang et al. may achieve even better results.

11.2.2 An Artificial Cephalopod Eye

As alluded to in the previous section, there is a second method nature has used to deal with spherical aberration (Land, 1988). This involves the use of a ball lens with a spherically-symmetric refractive index gradient that decreases from the center outwards (Figure 11.4). This is a particularly appropriate adaptation to the watery habitat of cephalopods, such as octopi and squid. Such an environment necessitates that the entire focusing power of the eye lie within the lens itself, as both sides of the cornea consist of essentially the same medium. In this arrangement, a spherical lens provides the shortest possible focal length. The result is a wide field of view from a relatively compact apparatus.

The theory that cephalopods use a spherical lens with a refractive index gradient was initially postulated in the latter half of the 1800s, first by Maxwell and later by Matthiessen (Land, 1988). Indeed, it was Matthiessen who determined that the ratio of the focal length to the lens radius is approximately 2.5 ("Matthiessen's ratio") in animals with lenses of this design. A precise mathematical description of the gradient was not established until 1944 (Luneberg, 1944), followed by a numerical solution in 1953 (Fletcher, 1953). Still, it was not until 33 years later that Koike et al. created an artificial ball lens with the required index of refraction gradient (Koike et al., 1986).

Besides the lens, construction of an artificial cephalopod eye involves a critical design issue. The retinas of many animals including cephalopods are curved structures, whereas man-made photodetector arrays are flat. This has much to do with the way electronics are manufactured in general, on flat semiconductor surfaces. Hung et al. (2004) have overcome this limitation by

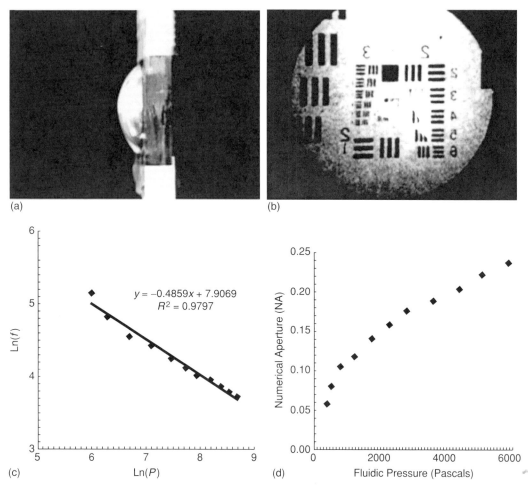

Figure 11.3   (a) Two centimeter aperture fluidic adaptive lens; (b) a picture of resolution measurements of fluidic adaptive lenses using positive standard; (c) dependence of focal length on fluidic pressure in a spherical fluidic adaptive lens. The solid line is a line fit of the data, indicating that the focal length is approximately proportional to the inverse square root of the pressure. (d) Numerical aperture versus fluidic pressure in spherical fluidic adaptive lens. (From Zhang, D., Lien, V., Berdichevsky, Y., Choi, J., and Lo, Y. *Applied Physics Letters* 2003: 82(19), 3171–3172. With permission.)

fabricating a unique array where individual photodetectors are connected by flexible structures on top of a PDMS polymer substrate. These S-flexures are a key requirement in the development of an artificial cephalopod eye, as each pixel remains connected to those around it while allowing a flexible, curve retina (Figure 11.5).

While research is ongoing, it is apparent that the development of this type of optical system will allow for a much wider field of view (180 to 200°) than conventional cameras. At the same time, the device will maintain a compact arrangement, allowing for space-efficient implementation in various applications.

11.2.3   A Foveated Imaging System

A final example of a camera-type system borrows more from the strategy of certain living organisms than from the design. A commonly observed trait in animals is the ability to scan a scene in order to increase their field of view. Many animals, including humans, have a much higher resolution in the

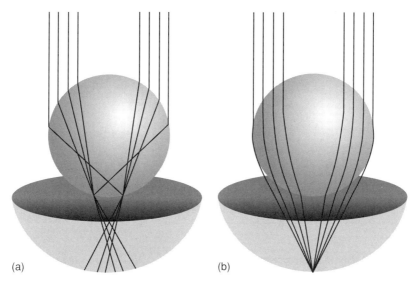

Figure 11.4   (a) Spherical aberration from a spherical lens with homogenous index of refraction; (b) properly focused rays from a spherical lens with a refractive index gradient.

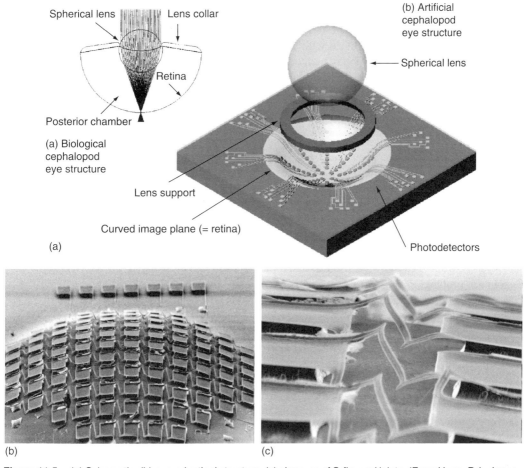

Figure 11.5   (a) Schematic, (b) curved retinal structure, (c) close-up of S-flexural joints. (From Hung, P.J., Jeong, K., Liu, G.L., and Lee, L.P. *Applied Physical Letters* 2004: 85(24), 6051–6053. With permission.)

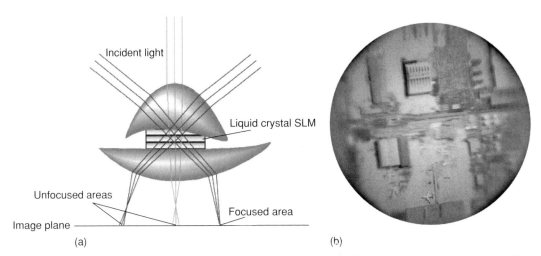

Figure 11.6 (a) A foveated imaging system where incident light rays are directed to a single imaging plane. The SLM changes the index of refraction to focus light from a specific direction. (b) Sample image from such a system. (From Martinez, T., Wick, D., and Restaino, S. *Optics Express* 2001: 8(10), 555–560. With permission.)

center of their field of view than towards the periphery. This is known as foveated imaging and allows for a relatively wide angle of view with the option of detailed resolution by scanning.

Optical engineers are often faced with a similar challenge of increasing the field of view while maintaining resolution. Traditional methods include decreasing the entrance pupil size (which increases the f/# at the expense of resolving power and illumination) and adding optical elements (increasing the complexity, size, and weight of the system). Martinez et al. (2001) have devised an artificial foveated viewing system as a unique solution.

Their design uses liquid crystal spatial light modulators (SLM), which are used to manipulate optical wavefronts. Voltages applied to liquid crystals alter the index of refraction such that aberrations are corrected. However, only aberrations from a limited range of angles can be corrected at one time (Figure 11.6). Rays of light from different field angles are not corrected, and the result is a region of high resolution surrounded by areas of low resolution.

By appropriately varying the SLM, the optical system effectively scans with a narrow field of view of high resolution while maintaining peripheral vision much like the human eye. An additional benefit of this optical system is a decreased bandwidth requirement for transmitting digital images, as only a portion of the entire field of view has high resolution. The low resolution areas may serve a purpose as well; they may be used as an initial assessment of whether an area warrants high resolution probing.

11.3   COMPOUND EYES

The appeal of insect compound eyes may be due, in part, to their being so different than our own. On the surface they also appear to be more diverse and complex with anywhere from a single ommatidium (individual eye unit) in the ant species *Pomera punctatissima* to over 10,000 per eye in some species of dragonflies. Again, the various manifestations of compound eyes are customized to the needs of their users. In general, compound eyes are broadly divided into two categories, superpositional and appositional.

The individual facets of appositional compound eyes are optically isolated, and each of them provides part of a scene. The result is a series of images slightly offset from one another (see Figure 11.7). The advantage of this arrangement is that the images are processed in parallel, leading to

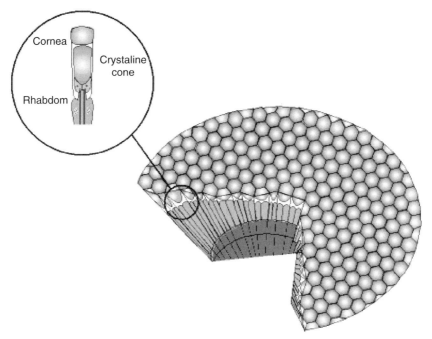

Figure 11.7   Appositional compound eye: (a) schematic, (b) electron micrograph, and (c) image projection.

very fast image recognition and motion detection. However, this comes at the expense of the brightness of the image, as only a limited amount of light is captured.

By contrast, the facets of superpositional compound eyes are not optically isolated, and, as their name would indicate, form images from the superposition of their respective light rays (see Figure 11.8). The light rays fall from multiple lenses onto a shared retina, with a concomitant increase in photosensitivity. Surprisingly, these eyes form a single erect image, which can be accomplished by cleverly designed refractive index gradients or mirrors.

In terms of biomimetic engineering, it seems that the appositional arrangement is vastly more popular than the superpositional. This may be due to the relative conceptual simplicity in setting up multiple cameras versus coordinating multiple lenses on a shared retina. Still, as will be discussed later, the superpositional compound eye is well adapted for certain fields and successes there may spur increased interest.

11.3.1 Appositional Compound Eyes

In the early 1990s, Ogata et al. (1994) fabricated an artificial compound eye and integrated electronic retina by taking advantage of integrated circuit technologies at the time. Their design used planar arrays (16×16) of gradient index of refraction rods (GRIN rods) which focused light through pinholes onto a photodetector array (with eight shades of gray scale). The pinholes acted as lens-isolating elements and the arrangement of the arrays in a circle increased the available field of view. The image output was decidedly pixelated, but this modest beginning has led to modern versions with increased resolution and color capabilities (Figure 11.9) (Tanida et al., 2003; Duparre et al., 2004).

Continuing with the theme of nonflat surfaces from their cephalopod eye design, Jeong et al. (2005) have also developed a number of appositional compound eyes arranged spherically. More faithful to biological designs than previous work, these eyes contain separate ommatidia, each with its own lens, waveguide, and photoreceptor. The difficulties lie primarily with shaping the polymers appropriately. Hemispherical substrates are produced by natural surface tension forces on droplets of liquid polymer which are then allowed to cure onto a flexible membrane. A micromolded lenslet

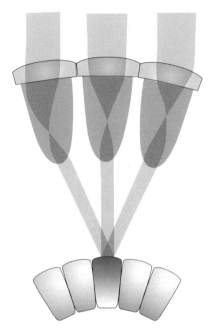

Figure 11.8   Superpositional compound eye: (a) electron micrograph and (b) schematic.

array is then superimposed upon this substrate, allowing for each lenslet to be facing a different radial direction. Subsequent exposure to light through the lenslets provides a method for defining the corresponding waveguides (Figure 11.10). The individual lenslets have been shown to have comparable fields of view to natural compound eyes.

Researchers in Israel have recently been developing an artificial version of a subset of the appositional compound eye. To biologists, it is known somewhat confusingly, as the neural super-positional eye. While the ommatidia are optically isolated from one another, the information from each one is superimposed through neural processing (Figure 11.11). In this way, the images obtained from multiple facets are compared in order to increase sensitivity. The structure is slightly different than the pure appositional eye described previously.

Rosen and Abookasis (2003) have taken this approach to develop a sensor capable of imaging through scattering media such as biological tissue. Coherent light is projected through a biological sample to a microlens array. Each microlens detects a unique speckle pattern and summation of these patterns reveal an outline of interior structures of the sample. Rosen et al. have demonstrated the ability to image bony arrangements placed between two slices of muscle tissue (Figure 11.12). While still in its initial stages, this technology has the potential for low-cost, safe, and noninvasive medical diagnoses.

11.3.2   Superpositional Compound Eyes

One of the most famous examples of biologically inspired optics comes from a paper written in 1979, entitled "Lobster eyes as x-ray telescopes" (Angel, 1979). As described in various papers of the time, lobster eyes consist of tapered tubes having rectangular cross-sections and reflective interior surfaces (Figure 11.13). These tubes are arranged spherically, and light reflecting twice from the interior surfaces behaves identically to a spherical mirror. The tubes share a common retina making the lobster eye a superpositional compound eye (Figure 11.14).

Roger Angel theorized that such a mechanism could be adapted for x-ray telescopes if each cell could be made with an aspect ratio of approximately 100. Conventional x-ray telescopes

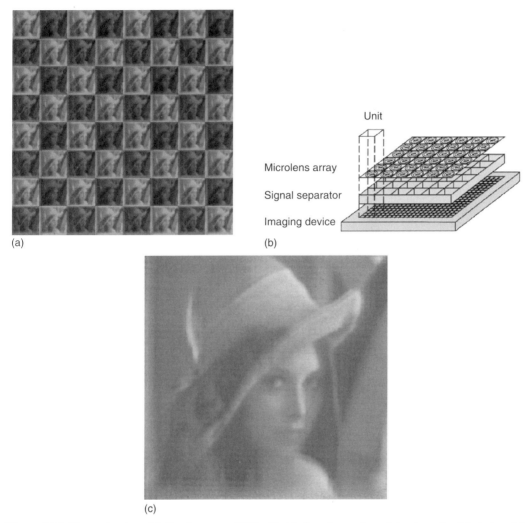

(a)

Unit

Microlens array

Signal separator

Imaging device

(b)

(c)

Figure 11.9 (See color insert following page 302) Tanida imaging system: (a) schematic, (b) device, (c) sample images. (From Tanida, J., Shogenji, R., Kitamura, Y., Yamada, K., Miyamoto, M., and Miyatake, S. *Optics Express* 2003: 11(18), 2109–2117. With permission.)

use concave mirrors, but the majority of x-rays pass through unaffected. Only at glancing angles are x-rays reflected in this manner, leaving conventional x-ray telescopes with only a $1°$ field of view. A lobster eye design, on the other hand, has the potential for a nearly unlimited field of view.

However, interest waned until Australian scientists revived the idea in 1989 and research began in earnest for a working device. While these attempts have not yet come to fruition, support for the concept of using lobster eyes has spread internationally and expanded to include satellite astronomy (Peele et al., 1996) and computer microchip processing (Chown, 1996a,b). Peele et al. have demonstrated x-ray focusing using microchannel plates manufactured by LIGA (lithographie, galvanoformung, abformung) processing of poly-methyl-methacrylate (PMMA). In the case of microchip technology, the lobster eye design is used in reverse to produce parallel x-rays (Figure 11.15). Traditionally, circuit linewidths are defined by light, with shorter wavelengths allowing smaller electronics. The wavelengths of x-rays are well-suited to this application, but it has proven difficult to generate parallel x-rays. The lobster eye design may be a feasible solution.

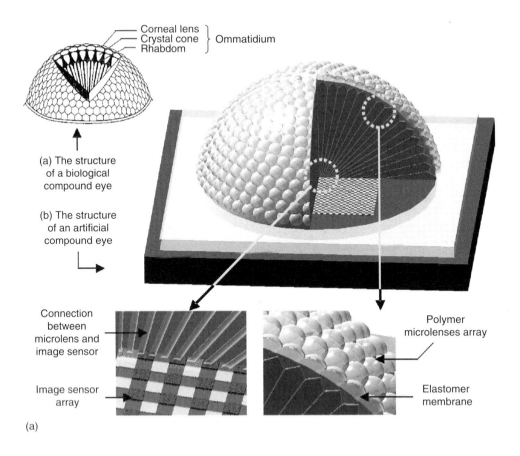

(a) The structure of a biological compound eye

(b) The structure of an artificial compound eye

Corneal lens
Crystal cone
Rhabdom
} Ommatidium

Connection between microlens and image sensor

Image sensor array

(a)

Polymer microlenses array

Elastomer membrane

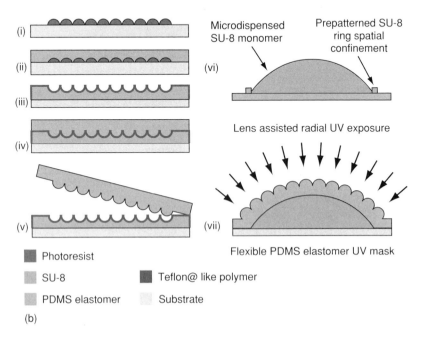

(i)

(ii)

(iii)

(iv)

(v)

(vi)

Microdispensed SU-8 monomer

Prepatterned SU-8 ring spatial confinement

Lens assisted radial UV exposure

(vii)

Flexible PDMS elastomer UV mask

Photoresist

SU-8

PDMS elastomer

Teflon@ like polymer

Substrate

(b)

Figure 11.10 (See color insert following page 302)   (*Continued*)

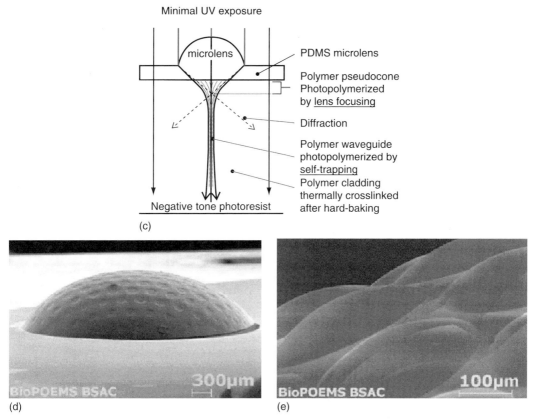

Figure 11.10 (See color insert following page 302) Artificial appositional eye approach: (a) schematic, (b) lens array process flow, (c) waveguide illustration, (d) picture of device, and (e) close-up picture of lenslets. (From Jeong, K., Kim, J., Nevill, J., and Lee, L.P. (2005). With permission.)

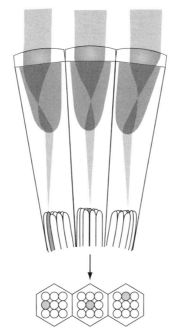

Figure 11.11 Neural superpositional eye schematic: The neural superpositional eye combines the sensitivities of all facets by comparing all retrieved images.

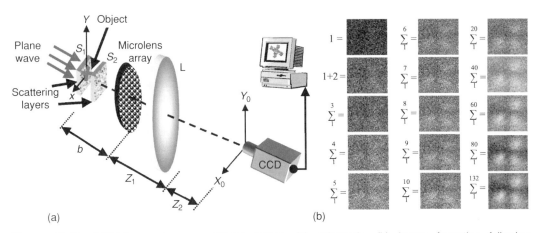

Figure 11.12 Artificial neural superpositional system (a) schematic, (b) image formation following summation over many images. (From Rosen, J. and Abookasis, D. *Optics Express* 2003: 11(26), 3605–3611. With permission.)

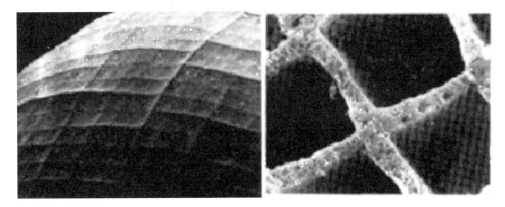

Figure 11.13   Electron micrograph of the lobster eye.

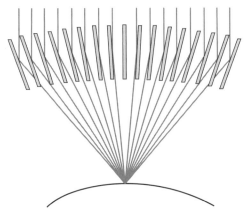

Figure 11.14   Cross-sectional schematic of lobster eye optics. Rays reflected by channel walls are focused onto a photodetector.

Figure 1.3 Bug-eating plants with traps that developed from their leaf.

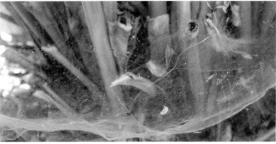

Figure 1.14 The spider constructs an amazing web made of silk material that for a given weight it is five times stronger than steel.

Figure 1.18 MACS crawling on a wall using suction cups.

Figure 1.19 JPL's Lemur, six-legged robots, in a staged operation. (Courtesy of Brett Kennedy, JPL.)

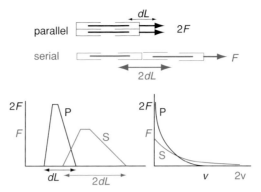

Figure 2.5 Functional effects of parallel (P) and serial (S) arrangement of sarcomeres. *F* represents force, *v* represents velocity, and *dL* represents the length ranges over which the muscle can generate force.

Figure 4.2 Evolving a controller for physical dynamic legged machine. (a) The nine-legged machine is powered by 12 pneumatic linear actuators arranged in two Stewart platforms. The controller for this machine is an open-loop pattern generator that determines when to open and close pneumatic valves. (b) Candidate controllers are evaluated by trying them out on the robot in a cage, and measuring fitness using a camera that tracks the red foot (see inset). (c) Snapshots from one of the best evolved gates. (From Zykov, V., Bongard, J., Lipson, H., (2004) Evolving dynamic gaits on a physical robot, *Proceedings of Genetic and Evolutionary Computation Conference*, Late Breaking Paper, GECCO'04. With permission.)

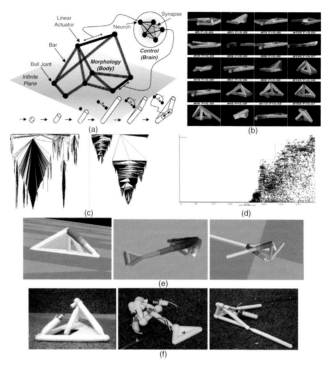

Figure 4.4 Evolving bodies and brains: (a) schematic illustration of an evolvable robot, (b) an arbitrarily sampled instance of an entire generation, thinned down to show only significantly different individuals, (c) phylogenetic trees of two different evolutionary runs, showing instances of speciation and massive extinctions from generation 0 (top) to approximately 500 (bottom), (d) progress of fitness versus generation for one of the runs. Each dot represents a robot (morphology and control), (e) three evolved robots, in simulation (f) the three robots from (e) reproduced in physical reality using rapid prototyping. (From Lipson, H., Pollack, J. B., (2000) *Nature*, 406, 974–978. With permission.)

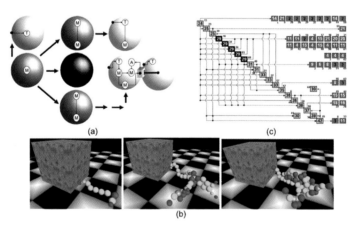

Figure 4.7 Artificial ontogeny: Growing machines using gene regulatory networks. (a) An example of cells that can differentiate into structural, passive cells (dark) or active cells (bright) which contains neurons responsible for sensing (T = touch, A = angle) and motor actuation (M). The connectivity of the neurons is determined by propagation of 'chemicals' expressed by genes and sensors, who are themselves expressed in response to chemicals in a regulatory network. (b) Three machines evolved to be able to push a block, (c) The distribution of genes responsible for neurogenesis (red) and morphogenesis (blue) shows a clear separation that suggests an emergence of a 'body' and a 'brain'. (From Bongard, J. C., Pfeifer, R., (2003) Evolving complete agents using artificial ontogeny, In: Hara, F., Pfeifer, R., (eds), *Morpho-functional Machines: the New Species (Designing Embodied Intelligence)*, Springer-Verlag, New York, New York. With permission.)

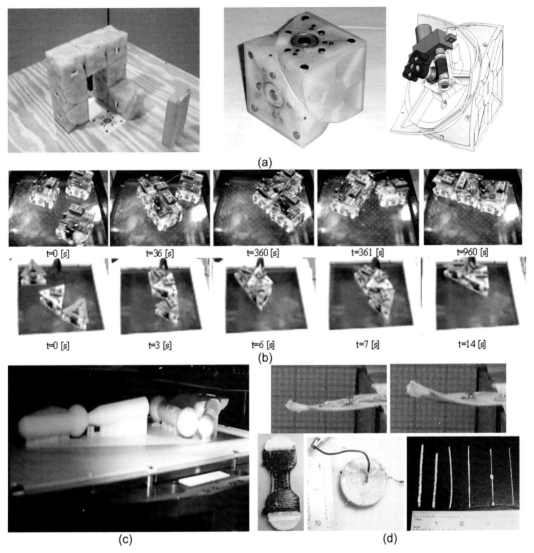

Figure 4.9 (a) Reconfigurable *molecube* robots. (From Zykov, V., Mytilinaios, E., Lipson, H., (2005) *Nature*, 435 (7038), 163–164. With permission.) (b) Stochastic modular robots reconfigure by exploiting Brownian motion, and may allow reconfiguration at a micro-scale in the future. (From White, P. J., Kopanski, K., Lipson, H. (2004) Stochastic self-reconfigurable cellular robotics, *IEEE International Conference on Robotics and Automation (ICRA04)*. With permission.) (c) Rapid prototyping. (d) Future rapid prototyping systems will allow deposition of multiple integrated materials, such as elastomers, conductive wires, batteries, and actuators, offering evolution of a larger design space of integrated structures, actuators, and sensors, not unlike biological tissue. (From Malone, E., Lipson, H. (2004) Functional freeform fabrication for physical artificial life, Ninth International Conference on Artificial Life (ALIFE IX), *Proceedings of the Ninth International Conference on Artificial Life (ALIFE IX)*. With permission.)

Figure 6.9 'Mask,' a 5 in. self-portrait by Ron Mueck, a graduate of Jim Henson Creature Shop, a leading animatronics studio. (Photo by Mark Feldman [Feldman, 2002 website]. With permission.)

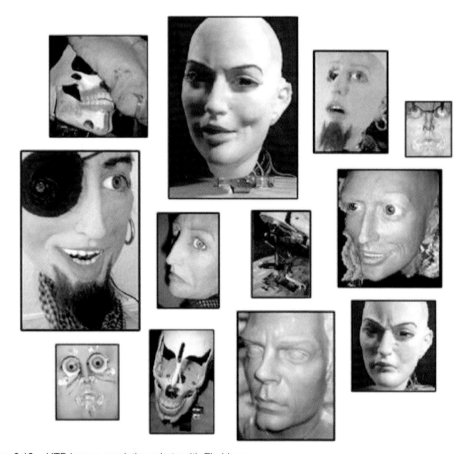

Figure 6.13 UTD human emulation robots with F'rubber.

Figure 6.17 Author's latest robot EVA. Because SPEM silicone requires little force to move, this robot's 36 DOF run for hours on four AA batteries.

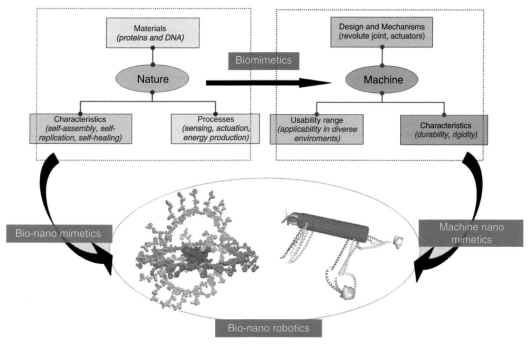

Figure 7.1 Biomimetics — bio-nano robotics, inspired by nature and machine.

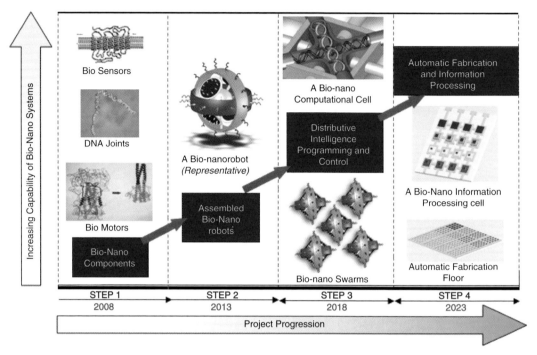

Figure 7.6   The roadmap illustrating the system capability targeted as the project progresses.

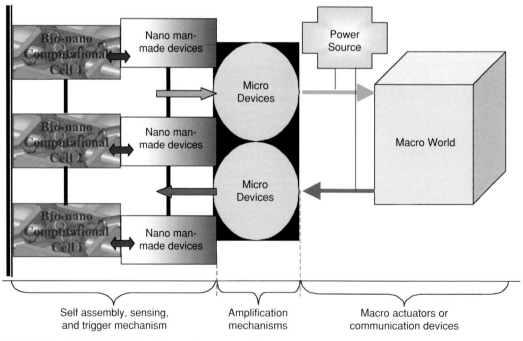

Figure 7.10   Feedback path from nano- to macro-world route.

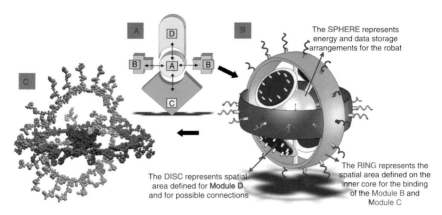

Figure 7.12 (A) A Bio-nano robotic entity 'ABCD', where A, B, C and D are the various bio-modules constituting the bio-nanorobot. In our case these bio-modules will be set of stable configurations of various proteins and DNAs. (B) A bio-nanorobot (representative), as a result of the concept of modular organization. All the modules will be integrated in such a way so as to preserve the basic behavior (of self-assembly and self-organization) of the biocomponents at all the hierarchies. The number of modules employed is not limited to four or any number. It is a function of the various capabilities required for a particular mission. (C) A molecular representation of the figure in part B. It shows the red core and green and blue sensory and actuation bio-modules.

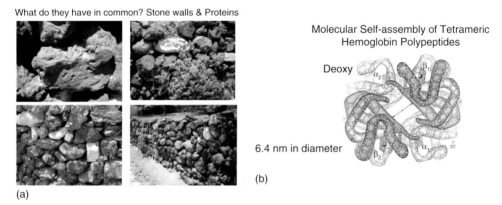

Figure 8.3 (a) The stone wall is built one stone at a time with different sizes and colors of stones. It has a defined function. (b) The protein — hemoglobin consisting of four chains — is built one amino acid at a time with 20 amino acids of all shapes and chemical properties. It also has a defined function to carry oxygen.

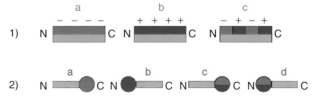

Figure 8.4 Two distinct classes of self-assembling peptide construction motifs are shown here. (a) The first class belongs to amphiphilic peptides that form well-ordered nanofibers. These peptides have two distinctive sides, one hydrophobic and the other hydrophilic. The hydrophobic side forms a double sheet inside of the fiber and hydrophilic side forms the outside of the nanofibers that interact with water molecules that they can form extremely high water content hydrogel, containing as high as 99.9% water. At least three types of molecules can be made, with $-$, $+$, $-/+$ on the hydrophilic side. (b) The second class of self-assembling peptide belongs to surfactant-like molecules. These peptides have a hydrophilic head and a hydrophobic tail, much like lipids or detergents. They sequester their hydrophobic tail inside of micelle, vesicles or nanotube structures and expose their hydrophilic heads to water. At least three kinds molecules can be made, with $-$, $+$, $-/+$ heads.

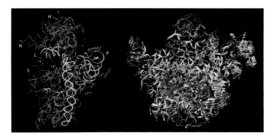

Figure 8.5 The bacterial ribosome. 30S ribosome (left panel) and 50S ribosome (right panel). The ribosome is one of the most sophisticated molecular machines nature has ever self-assembled. It has more than 50 different kinds of proteins and 3 different size and functional RNAs, all through weak interactions to form the remarkable assembly line. (*Source*: http://www.molgen.mpg.de/~ag_ribo/ag_franceschi/.)

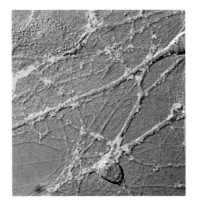

Figure 8.8 When primary rat hippocampal neuron cells are allowed to attach to the peptide scaffolds, the neuron cells not only project lengthy axons that follow the contours of the scaffold surface, but also form active and functional synaptic connections, each green dot is a functional neuronal connection (upper panel). Furthermore, when the peptide scaffold was injected into brain of animals, it bridged the gap and facilitated the neural cells to migrate cross the deep canyon (lower panel). The animals regained their visual function. Without the peptide scaffold, the gap remains, and the animals did not regain visual function.

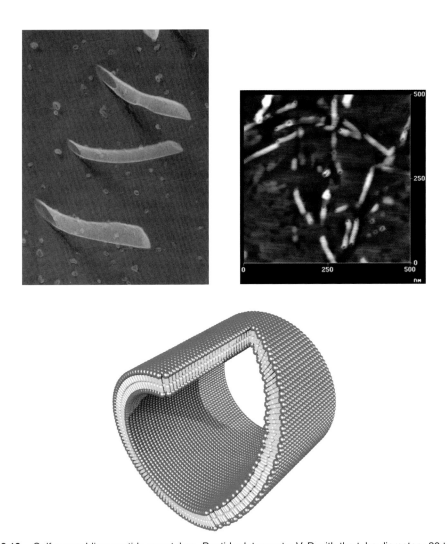

Figure 8.10 Self-assembling peptide nanotubes. Peptide detergents: V_6D with the tube diameter ~30 to 50 nm (left panel), A_6K with the tube diameter ~20 to 30 nm (middle panel) and the model for V_6D. The openings of the nanotubes are clearly visible. The wall of the tube has been determined using neutron scattering as ~5 nm, suggestive a bi-layer structure modeled here.

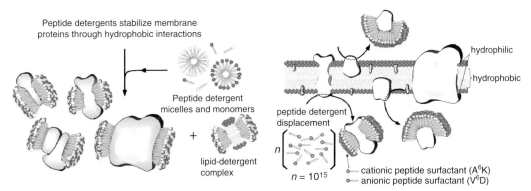

Figure 8.11 Schematic illustration of designed peptide detergents used to solubilize and stabilize membrane proteins. When mixed with membrane proteins, they solubilize and stabilize them, presumably at the belt domain where the membrane proteins are embedded in lipid membranes.

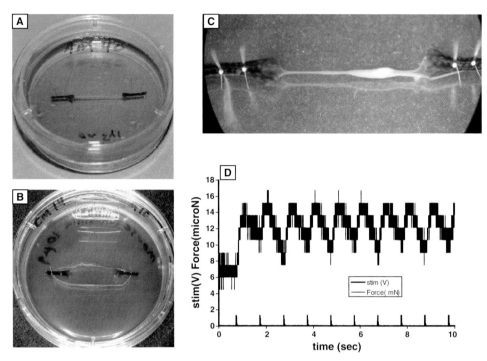

Figure 9.1 (A) Self-organized skeletal muscle construct after 3 months in culture, length ~12 mm. (B) Rat cardiac myocyte + fibroblast monolayer in the process of delaminating and self-organizing into a functional cardiac muscle construct, 340 h in culture. (C) Self-organized cardiac muscle construct, attached to laminin-coated suture anchors, 380 h in culture. (D) Electrically-elicited force trace from the cardiac muscle construct shown in C, stimulation pulses shown below, contractile force trace shown above (raw data, unfiltered).

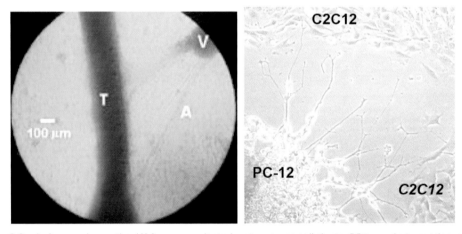

Figure 9.2 Left: axonal sprouting (A) from an explanted motor neuron cell cluster (V) toward a target tissue (T), in this case, an aneural cultured skeletal muscle 'myooid.' Right: a simple cell culture system demonstrating axonal sprouting between neural (PC–12) and myogenic (C2C12) cell lines. This co-culture system allows the study of synaptogenesis in culture. (Photographs taken by members of the Functional Tissue Engineering Laboratory at the University of Michigan: Calderon, Dow, Borschel, Dennis.)

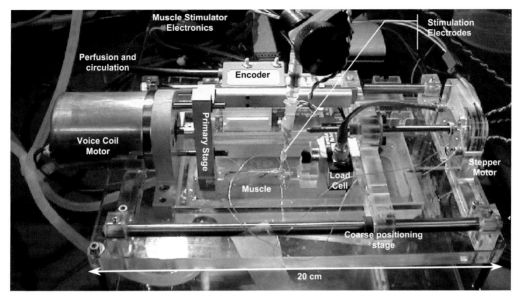

Figure 9.3 Muscle Bioreactor Technology. Muscle identification, control and maintenance apparatus is shown with the primary sensors and actuators noted. The coarse positioning stage is adjusted at the beginning of the experiment to accommodate different tissue lengths, but is typically kept at a constant position during a particular contraction. The primary stage provides the motion that simulates the boundary condition force law with which the muscle specimen pulls against. The vertical syringe has a suction electrode at its tip that is connected to the stimulation electronics in the background. The encoder and load cell measure muscle displacement and force, respectively, and are employed as sensory control inputs during FES control experimentation. Silicone tubing recirculates solution via a peristaltic pump, while oxygen is injected in the loop.

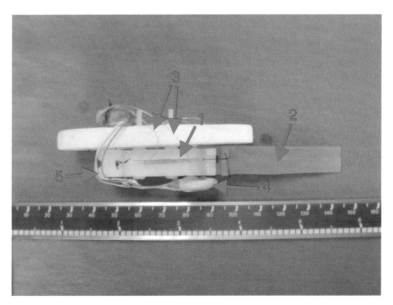

Figure 9.4 Biomechatronic swimming robot. To power robotic swimming, two frog semitendinosus muscles (1), attached to either side of elastomeric tail (2), alternately contract to move the tail back and forth through a surrounding fluid medium. Two electrodes per muscle (3), attached near the myotendonous junction, are used to stimulate the tissues and to elicit contractions. To depolarize the muscle actuators, two lithium ion batteries (4) are attached to the robot's frame (5). During performance evaluations, the robot swam through a glucose-filled ringer solution to fuel muscle contractions.

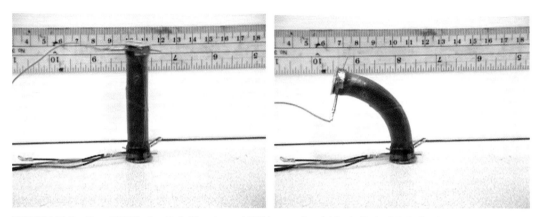

FIGURE 10.2   Two-DOF Spring Roll. (Courtesy of SRI International, Menlo Park, CA, U.S.A.)

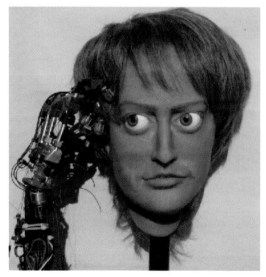

FIGURE 10.9   An android head and a robotic hand that serve as biomimetic platforms for the development of artificial muscles. (*Acknowledgement*: This photograph was taken at JPL where the head was sculptured and instrumented by D. Hanson, University of Texas, Dallas. The hand was made by G. Whiteley, Sheffield Hallam University, U.K.)

FIGURE 10.10   Four-finger EAP gripper lifting a rock.

FIGURE 10.11   A new class of multi-limbed robots called Limbed Excursion Mobile Utility Robot (LEMUR) is under development at JPL (Courtesy of Brett Kennedy, JPL).

FIGURE 10.17 The Virginia Tech students' arm being prepared for the match against Panna Felsen, the 17-year old student from San Diego.

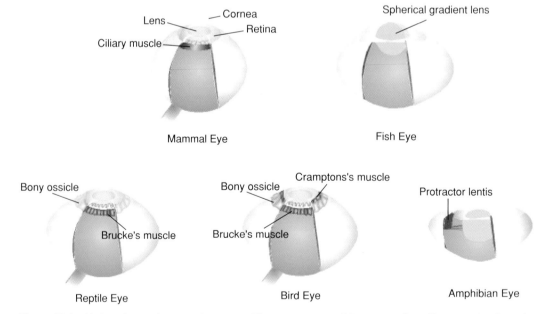

Figure 11.1 Various types of camera-type eyes. The arrangement of the mammalian ciliary muscle allows for passive changes in lens thickness. Brucke's muscles attached to bony ossicles in reptiles and birds, on the other hand, actively change the lens thickness. Birds have an additional muscle, Crampton's muscle, which can alter the shape of the cornea. The protractor lentis in some amphibian eyes moves a fixed-shape lens closer to or farther from the retina for accommodation.

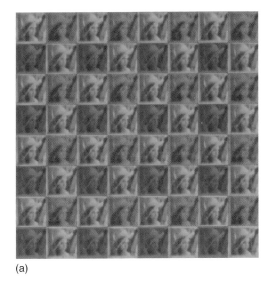

(a)

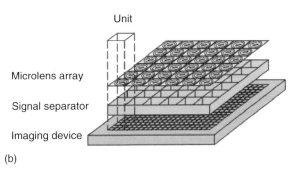

Unit

Microlens array

Signal separator

Imaging device

(b)

(c)

Figure 11.9 Tanida imaging system (a) Schematic, (b) device, (c) sample images (From Tanida, J., Shogenji, R., Kitamura, Y., Yamada, K., Miyamoto, M., and Miyatake, S. *Optics Express* 2003: 11(18), 2109–2117. With permission).

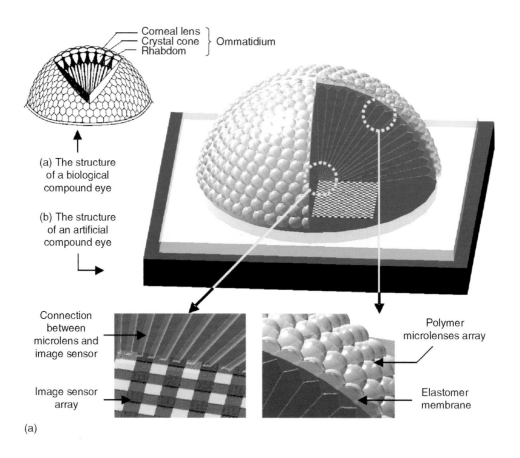

Corneal lens ⎤
Crystal cone ⎥ Ommatidium
Rhabdom ⎦

(a) The structure of a biological compound eye

(b) The structure of an artificial compound eye

Connection between microlens and image sensor

Image sensor array

Polymer microlenses array

Elastomer membrane

(a)

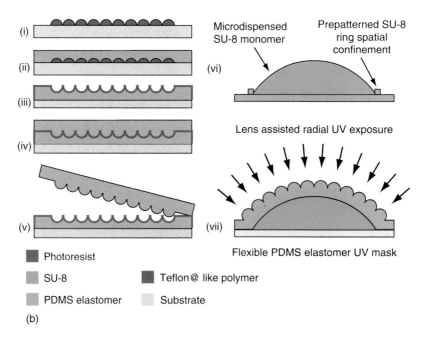

(i)

(ii)

(iii)

(iv)

(v)

Microdispensed SU-8 monomer

Prepatterned SU-8 ring spatial confinement

(vi)

Lens assisted radial UV exposure

(vii)

Flexible PDMS elastomer UV mask

■ Photoresist

▨ SU-8   ■ Teflon@ like polymer

▨ PDMS elastomer   ░ Substrate

(b)

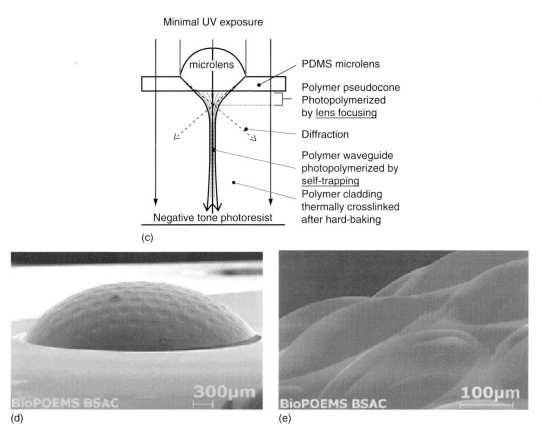

Minimal UV exposure

microlens

PDMS microlens

Polymer pseudocone
Photopolymerized
by <u>lens focusing</u>

Diffraction

Polymer waveguide
photopolymerized by
<u>self-trapping</u>
Polymer cladding
thermally crosslinked
after hard-baking

Negative tone photoresist

(c)

(d) 300μm   BioPOEMS BSAC

(e) 100μm   BioPOEMS BSAC

Figure 11.10 (*Continued*) Artificial apposition eye approach: (a) schematic, (b) lens array process flow, (c) waveguide illustration, (d) picture of device, and (e) close-up picture of lenslets (From Jeong, K., Kim, J., Nevill, J., and Lee, L.P. (2005) With permission).

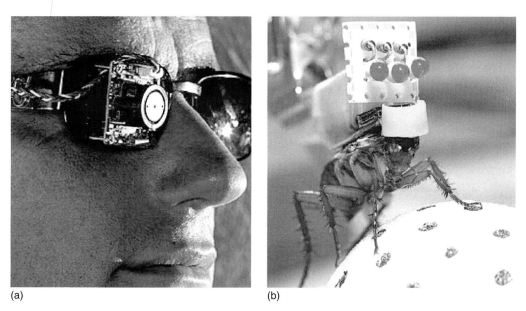

(a)

(b)

Figure 11.21 (a) An artificial eye directly interfaced with the brain and (b) an electronically controlled insect.

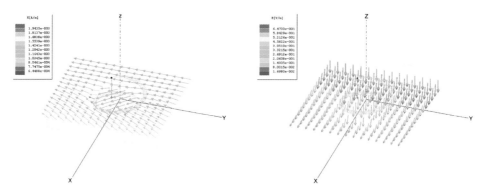

Figure 12.6 Electric field (left) and magnetic field (right) patterns calculated for a unit cell of a coiled medium using ANSOFT-HFSS. The wave is propagating in the *x*-direction and the fields on the two *yz* faces have 50° phase difference. The incoming wave (electric field) from the far *yz* face is at this time polarized parallel to the axis of the coil. However, the effect of the coil adds an out of phase normal component. Therefore, the field vectors of both electric and magnetic fields rotate as the wave travels through the cell.

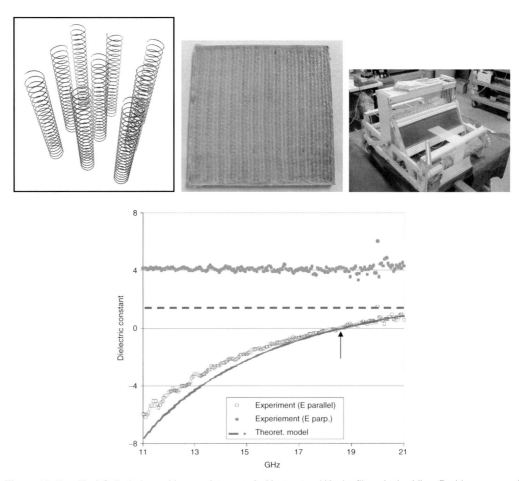

Figure 12.10 (Top) Coiled wire architecture integrated with structural Kevlar fibers by braiding. Braids woven and laminated into composite plates. (Bottom) EM characterization of the braided and woven composite showing typical plasmon media response when aligned parallel to the polarization of the EM radiation. Normal (nonplasma) dielectric response is observed when aligned in the perpendicular direction.

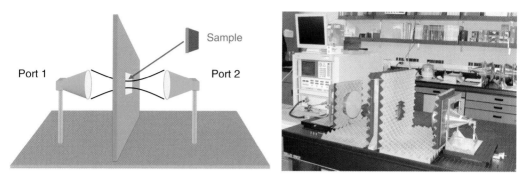

Figure 12.13 (Left) Schematic and (right) photo of Focused Beam system for EM characterization from 5 to 40 GHz at UCSD's CEAM.

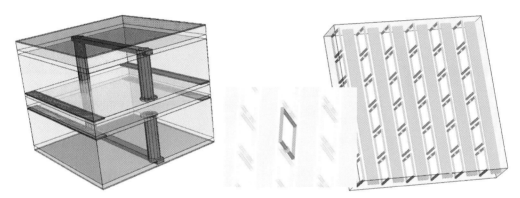

Figure 12.17 (Left) Unit cell of NIM. The negative permeability is achieved by ring resonators, formed from copper strips on the upper and lower surfaces, connected to vias that run through the structure, with one of the vias possessing a gap in the center to introduce capacitance. Copper strips are patterned on the central circuit board, giving rise to the negative permittivity of the structure. (Right) Views of conducting elements to be fabricated within a composite panel.

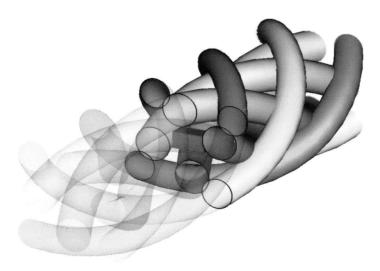

Figure 12.29 Illustration of a sensor embedded in a composite braid.

Figure 13.2 A cuttlefish (*Sepia officinalis*) can change its appearance according to the background. Here the animal changes its body pattern when moved from a sandy or gravel substrate to one with shells. (Courtesy of Roger T. Hanlon, Senior Scientist, Marine Biological Laboratory, Woods Hole, MA.)

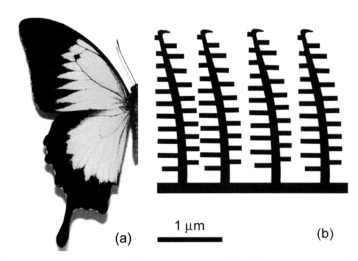

(a)

1 μm

(b)

Figure 15.10 Iridescence in butterflies. (a) Real color image of the iridescence from a *Papilio ulyssus* wing, (b) diagram based on the transmission electron micrograph showing wing-scale cross-section of the butterfly *Morpho rhetenor* (From Vukusic, P. and J.R. Sambles (2003) Photonic structures in biology. *Nature* 424: 852–855. With permission). The high density of structures and high layer number creates an intense reflectivity of wing scales.

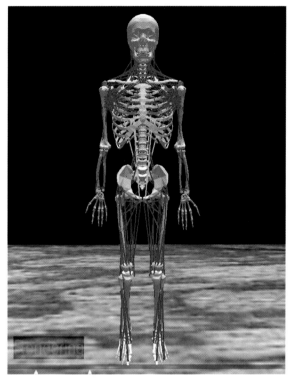

Figure 16.1 A 3D computer simulation model of whole body human dynamic musculo-skeletal system.

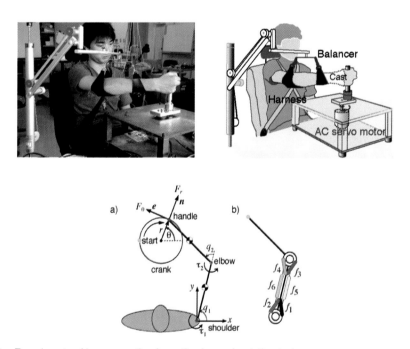

Figure 16.8 Experiments of human motion formation in crank rotation tasks.

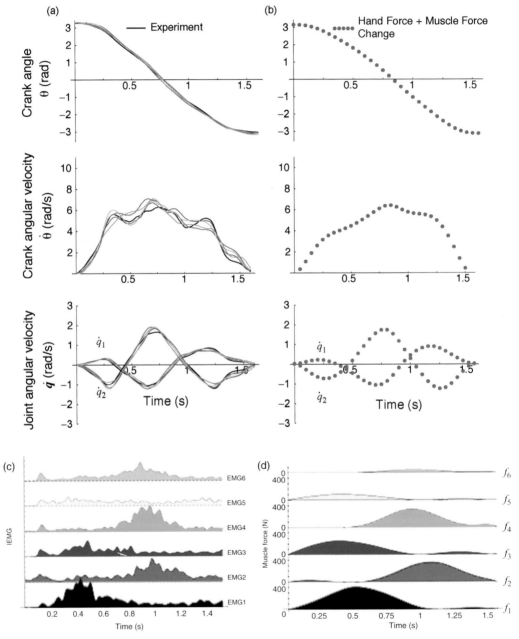

Figure 16.10 Comparison of the time responses of the human hand motions in (a), the EMG in (c) from the experiments and the numerical simulations in (b) and (d) with our combined criterion of the hand interaction force change and muscle force change.

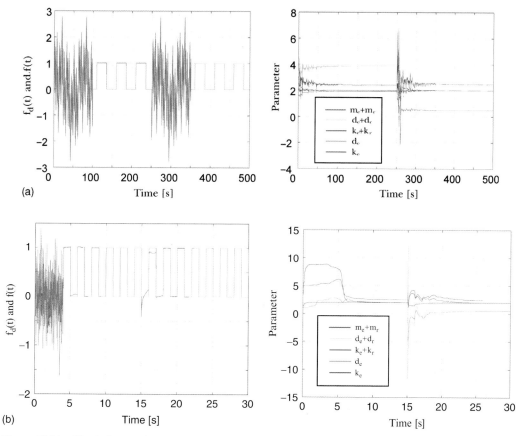

Figure 16.14 Simulation results on the time responses of the robot's force (the blue lines) tracking a time varying desired force (the green lines), and the parameter convergences of the feedforward controller. The environmental viscosity is set from 2 to 0.5 at the time $t = 250[s]$ in (a) and $t = 15[s]$ in (b). At the beginning 100(s), 250(s) to 350(s) in (a) and the beginning 4(s) in (b), the desired forces are set as noises.

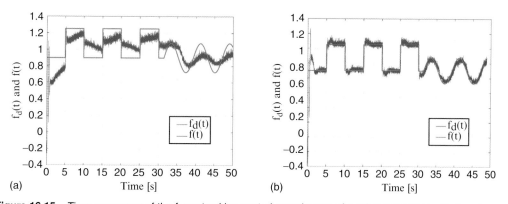

Figure 16.15 Time responses of the force tracking control experiments of a robot interacting with its unknown dynamic environment. The experimental setup is in Figure 16.11. Here, (a) is the result when using usual constant feedforward + PI feedback control (23), while (b) is the result of 2 D.O.F. adaptive control (41). The feedback control of (b) is used as same as that in (a). (a) Experimental result using PI control; (b) experimental result of adaptive control.

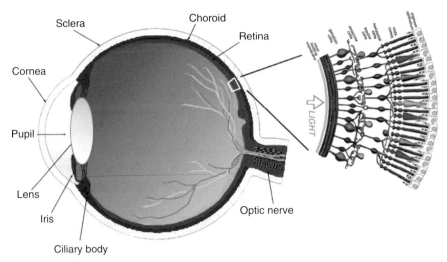

Figure 17.1 The human eye in cross section with an enlarged section of the retina (right). The light sensitive retina covers more than half the back of the eye. Over 100 million photoreceptors convert light into neural signals that are then transmitted to the proximal visual centers by the optic nerve. The optic nerve is composed of 1 million axons from the retinal ganglion cells, which are the output cells of the retina. In outer retinal diseases, the photoreceptors are degenerated, but the inner retina cells remain and can be electrically stimulated.

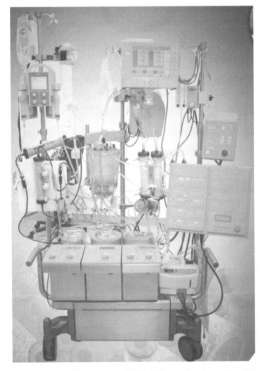

Figure 18.1 ELAD artificial liver system. (Courtesy of Vital Therapies Inc; San Diego, CA.)

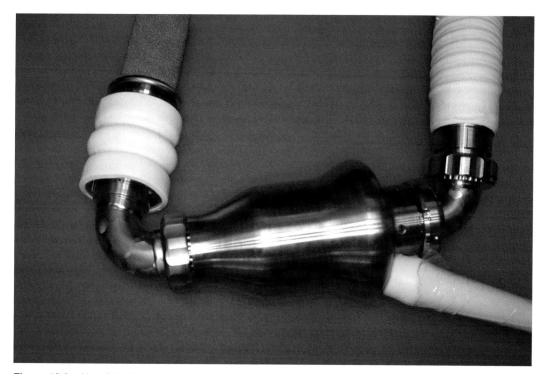

Figure 18.6   HeartMate II rotary axial pump device.

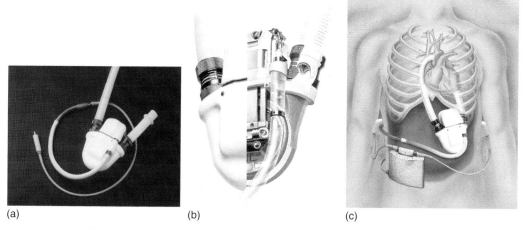

(a)   (b)   (c)

Figure 18.7   (a) Novacor VAD, (b) cross-section of Novacor, and (c) diagrammatic representation of Novacor, (With permission from WorldHeart Corporation, Ottawa, Canada.)

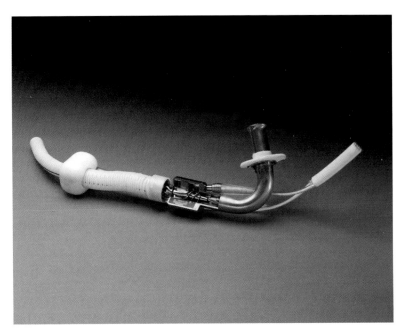

Figure 18.9 MicroMed DeBakey ventricular assist device. (With permission from MicroMed Technology, Inc; Huston, TX.)

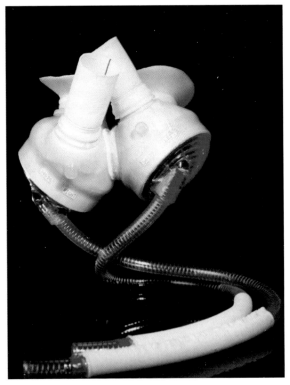

Figure 18.12 The SynCardia CardioWest total artificial heart (With permission from SynCardia Systems, Inc.; Tucson, AZ.)

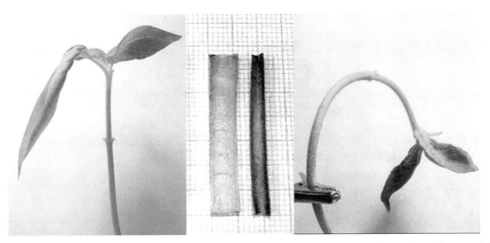

Figure 19.9 A stable, erect sunflower stem (left panel) depends on the pressure of internal, easily expandable hydrostat tissue (pith = transparent cells in center panel) that tensions the stronger-walled surface layers of the stem giving them rigidity and stability. The well-known limp shape of wilting young plant stems (right panel) occurs when the internal, thin-walled pith tissue dehydrates, shrinks (center panel) and ceases to exert radial pressure on the surface layer under tension. Left panel shows fully hydrated sunflower stem and right panel a dehydrated stem; the center panel demonstrates volume reduction in pith during dehydration of a segment slice. Pith parenchyma acts as a hydrostat motor that provides herbaceous stems with stability and the driving force for expansion.

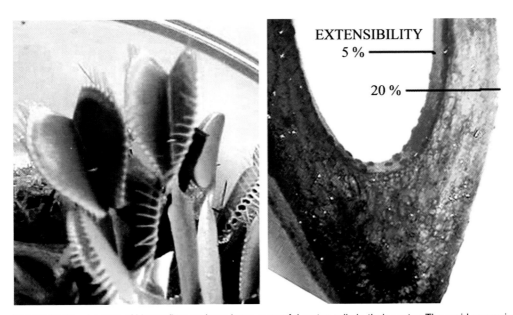

Figure 19.13 Leaves of Venus flytraps have large, powerful motor cells in their center. The rapid expansion of these cells is modified by different extensibilities of the two adjacent surface (epidermis) layers. These anisotropic restraints turn a linear expansion into the rapid curving of the leaf and closure of the trap. This mechanism is triggered when the sensory hairs at the upper surface are repeatedly touched.

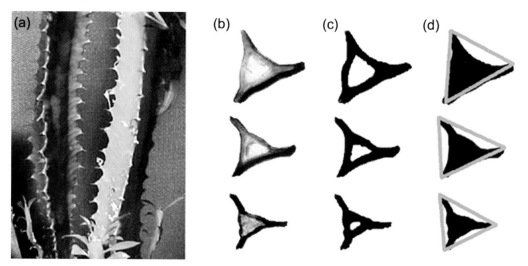

Figure 19.18 The shoots of desert plants in the *Cactus* and *Euphorbia* families often adopt the shape of pleated columns (a) a structure which allows the photosynthesizing periphery to maintain its area in spite of massive volume losses in the shoot center. Cross sections of a pleated-column shoot show increasing dehydration from upper to lower pictures (b) with a considerable volume loss of the shoot center (c) and remarkably constancy in the area of a surface that changes its shape from a convex to a concave outline.

Figure 20.2 A four-legged robot called STAR (Steep Terrain Access Rover) is under development at JPL. (Courtesy of Brett Kennedy, JPL.)

Figure 20.3 The Singaporean giant 'durians' building called the Esplanade Theater (left) has the shape of this fruit that is considered the King of fruits (right). (The photo on the right is the courtesy of Anand Krishna Asundi, Nanyang Technological University, Singapore.)

Figure 20.5 Grand challenge for the development of EAP-actuated robotics.

Figure 20.6 The chain of evolution of our mimicking nature is drawn into the artificial world that we create. (Top graphics is the courtesy of David Hanson, and Human Emulation Robotics, LLC. The bottom graphics is the modification that was made by Adi Marom, Research Artifacts Center Engineering, The University of Tokyo, Japan. The robotic arm in this figure was made by G. Whiteley, Sheffield Hallam U., U.K., and photographed in the author's lab.)

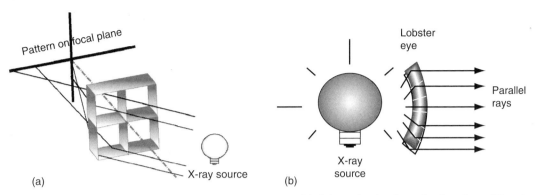

(a)   (b)

Figure 11.15   Practical applications of artificial lobster eyes. (a) Schematic showing reflections from different channels redirected to form different parts of a focal pattern. (b) Example of x-ray image by using a microchannel plate with a point source at 1.5 keV. (c) By using the lobster eye in reverse, parallel x-rays may be generated.

11.3.3   Hybrid Appositional or Superpositional Compound Eyes

In 2004, Szema et al. z proposed a new kind of compound eye based on both the superpositional and appositional arrangements. While not found in nature, the design attempts to combine the advantages of both types of systems. It begins with a standard superpositional compound eye modified by attaching optical shutters to each of the facets. In this way, images are acquired from individual ommatidia separately, as in the appositional arrangement. However, the use of a single retina simulates the superpositional compound eye.

Such a design not only reduces bandwidth and parallel processing requirements, but also has an additional advantage by recovering useful information that is lost in a typical superpositional arrangement. Namely, when images from multiple lenses are projected simultaneously onto a common retina, it is difficult to correlate image points to their originating facets (see Figure 11.16). This is significant because by knowing the relative position of the lenses and the objects in their respective fields of view, it is possible to derive three-dimensional information (Figure 11.17).

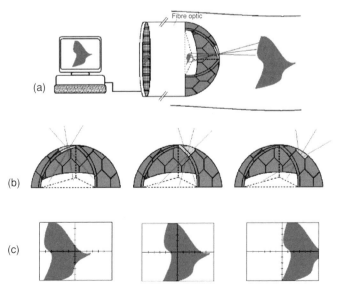

Figure 11.16   A hybrid appositional/superpositional eye (a) schematic of operation, (b) illustration of operation, (c) illustration of different points of view from each facet. (From Szema, R., Rastegar, J., and Lee, L. *Journal of Medical Engineering and Technology* 2004: 28(3), 117–240. With permission.)

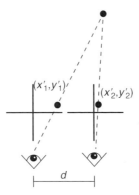

Figure 11.17   Depth perception through different visual fields.

As humans can triangulate depth from the fields of view of two eyes, a compound eye may prove to be even more adept at discerning distances and dimensional measurements. This is because the distances measured are based on calculations using multiple data points, decreasing the error of any pair of eyes alone. This has applications in robotic navigation systems, surveillance cameras, and borescopic devices.

11.4   OTHER BIOMIMETIC APPROACHES

This section deals with biomimetic engineering approaches to optics which do not fit easily into the categories of camera or compound eyes. While the first two examples of brittlestar microlenses and *Melanophila acuminate* beetle pit organs may seem simplistic relative to the other eye designs, they are no less suited for their applications. In addition, it is important that "simple" is not confused with "easily replicated" or even "easily understood." Indeed, the idea of image formation by projection from a lens onto a retina was noted by Johannes Kepler in 1604. But, as described below, it was not until 2001 that scientists even recognized that the brittlestar had eyes, let alone was covered with them.

11.4.1   Brittlestar Eyes

By all accounts, the brittlestar is a curious creature (see Figure 11.18). In particular, the species *Ophiocoma wendtii* demonstrates marked photosensitivity and the ability to evade predators, all without readily apparent eyes or a brain. In a letter published in 2001, Aizenberg et al. (2001) reported that, in fact, the calcite crystals embedded throughout the skeletal structure were the elements of an enveloping compound eye. The crystals, 40 to 50 μm in diameter, appear to form doublet lenses with the ability to correct for spherical aberration and birefringence. In addition, the focal point of the lenses is approximately 4 to 7 μm with a 3-μm spot size. This matches nerve bundle locations and sizes beneath the calcite crystal array, supplying evidence to the notion that the array forms a pixelated image of its surroundings.

Bell Laboratories or Lucent Technologies have used this discovery to design process flows for artificial lenses which also limit birefringence and spherical aberration effects. This may have applications in optical networking equipment and improved photolithography techniques. By organically modifying micropatterned templates, researchers are able to direct the growth of single calcite crystals into sub-10-μm patterns and defined crystallographic orientation (Aizenberg et al., 2003).

Self-assembled monolayers (SAMs) of ω-terminated alkanethiols on gold or silver supports are used as crystallization templates, and organic posts can be added without upsetting single-crystal growth. These posts allow for removing water and impurities during the transition from an

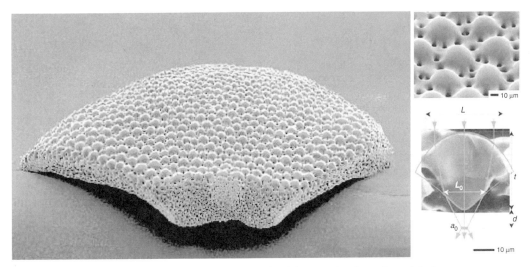

Figure 11.18   (a) The brittle star, (b) electron micrographs, (c) ray tracing through a calcite lens.

amorphous calcium carbonate phase to the crystalline phase. The result is a large, microporous single crystal. Light microscopy images of the crystallization process, as well as infrared spectral response of the crystals, are shown below (Figure 11.19).

11.4.2 *Melanophila acuminate* Beetle

Yet another interesting animal, the *M. acuminate* beetle, has the uncanny ability to find forest fire areas for breeding purposes. Females of this species lay their eggs in burnt wood, where the natural defense mechanisms of living trees are absent. This is the only place their larvae can survive. While *M. acuminate* is not the only creature with this practice, it is often the first to arrive, sometimes outpacing other insects by weeks to months (Sowards et al., 2001).

M. acuminate are guided by specialized structures, known as pit organs, which are more sensitive to the frequency of infrared light emitted by forest fires (Figure 11.20). These organs (450 μm × 200 μm × 108 μm deep) house 50 to 100 sensilla (each 15 μm in diameter). Following infrared light absorption, the expansion of a cuticular apparatus is detected by mechanoreceptors which direct the beetles to the fire (Sowards et al., 2001).

Equivalent man-made guidance systems, heat-seeking missiles, for example, rely on sensors that need to be cooled to freezing temperatures at significant cost. Schmitz et al. have been working on adapting insights gleaned from analyzing the *M. acuminate* beetle for use in similar applications (Roach, 2004). Their design uses a disc which expands in response to infrared radiation. The composition of the disc depends on the infrared source, with examples such as Teflon for heat from a human hand or polyethelene for fire detection. Currently, the mechanosensors used are able to detect a source 30 to 40 cm away. The sensitivity is expected to increase greatly with more sensitive mechanosensors and tailored disc materials.

11.5 CONCLUSION

When it comes to designing optical systems, humans have been unquestionably creative. Still, it has not been until relatively recently that there has been a major thrust towards biologically inspired optics. Of the multitude of natural vision systems that exist, only a relative few have been emulated using man-made materials. And it is probably safe to say that none of these artificial systems have matched the performance of their biological counterparts.

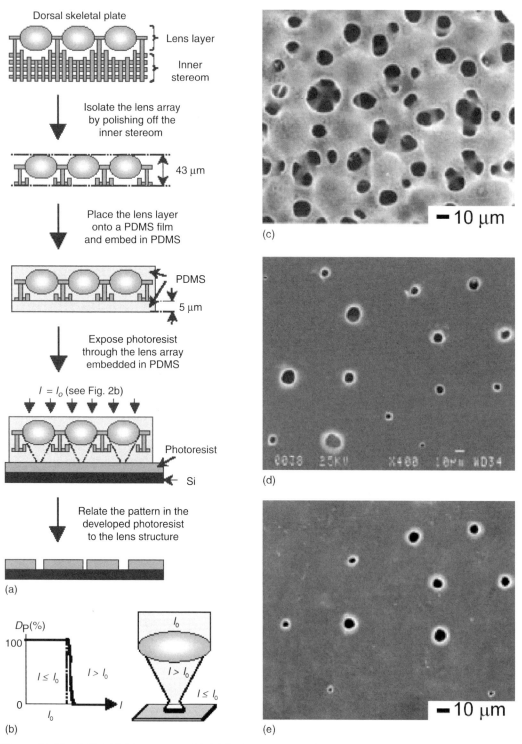

Figure 11.19 Artificial brittlestar array (a) fabrication process, (b) light microscopy images of crystallization process, (c) infrared spectral response of crystals. (From Aizenberg, J., Muller, D., Grazul, J., and Hamann, D. *Science* 2003: 299, 1205–1208. With permission.)

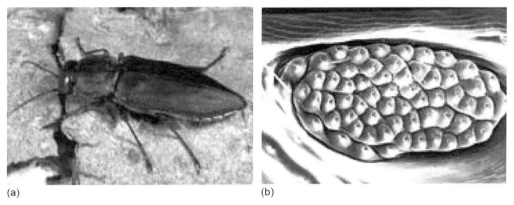

Figure 11.20   (a) A *Melanophila acuminate* beetle and (b) a close up of its pit organ.

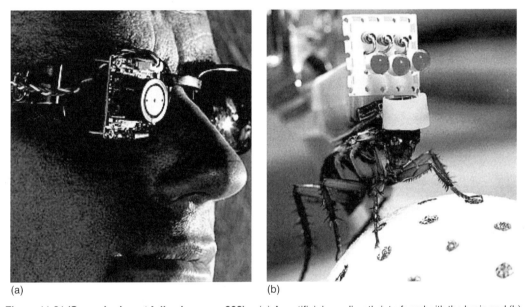

Figure 11.21 (See color insert following page 302)   (a) An artificial eye directly interfaced with the brain and (b) an electronically controlled insect.

The goal of biomimetic devices, however, should not be merely to imitate nature. Rather, devices should be designed according to their application with biology providing a valuable resource for consultation. Nature should be used to improve human designs. This will involve an increasing collaboration between biologists and engineers so that, eventually, we may improve upon nature's designs as well (Figure 11.21).

REFERENCES

Aizenberg, J., Tkachenko, A., Weiner, S., Addadi, L., and Hendler, G. Calcitic microlenses as part of the photoreceptor system in brittlestars. *Nature* 2001: 412, 819–22.

Aizenberg, J., Muller, D., Grazul, J., and Hamann, D. Direct fabrication of large micropatterned single crystals. *Science* 2003: 299, 1205–08.

Angel, J. Lobster eyes as x-ray telescopes. *Astrophysical Journal* 1979: 233(1), 364–73.

Chown, M. I spy with my lobster eye, *New Scientist* 1996a: 150(2025), 20.

Chown, M. X-ray lens brings finer chips into focus, *New Scientist* 1996b: 151(2037), 18.

Duparre, J., Schreiber, P., Dannberg, P., Scharf, T., Pelli, P., Reinhard, V., Herzig, H., and Brauer, A. Artificial compound eyes — different concepts and their application to ultra flat image acquisition sensors. *Proceedings of SPIE, Vol. 5346, MOEMS and Miniaturized Systems IV* 2004.

Fletcher, A., Murphy, T., and Young, A. Solutions of two optical problems. *Proceedings of the Royal Society, London A* 1953: 223, 216–25.

Hung, P.J., Jeong, K., Liu, G.L., and Lee, L.P. Suspensions for electrical connections on the tunable elastomer membrane, *Applied Physical Letters* 2004: 85(24), 6051–3.

Jeong, K., Kim, J., Nevill, J., and Lee, L.P. A microfabricated biomimetic compound eye (to be published 2005).

Koike, Y., Sumi, Y., and Ohtsuka, Y. Spherical gradient-index sphere lens. *Applied Optics* 1986: 25(19), 3356–63.

Land, M.F. The optics of animal eyes. *Contemporary Physics* 1988: 29(5), 435–55.

Land, M. and Nilsson, D. *Animal Eyes*. Oxford Press, Oxford, 2002.

Luneberg, R.K. *Mathematical Theory of Optics*. Brown University, Providence, RI, 1944, pp. 208–13.

Martinez, T., Wick, D., and Restaino, S. Foveated, wide field-of-view imaging system using a liquid crystal spatial light modulator. *Optics Express* 2001: 8(10), 555–60.

Ogata, S., Ishida, J., and Sasano, T. Optical sensor array in an artificial compound eye. *Optical Engineering* 1994: 33(11), 3649–55.

Peele, A., Nugent, K., Rode, A., Gabel, K., Richardson, M., Strack, R., and Siegmund, W. X-ray focusing with lobster-eye optics: a comparison of theory with experiment. *Applied Optics* 1996: 35(22), 4420–25.

Roach, J. U.S. military looks to beetles for new sensors. National Geographic News, available 8 March 2004.

Rosen, J. and Abookasis, D. Seeing through biological tissues using the fly eye principle. *Optics Express* 2003: 11(26), 3605–11.

Smythe, R. *Vision in the Animal World*. Macmillan Press, England, 1975.

Sowards, L., Schmitz, H., Tomlin, D., Naik, R., and Stone, M. Characterization of beetle *Melanophila acuminate* (Coleoptera: Buprestidae) infrared pit organs by high-performance liquid chromatography/Massspectrometry, scanning electron microscope, and fourier transform-infrared spectroscopy. *Annals of the Entomological Society of America* 2001: 94(5), 686–94.

Szema, R., Rastegar, J., and Lee, L. An artificial compound eye for stereoendoscopy. *Journal of Medical Engineering and Technology* 2004: 28(3), 117–240.

Tanida, J., Shogenji, R., Kitamura, Y., Yamada, K., Miyamoto, M., and Miyatake, S. Color imaging with an integrated compound imaging system. *Optics Express* 2003: 11(18), 2109–17.

Zhang, D., Lien, V., Berdichevsky, Y., Choi, J., and Lo, Y. Fluidic adaptive lens with high focal length tenability. *Applied Physics Letters* 2003: 82(19), 3171–2.

WEBSITE

http://news.nationalgeographic.com/news/2003/03/0314_030314_secretweapons3.html

12

Multifunctional Materials

Sia Nemat-Nasser, Syrus Nemat-Nasser, Thomas Plaisted,
Anthony Starr, and Alireza Vakil Amirkhizi

CONTENTS

12.1 INTRODUCTION

Multifunctional structural materials possess attributes beyond the basic strength and stiffness that typically drive the science and engineering of the material for structural systems. Structural materials can be designed to have integrated electrical, magnetic, optical, locomotive, power generative, and possibly other functionalities that work in synergy to provide advantages that

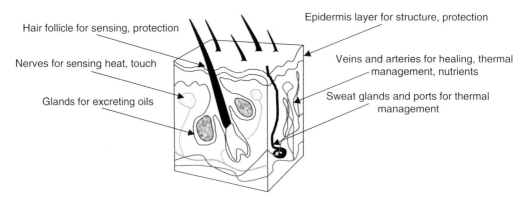

Figure 12.1   Illustration of the many integrated functions within the human skin.

reach beyond that of the sum of the individual capabilities. Materials of this kind have tremendous potential to impact future structural performance by reducing size, weight, cost, power consumption, and complexity while improving efficiency, safety, and versatility.

Nature offers numerous examples of materials that serve multiple functions. Biological materials routinely contain sensing, healing, actuation, and other functions built into the primary structures of an organism. The human skin, for instance (see Figure 12.1), consists of many layers of cells, each of which contains oil and perspiration glands, sensory receptors, hair follicles, blood vessels, and other components with functions other than providing the basic structure and protection for the internal organs. These structures have evolved in nature over eons to the level of seamless integration and perfection with which they serve their functions. Scientists now seek to mimic these material systems in designing synthetic multifunctional materials using physics, chemistry, and mathematics to their advantage in competing with the unlimited time frame of nature's evolutionary design process. The multifunctionality of these materials often occurs at scales that are nano through macro and on various temporal and compositional levels.

12.1.1 Multifunctional Concepts

In recent years, wide arrays of multifunctional material systems have been proposed. Each of these systems has sought to integrate at least one other function into a material that is capable of bearing mechanical loads and serves as a structural material element. Researchers at ITN Energy Systems and SRI International have integrated a power-generating function into fiber-reinforced composites (Christodoulou and Venables, 2003). Individual fibers are coated with cathodic, electrolytic, and anodic forming layers to create a battery. The use of the surface area of fibers as opposed to that of a foil in a thin film battery allows greater energy outputs, measured on the order of 50 Wh/kg in a carbon fiber-reinforced epoxy laminate. These batteries may be deposited on various substrates, including glass, carbon, and metallic fibers. This research and many others, including our own research, have been supported by DARPA under the first generation of Synthetic Multifunctional Materials Initiative (Figure 12.2).

Other power-generating schemes integrated into structural composites have been proposed, where the composite structure is consumed to generate power after its structural purpose is complete (Joshi et al., 2002; Thomas et al., 2002; Baucom et al., 2004; Qidwai et al., 2004). Physical Sciences Inc. have incorporated oxidizers into thermoplastic matrix composites and demonstrated significant energy output from directly burning the material (Joshi et al., 2002). Such a material would be useful for instance in a space application, where weight saving is critical, and structures required only for launch could provide an energy source once the structure is no longer needed.

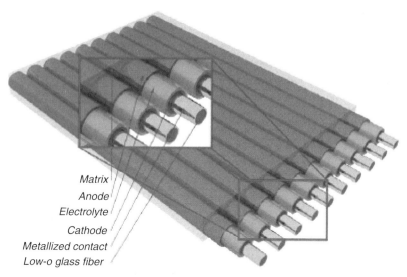

Matrix
Anode
Electrolyte
Cathode
Metallized contact
Low-o glass fiber

Figure 12.2   ITN Powerfiber with simultaneous battery and structure functionalities.

The integration of sensing into materials has made many advances in recent years. Much of the research has been conducted under the context of Structural Health Monitoring, or SHM. In line with the overall theme of this book, researchers seek to make a material sense its environment, feel internal damage, and signal an alert that repair is needed, essentially mimicking the behavior of biological organisms. Later in this chapter we provide an approach to integrating sensing into composite materials. For a comprehensive overview of the field, the reader is directed to a recent review paper on the subject (Mal, 2004).

This brief introduction to multifunctional materials only scratches the surface of the various multifunctional concepts developed to date. The remaining sections of this chapter will detail a further example of a multifunctional material under development at University of California, San Diego (UCSD), in the first author's laboratories. The functionalities of this material include integrated structural, electromagnetic, thermal, healing, and sensing capabilities. While this material in no way encompasses all of the possible functionalities that may be integrated into a material, it offers an example of how such an integration may be achieved while maintaining the structural integrity of the overall material. Particular attention is given to the interplay and resulting synergy between the various elements that contribute these functionalities.

12.2   MULTIFUNCTIONAL COMPOSITES

We focus on the issues that relate to integrating multiple functions into fiber-reinforced polymers to create composites with basic structural attributes that can also perform other functions. We discuss various methods that have been used to control the mechanical, electromagnetic (EM), and thermal properties of the material, while introducing self-healing, and environmental-sensing and prognostic capabilities into the material. The polymer matrix of these composites has the ability to covalently heal microcracks at rates that can be facilitated by moderate heating through thin conductors which are also used to control the EM properties of the material. The same conductors can also be used to create sensor-integrated electronic networks within the composite, capable of sensing, and local and global communications and decision making.

12.2.1 Electromagnetic Functionality

Recent advances in electromagnetic (EM) *metamaterials* have provided an opportunity to change and tune the dielectric constant as well as the index of refraction of the material over a range of useful frequencies. Electromagnetic metamaterials are artificially structured media with unique and distinct EM properties that are not observed in naturally occurring materials. A variety of meta-materials with striking EM properties have been introduced, most notably those with a negative refractive index (NRI). NRI is associated with a medium of simultaneously negative electric permittivity, ε and the magnetic permeability, μ. There are no known conventional materials with such exceptional properties. Recently, Smith et al. (2000a,b) at UCSD have produced a medium with effective ε and μ that are measured to be simultaneously negative. Later, Smith et al. performed a Snell's law experiment on a similar metamaterial wedge sample, and demon-strated the negative refraction of a microwave beam (Shelby et al., 2001). Thus they showed that their medium does indeed possess an NRI, that is, it is a negative index material (NIM). Such a property has been hypothesized by Veselago who termed the medium "left-handed" (Veselago, 1968). The work on controlling the dielectric constant and producing negative ε and μ has been discussed by Smith et al. (Smith et al., 2002, 2003, 2004a,b,c; Kolinko and Smith, 2003; Pendry et al., 2003). However, until recently, the NIMs produced have been experimental samples, suitable only for proof-of-concept demonstrations.

Based on the calculation of the effective EM properties of a medium containing period-ically distributed very thin conducting wires and electric resonators, the authors at UCSD have introduced into structural composites electromagnetic enhancements in the form of tunable index of refraction, radio frequency (RF) absorption, and when desired, a negative index of refraction (Starr et al. 2004). Such properties are the result of embedding periodic metal scattering elements into the material to create an *effective medium* response over desired RF frequency ranges. We have identified two wire architectures, namely thin straight wire arrays and coiled wire arrays, that are suitable for direct integration into fiber-reinforced composites (Nemat-Nasser et al., 2002). These arrays act as inductive media with a plasma-like response to control the electric permittivity. As a result, the dielectric constant may be tuned to negative or positive values. Such a medium may be used as a window to filter electromagnetic radiation. When the dielectric constant is negative, the material does not transmit incident radiation. As the dielectric constant approaches to and exceeds the *turn-on frequency*, the incident EM radiation is transmitted through the composite. Further-more, over a desired frequency range, the dielectric constant may be tuned to match that of the surrounding environment. For instance, the dielectric constant may be tuned to match that of air, with a dielectric constant of unity, such that incident radiation does not experience a difference when encountering the composite.

12.2.1.1 Thin-Wire Plasmonic Composites

The ionosphere is a dilute plasma. Many artificial dielectrics are plasma analogs. In 1996, Pendry et al. suggested an *artificial plasmon medium* composed of a periodic arrangement of very thin conducting wires, predicting a plasma frequency in the microwave regime, below the diffraction limit. Recently, other researchers have presented examples of artificial plasmon media at microwave frequencies (Smith et al., 1999). The dielectric constant κ of a dilute neutral plasma is given by

$$\kappa = 1 - \left(\frac{f_{\mathrm{p}}}{f}\right)^{2} \tag{12.1}$$

where f_{p} is the plasma frequency and f is the electromagnetic excitation frequency. This parameter must be evaluated empirically for any configuration, but analytical and numerical results can be easily used for design purposes. Pendry et al. provide the following analytical formula for thin wire media[1]:

$$f_\mathrm{p} = \frac{c_0}{d} \sqrt{\frac{1}{2\pi(\ln\left(\dfrac{d}{r}\right) - \dfrac{1}{2}(1 + \ln \pi))}} \tag{12.2}$$

where c_0 denotes the speed of light in vacuum, d is the lattice spacing, and r is the radius of the wires (Pendry et al., 1996). Straight wire arrays, such as those shown in Figure 12.3, are designed such that the radius of the wires is very small compared to the lattice spacing, so that the wavelength of the electromagnetic excitation frequency is large compared to the lattice size. For the medium to behave as a plasma at microwave frequencies, for instance, the wire radius must be on the order of tens of micrometers and spaced on the order of centimeters. To integrate such electromagnetic

(a)   (b)   (c)

Figure 12.3 (Top) Schematic of two-dimensional thin wire array. One hundred micrometer wires are periodically embedded between composite laminates with layup jig to yield a processed fiberglass/epoxy laminate with array visible inside. (Bottom) Laminating hot presses for processing composite panels.

designs into materials, one needs a periodic material that can accommodate the arrangement of the wire elements. Fiber-reinforced polymer composites facilitate such arrangements due to the natural periodicity of their fiber and laminate construction. The arrangement of fibers within each layer provides flexibility in orientation, spacing, and geometry of the conducting wire elements. Each layer may contain elements with orientation in only one direction, as in a uni-directional laminate, or the elements may be woven such that each layer has bi-directional elements. Variation of the spacing of these elements in the thickness (z) dimension of the material is controlled by the sequence in which laminates are stacked to form the laminate.

As an example, we have introduced arrays of thin, straight wires into various types of composite materials. Composite panels were made by hand-layup of preimpregnated woven fabric (prepreg). The samples varied in the type of host material, wire diameter, and number of electromagnetic layers. Host materials included E-glass fibers impregnated with epoxy resin, Spectra® (Honeywell UHMW polyethylene) fibers impregnated with vinyl ester resin, and quartz fibers impregnated with cyanate ester resin, chosen for their mechanical attributes and favorable dielectric characteristics. The dielectric constant of epoxy/E-glass was 4.44 at microwave frequencies with a loss tangent of 0.01, and that of vinyl ester/Spectra was 2.45 with a loss tangent of 0.002. Cyanate ester/quartz provided the best overall electromagnetic characteristics with a dielectric constant of 3.01 and a loss tangent of 0.001, where a low dielectric constant and loss tangent are preferable for optimal microwave transmission. The fiber volume fraction for each material was about 60%. The frequency at which the panels behave as plasma depends upon the dimensions of the embedded wire array. Numerical simulations were performed to predict the necessary array for plasma response in the microwave regime. In making each panel, copper wire of 75 or 50 μm diameter was strung across a frame to form the desired pattern and was subsequently encased in layers of prepreg. Panels were processed at elevated temperature and pressure to cure the resin and form the solid composite as shown in Figure 12.3. Electromagnetic characterization was performed to extract the effective material properties through measurements in an anechoic chamber that we developed in the Physics Department of UCSD. Additional characterizations have been performed on a focused beam electromagnetic system in the first author's laboratories, CEAM (Center of Excellence for Advanced Materials), as is discussed in connection with Figure 12.13 later on.

Representative dispersion relations of the dielectric constant in the microwave regime for each of these panels are given in Figure 12.4, comparing analytical and numerical predictions with the experimental results. The graphs in this figure show the characteristic trend of changing the dielectric constant from negative to positive values as a result of the plasmon media in a composite panel of each type. Results for the different host materials show similar behavior, though the turn-on frequency is shifted depending on the dielectric constant of the host material and the wire diameter and spacing. Moreover, the results show that a host material with a lower dielectric constant provides a wider bandwidth over which the dielectric constant of the free space can be matched (Plaisted et al., 2003b).

12.2.1.2 Coiled Wire Plasmon Media Composites

As an alternative to processing thin wires into composites, we may incorporate thicker, more robust wires in the form of coiled arrays. By proper design of the coil geometry, various degrees of inductance may be achieved with thicker wires as compared with the thin straight wires. Textile braiding of reinforcing fibers with wire is an ideal method to integrate the coil geometry into the composite. The braiding process interlaces two or more yarns to form a unified structure. Our process uses a two-dimensional tubular braiding machine, as shown in Figure 12.5, which operates in a maypole action, whereby half of the yarn carriers rotate in a clockwise direction, weaving in and out of the remaining counter-rotating carriers. This action results in a two-under two-over braid pattern. Each yarn makes a helical path around the axis of the braid to create a uniform coil. To integrate the wire coil into such a structure, we simply replace one of the fiber

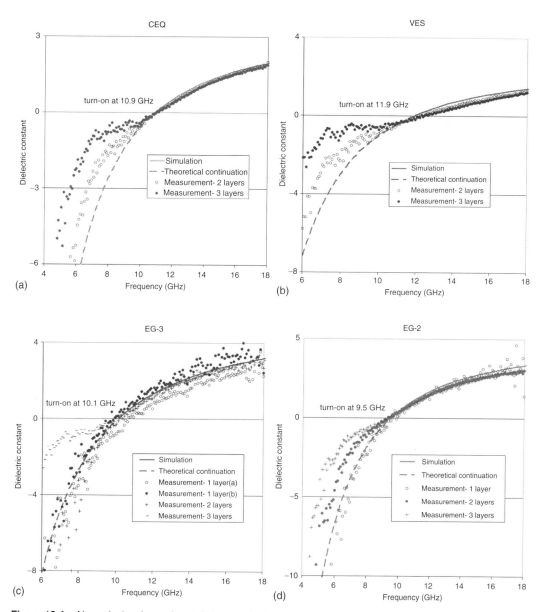

Figure 12.4 Numerical and experimental characterization of the thin-wire EM composite samples. Data for panels made of the same host composite material and wire diameter are displayed in one chart since their numerical simulations are identical. "Turn-on" indicates the transition between the stop-band and pass-band, or the frequency above which the material transmits electromagnetic radiation. (a) 50 μm (0.002 in.) diameter wires embedded in cyanate ester/quartz composite. (b) 50 μm diameter wires embedded in vinyl ester/Spectra composite. (c) 75 μm (0.003 in.) diameter wires embedded in epoxy/E-glass composite (two single layer samples were manufactured and measured for this case). (d) 50 μm diameter wires embedded in epoxy/E-glass composite.

carriers with a wire carrier. A comprehensive description of the textile braiding process is given by Ko et al. (1989) and Ko (2001). Modeling of the mechanical properties has also been developed for textile braids (see e.g., Cox et al., 1994; Naik, 1995; Xu et al., 1995; McGlockton et al., 2003; Yang et al., 2003).

Braiding wire with the reinforcing fibers results in an electromagnetic element with uniform geometry that maintains its shape under considerable handling and other processing conditions. The

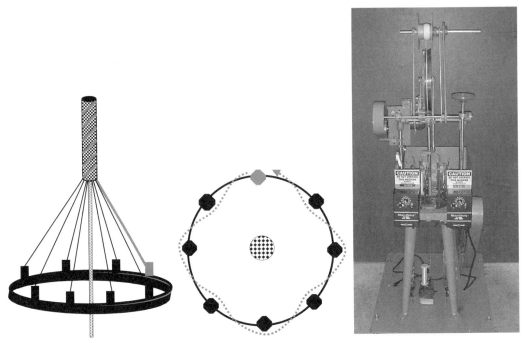

Figure 12.5   (Left) Schematic of tubular braiding machine. Fibers and wire (indicated in gray) are spooled from carriers that rotate on a circular track. Fibers may be braided around a center mandrel or other fibers in the core of the braid. (Center) Arrow indicates path taken by one yarn carrier in maypole braiding pattern. (Right) Photograph of tubular braiding machine at CEAM.

braid itself is a tough structure that protects elements woven into the outer sheath, as well as other elements in the core. Thus functional elements (wires and/or perhaps sensors) are truly integrated into the fibers of the host composite, rather than acting as inclusions in the matrix phase. Furthermore, braiding allows fine control of the pitch and diameter of the wire coil such that the electromagnetic properties may be tuned for desired performance. The sense of the coil, as left-handed or right-handed, may also be varied in this process to address issues of chirality, as discussed below (see Amirkhizi et al., 2003).

12.2.1.2.1   Chirality

The introduction of coil geometry not only affects the inductance of the medium and consequently the overall dielectric constant, but also introduces different capacitative response than mere straight wires. This capacitative response usually changes the overall magnetic properties of the medium, although the inductive response still remains the dominant effect. Part of the magnetic response is induced by the chirality effect which is discussed presently. However, a more careful and thorough study is needed since the techniques that can be used to eliminate chirality do not necessarily change the axial magnetic effects.

Of importance is the effect of the handedness of the coils on the EM field vectors. The geometry of the coils requires that the current density in the conductors has a circumferential component, in addition to the axial component which is the only component present in the case of the straight wires. The oscillating circumferential component of the current enhances the magnetic field of the propagating wave with a component parallel to the axis of the coils. Note that as the active component of the electric field is parallel to the axis of the coils, the accompanying magnetic field is normal to it. Therefore the enhanced magnetic field is normal to the external excitation.

Moreover the extra component is in phase with the current density and in turn with the external electric field, whereas the external magnetic field and electric field are out of phase by a quarter of a

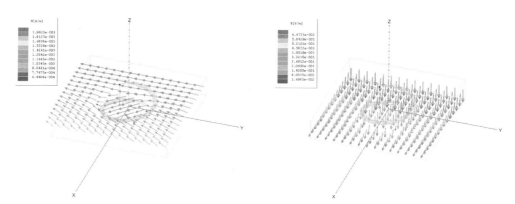

Figure 12.6 (See color insert following page 302)   Electric field (left) and magnetic field (right) patterns calculated for a unit cell of a coiled medium using ANSOFT-HFSS. The wave is propagating in the *x*-direction and the fields on the two *yz* faces have 50° phase difference. The incoming wave (electric field) from the far *yz* face is at this time polarized parallel to the axis of the coil. However, the effect of the coil adds an out of phase normal component. Therefore, the field vectors of both electric and magnetic fields rotate as the wave travels through the cell.

cycle. If the created magnetic component were in phase with the external excitation, the superposed field would be slightly skewed from the original field. This would have meant that one could still define principal axes for the material property tensors, although they are slightly angled compared to the structural axes. However, the phase incompatibility creates rotating magnetic fields which in turn create rotating electric fields. The principal propagating polarizations are not linear any more, but rather have elliptical polarization (see Figure 12.6).

The effect of chirality can be used to benefit some applications. However, in most cases, it may introduce unwanted complexity. In order to eliminate this behavior, two methods have been proposed. The first method is to include alternating coils in the array so that every right-handed coil should be adjacent to a left-handed coil. We considered this solution only for regular arrays as will be discussed, but we conjecture that since the wavelength is much larger than the spacing between coils for effective media, a randomly homogenous and statistically equal distribution of the right and left-handed coils should also have a similar effect. Note that for an irregular medium, the size of the volume that is randomly homogeneous must be considerably smaller than the wavelength as compared with a regular array. Another way to eliminate the chirality effect is to use double coils instead of simple single coils. If two concentric coils with the opposite handedness are together, most of the magnetic field created by the circumferential electric current is effectively canceled.

In the first method, one can stack alternating layers of right- and left-handed coils together. The traveling wave undergoes the opposite effects of the two layers and therefore the polarization of the fields will not be rotated. Another arrangement that has the same effect is to design each layer to have alternating coils. In other words, instead of having alternating layers in the thickness direction, one has alternating layers in the normal direction. Moreover by shifting these layers by one lattice spacing, one can achieve a 2-D checker board design. These three designs have similar behavior and do not significantly affect the plasmon frequency, compared to the original chiral medium. The design with alternating layers normal to the propagation direction is preferable, since the period length in the propagation direction is smallest and therefore the diffraction frequency limit is higher, as shown in Figure 12.7.

In the second method, the effect of clockwise or counter-clockwise circumferential current is not cancelled by adjacent coils, but by a local and concentric coil of the opposite handedness. The attraction of this method lies in the fact that no special ordering or arrangement at the time of manufacturing of the composite is required. The double coils can either be made by a two-stage braiding scheme or a similar design can even be achieved by braiding the conducting coils of insulated wires at the same time in opposite orientations. The double coils may have an advantage

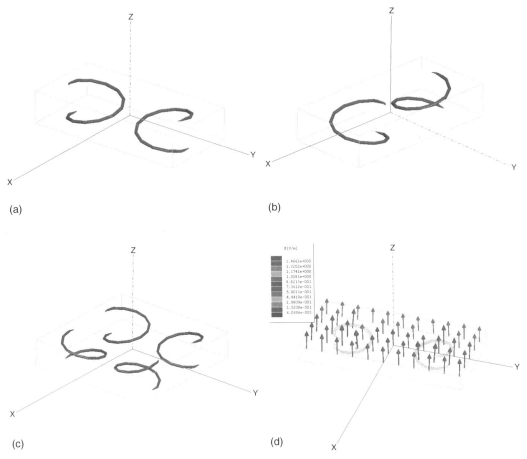

Figure 12.7 Alternating arrays of left-handed and right-handed coils. Considering an EM wave is propagating through the medium in the *x*-direction, each of the above sets can be used to cancel the polarization rotation effect. To envision the whole array, imagine these as blocks and fill the 3D space with similar blocks in each case (only translated by the size of the block in each direction). Top left: Each layer through the thickness consists of alternating coils. The layers are then stacked, such that normal to the thickness, the coils are similar. Top right: Layers of uniform right-handed and left-handed coils are stacked through the thickness. Bottom left: Checker board configuration. All four adjacent coils to any single one are of opposite handedness. Bottom right: The effect of the field rotation is canceled. However, the linear polarization of the electric field parallel to the axis of the coils is maintained through the medium. Note that the periodic length of the medium for the top right and bottom left cases is twice as much as it is for the top left case, hence providing a smaller diffraction frequency limit. The dispersion relation and plasmon frequency for the principal propagating modes remain essentially unaltered compared to the uniform arrays. However, the modes are dramatically different.

in mass production of composites. However, the additional inside loop increases the plasmon frequency and reduces the effective range of the pass band. Numerical studies show that higher pitch values can overcome this difficulty, as indicated in Figure 12.8. Simulation parameters for these results are given in Table 12.1.

12.2.1.3 Braided Composite Manufacturing

As an example, we have braided coil elements with para-aramid (DuPont Kevlar®) reinforcing fiber and polyamide (DuPont nylon 6,6) thermoplastic fiber. The outer braid consists of a single 30 gauge (0.254 mm diameter) copper wire, four ends of 200 denier Kevlar fiber, and three ends of 210

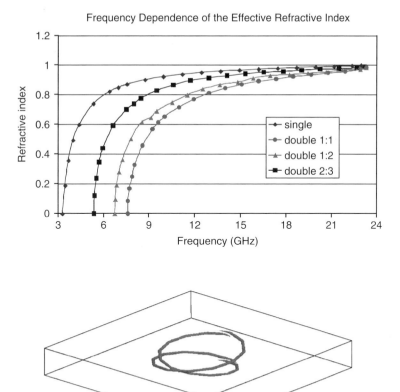

Figure 12.8 (Top) Frequency dependency of the effective refractive index for various coil geometries. Double coils (bottom) can also be used to cancel the effect of chirality. However, they also modify the plasma frequency of the medium as the effective inductance and capacitance per unit volume is changed.

denier nylon fiber. The core of the braid consists of one end of 1000 denier Kevlar fiber and three ends of 420 denier nylon fiber. An illustration is provided in Figure 12.9 showing the constituents of the braid architecture. Nylon is included in the braiding process since it will serve as the polymer matrix of the final composite, although it may not be the optimal choice in terms of mechanical strength of the resulting composite. Complete fiber wet-out can be a difficult processing challenge in braided composite materials, due to the inherent tight packing of fibers in the braiding process. We have initially addressed this issue by developing a commingled braid composite, which integrates the eventual matrix phase as a thermoplastic fiber that is braided along with the structural

Table 12.1 Parameters for Simulating Various Coil Geometries in HFSS Electromagnetic Simulations

		Single	Double 1:1	Double 1:2	Double 2:3
Outer cell	Spacing (mm)	6.35	6.35	6.35	6.35
	Cell height (mm)	1.1	1.1	1.1	1.1
	Inner diameter (mm)	2.6	2.6	2.6	2.6
Inner cell	Turns in one cell	1	1	2	3
	Inner diameter (mm)	—	2.2	2.2	2.2
	Turns in one cell	—	1	1	1
	Wire thickness (mm)	0.1	0.1	0.1	0.1
	Plasma frequency (GHz)	3.26	7.59	6.73	5.35

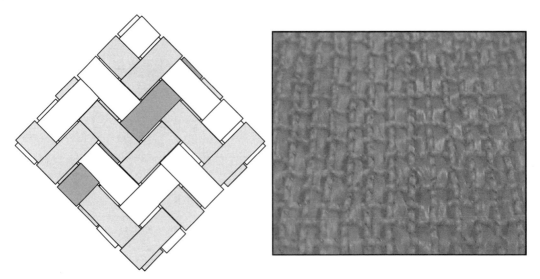

Figure 12.9 (Left) Schematic of outer braided architecture with 2 up 2 down braid pattern consisting of Kevlar fibers (light gray), nylon fibers (white) and copper wire (dark gray). (Right) Photograph of braids bi-directionally woven into fabric with additional Kevlar fibers. Coils with opposite sense are woven adjacent to one another.

fibers. Overall, the composite is designed to have a Kevlar fiber volume fraction of about 50%. Selection of the diameter of the core allows control of the diameter of the coil that is braided around it. The core may be composed of various other elements, including other electromagnetic elements, or perhaps sensors, though in this initial design we have incorporated only reinforcing fibers. The pitch of the braids is determined by the take-up and rotation speed of the carriers. The pitch of these coils was maintained at $60°$ from the axis of the braid.

The braided elements take the form of a laminate by weaving with other reinforcing fibers to form a cohesive fabric. The braids may be oriented in a single direction in each layer or may be woven together bi-directionally. Due to the inherent stiffness of the dry braid, tight weaving patterns in a bi-directional weave, such as plain weave and satin weave, may be restricted since the braid cannot be woven over small intervals without kinking, which compromises the braid structure. This factor is dependent on the braid and wire diameter, where smaller diameters are not subject to such limitations. This limitation is avoided when braids are woven uni-directionally since the fill yarns (weft direction) are able to accommodate such undulation while allowing the braid elements (warp direction) to remain straight. To achieve the desired spacing of the coil array, while maintaining a uniform composite fabric, blank braids may be woven into the layer or inserted between layers. The blank braid is identical to the electromagnetic braid element, however, the copper wire is replaced with an end of reinforcing fiber. Additionally, as mentioned above, chiral effects of the coil geometry can be eliminated by alternate placement of a left-handed coil next to a right-handed coil. Such an arrangement can be easily achieved in the braiding and weaving processes. Woven layers are stacked in accord with the electromagnetic design and processed with additional thermoplastic matrix at elevated temperature and pressure to form the consolidated composite.

These braided elements have been integrated into a composite panel and characterized electromagnetically. Figure 12.10 shows such a panel consisting of Kevlar braids woven into laminates and pressed into a nylon matrix composite. The coils were arranged in an alternating square matrix in one direction of the composite. Hence, the panel showed a plasmon response in one orientation and not in the other. The experimental results showed good agreement with our simulations. The dielectric constant of the structure is measured as a function of frequency

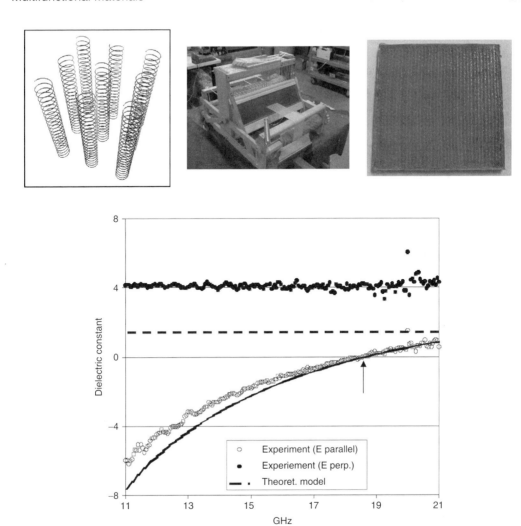

Figure 12.10 (See color insert following page 302) (Top) Coiled wire architecture integrated with structural Kevlar fibers by braiding. Braids woven and laminated into composite plates. (Bottom) EM characterization of the braided and woven composite showing typical plasmon media response when aligned parallel to the polarization of the EM radiation. Normal (nonplasma) dielectric response is observed when aligned in the perpendicular direction.

from 11 to 21 GHz, whereupon at around 18 GHz, the dielectric constant passes through zero. This dispersion relation follows the characteristic trend of the thin straight wire arrays studied previously. Between the plasma frequency and the upper limit of our frequency sweep, the dielectric constant of the composite array approaches unity. Since the index of refraction of the material is the square of the dielectric constant, we may also conclude that the index approaches unity.

12.2.1.4 Controlling the Effective Magnetic Permeability

Following Pendry et al. (1999), Smith et al. (2000a,b), and Shelby et al. (2001), we have shown that the effective magnetic permeability, μ, of free space can be rendered negative over a certain frequency range by suitably integrating the so called split-ring-resonators, as shown in Figure 12.11. The structure, however, cannot be integrated into a thin composite panel. To remedy this fundamental barrier, we considered collapsing the rings into nested folded plates, as shown in

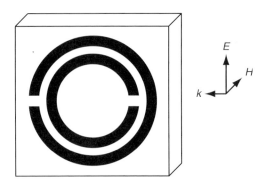

Figure 12.11 Original SRR design with wave vector *k*, electric *E*, and magnetic *H* fields indicated for effective negative permeability.

Figure 12.12, and called the construction folded-doubled-resonator (FDR). Measurements, using a focused beam EM characterization system, Figure 12.13, revealed that indeed the composite had a negative μ over a frequency range of about 8.5 to 9.5 GHz. Following this, a new design was conceived, numerically simulated, and constructed that had a more pronounced negative μ. This construction is shown in Figure 12.14, and the measured results are given in Figure 12.15. As is discussed in the next section, combining the negative ε and μ, it is possible to construct a composite panel with negative index of refraction.

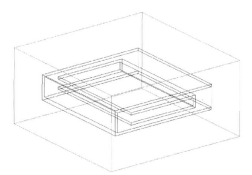

Figure 12.12 FDR design produces the required resonance with a thickness that lends itself to inclusion into an actual composite panel of reasonable thickness.

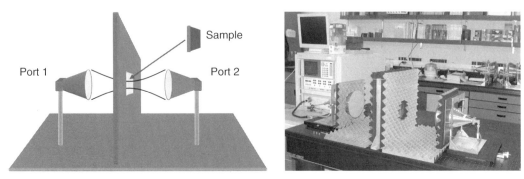

Figure 12.13 (See color insert following page 302) (Left) Schematic and (right) photo of Focused Beam system for EM characterization from 5 to 40 GHz at UCSD's CEAM.

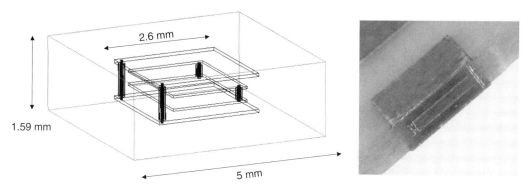

Figure 12.14 FDR unit cell and dimensions (left); and fabrication within a composite panel (right).

12.2.1.5 Negative Refractive Index Composites

As mentioned above, over the past several years, the authors at UCSD's Center of Excellence for Advanced Materials (CEAM) have developed methods to design, fabricate, and characterize NIMs, and have demonstrated these capabilities in illustrative microwave experiments. Composite panels of 2.7 mm thickness have been produced that possess through-the-thickness negative index that has been measured unambiguously by full S-parameters retrieval, as discussed below (Starr et al., 2004). Such samples are relatively easy to characterize, as both transmission and reflection measurements can be carried out on very thin samples.

Several views of the actual panel along with the dimensions of the elements within a unit cell of the CEAM NIM are shown in Figure 12.16–Figure 12.18. The elements that give rise to both electric and magnetic response are fabricated using multi-circuit board techniques. The composite is assembled from three laminated layers. The top and bottom layers consist of Rogers 4003 circuit board laminates ($\varepsilon = 3.38$, tan $\delta = 0.003$), with a prepreg layer of Gore SpeedBoard ($\varepsilon = 2.56$, tan $\delta = 0.004$). The measured (solid) and simulated (dashed) values of the real (black) and imaginary (gray) index of refraction are shown in Figure 12.19.

The layers are bound together by a layer of adhesive at the interfaces between the Gore and Rogers circuit boards. Both of the Rogers circuit boards initially have a thin layer of copper (half-ounce or approximately 1 μm in thickness) deposited on both sides from which the elements are patterned using conventional optical lithography. The wire elements are patterned on the sides of the Rogers boards that face the Gore SpeedBoard. This prototype was manufactured by Hughes Circuits.[2]

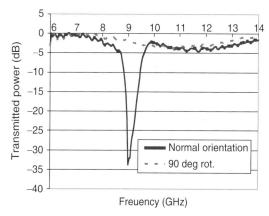

Figure 12.15 Negative magnetic permeability experimentally demonstrated from about 8.5 to 9.5 GHz for the FDR structure.

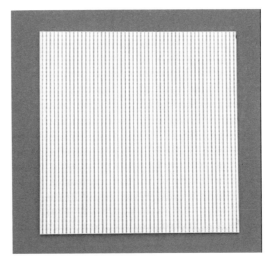

Figure 12.16   Planar view of the CEAM NIM.

12.2.2 Heating Functionality

Initial simulation and testing has been conducted to demonstrate the heating capabilities of our integrated thin wire arrays (Plaisted et al., 2003a,b; Santos et al., 2004). Using the same wire diameter and array dimensions as designed for EM functionality, we have applied direct current to resistively heat a composite sample. Embedded wires are currently used for resistive heating as a method of welding thermoplastic polymers and polymer composites (Eveno and Gillespie 1988; Jakobsen et al., 1989; Ageorges et al., 2000) Similarly, embedded heating elements have been used to cure the resin matrix in thermoset polymer composites (Sancaktar et al., 1993; Ramakrishnan et al., 2000).

12.2.2.1 Simulation and Testing

The thin copper wires in our composite can be connected to a DC electrical source and leveraged as heating elements, dissipating heat as a result of Ohm's Law:

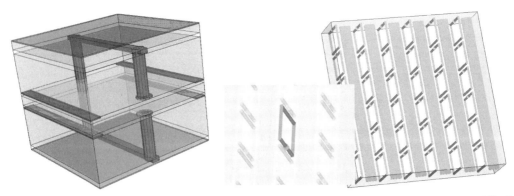

Figure 12.17 (See color insert following page 302)   (Left) Unit cell of NIM. The negative permeability is achieved by ring resonators, formed from copper strips on the upper and lower surfaces, connected to vias that run through the structure, with one of the vias possessing a gap in the center to introduce capacitance. Copper strips are patterned on the central circuit board, giving rise to the negative permittivity of the structure. (Right) Views of conducting elements as they are fabricated within a composite panel.

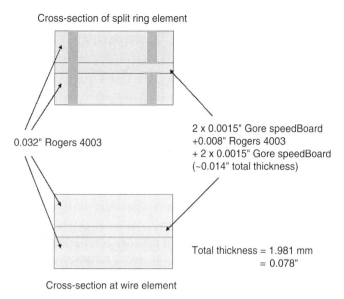

Cross-section of split ring element

0.032" Rogers 4003

2 x 0.0015" Gore speedBoard
+0.008" Rogers 4003
+ 2 x 0.0015" Gore speedBoard
(~0.014" total thickness)

Total thickness = 1.981 mm
= 0.078"

Cross-section at wire element

Figure 12.18   Dimensions for cross-sectioned view of NIM.

$$P = VI = I^2R = V^2/R \tag{12.3}$$

where P is power, V is voltage across the circuit element, I is direct current through the circuit, and R is the resistance of the circuit element. Finite element computer simulation software, NISA, is used to model the heating in conjunction with experimental testing. The heat transfer module of NISA, known as NISA/HEAT, uses finite element methods to solve the heat conduction equation for temperature based on a set of initial and boundary conditions.

Our thin wire arrayed composites typically have a spacing of 0.125 in. (3.175 mm) between copper wires of 100 μm diameter. To simulate this geometry in NISA's graphical interface, a unit cell of 0.125 in. by 0.125 in. is constructed to represent a cross-section of the composite as shown in Figure 12.20. To reduce calculations to a 2-D problem, the unit cell is assumed to have unit depth and a constant cross-section along the length of the wire.

A square element mesh is applied to the unit cell, with the circular cross-section of wire approximated by a square pattern of four elements. Boundary conditions are prescribed on the

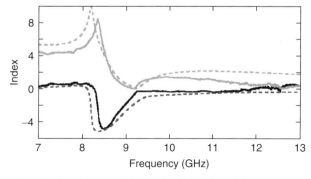

Figure 12.19   Recovered refractive index (n) from simulation data (dashed curves) and from measured S-parameters (solid curves). Black and gray curves represent the real and imaginary parts of the refractive index, respectively.

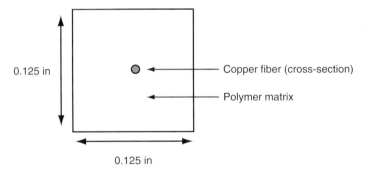

Figure 12.20   Unit cell geometry for NISA simulation of resistive heating scheme.

mesh based on: thermal conductivity, mass, density, and specific heat of the material; electrical power input and heat generation; and conditions at the edges of the unit cell. We approximate the thermal properties of the polymer matrix with those of epoxy commonly used in composites. These properties are prescribed on the polymer elements of the mesh, while the properties of annealed copper are prescribed on the wire elements. It is assumed that the electrical power input is constant over time and converted fully into heat, so a constant heat generation is prescribed on the wire elements. The conditions at the edges of the unit cell are either "insulated," implying that boundaries of zero heat flux are prescribed on all edges of the cell, or "exposed to air," where convection boundary conditions are prescribed on two opposite edges of the cell instead of zero heat flux. The insulated condition simulates a unit cell surrounded on all sides by identical material through periodic boundary conditions.

According to the results of our simulation, the temperature of the insulated unit cell increases linearly for a constant power input, while the temperature of the exposed unit cell holds constant after a period of time (Figure 12.21). Also, the temperatures at different locations in the exposed unit cell vary by as much as 15°C, as shown by the multiple lines on the graph. For the insulated unit cell, the temperature distribution differs by 4°C at the most. The power density value (W/cm^2) in these graphs denotes power distribution over the flat area of the composite panel, *not* the power distribution over the cross-section.

A sample composite panel was fabricated from glass–fiber-reinforced epoxy prepreg and 100-μm copper wire to test the resistive heating process. Copper wires were strung in a parallel arrangement in one direction and three thermocouple wires were included at various depths between the prepreg layers to monitor internal temperatures. The dimensions of the panel were

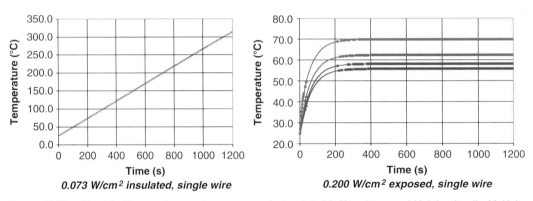

Figure 12.21   Simulated temperature vs. time response for insulated (left) and exposed (right) unit cells. Multiple lines indicate temperatures at various locations within the panel.

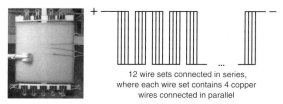

12 wire sets connected in series,
where each wire set contains 4 copper
wires connected in parallel

Figure 12.22   (Left) Composite panel in conductor frame. Embedded thermocouple wires protrude to the right of the panel. (Right) Abbreviated circuit diagram.

15 cm by 15 cm by 0.32 cm thick and its fiber volume fraction was around 60%. After curing, the copper wire strands that protruded from the edges of the panel were retained since they provided electrical connection.

The wires in the composite panel were combined into a single circuit by a custom apparatus we refer to as a conductor frame. The frame consists of conductor bars to which adhesive copper strips are attached. By clamping groups of wires to the conductor bars, a combined series–parallel circuit through the entire panel is created (Figure 12.22). The copper strips extend around the sides of the upper conductor bar so that they may be connected to the power source. The DC power source used in these tests was a voltage generator with maximum output of 36 V and 8 A. In addition, the thermocouple wires were connected to a multi-channel thermocouple monitor. To measure the electrical power input, two multimeters were included in the setup to measure total voltage across all wires in the composite and total current across the entire circuit.

The voltage for our initial tests was based on simulation and remained constant throughout each individual test. The voltage was then iteratively optimized in subsequent tests to achieve our target temperature. Prior to turning on the power source the initial temperature for all thermocouple channels was recorded. Once power was supplied to the composite panel, temperatures were recorded for each of the thermocouple wires at 30-s intervals. The voltage and current were also recorded every 30 seconds for a total duration of about 20 min.

Noninsulated test conditions were conducted with the panel configuration as shown in Figure 12.22. To test insulated conditions, sheets of cotton-like fiberglass were placed on both sides of the panel to minimize heat loss. The results of the resistive heating tests are qualitatively similar to the results of the finite element simulations. The temperature for an insulated composite rises almost linearly, while the temperature in the exposed composite rises quickly at first before holding constant (Figure 12.23). However, the quantitative results differ noticeably between simulation and experiment. For insulated conditions, the temperature after 1200 s is above 300°C in simulation, whereas the temperature in the actual test only exceeds 80°C. This error is less pronounced for the exposed case; the simulation predicts a maximum constant temperature of 70°C while the test results have a maximum temperature of 84°C. However, the simulation of exposed conditions

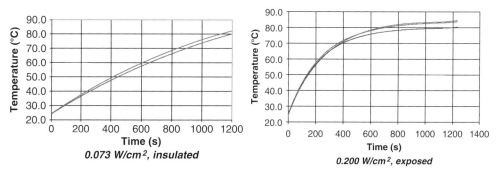

0.073 W/cm², insulated

0.200 W/cm², exposed

Figure 12.23   Experimental temperature vs. time for insulated (left) and exposed (right) panels.

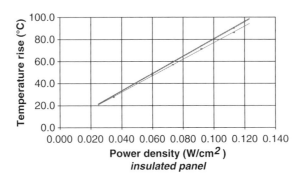

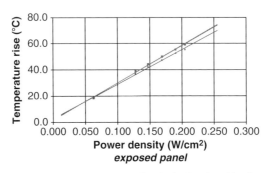

Figure 12.24   Experimental temperature rise vs. power density for insulated (top) and exposed (bottom) panels.

predicts a local temperature difference of 15°C between different areas in the composite, while this temperature difference is only about 5°C in the actual test.

The tests were carried out for different levels of electrical power, and a linear correlation was found between the final temperature and the power input (Figure 12.24). Power density over the face of the panel is used in place of total electrical power so that the value is normalized for any size of composite panel. Temperature rise is used in place of the actual final temperature so that temperature results are normalized for any initial temperature. According to the graphs, an insulated panel requires about 60% less power than the exposed panel to reach the same temperature. If an ambient temperature of 20°C is assumed, and our target temperature is 80°C, then a temperature increase of 60°C is desired, which corresponds to 0.073 W/cm^2 power input for the insulated panel, compared to the exposed panel's 0.20 W/cm^2.

Thermal management within the composite may be leveraged for a number of applications. In our multifunctional composite, we may utilize this heating function to induce a thermally activated healing process as detailed in the next section.

12.2.3   Healing Functionality

A material that can heal itself is of great utility where access for manual repair is limited or impossible, as in a biological implant or a material that is launched into orbit in the solar system. Structures made of such a material may have significantly prolonged service life in addition to improved safety if failure mechanisms such as cracking can be repaired *in situ*. Nature has long demonstrated this property in various biological materials, whereas, until recently, man-made healing materials have essentially not been demonstrated. However, interest in synthetic healing materials has recently gained significant attention with the creation of a truly autonomic healing polymer by White and other researchers at the University of Illinois (White et al., 2001). Since then other healing materials have been proposed (Bleay et al., 2001; Chen et al., 2002), one of which is a novel polymer that will be the focus of further research at UCSD. Wudl and his research group at

UCLA have created a strong, tough polymer that forms a high degree of thermally reversible, covalent cross-links. Mechanical failure of this polymer occurs preferentially along these cross-links, and due to the reversible nature of this bond it may be repaired by application of moderate pressure and heat.

12.2.3.1 Polymer Healing

Polymers offer many attributes that allow for healing of damage within the material. Polymers consist of long chain molecules with molecular weights ranging from 10^4 to 10^6. These long chains are made up of a string of monomers, which are the molecular repeat unit identifying that particular polymer. Linear, branched, and the other nonnetworked polymers generally form amorphous or semicrystalline polymers with thermoplastic character. Interaction between chains occurs through noncovalent bonding such as hydrogen bonding and chain entanglements. As a thermoplastic, the polymer may be heated to its melting temperature and solidified repeatedly with little change in the properties. In contrast, thermosets consist of cross-linked networks formed by covalent bonding and will degrade, rather than melt, upon heating.

Healing damage within a polymer is most often associated with the softening and flow of material across a damaged interface that occurs upon heating a thermoplastic polymer. This technique is commonly known as thermoplastic welding. Healing in thermoplastic polymers occurs largely due to the restoration of entanglements in the polymer interface. Secondary bonding between chains, in the form of van der Waals or London dispersion forces, is also critical to the healing process. Other bonding, such as hydrogen bonding and chemisorption, can play an important role. Chains are able to diffuse across the interface when heated above the glass transition temperature, T_g. The rate of crack healing is strongly time dependent, as the once separated molecular structures diffuse across the interface to form an equivalent bonding state to that of the virgin material. Crack healing in thermoplastic polymer surfaces seldom results from the reformation of broken bonds (primary bonds) in the polymer backbone. Typically the polymer chains at the crack interface have been irreversibly damaged through bond breakage which results in an average molecular weight significantly lower than that of the bulk polymer. Furthermore, the catalyst for polymerization (through addition or condensation reactions) is not present to repolymerize the material. The crack healing and welding in thermoplastics has been widely studied in the literature, particularly by Wool and Kausch and their co-workers (Wool, 1978, 1979, 1995; Jud and Kausch, 1979; Jud et al., 1981; Wool and O'Connor, 1981a,b, 1982; Kausch and Jud, 1982; Kausch, 1983; Kausch et al., 1987; Kausch and Tirrell, 1989; Wool et al., 1989)

Healing damage within thermoset polymers is typically not possible due to the cross-linked nature of these materials. The cross-link network prevents the polymer chains from diffusing through the material when heated above the T_g. Rather the material begins to thermally degrade when heated excessively, and in contrast to thermoplastics, cannot be returned to its original state. Until recently, there has been no evidence of repairing damage in highly cross-linked materials. An interesting repair scheme has been proposed by White et al. to embed a healing system within an epoxy polymer (White et al., 2001) Liquid monomer is microencapsulated and embedded with dispersed catalyst in an epoxy polymer, such that a propagating crack intersects a microcapsule and releases the healing agent to effectively glue the crack shut. A schematic of the healing process is given in Figure 12.25. Another system, proposed by Wudl and coworkers, uses a novel reversible bonding approach to repair cross-links (Chen et al., 2002). It is this polymer that has been utilized in the multifunctional material under development at UCSD.

12.2.3.2 Thermo-Reversibly Cross-Linked Polymer

In 2002, Wudl and coworkers published work on a polymer with the ability to repair internal cracking (Chen et al., 2002). Until that time, there had been no highly cross-linked polymers that

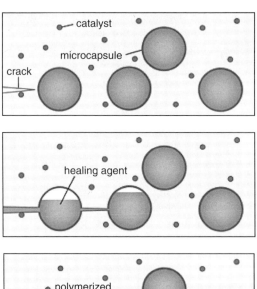

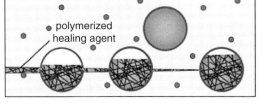

Figure 12.25   Healing concept of an autonomic healing polymer (From White, S.R., Sottos, N.R., Guebelle, P.H., Moore, J.S., Kessler, M.R., Sriram, S.R., Brown E.N. and Viswanathan S. *Nature* 2001: 409(6822), 794–797. With permission.).

could be repaired without the use of additional monomers or surface treatment to repair a cracked interface. Chen et al. accomplished this by synthesizing a polymer based on a thermally reversible Diels–Alder (DA) and retro-DA cycloaddition. The Diels–Alder cycloaddition is a widely used reaction in organic synthesis. Many polymers involving the DA cycloaddition have been synthesized, though in many cases the retro-DA reaction is not observed if the diene and dieneophile are not sufficiently stable on their own. Those polymers with suitable monomer combinations to exhibit the retro-DA reaction have incorporated the DA adduct into the backbone of the polymer (Chujo et al., 1990; Engle and Wagener, 1993; Imai et al., 2000) In contrast, the unique aspect of the polymer created by Chen et al. is that all of the monomer linkages, or cross-links, are formed by DA cycloaddition and furthermore exhibit the retro-DA reaction.

12.2.3.3 Healing Experiments

The weakest bond in the polymer structure is the polymerization or cross-linking bond of the DA adduct. While strong in comparison to other types of noncovalent chemical bonds, this is the first bond to break when the material is loaded to failure or heated above its transition temperature. However, because this bond is reversible, this is also the bond that reforms when the material is cooled below the transition temperature. To test the healing ability of this bond, quantitative testing of the fracture toughness was performed (Chen et al., 2003). Compact tension samples were notched with a razor blade and loaded in a direction perpendicular to the pre-crack. To arrest crack propagation in these tests, a hole was drilled into the middle of the specimen. In this way the cracks were arrested before fracturing the material into two halves and allowed more accurate alignment of the fracture surfaces during the healing treatment. Healing was carried out at 115°C for about 30 min with pressure applied by a clamp. Averaging over three tests, the material was able to recover 81% of its original fracture load. Furthermore, when the same healing procedure was applied a second time,

Furan Maleimide cyclo adduct

Figure 12.26 The polymer consists of a multifuran molecule combined in stoichiometric ratio with a multi-maleimide molecule.

the material recovered an average of 78% of the original fracture load, indicating that the material could be repaired multiple times. It was noted that the crack usually propagated along the same crack plane.

In addition, we have observed the healing mechanism in samples of this healable polymer. Our aim was to arrest crack growth prior to complete fracture of the sample. In this way we could improve the ability to match the severed interfaces back to their original location prior to heating. Due to the inherent high mechanical strength of this material, a controlled cracking procedure was devised. Samples were machined to 0.25 in. by 0.15 in. by 0.20 in. dimensions with a 0.08 in. diameter hole penetrating through the middle. Two notches were cut into the hole to initiate the crack on opposing sides of the hole. The samples were cooled in liquid nitrogen and immediately loaded in compression in the direction of the machined notch. The applied load caused cracks to grow from the notches in a controlled manner in the direction of the applied load. Cracked samples were then placed in a spring device that applied compression normal to the crack faces so as to put the crack faces in contact. Heat was then applied at various levels and durations. For samples treated for at least 6 h above 80°C under a nitrogen atmosphere with about 8 kPa of compression normal to the interface, the crack was observed to disappear, indicating healing. In these cases, no visible scar remained, apart from the initial starter notch. These tests were only qualitative in nature. However, it appeared that the crack had been completely repaired and visually the material had been restored to its original state. Figure 12.27 shows representative photographs before and after the healing event.

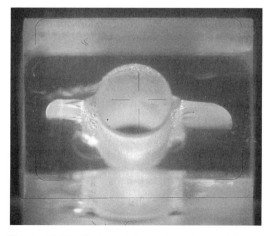 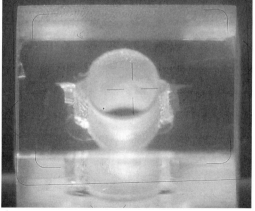

Figure 12.27 Optical photographs taken at 20 × magnification of representative healable polymer sample. Diagonal view of sample with predrilled hole and starter notches is shown. (Left) Polymer sample after controlled cracking. Note cracks have propagated from the starter notches to the left and right of the sample. (Right) Same sample after healing for 6 h under nitrogen and about 8 kPa of compression normal to the crack face. Crack faces have disappeared leaving only starter notches and predrilled hole visible.

12.2.3.4 Healing Summary

A cross-linked polymer with thermally reversible covalent bonds, such as that created by Wudl et al., offers many attractive attributes. When combined with a medium that distributes heat, such as a network of resistive heating wires, the healing mechanism may be initiated throughout the material. The effectiveness of such a system has yet to be fully determined, however, since healing occurs on the molecular level to reestablish broken covalent bond, there is considerable promise. Initial results on macro-cracked neat polymer samples show excellent potential that near full recovery of original strength is possible. Moreover, healing may be carried out multiple times on the same broken bond. This polymer requires outside intervention to initiate the heating (healing) process. We are developing self-sensing smart materials to embed in this material, where an integrated self-sensing, self-healing composite may act autonomously.

12.2.4 Sensing Functionality

The goal is to add *information-based* properties into multifunctional composites, mimicking nature's approach to local and global information acquisition, processing, and communication. Figure 12.28 identifies the necessary three interwoven challenges that must be successfully met in order to create intelligently sensing composite materials that are:

*   Aware of their environmental and internal changes; and
*   Can selectively acquire, process, and store or communicate information locally and globally.

As is suggested in this figure, integrating sensing functionality into structural materials begins with the challenge of composite fabrication that seamlessly integrates within the material the necessary sensing and electronic platforms, without sacrificing the structural attributes of the resulting system. The next challenge is that the structurally integrated micro-sensors must be able to monitor, interact with their neighboring sensors, make on-board decisions, and report on the local structural environment upon request, or in real-time as necessary. And, the final but equally vital challenge is to create an efficient data handling architecture with local–global processing and communication algorithms.

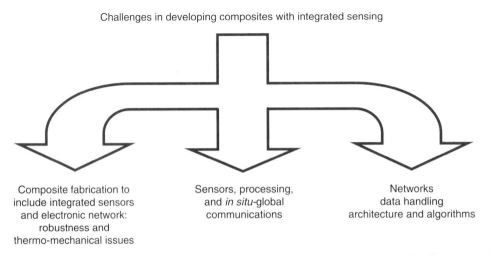

Figure 12.28 Three interwoven challenges that must be successfully met in order to create intelligently sensing composite materials.

12.2.4.1 Integrating Sensing into Composites

Here we focus on the development of a new type of composite with integrated high density of small, advanced sensors that would enable sensing without compromising the structural integrity. The volume of work in related areas is vast, and there have been a number of contributions aimed at incorporating novel nonstructural elements into composite systems (Varadan and Varadan, 2000; Lin and Fu-Kuo, 2002; Zhou and Sim, 2002). In this section we discuss several of the challenges associated with sensorized composites, including electronics, mechanical integration, and data management. We conclude by presenting results of some preliminary work.

12.2.4.2 Sensor Communications and Power

We seek to take advantage of advances in microelectronics, new capabilities in MEMS and sensor development, data feature extraction algorithms, multi-drop networking protocols, and composite fabrication technology to enable *in situ* sensing and damage detection at scalable, and potentially high areal density. Data bandwidth limitations require local data handling and efficient processing algorithms. Power management will be necessary to make high sensor densities possible. Embedded sensors require both power and a means of communication. Ideally, this would be done with the fewest number of conductors possible, and use a multi-drop network if feasible. The 1-Wire[R] network allows for distribution of power and two-way communications among many devices over a single pair of conductors.[3]

The 1-Wire network is specified for operation over a supply voltage range of 2.8 to 6 VDC. The positive, or data, line is held high by a nominal 5 kΩ impedance source. Multiple devices may be attached to the same network wire pair. Each device has a unique 64-bit address. The network protocol uses half duplex communications in master–slave architecture, where devices respond only upon command from the master. The protocol uses serial bit communications at 16.3 kbps. An overdrive mode is available that increases the data rate to 142 kbps. The network protocol not only allows individual device addressing, but also enables a novel search algorithm that allows a network master to discover all unique 64-bit addresses on any particular network. Power is drawn by the network device from the data line while it is high, and stored onboard the device in a nominal 800 pF capacitor for use during the intervals when the line is low.

The presence of multiple devices on a single network can increase the capacitive load on the network. Reflections at longer network distances and higher capacitance per unit length conductor pairs can cause slew rate problems, and contribute to limitations in the total fanout, or number devices that can be accommodated on a single network. Active pull-up drivers, careful attention to line termination, and proper transmission line design can compensate some of these effects. A well-designed system might accommodate several hundred devices, and has been demonstrated in a laboratory setting (Dallas, 2004).

12.2.4.3 Mechanical Integration

Previous efforts to integrate sensors and MEMS devices into fiber-reinforced composites have often required that such a device be placed between the fiber layers of a composite as it is being fabricated. As a result, the device is usually surrounded by the matrix phase, typically a polymer, which is generally the weak phase within the composite. Interlaminar regions are often the source of failure within a composite due to delamination, since often there is no reinforcement in the thickness direction (this problem may be minimized by through-the-thickness stitching). Depending on the size of an embedded device, this region is further weakened by the presence of sensors which can serve as stress concentrators. These sensors act as discontinuities in transferring stress within the material, leading to matrix cracking, debonding, delamination, and ultimately mechanical failure. Fiber optic sensors do not present a similar problem, since the sensor itself is a fiber that

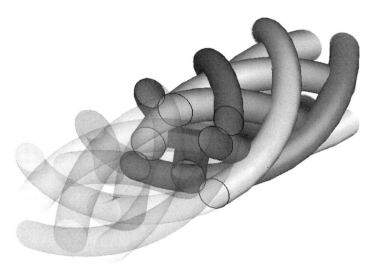

Figure 12.29 (See color insert following page 302)   Illustration of a sensor embedded in a composite braid.

commingles with the other reinforcing fibers of the material. However, fiber optic sensors do not offer the diverse sensing potential of MEMS and other microsensors. More importantly, they do not present a networking option within the structure itself, which is an essential ingredient of a truly *sensing* composite, as will be discussed in greater detail below.

To address the issues associated with structural integrity, we envisage a multicomponent braid, which consists of fibers for mechanical reinforcement, metallic (e.g., copper) wires for power and communications, and polymer matrix material impregnated around periodically spaced sensor or electronics packages, that can be integrated into the composite as a single fibrous braided element (see Figure 12.29). The multicomponent braid also acts to isolate the sensor or electronics elements within a protective environment that is commensurate with the composite structure as a whole. Rather than acting as an inclusion, the sensor network is integrated directly into the fiber phase of the composite. Braiding these sensors into and along with reinforcing fibers forms a protective casing around the discontinuity that mitigates the flaws and related failure modes of embedded designs.

The resulting sensor braid is suitable for inclusion in the composite panel in one of two ways. First, it may be directly laid into the composite layup, forming an integral vein within the composite structure, the woven fiber sheets forming the basis of the material in the usual manner of laminated composites. The spacing of these sensor braids can be such that they are isolated from each other while providing the desired degree of sensing within the material. Alternatively, the braided elements may be used in creating a woven fabric that commingles the braids with further reinforcing fibers, similar to the woven fabric used in traditional laminated composites. Depending upon the desired sensor density, the electronic sublattice braid can be included at appropriate spacing in one or multiple directions as needed.

A critical issue to be addressed in forming a multicomponent braid with sensing elements is the behavior of the sensor interconnects during normal composite processing. Both the high temperature and pressures can contribute to loss of interconnects during the processing; also, if not properly managed, stresses can build during the thermal equilibration that can result in both interconnect failure and incipient failure sites.

Three-dimensional braiding has been employed in the past as a method to integrate fiber optic sensors into 3-D woven preforms of composites (El-Sherif and Ko, 1993). Our method, however, uses two-dimensional tubular braiding to create continuous fibrous braids that protect the sensor and wiring within the core of the braid. In our tubular braiding process, carriers containing spools of reinforcing fiber weave in and out of each other in a radial pattern to form a protective sheath around the sensor and wiring that feeds from a central carrier. As mentioned previously, these braids may be woven with further reinforcing fibers to form composite fabric, or they may be placed between the layers of a laminate composite. Conventional resin transfer molding techniques such as VARTM may then be used to wet the composite fibers with thermoset polymer. Care must be taken in this step to avoid damage to the sensors due to excess pressure and in some cases temperature, if curing and/or postcuring of the polymer matrix at elevated temperature were necessary.

12.2.4.4 Data Management

As the number of sensor nodes increases, and as the data load at each sensor increases, data management becomes a key issue. If data were processed and analyzed external to the embedded network, the bandwidth needed to transport the data would grow to unrealizable levels for any reasonably sized structure. The computational requirements whether in- or ex-network become intractable as well. The size of the envisaged network of embedded sensors, even for a moderately, sized structure, will likely generate immense quantities of data that must be coordinated, interpreted, and acted upon. While optimal data fusion algorithms have been developed for small, typically unconstrained, networks of sensors (Waltz and Llinas, 1990; Goodman et al., 1997; Varshney, 1997), less is known about data fusion in networks of large numbers of sensors.

Localized processing algorithms provide a solution for the management of large data sets, and are appropriate for sensorized composites. In local processing algorithms, some primitive processing capability must be introduced to each sensor node, such that decision making in part becomes controlled by the local elements. The local processing scheme is entirely compatible with the low power, low bandwidth networks that can be *scalably* embedded in materials. Processing schemes from communication theory (Middleton, 1960; Gabrielle, 1966), specifically the FFT and decoding large block length codes, suggest that efficient and near optimal solutions can be achieved with local, hierarchical algorithms. The alternative to local processing, for example, extracting all data and performing external processing, is a computationally infeasible procedure (NP-complete); large-sized data sets cannot be used optimally (Tsitsiklis and Athans, 1985).

Any proposed method of data handling or processing will not be specifically applicable to all applications. The nature of the data or problem will generally dictate the type of solution needed.

12.2.4.5 Preliminary Results

We have undertaken some preliminary work to verify the feasibility of embedding sensors in composites. An early issue is whether the proposed sensors, as well as interconnects, can survive composite processing. As an initial test case, we embed sensors into a layup of thermally cured prepreg composite.

The DS18B20X 1-Wire digital thermometer was chosen as a demonstration sensor.[4] The DS18B20X has a 1.32 by 1.93 mm footprint, and is 0.6 mm high. A microprocessor was programmed to communicate over the micro-network. A planar substrate was chosen for the initial demonstration. The substrate was a conventional fiberglass-based printed circuit board material (FR4), 0.010 in. thick, 0.080 in. wide and 6 in. long with a $\frac{1}{2}$ oz. Cu foil overlayer. Strips with the device land pattern shown in Figure 12.30 were fabricated using a rapid PCB prototyping numerical controlled milling machine.[5] While 1-Wire devices can operate in a two conductor "parasitic" mode, a more robust three-wire configuration was selected to allow the power and communications to have separate lines, sharing a common ground. Convection hot gas solder reflow was used to connect to the substrate.

Following preliminary electronic testing of the device, it was embedded in a glass fiber-reinforced epoxy composite. Style 7781 E-glass fabric was preimpregnated with BT-250 epoxy resin supplied by Bryte Technologies, Inc. The substrate with attached sensor was embedded in the middle of eight layers of prepregged material. The layup was consolidated at 250°F for 1 h under 50 psi pressure according to the material processing specifications. The connecting wires were passed through a silicone sealing tape that protected them from excess resin during the curing

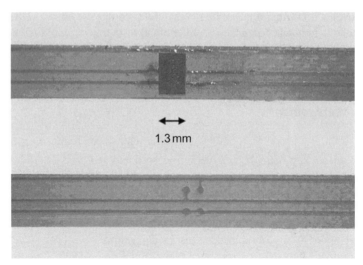

Figure 12.30 Photograph showing substrate with fabricated traces and land pattern (below) and with sensor attached (above).

process. The sensor as embedded was demonstrated to work successfully. The 1-Wire protocol was implemented in an SX48-based microcontroller that serves as the 1-Wire master. The total network length (master to device) was about 20 cm for this proof-of-principle work.

Any sensor with a temperature-dependent output may require control or calibration of thermal effects. Repetitive reads of the DS18B20X were observed to cause a temperature increase as a result of self-heating effects in the IC. The observed temperature rise after 5 min is shown in Figure 12.31 as a function of the reading rate. Obviously, the heat capacity and thermal conduction coefficients of the particular sample determine the dynamic and ultimate steady-state temperature increase.

A 10 by 10 array of the same sensors was also fabricated on similar FR4 circuit board material. After verification of the operation of the 10 by 10 array it was embedded in an aramid fiber (Kevlar) reinforced composite panel. The panel is 15 cm by 15 cm square, and is about 2.5 mm thick. It was formed from 16 layers of aramid fabric under vacuum-assisted resin transfer molding. A two-part epoxy resin was pulled through the fabric by the vacuum action to provide complete wetting of the fibers. In this case, the fabric was draped over the sensors and substrate. This panel is shown in Figure 12.32.

12.2.4.6 Sensors for Structural Health Monitoring

Continued research is necessary to successfully develop a sensor integration technology within a braided fiber component of a composite. It will be necessary to measure the mechanical properties

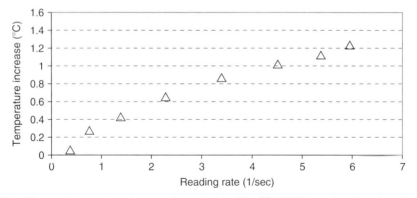

Figure 12.31 Observed temperature increase (over 5 min) of the DS18B20X as a function of reading rate.

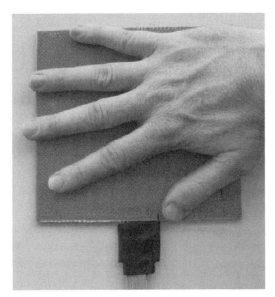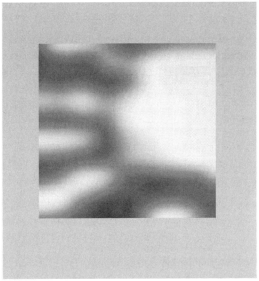

Figure 12.32 Composite panel with embedded network consisting of a 10 × 10 array of individually addressable thermal sensors. A hand is placed on the panel (left) generating a thermal image (right). The image shown is generated after about 20 sec, and represents about 3°C maximum increase over ambient.

of the embedded sensor composite in a variety of mechanical loading scenarios to determine and limit any adverse impacts on the strength of the composite. Additionally, detailed electrical design of the embedded network will need to be undertaken. If higher sensor densities are to be contemplated, the network architecture and data handling strategies will also need careful study.

The current work is focused on the implementation of a composite embedded network. As those problems are solved, it will be necessary to turn to specific structurally significant sensors to finally realize integrated structural health monitoring of composites. We expect continued progress in electronic miniaturization and power management. Smaller IC linewidths will drive the overall size of chip scale packages smaller. Current work in implementing 45° on chip interconnects also promise to reduce Si real estate demands. Higher levels of integration of MEMS-based sensors with standard IC processing will result in greater choices of microsensors to integrate into composite structures.

12.3 SUMMARY

The field of multifunctional materials is still in its infancy with regard to various functionalities that may be integrated into structural materials. With Nature as our guide the possibilities are limitless. We have presented an overview of a multifunctional composite material being developed at UCSD. This material incorporates electromagnetic, thermal management, healing and sensing functionalities into a structural composite. Integrated copper conductors resonate to provide a tuned dielectric constant and index of refraction, ranging from negative through positive values. These conductors may also serve as resistive heating elements to provide thermal management, which may be further utilized to activate a thermal repair mechanism in a healable polymer matrix. Integrated network sensors provide *in situ* sensing and damage detection at scalable and potentially high areal density with local processing and decision making. Future work will include the fabrication of smaller scale conductive element designs to achieve EM functionality in the terahertz frequency regime. The architecture of the braided elements is being tailored to obtain optimal mechanical properties of the composite structure. We are now studying other sensing technologies, such as piezoelectrics and MEMS devices, which will interact through a network similar to that which we have demonstrated with our thermal sensors.

ACKNOWLEDGMENTS

Part of the research summarized in this chapter has been conducted at the Center of Excellence for Advanced Materials (CEAM) at the University of California, San Diego under the support of ARO/ DARPA DAAD19-00-1-0525 contract to UCSD. The authors wish to thank Dr. Leo Christodoulou and Dr. John Venables of the DARPA Defense Science Office of for helpful guidance and encouragement. Research on the integration of sensing in composites is currently supported by the NSF under grant CMS-0330450. Dr. Shi-Chi Liu is gratefully acknowledged for his support and encouragement.

NOTES

1. We keep the additional numerical correction factor $-\frac{1}{2}(1 + \ln \pi)$ that Pendry usually drops because we typically employ this formula for values of d and r that do not follow the assumption that $\ln(d/r) \gg 1$.
2. Hughes Circuits, 540 S. Pacific St, San Marcos, CA 92069–4056.
3. 1-Wire is registered by its developer Dallas Semiconductor, who has subsequently been acquired by Maxim Integrated Products.
4. Maxim Integrated Products, Inc., 120 San Gabriel Drive, Sunnyvale, CA 94086.
5. CF100 from LPKF Lasers and Electronics, North America, 28880 SW Boberg Rd, Wilsonville, OR 97070.

REFERENCES

Ageorges, C., L. Ye and M. Hou (2000). Experimental investigation of the resitance welding of thermoplastic composite materials. *Composite Science and Technology* **60**: 1027–39.

Amirkhizi, A., T. Plaisted, S.C. Nemat-Nasser and S. Nemat-Nasser (2003). *Metallic Coil–Polymer Braid Composites: I. The Numerical Modeling and Chirality.* ICCM–14, Society of Manufacturing Engineers, San Diego, CA.

Baucom, J. N., J. P. Thomas, W. R. Pogue Iii and M. A. Qidwai (2004). Autophagous structure-power systems. Active materials: behaviour and mechanics, *Proceedings of SPIE — The International Society for Optical Engineering Smart Structures and Materials, San Diego, CA.*

Bleay, S. M., C. B. Loader, V. J. Hawyes, L. Humberstone and P. T. Curtis (2001). A smart repair system for polymer matrix composites. *Composites—Part A: Applied Science & Manufacturing* **32**(12): 1767–76.

Chen, X., M. A. Dam, K. Ono, A. Mal, S. Hongbin, S. R. Nutt, K. Sheran and F. Wudl (2002). A thermally re-mendable cross-linked polymeric material. *Science* **295**(5560): 1698–702.

Chen, X., F. Wudl, A. Mal, S. Hongbin and S.R. Nut (2003). New thermally remendable highly cross-linked polymeric materials. *Macromolecules* **36**: 1802–07.

Christodoulou, L. and J. D. Venables (2003). Multifunctional material systems: the first generation. *JOM* **55**(12): 39–45.

Chujo, Y., K. Sada and T. Saegusa (1990). Reversible Gelation of polyoxazoline by means of Diels–Alder reaction. *Macromolecules* **23**: 2636–4.

Cox, B. N., W. C. Carter and N. A. Fleck (1994). Binary model of textile composites. I. Formulation. *Acta Metallurgica et Materialia* **42**(10): 3463–79.

Dallas, S. (2004). *Tech Brief 1: MicroLAN Design Guide*, Dallas Semiconductor.

El-Sherif, M. A. and F. K. Ko (1993). *Co-braiding of Sensitive Optical Fiber Sensor in 3D Composite Fiber Network.* SPIE — the International Society for Optical Engineering, Bellingham, WA.

Engle, L. P. and K. B. Wagener (1993). A review of thermally controlled covalent bond formation in polymer chemistry. *Journal of Macromolecular Science* **C33**(3): 239–57.

Eveno, E. C. and J. W. Gillespie Jr. (1988). Resistance welding of graphite polyetheretherketone composites: an experimental investigation. *Journal of Thermoplastic Composite Materials* **1**: 322–38.

Gabrielle, T. L. (1966). Information criterion for threshold determination. *IEEE Transactions of Information Theory* **6**: 484–86.

Goodman, I. R., R. P. S. Mahler and H. T. Nguyen (1997). *Mathematics of Data Fusion.* Kluwer, Boston.

Imai, Y., H. Itoh, K. Naka and Y. Chujo (2000). Thermally reversible IPN organic–inorganic polymer hybrids utilizing the Diels–Alder reaction. *Macromolecules* **33**(12): 4343–6.

Jakobsen, T. B., R. C. Don and J. W. Gillespie Jr. (1989). Two-dimensional thermal analysis of resistance welded thermoplastic composites. *Polymer Engineering and Science* **29**: 1722–9.

Joshi, P. B., B. L. Upschulte, A. H. Gelb, D. M. Lester, I. A. Wallace, D. W. Starrett, D. M. Marshall and B. D. Green (2002). Autophagous spacecraft composite materials for orbital propulsion. *Proceedings of SPIE — The International Society for Optical Engineering, San Diego, CA.*

Jud, K. and H. H. Kausch (1979). Load transfer through chain molecules after interpenetration at interfaces. *Polymer Bulletin* 697–707.

Jud, K., H. H. Kausch and J. G. Williams (1981). Fracture mechanics studies of crack healing and welding of polymers. *Journal of Materials Science* **16**(1): 204–10.

Kausch, H. H. (1983). The nature of defects and their role in large deformation and fracture of engineering thermoplastics. *Pure and Applied Chemistry* **55**(5): 833–44.

Kausch, H. H. and K. Jud (1982). Molecular aspects of crack formation and healing in glassy polymers. *Plastics and Rubber Processing and Applications* **2**(3): 265–8.

Kausch, H. H. and M. Tirrell (1989). Polymer interdiffusion. *Annual review of material science. Annual Reviews.* **19**: 341–77.

Kausch, H. H., D. Petrovska, R. F. Landel and L. Monnerie (1987). Intermolecular interaction in polymer alloys as studied by crack healing. *Polymer Engineering and Science* **27**(2): 149–54.

Ko, F. (2001). Braiding. *ASM Handbook: Composites, ASM International* **21**: 69–77.

Ko, F. K., C. M. Pastore and A. Head (1989). *Atkins and Pearce Handbook of Industrial Braids.* Atkin and Pearce, Covington, KY.

Kolinko, P. and D. R. Smith (2003). Numerical study of electromagnetic waves interacting with negative index materials. *Optics Express* **11**(7): 640–8.

Lin, M. and C. Fu-Kuo (2002). The manufacture of composite structures with a built-in network of piezo-ceramics. *Composites Science and Technology* **62**: 919–39.

Mal, A. (2004). Structural health monitoring. *Mechanics* 33 (11–12): 6–16.

McGlockton, M. A., B. N. Cox and R. M. McMeeking (2003). A binary model of textile composites: III high failure strain and work of fracture in 3D weaves. *Journal of the Mechanics and Physics of Solids* **51**(8): 1573–600.

Middleton, D. (1960). *An Introduction to Statistical Communication Theory.* McGraw-Hill, New York.

Naik, R. A. (1995). Failure analysis of woven and braided fabric reinforced composites. *Journal of Composite Materials* **29**(17): 2334–63.

Nemat-Nasser, S. C., A. V. Amirkhizi, T. A. Plaisted, J. B. Isaacs and S. Nemat-Nasser (2002). Structural composites with integrated electromagnetic functionality. Smart Structures and Materials 2002: Industrial and Commercial Applications of Smart Structures Technologies, *Proceedings of SPIE — The International Society for Optical Engineering, San Diego, CA.*

Pendry, J. B., A. J. Holden, W. J. Stewart and I. Youngs (1996). Extremely low frequency plasmons in metallic mesostructures. *Physical Review Letters* **76**(25): 4773–6.

Pendry, J. B., A. J. Holden, D. J. Robbins and W. J. Stewart (1999). Magnetism from conductors and enhanced nonlinear phenomena. *IEEE Transactions on Microwave Theory & Techniques* **47**(11): 2075–84.

Pendry, J. B., D. R. Smith, P. M. Valanjui, R. M. Walser and A. P. Valanju (2003). Comment on "Wave refraction in negative-index media: always positive and very inhomogeneous" (and reply). *Physical Review Letters* **90**(2): 9703.

Plaisted, T., C. Santos, A. Amirkhizi, S. C. Nemat-Nasser and S. Nemat-Nasser (2003a). *Self-healing Structural Composites with Electromagnetic Functionality.* ICCM-14, Society of Manufacturing Engineers, San Diego, CA.

Plaisted, T. A., A. V. Amirkhizi, D. Arbelaez, S. C. Nemat-Nasser and S. Nemat-Nasser (2003b). Self-healing composites with electromagnetic functionality. Smart Structures and Materials 2003: Industrial and Commercial Applications of Smart Structures Technologies, *Proceedings of SPIE — The International Society for Optical Engineering, San Diego, CA.*

Qidwai, M. A., J. P. Thomas, J. C. Kellogg and J. Baucom (2004). Energy harvesting concepts for small electric unmanned systems. Active Materials: Behaviour and Mechanics, *Proceedings of SPIE — The International Society for Optical Engineering Smart Structures and Materials, San Diego, CA.*

Ramakrishnan, B., L. Zhu and R. Pitchumani (2000). Curing of composites using internal resistive heating. *ASME Journal of Manufacturing Science and Engineering* **122**(1): 124–31.

Sancaktar, E., W. Ma and S. W. Yugartis (1993). Electric resistive heat curing of the fiber–matrix interface in graphite/epoxy composites. *ASME Journal of Mechanical Design* **115**: 53–60.

Santos, C., T. A. Plaisted, D. Arbelaez and S. Nemat-Nasser (2004). Modeling and testing of temperature behavior and resistive heating in a multi-functional composite. Smart Structures and Materials 2004: Industrial and Commercial Applications of Smart Structures Technologies, *Proceedings of SPIE — The International Society for Optical Engineering, San Diego, CA.*

Shelby, R. A., D. R. Smith and S. Schultz (2001). Experimental verification of a negative index of refraction. *Science* **292**(5514): 77–79.

Smith, D. R., D. C. Vier, W. Padilla, S. C. Nemat-Nasser and S. Schultz (1999). Loop-wire medium for investigating plasmons at microwave frequencies. *Applied Physics Letters* **75**(10): 1425–7.

Smith, D. R., W. Padilla, D. C. Vier, S. C. Nemat-Nasser and S. Schultz (2000a). Negative permeability from split ring resonator arrays. *2000 Conference on Lasers and Electro-Optics Europe, Nice, France, IEEE 2000 Piscataway, NJ, USA.*

Smith, D. R., W. J. Padilla, D. C. Vier, S. C. Nemat-Nasser and S. Schultz (2000b). Composite medium with simultaneously negative permeability and permittivity. *Physical Review Letters* **84**(18): 4184–7.

Smith, D. R., D. Schurig and J. B. Pendry (2002). Negative refraction of modulated electromagnetic waves. *Applied Physics Letters* **81**(15): 2713–5.

Smith, D. R., D. Schurig, M. Rosenbluth, S. Schultz, S. A. Ramakrishna and J. B. Pendry (2003). Limitations on subdiffraction imaging with a negative refractive index slab. *Applied Physics Letters* **82**(10): 1506–8.

Smith, D. R., P. Kolinko and D. Schurig (2004a). Negative refraction in indefinite media. *Journal of the Optical Society of America B—Optical Physics* **21**(5): 1032–43.

Smith, D. R., P. Rye, D. C. Vier, A. F. Starr, J. J. Mock and T. Perram (2004b). Design and measurement of anisotropic metamaterials that exhibit negative refraction. *IEICE Transactions on Electronics* **E87C**(3): 359–70.

Smith, D. R., D. Schurig, J. J. Mock, P. Kolinko and P. Rye (2004c). Partial focusing of radiation by a slab of indefinite media. *Applied Physics Letters* **84**(13): 2244–2246.

Starr, A. F., P. M. Rye, D. R. Smith and S. Nemat-Nasser (2004). Fabrication and characterization of a negative-refractive-index composite metamaterial. *Physical Review B* **70**: 113102.

Thomas, J. P., M. A. Qidwai, P. Matic, R. K. Everett, A. S. Gozdz, M. T. Keennon and J. M. Grasmeyer (2002). Structure-power multifunctional materials for UAV's. *Proceedings of SPIE — The International Society for Optical Engineering, San Diego, CA.*

Tsitsiklis, J. and M. Athans (1985). On the complexity of decentralized decision making and detection problems. *IEEE Transactions on Automatic Control* **30**: 440–6.

Varadan, V. K. and V. V. Varadan (2000). Microsensors, microelectromechanical systems (MEMS), and electronics for smart structures and systems. *Smart Materials and Structures* **9**: 953–72.

Varshney, P. K. (1997). *Distributed Detection and Data Fusion.* Springer, New York.

Veselago, V. G. (1968). The electrodynamics of substances with simultaneously negative values of [permittivity] and [permeability]. *Soviet Physics USPEKI* **10**: 509.

Waltz, E. and J. Llinas (1990). *Multisensor Data Fusion.* Artech House, Boston.

White, S. R., N. R. Sottos, P. H. Guebelle, J. S. Moore, M. R. Kessler, S. R. Sriram, E. N. Brown and S. Viswanathan (2001). Autonomic healing of polymer composites. *Nature* **409**(6822): 794–7.

Wool, R. P. (1978). Material response and reversible cracks in viscoelastic polymers. *Polymer Engineering and Science* **18**(14): 1057–61.

Wool, R. P. (1979). Crack healing in semicrystalline polymers, block copolymers and filled elastomers. *Adhesion and adsorption of polymers.* Plenum, New York **12A**: 341–62.

Wool, R. P. (1995). *Polymer Interfaces: Structure and Strength,* Hanser, München.

Wool, R. P. and K. M. O'Connor (1981a). Craze healing in polymer glasses. *Polymer Engineering and Science* **21**(14): 970–77.

Wool, R. P. and K. M. O'Connor (1981b). A theory of crack healing in polymers. *Journal of Applied Physics* **52**(10): 5953–63.

Wool, R. P. and K. M. O'Connor (1982). Time dependence of crack healing. *Journal of Polymer Science: Polymer Letters Edition* **20**: 7–16.

Wool, R. P., B. L. Yuan and O. J. McGarel (1989). Welding of polymer interfaces. *Polymer Engineering and Science* **29**(19): 1340–1367.

Xu, J., B. N. Cox, M. A. McGlockton and W. C. Carter (1995). A binary model of textile composites. II. The elastic regime. *Acta Metallurgica et Materialia* **43**(9): 3511–24.

Yang, Q. D., B. Cox and Y. Qingda (2003). Spatially averaged local strains in textile composites via the binary model formulation. *Journal of Engineering Materials and Technology* **125**(4): 418–25.

Zhou, G. and L. M. Sim (2002). Damage detection and assessment in fibre-reinforced composite structures with embedded fibre optic sensors — review. *Smart Materials and Structures* **11**: 925–39.

13

Defense and Attack Strategies and Mechanisms in Biology

Julian F.V. Vincent

CONTENTS

13.1   INTRODUCTION

Whilst there are several proposed uses of biomimetics in defense or attack (martial, general law enforcement) systems, at present they seem to be mostly development of novel materials (occasionally novel mechanisms) in an established context. Examples are armor, personal or otherwise, made of analogs of silk, mother-of-pearl (nacre), or wood. I do not intend to rehearse this topic further. Camouflage is another area that has been examined, especially adaptive camouflage, but since there is still much to be learned about camouflage techniques in nature (which I take to include mimicry — camouflage is 'deception'), I have included it. In general, camouflage and armor are inimical; the tendency is for the more primitive (= evolutionarily older) animals of any particular phylum to be well armored but slow and relatively easily seen, whereas the more highly evolved ones are less well armored, or have no armor at all, but are fast-moving, or very well camouflaged, or both. Thus they rely on speed and behavioral adaptiveness and subtlety for their safety. The inevitable conclusion is that nature often employs guerrilla techniques rather than what we think of as "conventional" ones. This may be related to the perceived financial investment. In human warfare, an infantryman is seen as more expendable than the combination of a pilot and aircraft. Indeed a significant reason for having a pilot is as a hostage to the aircraft's expensive technology, so that it is brought back in one piece from a sortie.

The preparation of a chapter like this is especially difficult since I could not think of a suitable narrative to cover all the possibilities that exist in nature. Also, I have little understanding of the techniques that are available to, or desired by, the military and police (the obvious users of defense mechanisms). I decided, therefore, to adopt a classificatory approach, and to use an existing military classification as my template (Alexander et al., 1996). I have removed the obviously nonbiological techniques that involve explosives, lasers, etc., have retained others which, although biology does not present us with the same resource, are obvious functional analogs, and have included some that seemed to be missing from Alexander's list but are present in biology. These latter are presented without citations.

Man has many martial devices that have their reflections in nature, but the similarities have either not been recognized or have not been developed. And since the outcome in nature is, mostly for all parties, in an intraspecific encounter to live to fight another day (or at least live), perhaps we have still much to learn. As for the rest, I suspect we have an untapped resource for biomimicry; I have mostly left the extrapolation from biology to technology to the reader, otherwise this chapter would have been too long. But most of the examples quoted either have a technological counterpart or could be realized without much difficulty.

The Department of Defense defines (non-lethal) weapons as designed and deployed so as to incapacitate people or their weapons and other equipment, rather than destroying them; also to have minimal effects on the environment. Unlike conventional, lethal, weapons that destroy their targets principally through blast, penetration and fragmentation, non-lethal weapons have relatively reversible effects and affect objects differently (Alexander et al., 1996).

13.2 ACOUSTICS

13.2.1 Blast Wave Projector

Energy generation from a pulsed laser that will project a hot, high pressure plasma in the air in front of a target. It creates a blast wave with variable but controlled effects on hardware and troops (Alexander et al., 1996).

This could be akin to cavitation bubbles that are the loudest source of sound from ship propellers.

Snapping shrimps (Stomatopods or mantis shrimp) are very noisy; it has been long assumed that the noise was caused by their claws closing. In *Odontodactylus scyllarus*, the sound is caused by the collapse of cavitation bubbles due to the high speed at which the claw moves, powered by a highly elastic part of the exoskeleton. The shrimps appear to use cavitation to stun their prey (small crabs, fish, and worms); it certainly wreaks havoc with the shrimp's own exoskeleton. Although the claw is highly mineralized, its surface becomes pitted and damaged; stomatopods moult frequently and produce a new smashing surface every few months (Patek et al., 2004).

13.2.2 Infrasound

Very low-frequency sound that can travel long distances and easily penetrate most buildings and vehicles. Transmission of long wavelength sound creates biophysical effects; nausea, loss of bowels, disorientation, vomiting, potential internal organ damage or death may occur. Superior to ultrasound because it is "in band" meaning that its does not lose its properties when it changes mediums such as from air to tissue. By 1972 an infrasound generator had been built in France that generated waves at 7 Hz. When activated it made the people in range sick for hours (Alexander et al., 1996).

Whales are certainly able to generate low frequencies (15 to 30 Hz) which they use for communication over long distances (the capercaillie, a ground-living bird of the Scottish woodlands, uses low frequencies for the same reason) but they have not been tested for any damaging effects (Croll et al., 2002).

Although it does not really belong to "infrasound," animals (e.g., frogs, birds, and deer) advertize a false impression of exaggerated size by making low frequency sounds (Reby and McComb, 2003). The implication for other animals is that a low noise can only come from a large resonant cavity, so the animal producing the noise is probably large and therefore probably strong. Producing low frequency vibrations is therefore a premium especially if the animal cannot be seen and the assessment of size can be made only from the frequency range of the noise.

13.2.3 Squawk Box

Crowd dispersal weapon field tested by the British Army in Ireland in 1973. This directional device emits two ultrasonic frequencies which when mixed in the human ear become intolerable. It produces giddiness, nausea or fainting. The beam is so small that it can be directed at specific individuals (Alexander et al., 1996).

There are many reports of dolphins using a similar technique, either when hunting or when swearing at a human experimenter. In a U.K. radio programme some years ago, a researcher recounted playing back its own sounds to a dolphin to see what it would do, including listening to the dolphin's response with a hydrophone. The dolphin was quite amenable to this game and cooperated well. But by mistake the experimenter sent the dolphin a rather loud signal to which the dolphin obviously objected. The dolphin looked at the experimenter through the walls of the aquarium, then went to the hydrophone and blasted into it before the experimenter could rip off his earphones. The experimenter experienced much pain! The implication is that we could probably learn about physiologically damaging noise from dolphins and other cetaceans that are also much more experienced with the technique, having been using it for longer than we have.

13.3   ANTILETHAL DEVICES

13.3.1   Body Armor

Many animals have a hard outer covering that serves as armor, but there are many different ways in which the function is realized. Whereas the armor developed for individuals or vehicles is based on the inevitability of attack, and relies on resisting by strength, biological armor can come in many guises. Obvious ones are armadillo and tortoise, although nobody seems to have made any measurements of the protection that is given. The same is not true of ankylosaurs (Figure 13.1) and their relatives, herbivorous dinosaurs that grew to 10 m long during the late Jurassic and Cretaceous. They had centimeter-sized osteodermal plates that covered back, neck, head, and also protected the eyes. In polarized light, sections of the plates show where collagen — a normal precursor of bone and an essential component of skin — was incorporated. Comparing similar dermal bones from stegosaurus and crocodile, the polocanthids had extra collagen fibres that may have stabilized the edges of the bony plates. But in nodosaurids — which also had plates between 2 and 5 cm thick, the collagen fibres ran parallel and perpendicular to the surface, and then at 45° to each of these axes, providing reinforcement in all directions. Ankylosaurids had thinner plates that were 0.5 to 1.0 cm thick, convex shaped, which will have increased their stiffness in bending, and with the collagen fibres randomly arranged.

The dinosaur structure seems to be repeated in the bone-free collagenous skin of the white rhinoceros, which is three times thicker and contains a dense and highly ordered three-dimensional array of relatively straight and highly crosslinked collagen fibres. The skin of the back and sides of the animal is therefore relatively stiff (240 MPa) and strong (30 MPa), with high breaking energy (3 MJ m^{-3}) and work of fracture (78 kJ m^{-2}). These properties fall between those of tendon and skin as would be expected from a material with a large amount of collagen (Shadwick et al., 1992).

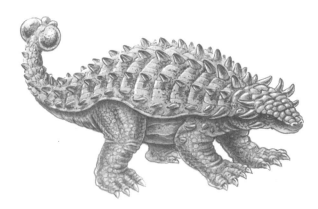

Figure 13.1   An ankylosaur.

Unfortunately the data on "soft" body armor (e.g., Kevlar) does not quote performance in these units, preferring to equate energy of an incoming threat to depth of penetration through the armor. Presumably one has to go to reports from the old big game hunters to get similar information about the rhinoceros. However, leather is still tougher than Kevlar, although nobody really understands why, since the collagen fibres are not dissimilar from Kevlar in general morphology.

A concept that is entirely alien to the current design of man-made armor is the porcupine quill, although the pikestaff of the medieval infantryman might be considered analogous, and parts of mediaeval armor and their weapons were equipped with spikes to keep the enemy at bay. The porcupine has several different types of quill; those with a length-to-diameter ratio greater than about 25 are mostly rattles to warn enemies that there are quills here. Those with a lower length-to-diameter ratio (15 or less) act as columns when they meet an end load, and with the sharp tip, can easily penetrate flesh. They are sometimes brittle and the tip can break off, but they also have weak roots in the porcupine's skin and so can easily be pulled out when the impaled attacker moves away. The quills are filled with a variety of reinforcing foams, struts, and stringers, so that they rarely break when buckled (Vincent and Owers, 1986). Quills are modified hairs and are made of keratin.

In general, plants have totally passive defense mechanisms, which is energetically probably much cheaper. They are thus built to survive a certain amount of damage due to grazing, and may even grow more vigorously in response. Many plants, especially those living under dry conditions, such as the acacia, have spines, thorns, or hooks that cause pain to the animals attacking them. Presumably the giraffe, which feeds on such plants, has a reinforced surface to its tongue so that it can cope with the abuse. Many of the grain-bearing plants (Graminae) have silica particles — sometimes as much as 15% of the dry weight — which wears down the teeth of the animals feeding on them. Indeed the performance of the teeth is frequently dependent on such wear, exposing a complex of self-sharpening cutting and grinding surfaces (Alexander, 1983). The literature on plant–animal interactions is large, mostly concerned with how plants control the ease with which they can be grazed, commonly by limiting crack propagation with inhomogeneities such as embedded fibres; and their chemical defenses which range from repulsive taste or smell, through manipulation of the digestion or behavior of the grazer (by psychoactive drugs) to lethal chemicals, mostly in those plants which cannot afford to be eaten since they grow so slowly.

In both plants and animals, spines and thorns are passive and are of use only at close quarters. The closest equivalent is barbed wire which many claim to be biomimetic.

Horns and antlers can be used for both attack and defense, an unusual concept for technology — the closest analogy is the sword, which can be used both to deliver a blow and to parry one. The utility of antlers (dead, made of bone, replaced each season, grown from the tip) and horns (living, made of a thick keratin sheath over a bone core, incremented each season, grown from the base) has been questioned by animal behaviorists who find difficulty coping with the wide range in sizes of horns and antlers, and the range in forces imposed on them during fighting. These problems were largely resolved by Kitchener, who showed that there is a linear relationship between the second moment of area at the base of the horn or antler and the body weight of the animal, and that this relationship is constant for any single style of fighting. Most styles are ritualistic and akin to wrestling; sheep and goats are far more agonistic, throwing themselves at each other resulting in more random forces being exerted on their horns (Kitchener, 1991).

Ever since their discovery in the 16th century, the enormous antlers of the extinct Irish elk or giant deer (*Megaloceros giganteus*) have attracted scientific attention. Mechanical analysis of the antlers of the Irish elk shows that they are massively over-designed for display (for which, as John Currey pointed out, they really only need to be made of waterproof cardboard) because the force exerted by gravity acting on the antlers is less than 1% of their strength. In contrast, the antlers seem to be optimally designed for taking the maximum estimated forces of fighting, that are more than 50% of the strength of the antler, as would be expected for a biological structure of this kind.

However, this analysis assumes that the mechanical properties of the bone of the Irish elk antlers and living deer are similar. It would be unwise to measure directly the mechanical properties of

Irish elk antler after more than 10,000 years in a peat bog. Instead, neutron diffraction, which measures the degree of preferred orientation of the hydroxyapatite crystals that comprise bone, showed that the orientation of the hydroxapatite is predictable from the presumed forces generated during fighting. Thus on the tensile faces of the antler, the orientation was along the length of the antler, whereas on the compressive faces, the orientation was more orthogonal to the long axis — exactly what the theory of fibrous composite materials predicts (Kitchener et al., 1994).

13.3.2 Passive Camouflage

Many American hunters recommend that more effort should be put into the research on camouflage, and that body armour should be a second priority to finding effective concealment. The logic is that what you can't see, you can't hit. Body armour is required only when you can be seen and identified.

Many animals and plants, especially insects, can look like inert objects such as bits of wood or stones (e.g., the succulent South American plant *Lithops*). Because of their colored wings, many moths can conceal themselves when placed against a suitable background such as the bark of a tree. The peppered moth (*Biston betularia*) in industrial areas of England has been held as a classic example of natural selection, with birds eating those moths that they could see only when they were sitting on an unsuitably colored bark. In this instance the moth was originally light with small black speckling, but pollution produced in the early industrial revolution blackened the trees, so an initially rare dark form of the moth was selected by being less easily seen and eaten (Kettlewell, 1955). Later, with reduced pollution and clearing of the woods, the bark was lighter and better lit and the lighter-colored form again predominated. Similarly many nesting birds are difficult to see; ground-nesting birds have camouflaged eggs and chicks. Many insects, especially grasshoppers, have bright hind wings which disappear when the insect stops flying, settles, and folds its wings thus becoming camouflaged. This sudden change makes it difficult to spot the insect.

Another basic component of passive camouflage, well known to technology, is countershading, in which, those parts of the body that are normally well illuminated are darkly colored, and those that are normally shaded lightly colored . This is seen in both terrestrial and aquatic animals; the corollary is the larva of the privet hawk moth (*Sphinx ligustri*) which is dark on the underside and light on the upperside, and habitually hangs inverted beneath its twig. The effect is to flatten the aspect of the animal, making it difficult to judge its size and how far away it is.

The literature of camouflage in biology is very large (Wickler, 1968).

13.3.3 Warning Coloration

The announcement that you are strong or dangerous is useful since it can deter an enemy from attacking, and gains its best effect by the strong making themselves easily seen. But one can also pretend strength. This is not novel, and has been used for hundreds of years with armies making themselves appear larger than they are with hats on sticks, unattended guns protruding through the battlements, and soldiers circulating past a small gap for the enemy to see ...

Many animals and plants (especially fruits) advertize that they are poisonous or that they have a very nasty sting or bite. Typical warning colors are bright, for instance red and yellow associated with black, mutually arranged to maximize contrast and visibility (aposematic coloration). There is a vast amount of literature on this aspect of coloration, which includes mimicking of an unpalatable animal by a palatable one (Batesian mimicry) and mimicry of palatable mimics of unpalatable animals (Müllerian mimicry). Such mimicry is probably commonest amongst butterflies, where the main selection agent is predatory birds and the habitat is thick forest or woodland (Wickler, 1968). Thus, the predatory bird probably only ever gets a fleeting glimpse, poorly illuminated of its prospective prey, and with this minimal information it has to decide whether or not to attack. It is imaginable that under these conditions even a slight resemblance to an unpleasant species is enough to convince a bird not to attack.

Most insects, in particular beetles, butterflies, and moths, get their noxious chemicals from the plants they feed on. The first bird to be discovered with warning coloration and toxic feathers is the Pitohui of New Guinea (Dumbacher et al., 2004). The source of the alkaloids, also found in poison-dart frogs, is Melyrid beetles.

13.3.4 Active Camouflage

Created by dynamically matching the object to be camouflaged to its background colors and light levels thus rendering it virtually invisible to the eye. This is conceptually the same camouflage process as that used by a chameleon. This is accomplished through a sophisticated color and light sensor array that detects an object's background color and brightness. This data is then computer matched and reproduced on a pixel array covering the viewing service of the object to be camouflaged.

Pattern control is achieved by flatfish such as the plaice (*Pleuronectes platessa*) that can change its shading and patterns to suit a variety of backgrounds — including a chequer board! However, it can manage only black and white, and then only slowly, over a matter of minutes, since its color-change cells (melanophores) are hormonally controlled. They change color by moving pigment around inside the cell going from "concentrated" (the pigment is centered making the cell white or translucent) to "dispersed" (the pigment is spread around the cell which now appears dark) (Fuji, 2000; Ramachandran et al., 1996).

Color control in octopus and squid (cephalopod — literally "head-footed" — molluscs) is managed by colored cells — chromatophores — that are found in the outer layers of the skin. Each comprises an elastic sac containing pigment to which is attached radial muscles. When the muscles contract, the chromatophore is expanded and the color is displayed; when they relax, the elastic sac retracts. The chromatophore muscles are controlled by the nervous system. Differently colored (red, orange, and yellow) chromatophores are arranged precisely with respect to each other, and to reflecting cells (iridophores producing structural greens, cyans and blues, and leucophores, reflect incident light of whatever wavelength over the entire spectrum) beneath them. Neural control of the chromatophores enables a cephalopod to change its appearance almost instantaneously (Hanlon et al., 1999), a key feature in some escape behaviors and during fighting signalling. Amazingly the entire system apparently operates without feedback from sight or touch (Messenger, 2001).

The primary function of the chromatophores is to match the brightness of the background and to help the animal resemble the substrate or break up the outline of the body. Because the chroma-tophores are neurally controlled, the animal can, at any moment, select and exhibit one particular body pattern out of many, which presumably makes it difficult for the predator to decide or recognize what it is looking at. When this is associated with changes in shape or behavior, the prey can become totally confusing. Consider this performance by an octopus found in Indo-Malaysian waters. It is seen on the seabed as a flatfish and swims away with characteristic "vertical" (remember the flatfish swims on its side) undulations. As it does so it changes into a poisonous zebra fish. It then dives into a hole and sends out two arms in opposite directions to mimic the front and back ends of a poisonous banded sea snake (videos of these behavior patterns are available to download with the paper by Norman et al.). It also sits on the sea bed with its arms raised, possibly in imitation of a large poisonous sea anemone. Or it can sink slowly through the water column apparently imitating a jellyfish (Norman et al., 2001). Each of these types of animal requires a different response on the part of the predator, which presumably is totally confused. Such dynamic mimicry is seen only in cephalopods and the films of the Marx Brothers.

Countershading in animals is widespread and cephalopods are no exception. On the ventral surface, the chromatophores are generally sparse, sometimes with iridophores to enhance reflec-tion; dorsally the chromatophores are much more numerous and tend to be maintained tonically

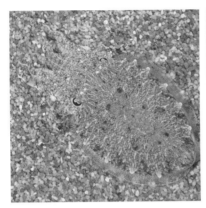 

Figure 13.2 (See color insert following page 302)   A cuttlefish (*Sepia officinalis*) can change its appearance according to the background. Here the animal changes its body pattern when moved from a sandy or gravel substrate to one with shells. (Courtesy of Roger T. Hanlon, Senior Scientist, Marine Biological Laboratory, Woods Hole, MA.)

expanded. More remarkably, however, cephalopods can maintain countershading when they become disorientated. The countershading reflex ensures that chromatophores on the ventral surface of the entire body expand when the animal rolls over on its back: a half-roll elicits expansion of the chromatophores only on the upper half of the ventral body. Such a response is, of course, possible only in an animal whose chromatophores are neurally controlled (Ferguson et al., 1994). When matching brightness, the chromatophores act like a half-tone screen; color matching is achieved with the chromatophores, iridophores, and leucophores (Hanlon and Messenger, 1988).

On variegated backgrounds, a cuttlefish will adopt the disruptive body pattern, whose effect is to break up the "wholeness" of the animal (Figure 13.2). Disruptive coloration is a concealment technique widespread among animals. *Octopus vulgaris* has conspicuous frontal white spots; loliginid squids show transverse dark bands around the mantle that probably render the animal less conspicuous, and the harlequin octopuses have bold black-and-white stripes and spots.

Although many animals use patterning for concealment, it is nearly always a fixed pattern. Because they control their chromatophores with nerves and muscles, cephalopods can select one of several body patterns to use on a particular background.

Cephalopods also produce threatening or frightening displays. In its extreme form, the animal spreads and flattens, becoming pale in the middle and dark around the edges, creating dark rings around the eyes and dilating the pupil, and in sepioids and squids, creating large dark eyespots on the mantle. This effect is extremely startling. The animal also seems to get bigger.

13.3.5 Translucent Camouflage

The best way to avoid being seen is to be invisible and so cast no shadow. The equivalent of translucence is to present the observer with the scene which the object is blocking out. In a technical world this can be done using a camera to film the scene that is blocked and presenting it to the observer in front of the object.

Whole animals (e.g. pelagic marine organisms such as jelly fish, sea gooseberries, and many larval forms) or parts of animals (e.g. the cornea of the eye) can be translucent and therefore nearly invisible. To be translucent, reflection of incident light must be kept to a minimum and light must be neither scattered nor absorbed as it passes through the body. Scattering is caused by variations in refractive index. Animal tissue normally has many variations in refractive index (cells, fibres, nuclei, nerves, and so on). The most important factors are the distribution and size of the components; refractive index is less important; the shape of the components is least important. For instance, if a cell requires a certain volume of fat to survive but must scatter as little light as possible, it is best to divide the fat into many very small droplets. Slightly worse is to divide it into a few large droplets, but the very worst is to divide it into drops about the size of the wavelength of light (Johnsen, 2001).

Variations in refractive index do not always cause scattering. If the refractive indices vary by less than half the wavelength of light, the scattered light is eliminated by destructive interference and the light waves overlap in such a way that they cancel each another. This happens in the cornea of the eye, which is constructed of an orthogonal array of collagen fibres.

Many organisms living in the deeper ocean, where there is little or no ambient light to be reflected or by which camouflage color can be seen, produce their own light. The organs that do this — photophores — can be mounted on mechanisms which rotate them so that they face the body and are effectively obscured, hence can be modulated and switched on and off (Johnsen et al., 2004).

13.3.6 Reflecting Camouflage

If an object can simply reflect the color and pattern of its surroundings, then it will be adaptive. But if it merely reflects the sky when looked at from above, or the ground when looked at from below, this will be ineffective. The geometry of the reflecting surface is crucial. In deep water, the laterally scattered light is equal in intensity from a range of angles. Looking up, one sees brightness; looking down there is dim blue-green. A perfect mirror suspended vertically in the water would be invisible since the light from the surface is reflected to a viewer below, making the mirror appear translucent. Many fish have platelets of guanine in their scales arranged vertically, thus generating such a mirror independently of the shape of the section of the body. The fish is also countershaded. Viewed laterally the fish is a reflector and therefore invisible. Viewed from the top, it is dark like the depths below it. Viewed from below it is silvery white like the surface.

The most difficult view to camouflage is that from directly below when the fish obscures light from above. Many clupeids, such as the threadfin shag *Dorosoma petense*, are thin and come to a sharp edge at the belly. This allows light from above to be reflected vertically downwards over the entire outline (Johnsen, 2002).

Another form of reflecting camouflage is provided by the cuticle of some scarab beetles. The cuticle is made of structures that look like liquid crystals, mainly nematic and cholesteric. Thus, of the incident light on the cuticle, the right circularly polarized component can be reflected and the left circularly polarized light can penetrate the helicoidally structured cuticle. However, at a certain depth, there is a layer of nematic structure that acts as a half-wave plate, reversing the sense of polarization of the light, which is then reflected when it reaches the next layer of helicoidal structure, has its sense of polarization reversed again by the nematic layer, and continues back out through the helicoidal cuticle with very little loss. The refractive index of the cuticle is increased by the addition of uric acid. Thus the cuticle is an almost perfect reflector, making the beetle appear the same green as its surroundings. This system will work only when the color and light intensity are the same in all directions (Caveney, 1971).

13.3.7 Motion Camouflage

This is included here since it is a way of observing and approaching an object without making it obvious to an observer or the object that it is being observed. The technique might have been unintentionally deployed by attacking fighter aircraft, and is currently in development for disguising the intended target of guided missiles. An everyday equivalent, converted to the acoustic environment, would be that if you are following someone closely, make sure that the noise of your footfall is in synchrony with that of your quarry.

This is a stealth shadowing technique used by, for instance, the dragonfly approaching its prey on the wing. The dragonfly follows a path such that it always lies on a line connecting itself and a fixed point. Then the only visual cue to the dragonfly's approach is its looming (i.e., the increase in the size of its image as it closes in on the object). The observer of the object thus sees no movement away from the direction of the fixed point. The fixed point could be a part of the background against which the dragonfly is camouflaged, or the initial position of the dragonfly, in which case the dragonfly appears not to have moved from its starting point (Anderson and McOwan, 2003).

13.3.8 False Target Generation

A device that creates and presents an image of a target that causes a weapon to aim at a false target. Used as a countermeasure to precision guided weapons (Alexander et al., 1996).

This is a common ploy in insects; for instance, butterflies have eye spots on the trailing edge of the hind wing. Predating birds tend to aim for the eyes rather than the body of the insect, and so the insect escapes with relatively slight damage to the hind wing. Similarly fish can have an eyespot on the tail fin with the true eye concealed in a dark marking across the head. A number of moth larvae have a false "head" at the tail end which can simply be eye spots or an image of the head of another animal such as a snake. The advantage then is not just that the attack will be at the "wrong" end of the animal, thus protecting the nervous system, but that the animal will apparently move backwards in order to escape.

A more sophisticated false target is generated by autotomy of part of the animal. A well-known example is the salamander which leaves the end of its tail behind. A more sophisticated example is provided by certain opilionids (harvestmen), which can autotomize a leg which will continue to move and thus confuse and divert the predator whilst the putative prey makes its escape (Gnaspini and Cavalheiro, 1998). Since the opilionid has eight legs (at least at the start of the chase) it can employ this subterfuge a number of times. However, studies on wolf spiders (which play a similar trick) show that the loss of a leg slows them down (Amaya et al., 1998).

13.4 BARRIERS

13.4.1 Slick Coating

Teflon lubricants that create a slippery surface because of their chemical properties. These chemical agents reduce friction with the intent to inhibit the free movement of the target. In the 1960s Riotril ("Instant Banana Peel") was applied as an ostensibly inert white powder to a hard surface and wetted down. It then became like an ice slick. It is virtually impossible for an individual to move or stand up on a hard surface so treated; tyres skid. Riotril, if allowed to dry, can easily be peeled away or, because it's water-soluble, can be washed away (Alexander et al., 1996).

A similar phenomenon is found in the carnivorous pitcher plants (Figure 13.3). Several mechanisms have been proposed for the way they capture insects, mostly slippery surface wax crystals. But the important capture mechanism is due to the surface properties of the rim of the pitcher, which has smooth radial ridges. This surface is completely wettable by nectar secreted by the rim, and by rain water, so that a film of liquid covers the surface when the weather is humid. The rim is then slippery both for soft adhesive pads (the liquid sees to that) and for the claws, due to the surface topography. This dual system starts sliding ants down the slippery slope (Bohn and Federle, 2004).

13.4.2 Sticky Coating

Polymer adhesives used to bond down equipment and human targets. Also known as stick'ems' and superadhesives (Alexander et al., 1996).

The best known biological adhesives are those occurring in spiders' webs and those on the leaves of the sundew, *Drosera*. Neither adhesive has yet been characterized.

Figure 13.3   A pitcher plant trap, which is a modified leaf. The rim of the trap is curled over, forming a slippery platform onto which insects can walk.

The Peripatus (the velvet worm, Figure 13.4) shoots out sticky adhesive threads that entangle its prey. The threads contain protein, sugar, lipid, and a surfactant, nonylphenol. The proteins are the principal component of the slime; the amino acid composition suggests collagen. The original function of the secretion was probably defense, developing into attack as the viscosity, amount, and distance that the substance could be expelled all increased. This defensive substance would in turn be also useful for hunting, if the original condition consisted of capturing prey directly using mandibles, as when onychophorans handle small prey. The adhesive substance probably allows the entanglement of larger and therefore more nutritious prey (Benkendorff et al., 1999).

When in danger, some species discharge sticky threads that can entangle predators. Some like the sea cucumber can even expel their internal organs, which they regrow causing it no harm at all. Although the mechanical properties of the threads have not been measured, they are obviously very

Figure 13.4   The velvet worm, *Peripatus capensis*. It lives in damp places and has no external armor. However, it can shoot sticky threads several times its body length.

tough since the Palauan people of the south Pacific squeeze the sea cucumber until it squirts out its sticky threads, which they put on their feet to protect them when they walk around the reef.

When attacked, the centipede *Henia* rolls itself up with its ventral surface facing outward. This is the opposite to most centipedes, which either attack with their large mandibles or roll up with their dorsal surface — the most armored — facing outward. However, *Henia* has a large gland on the underside of each segment which secretes an adhesive. The amount of adhesive is more than 10% of the body weight. The adhesive sticks to the mouthparts, etc., of the assailant preventing the parts from working. While, the assailant retires to clean itself, the centipede escapes. The glue seems to be made of two components: a fibrous protein (possibly silk-like) and a globular protein, which is the actual adhesive. At high magnification, the adhesive appears as a large number of fine fibres stuck firmly at each end. Thus removing the adhesive is not as simple as initiating a crack and propagating it; each fibre has to be broken separately, taking a lot of time and effort (Hopkin et al., 1990). The adhesive can stick to dirty wet surfaces, desirable for any technical adhesive. When sticking two glass plates together it is as effective as a cyanoacrylate adhesive.

13.4.3 Sticky Foam

A name given to a polymer-based superadhesive agent. The technology first began appearing in commercial applications such as "super glue" and quick setting foam insulation. It is extremely persistent and is virtually impossible to remove. Sticky foam came to public attention on February 28, 1995 when U.S. Marines used it in Mogadishu, Somalia, to prevent armed intruders from impeding efforts to extricate United Nation forces from that country (Alexander et al., 1996).

A foam allows a limited amount of material to occupy a greater volume, and since the intent is to impede rather than to entrap, the greater difficulty of breaking a structure that can accommodate higher strains, and is made of multiple threads, contributes to the effectiveness of the mechanism. This is probably why it occurs in the adhesive plaque which sticks the byssus thread of the mussel onto the rock. Otherwise, foams in biology are more used for protection than for attack and are an integral part of many egg cases, especially in snails and insects (e.g., *Mantis*, *Locusta*). They are commonly made of protein, often phenolically tanned and waterproofed, although their primary stability comes from their liquid crystalline structure (Neville, 1993)

13.4.4 Rope

Nylon rope dispersed by a compressed air launcher mounted on a truck (Alexander et al., 1996).

With animals the rope can become part of an entrapment mechanism — basically with an adhesive device on the end of the rope. Examples are the ballistic snares of the chameleon and the squid.

In the arms of the squid, transverse muscle provides the support required for the relatively slow bending movements while in the tentacles the transverse muscle is responsible for the extremely rapid elongation that occurs during prey capture. In the squid *Loligo pealei*, the thick filaments of the obliquely striated muscle fibres of the arms are approximately 7.4 mm long while those in the cross-striated fibres of the tentacle are approximately 0.8 μm long. This results in more series elements per unit length of fibre. Since shortening velocities of elements in series are additive, this results in the shortening velocity of the tentacle fibres to be approximately $15\ L_0\ s^{-1}$ compared with the arm transverse muscle $1.5\ L_0\ s^{-1}$ at 19°C.

The strike of *L. pealei* when it is capturing its prey takes as little as 20 ms. During the strike, the proximal portion of the tentacle, the stalk, elongates. The nonextensible distal portion of the tentacle, the club, contacts the prey and attaches using suckers. Extension takes 20 to 40 ms with peak strains in the stalk of 0.43 to 0.8. Peak longitudinal strain rates vary from 23 to 45 s^{-1}. The stalk can extend at over 2 ms^{-1} at an acceleration of 250 ms^{-2}. Once the tentacular clubs have contacted the prey, the stalks often buckle (Kier and Thompson, 2003).

13.4.5 Smoke

A thick, disorienting "cold smoke" that can be generated in areas from 2,000 to 50,000 cubic feet. It restricts an intruders eye–hand coordination and interactions among members of an intruding group. White obscuring smoke can be delivered by grenades or smoke pots. Relatively inexpensive, noncontaminating and tactically ideal for police use. Obscuring smokes are temporarily irritating to the nose and throat and cause those affected to lose their senses of purpose and direction (Alexander et al., 1996).

Compared to smell, all the intricate color and shape changes of the octopus are ineffective. One way to counter this threat is to block the predator's sense of smell, which has been shown to be one way in which the ink is used, though in large quantities. Obviously ink can be used to cover the animal's hasty departure, but it can also be used as a decoy, since the octopus or cuttlefish can produce a coherent plume of ink that is more or less of its own size and shape.

13.4.6 Stakes

A sharp stake, often of wood or bamboo, that is concealed in high grass, deep mud or pits. It is often coated with excrement, and intended to wound and infect the feet of enemy soldiers. Can be utilized both as a booby trap and as a barrier. Commonly known as punji stick or punji stakes (Alexander et al., 1996).

The Komodo Dragon, *Varanus komodoensis*, the largest land-living lizard, feeds mainly on carrion. Even though it is large and strong, mostly when it attacks living animals, it only wounds rather than kills them. But even minor wounds often become septic, so septicemia seems to be a significant mechanism for weakening and eventually killing prey. However, when the dragons fight each other, they appear to suffer no ill effects, even though their fights are frequent and often result in deep puncture wounds. If one could identify the bacteria in the dragon's saliva, including those capable of killing its mammalian prey, then one might have not only a chemical weapon but also its antidote. Additionally the wounds made by the dragon bleed profusely and it takes longer for the blood to clot, so the saliva also contains an anticoagulant.

13.5 BIOTECHNICALS

13.5.1 Hypodermic Syringe or Dart

Modified shotgun or handgun in which the projectile is a drug-filled syringe activated by a small charge on impact. Wide variety of drugs available including emetics (Alexander et al., 1996).

Organisms have two methods of delivering poison: externally (on being attacked) and internally (on being eaten). Since plants can usually afford to lose a leaf or two, they tend to have the poisons internally and are not necessarily brightly colored as warning. Animals are either brightly colored (for instance, the poison-dart frogs, *Dendrobates* spp. or poisonous nudibranchs or insects, q.v.) or carry their poisons in spines or stings. Bees, wasps, and scorpions are obvious examples of the latter; the sting is deployed, penetrates the victim with effort from the stinger, and poison is injected from a sac which contracts. In hive bees and presumably others, the sting sac also releases a pheromone which attracts other bees and encourages them to sting — rather like a beacon or marker used in bombing raids. The urticaceous hair found on stinging nettles (*Urtica* spp.) and many caterpillars is a passive mechanism. On the stinging nettle there are hollow hairs (Figure 13.5) containing several irritating substances such as histamine (the mediator of some allergic reactions), serotonin, acetylcholine, and formic acid. When lightly brushed against, the tip of the hair (made of brittle silica) snaps off at an angle leaving a sharp tip that pierces the skin and delivers the cocktail. A similar system operates in caterpillars. The urticating hairs or spines of the larva of the moth

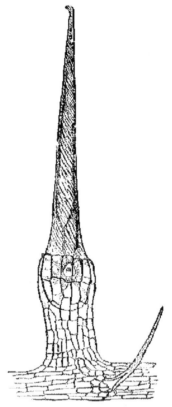

Figure 13.5 A nettle sting, about 1-mm long. The tip is highly silicious and brittle, so that when it breaks off it leaves a sharp end like a syringe needle. It contains an irritant poison.

Automeris io (which is related to silkworms) are of two types, both having a poison gland (Gilmer, 1925). The chemical nature of the poisons is not fully known, though they can contain formic acid, histamines, and enzymes which can dissolve human tissues and cause dermatitis. The spines work very much like nettle stings. Severe allergic reaction can cause death. The skin bleeds after contact with caterpillars of the Venezuelan *Lonomia achelous* which have poison spines containing an anticoagulant.

13.5.2 Neuro-Implant

Computer implants into the brain that allow for behavioural modification and control. Current research is experimental in nature and focuses on lab animals such as mice (Alexander et al., 1996).

There are several (probably many) parasites which affect the behavior of the host to the benefit of the parasite. The parasite can therefore be thought of reprogramming its host, though of course the effective agent, being chemical, is far more subtle and would be much easier to administer. Consider *Dicrocoelium dendriticum*, a parasitic worm; its main or primary host is sheep. The eggs are released in the dung of the sheep and are eaten by the snail *Cionella lubrica*. The eggs develop and the next stage (cercaria) is released into the snails mucus slime balls (which form in its respiratory chamber) and deposited on vegetation. Ants (*Formica fusca*) then eat the slime balls. Most of the cercaria become dormant in the ant's abdomen. However, some of them migrate into the ant's head where they enter the nervous system of the ant and affect its behavior. As evening approaches and the air cools, the infected ants, instead of returning to their nest, climb to the top of the vegetation and clamp on to the leaves with their mandibles. They stay there immobile until the next morning. The ants are thus likely to be eaten by passing sheep, thus completing the life cycle of the parasite. Although the parasite is obviously far more complex than a computer chip, the change in the ant's behavior is minimal: the interaction of the insect's temperature response with its response to gravity.

13.5.3 Pheromones

The chemical substances released by animals to influence physiology or behavior of other members of the same species. One use of pheromones, at the most elemental level, could be to mark target individuals and then release bees to attack them. This would result in forcing them to exit an area or abandon resistance (Alexander et al., 1996).

Lima beans (*Phaseolus lunatus*) infested with spider mites release chemicals that attract predatory mites that then prey on the spider mites. The uninfected plants downwind also attract predatory mites. Jasmonic acid sprayed onto tomato plants may regulate volatiles that attract parasitoid wasps that prey on caterpillars feeding on the tomato plants. Such indirect defenses may be even more complex. This may then be why some plants house and feed the predators as has happened in ant plants. The ants can be considered to be an induced biotic defense because the number of ants that patrol the leaves increases severalfold as a result of attraction by volatiles emitted from the damaged tissue when a herbivore chews a leaf. The ants are acting as a Praetorian body guard.

13.6 ELASTIC MECHANISMS

Human technology used elastic mechanisms as power amplification of human or animal energy to launch arrows and other projectiles; this approach is used in nature but man has replaced elastic mechanisms with explosives.

The ability to escape quickly from a predator is vital for most prey, while predators have obvious advantages if they are able to outrun fast prey and overpower it using even faster weapons. The speed of running, jumping, predatory strikes, etc. is generally correlated with the animal's size. In order to achieve velocities comparable to those of larger animals, small ones such as most arthropods have to rely on very high accelerations (Alexander and Bennet-Clark, 1977). Therefore, in many insects, the speed of action reaches or even surpasses the velocity limitations inherent in muscle contraction. Irrespective of phylogenetic relationships, convergent evolution has resulted in special mechanical designs (e.g., springs or catapults) that overcome the constraints of muscle action in many arthropods (Bennet-Clark and Lucey, 1967).

In addition to fast mechanics, both prey and predators rely on rapid neuronal and muscular systems to initiate and control their swift escape or predatory actions. Among the ants, several species employ particularly fast mandible strikes in order to catch swift prey or to defend themselves. This so-called trap-jaw mechanism (a mandible strike which far exceeds the speed allowed for by muscular contraction) has evolved independently in three ant species (Gronenberg, 1996). These studies reveal that the fast strike results from energy storage in a catapult design, and its control relies on fast neurones and on a high velocity trigger muscle.

In biological elastic mechanisms, strain energy is stored only when the spring mechanism is in the position from which the energy will be released — its loaded configuration. This is in contradistinction to most man-made systems, where the assumption of the loaded configuration is also the means by which the energy is stored (e.g., drawing a bow). For instance, the locust brings its legs into the jumping position, then loads the main jumping tendon using muscle power. This probably makes the system safer and allows a lower safety factor in the strength of the components (Bennet-Clark, 1975).

Nature commonly uses bistable mechanisms. This is intimately associated with the separation of the assumption of the loaded configuration from the storage of strain energy. The mechanism is drawn over center by the main spring, and then the spring is loaded. The main spring has a low mechanical advantage and can store a large amount of strain energy, generating high forces. When the system is "fired" the trigger, which can generate only a low force but has a high mechanical advantage, allows the mechanism to move back over center and the energy from the main spring is fed into the system (Bennet-Clark and Lucey, 1967). This has the advantage that there are no firing pins or hooks to jam or break. Thus, control is smoother and reliability improved. In the snap-jaw ant, the mandibles are clicked against each other, rather like snapping finger and thumb over each other. The ant can then move comparatively massive objects. The mandibles are first held with the tips just touching, then loaded. Large muscles contract against the closed mandibles that are thus bent and store some elastic energy. However, most of the muscular energy is transformed and elastically stored within the apodeme and its cuticular threads, within the muscle fibres and probably also within the entire head capsule. Slight rotation of one of the mandibles then causes its lower edge to bend slightly inwards and lets the other mandible slide above it, powered by the strain energy stored within the contracted muscles and the mandible shaft (Gronenberg et al., 1998). Immediately afterwards the unstimulated mandible hits the object and bounces it away. The stored energy thus is spent and the mandibles are decelerated during the second half of their trajectory and come to a hold before they could bump into the front of the head.

The Venus fly trap (*Dionaea muscipula*) preys on insects and other small animals that venture onto its trap leaves and trigger their closure by disturbing certain sensitive hairs. The leaves routinely shut in 1/25 s. Such speed of movement is uncommon amongst plants and so has attracted attention and theories for many years. The mechanism is based on a turgor-driven elastic instability of the leaf, which is in effect a prestressed mechanical bistable structure (Forterre et al., 2005; Thom, 1975). A better understanding of this mechanism and the way in which it is designed and actuated would not only solve a long-standing conundrum, but could also give rise to a series of novel hydraulic actuators and switches.

Nature does use explosives, in the sense that an explosive chemical reaction proceeds at very high speed, is exothermic, and produces large amounts of hot gas that do the damage. The insect in question is the bombardier beetle, of which there are many species, for example *Brachinus explodens*, which produces a jet of steam and hydroquinone at a temperature probably in excess of 100°C. The propellant is oxygen produced from the breakdown of hydrogen peroxide. The jet is pulsed (at about 500 Hz) and can, depending on the species of beetle, be aimed very accurately (Dean et al., 1990).

13.7 ELECTRICAL

13.7.1 Stun Gun

A small, two-pronged, hand held electrical discharge weapon. Effective range is less than an arm length. It works by affecting the muscle signal paths, disturbing the nervous system (Alexander et al., 1996).

The electric eel is different from other electric fish in its ability to generate a stunning or even a killing electrical discharge. The electric eel can produce up to 600 V in a single discharge. The electric organ, which consists of a series of modified tail muscles, is similar to a row of batteries connected in a series. It is subdivided into three sections: two small and one large. One small battery is used for navigational signals. The large battery and the other small one are used to generate the stunning discharge. After delivering a strong shock, the electric eel must then allow the electric organ to recharge (Heiligenberg, 1977).

A discharge from an electric eel can kill the small fish that are its primary food, but electric eels can also shock potential predators. A touch from the electric eel's tail can effectively disable a human or a large animal with a stunning shock, although a single discharge is usually not enough to kill. However, repeated shocks could kill.

13.8 ENTANGLERS

13.8.1 Bola

Device consisting of two or three heavy balls attached by one or two ropes or cords and used for entanglement purposes. It is twirled overhead in one hand and hurled or cast at the intended target. Designed to entangle legs to retard or stop movement. Probably an ancient weapon, but made famous by the gauchos of South America, who used them to catch cattle and ostriches (Alexander et al., 1996).

Ordgarius magnificus, the Australian bola spider, hides in a silk-lined retreat among the leaves of native trees such as eucalypts. At night it hangs, head down, from a horizontal silk strand, and using an extended front leg, suspends a silk thread about 4 cm long with a sticky blob on the end (Figure 13.6). Thread + blob = bola. The blob contains an attractant moth pheromone. When the spider detects the vibrations in the air made by an attracted moth flying close, it begins to jerk its body so as to swing the bola around in a circle. When the moth is close enough, she lets the thread run then flicks it to hit the moth. The moth is then entangled, the spider reels it in, wraps it in silk and sucks it dry. Different pheromones are used for different seasons or growth stages to capture the moth species that are available or are of best size. The difference is that whilst in technology the target is probably running away, in nature it is flying towards you, with friendly intent!

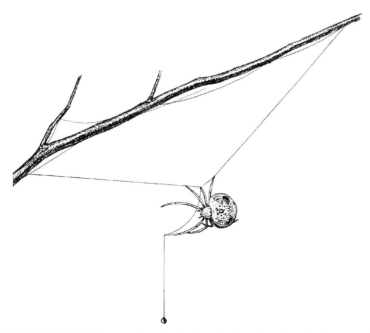

Figure 13.6 A bola spider (an American species, *Mastophora*, is shown here), waiting for a prey insect to fly past.

13.8.2 Cloggers

Polymer agents, sticky-soft plastics, used in burst munitions to clog up jet and tank engine intakes (Alexander et al., 1996).

Hagfish slime is a mixture of mucus and threadlike fibres, secreted in concentrated form from pores on the side of the hagfish's body. Upon contact with the seawater, the slime absorbs water rapidly, expands into a sticky gel that can ensnare and sometimes suffocate an attacker. Up to 5 l of gel can be produced within seconds (Koch et al., 1991). We are probably missing a trick by producing only single-phase "cloggers." The addition of fibres would greatly increase the coherence of the clogging substrate, generating a compliant fibrous composite material.

The hagfish rids itself of the mucus by tying itself into a knot that it runs down its body pushing the mucus ahead of it (Fernholm, 1981). Over the ranges of temperature encountered by the hagfish, the gel strength is relatively independent of temperature, which perhaps ensures that slime is an effective defense in a variety of conditions.

13.9   PROJECTILE

13.9.1 Water Stream

Mobile unit that projects a continuing stream of water for riot control purposes (Alexander et al., 1996).

The archer fish (*Toxotes jaculator*) is the best-known analog, though it conserves its energy by aiming the jet of water very carefully and bringing down one object at a time. The object is usually an insect or other small animal sitting on a plant overhanging the water. With the tongue against a groove on the roof of the mouth, the fish forms a tube, and forces water out by snapping the gills shut. The jet of water is directed with the tip of the tongue. The fish can squirt up to seven times in quick succession, and the jet can reach 2 to 3 m, but it is accurate to only 1 to 1.5 m. Fish as small as 2 to 3 cm long can already spit, but their jets reach only 10 to 20 cm (Rossel et al., 2002). The disadvantage of this technique is obviously that when the dislodged prey falls into the water, it can be taken by any of the other fish. So the archer tends to position itself below the prey, and also knows how to catch an object falling on a curved trajectory, a skill that would make it a good ball player!

13.10   RIOT CONTROL AGENT

13.10.1 Chemical Mace

Small spray can containing a 0.9% solution of agent CN in a variety of petroleum-based carriers including a mixed freon/hydrocarbon solvent. First introduced in 1966. CSMace then developed in 1968 by suggestion of the U.S. Army (Alexander et al., 1996).

Chemical agents produced by animals or plants tend to be for defense, sometimes against a single individual and sometimes against large numbers. Carnivorous ground- and water-beetles (Adephaga) are some of the better known animals that deliver compounds in one of three ways:

(1)   *Oozing:* The glands of many beetles do not have muscles for discharging large amounts of substance and so the material only oozes out from the openings. This is helped by internal pressure.

(2)   *Forceful spraying:* Many ground beetles have intrinsic muscles with the glands. The beetle *Pasimachus subsulcatus* can forcibly discharge a spray up to several centimetres that is irritating to the eyes and hurts abraded skin.

(3)   *Crepitation or squirting* is characteristic of bombardier beetles (q.v.). Hydroquinones are stored with hydrogen peroxide in the major gland chambers and the ezymes catalase (which converts

hydrogen peroxide to water and oxygen) and peroxidase are stored in an accessory chamber. When the beetle is disturbed, these compounds are mixed. This produces a strongly exothermic reaction that generates quinines, discharged as a vapour of about 100°C, an effective deterrent against predators. The emission occurs as a pulsed jet rather than as a steady stream that allows for a higher discharge velocity due to increased pressure in the reaction chamber (Dean et al., 1990).

Many compounds in these beetles have been implicated as toxins or feeding deterrents against predators; the secretions are usually mixtures of a number of components. Some surfactant components may help the toxic compounds penetrate the skin of a predator.

Amongst sea birds, fulmars are well-known masters of the art of projectile vomiting. Also they are exceptionally courageous and will stay by their single egg if people come close. They are mostly silent apart from a low cackling noise made to other fulmars. So the first a person may know of the presence of a fulmar is a stream of foul and evil-smelling orange vomit spewing straight into their eyes from a few feet away. Even a young fulmar chick can do this.

Spitting cobras also figure here among projectile vomitters.

13.11   OPERATIONAL

13.11.1   Long-Term Disablement

The outcome of the application of nonlethal force that affects the opponent beyond duration of the confrontation or conflict. Blinding, maiming or psychologically deranging the opponent represent forms of long-term disablement. This form of disablement burdens a society and is anathema to the Western definition of nonlethality (Alexander et al., 1996).

Few animal encounters end with disablement — the tendency is for the victor to eat the vanquished. This is not true of herbivores, of course, when the fight will be in dispute of territory or reproductive access to a harem. Deer, whose antlers grow afresh every year, can cope with the 30% breakage which results from fights in the rutting season (Kitchener, 1987), although they may get wounded on the flank. On the other hand, sheep, goats, and antelope, whose horns grow from the root and are not renewed annually, lose the ability to fight if the horn is broken and thus cease to be reproductively active. It is likely that only amongst elephants and cetaceans is there any care for the disabled; social carnivores (wolves, lions, wild dogs) also show some concern. Otherwise the injured die and are no further burden.

13.11.2   Passive Deterrents

Non-lethal weapons that do not affect the physiology of the target individual. Includes dyes, personal alarms, and scent sprays (Alexander et al., 1996).

The best-known animal using a similar technique is the striped skunk that has about a table-spoonful of oily yellow musk in its scent glands located at its anus. This will produce five or six sprays, each of which is accurate and can travel up to 5 m. The mist from the spray can travel 10 to 15 m with the smell carrying up to 2 km. A great many insects produce repellant chemicals (q.v.).

13.12   PHYSIOLOGICAL

13.12.1   Neurochemical

There are many neurotoxins. For instance, a sea anemone uses its tentacles to capture prey and defend itself against predators. Every tentacle is covered with thousands of tiny stinging capsules

Figure 13.7   The South American peacock butterfly *Automeris memusae*, showing its cryptic- (above) and warning- or flash- (below) wing positions.

called nematocysts about 5 μm across. Each capsule contains a coiled hollow thread with a barb at the end. The capsules contain a poison capable of paralysing or killing small animals. When a small animal contacts the tentacles, the capsules are triggered and fire their barbed threads like harpoons, which pierce the skin of the animal and inject their poison. In the fresh-water polyp *Hydra vulgaris*, the capsule contains a 2 M salt solution and so reaches a turgor pressure of 150 atmospheres (15 MPa) before it shoots out the dart at an acceleration of 40,000 g (Holstein et al., 1994). The sea anemone uses its stinging cells for defense, as do other animals. The digestive tract of the nudibranch sea slug *Aeolidia* (a sort of snail) is lined with a protective coating to prevent injury from any unactivated nematocysts it consumes, which it then transports into its skin to use for its own defense. Sea anemones also use their poisonous stings against their own kind, usually while competing for territory. Some species even possess special club-like structures, packed with potent stinging capsules, that they use to battle other anemones. Territorial fights often result in serious injury and even death to one or both anemones.

13.12.2 Diversion

A diversion that acts directly by affecting one or more of the five senses. Noise that lasts less than one second (Alexander et al., 1996).

An obvious example from biology is flash coloration and its commonest manifestation is in the hind wings of cryptically colored moths with brightly colored hind wings that are revealed suddenly when the insect is threatened. The hind wings can be of one or two colors in well-defined patches (often red or yellow with black) or have large and colorful eyespots (Figure 13.7). Another well-known example is feigning injury; the lapwing nests on the ground and will lure a potential predator away from its nest by dragging one wing on the ground with the pretence that it is broken. When the predator is safely away from the nest, the lapwing flies away.

13.13 SURVEILLANCE

13.13.1 Electrosensing

A 'sixth sense' based on reception of electrical signals in the environment. Akin to electronic eavesdropping.

Teleost (bony) fish, elasmobranchs (sharks and rays) and the duckbilled platypus (and probably many more types of animals) have an electric sense. It is best developed in the elasmobranchs, which have rows of pit organs (ampullae of Lorenzini) that can detect electric fields as weak as

5 nV cm^{-1} and so detect the fields induced through their bodies as they swim through the earth's magnetic field. They can use this sense to detect the presence of prey and there is evidence that they also use it in navigation. However, the electroreceptors cannot measure DC voltages so that a voltage due to water flow in the ocean is not uniquely interpretable in terms of the speed and direction of flow at the point where the electrical measurement is made. Perhaps the cue is the directional asymmetry of the change in induced electroreceptor voltage during turns. A neural network could use this cue to determine swimming direction by comparing electrosensory signals and signals from the semicircular canals of the inner ear, which function as an accelerometer (Kajiura and Holland, 2002). Weakly electric fish such as nocturnal fish and the gymnotids and mormyrids of the murky waters of the Amazon use active electrolocation — the generation and detection of electric currents — to explore their surroundings. Although electrosensory systems include some of the most extensively understood circuits in the vertebrate central nervous system, relatively little is known quantitatively about how fish electrolocate objects (Assad et al., 1999).

13.14 CONCLUSIONS

Biomimetics is not a particularly new study, but it seems to be generating success. It is possible to calculate the number of functions of biology that appear in a technical environment, which is about a 10% overlap. This does not mean that the route has been biomimetic, but it does mean that the technology has been transferred. The transfer has so far been somewhat adventitious, so a chapter like this one can perhaps not so much report what *has* been successful (not a lot), but what *might* be successful. The difference is: how many of the possibilities have been tried? This chapter is more a list of things to do than of things done.

REFERENCES

Alexander R.M., *Animal Mechanics*, Blackwell Scientific Publications, Oxford (1983), pp. 1–10, 1–301.

Alexander R.M. and H.C. Bennet-Clark, Storage of elastic strain energy in muscle and other tissues, *Nature*, Vol. 265 (1977), pp. 114–117.

Alexander J. B., R. Applegate, J. B. Becker, M. Begert, J. H. Cuadros, A. Flatau, and C. Heal, *Nonlethal Weapons: Terms and References*, INSS Occasional Paper 15 (ed. Bunker R.J.), United States Air Force Institute for National Security Studies, USAF Academy, Colorado (1996), pp. 1–72.

Amaya C.C., P.D. Klawinski, and J. D.R. Formanowicz, The effects of leg autonomy on running speed and foraging ability in two species of wolf spiders, *American Midland Naturalist*, Vol. 145 (2001), pp. 201–205.

Anderson A.J. and P.W. McOwan, Model of a predatory stealth behaviour camouflaging motion, *Proceedings of the Royal Society B*, Vol. 270 (2003), pp. 489–495.

Assad C., B. Rasnow, and P.K. Stoddard, Electric organ discharges and electric images during electrolocation, *Journal of Experimental Biology*, Vol. 202 (1999), pp. 1185–1193.

Benkendorff K., K. Beardmore, A. Gooley, N. Packer, and N. Tait, Characterisation of the slime gland secretion from the Peripatus *Euperipatoides kanangrensisi* (Onychophora: Peripatopsidae), *Comparative Biochemistry and Physiology B*, Vol. 124 (1999), pp. 457–465.

Bennet-Clark H.C., The energetics of the jump of the locust, *Schistocerca gregaria*, *Journal of Experimental Biology*, Vol. 63 (1975), pp. 53–83.

Bennet-Clark H.C. and E.C. A. Lucey, The jump of the flea, *Journal of Experimental Biology*, Vol. 47 (1967), pp. 59–76.

Bohn H.F. and W. Federle, Insect aquaplaning: nepenthes pitcher plants capture prey with the peristome, a fully wettable water-lubricated anisotropic surface, *Proceedings of the National Academy of Sciences of the United States of America*, Vol. 101 (2004), pp. 14138–14143.

Caveney S., Cuticle reflectivity and optical activity in scarab beetles: the role of uric acid, *Proceedings of the Royal Society B*, Vol. 178 (1971), pp. 205–225.

Croll D.A., C.W. Clark, A. Acevedo, B. Tershy, S. Flores, J. Gedamke, and J. Urban, Only male fin whales sing loud songs, *Nature*, Vol. 417 (2002), pp. 809.

Dean J., D.J. Aneshansley, H.E. Edgerton, and T. Eisner, Defensive spray of the bombardier beetle: a biological pulse jet, *Science*, Vol. 248 (1990), pp. 1219–1221.

Dumbacher J. P., Wako A, S.R. Derrickson, A. Samuelson, T.F. Spande, and J. W. Daly, Melyrid beetles (Choresine): a putative source for the batrachotoxin alkaloids found in poison-dart frogs and toxic passerine birds, *Proceedings of the National Academy of Science*, Vol. 101 (2004), pp. 15857–15860.

Ferguson G.P., J. B. Messenger, and B.U. Budelmann, Gravity and light influence the countershading reflexes of the cuttlefish *Sepia officinalis*, *Journal of Experimental Biology*, Vol. 191 (1994), pp. 247–256.

Fernholm B., Thread cells from the slime glands of hagfish (Myxinidae), *Acta Zoologica*, Vol. 62 (1981), pp. 137–145.

Forterre Y., J. M. Skotheim, J. Dumais, and L. Mahadevan, How the Venus flytrap snaps, *Nature*, Vol. 433 (2005), pp. 421–425.

Fuji R., The regulation of motile activity in fish chromatophores, *Pigment Cell Research*, Vol. 13 (2000), pp. 300–319.

Gilmer P.M., A comparative study of the poison apparatus of certain lepidoterous larvae, *Annals of the Entomological Society of America*, Vol. 18 (1925), pp. 203–239.

Gnaspini P. and A.J. Cavalheiro, Chemical and behavioural defenses of a neotropical cavernicolous harvestman: *Goniosoma spelaeum* (Opiliones, Laniatores, Gonyleptidae), *Journal of Arachnology*, Vol. 26 (1998), pp. 81–90.

Gronenberg W., The trap-jaw mechanism in the Dacetine ants *Daceton armigerum* and *Strumigenys* sp., *Journal of Experimental Biology*, Vol. 199 (1996), pp. 2021–2033.

Gronenberg W., B. Hölldobler, and G.D. Alpert, Jaws that snap: control of mandible movements in the ant *Mystrium*, *Journal of Insect Physiology*, Vol. 44 (1998), pp. 241–253.

Hanlon R.T. and J. B. Messenger, Adaptive coloration in young cuttlefish (*Sepia officinalis* L) — the morphology and development of body patterns and their relation to behaviour, *Philosophical Transactions of the Royal Society B*, Vol. 320 (1988), pp. 437–487.

Hanlon R.T., R.T. Forsythe, and D.E. Joneschild, Crypsis, conspicuousness, mimicry and polyphenism as antipredator defences of foraging octopuses on Indo-Pacific coral reefs, with a method of quantifying crypsis from video tapes, *Biological Journal of the Linnean Society*, Vol. 66 (1999), pp. 1–22.

Heiligenberg W., Principles of electrolocation and jamming avoidance in electric fish: a neuroethological approach, in: Braitenberg V. (ed.), *Studies of Brain Function*, Springer, Berlin (1977), pp. 1–85.

Holstein T.W., M. Benoit, G.V. Herder, G. Wanner, C.N. David, and H.E. Gaub, Fibrous mini-collagens in *Hydra* nematocysts, *Science*, Vol. 265 (1994), pp. 402–404.

Hopkin S.P., M.J. Gaywood, J. F.V. Vincent, and E.L. V. Mayes-Harris, Defensive secretion of proteinaceous glues by *Henia* (= *Chaetechelyne*) *vesuviana* (Chilopoda, Geophilomorpha), *Proceedings of the 7th International Congress of Myriapodology, Leiden* (1990), pp. 175–181.

Johnsen S., Cryptic and conspicuous coloration in the pelagic environment, *Proceedings of the Royal Society B*, Vol. 296. (2002), pp. 243–256.

Johnsen S., Hidden in plain sight: the ecology and physiology of organismal transparency, *Biological Bulletin*, Vol. 201 (2001), pp. 301–318.

Johnsen S., E.A. Widder, and C.D. Morley, Propagation and perception of bioluminescence: factors affecting counterillumination as a cryptic strategy, *Biological Bulletin*, Vol. 207 (2004), pp. 1–16.

Kajiura S.M. and K.N. Holland, Electroreception in juvenile scalloped hammerhead and sandbar sharks, *Journal of Experimental Biology*, Vol. 205 (2002), pp. 3609–3621.

Kettlewell H.B. D., Selection experiments on industrial melanism in the Lepidoptera, *Heredity*, Vol. 9 (1955), pp. 323–342.

Kier W.M. and J. T. Thompson, Muscle arrangement, function and specialisation in recent coleoids, *Berliner Paläobiologie*, Vol. 3 (2003), pp. 141–162.

Kitchener A.C., Fracture toughness of horns and a reinterpretation of the horning behaviour of bovids, *Journal of Zoology*, Vol. 213 (1987), pp. 621–639.

Kitchener A.C., The evolution and mechanical design of horns and antlers, in: Rayner J. M.V. and R. J. Wootton (eds.) *Biomechanics in Evolution*, Cambridge University Press, Cambridge, MA (1991), pp. 229–253.

Kitchener A.C., G.E. Bacon, and J. F.V. Vincent, Orientation in antler bone and the expected stress distribution, studied by neutron diffraction, *Biomimetics*, Vol. 2 (1994), pp. 297–307.

Koch E.A., R.H. Spitzer, and R.B. Pithawalla, Structural forms and possible roles of aligned cytoskeletal biopolymers in hagfish (slime eel) mucus, *Journal of Structural Biology*, Vol. 106 (1991), pp. 205–210.

Messenger J. B., Cephalopod chromatophores: neurobiology and natural history, *Biological Reviews*, Vol. 76 (2001), pp. 473–528.

Neville A.C., *Biology of Fibrous Composites; Development Beyond the Cell Membrane*, Cambridge University Press, Cambridge (1993), pp. i–vii, 1–214.

Norman M.D., J. Finn, and T. Tregenza, Dynamic mimicry in an Indo-Malayan octopus, *Proceedings of the Royal Society of London Series B*, Vol. 268 (2001), pp. 1755–1758.

Patek S.N., W.L. Korff, and R.L. Caldwell, Deadly strike mechanism of a mantis shrimp, *Nature*, Vol. 428 (2004), pp. 819–890.

Ramachandran V.S., C.W. Tyler, R.L. Gregory, D. Rogers-Ramachandran, S. Duensing, C. Pillsbury, and C. Ramachandran, Rapid adaptive camouflage in tropical flounders, *Nature*, Vol. 379 (1996), pp. 815–818.

Reby D. and K. McComb, Anatomical constraints generate honesty: acoustic cues to age and weight in the roars of red deer stags, *Animal Behaviour*, Vol. 65 (2003), pp. 519–530.

Rossel S., J. Corlija, and S. Schuster, Predicting three-dimensional target motion: how archer fish determine where to catch their dislodged prey, *Journal of Experimental Biology*, Vol. 205 (2002), pp. 3321–3326.

Shadwick R.E., A.P. Russell, and R.F. Lauff, The structure and mechanical design of rhinoceros dermal armor, *Philosophical Transactions of the Royal Society B*, Vol. 337 (1992), pp. 419–428.

Thom R., *Structural Stability and Morphogenesis, an Outline of a General Theory of Models*, WA Benjamin, Inc., Reading, MA, 0 8053 9277 7 (1975), pp. i–xxv, 1–348.

Vincent J. F.V. and P. Owers, Mechanical design of hedgehog spines and porcupine quills, *Journal of Zoology*, Vol. 210 (1986), pp. 55–75.

Wickler W., *Mimicry*, Weidenfeld and Nicolson, London (1968), pp. 1–225.

14

Biological Materials in Engineering Mechanisms

Justin Carlson, Shail Ghaey, Sean Moran,
Cam Anh Tran, and David L. Kaplan

CONTENTS

14.1 INTRODUCTION

The biological world utilizes an amazing range of materials that provide function and survival to organisms faced with a wide range of environmental threats. The biosynthesis, processing, and assembly of these materials provide insight into design rules and strategies that can serve as useful templates for broader materials science and engineering needs. High strength fibers, toughened organic–inorganic composites, designs for efficient fluid flow, adhesion mechanisms, and actuators are examples reviewed herein. The knowledge gained from the study of these types of complex high performance materials systems should continue to stimulate new directions in materials science, including new hybrid systems to exploit the strengths and utility of both biological and synthetic versions of future materials designs.

The field of biomimetics encompasses a broad range of topics, generally based on the concept of "learning from Nature" in areas of materials science and engineering. This "learning" may be through inspiration in design, function, or a combination of both. Usually, this inspiration derives from a novel attribute of a biological system that suggests new and important insights into structure and function for materials science applications. Examples used in this chapter illustrate features of unique materials from Nature to inspire designs and functions for new materials: (a) silk proteins used by spiders and silkworms to construct composite encasements (cocoons) or strong and functional webs to entrap prey; (b) organic–inorganic composite structures found in sea shells to form highly engineered, hard, tough materials; (c) surfaces such as shark skin to reduce hydro-dynamic friction; (d) modes of "sticking" to surfaces used by the gecko to produce strong adhesives; and (e) muscles in the human body to create highly engineered actuators. These examples, optimized through evolution, provide a range of topics for inspiration in materials designs and functions. They also provide generic insight into the underlying principles employed by biological systems to achieve remarkable plasticity in materials structure and function.

Each of the topics listed above is reviewed with a focus on what is currently understood in terms of structure and function, a mechanistic view of the system, and the current state of the art in mimicking these systems. It will be obvious at the end of this chapter that the learning curve is barely past the lag phase (microbial growth curve perspective). It is also worth considering that we are inherently limited in gaining additional insight into these systems due to the complexity of the biological systems of interest, our current limited understanding of their structure and function, and our preconceived bias of how to understand these systems due to training or perspective from more traditional materials science and engineering approaches. The excitement with Nature as a guide to materials science and engineering is that this is only the beginning and there is a lot to be learned in order to elucidate the "rules" that govern the processes involved.

14.2   COMPARISONS: BIOLOGICAL MATERIALS
AND SYNTHETIC MATERIALS: SYNTHESIS AND ASSEMBLY

At the core of this chapter are the novel "rules" that govern materials formation in Nature. These rules originate from the template-based synthesis driven by genetic blueprints. Furthermore, the building blocks (e.g., amino acids, sugars, nucleic acids) are linked (via enzymatic coupling reactions) into polymers with control of stereochemistry to affect regularity in chemistry and thus higher order interactions (intra and interchain). These polymeric building blocks (proteins, poly-saccharides, nucleic acids, and other biological macromolecules) are therefore "programmed" (chemically and physically) to self-organize into more complex materials through hierarchical structural complexity that gives rise to novel materials performance. The control of this structural hierarchy initiates with the regularity in structure at the individual monomer and chain levels, and is propagated up length scales from the molecular (chains), through the mesoscopic (mesophases), and finally to the macroscopic (material ultrastructure) level. Remarkably, these processes occur within a complex mixture of small and large molecules inside and outside of the sites of synthesis (cells). Compartmentalization helps in these processes, along with membrane interfaces. Most of the details involved in these processes are largely unexplored territory scientifically. The entire materials assembly process is governed by the interplay between genetic programs, environmental conditions inside and outside of the cells, and the remarkable specificity and control achieved through enzym-atic processes. Historically, these hierarchical interactions have been studied from the "top-down" or at the macro-scale, using electron microscopy to interrogate ultrastructure, or by testing mechan-ical properties of the materials and using this to interpret structural organization. In recent years, the focus of inquiry has shifted to the "bottom-up" paradigm, molecular-level interactions.

Polymer assembly as the basis for structural hierarchy and function in biology is most often governed by many weak bonds (hydrogen bonding, van der Waal). It is the high frequency and location of these types of bonds that allow assembly or disassembly of these material systems within reasonable energy demands to permit functions (e.g., such as denaturation and renaturation (replication fork) of DNA during semiconservative replication). These processes are mediated by water, structure, and location, with respect to the organic components and features such as hydrophobic hydration play a major role in the processes. General themes to consider that contrast the process of materials formation and assembly in Nature vs. in the laboratory via synthetic approaches are listed in Table 14.1.

Table 14.1 Comparison between Biological and Nonbiological Polymer or Materials Synthesis and Assembly

Feature	Laboratory	Nature
Synthesis		
Monomer	Usually racemates	Stereochemically pure
Blocks or domains	Usually mono or diblocks	Mono to highly diverse blocks
Polymerization	Comparatively rapid — mostly polydisperse	Comparatively slow template control — monodisperse (proteins, nucleic acids); others polydisperse (polysaccharides)
Processing or assembly		
Plasticizers	Varied, mostly organic	Water
Polymer interactions	Chain entanglements, fringed micelle model	Less chain entanglements, extensive hydrogen bonding and other weak interactions
Higher order structures features	Varied, rare	Common, controlled by chain interactions
Organic–inorganic composites	Usually mixtures, composites	Molecular-level interfaces controlled by weak bonds
Fate		
Environmental stability	Wide range of temperature	Narrower range of temperature
Degradability	Varies with polymer, most nondegradable	Universally degradable, rate matches function

14.2.1 Silk Processing and Assembly by Insects and Spiders — High Performance Fibers from Nature

Background — Silks are externally spun protein fibers generated by spiders and insects (Kaplan et al., 1994). Reeled silkworm silk (*Bombyx mori*) has been used in the textile industry for over 5,000 years. Unlike silkworm silks, spider silk production has not been domesticated because spiders are more difficult to raise in large numbers due to their solitary and predatory nature. In addition, orb webs are not reelable as a single fiber and they generate only small quantities of silk. Silkworms can be raised in large numbers and generate one type of silk at one stage in their lifecycle, forming the basis for the sericulture industry. Many spiders have evolved families of silk proteins (different polymer chain chemistries — primary amino acid sequences) with different functions. For example, the spider, *Nephila clavipes*, generates at least six different silks from sets of different glands, each silk specifically matched to function — such as for environmental glues, strong or flexible web components, prey capture, and encapsulation (cocoons) for offspring development.

Silks are of interest for their remarkable mechanical properties as well as their durability, luster, and "feel." Silk fibers generated by spiders and silkworms represent the strongest natural fibers known, even rivaling synthetic high performance fibers in terms of mechanical properties (Gosline et al., 1986). The best properties of *N. clavipes* native dragline fibers collected and tested at quasi static rates were 60 and 2.9 GPa for initial modulus and ultimate tensile strength, respectively. In addition, these fibers display resistance to mechanical compression that distinguishes them from other high performance fibers (Cunniff et al., 1994). Based on microscopic evaluations of knotted single fibers, no evidence of kink-band failure on the compressive side of a knot curve was observed. Synthetic high performance fibers fail by this mode even at relatively low stress levels. Silks are mechanically stable up to almost 200°C (Cunniff et al., 1994).

Spider dragline and silkworm cocoon silks are considered semicrystalline materials with the crystalline components termed β-sheets (Gosline et al., 1986). Most silks assume a range of different secondary structures during processing from water-soluble protein in the glands to water-insoluble spun fibers. Marsh et al. (1955) first described the crystalline structure of silk as an antiparallel hydrogen bonded β-sheet. The unit cell parameters in the silk II structure (the spun form of silk that is insoluble in water) are: 0.94 nm (interchain), 0.697 nm (fiber axis), 0.92 nm (intersheet). These unit cell dimensions are consistent with a crystalline structure in which the protein chains run antiparallel with interchain hydrogen bonds perpendicular to the chain axis between carbonyl and amine groups, and van der Waal forces stabilizing the intersheet interactions (based on the predominance of short side chain amino acids such as glycine, alanine, and serine in β-sheet regions). Generally, silkworm fibroin in cocoons contains a higher content of crystallinity (β-sheet content) than spider dragline silks such as from *N. clavipes*.

As more protein sequence data from various spiders and silkworms has been elucidated, it is clear that these families of silk proteins are similar but also encompass a range of sequence variations that reflect their functional properties. Silkworm fibroin is the protein that forms the structural aspects of the fibers. These fibers are encased in a family of glue-like sericin proteins. The primary sequence of amino acids of these proteins is found to be highly repeated, thus, small regions of sequence chemistry in the protein chain are found elsewhere in other regions of the chains. This design feature is critical to the function of this group of proteins as structural materials. This design feature also allows the sequences of these large proteins to be represented in relatively short sequences in terms of polymer design. This has led to the option of forming synthetic genetic variants to represent the larger proteins, a useful laboratory technique to enhance the ability to understand these proteins in a simplified form. These shorter genetic pieces can be polymerized (multimerized) into longer genes to explore sequence and size relationships. These powerful tools in molecular biology facilitate direct insight into the role of sequence chemistry, protein block sizes and distributions, and protein polymer chain length on materials structure and function. Native and synthetic silk clones have been generated in a variety of heterologous expression systems, including bacteria, yeast, insect cells, plants, and mammalian cells (Wong and Kaplan, 2002).

Mechanism — Silk proteins are hydrophobic based on the predominance of glycine and alanine amino acids. The chains self-assemble into insoluble β-sheets. Spiders and silkworm keep these hydrophobic proteins soluble in water during their processing into fibers at concentrations up to 30 weight percent. The formation of liquid crystalline phases during silk processing has been reported as part of this process (Vollrath and Knight, 2001). This process is accomplished *in vivo* without premature crystallization into the insoluble β-sheets. Premature crystallization would be catastrophic for the animal as it would clog the spinning device. In recent studies, micelle and gel states were identified that suggest these are important steps governing chain silk protein interactions toward organized silk structures. These features can be duplicated in part using regenerated silkworm silk and the control of water removal from aqueous solutions of these proteins via osmotic stress (Jin and Kaplan, 2003). These "soft" micelles consist of flexible molecules and the structures can grow and change shape in response to changes in protein concentration. This is typical behavior of amphiphilic flexible surfactants that assemble into small spherical micelles and evolve morphologically into cylinders and lamellar-layered structures. Lyotropic liquid crystalline phase behavior can help explain the assembly of silk protein polymer chains into domains of high concentration (Vollrath and Knight, 2001), with orientation driven by micellar behavior and water efflux through channels in these structures.

Based on the micellar behavior and subsequent morphological features generated during silk processing, critical design rules (chemistry of the amino acid sequence and blocks or regions of the sequence chemistry) to match the processing environment have been described (Bini et al., 2004). These design rules demonstrate a modified triblock design for these proteins that also contrasts in important ways with traditional synthetic triblock co-polymers. In silks, the large (dominating in size and chemical influence) internal hydrophobic blocks (crystallizable domains which promote intra and interchain folding) are interrupted with very short hydrophilic blocks or spacers, putatively to control water content in micellar states to prevent premature crystallization into β-sheets. The large N- and C-terminal hydrophilic blocks interact with water and define micellar partitioning. All protein sequences in silks adopt these general block "design rules" (Bini et al., 2004) presumably to match the limits of the all aqueous processing environment. There is, however, significant sequence (chemistry) variability permitted in the hydrophobic or dominant regions of the protein which give rise to the variations in functional properties of the different silks. In the final step of the process of formation of silk fibers, the assembled proteins are spun through orifices in the abdomen of spiders or mouth of silkworms, inducing the structural and morphological transition of the gel state from the gland into the β-sheet structure and fiber morphology. The spun fibers are insoluble in water.

In general, the combined control of sequence chemistry and processing conditions is central to the successful formation of high-performance fibers based on silk proteins. Included in this is the presence of suitable blocks and chemical features of these blocks to deal with the processing environment and the appropriate mechanical properties. To date, no successful example has been reported of spinning recombinant or reconstituted silk proteins that emulate the full range of novel mechanical properties of silkworm or spider silk fibers.

Mimetic Systems — Many aspects of the silk spinning process can be mimicked *in vitro* and the all-aqueous environment used is instructive as a model for polymer processing, in general. Employing new insights into silk fibroin solubility and assembly has permitted new forms of these protein-based materials to be generated. For example, porous 3-D sponges have been formed from regenerated silk fibroin (Nazarov et al., 2004). Blending silk fibroin and polyethylene oxide to obtain sufficient solution viscosity suitable for electrospinning to generate nanoscale-diameter fibers has also been reported (Jin et al., 2002). The insights into silk protein assembly have been utilized toward the formation of patterned peptide multilayer thin films with nanoscale order (Valluzzi et al., 2003). This engineered liquid crystallinity to form smectic layers was generated with thin films and bulk materials, with tunability of the layer thickness and patterning based on sequence design and chain length. In addition to recapitulation of the silk assembly process *in vitro* as outlined above based on new insight into the process of silk protein–protein interactions, efforts are underway in many laboratories to develop synthetic analogs of silk. For example, copolymers of glycine–alanine repeats (hard segments, β-sheet formation) with polyethylene glycol (soft segments) have been studied to provide fundamental insight into control of polymer assembly and structure (Rathore and Sogah, 2001).

With continuing advances in genetic engineering, improved quantities and the availability of additional sequence variants of silk proteins can be anticipated. These materials, combined with new understanding of aqueous processing of these polymers, should help with continued improvements in the ability to generate silk-based materials for a wide range of potential utility. At the same time, novel polymer mimics of these systems in which the key design rules are considered will continue to emerge. Finally, with the wealth of silk protein sequence–structure–function data being generated from the study of different silks, this family of unusual protein polymers can serve as a blueprint for future designs and engineering to develop synthetic analogs using more traditional synthetic polymer approaches.

14.2.2 Seashells — High Performance Organic–Inorganic Composites from Nature

Background — The shells of mollusks exhibit exceptional toughness, despite compositions of predominately brittle inorganic salts. Calcium carbonate ($CaCO_3$), or "chalk" comprises up to 98% of the content of many of these shells, with the remaining few percent consisting of proteins and glycoproteins (Levi-Kalisman et al., 2001). The organic components create a molecular-level template for control of nucleation and crystal growth. The mechanical strength of these shells can be attributed to these molecular interfaces as well as the complex hierarchal structure consisting of ordered lamellae or layers (Song et al., 2003). The nacreous layer, the interior lining of seashells, exhibits three orders of magnitude increased mechanical strength when compared to pure aragonite crystals (Almqvist et al., 1999). Aragonite, calcite, and vaterite are polymorphs of these calcium salts with the same chemical composition ($CaCO_3$), but different crystal shape, size, and symmetry. In the nacreous layer, these inorganic "bricks" are polygonal aragonite tablets separated by a thin layer of soluble and insoluble proteins (predominately) that act as the "mortar." Soluble, aspartic acid-rich glycoproteins direct crystal formation by binding with calcium leading to the formation of aragonite crystals. Depending on the species, the organic matrix also consists of insoluble β-chitin and silk-like proteins that do not contribute to the mineralization process directly, but function by establishing a more rigid scaffold in part responsible for the superior fracture toughness and strength associated with these shells. In recent studies, some of these concepts were studied through the isolation of specific proteins or groups of proteins from native mollusks and utilized in studies of controlled mineralization, such as for the formation of biopearls (Zaremba et al., 1996). It is also worth noting that other organic–inorganic composites in nature, such as bone and tooth enamel, are similarly organized in terms of organic templates and molecular scale interactions, although the specific organic components are different. For example, collagens represent the bulk of the organic matrix in these composites and hydroxyapatite is the major inorganic component.

Mechanism — Unlike the mortar in a brick wall, the organic matrix of the nacreous layer in mollusk shells is flexible and contributes to the strength and toughness of the shell by absorbing and displacing stress applied to the aragonite tablets. Insoluble fibers bind to the aragonite tablets at the optimized inorganic interface, acting as a natural adhesive between the layers. In addition, the organic matrix acts to dissipate crack propagation (Pokroy and Zolotoyabko, 2003). The fracture toughness of these types of structures is directly related to the presence of specific proteins in the organic matrix with domains characteristic of elastic behavior based on the amino acid sequence chemistry. For example, studies of nacre with Atomic Force Microscopy illustrated stepwise unfolding of the associated proteins, reflective of this elastic behavior (Smith et al., 1999). Specific proteins have been isolated and ascribed with these features, such as Lustrin A, which contains cysteine- and porline-rich domains (Zhang et al., 2002). Other domains in this protein also appear to provide regions with direct interactions with the aragonite component of nacre (Wustman et al., 2002).

Biomimetics — To form relatively inexpensive materials that mimic the mechanical features of nacre, weak interfaces have been layered between sheets of ceramics. In this method each layer of ceramic is approximately 200 μm thick and is cut from a larger sheet made by treating silicon carbide powder with boron to create a pliable material. The silicon carbide layers are coated with graphite and subsequently pressed together and sintered at 2000°C. The resulting material has a fourfold increase in fracture toughness when compared to monolithic silicon carbide, requiring 100 times the amount of work to break the layered ceramic (Clegg et al., 1990). This ceramic material also offers increased heat resistance, which contributed to its successful testing as a combustion liner for gas turbine engines. Although this ceramic mimics nacre, it does not harness the strength and toughness associated with the nano-scale architecture found in biocomposites such as nacre, bone, and coral. The ability to manufacture materials that resemble nacre structure and properties has been approached using alternating layers of clay and polymer. Unlike the self-assembling components of nacre, this artificial nacre is prepared by physically applying sequential layers of negatively charged clay, montmorillonite, and positively charged polyelectrolytes, poly(diallydi-methylammonium) chloride. The high affinity between the two components induces a strong inorganic/organic interface. Under stress, these sacrificial bonds are broken to allow platelet movement, which results in displacement of force much like the organic matrix of the nacreous layer. Two hundred sequential clay and polymer layers resulted in the formation of a film with a thickness of 4.9 μm, thus each clay or polymer layer had an average thickness of 250 nm (Tang et al., 2003). This film exhibited similar mechanical properties to nacreous layers.

The self-assembly of the shell components is an attractive feature because this ensures highly specific spacing, alignment, and placement of material components at small length scales, a feature more difficult to attain by current synthetic fabrication methods. Molecular erector sets have been proposed to mimic this self-assembly process by employing cell-surface and phage display technologies, which can produce polypeptide sequences that specifically interact with an inorganic surface with high affinity (Sarikaya et al., 2003). The development of molecular erector sets is under study in a number of laboratories and uniformly thin layers over inorganic surfaces have been generated (Tamerler et al., 2003).

14.2.3 Shark Skin — Biological Approaches to Efficient Swimming Via Control of Fluid Dynamics

Background — Sharks are in the class Chondrichthyes, or cartilaginous fish that includes rays, skates, and others. The dermis is composed of collagen type I fibers organized in helices around the shark's body in alternating layers that form 50 to 70° angles with each layer between the pectoral and anal fins and 45 to 50° angles in the thin caudal peduncle just in front of the tail. The epidermis is covered with placoid scales called dermal denticles, which are like thousands of teeth embedded in the skin. Unlike the scales of fish, which in most species tend to be broad and flat, placoid scales of sharks are pointed with a basal plate, a pedicel, and a crown enclosure. The denticles vary among species and as sharks age, the number of scales increases. The scales are compared to teeth, given that each is covered by dentine and composed of enamel and have a pulp cavity. Denticles vary widely in size among species. For example, the nurse shark has denticles that are so large and so closely spaced that they can form a barrier against even harpoons. The morphologies vary including blunt, scalloped, spade-shaped, thorn-like, geometric, and heart-shaped. Occasionally, denticles develop independently and become comparatively gigantic structures as in the fin spine, a thorn-like quill, in the spiny dogfish and Port Jackson shark, or the tooth found in the sawfish and the stinger of stingrays.

Mechanism — The roughness of shark skin is paradoxical to principles of fluid dynamics since rough surfaces increase drag, and shark skin is considered rough due to the denticles. However, the rough texture of shark skin reduces drag due to the presence of microscopic riblets on the surface of the skin. Riblets channel the laminar flow over the skin to further reduce drag after the larger structures, the denticles, create a boundary against turbulent flow. The water is channeled through the small valleys created by the microscopic ridges, speeding up the flow of water over the surface of skin. Without the riblets and denticles, the water would flow over smooth skin and suffer the full effects of friction. The ridges on the denticles, like the ridge that runs longitudinally along the shark's body, help in drag reduction and in the smoothing boundary layer turbulence. The efficiency of shark skin and shark swimming in water originates in principles of fluid dynamics. Body geometries, movements, and wake evolution have been modeled (Cheng and Chahine, 2001). The structure and dynamic behavior of the vortex wakes generated by a swimming body are responsible for the highly efficiency propulsion and maneuverability.

Hydrostatic pressure under the skin of sharks varies with the swimming speed (Wakling, 2001). The stress in the skin varies with internal pressure and this stress controls skin stiffness. The inertial pressure on sharks increases tenfold between slow and fast swimming. The skin acts as an external tendon by transmitting muscular force and displacement to the tail. Hydrostatic pressures of 7 to 14 kN m^2 occur just under the skin when swimming slowly and with bending pressure vary between 20 and 35 kN m^2. During bursts of swimming, tighter bends generate pressures up to 200 kN m^2. To bend sharply as in fast swimming the muscles on one side shorten and increase in cross-sectional area, causing the fibers in the skin overlying the contracting muscles to increase their angle. The changes in fiber angle cause the skin to remain taut in and avoid wrinkling or loss of tension.

As a variation, some sharks have special arrangements of riblets that converge or diverge in a V pattern on the skin surrounding the shark's sensory organs. One set angles in toward the shark's pit organ and others angle away from the lateral-line organ. The function of the pit organ is unclear but the lateral-line organ functions similarly to the human ear (Koeltzsch et al., 2002). It is suspected that the diverging riblets draw water away from a shark's "ears" to prevent the noisy sound of rushing water, which would otherwise inhibit hearing. At the rostrum and on the leading edges of the fins, the skin is almost totally devoid of riblets. This arrangement promotes smooth water flow to each side. If the leading edges of the body and fins were covered by the same denticles as the majority of the body, a swimming shark would deflect the boundary layer away from the body and increase drag. The posterior edges of fins are flexible and denticle-free. This may help to reduce turbulence, saving energy lost to the vortices occurring immediately behind the fins (Bargar and Thorson, 1995). Three factors affect the drag reducing properties of riblets, sharp-edged riblets, riblet protrusion height as there is an optimal height for riblets to protrude into the boundary layer beyond which they would interfere with the flow of seawater, and the lateral spacing of the riblets to affect the dynamics of the water passing over the skin (Bechert et al., 1986).

Biomimetics — The structure of shark skin has prompted swimsuit and wetsuit manufacturers to develop new designs to reduce drag in water to improve times for competitive swimmers or to improve navigation by scuba divers. Properties of shark skin have also been used as models for movements of submersible and surface vessels in order to reduce the drag created by the speed of solid boat structures through water. Finally, aeronautics research has keyed into the structures of shark skin to reduce air resistance for planes. The human's body with smooth skin covered with hair creates a great deal of drag. Speedo, Inc. has developed a swimsuit for competitive swimmers based on shark skin designs. The Speedo Fastskin FSII suit reduces drag in water by as much as 4%. Passive drag affects a swimmer in the streamline position, usually after the initial dive into the water and following a turn. During a 50-m race, a swimmer is likely to be in the streamline position for up to 15 m. Swimmers from more than 130 countries wore this biomimetic suit at the Sydney Olympics and over 80% of the swimming medals and 13 out of the 15 world records set were with swimmers in this new suit. Computational fluid dynamics were used to design the swimsuit which directs water along grooves in the fabric, allowing the water to swirl in microscopic vortices, reducing drag. This control of fluid flow creates greater efficiency in movement and up to 3% improvement in overall speed. A similar design could be applicable to wetsuits to reduce transit time to great depths. Other applications include the design of highly efficient, fast, and maneuverable underwater craft, and options for pipes in water distribution systems. Lining a pipe with riblet-like grooves speeds flow by up to 10% (Koeltzsch et al., 2002).

Interest in these general features has also been seen in the aerospace industry for airplane design. In 1997, two Airbus Industry A30 planes were designed to test a specially ribbed plastic film that cuts aerodynamic drag when attached to aircraft surfaces and is expected to decrease fuel consumption by 1% (Ball, 1999). The riblets are barely perceptible to the touch, and they appear like a matte finish on the aircraft skin. Cathay Pacific and Lufthansa have already begun flying planes with small percentages of their surfaces covered in riblets to test durability.

14.2.4 Gecko and Burrs — Biological Solutions to Sticking to Surfaces

Background — Gecko lizards do not have little suction cups on their feet but are able to climb up walls and stick to ceilings. The feet of these animals have toe pads consisting of tiny hair-like structures called setae, made of keratin (Autumn et al., 2000, 2002). The setae are arranged in lamellar patterns and each seta has 400 to 1,000 microhair structures, called spatulae. These tiny structures allow geckos to climb vertical walls or across ceilings. Lizards can cling to hydrophilic or hydrophobic surfaces, although adhesion strength is related to the polarity of the substrate with the more polar the better (Autumn et al., 2002). Setae range from 30 to 130 μm, and there are 5,000 setae per mm^2, thus the total number of setae per gecko foot is greater than half a million (Autumn et al., 2000). The size of the spatulae that are attached to the setae ranges from 0.2 to 0.5 μm, distances in which molecular interactions can occur and accounting for van der Waals interactions (Autumn et al., 2002). The average adhesive force of a seta is ~194 ± 25 μN (Autumn et al., 2000). If the average lizard foot is 100 mm^2, the total adhesive force by a lizard is ~400 N. If a human hand were covered in setae, similar to a gecko lizard, the total adhesive force created from just human hands would be over 30,000 N (equivalent to 6,744 pound-force or 3,059 kg-force).

Mechanism — Geckos use van der Waal forces to adhere to different surfaces and do not use secreted sticky materials for this function (Autumn et al., 2002). With such strong adhesive forces the movement of a lizard could be problematic. However, to "unstick" their feet, lizards curl their toes. This movement breaks the van der Waal forces and allows movement across substrates. The critical angle needed to break the forces is $30.6° \pm 1.8$ (Autumn et al., 2000).

Biomimetics — New tapes, similar in design to lizard feet, have been designed and developed as a dry adhesive to eliminate the sticky residue normally left from more traditional tapes and glues. The adhesive sticks to most surfaces. Various microfibers have been synthesized from polyester, silicone, and polyimide materials. Their adhesive forces are 294 ± 21 nN per spatula, 181 ± 9 nN per spatula, and 70 nN per hair, respectively (Geim et al., 2003). These microfibers adhere in a similar fashion to natural keratinous setae, but after repeated use, the microfibers lose their stickiness due to bunching of the fibers (Geim et al., 2003). Further research studying the traits of chitin microfibers, similar to those found in insects, should help resolve this problem. This type of Gecko-like adhesive could eventually be used for a wide range of applications, from hanging tapestries and rock climbing to wound closure in surgery.

14.2.5 Muscles — Efficient Biological Conversion of Chemical Energy into Mechanical Energy

Background — Muscles are classified into (a) skeletal — often associated with bone, (b) cardiac — found in the heart, and (c) smooth — lining of hollow organs in the gastrointestinal, urogenital, and cardiovascular systems. Skeletal muscle functions in generating force and movement but also in the support and maintenance of body structure. Cardiac muscle pumps and regulates flow of blood throughout the body. Smooth muscle contractions function in general housekeeping activities, such as the movement of food through the gastrointestinal system. The scaling and interactions of proteins to form contractile muscle systems provide a classic example of the structural hierarchy in biological materials to achieve remarkable function. Skeletal muscle contains myofibrils which consist of a series of sarcomeres, the functional units of muscle contraction, composed of contractile proteins, actin, and myosin. The thick myosin and thin actin filaments are arranged in a repeating interlocking mesh structure, generating a striated appearance. In addition to contractile proteins, there are also energy stores and signaling mechanisms associated with the myofibrils.

Mechanism — The mechanism of muscle contraction is based on the crossbridge cycle (sliding filament model) and Lymn Taylor actomyosin ATPase hydrolysis kinetic scheme. Myosin consists of a globular head with ATP and actin binding sites and a long tail that is involved in polymerization into myosin filaments. Actin is polymerized into filaments and forms the "ladder" along which the myosin filament climb to generate motion. Contraction and force generation result from the shortening of the sarcomere as a result of thin actin filaments sliding over thick myosin filaments. The formation of crossbridges between the filaments is ATP dependent. In sequence, (a) ATP binds to myosin, which results in the dissociation of the myosin head, the crossbridge between the filaments, from the actin filament, (b) the free myosin bridge moves into position to reattach to the actin and ATP is hydrolyzed, (c) the free myosin bridge rebinds to the actin filament and the working stroke, and (d) force is generated and the products of ATP hydrolysis are released from the myosin crossbridge.

Skeletal muscles are attached to two or more bones by tendons. They work in antagonistic groups such as flexing or bending vs. extending or straightening. Antagonist groups can also contract together to stabilize joints or maintain body posture. Muscle arrangement is based on the orientation of the fibers relative to the axis of force generation. The physiological cross-sectional area (PCSA) of a muscle is the estimated sum of the cross-sectional areas of all the myofibers. The PCSA is proportional to muscle force generation, increased PCSA results in increased force. The extent and velocity of muscle contraction increases with myofiber length. Two other relationships used to describe muscle properties are based on length–tension and force–velocity. The length–tension relationship illustrates that the isometric tension generated in skeletal muscle is a function of the magnitude of overlap between actin and myosin filaments. The greatest force is generated at an intermediate "optimal" sarcomere length. Long lengths result in decreased myosin cross-bridges with actin filaments and short lengths create actin filament overlaps, which interfere with crossbridges. Passive tension increases as a muscle is stretched, providing a resistive force in the absence of muscle activation. The force–velocity relationship demonstrates that the maximum force generated by a muscle is a function of its velocity and is used to define the kinetic properties of the cross-bridge cycle during contraction.

Cardiac muscle has actin and myosin assemblies similar to that of skeletal muscle, but the muscle cells are mononucleated and arranged in a continuous network of branching and anastomosing cells. Cardiac muscles maintain constant, continuous contractions to pump blood throughout the body. Smooth muscles also contain actin and myosin filaments that produce contractions but not in a striated pattern. The cells are spindle shaped and mononucleated. Smooth muscle contractions are slower and can be sustained for longer periods compared with skeletal muscle. Smooth muscles are responsible for the contractability of hollow organs and have a role in various functions such as regulating flow through blood vessels. Smooth muscle fibers are organized into sheets that are arranged into two layers, which alternately contract. The fibers have a longitudinal arrangement in the outer layer and fibers of the inner layer are organized in a circular pattern. Unlike skeletal muscle, cardiac muscle and smooth muscle control is involuntary.

Biomimetics — Muscles are natural actuators. Actuators are controllable machines that convert energy to mechanical work by means of a shape or dimensional change. Braided pneumatic muscle actuators, also known as McKibben artificial muscles, were first developed in the 1950s specifically for use in artificial limbs. The device consists of an expandable fiber mesh wrapped around an inflatable bladder. When the bladder is filled with pressurized air, its volume increases. A muscle-like contraction is produced as the actuator expands radially while contracting along its axis due to the constant length of the mesh fibers. Pneumatic muscles produce a large force when shortened and can achieve greater power or weight ratios than natural muscle. They can operate in antagonistic modes like natural muscles and are capable of more natural motion and control. The system is lightweight, adaptable, and easily powered. Pneumatic actuators have high power efficiency and are capable of amplifiable force output. Precise control of pneumatic muscles can be difficult because of their nonlinear and time varying characteristics. Actuator properties can also vary with temperature and use. Another limitation is that current pneumatic actuators have a short fatigue life in the order of 10,000 cycles, but a new pneumatic device in which the fiber mesh is impregnated inside the bladder has demonstrated a fatigue life of 10,000,000 cycles. Force output is dependent on the thickness of the actuator and the speed of the McKibbean muscle is slow relative to natural muscle response.

Other examples of mimetic structures include shape memory alloys, such as Ni–Ti (Nitinol). These are a class of metallic materials capable of returning to a predetermined size and shape when subjected to a thermomechanical load. Macroscopic deformations of the material are caused by microscopic crystalline structure changes between the hard, high temperature austenite phase and soft, low temperature martensite phase. Shape memory alloys exhibit typical thermomechanical properties of psuedoelasticity and shape memory effects. Nitinol wires have been used as artificial muscles for robotic and prosthetic applications (De Laurentis and Mavroidis, 2002). Actuation occurs by a voltage drop across the wire causing current flow through the material, which results in heating that causes a crystalline transformation and accompanying shape change such as bending. The advantages of using shape memory alloys are ease of actuation, small size and weight, noiseless operation, and low cost. Shape memory alloys have high strength to area ratios and are capable of high grasping strength. Disadvantages include slow response, high power consumption, dependence of attainable motion and force on wire length and thickness, and heat generation. Actuation effects are also nonlinear and have short life cycles.

Electroactive polymers often exhibit the closest performance resemblance to biological muscle. They are considered to be electrically hard but mechanically soft materials. There are many types of actuators which fall into this category with varying characteristics and properties. Ionic polymer–metal composites are capable of large deformations in response to low applied voltage (Jung et al., 2003). These structures are composed of an ion exchange polymer such as a perfluorinated ion membrane chemically deposited with a metal such as gold or platinum. The metal ions are dispersed throughout the hydrophilic regions of the polymer. Bending is under the control of diffusion and Coulomb forces, occurring as a result of mobile ions migrating within the polymer network due to the application of an electric field. These structures do not function in a dry environment, necessitating a protective coating to retain moisture for function. Electrolysis occurs at high voltages, which result in degradation of the material and the release of heat and gases, restricting allowable voltage. Force generation is proportional to the thickness of the material. Ionic polymer–metal composites have a fast response time, are tough, and can achieve large actuation strain. They are light, compact, soft, and can be miniaturized. Conducting polymers such as polypyrrole have been evaluated as artificial muscles (Otero et al., 1996). Volume changes are induced by the movement of charge compensating ions to or from the polymer layers during oxidation and reduction reactions. During oxidation electrons are extracted from the polymer chains, resulting in the rearrangement of bonds and storage of positive charges. Conformational movements of the chains in order to maintain electroneutrality generate free volume, which are occupied by counterions and water molecules from the solution, resulting in swelling of the film. During reduction, these series of events are reversed, promoting shrinkage of the film. A polymeric triple layer muscle made of polypyrrole has demonstrated smooth, uniform movement comparable to that of natural muscle. In addition to reproducible actuation, the rate and direction of movement can be controlled by applied current conditions. Polymer actuators exhibit large strain and high strength, are lightweight, and can operate at room or physiological temperatures. Actuation of conducting polymers can be achieved at low voltages. Development of conjugated polymer actuators containing a polymer electrolyte, which functions as the electron source and sink, between polypyrrole layers has allowed for "dry" actuator function outside of an aqueous solution.

Carbon single-walled nanotube sheets exhibit properties such as mechanical flexibility, high toughness, and electrochemical behavior that have potential for development as artificial muscles (Baughman et al., 1999). Unlike conducting polymers in which actuation is based on electrochemical dopant intercalation, carbon nanotube actuation results from quantum chemical and double-layer electrostatic effects. Dimensional changes occur in covalently bonded directions caused by a double charge injection. Applied voltage change injects an electronic charge into the carbon nanotube sheet that is compensated at the nanotube electrolyte interface by electrolyte ions, the double-layer. Carbon nanotube (CNT) actuator sheets are nanoscale actuators that are organized into nanotube arrays in a manner similar to natural muscle. CNTs exhibit high work density and good work capacity per cycle. Compared to natural muscle, CNT-actuators are stronger and more durable. Activation only requires low voltages and the material is capable of operation at high temperatures. The material is light, shows fast response, has a long life cycle, and is capable of large displacements. Like most electroactive polymers, actuation requires an electrolyte solution.

Other examples include vanadium oxides which can be used in various redox-dependent applications such as the conversion of electrical energy into mechanical energy (Gu et al., 2003). Also, hydrogels are an intermediate between a liquid and solid state, consisting of a polymer network and interstitial fluid. Gel actuators made of materials such as poly(vinyl alcohol), poly-acrylonitrile, polyacid acrylic, and polyacrylamide can undergo large volume changes in response to stimuli such as pH, temperature, or electricity. Bending action can be induced by controlled swelling. Conformational changes are dependent on the diffusion of solvent through the gel and as a result of the amorphous nature of many gels, response time can be slow. The size of the gel and distance for fluid flow are other factors that can affect response time. Hydrogels also tend to lack mechanical strength or are unable to hold a load. A greater degree of cross-linking would improve response time and mechanical strength but result in reducing the extent of deformation. Development of copolymer gels made of cross-linked hydrogels have resulted in actuators with improved response time and good mechanical properties (Galaev and Mattiason, 1999). Cross-linked hydrogel actuators undergo linear contraction and expansion without volume change, and show faster response time and improved mechanical strength compared to the individual components (Calvert and Liu, 1999). Gels require an aqueous environment for operation, so a sealed environment or coating is required for "dry" operation. Hydrogel actuators are light, compact, and flexible. A natural polymer hydrogel made of a semi-interpenetrating polymer network (semiIPN) composed of glutaraldehyde cross-linked chitosan and interpenetrating silk fibroin has demonstrated reversible swelling–shrinking behavior and may have applications as an artificial muscle (Chen et al., 1997).

An emerging area of research has been the development of "molecular actuators," molecules, or molecular assemblies that undergo conformational change in response to stimuli and perform work. Rotaxane is a molecular system that consists of a ring threaded on a rod-like structure with blocking elements at the ends. The ring shuttles along the rod between two points and can be driven by chemical, photochemical, or electrochemical forces. A multicomponent rotaxane system consisting of two string and ring units threaded together that contracts and stretches in response to a metal exchange reaction has been developed (Collin et al., 2001). The string contains both bidentate and terdentate ligands and the ring contains a bidentate ligand. This synthetic molecular muscle unit adopts a stretched conformation in the presence of copper ions, which prefer to bind to two bidentate ligands. Zinc ions prefer binding to a bidentate and terdentate ligand. When zinc ions replace the copper, the rings slide along the string from the bidentate ligand to the terdentate ligand, resulting in a contracted conformation. This synthetic molecule replicates the sliding motion of natural muscle actin and myosin filaments.

In addition to the synthetic replication of molecular motors, there has been research into developing actuators using natural proteins. An artificial muscle from real muscle components has been synthesized *in vitro* (Kakugo et al., 2002). Isolated myosin molecules cross-linked under stretching showed self-organization capabilities and orientated to form hierarchical structures. ATPase activity comparable to native myosin was seen in the presence of actin and addition of ATP resulted in the motion of F-actin along the axis of oriented myosin gel. Actin gels formed from cross-linked actin–polymer complexes also showed preferential motility on the orientated myosin gel in the presence of ATP. NonATP based molecular motors have been isolated from sieve elements of legumes (Knoblauch et al., 2003). The crystalloid protein bodies, dubbed forisomes, are part of the microfluidic system for transport of water and minerals throughout the plant. Forisomes have a disordered (extended) shape in the absence of calcium ions. In the presence of calcium ions forisomes take on an ordered (swollen) conformation, acting as a cellular stopcock to block fluid flow. Isolated forisomes were observed to swell radially and contract longitudinally in the presence of calcium ions. This anisotropic deformation response was reversible in the absence of calcium ions and multiple expansion–contraction cycles were induced without causing a decrease in responsiveness.

Tissue engineering is a means of creating a biological substitute that is capable of restoring, maintaining, and improving function. Smooth muscle tissue has been successfully engineered *in vitro* on tubular scaffolds of poly(lactic-glycolic acid) seeded with bone marrow derived mesenchymal stem cells (Cho et al., 2004). The differentiated cells exhibited smooth muscle-like morphology and expressed smooth muscle cell specific markers, SM a-actin and SM myosin heavy chain. Bone marrow cells also have the capacity to differentiate into cardiac tissue both *in vivo* and *in vitro*. Stem cells injected into the myocardium develop a cardiomyogenic phenotype and BMSC transplant experiments have been shown to be effective in treating infarcted myocardium by generating de novo myocardium (Orlic et al., 2001). Differentiation of stem cells treated with 5-azacytidine, a cytosine analog, that regulates differentiation into cardiomyocytes has also been demonstrated *in vitro* (Makino et al., 1999). The cultured cells beat spontaneously, expressed cardiomyocyte-specific genes, and exhibited electrophysiological characteristics similar to *in vivo* cardiomyocytes. Another progenitor cell that exhibits developmental plasticity is the hematopoietic stem cell. Transplanted hematopoietic stem cells have been shown to be involved in skeletal muscle repair and regeneration (Camargo et al., 2003). However, it is still unclear whether the hematopoietic stem cells switches to a myogenic cell fate in response to microenvironmental cues as in the case of bone marrow derived mesenchymal stem cells or the generation of myogenic cells result from the fusion of hemopoietic stem cells with the muscle myofibers. In the later case, it is believed muscle nuclear factors may play a role in activating a myogenic program in the fused hemopoietic stem cells. A variety of specialized bioreactors have been used to optimize tissue outcomes (Kim et al., 1998; Carrier et al., 1999; Radisic et al., 2004).

14.3 CONCLUSIONS

The examples provided from Nature illustrate the diverse and functional (high performance) materials that are available as blueprints for exploitation in the broader field of materials science and engineering. The gap between synthetic and natural polymers in terms of diverse yet controlled sequence chemistry, coupled with control of regioselective and stereoselective chemical features, suggests that a significant hurdle will remain for some time for synthetic systems to fully emulate the novel features of natural materials. As control of synthetic processes for polymers continues to improve, and as models from biology continue to be understood via reverse engineering, more crossover among these systems will be realized. The marriage of both biological and synthetic approaches may provide a useful bridge toward the future such that new materials, new processing paradigms, and new assembly controls can be studied and technologically exploited.

ACKNOWLEDGMENTS

We thank various agencies for funding various background aspects used in this chapter, including the NIH, NSF and NASA. We also wish to acknowledge the opportunity to organize this chapter as a class project — Biotechnology Engineering Seminar. We also greatly appreciate the input and comments from the reviewers that helped the class improve the focus and refine the details.

REFERENCES

Almqvist N, Thomson NH, Smith BL, Stucky GD, Morse DE, and Hansma PK, Methods for fabricating and characterizing a new generation of biomimetic materials. *Materials Science and Engineering*, C7: 37–43, 1999.

Autumn K, Liang YA, Hsieh ST, Zesch W, Chan WP, Kenny TW, Fearing R, and Full RJ, Adhesive force of a single gecko foot-hair. *Nature*, 405: 681–685, 2000.

Autumn K, Sitti M, Liang YA, Peattie AM, Hansen WR, Sponberg S, Kenny TW, Fearing R, Israelachvili JN, and Full RJ, Evidence for van der Waals adhesion in gecko setae. *PNAS*, 99(19): September 17, 12252–12256, 2002.

Ball P, Engineering shark skin and other solutions, *Nature*, 400: 507–509, 1999.

Barger TW and Thorson TB, A scanning electron microscopy study of the dermal denticles of the bull shark, *Carcharhimus leucas. Journal of Aquariculture Acquatic Science*, 7: 120–137, 1995.

Baughman RH, Cui C, Zakhidov AA, Iqbal Z, Barisci JN, Spinks GM, Wallace GG, Mazzoldi A, De Rossi D, Rinzler AG, Jaschinski O, Roth S, and Kertesz M, Carbon nanotube actuators. *Science*, 24: 1340–1344, 1999.

Bechert DW, Bartenwerfer M, Hoppe G, and Reif WE, Drag reduction mechanisms derived from shark skin. Presented at the 15th ICAS Congress, 1986. London 86–1.8.3. Distributed as AIAA paper.

Bini E, Knight DP, and Kaplan DL, Mapping domain structures in silks from insects and spiders related to protein assembly. *Journal of Molecular Biology*, 335: 27–40, 2004.

Camargo FD, Green R, Capetenaki Y, Jackson KA, and Goodell MA, Single hematopoietic stem cells generate skeletal muscle though myeloid intermediates. *Nature Medecine*, 9: 1520–1527, 2003.

Carrier RL, Papadaki M, Rupnick M, Schoen FJ, Bursac N, Langer R, Freed LE, and Vunjak-Novakovic G, Cardiac tissue engineering: cell seeding, cultivation parameters, and tissue construct characterization. *Biotechnology and Bioengineering*, 64: 580–589, 1999.

Chen X, Li W, Zhong W, Lu Y, and Yu T, pH sensitivity of hydrogels based on complex-forming chitosan/silk fibroin interpenetrating polymer network. *Journal of Applied Polymer Science*, 65: 2257–2262, 1997.

Cheng JY and Chahine GL, Computational hydrodynamics of animal swimming: boundary element method and three-dimensional vortex wake structure. *Comparative Biochemistry and Physiology Part A*, 131 December: 51–60, 2001.

Cho SW, Kim IK, Lim SH, Kim DI, Kang SW, Kim SH, Kim YH, Lee EY, Choi CY, and Kim BS, Smooth muscle-like tissues engineered with bone marrow stromal cells. *Biomaterials*, 25: 2979–2986, 2004.

Clegg WJ, Kendall K, Alford NM, Button TW, and Birchall JD, A simple way to make tough ceramics. *Nature*, 347: 455–457, 1990.

Collin JP, Dietrich-Buchecker C, Gavina P, Jimenez-Molero MC, and Sauvage JP, Shuttles and muscles: linear molecular machines based on transition metals. *Accounts of Chemical Research*, 34: 47–87, 2001.

Cunniff PM, Fossey SA, Auerbach MA, Song JW, Kaplan DL, Adams WW, Eby RK, Mahoney D, and Vezie DL, Mechanical and thermal properties of dragline silk from the spider *Nephila clavipes. Polymers for Advanced Technologies*, 5: 401–410, 1994.

De Laurentis KJ and Mavroidis C, Mechanical design of a shape memory alloy actuated prosthetic hand, *Technology and Health Care*, 10: 91–106, 2002.

Ferrari, Andrea, and Antonella, *Sharks* 2002 Firefly Books.

Galaev IY and Mattiasson B, 'Smart' polymers and what they could do in biotechnology and medecine. *Trends in Biotechology*, 17: 335–340, 1999.

Geim AK, Dubonos SV, Grigorieva IV, Novoselov KS, Zhukov AA, and Shapoval SY, Microfabricated adhesive mimicking gecko foot-hair. *Nature Materials*, 2 July: 461–463, 2003.

Gosline JM, DeMont ME, and Denny MW, The structure and properties of spider silk. *Endeavour*, 10: 37–43, 1986.

Gu G, Schmid M, Chiu PW, Minett A, Fraysse J, Kim GT, Roth S, Kozlov M, and Munoz E, V_2O_5 nanofibre sheet actuators. *Nature Materials*, 2: 316–318, 2003.

Jin H-J, Fridrikh SV, Rutledge GC, and Kaplan DL, Electrospinning *Bombxy mori* silk with poly(ethylene oxide). *Biomacromolecules*, 3: 1233–1239, 2002.

Jin H-J and Kaplan DL, Mechanism of processing silk in insects and spiders. *Nature*, 424: 1057–1061, 2003.

Jung K, Nam J, and Choi H, Investigatios on actuation characteristics of IPMC Artificial muscle actuator. *Sensors and Actuators A*, 107: 183–192, 2003.

Kakugo A, Sugimoto S, Gong JP, and Osada Y, Gel machines constructed from chemically cross-linked actins and myosins. *Advanced Materials*, 12: 1124–1126, 2002.

Kaplan DL, Adams WW, Viney C, and Farmer B, Silks: materials science and biotechnology, *American Chemical Society Symposium Series*, 544: 1994.

Kim BS, Putnam AJ, Kulik TJ, and Mooney DJ, Optimizing seeding and culture methods to engineer smooth muscle tissue on biodegradable polymer matrices. *Biotechnology and Bioengineering*, 57: 46–54, 1998.

Knoblauch M, Noll GA, Muller T, Prufer D, Schneider-Huther I, Scharner D, Van Bel AJE, and Peter WS, ATP-independent contractile proteins from plants. *Nature Materials*, 2: 600–603, 2003.

Koeltzsch K, Dinkelacker A, and Grundmann R, Flow over convergent and divergent wall riblets. *Experiments in Fluids*, 33(2): 346–350, August 2002.

Levi-Kalisman Y, Falini G, Addadi L, and Weiner S, Structure of the nacreous organic matrix of a bivalve mollusk shell examined in the hydrate state using Cryo-TEM. *Journal of Structural Biology*, 135: 8–17, 2001.

Makino S, Fukuda K, Miyoshi S, Konishi F, Kodama H, Pan J, Sano M, Takahashi T, Hori S, Abe H, Hata J, Umezawa A, and Ogawa S, Cardiomyocytes can be generated from marrow stromal cells *in vitro*. *Journal of Clinical Investigation*, 103: 697–705, 1999.

Marsh RE, Corey RB, and Pauling L, The crystal structure of silk fibroin. *Acta Crystallographica*, 8(1): 62–62, 1955.

Nazarov R, Jin H-J, and Kaplan DL, Porous 3-D scaffolds from regenerated silk fibroin. *Biomacromolecules*, 5: 718–726, 2004.

Orlic D, Kajstura J, Chimenti S, Jakoniuk I, Anderson SM, Li B, Pickel J, McKay R, Nadal-Ginard B, Bodine DM, Leri A, and Anversa P, Bone marrow cells regenerate infarcted myocardium. *Nature*, 410: 701–705, 2001.

Otero TF, Grande H, and Rodriguez J, Reversible electrochemical reactions in conducting polymers: a molecular approach to artificial muscles. *Journal of Physical Organic Chemistry*, 9: 381–386, 1996.

Pokroy B and Zolotoyabko E., Microstructure of a natural plywood-like ceramics: a study by high resolution electron microscopy and energy-variable x-ray diffraction. *Journal of Materials Chemistry*, 13(4): 682–688, 2003.

Radisic M, Yang L, Boublik J, Cohen RJ, Langer R, Freed LE, and Vunjak-Novakovic G, Medium perfusion enables engineering of compact and contractile cardiac tissue. *American Journal of Physiology: Heart Circulatory Physiology*, 286: H507–H516, 2004.

Rathore O and Sogah DY, Self-assembly of beta-sheets into nanostructures by poly(alanine) segments incorporated in multiblock copolymers inspired by spider silk. *JACS*, 123(22): 5231–5239, 2001.

Sarikaya M, Tamerler C, Jen AKY, Schulten K, and Baneyx F, Molecular biomimetics: nanotechnology through biology. *Nature Materials*, 2(9): 577–585, 2003.

Smith BL, Schaffer TE, Via M, Thompson JB, Frederick NA, Kindt J, Belcher A, Stucky GD, Morse DE, and Hansma PK, Molecular mechanistic origin of the toughness of natural adhesives, fibres and composites, *Nature*, 399: 761–763, 1999.

Song F, Soh AK, and Bai YL, Structural and mechanical properties of the organic matrix layers of nacre. *Biomaterials*, 24: 3623–3631, 2003.

Tamerler C, Dincer S, Heidel D, Zareie MH, and Sarikaya M, Biomimetic multifunctional molecular coatings using engineered proteins. *Progress in Organic Coatings*, 47: 267–274, 2003.

Tang ZY, Kotov NA, Magonov S, and Ozturk B, Nanostructured artificial nacre. *Nature Materials*, 2: 413–419, 2003.

Valluzzi R, Probst W, Jacksch H, Zellmann E, and Kaplan DL, Patterned peptide multilayer thin films with nanoscale order through engineered liquid crystallinity. *Soft Materials*, 1: 245–262, 2003.

Vollrath F and Knight DP, Liquid crystalline spinning of spider silk. *Nature*, 410: 541–548, 2001.

Wakling JM, Biomechanics of fast-start swimming in fish. *Comparative Biochemistry and Physiology*, 131: 31–40, 2001.

Wong C and Kaplan DL, Genetic engineering of fibrous proteins: spider dragline silk and collagen. *Advances of Drug Delievery Reviews*, 54: 1131–1143, 2002.

Wustman BA, Morese DE, and Evans JS, Structural analyses of polyelectrolyte sequence domains with the adhesive elsastomeric biomeralization protein Lustrin A. *Langmuir*, 18: 9901–9906, 2002.

Zaremba CM, Belcher AM, Fritz M, et al. Critical transitions in the biofabrication of abalone shells and flat pearls. *Chemical Materials*, 8(3): 679–690, 1996.

Zhang B, Wustman BA, Morse D, and Evans JS, Model peptide studies of sequence regions in the elastomeric biomeralization protein, Lustrin A.I. The C-domain consensus–PG-, -NVNC, 2002.

15

Functional Surfaces in Biology: Mechanisms and Applications

Stanislav N. Gorb

CONTENTS

15.1   INTRODUCTION: FUNCTIONS OF BIOLOGICAL SURFACES

Biological surfaces represent the interface between living organisms and the environment and serve many different functions: (1) They delimit dimensions, often give shape to the organism, and provide mechanical stability to the body. (2) They are barriers against dry, wet, cold, or hot environments. (3) They take part in respiration and in the transport of diverse secretions, and serve as a chemical reservoir for the storage of metabolic waste products. (4) A variety of specialized surface structures are parts of mechano- and chemoreceptors. (5) The coloration and chemical components of surfaces are important components for thermoregulation, and are often involved in diverse communication systems. (6) A number of specialized surface structures may serve a variety of other functions, such as air retention, food grinding, body cleaning, etc.

There are numerous publications describing biological surfaces by the use of light and electron microscopy. Because of the structural and chemical complexity of biological surfaces, exact working mechanisms have been clarified only for a few systems. However, biological surfaces hide a virtually endless potential of technological ideas for the development of new materials and systems. Because of the broad diversity of functions, inspirations from biological surfaces may be interesting for a broad range of topics in engineering sciences: adhesion, friction, wear, lubrication, filtering, sensorics, wetting phenomena, self-cleaning, antifouling, thermoregulation, optics, and so on. Since all biological surfaces are multifunctional, it makes them even more interesting from the

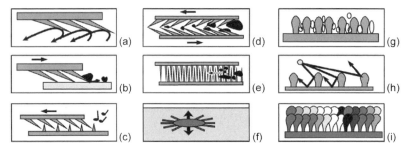

Figure 15.1   Diagram of functions of cuticular microstructures (a) aerodynamically active surfaces, (b) grooming, (c) sound generation, (d) food grinding, (e) filtration devices, (f) hydrodynamically active surfaces, (g) air retention, (h) thermoregulation, and (i) body coloration pattern. (With permission of Springer Science + Business Media B.V.)

point of view of biomimetics. It is also important to mention here their rather special properties as living structures, such as growth without interruption of function, ability to adjust to changing environment, and ability of self-repair. These functions are still unavailable to engineers using nonliving materials, but clearly represent a challenge for the future developments.

In the present chapter, we discuss some functions of biological surfaces (Figure 15.1) that are potentially interesting for biomimetics, and demonstrate several examples of materials and systems that were developed based on inspirations from biology.

15.2 SURFACES OF JOINTS AND SKIN: ANTIFRICTION AND DRAG REDUCTION

One of the challenges in designing moving parts of microelectromechanical systems (MEMS) is fabrication of joints allowing precise motion of parts about one rotational axis or multiple axes. One problem is the high friction, stiction, and wear rate of joints (Scherge et al., 1999; Komvopoulos, 2003). Wear of the interacting surfaces is a consequence of friction, affecting the material's contact points by becoming deformed or being torn away. Friction and wear are strongly correlated processes by which the points of the surfaces in contact change their topography continuously. Capillary adhesion, due to the presence of a water layer in contact, can account for a great part of the measured friction or lead to the stiction between a contact pair (Scherge et al., 1999). These are critical issues limiting the operational lifetime and negatively influencing the technological potential of MEMS. Conventional methods of lubrication cannot always be used, especially in devices with medical applications. Friction reduction in some man-made mechanical systems is based on the different hardnesses of elements in contact (Miyoshi, 2001; Li et al., 2004), the use of hydrophobized surfaces, and the application of surface texture, which minimizes the real contact area between two solid surfaces. Research on optimization of surface texture has been done using a "trial and error" approach (Scherge and Schaefer, 1998; Etsion, 2004). However, ideas from the studies of surface properties of biological micro-joints might represent a shortcut towards a solution.

The biological world is part of the physical world, and therefore, the rules of mechanics also apply to living systems. Living creatures move on land, in air, and in water. Complex motions inside their bodies provide fluid circulation or generate forces for locomotion. The resistance against motion mediated by surrounding media and by the mechanical contact with various substrates was an evolutionary factor, which contributed to the appearance of many surfaces adapted to reduce such resistance. But, one always needs friction to generate force to move on a substrate or to overcome the drag caused by friction elsewhere. A living motion system becomes optimized when it is capable of minimizing friction at one end of the system while maximizing it at the other end (Radhakrishnan, 1998). In other words, a living device needs a combination of maximum friction required for acceleration, deceleration, and maneuvering combined with minimum friction in joints for economic energy expenditure. Adhesion phenomena can also contribute to the functionality of such a system.

Vertebrate bones, that are joined with each other, are covered by cartilage, which is the gliding surface of the joint. The coefficient of friction is very low (0.0026) (Fung, 1981). Cartilage is a fibrous composite material of collagen fibers embedded in a highly hydrated proteoglycan gel (Buckwalter, 1983; Aspden, 1994). The so-called white fibro-cartilage is responsible for joint mobility. It provides lubrication of surfaces in contact (Ateshian, 1997) and serves as a kind of damper under dynamic loads. Four theories explaining the cartilage lubrication mechanism have been previously reviewed. These are fluid transport theory, lubrication layer theory, roller-bearing theory, and cartilage material theory (Fung, 1981; Scherge and Gorb, 2001). In insect joints, which work under lower loading forces, but much higher frequencies than vertebrate joints (Wootton and Newman, 1979; Gronenberg, 1996), the joint surfaces are usually smooth or present a combination of wavy and smooth counterparts (Figure 15.2). Underlying tissues are penetrated with canals, which are presumably responsible for delivering lubricants in the contact area. The specialized surface structures in the insect joints have been shown to confer friction-reducing properties in certain insect surfaces (Perez Goodwyn and Gorb, 2004). The next step is to transfer the structural and functional solutions found in biological joints to industrial systems.

Evolutionary processes have adapted swimming and flying organisms to interact efficiently with the surrounding medium. Reduction of drag due to friction in the boundary-layer close to the body surface is one of these adaptations. Skin secretion (mucus), compliant material of skin, scales, riblets and the degree of roughness may influence the flow velocity gradient, the type of flow, and the thickness of the boundary-layer around animals, and may seriously affect their drag in a positive or negative way. Boundary-layer damping results from a combination of elastic and viscoelastic structures in the skin of some animals. Dolphin skin has a very special design (Nachtigall, 1977). It is very smooth and relatively soft. Under the pressure of microturbulence, the rubber-like outer

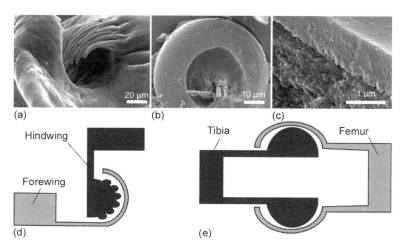

Figure 15.2   Examples of micro-joints in insects. (a) Lateral view of the wing double wave locking mechanism in the bug *Coreus marginatus* (forewing part). This joint provides interlocking between both wings on the same side of the body in the anterior direction allowing them to slide in the medial and lateral directions. (From Perez Goodwyn, P.J. and S.N. Gorb (2004) *J. Comp. Physiol. A* 190: 575–580. With permission of Springer Verlag.) (b) Medial aspect of the femoro-tibial joint (femoral part) of the leg in the beetle *Melolontha melolontha*. (c) Fracture of the material of the joint in the beetle *M. melolontha*. (d) Diagram of the wing locking mechanism shown in (a). (e) Diagram of the femoro-tibial joint shown in (b) and (c). Constructional principles and mechanical principles found in such joints can be used to design joints in technical actuators (a)–(c).

layer deforms and changes the local shape of the skin surface due to a shift of the damping fluid located under the rubber-like layer. Such a material design seems to be able to damp turbulences.

Many fish have developed another mechanism to reduce friction (by up to 60% in some species) in the boundary-layer. They produce skin secretions, which are usually slightly soluble in water. In areas, where microturbulence is strong, these substances can be locally dissolved. This results in the damping of microturbulence (Nachtigall, 1977). Due to skin secretions, fish can reach an extremely high speed in a short time.

In aquatic vertebrates, skin, specialized for increased friction, often contains patterns with microridges and micro-outgrowths (Fahrenbach and Knutson, 1975). Friction in the boundary-layer of the body moving in the medium at high Reynolds numbers may be decreased due to such a sculpturing of the surface. The grooved scales of the shark skin is an example of such a system. Their size ranges from 200 to 500 μm. The surface of each scale contains parallel grooves between so-called riblets directed almost parallel to the longitudinal body axis. Interestingly, grooves and riblets of neighboring scales correspond exactly to each other so that the shark surface looks like a pattern of parallel stripes (Reif and Dinkelacker, 1982). Experiments on flow resistance have been carried out with smooth-bodied models and with those covered with grooves of dimensions similar to original shark skin. The flow resistance measured in the grooved model was about 5 to 10% lower than the resistance in the smooth model at a Reynolds number of 1.5×10^6. The geometry of the grooves and riblets can also influence results. Small bristles, scales, and microtrichia of the wings of flying insects (Bocharova-Messner and Dmitriev, 1984; D'Andrea and Carfi, 1988) have similar function (Figure 15.3). The microturbulences, generated around such structures in flight, presumably build a kind of a lubricating layer of air between an air stream and the insect surface. This can possibly decrease friction during high-speed flight. A foil covered by tiny riblet-like structures, inspired by biological surfaces, has been suggested for aeroplane surfaces (Bechert et al., 2000).

Terrestrial animals, such as snakes, must overcome problems related to friction in contact with solid or friable media. Friction-modifying nanostructures of the scaly surface of snakes have recently been described (Hazel et al., 1999). These include an ordered microstructure array (Figure 15.4), presumably to achieve adaptable friction characteristics. Significant reduction of adhesive forces in the contact areas caused by the double-ridge microfibrillar geometry provides ideal conditions for sliding in a forward direction with minimum adhesive forces. Low surface adhesion in these local contact points may reduce local wear and skin contamination by environmental debris. The highly asymmetric profile of the microfibrillar ending with a radius of curvature of 20 to 40 nm may induce friction anisotropy along the longitudinal body axis and functions as a kind of stopper for backward motion, while providing low friction for forward motion. Additionally, the system of micropores penetrating the snakeskin may serve as a delivery system for a lubrication or anti-adhesive lipid mixture that provides boundary lubrication of the skin.

15.3 ATTACHMENT SYSTEMS

Materials and systems preventing the separation of two surfaces may be defined as adhesives. There are a variety of natural attachment devices based on entirely mechanical principles, while others additionally rely on the chemistry of polymers and colloids (Gorb, 2001; Scherge and Gorb, 2001; Habenicht, 2002). There are at least three reasons for using adhesives: (1) they join dissimilar materials; (2) they show improved stress distribution in the joint; and (3) they increase design flexibility (Waite, 1983). These reasons are relevant both to the evolution of natural attachment systems and to the design of man-made joining materials.

In general, adhesive-bond formation consists of two phases: contact formation and generation of intrinsic adhesion forces across the joint (Naldrett, 1992). The action of the adhesive can be supported by mechanical interlocking between irregularities of the surfaces in contact. Increased surface roughness usually results in an increased strength of the adhesive joint, due to the increased

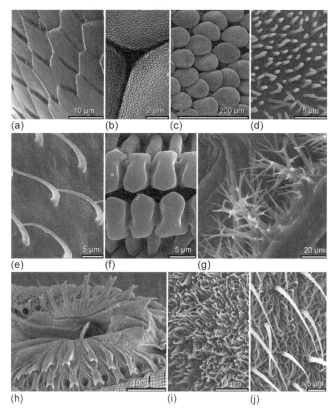

Figure 15.3 Functional diversity of noninnervated cuticular protuberances in insects. (a) Unspecialized polygonal surface on the tarsus of the scarabaeid beetle, *Melolontha melolontha*, (b) Ommatidia surface in the calliphorid fly, *Calliphora vicina*, (c) scales on the dorsal surface of the elytron in the scarabaeid beetle, *Hoplia* sp., (d) the surface of a single scale in the scarabaeid beetle, *Hoplia* sp., (e) wing surface in the bibionid dipteran, *Bibio ferruginatus*, (f) pseudotracheae of the labellum in the *C. vicina*, (g) filter system of the spiracle in the tenebrionid beetle, *Tenebrio molitor*, (h) prestomal teeth in the *C. vicina*, (i) plastron in the nepid bug, *Ranatra linearis*, and (j) antiwetting surface in the water-strider, *Gerris lacustris*. (With permission of Springer Science + Business Media B.V.)

contact area between the contacting surfaces and the adhesive substance. Strong adhesion is also possible between two ideally smooth surfaces. If sufficient contact between the substrate and adhesive interface is reached, forces will be set up between atoms and molecules of both contacting materials, and they will adhere. Van der Waals forces are the most common of such forces, together with the hydrogen bond. Electrostatic forces may be also involved. However, for typical biological adhesion, their contribution is not significant.

Adhesive organs, which may be used for attachment to substrates as well as being involved in catching prey, demonstrate a huge diversity among living organisms due to their structural and chemical properties. Biological adhesion underlies the organization of all living tissues. Cell contact phenomena have been extensively reviewed in the biomedical and biophysical literature (Weiss, 1970; Steinberg, 1996; Strange, 1997). The function of attachment appeared very early in evolution; early unicellular organisms have a variety of cellular adaptations for adhesion. Many multicellular organisms often bear adhesive organs composed of single cells. In many cases, however, cells are specialized into glands, which may be composed of several cell types.

Attachment devices are functional systems, the purpose of which is either temporary or permanent, attachment of an organism to the substrate surface of another organism or temporary interconnection of body parts within an organism. Their design varies enormously and is subject to

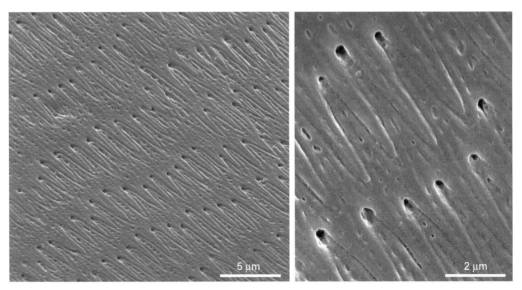

Figure 15.4   SEM micrographs of the ventral skin surface of the rattlesnake.

different functional loads (Nachtigall, 1974; Gorb, 2001). There is no doubt that many functional solutions have evolved independently in different lineages. Many species of animals and plants are supplied with diverse attachment devices, the morphology depending on the species biology and the particular function in which the attachment device is involved. Evolutionary background and behaviour influence the specific composition of attachment systems in each particular species. There are eight fundamental classes of attachment principles: (1) hooks, (2) lock or snap, (3) clamp, (4) spacer, (5) suction, (6) expansion anchor, (7) adhesive secretions (glue), and (8) friction (Gorb, 2001). However, different combinations of these principles also occur in existing attachment structures. Three types of adhesion at the organism level are known: (1) temporary adhesion allowing an organism to attach strongly to the substrate and detach quickly when necessary (see below the subsection about locomotory attachment devices); (2) transitory adhesion permitting simultaneous attachment and movement along the substrate; (3) permanent adhesion involving the secretion of a cement. These three types of adhesion do not have the same purpose and use different adhesive systems.

The industry of adhesives presently follows three main goals (Hennemann, 2000): (1) an increase in the reliability of glued contact; (2) mimicking of natural, environment-friendly glues; (3) development of mechanisms for application of a minute amount of glue to the surface. An additional challenge is the use of substances or mechanisms which allow multiple attachment and detachment, and enable attachment to a variety of surfaces.

Many biological attachment devices correspond to some of these requirements. One such example is the hairy surface of the leg pulvillus in flies. This system uses a secretion enabling hairs to attach and detach to diverse substrata very quickly. The hair design includes a mechanism that delivers the secretion, in extremely small amounts, directly to the contact area, and only then when contact to the substrate is achieved (Figure 15.5).

Walking machines usually use suckers to hold onto vertical and overhanging surfaces. A primary disadvantage of this attachment principle is that a very smooth substrate surface is required. The future goal should be to make robots that are able to walk on a variety of surfaces. Insects can walk rather well on smooth and structured substrata, on inclines, vertical surfaces, and some of them even on the ceiling.

In their evolution, animals have developed two distinctly different mechanisms to attach themselves to a variety of substrates: with smooth pads or with setose, or hairy surfaces. Due to

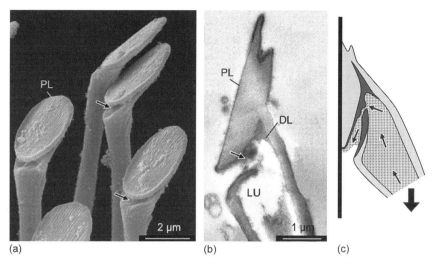

Figure 15.5 Dispensing system of the tenent seta in the syrphid fly *Episyrphus balteatus*. (From Gorb, S.N. (2001) *Attachment Devices of Insect Cuticle*. Dordrecht, Boston, London: Kluwer Academic Publishers. With permission of Springer Science + Business Media B.V.) (a, b) SEM (a) and TEM (b) micrographs of the tenent setae, (c) diagram of position of the seta on the substratum. Dotted area indicates lipid-containing secretion. Small arrows indicate the route of secretion release. Large arrow indicates direction of pulling force. DL, dense layer; LU, lumen; PL, end plate.

the flexibility of the material of the attachment structures, both mechanisms can maximize the possible contact area with the substrate, regardless of their microsculpture (Figure 15.6a,b,e,f). Tenent setae are relatively soft structures (Figure 15.6c,d). In *Calliphora* flies, their tips are usually compressed, widened, and bent at an angle of 60° to the hair shaft (Bauchhenss and Renner, 1977). When walking on smooth surfaces, these hairs in flies and beetles produce a secretion, which is essential for attachment (Ishii, 1987; Gorb, 1998).

Different forces may contribute to the resulting attachment force: capillary adhesion and intermolecular van der Waals forces. Geckos, which possess hairy attachment system, do not produce any secretory fluid in contact area. Different authors have carried out force measurements of gecko attachment system, at the global and local scales, and found evidences for contribution of both van der Waals and capillary forces, generated by the layer of absorbed water, to the overall adhesion (Hiller, 1968; Autumn et al., 2000; Autumn and Peattie, 2002; Huber et al., 2005). The action of intermolecular forces is possible only at very close contact between surfaces. The forces increase, when the contacting surfaces slide against each other. This may explain, why flies placed on a smooth undersurface always move their legs in a lateral–medial direction (Wigglesworth, 1987; Niederegger and Gorb, 2003). During these movements, pulvilli slide over the surface obtaining optimal contact. A contribution of intermolecular interaction to the overall adhesion has been shown in experiments on the adherence of beetles (Stork, 1980) and beetle setae (Stork, 1983) on a glass surface. The presence of claws, decrease of air pressure, decrease of relative humidity, or electrostatic forces do not influence beetle attachment on the smooth substrata. In the beetle *Chrysolina polita* (Chrysomelidae), the resulting attachment force directly depends on the number of single hairs contacting the surface. Recently, the contribution of intermolecular interaction and capillary force has been demonstrated for the fly, *Calliphora vicina*, in a nanoscale experiment with the use of atomic force microscopy (Langer et al., 2004). Smooth systems are composed of cuticles of unusual design (Figure 15.6g,h). The key property of smooth attachment devices is deformability and the softness of the pad material. Viscoelastic properties have recently been demonstrated (Gorb et al., 2000; Jiao et al., 1999).

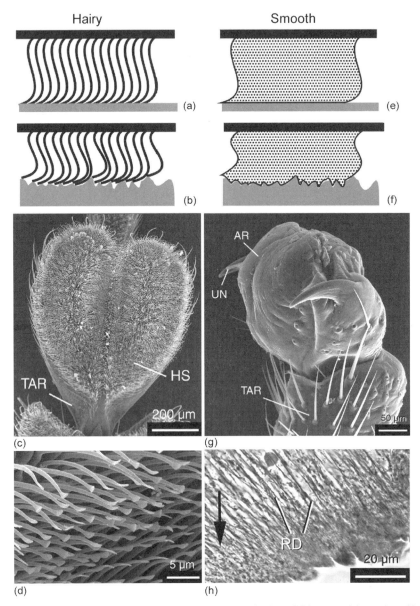

Figure 15.6   Hairy (a–d) and smooth (e–h) attachment systems. (a–b, e–f) Diagram of the action of both types of systems on the smooth (a, e), and structured (b, f) substrata. (From Gorb, S.N. (2001) *Attachment Devices of Insect Cuticle*. Dordrecht, Boston, London: Kluwer Academic Publishers. With permission of Springer Science + Business Media B.V.) Both systems are able to adapt to the surface profile. (c) SEM micrograph of the ventral surface of the tarsus in the beetle *Cantharis fusca* (Coleoptera, Polyphaga). (d) SEM micrograph of the ventral surface of the tarsus in the dobsonfly *Sialis lutaria* (Megaloptera). (g) SEM micrograph of the pretarsus of the cicada *Cercopis vulnerata* (Auchenorrhyncha). (h) Sagittal section of the arolium of the locust *Schistocerca gregaria* (Caelifera). AR, arolium; *black arrow*, ventral direction; HS, hairy soles; RD, filaments; TAR, tarsal segments; UN, claw.

Hairy and smooth leg attachment pads are promising candidates for biomimetics of robot soles adapted for locomotion. Similar principles can be applied to the design of microgripper mechanisms with an ability to adapt to a variety of surface profiles. Most recent data on hairy systems demonstrated their excellent adhesion and high reliability of contact. In contrast to smooth systems, some hairy systems seem to operate with dry adhesion, and do not require supplementary fluids in the contact area. Contacting surfaces in such devices are subdivided into patterns of micro or nanostructures with a high aspect ratio (setae, hairs, pins). The size of single points gets smaller and their density higher as the body mass increases (Scherge and Gorb, 2001). We have explained this general trend by applying the Johnson-Kendall, Roberts (JKR) contact theory, according to which splitting up the contact into finer subcontacts increases adhesion (Arzt et al., 2003). The fundamental importance of contact splitting for adhesion on smooth and rough substrata has been explained by a very small effective elastic modulus of the fibre array (Persson, 2003). Adhesion enhancement by division of the contact area has also been demonstrated experimentally (Peressadko and Gorb, 2004). A patterned surface, made out of polyvinylsiloxane (PVS), has significantly higher adhesion on a glass surface than a smooth sample made out of the same material (Figure 15.7). This effect is even more pronounced on curved substrata. An additional advantage of patterned surfaces is the reliability of contact on various surface profiles and the increased tolerance to defects of individual contacts.

Recently, *Continental* has developed a winter tyre with honeycomb profiles similar to those existing on the attachment pads of the grasshopper *Tettigonia viridissima* (Gorb et al., 2000)

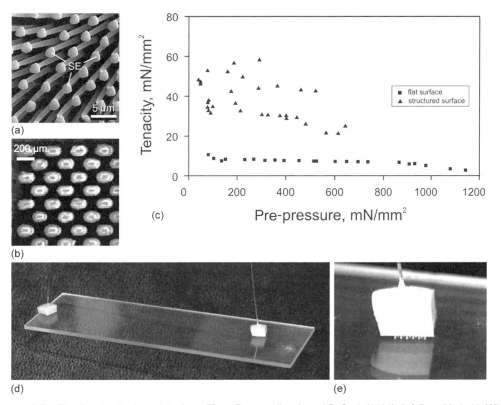

Figure 15.7 Biomimetic attachment devices. (From Peressadko, A. and S. Gorb (2004) *J. Adhes.* 80: 1–15. With permission of Taylor and Francis.) (a) Surface of the attachment organ in the fly *Calliphora vicina* (SEM micrograph), (b) prototype made of polyvinylsiloxane (PVS), (c) dependence of tenacity (adhesion per unit area) of the structured and flat samples on the prepressure, and (d, e) structured PVS surfaces used to hold a glass slide.

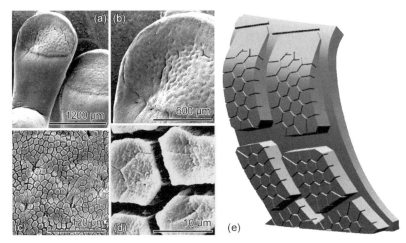

Figure 15.8 Surfaces generating grip on the wet substrata. (a–d) Scanning electron micrographs of the toe pads of the tree frog *Phyllomedusa trinitatis*. (Courtesy of J. Barnes, University of Glasgow.) (a) Low power view of the terminal portions of two toes, with toe pad epithelial cells just visible, (b) expanded view of a single toe pad, (c) medium power view of toe pad epithelium with mucous pores, (d) high power view to show detailed structure of the columnar epithelial cells separated from each other by grooves which, in life, would be filled with mucus, and (e) hexagonal sipes of Conti Winter Contact TS780. (Courtesy of R. Mundl, Continental AG.)

and tree frogs (Hanna and Barnes, 1990) (Figure 15.8a–d). The company promises enhanced wear performance on dry roads, less aquaplaning and better braking on wet roads, substantially improved lateral guidance, better grip, and more traction on ice. Fine grooves and longitudinal sipes in the individual tread blocks, provide lateral guidance. These are intersected by lateral sipes that provide maximum traction. This optimization ensures maximum safety on wintry roads. As lateral forces are generated in bends, the honeycomb sipes provide more gripping edges than conventional sipes, thus considerably improving trackholding when cornering (Barnes et al., 2002) (Figure 15.8e).

15.4 ANTI-ADHESIVE AND SELF-CLEANING SURFACES

Some biological systems have developed surfaces covered with micro and nanostructures, having antiwetting, anti-adhesive, and self-cleaning properties. The most prominent example is the so-called lotus-effect recently described for plant surfaces and successfully applied in numerous industrial materials, such as paints, roof tiles, spoons, and sinks.

The majority of surfaces of vascular plants are covered by a hydrophobic cuticle, which has an external layer consisting of so-called epicuticular waxes. The layer often contains wax crystalloids, with dimensions ranging from hundreds of nanometers to micrometers (Figure 15.9). The roughness of the hydrophobic plant surface decreases wettability (Holloway, 1969a,b, 1994), which is reflected in a greater contact-angle of water droplets on such surfaces, compared to smooth surfaces of the same chemical composition. This property of structured hydrophobic plant surfaces results in their ability to be cleaned by rolling drops of water (Barthlott and Neinhuis, 1997, 1998). Particles contaminating plant surfaces consist, in most cases, of material that is more readily wetted than hydrophobic wax components. Contaminants usually rest on the tips of the surface structures, so that the real contact area between the particles and plant surface is minimized. Thus, these particles can be easily removed by water droplets rolling over the surface. In this case, adhesion between particles and water droplets is greater than between particles and plant surfaces due to the reduced contact between the particles and plant surfaces. In the case of a smooth plant surface, the real contact area between the contaminating particles and the surface is large enough to avoid particle adherence

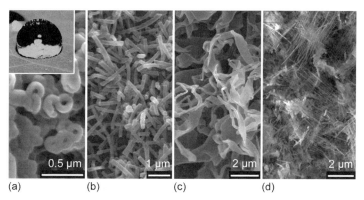

Figure 15.9 Plant antiwetting surfaces mediated by wax crystalloids. Tubules in *Prunus domestica* (a) and *Chelidonium majus* (b), polygonal rodlets in *Acer negundo* (c), terete rodlets in *Brassica oleracea* (d). (From Gorb, E.V. and S.N. Gorb (2002) *Entomologia Experimentalis et Applicata* 105: 13–28. With permission of Springer Science + Business Media B.V.) Inset shows the water droplet on the plant surface covered with wax-crystallites.

to rolling water droplets. A series of experiments revealed that surfaces of such plants as *Nelumbo*, *Colocasia,* and *Brassica* are richly covered by wax crystallites, which are responsible for the cleaning effect of their surfaces (Barthlott and Neinhuis, 1997). Similar adaptations have also been described for animal surfaces. Wings of insects (Odonata, Ephemeroptera, and Neuroptera) are covered with wax crystallites with dimensions that are comparable to those found in plants. These surfaces have been experimentally proven to be extremely nonwettable (Wagner et al., 1996).

Wax crystalloids on the flowering shoots of plants, such as *Salix* spp, *Hypenia*, and *Eriope* are adaptations to prevent crawling insects from robbing nectar and other resources (Eigenbrode, 1996). The wax blooms of ant-plants from the genus *Macaranga* seem to be an ecological isolation mechanism for the symbiotic ants. This mechanism is based exclusively on the influence of the ant attachment abilities, but not on the repellent effects of the wax. The comparison of surfaces in different species of the *Macaranga* revealed a high correspondence between the occurrence of wax coverage and obligatory ant associations (Federle et al., 1997). To explain the anti-adhesive properties of plant surfaces covered with waxes, several hypotheses are proposed: the roughness-hypothesis, the contamination-hypothesis, the wax-dissolving-hypothesis, and the fluid-absorption-hypothesis (Gorb and Gorb, 2002).

Many aquatic and semiaquatic arthropods have sculptured surfaces involved in holding air underwater for respiration. Such surfaces, called plastrons usually contain fields of microtrichia (Heckmann, 1983). These structures appear convergently in various arthropod taxa, as an adaptation to aquatic environments: Collembola, Lepidoptera, Coleoptera, Heteroptera, Diptera, Araneae, and Diplopoda (Thorpe and Crisp, 1947; Hinton, 1976; Messner, 1988). Some terrestrial insects, such as Aphididae (Auchenorrhyncha), also bear similar structures in the form of bristles, mush-room-like spines, or stigmal plates, which can protect their surfaces from moisture (Heie, 1987). In water striders and some spiders, antiwetting surfaces of legs and ventral body side are involved in the locomotion mechanism of walking on the water surface (Figure 15.3i,j).

15.5   OPTICS

Structural coloration, due to the presence of scales and bristles, is well-known in insects such as butterflies (Ghiradella, 1989) and beetles (Schultz and Hadley, 1987). For example, scales of the scarabaeid beetles from the genus *Hoplia* bear additional microtrichia on their surfaces responsible

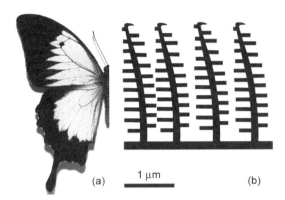

Figure 15.10 (See color insert following page 302)   Iridescence in butterflies. (a) Real color image of the iridescence from a *Papilio ulyssus* wing, and (b) diagram based on the transmission electron micrograph showing wing-scale cross-section of the butterfly *Morpho rhetenor.* (From Vukusic, P. and J.R. Sambles (2003) *Nature* 424: 852–855. With permission.) The high density of structures and high layer number creates an intense reflectivity of wing scales.

for lustreless appearance of the elytra surface (Figure 15.3c,d). The coloration pattern serves for species and sex recognition, and also for camouflage and mimicry. The most interesting type of structural coloration is called iridescence, which is well known in insects and birds, and has been characterized for many different species (Ghiradella et al., 1972; Huxley, 1975). The iridescence is a result of optical interference within multilayer structures (Ghiradella, 1991) which are rather complex in their architecture (Figure 15.10) and may be incorporated into systems that can produce several different optical effects. Such effects include diffraction-assisted reflection angle broadening (Vukusic et al., 1999, 2000a), all-structural color mixing and strong polarization effects (Vukusic et al., 2000b).

Another interesting optical property of surface structures has been described from the eye surface of insects. Ommatidial gratings are antireflective structures on the eyes of insects, especially those which are nocturnally active (Figure 15.3b). These protuberances are very small microtrichia (200 nm in diameter), which increase visual efficiency through decreased surface reflection in their density, and increased photon capture for a given stimulus condition (Parker et al., 1998; Vukusic and Sambles, 2003). Such a grating is particularly useful on a curved corneal surface, as it would increase the transmission of incident light through the cornea, compared with a smooth surface. For an increase in transmission and reduced reflection, a continuous matching of the refractive index n_1 and n_2 at the boundary of both adjacent materials is very critical. If the periodicity l_1 of the grating is smaller than wavelength l_2 of transmitting light, only light of zero order can be reflected or transmitted. For a constant ratio of both materials (cuticle-air) at the boundary between media, the electromagnetic field strength of incoming light is nearly constant (Bernhard et al., 1965). This region, therefore, can be considered as homogenous and an effective refractive index can be given. Such structured surfaces may also be self-cleaning utilizing the lotus-effect mechanism described above.

15.6 THERMOREGULATION AND PREVENTION OF DRYING

Surface outgrowths may provide multi-level reflection of sunlight. Such an ability of wing scales is suggested to be an adaptation for cooling in butterflies (Grodnicky, 1988). Body coverage by bristles, scales, and hairs in the honey-bee, *Apis mellifera*, may be used for warming up and influencing metabolism at low temperatures (Southwick, 1985). In species of curculionid beetles of the genus *Tychius* inhabiting arid areas, cuticular scales have been suggested to be a system responsible for maintaining thermal balance (Karasev, 1989). An additional function of such coverage is preventing water loss. This function was suggested for leaf-like bristles at the body margins in Aphididae (Auchenorrhyncha) (Heie, 1987). Surfaces of desert plants are covered with hydrophobic wax-crystalloids decreasing water loss through the cuticle.

15.7 SOUND GENERATION

Highly specialized areas of cuticle, responsible for sound generation, usually consist of patterns of cuticular plates, seldom of microtrichia fields (Hinton, 1970). When these surfaces slide over each other, sound is generated. Such structures have been previously described from elytra, abdomina, and coxae in phylogenetically distantly related arthropod taxa: spiders (Aranei), crustaceans *Trizopagurus*, bugs Cimicomorpha, Pentatomorpha, and beetles *Geotrupes* (Scarabaeoidea) (Gogala, 1984; Starck, 1985; Field et al., 1987; Palestrini et al., 1987). The sound frequency corresponds to the periodicity of these structures contacting functionally corresponding surfaces, and the speed of sliding. These systems should be very wear-resistant. However, rigorous experiments supporting or rejecting this statement are absent in the literature.

15.8 DEFENSE, GROOMING, SAMPLING, FILTRATING, GRINDING

Long, stiff, sclerotized cuticular spines are often used as defense mechanisms against predators. These structures are particularly widespread in the marine arthropods, such as zoea of crabs (Morgan, 1989). A similar function was described for long setae of certain beetle larvae from the family Dermestidae (Nutting, 1963).

Grooming is a very important function for insects, which sometimes live in extremely dirty or dusty environments. Their rich sensory equipment has to be kept clean in order to respond adequately to external signals. Many Hymenoptera bear modified leg spines specialized for the cleaning of antennae (Schönitzer and Lawitzky, 1987; Francouer and Loiselle, 1988). In aquatic environments, the problem of grooming is even more important because of biofouling. Pinnoterid crabs (Decapoda, Brachiura) use epipodite lobes covered by bristles for cleaning the gills (Pohle, 1989).

The collection of pollen grains and food sampling are functionally similar to grooming. They require similar surface structures and similar motorics. In bees (Apoidea), systems responsible for collecting pollen grains are usually equipped with urticating bristles (Pasteels and Pasteels, 1972; Hesse, 1981). Mouthpart surfaces used for scratching food particles from substrates appeared independently in the evolution of such phylogenetically far-related animal groups as insects (Figure 15.3h), crustaceans, molluscs and even some vertebrates (Arens, 1989). When food is sampled from rough hard substrata, surfaces of sampling devices show traces of wear. Cuticular protuberances that grind food items occur not only in the insect mouthparts, but also in one part of the insect digestive system, called the proventriculus. Specialized spines on the inner surface of the proventriculus have been reported in insects (Richards and Richards, 1969). Similar teeth are found in crustacean stomachs.

Filtration systems are usually equipped with long bristles too. Such systems are well-known from mouthparts of aquatic insect larvae. The filtering system of insect spiracles that keeps the tracheae free of dirt particles is composed of branched acanthae (Figure 15.3g). However, in some cases, completely different principles may be involved in the design of these systems. The labellum of many flies (Brachycera, Diptera) bears so-called pseudotracheae (Gracham-Smith, 1930; Elzinga and Broce, 1986). Outgrowths of the pseudotracheae have a complex material design. Labellum and pseudotracheae are driven by muscles, resilin springs, and hydraulic pressure, which enable a change in the diameter of the filtration sieve, depending on the size of the particles in the food (Figure 15.3f).

15.9 BIOMIMETICS OF SURFACES: WHAT CAN WE LEARN FROM EVOLUTION?

Throughout evolution, nature has constantly been called upon to act as an engineer in solving surface-related technical problems. Organisms have evolved an immense variety of shapes and structures. Although often intricate and fragile, they can nonetheless deal with extreme mechanical loads. This chapter has demonstrated that many functions are based on a variety of ingenious surface-related solutions. The success of biologically inspired technological surfaces is an indication that knowledge from biology is also highly relevant for technical applications.

One of the greatest challenges for today's engineering science is miniaturization. Many organisms have solved many problems correlated with extremely small size, during their evolution. Zoologists and botanists have collected a huge amount of information about the structure of such living micromechanical systems. This information can be utilized to mimic them for further industrial developments. An important feature of the evolution of functional surfaces is the multiple origins of similar solutions in different lineages of living organisms, and in different functional systems. For example, there is no doubt that attachment systems consisting of a pair of microtrichia-covered surfaces appeared independently in the head-arresting system of Odonata, coxal-locking devices of Neuroptera, and the elytra-locking system of Coleoptera (Gorb, 1999, 2001). In the case of locomotory attachment devices, hairy pads appeared three times independently within insects and additionally at least two times within other animals (Gorb and Beutel, 2001). Also, similar optical structures, made of different materials, have been described for insects, birds, and plants (Vukusic and Sambles, 2003). Such examples of convergence are the most interesting from the engineering point of view, because they indicate a kind of optimal solution for a particular problem. These examples should be the very first candidates for biomimetics.

An engineering approach, applied after detailed studies on the natural system, involving modeling and then prototyping, would be most promising. However, in some cases, engineers can also, simply copy the surface shape and replicate it at a variety of scales and materials using available technologies of chemistry and processing. Both approaches may run parallel for some time and possibly converge later.

ACKNOWLEDGMENTS

The author is very thankful to the members of the Biological Microtribology Group at MPI for Developmental Biology (Tübingen, Germany) and Evolutionary Biomaterials Group at MPI for Metals Research (Stuttgart, Germany) for the fruitful joint work on various surface-related projects. Valuable suggestions of John Barnes (University of Glasgow, Glasgow, UK), the Editor, and four reviewers on the text of this review are greatly acknowledged.

REFERENCES

Arens, W. (1989) Comparative functional morphology of the mouthparts of stream animals feeding on epilithic algae. *Arch. Hydrobiol. Suppl.* 83 (3): 253–354.

Arzt, E., Gorb, S., and R. Spolenak (2003) From micro to nano contacts in biological attachment devices. *Proc. Natl Acad. Sci. USA* 100 (19): 10603–10606.

Aspden, R.M. (1994) Fibre reinforcing by collagen in cartilage and soft connective tissues. *Proc. R. Soc. Lond. B* 258 (1352): 195–200.

Ateshian, G.A. (1997) A theoretical formulation for boundary friction in articular cartilage. *J. Biomech. Eng. — Transact. ASME* 119 (1): 81–86.

Autumn, K. and A.M. Peattie (2002) Mechanisms of adhesion in geckos. *Integr. Comp. Biol.* 42: 1081–1090.

Autumn, K. Liang, Y.A., Hsieh, S.T., Zesch, W., Chan, W.P., Kenny, T.W., Fearing, R., and R.J. Full (2000) Adhesive force of a single gecko foot-hair. *Nature* 405: 681–685.

Barnes, J., Smith, J., Oines, C., and R. Mundl (2002) Bionics and Wet Grip. Tire Technology International, December 2002: 56–60.

Barthlott, W. and C. Neinhuis (1997) Purity of the sacred lotus or escape from contamination in biological surfaces. *Planta* 202: 1–8.

Barthlott, W. and C. Neinhuis (1998) Lotusblumen und Autolacke: Ultrastruktur pflanzlicher Grenzflächen und biomimetische unverschmutzbare Werkstoffe. In: 4. Bionik — Kongress, München 1998, Nachtigall, W. and A. Wisser, Stuttgart, Jena, Lübeck (eds). Ulm: Gustav Fischer Verlag, pp. 281–293.

Bauchhenss, E. and M. Renner (1977) Pulvillus of *Calliphora erythrocephala* Meig. (Diptera; Calliphoridae). *Int. J. Insect Morphol. Embryol.* 6 (3/4): 225–227.

Bechert, D.W., Bruse, M., Hage, W., and R. Meyer (2000) Fluid mechanics of biological surfaces and their technological application. *Naturwissenschaften* 87: 157–171.

Bernhard, C.G., Miller, W.H., and A.R. Moller (1965) Insect corneal nipple array — a biological broad-band impedance transformer that acts as an antireflection coating. *Acta Physiol. Scand.* S63 (Suppl. 243): 5.

Bocharova-Messner, O.M. and A.Z. Dmitriev (1984) Morphological and functional analysis of the wing venation in Odonata according to the data of the scanning electron microscopy. In: IX Congress of All-Union Entomological Society. Abstr. Theses. Kiev: Naukova Dumka, p. 65 (In Russian).

Buckwalter, J.A. (1983) Articular cartilage. *Instr. Course Lect.* 32: 349–370.

D'Andrea, M. and S. Carfi (1988) Spines on the wing veins in Odonata. 1. Zygoptera. *Odonatologica* 17 (4): 313–335.

Eigenbrode, S.D. (1996) Plant surface waxes and insect behaviour. In: *Plant Cuticles and Integral Functional Approach*, Kerstiens, G. (ed.). Oxford: BIOS, pp. 201–222.

Elzinga, R.J. and A.B. Broce (1986) Labellar modifications of Muscomorpha flies (Diptera). *Ann. Entomol. Soc. Am.* 79 (1): 150–209.

Etsion, I. (2004) Laser surface texturing — Measure to reduce friction. 14th International Colloquium Tribology, January 2004, Esslingen, Germany. 1: 329–333.

Fahrenbach, W.H. and D.D. Knutson (1975) Surface adaptations of the vertebrate epidermis to friction. *J. Invest. Dermatol.* 65 (1): 39–44.

Federle, W., Maschwitz, U., Fiala, B., Riederer, M., and B. Hölldobler (1997) Slippery ant-plants and skilful climbers: selection and protection of specific ant partners by epicuticular wax blooms in *Macaranga* (Euphorbiaceae). *Oecologia* 112: 217–224.

Field, L.H., Evens, A., and D.L. Macmillan (1987) Sound production and stridulatory structures in hermit crabs of the genus *Trizopagurus. J. Mar. Biol. Assoc. U.K.* 67 (1): 89–110.

Francouer, A. and R. Loiselle (1988) Evolution du strigile chez les formicides (Hymenopteres). *Nat. Can.* 115 (3–4): 333–335.

Fung, Y.C. (1981) *Biomechanics: Mechanical Properties of Living Tissues*. Berlin: Springer Verlag.

Ghiradella, H. (1989) Structure and development of iridescent butterfly scales: lattices and laminae. *J. Morphol.* 202: 69–88.

Ghiradella, H. (1991) Light and colour on the wing: structural colours in butterflies and moths. *Appl. Optics* 30: 3492–3500.

Ghiradella, H., Aneshansley, D., Eisner, T., Silbergleid, R. E., and H.E. Hinton (1972) Ultra-violet reflection of a male butterfly: interference colour caused by thin layer elaboration of wing scales. *Science* 178: 1214–1217.

Gogala, M. (1984) Vibration producing structures and songs of terrestrial Heteroptera as systematic character. *Biol. Vestn.* 32 (1): 19–36.

Gorb, S.N. (1998) The design of the fly adhesive pad: distal tenent setae are adapted to the delivery of an adhesive secretion. *Proc. R. Soc. Lond. B* 265: 747–752.

Gorb, S.N. (1999) Evolution of the dragonfly head-arresting system. *Proc. R. Soc. Lond.* B 266: 525–535.

Gorb, S.N. (2001) *Attachment Devices of Insect Cuticle*. Dordrecht, Boston, London: Kluwer Academic Publishers.

Gorb S.N. and R.G. Beutel (2001) Evolution of locomotory attachment pads of hexapods. *Naturwissenschaften* 88: 530–534.

Gorb, E.V. and S.N. Gorb (2002) Attachment ability of the beetle *Chrysolina fastuosa* on various plant surfaces. *Entomologia Experimentalis et Applicata* 105: 13–28.

Gorb, S., Jiao, Y., and M. Scherge (2000) Ultrastructural architecture and mechanical properties of attachment pads in *Tettigonia viridissima* (Orthoptera Tettigoniidae). *J. Comp. Physiol. A* 186: 821–831.

Gracham-Smith, G.S. (1930) Further observations on the anatomy and function of the proboscis of the blow-fly, *Calliphora erythrocephala* L. *Parasitology* 22: 47–115.

Grodnicky, D.L. (1988) Structure and function of the scale coverage of the wings in butterflies (Lepidoptera/Hesperioidea, Papilionoidea). *Entomol. Rev.* 67 (2): 251–256.

Gronenberg, W. (1996) Fast actions in small animals: springs and click mechanisms. *J. Comp. Physiol.* A 178: 727–734.

Habenicht, G. (2002) *Kleben: Grundlagen, Technologien, Anwendung.* Berlin: Springer Verlag.

Hanna, G. and W.J.P. Barnes (1990) Adhesion and detachment of the toe pads of tree frogs. *J. Exp. Biol.* 155: 103–125.

Hazel, J., Stone, M., Grace, M.S., and V.V. Tsukruk (1999) Nanoscale design of snake skin for reptation locomotions via friction anisotropy. *J. Biomech.* 32 (5): 477–484.

Heckmann, C.W. (1983) Comparative morphology of arthropod exterior surfaces with capability of binding a film of air underwater. *Int. Rev. Ges. Hydrobiol.* 68 (5): 715–736.

Heie, O.E. (1987) Morphological structure and adaptations. In: *Aphids: Biology, Natural Enemies and Control.* Amsterdam, Minks, A.K. and P. Harrewijn (eds). Elsevier, pp. 393–400.

Hennemann, O.-D. (2000) Kleben von Kunststoffen. Anwendung, Ausbildung. *Trend. Kunststoffe* 90: 184–188.

Hesse, M. (1981) Auf welche Weise transportieren Insekten den Blütenstaub? *Linz. Biol. Beitr.* 13 (1): 50.

Hiller, U. (1968) Untersuchungen zum Feinbau und zur Funktion der Haftborsten von Reptilien. *Z. Morphol. Tiere* 62: 307–362.

Hinton, H.E. (1970) Some little-known surface structures. In: *Insect Ultrastructure*, Neville, A.C. (ed.). Oxford and Edinburgh: Blackwell Scientific Publishers, pp. 41–58.

Hinton, H.E. (1976) The fine structure of the pupal plastron of simulid flies. *J. Insect Physiol.* 22: 1061–1070.

Holloway, P.J. (1969a) Chemistry of leaf waxes in relation to wetting. *J. Sci. Food Agric.* 20: 124–128.

Holloway, P.J. (1969b) The effects of superficial wax on leaf wettability. *Ann. Appl. Biol.* 63: 145–153.

Holloway, P.J. (1994) Plant cuticles: physicochemical characteristics and biosynthesis. In: *Air Pollution and the Leaf Cuticle*, Percy, K.E., Cape, J.N., Jagels, R., and C.J. Simpson (eds). Berlin: Springer Verlag, pp. 1–13.

Huber, G., Gorb, S., Spolenak, R., and E. Arzt (2005) Adhesion measurements for single gecko spatulae by atomic force microscopy. *Biol. Lett.* 1:24. doi:10.1098/rsbl.2004.0254.

Huxley, J. (1975) The basis of structural colour variation in two species of *Papilio*. *J. Entomol.* A 50: 9–22.

Ishii, S. (1987) Adhesion of a leaf feeding ladybird *Epilachna vigintioctomaculata* (Coleoptera: Coccinellidae) on a vertically smooth surface. *Appl. Ent. Zool.* 22: 222–228.

Jiao, Y., Gorb, S.N., and M. Scherge (1999) Adhesion measured on the attachment pads of *Tettigonia viridissima* (Orthoptera, Insecta). *J. Exp. Biol.* 203 (12): 1887–1895.

Karasev, V.P. (1989) Scale coverage of the curculionid beetles of the genus *Tichius* Germar (Coleoptera, Curculionidae). In: *Dynamics of Zoocoenozes and Animal Conservation in Belorussia, Minsk*, p. 85 (In Russian).

Komvopoulos, K. (2003) Adhesion and friction forces in microelectromechanical systems: Mechanisms measurement, surface modification techniques and adhesion theory. *J. Adhes. Sci. Technol.* 17: 477–517.

Langer, M., Ruppersberg, P., and S. Gorb (2004) Adhesion forces measured at the level of a terminal plate of the fly's seta. *Proc. R. Soc. Lond.* B 271: 2209–2215.

Li, X., Madan, I., Bozkurt, H., and H. Birkhofer (2004) Lifetime of solid lubricated roller bearings. 14th International Colloquium Tribology, January 2004, Esslingen, Germany. 3: 1361–1364.

Messner, B. (1988) Funktionelle Morphologie der Insektenkutikula am Beispiel der Plastronatmer. *Wiss. Z. E. M. Arndt-Univ. Greifswald. Math Naturwiss. R.* 37 (2–3): 27–30.

Miyoshi, K. (2001) *Solid Lubrication: Fundamentals and Applications.* New York, Basel: Marcel Decker.

Morgan, S.G. (1989) Adaptive significance of spination in estuarine crab zoeae. *Ecology* 70 (2): 464–482.

Nachtigall, W. (1974) *Biological Mechanisms of Attachment.* Berlin, Heidelberg, New York: Springer-Verlag.

Nachtigall, W. (1977) Biophysik des Schwimmens. In: *Biophysik*, Hoppe, W., Lohmann, W., Markl, H., and H. Iegler (eds). Berlin: Springer Verlag, pp. 525–536.

Naldrett, M.J. (1992) Adhesives. In: *Biomechanics-Materials: A Practical Approach*, Vincent, J.F.V. (ed). Oxford: IRL Press, pp. 219–240.

Niederegger, S. and S. Gorb (2003) Tarsal movements in flies during leg attachment and detachment on a smoothe substrate. *J. Insect Physiol.* 49: 611–620.

Nutting, W.L. (1963) The hastate setae of certain dermestid beetle larvae: an unusual entangling defence mechanism. *J. Arizona Acad. Sci.* 2: 189.

Palestrini, C., Piazza, R., and M. Zunino (1987) Segnali sonori in the species di Geotrupini (Coleoptera, Scarabaeoidea, Geotrupidae). *Boll. Soc. Entomol. Ital.* 119: 139–151.

Parker, A.R., Hegedus, Z., and R.A. Watts (1998) Solar-absorber antireflector on the eye of an eocene fly. *Proc. R. Soc. Lond. B* 265 (1398): 811–815.

Pasteels, J.J. and J.M. Pasteels (1972) Les soies cuticulaires des Megachilidae (Hymenoptera; Apoidea) vues au microscope electronique a balayage. *Bruxeles.* 27 p.

Peressadko, A. and S. Gorb (2004) When less is more: experimental evidence for tenacity enhancement by division of contact area. *J. Adhes.* 80: 1–15.

Perez Goodwyn, P.J. and S.N. Gorb (2004) Frictional properties of contacting surfaces in the hemelytra-hindwing locking mechanism in the bug *Coreus marginatus* (Heteroptera, Coreidae). *J. Comp. Physiol. A* 190: 575–580.

Persson, B.N.J. (2003) On the mechanism of adhesion in biological systems. *J. Chem. Phys.* 118: 7614–7621.

Pohle, G. (1989) Structure, function and development of setae on gill-grooming appendages and associated mouthparts of pinnoterid crabs (Decapoda: Brachiura). *Can. J. Zool.* 67 (7): 1690–1707.

Radhakrishnan, V. (1998) Locomotion: dealing with friction. *Proc. Natl Acad. Sci. USA* 95: 5448–5455.

Reif, E. and A. Dinkelacker (1982) Hydrodynamics of the squamation in fast swimming sharks. *Neues Jahrbuch für Geologie und Paläontologie* 164: 184–187.

Richards, P.A. and A.G. Richards (1969) Acanthae: a new type of cuticular process in the proventriculus of Mecoptera and Siphonaptera. *Zool. Jb. Anat.* 86: 158–176.

Scherge, M. and S.N. Gorb (2001) *Biological Micro- and Nanotribology.* Berlin: Springer Verlag.

Scherge, M. and J.A. Schaefer (1998) Surface modification and mechanical properties of bulk silicon. In: *Tribology Issues and Opportunities in MEMS.* Dordrecht: Kluwer Academic Publishers, pp. 529–538.

Scherge, M., Li, X., and A. Schaefer (1999) The effect of water on friction of MEMS. *Tribol. Lett.* 6 (3–4): 215–220.

Schönitzer, K. and G. Lawitzky (1987) A phylogenetic study of the antenna cleaner in Formicidae, Mutillidae and Tiphiidae (Insecta, Hymenoptera). *Zoomorphology* 107 (5): 273–285.

Schultz, T.D. and N.F. Hadley (1987) Structural colors of tiger beetles and their role in heat transfer through the integument. *Physiol. Zool.* 60 (6): 737–745.

Southwick, E.E. (1985) Bee hair structure and the effect of hair on metabolism at low temperature. *J. Apic. Res.* 24 (3): 144–149.

Starck, J.M. (1985) Stridulationsapparate einiger Spinnen — Morphologie und evolutionsbiologische Aspekte. *Z. Zool. Syst. Evolutionsforsch.* 23 (2): 115–135.

Steinberg, M.S. (1996) Adhesion in development: an historical overview. *Dev. Biol.* 180: 377–388.

Stork, N.E. (1980) Experimental analysis of adhesion of *Chrysolina polita* (Chrysomelidae, Coleoptera) on a variety of surfaces. *J. Exp. Biol.* 88: 91–107.

Stork, N.E. (1983) The adherence of beetle tarsal setae to glass. *J. Nat. Hist.* 17: 583–597.

Strange, C.J. (1997) Biological ties that bind. In multicellular organisms, much of the secret of life lies outside the cell. *Bioscience* 47 (1): 5–8.

Thorpe, W.H. and D.J. Crisp (1947) Studies on plastron respiration. I. The biology of *Aphelocheirus* and the mechanism of plastron retention. *J. Exp. Biol.* 24: 227–269.

Vukusic, P. and J.R. Sambles (2003) Photonic structures in biology. *Nature* 424: 852–855.

Vukusic, P., Sambles, J.R., Lawrence, C.R., and R.J. Wootton (1999) Quantified interference and diffraction in single *Morpho* butterfly scales. *Proc. R. Soc. Lond. B* 266: 1403–1411.

Vukusic, P., Sambles, J.R., and H. Ghiradella (2000a) Optical classification of microstructure in butterfly wing-scales. *Photonics Sci. News* 6: 61–66.

Vukusic, P., Sambles, J.R., and C.R. Lawrence (2000b) Structural colour: colour mixing in wing scales of a butterfly. *Nature* 404: 457.

Wagner, T., Neinhuis, C., and W. Barthlott (1996) Wettability and contaminability of insect wings as a function of their surface sculpture. *Acta Zoologica* 77 (3): 213–225.

Waite, J.H. (1983) Adhesion in byssally attached bivalves. *Biol. Rev.* 58 (2): 209–231.

Weiss, L. (1970) A biophysical consideration of cell contact phenomena. In: *Adhesion in Biological Systems,* Manly, R.S. (ed.). New York and London: Academic Press, pp. 1–12.

Wigglesworth, V.B. (1987) How does a fly cling to the under surface of a glass sheet? *J. Exp. Biol.* 129: 363–367.

Wootton, R.J. and D.J.S. Newman (1979) Whitefly have the highest contraction frequencies yet recorded in non-fibrillar flight muscles. *Nature* 280: 402–403.

Biomimetic and Biologically Inspired Control

Zhiwei Luo, Shigeyuki Hosoe, and Masami Ito

CONTENTS

16.1 REVIEW OF THE DEVELOPMENT OF SYSTEM CONTROL

From the micro-level of molecular biochemical reactions to the macro-level of ecosystems, *control* exists everywhere in complex biological systems. System control engineering, which is strongly motivated by mimicking biological systems, affects powerfully the late development of overall engineering fields including bioengineering. Recent rapid development of biological science and technologies will further improve the active applications of control engineering. Meanwhile, system control theory itself will also be promoted by advanced biomimetic researches.

The aim of system control engineering is to abstract the essences of *control* existing in biological and physical systems of nature and to apply them to create artificial systems. In 1948, N. Wiener published his book *Cybernetics*, in which he defined cybernetics as the science that deals with the control and communication of systems that have an organized structure (Wiener, 1948). By considering the similarity between control in biological systems and in artificial systems, he tried to unify the interpretation of living organisms and machines from the viewpoint of organized systems. Several theoretical concepts have been evolved in control theory, typified by feedback control, optimal control, sequence control, and so on. The main roles of feedback control are regulation and adjustment, whereas optimal control involves planning and supervision with a higher level of control state than feedback control. Sequence control, on the other hand, has the objective of rationalizing logical procedures, scheduling, and decision.

Basically, the objective of feedback control is to follow up a desired command input (set point) by controlled variables subjected to unpredictable external disturbance. For the constant set point, the control problem is referred to as a constant value control (regulator problem), and if the command changes over time, it is referred to as a follow-up control (servomechanism problem). The linear feedback control theory using transfer functions has already reached its maturity in the 1940s. The similarity of biological and mechanical systems in terms of homeostasis was considered within the framework of feedback control in that time. For example, tremor can be seen in many neurological disorders, their hand trembles when they attempt to grasp an object. Some tremors may be related to feedback instabilities. Since the feedback control is used to suppress disturbances and to maintain a set point in constant value control, it corresponds to homeostasis in living organisms.

In the 1960s, modern control theory based on the foundation of state space methods has been established. This leads to the rapid appearance and development of the pole assignment principle, observer theory, optimal regulator, Kalman filter, the internal model principle, and their extension to multiple-input–multiple-output systems. While the modern control theory contributed to the refinement of the linear feedback theory, an even more important development was the clarification of the principles related to duality of control and observation. Specifically, Kalman (1960) introduced the concepts of controllability and observability and demonstrated the duality of control and observation for the linear systems. He showed that it is possible to construct the pole-assignment of a closed-loop system and the state-estimation of a controlled system in the same framework. Furthermore, optimal regulator and Kalman filter can also be constructed in the same way. Therefore, if a system (the controlled system) is controllable and observable, then it is possible to construct any kind of desired dynamics (poles) for a closed-loop system.

Although the similarity of biological and mechanical systems was studied in terms of feedback control, often the subject was limited to the range in the vicinity of operating point in which the linear theory holds. In the 1980s, the control design theory has returned to the frequency response methods and robust control theories based on $H\infty$ norm were developed (Doyle et al., 1989). $H\infty$ control theory makes it possible to quantitatively handle the influence from the variations of the controlled object's dynamic characteristics, which had been previously treated only qualitatively. However, the control concept remains unchanged within feedback control.

On the other hand, adaptive control is a more direct strategy than robust control to handle the influence of the object's variations. In robust control, the controller itself does not change even if the controlled object was fluctuated. However, adaptive control has the function of identification to monitor the fluctuation in the controlled object, and based on the result of identification, the controller is adjusted. Accordingly, identification and adjustment are the two basic functions of adaptive control. Recently, robust adaptive control has also been developed to increase the system's robustness against the fluctuations of the controlled object. Adaptive control system can be regarded as a nonlinear system in the sense that the compensation is adjusted. However, since the controlled system and the compensation element are simply linear, and the purposes of the control are still limited to suppress disturbances and to track command input as in feedback control, present adaptive control does not have the ability to self-organize the system's internal states with respect to the environmental dynamic variations as seen in biological systems. It left far from the goal of automatic system control design. The meaning of adaptation in the adaptive control systems that has been developed so far is still quite different from the one featured in living systems.

System control is not simply limited to regulation or adjustment as that in the feedback control, it also includes management. In the 1960s, system optimization was extensively discussed in modern control theory. This brought forth the optimal control theory typified by Pontryagin's maximum principle (Boltyanski et al., 1960) and Bellman's dynamic programming (Bellman and Kalaba, 1960). Optimal control is related to the optimization of a dynamic system while the theory to optimize a static system is referred to as optimalizing control. Both concepts correspond to planning and management or supervision. They are different from regulation. An example of optimal control problem is a terminal control problem, such as to design trajectories of a rocket to send a satellite accurately onto a preprogrammed orbit. This can be regarded as the problem of obtaining a desired command input for a follow-up control in the feedback control problem. Therefore, it is recognized that this control strategy is on the planning level, which is generally one level higher than the strategy of feedback control. Mathematically, Pontryagin's maximum principle and Bellman's dynamic programming can be regarded as an extension of the classical calculus of variations to control problems, that is, the Hamilton's canonical equations and the Hamilton–Jacobi formula. On the other hand, optimalizing control optimizes a performance function by considering only the static input–output relationship in steady state as opposed to optimizing dynamic processes in optimal control. In early 1951, Draper and Li first considered the optimalizing control problem in order to keep the internal combustion engine running at an optimum operating condition regardless of variations in load (Draper and Li, 1951). Recent electronic fuel injection (EFI) system, in which a microprocessor adjusts the air–fuel ratio, has replaced the carburetor in some automobiles. Although it was not exactly what Draper and Li intended to do, optimalizing control has been implemented in various fields. In this way, optimalizing control determines the optimum set point (i.e., target value) of the constant value control, it can also be regarded as an upper level above feedback control.

Though half a century has passed since Wiener's Cybernetics, machines are not even close to resembling animals and their abilities. When we dreamed of flying like a bird, the airplane was born, far outdoing the birds in speed and size. However, the airplane cannot realize the bird's agility to move from branch to branch. The same thing can be found for the present robots and computers. Nowadays, industrial robots can only perform predefined operations in a well-structured task space but do not have full capability to deal with unexpected situations in natural complex environment. An autonomous robot, which is able to identify its own environment and determine its own voluntary actions, is yet something of the future. Similarly, though the computer has increased enormous computational capabilities, it is nevertheless a serial sequential machine that can perform only preprogrammed actions. Therefore, although machines have been made to imitate and to amplify special functions of human and animals, they are far from achieving the level of the autonomy, flexibility, environmental adaptability, and functional variety of biological systems.

Recently, as marked by the rapid spread of computers, internet, and mobile communication, the developments of the information science and technologies make it possible for us to process higher capacity of information much faster and more intelligently in the worldwide scale. It provides us with enormous challenging realms. Meanwhile, the systems around us are becoming larger and more complex. It becomes more and more important for the artificial systems to have high flexibility, diversity, reliability, and affinity. System control theory, which forms the core foundation for understanding, designing, and operating of systems, is still limited and insufficient to handle complex large-scale systems and to process spatial temporal information in real time as biological systems. Under this background, biomimetic and biologically inspired control research is becoming a very important subject. This subject is widely expected to breakthrough the next information and system control theories.

Animals acquire and develop their extremely sophisticated movement through active interaction with the environment using parallel decentralized processing of spatial temporal information in real time. They also use logical recognition together with dynamic physical motion. The analysis and clarification of these functions mathematically at the system level, and imitation of them in engineering, will lead to a deeper understanding of ourselves and will be significant for constructing the next generation of advanced artificial systems such as human friendly robots.

The following sections of this chapter will introduce the recent biomimetic system control researches. From the point of view of the system's self-organization, we will describe in Section 16.2 the nonlinear and redundant sensory-motor learning problem. We will introduce the problem of optimal motion formation under environmental constrains in Section 16.3. In Section 16.4, we will study the system's mechanical interaction and environmental adaptation, and show a novel biologically inspired two degree of freedom adaptive control theory with its application to a robot's force tracking control. The conclusion will be given in Section 16.5.

16.2 SENSORY-MOTOR ORGANIZATION

Animals survive in complex natural environment using their sensory-motor behavior. The organization and development of brain nervous system's motor control functions largely depend on the physical interaction with the external environment. Self-organization of the environmental adaptive motor function is one of the most interesting characteristics that we should learn in biomimetic control research.

Charles T. Snowdon, who was a President of the Animal Behavior Society, described animal's behavior as follows: "Animal behavior is the bridge between the molecular and physiological aspects of biology and the ecological. Behavior is the link between organisms and environment, and between the nervous system and the ecosystem. Behavior is one of the most important properties of animal life. Behavior plays a critical role in biological adaptations. Behavior is how we humans define our own lives. Behavior is that part of an organism by which it interacts with its environment. Behavior is as much a part of an organisms as its coat, wings, etc. The beauty of an animal includes its behavioral attributes" (Snowdon).

Historically, there are two broad approaches to studying animal behaviors: (1) ethological approach and (2) experimental physiological approach (http://salmon.psy.plym.ac.uk/). Ethologists mainly concern with the problems of how to identify and describe species-specific behaviors. They try to understand the evolutionary pathway through which the genetic basis for the behavior came about. They use field experiments and make observations of animal behavior under natural conditions. On the other hand, behaviorists and comparative psychologists concentrate on how we learn new behaviors by using statistical methods and carefully controlled experimental variables for a restricted number of species, principally rats and pigeons, under laboratory conditions.

The famous Russian physiologist Pavlov, who is recognized as the founder of behaviorism, trained a dog by ringing a bell before mealtime. Through the course of time, he discovered that simply by ringing the bell, the dog would salivate. It is now known as the concept of conditioned reflex (Pavlov, 1923). A similar conceptual approach was also developed by him to study human behavior. Sherrington, on the other hand, studied spinal reflexes and gave out his theory of the reciprocal innervation of agonist and antagonist skeletal muscle innervation, which is known as Sherrington's Law (Sherrington, 1906). Bernstein, another Russian scientist, further pointed out different important problems in motor learning and organization. Dealing with the redundancy problem in motor behavior, he proposed the concept of *synergy* in muscles' coordinative actions so as to constraint the motion D.O.F. with respect to the required tasks. He suggested that it is such a synergy that results in the reflex motions between each D.O.F. Moreover, this synergy changes with respect to the environmental variations, which is beyond the philosophy of Pavlov's conditioned reflex (Bernstein, 1967).

Today, with the development of information science, robotics, and control engineering, it becomes easier to study the motor behaviors and the principal control mechanisms of the brain nervous system more quantitatively and systematically.

16.2.1 Nonlinear and Redundant Sensory-Motor Organization

Even simple reaching movements that can be performed by a 5-month-old baby are never simple in cybernetics. At least, it requires solving several *nonlinear coordination transformations* from the object space to the muscle space. The transformations may also contain the problem of *redundancy*.

Figure 16.1 shows a 3D computer simulation model of whole body dynamic musculo-skeletal system of human developed in RIKEN BMC. As seen from this model, in order to realize the natural human motions, it is necessary to control more than 105 D.O.F by over 300 muscles. Figure 16.2 shows another research example on how to control the 3 D.O.F. position of an object by whole arm cooperative manipulation under the influence from the external forces. Here, each arm has 4 D.O.F. and interacts with the object by all links not the end-effectors. Human body is such a super-redundant system, and the redundancy exists in a lot of levels of the motor coordinates. The inverse solution of the redundancy problem generally forms a solution manifold in the motor control space, the solution is not unique and thus is not easy to define. Therefore, although the redundant D.O.F. provides the biological system with powerful hardware foundation to realize various smooth and delicate motions that have high *tolerance* (fault-tolerance to the functional disability in some parts of the system) and *adaptability* (adapt to the environmental uncertainties, variations, and different objectives), in order to enjoy these benefits, during organizing the sensory-motor coordination, we have to overcome the *ill-posed* nonlinear problems. These problems come not only from the kinematics but also from the dynamics. By now, there are many researches proposed from the viewpoints of robotic engineering as well as biologically inspired learning theory. The proposed approaches can be largely summarized as: (1) learning approach based on neural network; and (2) Jacobian approach from robotic engineering.

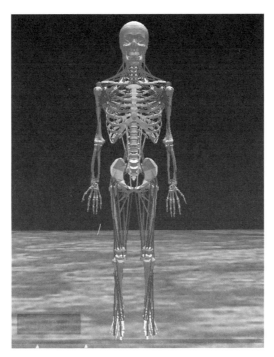

Figure 16.1 (See color insert following page 302)   A 3D computer simulation model of whole body human dynamic musculo-skeletal system.

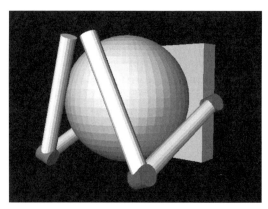

Figure 16.2   Whole body cooperative manipulation of an object.

16.2.2 Motor Learning Using Neural Network

In the learning-based approach, the main efforts have been made through: (1) supervised learning; and (2) self-organization.

Fundamentally, supervised learning depends closely on the availability of an external teacher. In this approach, we first construct a neural network and define a smooth nonlinear function for a set of neurons. Then, for a given set of inputs, we use the error between the desired response from the teacher and the network's actual output to adjust the interconnection weights between each neuron. Researches of supervised learning resulted in the later biological discovery of long-term depression (LTD) in cerebellum (Rosenblatt, 1962; Ito, 1984), which in turn clarified one of the basic functions of cerebellum in motor learning and adaptation. However, the later developments of supervised learning in artificial neural network may not match in detail with the real neural networks (Rumelhart et al., 1986).

One of the important abilities of supervised learning is the so-called generality, which means that, after sufficient learning, for a new input that was not learned before, the network can generate proper output. It is proved for the multi-layered artificial neural networks that, with sufficient numbers of neurons in the hidden layer, the network can approximate any continuous mapping from input to output (Funahashi, 1989). For motor learning, however, the condition of sufficient learning indicates that we have to perform sufficient physical trial motions by the body. This is necessary in supervised learning but is not efficient for motor learning in biological systems. For motor learning, the main target is rather to realize the generality of motion with limited physical trials.

By modifying the supervised learning, three models: (1) direct inverse (Kuperstein, 1988); (2) distal supervised learning (Jordan and Rumelhart, 1992); and (3) feedback error learning (Kawato et al., 1987; Miyamoto et al., 1988) have been proposed for the specific problem of motor learning. The main considerations of the modifications are about the selection of the suitable teacher signal and the concave property of the nonlinear transformation. However, these three models have two common disadvantages derived more or less from supervised learning. Firstly, in applying an algorithm such as backpropagation, global information of the network's output error is used to adjust all weights between nerve cells. It requires massive connections among all neurons, which is difficult to realize artificially. Secondly, the resultant motor output may not have topology conserving property with respect to the sensory input, or even no spatial optimality as we will show in the next subsection. Because of these disadvantages, in the tasks such as to move the hand smoothly in the task space, there may exist a dramatic change in the joint angles (Guez and Ahmad, 1988; Gorinevsky, 1993).

Comparing with supervised learning, the self-organization approach does not depend on any external teacher. It focuses on the spatial order of the input data and organizes the learning system so that the neighbor nodes have the similar outputs (Amari, 1980; Kohonen, 1982). By considering spatial characteristics of motor learning, self-organization algorithm has also been extended to generate the topology conserving sensory-motor map (Ritter et al., 1989). In this approach, we first construct a three-dimensional lattice, and specify the sensory input vectors, the corresponding inverse Jacobian matrixes, and the joint angle vectors to each node within the lattice. The lattice then outputs desired joint angles for the arm to perform many physical trial motions. For each trial motion, a visual system is used to input the end-effector position of the arm in the task space. The algorithm is then used to search for a winner node with its sensory vector closest to the visual input. After that, the sensory vector, the inverse Jacobian matrix as well as the joint angle vector of the winner node, together with that in its neighbor nodes, are adjusted, respectively. The neighbor region of adjustment decreases as the learning proceeds. As a result, the vectors (or a matrix) in one node are similar to that of its neighbor nodes. That is, a topology conserving map is self-organized without any supervisor's command. In this algorithm, for every adjustment step, the arm has to perform the real physical trial motions. Since it is still within the learning process, sometimes these trial motions are dangerous or may be impossible due to the incorrectness of the map. In addition, both in searching the winner node as well as when adjusting the neighbor nodes, the approach requires a centralized gating network to interact with all nodes, which makes the learning algorithm centralized and not parallel as seen from the computation point of view. Finally, besides the fact of topology conserving, we could not obtain any information about the map's spatial optimality.

16.2.3 Diffusion-Based Learning

Researches on motor learning of biological system are not limited to the two learning approaches in above subsection. In order to overcome their drawbacks, we presented a diffusion-based motor learning approach, in which each neuron only interacts with its neighbor neurons and generates a sensory-motor map with some spatial optimality.

In detail, we consider the spatial optimality of the coordination: to minimize the motor control error of the system as well as the differentiation of the motor control with respect to the sensory input overall the bounded task space. By using variational calculus, we derive a partial differential equation (PDE) of the motor control with respect to the task space. The equation includes a diffusion term. For the given boundary conditions and the initial conditions, this PDE can be solved uniquely and the solution is a well-coordinated map (Luo and Ito, 1998).

From the motor learning point of view, our approach contains both the aspects of supervised learning and self-organization. Firstly, we assumed that the forward many-to-one relation from the hand system's motor control to the task space sensory input can be obtained using supervised learning, and at the boundary, the supervisor can provide correct motor teacher information. Then, by evolving the diffusion equation, we can obtain the sensory-motor coordination overall the bounded task space.

16.2.3.1 Robotic Researches of Kinematic Redundancy

Before describing diffusion-based learning, we first briefly review the redundancy problem and summarize previous robotic approaches. Without losing generality, we only consider the kinematic nonlinear relation between the work space and the joint space which is represented as

$$\mathbf{x} = f(\theta) \tag{16.1}$$

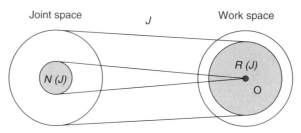

Figure 16.3 Nonlinear and redundant mapping.

where $\theta = [\theta_1, \theta_2, \ldots, \theta_m]^T$, $\mathbf{x} = [x_1, x_2, \ldots, x_n]^T$, $m > n$, and

$$d\mathbf{x} = J \, d\theta \tag{16.2}$$

where J is a $n \times m$ matrix. As shown in Figure 16.3, the range and null spaces of J are

$$R(J) = \left\{ \dot{\mathbf{x}} \in R^n : \dot{\mathbf{x}} = J(\theta)\dot{\theta} \text{ for } \forall \dot{\theta} \in R^m \right\}, \; N(J) = \left\{ \dot{\theta} \in R^m : J(\theta)\dot{\theta} = 0 \right\} \tag{16.3}$$

and dim $R(J) +$ dim $N(J) = m$.

Assuming the Jacobian J is known, we summarize five typical inverse kinematics approaches:

1.   By using the transpose of the matrix J, we calculate

$$\dot{\theta} = J^T(\mathbf{x}^d - \mathbf{x}) \tag{16.4}$$

where $\mathbf{x}^d$ is the desired end-effector position (Chiacchio et al., 1991).

2.   For the case when rank $(J) = n$, we use J^+, the pseudo-inverse of J, to obtain

$$\dot{\theta} = J^+ \dot{\mathbf{x}} \quad \text{or} \quad \dot{\theta} = J^+ \dot{\mathbf{x}} + (I - J^+ J)\eta \tag{16.5}$$

where $JJ^+ = I$ and vector $(I - J^+ J)\eta \in N(J)$. When rank$(J(\theta)) < n$, then J is singular, the joint θ is the singular configuration (Klein and Huang, 1983).

3.   By specifying additional task constraints to extend J as a full rank square matrix J_e, we have (Baillieul, 1985)

$$\dot{\theta} = J_e^{-1} \dot{\mathbf{x}} \tag{16.6}$$

4.   The regularization method to minimize the cost function $\| \, d\mathbf{x} - J d\theta \, \| + \lambda \, \| \, d\theta \, \|$.

5.   Based on compliance control, by using the relations:

$$\tau = K_\theta \, d\theta \Leftrightarrow F = K_x \, d\mathbf{x}; \quad \text{and} \quad \tau = J^T F, \quad d\mathbf{x} = J \, d\theta \tag{16.7}$$

then we have $K_\theta = J^T K_x J$, and therefore $d\theta = (J^T K_x J)^{-1} J^T K_x \, d\mathbf{x}$.

In approach 3, the specification of the additional task constraints may be closely related to the Bernstein's concept of synergy. However, from the biological point of view, the main problem inherent in all the above approaches is the assumption that the system's Jacobian is known a priori, which seems unlikely in biological system. In addition, the cost functions and task constraints considered in some of these approaches may not really be applied by the biological system. There are also several drawbacks such as: (1) all approaches need numerous computation of the Jacobian matrix and/or its pseudo-inverse; (2) the approaches 2 and 3 may be numerically unstable; and (3) the so-called quasicyclic problem (Lee and Kil, 1994). Therefore, research on how the biological system organizes its sensory-motor coordination should reflect not only the mathematical aspects of the algorithm and its computational efficiency, but also the biological reality.

16.2.3.2 Diffusion-Based Learning Algorithm

Consider again the nonlinear and redundant relation represented by an unknown function

$$\mathbf{x} = \mathbf{g}(\mathbf{y}); \quad \mathbf{x} \in R^n, \quad \mathbf{y} \in R^m, \ m \geq n \tag{16.8}$$

We try to obtain the inverse $\mathbf{y} = \mathbf{g}^{-1}(\mathbf{x})$ that minimizes a spatial criterion in a bounded task space:

$$V(\mathbf{y}) = \frac{1}{2} \int_x \left\{ \alpha(t) tr \left(\frac{\partial \mathbf{y}^T}{\partial \mathbf{x}} \frac{\partial \mathbf{y}}{\partial \mathbf{x}} \right) + \beta(t) \parallel \mathbf{A}[\mathbf{x} - \mathbf{g}(\mathbf{y})] \parallel^2 \right\} d\mathbf{x} \tag{16.9}$$

where $\alpha(t)$ and $\beta(t)$ are two adjustment coefficients, and $\mathbf{A}$ is the inverse Jacobian that will be mentioned later. Using variational method, it can be proved that the optimal solution of the inverse $\mathbf{y} = \mathbf{g}^{-1}(\mathbf{x})$ follows the PDE:

$$\frac{\partial \mathbf{y}}{\partial t} = \alpha(t)\nabla^2 \mathbf{y} + \beta(t)\mathbf{A}[\mathbf{x} - g(\mathbf{y})] \tag{16.10}$$

This PDE has two terms. The first term is a diffusion term that has the effect to interpolate the solutions of the $\mathbf{y}$ in the task space $\mathbf{x}$, while the second term acts for reducing the position errors. The discrete version of the equation is

$$\mathbf{y}_{i,j}^{t+1} = \tfrac{1}{4}\alpha(t)(\mathbf{y}_{i,j-1}^t + \mathbf{y}_{i-1,j}^t + \mathbf{y}_{i,j+1}^t + \mathbf{y}_{i+1,j}^t) + \beta(t)\mathbf{A}_{ij}^t(\mathbf{x}_{i,j}^d - \mathbf{g}(\mathbf{y}_{i,j}^t)) \tag{16.11}$$

where t is the evolution step, (i, j) are position in task space. As shown in Figure 16.4, $\mathbf{y}_{i,j}^{t+1}$ is adjusted by its four neighbor sides. Here, in order to represent all configurations of a 3 D.O.F. robot reaching its end-effector to all discrete positions of the $\mathbf{x}$ space, we reduce the scale of the robot and shift its origin to each discrete points of $\mathbf{x}$.

One of the main points in this approach is how to set the adjustment coefficients $\alpha(t)$ and $\beta(t)$ in the learning process. In our study, in order to learn the inverse Jacobian matrix $\mathbf{A}$, we set the time functions $\alpha(t)$ and $\beta(t)$, so that $\beta(t) = 1 - \alpha(t)$. For example, initially we select coefficients $\alpha = 1$ and $\beta = 1$ for only diffusion, after that, set $\alpha = 0$ and $\beta = 1$ for error correction. Therefore, during the diffusion process, the inverse matrix $\mathbf{A}$ can be obtained by

$$\mathbf{A}_{i,j}^{t+1} = \mathbf{A}_{i,j}^t + \frac{1}{\parallel \Delta \mathbf{x}_{ij}^t \parallel^2}(\Delta \mathbf{y}_{ij}^t - \mathbf{A}_{ij}^t \Delta \mathbf{x}_{ij}^t)\Delta \mathbf{x}_{ij}^{t^T} \tag{16.12}$$

considering the minimization of the cost function

$$E_{i,j} = \tfrac{1}{2} \parallel \Delta \mathbf{y}_{i,j}^t - \mathbf{A}_{i,j}^t \Delta \mathbf{x}_{i,j}^t \parallel^2 \tag{16.13}$$

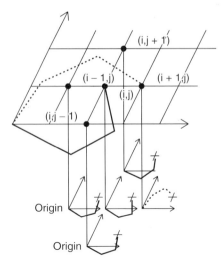

Figure 16.4   Adjusts of $\mathbf{y}_{i,j}^{t+1}$ by its four neighbor sides. Here, in order to represent all configurations of a 3 D.O.F. robot reaching its end-effector on all discrete positions of the $\mathbf{x}$ space, we give four solid line cases and one dotted line case of the scale-reduced robot's configurations and shift their origins to each discrete points of $\mathbf{x}$. "+" is then used to show the target positions that the robot should reached.

where $\Delta \mathbf{y}_{i,j}^{t} = \mathbf{y}_{i,j}^{t} - \mathbf{y}_{i,j}^{t-1}$ and $\Delta \mathbf{x}_{i,j}^{t} = \mathbf{x}_{i,j}^{t} - \mathbf{x}_{i,j}^{t-1}$ are calculated during the two learning steps using forward relation of Equation (16.8).

The final learning algorithm is summarized as follows:

1.   Use supervised learning to learn forward $\mathbf{x} = \mathbf{g}(\mathbf{y})$.
2.   Select a boundary range in the task space $\mathbf{x}$ and divide it into a $N \times N$ lattice.
3.   Perform trial motions on the boundary and remember the corresponding $\mathbf{y}$.
4.   Set the initial condition $\mathbf{y}_{i,j}^{0}$ and the initial inverse Jacobian $A_{i,j}^{0}$, respectively, for all $i, j = 1, 2, \ldots, N$, and set the time functions $\alpha(t)$ and $\beta(t)$ initially as $\alpha = 1$ and $\beta = 0$ for only diffusion, after that, set $\alpha = 0$ and $\beta = 1$ for error correction.
5.   Calculate $\Delta \mathbf{y}_{i,j}^{t}, \Delta \mathbf{x}_{i,j}^{t}, \frac{\partial E_{i,j}}{\partial A_{i,j}^{t}} = -(\Delta \mathbf{y}_{i,j}^{t} - \mathbf{A}_{i,j}^{t} \Delta \mathbf{x}_{i,j}^{t}) \Delta \mathbf{x}_{i,j}^{t\mathrm{T}}$ for $E_{i,j} = \frac{1}{2} \| \Delta \mathbf{y}_{i,j}^{t} - \mathbf{A}_{i,j}^{t} \Delta \mathbf{x}_{i,j}^{t} \|^{2}$.
6.   Adjust $\mathbf{y}_{i,j}^{t+1}$ and the inverse Jacobian matrix $A_{i,j}^{t+1}$ as in Equations (16.11) and (16.12).

Note that, for the *step 1*, since $\mathbf{x} = \mathbf{g}(\mathbf{y})$ is a function from high to lower dimension, it is possible to learn it using the general supervised learning. If we already learned the system's forward relation in *step 1*, then during performing the learning *steps* of *5* and *6*, the motor system is not necessary to perform the physical trial motions. Figure 16.5 shows the resultant map for a three-link robot arm using above learning approach. It is clear that the arm not only reaches its desired positions in all of the task space, but also the joints change smoothly with respect to the change of the arm's end-effector.

This approach has three advantages:

1.   It does not require too many trial motions for the sensory-motor system.
2.   During the map formation process, it requires only the local interactions between each node.
3.   It guarantees the final map's spatial optimality overall the bounded task space.

The detailed proof of the above diffusion-based learning algorithm using variational technique is given in Luo and Ito (1998).

It should be noted that the redundancy considered here only involves the kinematic aspect. For the redundancy problem considering the system's dynamics, refer to Arimoto's recent research (Arimoto, 2004).

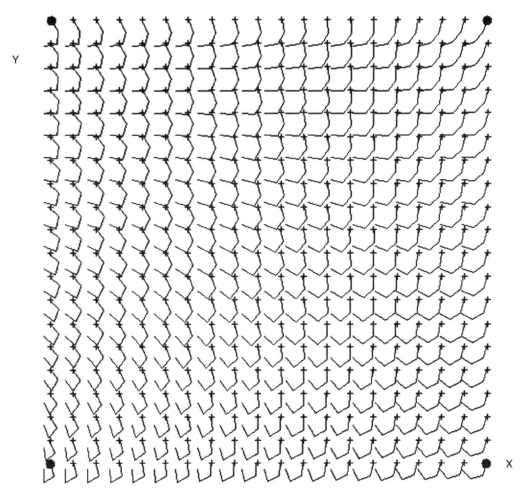

Figure 16.5   Resultant map of the 3 D.O.F. robot reaching its end-effector onto different positions of **x** space with different configurations. For the smooth change of the end-effector's position, the robot's joint angles are also changed smoothly.

16.2.3.3 Diffusion-Based Generalization of Optimal Control

Diffusion-based learning can also be effectively applied to generalize an optimal control for a robot manipulator (Luo et al., 2001).

Generally, in optimal control we have to solve a two-point boundary value problem with respect to increase and decrease of time. However, it is very difficult to solve it analytically, especially for a nonlinear system like a robot.

By now, there are many numerical approaches to solving the optimal control problem for a given set of initial and terminal conditions. However, these approaches require enormous computations. For every change in the initial and terminal conditions, they have to perform the complex computation again, which make it difficult to realize the optimal control for the robot in real time.

In our approach, we assume that, for some initial and terminal conditions, we already obtained the optimal solutions. Then, by using the diffusion-based learning algorithm, these optimal solutions can be generalized overall the bounded task space. For example, as shown in Figure 16.6, we assume that if for the initial S and four terminal conditions of T_1 to T_4, the optimal control inputs

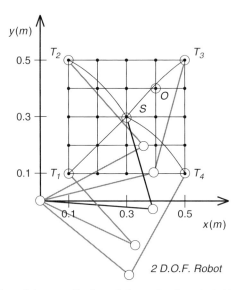

Figure 16.6   Diffusion-based spatial generalization of the optimal control. Here **S** is an initial position and **T₁** to **T₄** are four terminal positions for which we already have the optimal controls. We can then obtain the semioptimal controls from **S** to any terminal positions such as **O** without solving the complex two-point boundary value problems.

are already obtained, then, by using diffusion-based algorithm, we can obtain all semioptimal control solutions for all the initial and terminal conditions within a bounded work space as shown in Figure 16.7 without solving the nonlinear two-point boundary value problem.

Our approach greatly reduces the computational cost. In addition, since the diffusion-based learning process is completely parallel distributed, it only requires local interaction between the nodes of a learning network (a lattice) and therefore can be realized by the modern integrated circuit technology easily.

Recent neuron scientific discoveries show that, nitric oxide (NO), a gas that diffuses between neuron cells locally, can modulate the local synaptic plasticity and thus plays an important rule in motor learning and generalization (Yanagihara and Kondo, 1996). We expect that our diffusion-based learning theory may provide some mathematical understanding of the function of NO in the neural information processing and motor learning.

16.3 OPTIMAL MOTION FORMATION

In the previous section, we described on how to solve the sensory-motor organization from the redundant sensory space input to the motor control output. In this section, we consider the optimal motion formation problem for the arm to move from one position to another in the task space.

16.3.1 Optimal Free Motion Formation

For a simple human arm's point-to-point (PTP) reaching movement in free motion space, it is found experimentally that the path of human arm tends to be straight, and the velocity profile of the arm trajectory is smooth and bell-shaped (Morasso, 1981; Abend et al., 1982). These invariant features give us hints about the internal representation of motor control in the central nervous system (CNS).

One of the main approaches adopted in computational neuroscience is to account for these invariant features via optimization theory. Specifically, Flash and Hogan (1985) proposed the minimum jerk criterion

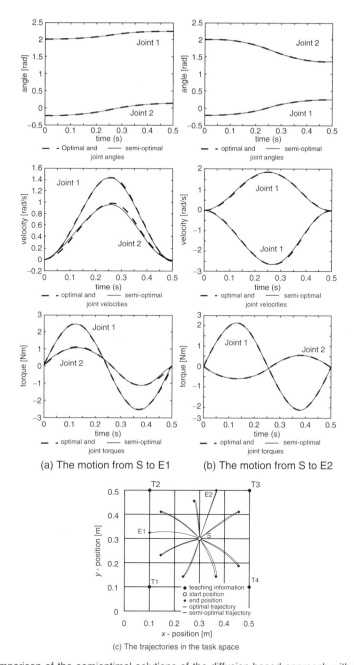

Figure 16.7 Comparison of the semioptimal solutions of the diffusion-based approach with the optimal ones. Here, (c) shows the robot's end-effector trajectories in the task space, while (a) and (b) show two examples of the time responses for the motions from **S** point to **E**1 and **E**2 points as given in (c), respectively. It is clear that the solutions of our diffusion-based approach are almost the same as those that are obtained by solving the complex two-point boundary problem.

$$J = \frac{1}{2} \int_0^{T_f} \ddot{\boldsymbol{x}}^T \ddot{\boldsymbol{x}} \, dt \qquad (16.14)$$

which shows that human implicitly plans the PTP movements in the task space. Here $\boldsymbol{x}$ is the position vector of the human arm's end-point. The optimal trajectory with zero boundary velocities and accelerations can be obtained based only on the arm's kinematic model as

$$\mathbf{x}(t) = \mathbf{x}(0) + (\mathbf{x}(T_f) - x(0))(10s^3 - 15s^4 + 6s^5) \qquad (16.15)$$

where $s = t/T_f$.

Uno et al., on the other hand, proposed to take into account about the arm's dynamics as a constraint condition when performing optimal motion planning. Based on this idea, the minimum joint torque–change criterion

$$J = \frac{1}{2} \int_0^{T_f} \dot{\tau}^T \dot{\tau} \; dt \qquad (16.16)$$

is presented (Uno et al., 1989a), which implies that human implicitly plans the PTP reaching movements in the human body space based on the arm's dynamic model. Here τ is the combined vector of the joint torques. They also expanded this model to a muscle model (Uno et al., 1989b) and proposed the minimum muscle force change criterion to show that CNS may generate unique hand trajectory by minimizing a global performance criterion of

$$J = \int_0^{T_f} \dot{\mathbf{f}}^T \dot{\mathbf{f}} \; dt \qquad (16.17)$$

where $\mathbf{f}$ is the combined vector of the muscle forces.

Kawato et al. also presented a cascade neural network model that may be possible for the nervous system to solve such a minimizing torque–change problem (Kawato et al., 1987; Miyamoto et al., 1988).

16.3.2 Optimal Motion Formation under Environmental Constraints

Studies of above section considered only the simple PTP human arm movements in the free motion space. However, how about the optimal criterion for the more complex constraint motions such as opening a door, turning a steering wheel, rotating a coffee mill, et al.?

To ask this question, we performed experiments of crank rotation task. As shown in Figure 16.8, rotating a crank requires only one degree of freedom force, however, we have to define the torques for the two joints of the arm. This is also a force redundant problem.

At the same time, we have performed many optimum calculations for the different kinds of criterions including the minimum jerk, minimum torque change, the minimum muscle force change, the minimum end-effector's interaction force change as well as our proposed criterion to minimize the combination of end-effector's interaction force change and muscle force change.

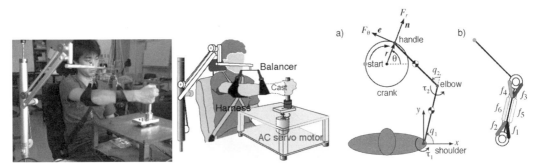

Figure 16.8 (See color insert following page 302)   Experiments of human motion formation in crank rotation tasks.

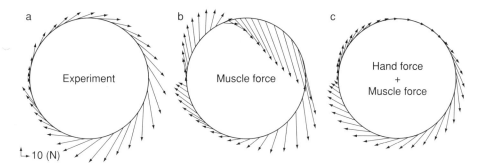

Figure 16.9 Comparison of the interaction force vectors between the human hand and the crank in experiment (a), and numerical simulations of (b) that use the minimum muscle force change criterion, (c) the combination of the hand interaction force change and the muscle force change criterion.

Compared with the experimental result of the measured force shown in Figure 16.9(a), the computational results for the cases of the minimum muscle force change criterion and the combination of the hand interaction force change and muscle force change criterion are, respectively, shown in Figure 16.9(b) and (c) (Ohta et al., 2004).

From Figure 16.9, it is observed that the predicted numerical result of the contact force vectors when using the minimum muscle force change criterion (which was proposed for P.T.P. motion in the free motion space) is inadequate here. Instead, the human arm tends to minimize

$$J = \int_0^{T_f} (\dot{\mathbf{F}}^{\mathrm{T}}\dot{\mathbf{F}} + w\dot{\mathbf{f}}^{\mathrm{T}}\dot{\mathbf{f}})\mathrm{d}t$$

the combination of the hand interaction force $\mathbf{F}$ change and the muscle force $\mathbf{f}$ change as in Figure 16.9(c) and Figure 16.10. Therefore, we strongly suggest that human arm movement is realized by different optimal criterions according to different task conditions as well as task requirements.

The combined criterion also captures well the muscle activities in the constrained multi-joint motions. It covers both the motions in the free motion space and the constrained motion space, since in the free motion space the interaction force at the end-effector is zero. Therefore, the combined criterion reduces merely to the minimum muscle force change criterion in the free motion space. How can the central nervous system measure the hand contact force and how can it solve the optimal constraint dynamic motion control are left as open questions.

16.4 MECHANICAL INTERACTION AND ENVIRONMENTAL ADAPTATION

This section further goes to describe the motor control functions on the mechanical interaction with dynamic environment. It is well known that human can perform physical interactions with uncertain dynamic environment skillfully. In fact, through force feedback from tendon and the co-activation of antagonist muscles, human can control the arm's mechanical impedance adaptively with respect to the environmental dynamics so as to realize the desired time response of the motion as well as the contact force (Hogan, 1984).

In order to realize such adaptive motor functions by a robot, we should not only search for the soft artificial actuators such as biological muscles, but also discover the control principles of the motor functions. Technically, according to the task requirements, the contact tasks can be specified

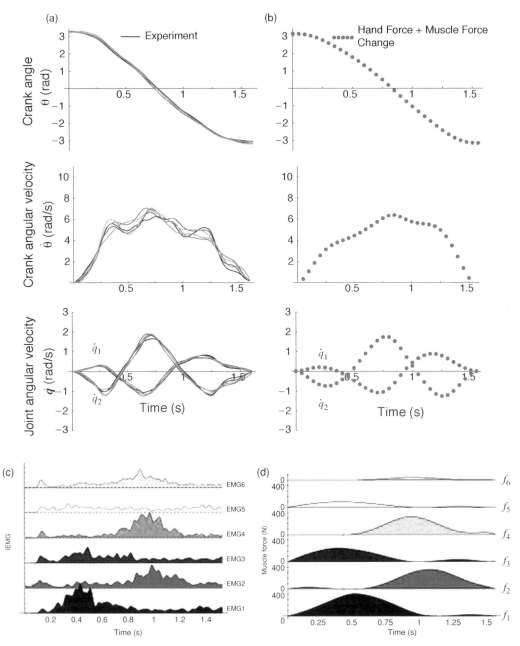

Figure 16.10 (See color insert following page 302) Comparison of the time responses of the human hand motions in (a), the EMG in (c) from the experiments, and the numerical simulations in (b) and (d) with our combined criterion of the hand interaction force change and muscle force change.

into two classes: those that require compliant interaction with the environment such as to push and open a door, and those that require to impose some exact force to the environment. With respect to these two different contact requirements, impedance control and explicit force control have been proposed, respectively. In this section, we summarize these two control approaches and introduce their recent developments.

16.4.1 Impedance Control

Let us formulate a robot's dynamic equation in contact task space as

$$\mathbf{I}_r(\mathbf{x})\ddot{\mathbf{x}} + \mathbf{C}_r(\mathbf{x}, \dot{\mathbf{x}})\dot{\mathbf{x}} = \mathbf{u}_r - \mathbf{f} \qquad (16.18)$$

where $\mathbf{I}_r(\mathbf{x})$ is the robot's inertia matrix, $\mathbf{c}_r(\mathbf{x},\dot{\mathbf{x}})\dot{\mathbf{x}}$ is the centrifugal and Coriolis force vector, respectively. $\mathbf{u}_r$ is the robot's control input vector and $\mathbf{f}$ is the contact force from the environment. For simplicity, let us consider the environmental dynamics as

$$\mathbf{M}_e\ddot{\mathbf{x}} + \mathbf{D}_e\dot{\mathbf{x}} + \mathbf{K}_e\mathbf{x} = \mathbf{f} \qquad (16.19)$$

where $\mathbf{M}_e$, $\mathbf{D}_e$, and $\mathbf{K}_e$ are the environmental inertia, viscosity, and stiffness, respectively, and is assumed unknown. It is well known that by specifying nonlinear compensation of

$$\mathbf{u}_r = \mathbf{f} + \mathbf{M}_r^{-1}\mathbf{I}_r(\mathbf{x})[\mathbf{u} - \mathbf{D}_r\dot{\mathbf{x}} - \mathbf{k}_r\mathbf{x}] + \mathbf{c}_r(\mathbf{x},\dot{\mathbf{x}})\dot{\mathbf{x}} \qquad (16.20)$$

the robot's dynamics becomes

$$\mathbf{M}_r\ddot{\mathbf{x}} + \mathbf{D}_r\dot{\mathbf{x}} + \mathbf{K}_r\mathbf{x} = \mathbf{u} - \mathbf{f} \qquad (16.21)$$

where $\mathbf{M}_r$, $\mathbf{D}_r$, and $\mathbf{K}_r$ are the robot's inertia, viscosity, and stiffness, respectively, and $\mathbf{u}(t)$ is a new control input of the robot that we should design late.

In impedance control, we usually control the robot's position and use the force feedback to adjust the robot's mechanical impedance as seen from the environment so as to keep a compliant contact with the environment (Hogan, 1985; Luo and Ito, 1993).

In detail, the control input is designed as

$$\mathbf{u} = C_x(\mathbf{x}_d - \mathbf{x}) + C_f\mathbf{f}$$

so that the robot dynamics as seen from the environment be as

$$\mathbf{M}_{rd}\mathbf{x} + \mathbf{D}_{rd}\dot{\mathbf{x}} + \mathbf{K}_{rd}(\mathbf{x} - \mathbf{x}_d) = -\mathbf{f}$$

where C_x and C_f are the robot's position and force feedback controllers, respectively.

From the stability point of view, we usually require the robot to be passive with respect to the environmental interactions. Passivity is defined as the property that the system does not flow energy to outside. The robot's passivity as seen from its environment or the manipulated object is very useful for the stable and safely mechanical interaction. When applying impedance control, if the desired position $\mathbf{x}_d$ is constant, then the robot is passive. However, if the $\mathbf{x}_d$ changes with respect to time, then the robot may lose the passivity as seen from the environment.

In order for the robot to realize the passivity while performing the time varying interactions, Li and Horowitz (1999) proposed a passive velocity field control (PVFC), they also suggested to apply PVFC to control a human interactive robot and smart exercise machines. Unlike the passivity based control scheme by Slotine and Li (1991), in which they considered the passivity of a tracking error system, PVFC remains passive of the robot with respect to the external environment by adding a virtual flywheel to exchange the mechanical energy with the real robot. However, PVFC has the following two main problems. Firstly, when specifying desired velocity vector field, PVFC does not consider the uncertainties of the environmental geometric constraints. Secondly, although PVFC maintains the passivity, the contact task performance cannot be adjusted with respect to the environment dynamics. In most contact tasks, we do not know the environmental constraints before performing the tasks; specifically, its shape, size, location as well as its mechanical dynamics. Therefore, the environmental uncertainties will influence the robot's task performance even using PVFC.

Mussa-Ivaldi et al., on the other hand, investigated the organization of motor output of spinally dissected frogs. They stimulated the spinal cord with microelectrodes and observed isometric forces produced by the muscles of the legs. It is reported that, elicited by a single stimulation, the ankle position converges to a single equilibrium point of a force vector field, and the point of convergence is shifted by superposing several vector fields resulting from multiple stimulations (Mussa-Ivaldi and Gister, 1992). Inspired by the biological studies on primitive motor behavior, we proposed an adaptive PVFC approach where learning of the unknown environmental geometry is based on the vector field interpolation theory (Luo et al., 1995; Saitoh et al., 2004). In our approach, we first parameterize the desired velocity vector field by the weighted combination of a set of basis vector fields according to the environmental model. Then, in order to overcome the influences from the model uncertainties of the environment, we use force feedback to adjust the weight parameters of the desired velocity field for the robot to approach the real environment.

For the dynamic environment, in order to maintain the passivity of the impedance control with time varying impedance center, a sufficient condition to adjust the impedance center is derived as follows (Kishi et al., 2003).

Theorem: *Define* $V(\mathbf{x})$ *as the robot's desired velocity vector of impedance center in the task space without considering the environment uncertainties,* $\dot{\mathbf{x}}_0 = \alpha V(\mathbf{x})$*, the adjustable robot impedance center, where* α *is an adjustment parameter. For a given constant* $r > 0$*, if we adjust*

$$\alpha \leq -\frac{rS + \dot{\mathbf{x}}^{T}\mathbf{D}_{rd}\dot{\mathbf{x}}}{z}, \quad when \quad z < 0$$

$$\alpha \geq -\frac{rS + \dot{\mathbf{x}}^{T}\mathbf{D}_{rd}\dot{\mathbf{x}}}{z}, \quad when \quad z > 0$$

then the impedance controlled robot with time varying impedance center remains passive with respect to the supply rate $\dot{\mathbf{x}}^{T}\mathbf{f}$ *for any initial* $S_{O} > 0$ *and any velocity vector* $V(\mathbf{x})$*. Where*

$$z \equiv V(\mathbf{x})^{T}\mathbf{K}_{rd}(\mathbf{x} - \mathbf{x}_0), \quad and \quad \dot{S} \equiv \dot{\mathbf{x}}^{T}\mathbf{D}_{rd}\dot{\mathbf{x}} + \dot{\mathbf{x}}_0^{T}\mathbf{K}_{rd}(\mathbf{x} - \mathbf{x}_0)$$

$\mathbf{x}$ *is the robot's position and* $\mathbf{f}$ *is contact force between the robot and its environment.*

16.4.2 Force Control

Here we consider another contact task requirement to realize a specific desired interaction force to the environment at the end-effector of a robot manipulator as shown in Figure 16.11.

For above robot's dynamics (16.18), the basic control objective considered here is to design the robot's control input $\mathbf{u}_r$ such that the robot realizes the desired contact force $\mathbf{f}_d$.

By Equations (16.19) and (16.21), the force controlled object from $\mathbf{u}$ to $\mathbf{f}$ becomes

$$\mathbf{P}(s) = [\mathbf{I} + \mathbf{E}(s)\mathbf{R}(s)]^{-1}\mathbf{E}(s)\mathbf{R}(s) \qquad (16.22)$$

which includes the robot dynamics $\mathbf{R}(s) = [\mathbf{M}_r s^2 + \mathbf{D}_r s + \mathbf{K}_r]^{-1}$ and the dynamic environment $\mathbf{E}(s) = \mathbf{M}_e s^2 + \mathbf{D}_e s + \mathbf{K}_e$, $\mathbf{P}(s)$ is usually assumed to be biproper.

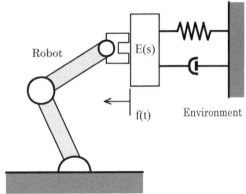

Figure 16.11   Interaction between a robot and its unknown dynamic environment.

By now, there are many approaches to designing the force control loop. Among them, the simplest one is the constant feedforward + PI feedback control

$$\mathbf{u}(t) = \mathbf{f}_d + K_P(\mathbf{f}_d - \mathbf{f}(t)) + K_I \int_0^t (\mathbf{f}_d - \mathbf{f}(\tau))d\tau \qquad (16.23)$$

that uses the force feedback error to generate the control input (Raibert and Craig, 1981). However, since the feedback control always has loop delay, the robot is impossible to realize the exact force tracking.

In what follows, we simply consider the scalar case and introduce a recent two D.O.F. adaptive tracking control theory. We then study its application to the robot force tracking control. We will show by computer simulations and experiments that, if we perform this adaptive control for the complex time varying force reference (we call it a PE condition and will explain it later), then after the adaptation process, the robot can realize exact force tracking without any loop delay.

16.4.2.1 Two D.O.F. Adaptive Tracking Control

Referenced from the biomimetic research of biological motor control functions of brain, Ito and Kawato proposed a feedback error learning approach to show that brain may learn the inverse dynamics model of the controlled object within the cerebellum, and uses this model in the feedforward loop so as to realize exactly tracking of the desired motions (Ito, 1984; Miyamoto, et al., 1988). Artificial neural network was also applied to learn the inverse dynamics model.

Inspired by these biomimetic studies of the feedback error learning, Miyamura et al. (2002) proposed a two D.O.F. adaptive control limited to linear systems. By assuming that the controlled plant is biproper together with a condition on the strictly positive realness of a specific transfer function (see Miyamura and Kimura, 2002 for the details), they proposed an adaptation algorithm and proved its convergence. However, since the specific transfer function in their condition contains both the feedback controller and the unknown parameters in the inverse model, the strictly positive realness condition requires a high gain in the feedback controller with respect to the unknown parameters of the inverse model. This makes it difficult to design the feedback controller in advance. To overcome this problem, Muramatsu and Watanabe (2004) proposed a new two D.O.F. adaptation algorithm without the assumption on the strictly positive realness condition. Here, we briefly introduce this algorithm.

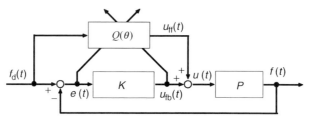

Figure 16.12 Two D.O.F. adaptive control.

As shown in Figure 16.12, P is an unknown transfer function from the control input $u(t)$ to the control output $f(t)$, K, and $Q(\theta)$ are the feedback and feedforward controllers, respectively, with the adjustable parameter θ. P is assumed unknown and is stable and has stable inverse. The objective here is not only limited to the convergence of the control error $e(t) \to 0$ as in the previous adaptive control or learning control researches, but also to realize the inverse of P by $Q(\theta)$. It is clear that, once $Q(\theta) \to P^{-1}$, then the control system can track any other types of desired signals without any loop delay. The only condition here is to perform an adaptation process with respect to a desired input that satisfies a PE condition that will be explained later in the theorem.

Firstly, let us describe the state space equation of the unknown P^{-1} as

$$\frac{d\eta_1(t)}{dt} = F\eta_1(t) + gf_d(t) \tag{16.24}$$

$$\frac{d\eta_2(t)}{dt} = F\eta_2(t) + gu_0(t) \tag{16.25}$$

$$u_0(t) = c_0^T \eta_1(t) + d_0^T \eta_2(t) + k_0 f_d(t) = \theta_0^T \eta(t) \tag{16.26}$$

$$\theta_0 = [\, c_0^T \quad d_0^T \quad k_0 \,]^T, \; \eta(t) = [\, \eta_1^T(t) \quad \eta_2^T(t) \quad f_d(t) \,]^T$$

Here, F is any stable matrix and g is any vector with (F, g) being controllable. In Equation (16.26),

$$c_0 = [\, c_1 \quad c_2 \quad \cdots \quad c_n \,]^T$$

$$d_0 = [\, d_1 \quad d_2 \quad \cdots \quad d_n \,]^T$$

and k_0 are unknown parameters to be estimated. The parameterization of (16.24)–(16.26) can yield an arbitrary transfer function from f_d to $u_0(t)$. If the matrices F and g are represented in the controllable canonical form

$$F = \begin{bmatrix} 0 & 1 & 0 & \cdots & 0 \\ \vdots & \ddots & \ddots & \ddots & \vdots \\ \vdots & & \ddots & \ddots & 0 \\ 0 & \cdots & \cdots & 0 & 1 \\ -f_1 & -f_2 & \cdots & \cdots & -f_2 \end{bmatrix}, \quad g = [\, 0 \quad \cdots \quad 0 \quad 1 \,]^T$$

then the transfer function from f_d to u_0 can be calculated as

$$T_{f_d, u_0} = P^{-1}(s) = \frac{k_0 + c_0^T(sI - F)^{-1}g}{1 - d_0^T(sI - F)^{-1}g} = \frac{k_0 s^n + (f_n k_0 + c_n)s^{n-1} + \cdots + (f_1 k_0 + c_1)}{s^n + (f_n - d_n)s^{n-1} + \cdots + (f_1 - d_1)} \tag{16.27}$$

which is parameterized by $c_0 = \begin{bmatrix} c_1 & c_2 & \cdots & c_n \end{bmatrix}^T$, $d_0 = \begin{bmatrix} d_1 & d_2 & \cdots & d_n \end{bmatrix}^T$, and k_0.

With respect to the unknown system of (16.24)–(16.26), the feedforward controller $Q(\theta)$ is described by

$$\frac{d\xi_1(t)}{dt} = F\xi_1(t) + gf_d(t) \tag{16.28}$$

$$\frac{d\xi_2(t)}{dt} = F\xi_2(t) + gu(t) \tag{16.29}$$

$$u_{ff}(t) = c^T\xi_1(t) + d^T\xi_2(t) + kf_d(t) = \theta^T\xi(t) \tag{16.30}$$

so as to generate the feedforward input u_{ff} from the desired force signal $f_d(t)$. Here

$$\theta = \begin{bmatrix} c^T & d^T & k \end{bmatrix}^T, \quad \xi(t) = \begin{bmatrix} \xi_1(t)^T & \xi_2(t)^T & f_d(t) \end{bmatrix}^T \tag{16.31}$$

and the robot's control input $u(t)$ *is then given by*

$$u(t) = u_{ff} + u_{fb} = \theta^T(t)\xi(t) + K(s)e(t) \tag{16.32}$$

Next, introducing a new state equation for the vector $\xi_e(t)$ with respect to the control error input $e(t) = f_d(t) - f(t)$ as

$$\frac{d\xi_e(t)}{dt} = F\xi_e(t) + ge(t) \tag{16.33}$$

and defining

$$\tilde{\xi}(t) := \begin{bmatrix} \xi_1(t)^T - \xi_e(t)^T & \xi_2(t)^T & f_d(t) - e(t) \end{bmatrix}^T \tag{16.34}$$

we can express the total control input $u(t)$ as

$$u(t) = \theta_0^T\tilde{\xi}(t) \tag{16.35}$$

which is linearly parameterized by θ_0, see Muramatsu and Watanabe (2004) for the details of the derivation.

Finally, to derive an adaptive rule, let us define

$$\hat{u}(t) := \theta^T(t)\tilde{\xi}(t) \tag{16.36}$$

by replacing θ_0 in Equation (16.35) with $\theta(t)$, and defining an error signal $\varepsilon(t)$ as

$$\varepsilon(t) := u(t) - \hat{u}(t) = \{\theta_0 - \theta(t)\}^T\tilde{\xi}(t) = -\psi(t)^T\tilde{\xi}(t) \tag{16.37}$$

where

$$\psi(t) := \theta(t) - \theta_0 \tag{16.38}$$

Substituting Equations (16.32) and (16.36) into (16.37), the signal $\varepsilon(t)$ is transformed to

$$\varepsilon(t) = u_{\text{fb}}(t) + c(t)^{\text{T}}\xi_{\text{e}}(t) + k(t)e(t) \tag{16.39}$$

Using these signals, we have then the following two D.O.F. adaptive control theorem:

Theorem: *For the force controlled object P(s) with its unknown inverse dynamics given as (16.24)–(16.26), if we adjust the parameter of the feedforward controller (16.28)–(16.30) as*

$$\frac{d\theta(t)}{dt} = \alpha\tilde{\xi}(t)\left\{u_{\text{fb}}(t) + c(t)^{\text{T}}\xi_{\text{e}}(t) + k(t)e(t)\right\} \tag{16.40}$$

then the force control error e(t) → 0. In addition, if $\tilde{\xi}(t)$ satisfies the PE condition such that $\tilde{\xi}(t)\tilde{\xi}(t)^{T}$ be positive, then the feedforward controller Q(θ) described by (16.28)–(16.30) tends to P^{-1}

The detailed prove of this theorem is given in Muramatsu and Watanabe (2004).

The adaptation law (16.40) can be interpreted as a combination of the feedback error learning and the learning control, since $\theta(t)$ is adjusted by both the feedback input $u_{\text{fb}}(t)$ and the feedback error $e(t)$. In addition, $\xi_{\text{e}}(t)$ is also generated by $e(t)$.

Note that the convergence of $Q(\theta) \rightarrow P^{-1}$ means that we can realize the time response $f(t)$ exactly as the desired $f_{\text{d}}(t)$ without any feedback loop delay. However, in order to realize this convergence, it is necessary for the desired $f_{\text{d}}(t)$ to satisfy the PE condition during adaptation process.

The convergence of adaptation can be increased further by the following modification:

$$\frac{d\theta(t)}{dt} = \Gamma\tilde{\xi}\varepsilon(t) \quad \text{and} \quad \frac{d\Gamma}{dt} = -\Gamma\tilde{\xi}\tilde{\xi}^{T}\Gamma \tag{16.41}$$

instead of Equation (16.40).

16.4.2.2 Application to a Robot's Force Tracking Control

To evaluate the effectiveness of above two D.O.F. adaptive tracking control, we performed computer simulations and robotic experiments.

In the simulations, as shown in Figure 16.13, we set the robot's parameters as $m_{\text{r}} = 1$, $d_{\text{r}} = 2$, and $k_{\text{r}} = 0.5$; and the unknown dynamic environmental parameters as $m_{\text{e}} = 1$, $d_{\text{e}} = 2$, and $k_{\text{r}} = 2$, respectively, at the beginning of the simulation. The simulation results are given in Figure 16.14, where Figure 16.14(a) shows the result for the adaptation law (16.40), and (b) is the fast convergence result when using (16.41). We change the environmental viscosity from $d_{\text{e}} = 2$ to 0.5 at the simulation time $t = 250[s]$ in Figure 16.14(a) and at $t = 15[s]$ in (b). In order for the feedforward controller ($Q(\theta)$ in Figure 16.12) to converge to the inverse of the force controlled object

$$P(s) = \frac{m_{\text{e}}s^2 + d_{\text{e}}s + k_{\text{e}}}{(m_{\text{e}} + m_{\text{r}})s^2 + (d_{\text{e}} + d_{\text{r}})s + (k_{\text{e}} + k_{\text{r}})},$$

we set the desired force $f_{\text{d}}(t)$ as noise at the beginning 100[s] and during the time of 250[s] to 350[s] in Figure 16.14(a) and at the beginning 4[s] in (b).

In both cases, since the force tracking error converged very fast, it is hard to distinguish between the desired and the reaction forces in these figures. Figure 16.14 also shows that the unknown parameters of the robot and environment are converged to the real parameters, which means that the feedforward compensation realizes the exact inverse of the force controlled object $P(s)$. Therefore, even for the rectangle type of desired forces, the control system can realize exactly

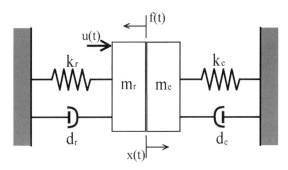

Figure 16.13   Simulation of a robot's force control while interacting with its unknown dynamic environment.

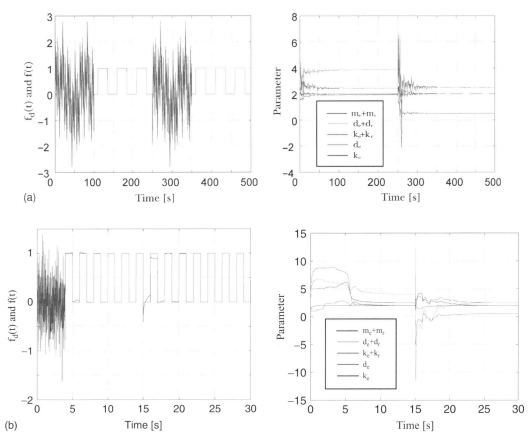

Figure 16.14 (See color insert following page 302)   Simulation results on the time responses of the robot's force (the blue lines) tracking a time varying desired force (the green lines), and the parameter convergences of the feedforward controller. The environmental viscosity is set from 2 to 0.5 at the time $t = 250[s]$ in (a) and $t = 15[s]$ in (b). At the beginning 100[s], 250[s] to 350[s] in [a] and the beginning 4[s] in (b), the desired forces are set as noises.

the same reaction force. It is also clear that when the environment changes its dynamics (at simulation time $t = 250[s]$ in Figure 16.14(a) or $t = 15[s]$ in (b)), only if the desired forces satisfy the PE condition, we can obtain the new inverse of the force controlled object. Therefore, the adaptation (16.40) or (16.41) has the so-called generality as we discussed in the learning theory of artificial neural networks.

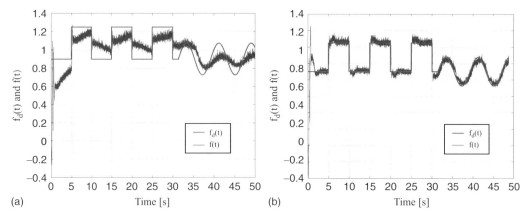

Figure 16.15 (See color insert following page 302) Time responses of the force tracking control experiments of a robot interacting with its unknown dynamic environment. The experimental setup is in Figure 16.11. Here, (a) is the result when using usual constant feedforward + PI feedback control (23), while (b) is the result of 2 D.O.F. adaptive control (16.41). The feedback control of (b) is used as same as that in (a). (a) Experimental result using PI control; (b) experimental result of adaptive control.

Experiments of the robot force tracking control are also performed on the robot system shown in Figure 16.11. The two-link robot is firstly controlled using a high gain position feedback so that it can only move along the force control direction. The environment is set as a second-order mass-spring-damper system as that in Figure 16.13.

Figure 16.15(a) shows the time response of the reaction force when using the usual constant feedforward + PI control as in Equation (16.23), while (b) is the result of the two D.O.F. adaptive tracking control that uses the same PI feedback controller as in (a) but the feedforward controller of (16.28)–(16.30) and (16.33) with the adaptation law (16.41). The matrices F and g in (16.28), (16.29), and (16.33) are set as

$$F = \begin{bmatrix} 0 & 1 \\ -5 & -6 \end{bmatrix}, \quad g = \begin{bmatrix} 0 \\ 1 \end{bmatrix}.$$

Technically, in order to overcome the influence of force measurement noise in the experiments, we simply choice to stop the adaptation if the signal $\varepsilon(t)$ in Equation (16.41) is small.

From the experiments, it is clear that the adaptive tacking control approach can lead to better tracking performance with almost no transitional delay, even if we change the desired force to the sine wave (at the time 30[s] in the experiments here).

16.5   CONCLUSIONS AND FURTHER RESEARCHES

Traditional biomimetic researches mainly abstract the specific functions of specific biological systems under specific environmental conditions. As a result, many artificial systems such as airplanes, computers as well as present robots were generated. These efforts not only realized but also greatly improved the specific functions within artificial systems, sometimes far beyond the biological systems. However, as seen from the system flexibility, diversity, and the environmental adaptability, there still exists too large a gap between the biological and artificial systems. This gap exists not only at the material levels such as muscle and skeletal structures but also at system control level.

In this chapter, we have introduced some of the recent researches from the point of view of biomimetic control. Especially we have concentrated our discussions on how to organize the system redundancy, the optimal motion formation, and the environmental adaptive control.

There are also many other issues, such as autonomous decentralized system control and hybrid system control, that is not mentioned here but are basically important in biomimetic control.

An autonomous decentralized system is a system in which the functional order of the entire system is generated by cooperative interactions among its subsystems (Yuasa and Ito, 1990). Each subsystem has the autonomy to control a part of states of the system. It is well known that biological systems possess this autonomous decentralized characteristic. For example, animal movements are generated by the cooperation of many motor neurons. These motor neurons control muscular fibers to generate different movements. Rhythmic movements, such as walking, flying and swimming, are generated by the control from the mutually coupled endogenous neural oscillators. The rhythmic moving patterns can be changed with respect to the environmental conditions as well as various objectives. The walking movement of a cat is a typical example. Here, the periodic motions of each limb cooperate with each other to generate stable gait patterns. As the cat moves faster, the gait pattern changes from "walk" to "trot," and finally to "gallop." Although the gait changes discontinuously, the speed of the body movement varies continuously. This pattern switching behavior can be formulated based on the bifurcation theory of nonlinear dynamic systems. In addition, the graph structure of the interaction network between all subsystems and the time delay of the interactions are important to determine the over system's performance such as synchronization (Wu and Chua, 2002; Amano et al., 2004).

On the other hand, if the system contains both continuous time-driven and discrete event-driven dynamics, it is called a hybrid system. Human brain has a high ability to mix discrete logical thinking with dynamic body movements control simultaneously. It realizes the diversity of the movements with respect to the environmental conditions as well as task requirements. The biological research on *basal ganglia* suggests that such kind of a skillful optimal motion pattern scheduling function is generated through the interaction between the *basal ganglia* and the high level *motor cortex* (Hikosaka et al., 1996). Nowadays, hybrid system arises in a large number of application areas. However, the problem of the hybrid system is inherently difficult because of its combinatorial nature. A straightforward application of the available frameworks faces the limitation of computational complexity and lacks the theoretical prediction of system properties. Hybrid systems can be formulated by many kinds of models, such as piecewise affine (PWA) system, linear complementary (LC) system, and mixed logical dynamical (MLD), system, etc. The equivalence between each model formulation was studied. For the MLD system, several powerful mixed integer quadric programming (MIQP) algorithms have been proposed to solve the on-line optimization procedures (Bemporad, 1999). In robotic applications, recently, modeling and control of dexterous multi-fingered hand operations and multi-legged dynamic walking movements are studied from the hybrid system point of view (Yin et al., 2003). But still few theoretic results have been achieved, and there remain many challenging problems in biomimetic control research. It is expected that theoretical study of hybrid system control may also lead to a better understanding of biological motor control functions at the high level of brain motor cortex and basal ganglia.

ACKNOWLEDGMENT

The authors are pleased to thank Professor Neville Hogan, Dr Yoseph Bar-Cohen, Dr Mikhail Svinin, and all reviewers for their very contributive comments and help.

REFERENCES

Abend W., Bizzi E., and Morasso P., Human arm trajectory formation, *Brain*, 105 (1982), pp. 331–348.

Amano M., Luo Z.W., and Hosoe S., Graph dependant sufficient conditions for synchronization of dynamic network system with time-delay, *Trans. SICE*, 40 (2004).

Amari S.I., Topographic organization of nerve fields, *Bull. Math. Biol.*, 42 (1980), pp. 339–364.

Arimoto S., A natural resolution of Bernstein's degrees-of-freedom problem in case of multi-joint reaching, *Proceedings of the IEEE International Conference on Robotics and Biomimetics* 445 (2004).

Baillieul J., Kinematic programming alternatives for redundant manipulators, *Proceedings of IEEE International Confernce on Robotics and Automation* (1985), pp. 722–728.

Bellman R. and Kalaba R., Dynamic programming and feedback control, *Proceedings of 1st IFAC* (1960), pp. 460–464.

Bemporad A. and Morari M., Control of systems integrating logic, dynamics, and constraints, *Automatica*, 35(3) (1999), pp. 407–427.

Bernstein N., *The Coordination and Regulation of Movements*, Pergamon Press, London (1967).

Boltyanski V.G., Gamkrelidge R.V., Mishchenko E.F., and Pontryagin L.S., The maximum principle in the theory of optimal processes of control, *Proceedings of 1st IFAC* (1960), pp. 454–459.

Chiacchio P., Chiaverini S., Sciavicco L., and Siciliano B., Closed-loop inverse kinematic schemes for constrained redundant manipulators with task space augmentation and task priority strategy, *Int. J. Robotics Res.*, 10 (1991), pp. 410–425.

Draper C.S. and Li Y.T., *Principles of Optimalizing Control Systems and an Application to Internal Combustion Engine*, ASME Publications, Mawson, ACT (1951).

Doyle J.C., Glver K., Khargonekar P.P., and Francis B., State-space solutions to standard H2 and H8 control problems, *IEEE Trans. Automatic Control*, 34 (1989), pp. 831–847.

Flash T. and Hogan N., The coordination of arm movements, *J. Neurosci.*, 5 (1985), pp. 1688–1703.

Funahashi K.I., On the approximate realization of continuous mapping by neural networks, *Neural Networks*, 2 (1989), pp. 183–192.

Gorinevsky D., Modeling of direct motor program learning in fast human motions, *Biol. Cyber.*, 69 (1993), pp. 219–228.

Guez A. and Ahmad Z., Solution to the inverse kinematics problem in robotics by neural networks, *Proceedings of IEEE International Joint Conference on Neural Networks*, 2 (1988), pp. 617–624.

Hikosaka O., Sakai K., Miyauchi S., Takino R., Sakai Y., and Putz B., Activation of human pre-SMA in learning of sequential procedures — a functional MRI study, *J. Neurophysiol.* 76(1) (1996), pp. 617–621.

Hogan N., Adaptive control of mechanical impedance by coactivation of antagonist muscles, *IEEE Trans. Automatic Control*, 29(8) (1984), pp. 681–690.

Hogan N., Impedance control, an approach to manipulation: Part I, II, *Int. J. of Robotics Res.*, 107 (1985), pp. 1–24.

Ito M., *The Cerebellum and Neural Control*, Raven Press, New York (1984).

Jordan M.I. and Rumelhart D.E., Forward models: supervised learning with a distal teacher, *Cogn. Sci.*, 16 (1992), pp. 307–354.

Kalman P.E., On the general theory of control systems, *Proceedings of 1st IFAC* (1960), pp. 481–492.

Kawato M., Furukawa K., and Suzuki R., A hierarchical neural-network model for control and learning of voluntary movement, *Biol. Cyber.*, 57 (1987), pp. 169–185.

Kishi Y., Luo Z.W., Asano F., and Hosoe S., Passive impedance control with time-varying impedance center, *The 5th IEEE International Symposium on Computational Intelligence in Robotics and Automation* (2003).

Klein C.A. and Huang C.H., Review of pseudo-inverse control for use with kinematically redundant manipulators, *IEEE Trans. Syst. Man Cyber.*, 13 (1983), pp. 245–250.

Kohonen T., Self-organized formation of topographically correct feature maps, *Biol. Cyber.*, 43 (1982), pp. 59–69.

Kuperstein M., Neural model of adaptive hand-eye coordination for single postures, *Science*, 239 (1988), pp. 1308–1311.

Lee S. and Kil R.M., Redundant arm kinematic control with recurrent loop, *Neural Networks*, 7 (1994), pp. 643–659.

Li P.Y. and Horowitz R., Passive velocity field control of mechanical manipulators, *IEEE Trans. Robotics Automation*, 15(4) (1999), pp. 751–763.

Luo Z.W. and Ito M., Control design of robot for compliant manipulation on dynamic environments, *IEEE Trans. Robotics Automation*, 9(3) (1993), pp. 286–296.

Luo Z.W. and Ito M., Diffusion-based learning theory for organizing visuo-motor coordination, *Biol. Cyber.*, 79 (1998), pp. 279–289.

Luo Z. W., Ito M., and Yamakita M., Estimation of environment models using vector field and its application to robot's contact tasks, *Proceedings of IEEE International Conference on Neural Networks* (1995), pp. 2546–2549.

Luo Z.W., Ando H., Hosoe S., Watanabe K., and Kato A., Spatial generalization of optimal control for robot manipulators, *J. Robotics Mechatronics*, 12(5) (2001), pp. 533–539.

Miyamoto H., Kawato M., Setoyama T., and Suzuki R., Feedback-error-learning neural network for trajectory control of a robotic manipulator, *Neural Networks*, 1 (1988), pp. 251–265.

Miyamura A. and Kimura H., Stability of feedback error learning scheme, *Syst. Control Lett.*, 45 (2002), pp. 303–316.

Morasso P., Spatial control of arm movements, *Exp. Brain Res.*, 42 (1981), pp. 223–227.

Muramatsu E. and Watanabe K., Feedback error learning control without recourse to positive realness, *IEEE Trans. Automatic Control*, 10 (2004), to appear.

Mussa-Ivaldi F.A. and Gister S.F., Vector field approximation: a computational paradigm for motor control and learning, *Biol. Cyber.*, 67 (1992), pp. 491–500.

Ohta K., Svinin M.M., Luo Z.W., Hosoe S., and Laboissiere, R., Optimal trajectory formation of constrained human arm reaching movements, *Biol. Cyber.*, 91(1) (2004), pp. 23–36.

Pavlov I.P., *Twenty Years of Experience in the Objective Study of Higher Nervous Activity in Animals*, Moscow (1923).

Raibert M.H. and Craig J.J., Hybrid position/force control of manipulators, *Trans. ASME J. DSMC*, 102 (1981), pp. 126–133.

Ritter H.J., Martinetz T.M., and Schulten K.J., Topology-conserving maps for learning visuo-motor-coordination, *Neural Networks*, 2 (1989), pp. 159–168.

Rosenblatt F., *Principles of Neurodynamics*, Spartan, New York (1962).

Rumelhart D.E., Hinton G.E., and Williams R.J., Learning internal representations by error propagation, in Rumelhart D.E. and McClelland J.L. (eds), *Parallel Distributed Processing*, 1, MIT Press, Cambridge, MA (1986).

Saitoh Y., Luo Z.W., and Watanabe K., Adaptive modular vector field control for robot contact tasks in uncertain environments, *J. Robotics Mechatronics*, 16(4) (2004), pp. 374–380.

Sherrington Sir C. S., *The Integrative Action of the Nervous System*, Cambridge University Press, Cambridge, MA (1906).

Slotine J.E. and Li W., *Applied Nonlinear Control*, Prentice Hall, Englewood cliffs, NJ (1991).

Snowdon C.T., *Significance of Animal Behavior Research*, in http://www.animalbehavior.org

Uno Y., Kawato M., and Suzuki R., Formation and control of optimal trajectory in human multi-joint arm movement minimum torque change model, *Biol. Cyber.*, 61 (1989a), pp. 89–101.

Uno Y., Suzuki R., and Kawato M., Minimum muscle tension change model which reproduces human arm movement, *Proceedings of the 4th Symposium on Biological and Physiological Engineering* (1989b), pp. 299–302 (in Japanese).

Wiener N., *CYBERNETICS, or Control and Communication in the Animal and the Machine*, John Wiley and Sons, Inc., New York (1948).

Wu C.W. and Chua L.O., Synchronization in an array of linearly coupled dynamical systems, *IEEE Trans. Circuit Syst. I: Fundam. Theory Appl.*, 42 (2002), pp. 430–447.

Yanagihara D. and Kondo I., Nitric oxide plays a key role in adaptive control of locomotion in cat," *Proc. Natl Acad. Sci. U.S.A.*, 93 (1996), pp. 13292–13297.

Yin Y., Luo Z.W., Svinin M., and Hosoe S., Hybrid control of multi-fingered robot hand for dexterous manipulation, *Proceedings of 2003 IEEE International Conference on Systems, Man and Cybernetics* (2003), pp. 3639–3644.

Yuasa H. and Ito M., Coordination of many oscillators and generation of locomotory patterns, *Biol. Cyber.*, 63, (1990), pp. 177–184.

WEBSITES

http://www.bmc.riken.go.jp/~robot/index-e.html
http://salmon.psy.plym.ac.uk/

Interfacing Microelectronics and the Human Visual System

Rajat N. Agrawal, Mark S. Humayun, James Weiland,
Gianluca Lazzi, and Keyoor Chetan Gosalia

CONTENTS

17.1   INTRODUCTION

Vision is an enormously complex form of information processing that depends on a remarkable neuroprocessor at the back of the eye called the retina (Figure 17.1). For an intelligent living being to see, every component of this complex system has to work in tandem. Blindness can result when any step of the optical pathway — the optics, the retina, the optic nerve, visual cortex, or other cortical areas involved in the processing of vision — sustains damage (Zrenner, 2002).

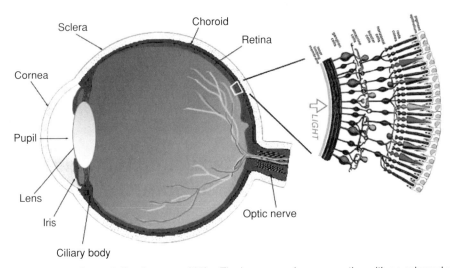

Figure 17.1 (See color insert following page 302)   The human eye in cross section with an enlarged section of the retina (right). The light sensitive retina covers more than half the back of the eye. Over 100 million photoreceptors convert light into neural signals that are then transmitted to the proximal visual centers by the optic nerve. The optic nerve is composed of 1 million axons from the retinal ganglion cells, which are the output cells of the retina. In outer retinal diseases, the photoreceptors are degenerated, but the inner retina cells remain and can be electrically stimulated.

Blindness afflicts more than 1 million Americans; approximately 10% amongst them have no light perception (Chiang et al., 1992). Some of these patients can be considered for experimental approaches, such as gene therapy and drugs, to prevent the development of blindness. But there are very few approaches to treatment of total loss of vision; one of them considered to be a serious contender is visual prosthesis.

During the 18th century, scientists began to understand that electricity could elicit a response in biological tissues. This knowledge has been applied to medicine with electronic implants such as the cardiac pacemakers and cochlear implants. Success of these implantable devices and improved electronic technology aroused interest in the development of a visual prosthesis that could help blind people.

17.1.1 Cortical Prosthesis

Initial work in the field of visual prostheses started with electrical stimulation of the visual cortex. Direct electrical stimulation of the cortical surface under local anesthesia of a sighted human subject resulted in seeing a spot of light, described as phosphene. The position of the light in space corresponded correctly to the stimulated anatomical region (Foerster, 1929; Greenberg et al., 1999). Subsequently, similar results were obtained in blind patients (Maynard, 2001; Ranck, 1975). Also, further experiments showed that a chronically activated electrical stimulation device could be safe, and the patient could read random letters (Button and Putnam, 1962; Brindley and Lewin, 1968; Brindley and Rushton, 1974).

In spite of the success with the experiments (Pollen, 1977; Dobelle and Mladejovsky, 1974), there were major issues that prevented widespread use: high currents and large electrodes were required to induce phosphenes (Karny, 1975); a single stimulating electrode induced multiple phosphenes; those produced were sometimes inconsistent. Other drawbacks included limited two-point discrimination, local heating and electrolysis (Bak et al., 1990), phosphene persistence following cessation of electrical stimulation (Dobelle et al., 1976), and phosphene flickering during surface stimulation (Brindley and Lewin, 1968; Pollen, 1977; Karny, 1975). These problems forced development of intracortical electrodes (Dobelle et al., 1976; Uematsu et al., 1974; Schmidt et al.,

1996; Normann et al., 1999; Jones and Normann, 1997; Maynard et al., 1997; Nordhausen et al., 1996). Experiments done with intracortical electrodes confirmed that the stimulus current threshold was 10 to 100 times lower than that for stimulation using surface electrodes (Dobelle et al., 1976). Subjects were also able to perceive phosphenes at a predictable and reproducible location in the visual space (Schmidt et al., 1996). Separate patterned perceptions could be evoked by electrical stimulation via electrodes spaced as close as 500 µm apart. The preferable stimulation location was probably the fourth layer of the visual cortex (Dobelle et al., 1976).

Undoubtedly, the lower current threshold of the intracortical microstimulation, the predictable forms of generated phosphenes, the capability of increasing the number of electrodes, power requirement improvement, and the current reduction per microelectrode are the main advantages of the intracortical microstimulation approach (Dobelle et al., 1976; Schmidt et al., 1996). Also, the skull protects both the electronics and the electrode array. The cortical prosthesis will bypass all diseased neurons distal to the primary visual cortex, and hence, has the potential to restore vision to the largest number of blind patients. However, there are problems in this approach. Spatial organization is more complex at the cortical level and two adjacent cortical loci do not necessarily map out to two adjacent areas in visual space, so that patterned electrical stimulation may not produce the desired patterned perception. In addition, the convoluted cortical surface makes it difficult for implantation, and surgical complications can have serious and devastating complications, including death.

17.1.2 Concept of Retinal Prostheses

A patent for a subretinal microphotodiode was issued to Graham Tassicker in the 1950s. This is the first written record of a retinal prosthesis. The device was never realized as described in the patent. During the early 1970s, it was found that blind humans can also perceive electrically elicited phosphenes in response to ocular stimulation, with a contact lens on the cornea as the stimulating electrode (Potts et al., 1968; Potts and Inoue, 1969, 1970). When obtainable, these electrically elicited responses indicated the presence of at least some functioning inner retinal cells. Clearly, stimulation through a single channel on the cornea could not produce form vision, yet it did demonstrate that electrical stimulation at the level of the eye could evoke phosphenes in blind subjects.

The idea of stimulating the remaining inner retina with an electrode array on the retinal surface came about due to the fact that a number of blinding retinal diseases are predominantly outer retinal or photoreceptor degeneration (Stone et al., 1992; Santos et al., 1997; Humayun et al., 1999a,b). Two of the more common outer retinal degeneration diseases are retinitis pigmentosa (RP) and age-related macular degeneration (AMD). Incidence of RP is 1 in 4000 live births; there are approximately 1.5 million people affected worldwide and it is the leading cause of inherited blindness (Berson et al., 1993). AMD is the main cause of visual loss amongst older adults (>65 years old) in Western countries. Annually, there are approximately 700,000 new patients in the U.S. who lose vision due to AMD; 10% of these become legally blind each year (Curcio et al., 1996).

Analysis of eyes with outer retinal degeneration suggests that cells are present, but the retinal circuitry is disrupted. Morphometric analysis of the RP retina has revealed that many more inner nuclear layer cells (bipolar cells and others — 78.4%) are retained compared to outer nuclear layer (photoreceptors — 4.9%), and ganglion cell layer — 29.7% (Potts and Inoue, 1970; Stone et al., 1992). Similar results were obtained from AMD patients (Kim et al., 2002). Thus, it appeared feasible to stimulate remaining retinal neurons. However, more recent studies of animal and human retina with outer retina disease have shown that the retinal circuitry and structure undergoes significant changes after photoreceptor loss.

Stimulation in blind humans with temporarily implanted electrodes demonstrated the principle of electrical stimulation with epiretinal electrodes. Focal electrical stimulation elicited phosphenes in all patients; four out of five patients were able to describe spatial and temporal aspects of the stimuli. The resolution could be estimated to be 4.5/2000, a crude ambulatory vision (Humayun et al., 1996). The electrical stimulation threshold has been found to be dependent on the location of the electrode; the macular region required higher threshold currents than the peripheral retina to elicit visual perceptions. Also, patients with less advanced RP or AMD required lower threshold currents than those with more advanced disease stages. These findings are important because lower thresholds would allow for smaller electrodes and greater resolution. Lower threshold values in healthier retinae were later confirmed by other experiments (Suzuki et al., 1999; Chen et al., 1999; Weiland et al., 1999; Rizzo et al., 2000; Majji et al., 1999). Perhaps the most important result of these studies was associated with form recognition. Patients were able to identify crude forms such as letters or a box shape during the short period of electrical stimulation testing. When the electrical stimulation ended, there was no persistence of the image. Another group later confirmed this result of form recognition by electrical stimulation of a healthy retina of a sighted volunteer (Rizzo et al., 2000). Other important psychophysical perceptions of this study included flicker fusion (at a frequency of 40 to 50 Hz) and different color perceptions (Humayun et al., 1999a,b).

17.1.3 Optic Nerve Prostheses

Attempts have been made to stimulate the optic nerve to generate optical impulse. But the high density of the axons within the optic nerve (1.2×10^6 within a 2-mm diameter cylindrical structure) could make it difficult to achieve focal stimulation and detailed perception. In addition, any surgical approach to the optic nerve requires dissection of the dura and can have harmful side-effects. Similar to the retina prosthesis approach, optic nerve stimulation requires intact RGCs and is limited to outer retinal pathologies. A volunteer with retinitis pigmentosa and no residual vision was chronically implanted with an optic nerve electrode connected to an implanted neurostimulator and antenna. An external controller with telemetry was used for electrical activation of the nerve which resulted in phosphene perception. Low perception thresholds allowed for large current intensity range within safety limits. In a closed-loop paradigm, the volunteer was using a video camera to explore a projection screen. The volunteer underwent performance evaluation during the course of a training program with 45 simple patterns. After learning, the volunteer reached a recognition score of 63% with a processing time of 60 sec. Mean performance in orientation discrimination reached 100% with a processing time of 8 sec (Veraart et al., 2003).

17.1.4 Sensory Substitution Devices

An alternative to direct stimulation of the visual system has been tried and is being developed to aid visually handicapped people. This method attempts to convert visual information into vibro-tactil or auditory signals (i.e., sensory substitution devices) (Margalit et al., 2004; Brabyn, 1982; Rita et al., 1998).

17.2   CURRENT CONCEPTS IN RETINAL PROSTHETIC DEVICES

Worldwide, there are currently several groups involved in the process of development of retinal prostheses. Their respective work is classified according to the location of the device in relation to the retina: on the retinal surface (*epiretinal*), or behind the retina (*subretinal*). There is also another group in Japan developing a new approach, suprachoroidal–transretinal stimulation (STS), where the array is to be placed in the suprachoroidal space (Kanda et al., 2004).

Epiretinal implantation has the advantage of keeping the majority of electronics outside the eye, minimizing the risk of failure, and optimizing the ease of replacement or upgrading the electronics. Additionally, the majority of the intraocular electronics could be placed in the vitreous cavity, a naturally existing space, which is fluid filled postsurgery. This greatly helps in dissipating the heat generated by the electronics (Majji et al., 1999; Eckmiller, 1997; Rizzo and Wyatt, 1997). However, an epiretinal prosthesis will be exposed to ocular rotational movements, making the need for a nontraumatic and yet sturdy attachment method paramount. Also, by physically being closer to the retinal ganglion cells (RGC), it may be difficult to stimulate bipolar cells and therefore one may lose visual processing that takes place in this layer.

The subretinal positioning of the retinal prosthesis has the advantage of placing the stimulating electrodes close to the bipolar cells, which may also permit lower stimulus thresholds (Zrenner et al., 1997, 1999; Chow and Peachey, 1998; Tassicker, 1956; Chow and Chow, 1997; Peyman et al., 1998; Guenther et al., 1999). However, the placement of any object between the choroid and the retina can be more disruptive to the nutritional supply of the retina derived from the choroids (Zrenner et al., 1999). Another drawback of this method is the limited amount of light that can reach the array coupled with the inefficiency of modern day photovoltaic or solar cells. This translates into using a very bright image intensifier (10 suns) in order for the stimulator chip to generate the level of currents that have resulted in visual perceptions in the blind.

There are advantages and disadvantages of the retinal stimulation approaches. The advantages include the ability to use the retinotopic organization of the eye in addition to the natural processing ability along the proximal visual pathways. Furthermore, the vitreous cavity fluid can be utilized as a heat sink and the prosthesis can be visualized by dilating the pupil in an out-patient setting. Less surgical morbidity and mortality are expected in comparison to any of the cortical prostheses implantation methods. But there are disadvantages as well. These include possible disruption of retinotopic organization due to stimulation of ganglion cells' axons along their path to the optic nerve, as axons from different retinal locations pass within short distances from each other. Ganglion cells encode many properties of the visible light (color, intensity, onset of light, extinction of light, etc.) and the question is which property will be encoded during electrical stimulation if many ganglion cells are activated simultaneously. Also, this approach will probably be limited to outer retinal pathologies.

Suprachoroidal–transretinal stimulation approach has the electrodes away from the retina. The anodic stimulating electrode is positioned on the choroidal membrane, while the cathode is placed in the vitreous body. STS is expected to activate the retinal network on the basis that various types of transretinal stimulation induce field responses in central visual areas (Lederman and Noell, 1969; Potts et al., 1970; Crapper and Noell, 1963; Shimazu et al., 1999). Since the STS is not in contact with the retina, it may require a high threshold for stimulation of the retinal circuitry. The strong electrical stimulation may damage the retina and increase the power load on implanted electronic devices. It may also stimulate a broader area of the retina, resulting in low resolution.

A different approach to retinal prostheses is the Artificial Synapse Chip (Peterman et al., 2003). This group is working on a high-resolution, physiologic flexible retinal interface, incorporating cell micropatterning and localized chemical stimulation with flexible biocompatible materials. This device aims to improve visual resolution of an electronic retinal prosthesis by addressing cells individually and mimicking the physiological stimulation achieved in synaptic transmission. By patterning the growth of cells to individual stimulation sites, the selectivity of stimulation is increased by selective stimulation of individual cells amongst closely packed neighbors, while decreasing the distance of cells from the stimulation sites and therefore the associated power requirements. This technology is to be coupled to a prosthesis based on localized, microfluidic neurotransmitter delivery.

Another approach to a retinal implant has been dubbed "Hybrid Implant." This approach proposes to develop an integrated circuit, which would include both electronic and cellular components. The electronics will perform image recognition and the neurons on the device will extend their axons to synapse with the lateral geniculate body and thus create a device–CNS interface and restore vision (Yagi and Watanabe, 1998). The advantage of this approach would be its capability of reconstructing an eye with total or inner retinal degeneration. Disadvantages include difficulties in precisely directing axons to the lateral geniculate body, developing the interface between the electronics and neurons, and the matrix to enable survival of the cellular components while being housed in microelectronics.

17.2.1 Simulations of Prosthetic Vision

One of the major arguments supporting the concept of a retinal prosthesis is the fact that cochlear implant patients can understand speech with only six input channels. Simulations of cochlear implant audition have shown that speech reduced to as few as four frequencies provides enough information for the human brain to understand language. Similarly, it is hoped that visual prostheses will be able to transmit useful information without replacing the input from all 100 million photoreceptors. Several experiments were done to define the minimum acceptable resolution for useful vision. Early studies in this area focused on simulating prosthetic vision from a cortical implant. The points of stimulation (pixels) required for specific activities varied from 80 to more than 600, depending on the activity being performed (Brindley, 1965). Most recent studies show that 625 pixels is a better estimate for certain tasks. It was concluded that 625 electrodes implanted in a 1 cm^2 area near the foveal representative of the visual cortex could produce a phosphene image with a visual acuity of approximately 20/30. Such acuity could provide useful restoration of functional vision for the profoundly blind (Cha et al., 1992a–c).

Although these studies began to delineate the number of electrodes needed, the fact that all the pixels were projected on a very small area of the retina, made it impractical to translate to the design of a retinal prosthesis, in which the electrodes would be spread over the entire macular region. Thus, a low vision enhancement system (LVES) has been modified to filter images on a head mounted display in order to simulate pixelized prosthetic vision and to produce an array of dots. The results suggested that a fair level of visual function can be achieved for facial recognition and reading large print text using pixelized vision parameters such as a 25×25 grid in a $10°$ field, with high contrast imaging and four or more gray levels.

17.3 MECHANICAL EFFECTS OF IMPLANTATION OF RETINAL PROSTHESIS

Retinal tissue is delicate and can easily tear or detach from the back of the eye. The delicate nature of the retinal tissue can also predispose it to pressure necrosis by a chronic implant being placed on it. Increased intraocular pressure, typical in glaucoma, can lead to damage to retinal ganglion cells and significant visual loss. Also, there is an abundant blood supply within and underneath the retina. Disruption of this vasculature can lead to chronic inflammation or new blood vessel formation, both of which can lead to retinal damage. Studies have shown that an epiretinal array can be secured to the inner retinal surface in a safe and secure manner, is mechanically stable, and biologically tolerated over a 6-month period (Majji et al., 1999).

Any intraocular implantable device has to be tested for biocompatibility. Since these devices are to remain within the intraocular environment for many years, they have to continue to be electrically effective, and also not cause mechanical damage over time. Moreover, the device should also not undergo long term degradation, like corrosion, in the ocular environment.

17.3.1 Infection and Inflammation

The eye, as is the central nervous system, has been described as immunological or partially immunological privileged (Rocha et al., 1992). Despite this fact, the inflammatory course is identical to that occurring elsewhere in the body once an incitement for inflammation has occurred (Oehmichen, 1983). Mere surgical manipulation, any infection, biodegradation or any release of toxic substances from a foreign body can provoke a severe inflammatory response. Bacterial infections are often delayed and appear to be due in part to the host's inability to respond properly to infections. Their origins are frequently distant infected sites in the body or skin flora (Dougherty and Simmons, 1982).

17.3.2 Ocular Side-Effects of Long Term Implantation

Since the field of retinal implants is relatively new, there are few reports available on the long-term side-effects or complications related to implantation of a device. Sham surgeries have been done, with no electrical stimulation, to simulate prosthetic implantation, to study the mechanical damage to the eye. In one such study, performed in four dogs, mild retinal folds were noticed at one edge of the array, which did not progress over time; there was no retinal detachment (RD) seen in any of the dogs. Retinal pigment epithelium (RPE) changes were noted near the retinal tacks which are used to fix the epiretinal implant (Majji et al., 1999). In another study (Walter et al., 1999), nine out of ten rabbits were implanted without serious complications. The implant was found to be stable at the original fixation site and there was no change noted in retinal architecture underneath the implant by light microscopy. In three cases, mild cataract formation was observed, while in one case, a total RD was found after a 6-month follow-up. In another study, three rabbits were implanted with an electrode array in the subretinal space. No side-effects were reported (Chow and Chow, 1997).

The anatomy and physiology of the retina evaluated after implantation of a retinal implant. Vascular integrity was evaluated by injection of fluorescent dye into the blood stream and subsequent imaging of the dye's presence in the ocular blood flow (a technique called fluorescein angiography). Good vascular perfusion was noted during the entire follow-up period of more than 6 months (Majji et al., 1999). Also, in the same study, electroretinogram (ERG) findings were found to be within reasonable limits after the surgery. There is histopathological confirmation that the retina underneath an epiretinal array does not undergo any damage over 6 months of follow-up. Light microscopy and electron microscopy have proved that the retinal microstructure does not show any signs of degradation over this time, though the area around the tack showed localized loss of retinal and RPE layers.

A single volunteer with end-stage RP has been chronically implanted with an optic nerve cuff electrode connected to an implanted neurostimulator and antenna in February 1998. Chronic follow-up of this patient has not shown any side-effects to the surgery or the presence of electrodes around the optic nerve.

17.3.3 Attachment of the Implant to the Retina

Any implanted device will be exposed to the ocular movements, especially in cases where vitreous surgery replaces the vitreous gel with fluid-filled cavity, where counter-currents from the fluid can generate forces on the epiretinal implant; hence, it requires a stable fixation to its intended anatomic location. Ocular rotational movements have been recorded to reach $700°$ visual angle/sec. These extreme movements can certainly dislodge the epiretinal device and move it away from the required location. The subretinal implant will not face the same counter-current movements as an epiretinal implant would, since it is expected to stay within the confines of the subretinal space taking the advantage of the adherence forces between the sensory retina and the retinal pigment epithelium. Even though the likelihood of displacement of such devices is low, they have been known to be displaced after implantation (Peyman et al., 1998). Surgical implantation of such a device can be either through the sclera (*ab externo*) or intraocularly through a retinotomy site after a vitrectomy procedure.

There have been various approaches to the attachment of the epiretinal implant or device to the retina. Bioadhesives, retinal tacks, and magnets have been considered and tested as some of the methods for the array attachment. Retinal tacks and the electrode array have been shown to be firmly attached to the retina for up to 1 year of follow-up with no significant clinical or histological side-effects (Majji et al., 1999). Similar results were seen in rabbits (Walter et al., 1999).

There have been studies on the use of commercially available compounds for their suitability as intraocular adhesives in rabbits. One type of adhesive (SS-PEG hydrogel, Shearwater Polymers, Inc.) proved to be strongly adherent and nontoxic to the retina (Margalit et al., 2000). Other groups have done similar experiments (Lowenstein et al., 1999).

The preferable fixation site for the intracortical microstimulation arrays is the cortex itself; skull will not be a good site due to the brain's constant movement in relation to the skull. These arrays are currently inserted either manually in an individual fashion or in a group of 2 to 3 electrodes normal to the cortical surface to a depth of 2 mm or by a pneumatic system that inserts 100-electrode arrays into the cortex in about 200 msec.

17.3.4 Hermetic Sealing of the Electronics

Prostheses will be composed of electronic parts within the eye. These components will be exposed to the chemical environment in the eye. These implanted parts will have to be sealed, such that they are not exposed to corrosion of the ocular fluids. Also, this protective coat will have to last for some years or decades for the continued functioning of the implant. The requirement of hermetically sealing a circuit in the case of neural stimulating devices is complicated by the demand that multiple conductors (feedthroughs) must penetrate the hermetic package so that the stimulation circuit can be electrically connected to each electrode site in the array. These connections are the most vulnerable leakage points in the system (Margalit et al., 2004).

17.4 ELECTRICAL CONSIDERATIONS IN RETINAL PROSTHETIC DEVICES

The effectiveness of an electrical stimulation for an intraocular retinal prosthesis, whether epiretinal or subretinal, is governed by a number of parameters characteristic of the electrode array, including shape and size of the electrodes, spacing between electrodes, electrode materials, current return positions, and stimulating current waveform, to name a few. Optimal electrode array type and characteristics must also take into account other factors that can influence the one or more parameters, including thermal or electrical safety or ease of surgical implantation.

17.4.1 Stimulating Electrodes: General Considerations with Regard to Electrical Stimulation of the Retina

The characteristics of the stimulating electrode array are often of competing nature: for example, it might be desirable to mechanically position the electrodes as close as possible to the ganglion and bipolar cells, but that would then result in penetrating electrodes that could harm the fragile structure of the retina. Similarly, it may appear natural to develop small electrodes to achieve high-resolution electrical stimulation of the retina; however, current densities needed to elicit phosphenes may exceed safety limits and potentially cause damage to the retina. Further, it is not completely clear, to say the least, the relation between size of the electrode and size of the visual spot induced by that electrode.

The problem is phenomenally complex, as it simultaneously involves neural activation at the microscopic level and control of the spread of the current in retinal tissue at the macroscopic level. Both problems are strongly coupled and involve very different scales and methods of analysis, which increases the complexity of solving the problem of optimal stimulation of retinal tissue and, indirectly, the problem of optimal physical characteristics of the stimulating electrode arrays.

Besides geometrical considerations that can affect the effectiveness of the electrical stimulation of the retinal tissue, other aspects of the system design can have a significant impact on the induced stimulation. Among the challenges that must be considered to achieve optimal electrical stimulation, in the sense of an electrical stimulation which uses as little current as possible to elicit visual perception, there are the actual characteristics of the "contact" between retina and electrode, which strongly impact the current magnitude and direction in retinal tissue. In fact, even though each layer of the retina is characterized by a different conductivity, the vitreous humor is in general significantly more conductive than each of the layers of the retinal tissue. The consequences of this can easily be understood by thinking of the vitreous humor as the "preferred path" of the electrical current as opposed to the retina, if the conditions are such to make this possible. Therefore, if a stimulating electrode has its surface in contact with the vitreous humor, and not only with the retina as it may happen for example with dome-shaped electrodes with only the tip in actual contact with the retina, most of the current will tend to flow through the vitreous humor without passing through the retina when the current return is located in the eyeball. This, in turn, may result in higher currents needed to stimulate the retina and therefore elicit vision. It is therefore clear that the choice of stimulating electrodes in terms of shape, size, and characteristics, as well as the system design in its entirety, including the choice of the current return location for the electrodes, can have a substantial impact on the effectiveness of the electrical stimulation of the retina. This, in turn, has a significant impact on the feasibility of the entire system, since a more effective stimulation will require less current, which will result in less power dissipation by the stimulating microchip, leading to a lower temperature increase in the eye and surrounding tissue due to the operation of the retinal prosthesis.

17.4.2   The Impedance Method for the Solution of Quasi-Static Electromagnetic Problems

The problem of characterizing the current spread in retinal tissue, which can also lead to a better understanding of the neural activation once coupled with models of the neural cells, can be solved through quasi-static electromagnetic methods. A very versatile method that has a number of benefits in the modeling of the system is the impedance method (Gandhi et al., 1984) (or admittance method [Armitage et al., 1983]), but other methods based on the solution of the quasi-static electromagnetic problem can be used as well (finite-element method, finite-difference method, scalar potential finite-difference method [Dawson et al., 1996], to name a few). The impedance method is based on the discretization of the physical model that must be modeled into computational cells. The edges of these computational cells are impedances (or admittances) which are computed using the electrical conductivity of the material in the cell and the width, length, and height of the computational cell. Therefore, the physical model is represented by means of an electrical network with resistance or admittances derived from the physical properties of the physical model itself. In its basic formulation, the impedance method uses uniform cells to discretize the physical model; however, nonuniform cells, leading to a multiresolution impedance method, can be used to reduce the computational time and computer memory needed to solve the problem (Eberdt et al., 2003).

The problem of characterizing the current spread in the retina translates, therefore, into the problem of developing an accurate model of the eye and the retina, with a geometrical resolution sufficiently high to describe current variations on the geometrical scale of interest (DeMarco et al., 2003). Even with the multiresolution impedance method, however, it is extremely challenging to develop a model that reaches cellular scales in the retinal tissue and at the same time covers an extended area such as the entire eyeball. Therefore, some compromise must be reached in terms of resolutions vs. geometrical scales of interest for the complete characterization of the system. A possible approach is to discretize the fine retinal structure and electrode geometries with resolutions as low as 5 μm, for example, and subsequently use neural models with the current levels found in the neural layers in order to model the response to electrical signals. Another approach would be the direct coupling of the macro-scale current spread modeling with electrical circuits to model the neural interaction. This is because in methods such as the impedance or admittance methods, there is no restriction on the circuit element used between two nodes. In the

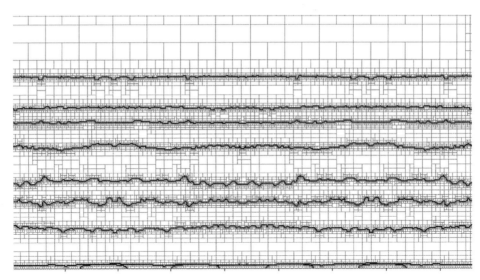

Figure 17.2   Example of a multiresolution computational mesh of a frog retina.

simplest case this is impedance related to the electrical properties of the biological tissue or electrodes: in more complex cases it can be an arbitrarily complex circuit that can be solved with circuit simulators such as SPICE®. In fact, the entire impedance or admittance network can be solved with such circuit simulators, with subcircuits describing specific functions or particular behaviors related to the electrical stimulation.

Figure 17.2 shows an example of a multiresolution computational mesh of a retinal section, with its various layers classified and associated to a conductivity specific for each of them (Eberdt et al., 2003). Figure 17.3 shows instead the current spread in this classified model of the retina for two types of electrodes, coaxial electrodes and dome electrodes with side current return, respectively, as obtained by two-dimensional multiresolution impedance method simulations. It can be qualitatively seen that the current magnitudes in various layers of the retina depend upon the type of electrode. Higher resolution and coupling with neural models can also be incorporated in these models. It should be noted, however, that there is a degree of uncertainty with respect to a number of parameters, such as the conductivity of each layer, which is estimated based on water content and affinity with other tissues, and actual retinal geometric features, which can be significantly distorted in diseased retinas.

17.5   RETINAL PROSTHESIS AND RELATED THERMAL EFFECTS

An implantable device for neural stimulation should generally receive power and data wirelessly (Rucker and Lossinsky, 1999) — through a telemetry link — process the received data, and inject currents in the neural tissue by means of a number of stimulating electrodes that in general need to accommodate desired waveforms, frequency of stimulation, and amplitudes of stimulating signals. Each of these characteristics is generally responsible for power dissipation, which may result in thermal increase in the human body in proximity of the implanted device.

A dual-unit epiretinal prosthesis (DeMarco et al., 1999; Liu et al., 2000), consisting of an extraocular unit with an external camera for image collection, a data encoding chip, and the primary coil for inductive power and data transfer and an intraocular unit with the secondary coil, data processing chips, an electrode stimulator chip, and the electrode array for epiretinal stimulation, could potentially lead to significant temperature increase in the eye and surrounding tissues.

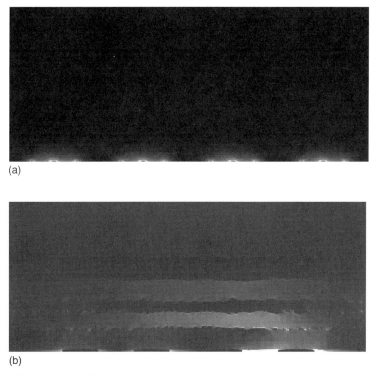

(a)

(b)

Figure 17.3 Qualitative image of the current spread in the frog retina due to (a) coaxial electrodes and (b) disc electrodes. Current density values range from white (max) to black (zero).

The wireless link causes electromagnetic power deposition in the head and eye tissues, which could lead to indirect thermal rise in the tissue, known to be the dominant physiological hazard due to power deposition in human tissues (Adair and Petersen, 2002). Moreover, the implanted electronic IC chips will dissipate power in the form of heat, which will directly lead to the thermal elevation in the surrounding tissues. It is therefore necessary to quantify these thermal effects in order to determine the safe limits of operation of the prosthetic system.

The temperature rise in the head and eye tissues due to the operation of the prosthesis can be experimentally determined with *in vivo* experiments or computationally evaluated by means of a computer code for the solution of the bio-heat equation. Preliminary computational predictions have been performed to evaluate the thermal influence of a dual-unit epiretinal prosthesis system on the human head and eye tissues and, therefore, provide a quantitative measure of the temperature rise in human body as a result of the operation of an implantable neurostimulator. As an example of typical methods and results, the following paragraphs and subsections provide a brief account of the methods and model used in such bio-engineering computations.

To quantify the thermal impact of the dual-unit epiretinal prosthesis system, the bio-heat equation can be numerically discretized both spatially and temporally using the well-known finite-difference time domain (FDTD) method (Sullivan, 2000; Wang and Fujiwara, 1999). In this example, the computational prediction was performed on a very high-resolution anatomically accurate three-dimensional human head model obtained from the National Library of Medicine (The National Library of Medicine, The Visible Human Project, 2000). For the computational study, the different tissues in the head model were modeled by their dielectric and thermal properties (DeMarco et al., 2003). Figure 17.4 shows the head model, which was utilized in the computational domain to evaluate the natural steady state (or basal, initial) temperature distribution in the model (due to the internal tissue metabolism with no implanted heat sources).

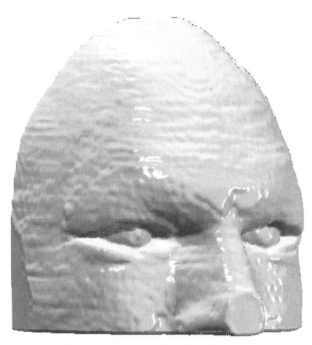

Figure 17.4 Example of a three-dimensional computational head model used for numerical simulation of the temperature increase in the tissue due to the operation of an implantable neurostimulator.

The bio-heat equation is developed from the well-known heat equation (Necati, 1985) by considering the additional sources of thermal influence for computations involving the human body (DeMarco et al., 2003; Bernardi et al., 2003; Gosalia et al., 2004). In the presence of implantable devices and sources of electromagnetic power deposition, the bio-heat equation is given as:

$$C\rho\frac{\partial T}{\partial t} = \nabla \cdot (K\nabla T) + A - B(T - T_\mathrm{B}) + \underbrace{\rho SAR + P_\mathrm{chip}^\mathrm{density}}_{\text{External heat sources}}\left[\frac{W}{m^3}\right] \qquad (17.1)$$

which equates the product of thermal capacitance ($C\rho$) and temperature rise per unit time to the different ways of accumulation of heat energy in the tissues. In Equation (17.1), the following notations have been used:

- $\nabla \cdot (K\nabla T)$: thermal spatial diffusion term, which leads to heat transfer through conduction (K [J/m · sec · °C]);
- A: tissue specific internal metabolic heat production, which will lead to an initial natural steady state temperature distribution (J/m^3 · sec);
- B: tissue specific capillary blood perfusion coefficient (J/m^3 · sec · °C). This has a cooling influence proportional to the difference in tissue temperature (T) and blood temperature (T_B);
- ρSAR and $P_\mathrm{chip}^\mathrm{density}$: external heat sources due to electromagnetic power deposition and power dissipated by the implanted electronics, which will lead to a thermal rise beyond the initial natural steady state temperature distribution in the head model.

Besides the bio-heat equation, the heat exchange at the tissue interface with the external environment has to be modeled accurately. At this interface, a boundary condition to model the heat exchange with the surrounding environment is imposed on the computations,

$$K\frac{\partial T}{\partial n}(x, y, z) = -H_\mathrm{a}\big(T_{(x, y, z)} - T_\mathrm{a}\big)\left[\frac{W}{m^2}\right] \qquad (17.2)$$

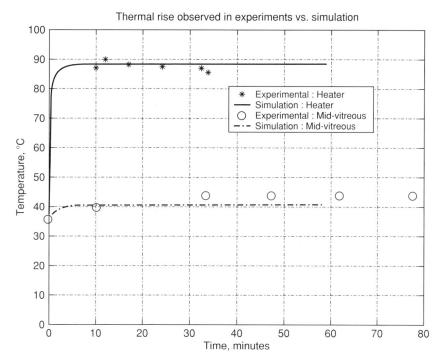

Figure 17.5 Comparison between observed experimental results and computationally derived results for an experiment designed to validate the computational models. (From Gosalia K, Weiland J, Humayun M, and Lazzi G. *IEEE Transactions on Biomedical Engineering*, 51(8): 1469–1477, 2004. With permission.)

where n is perpendicular to the skin surface and the right hand expression models the heat losses from the surface of the skin due to convection and radiation, which is proportional to the difference between skin temperature ($T_{(x,y,z)}$) and external environmental temperature (T_a).

For all the computations performed in the example above, the temperature of blood was assumed to be constant at 37°C, while H_a is the heat convection coefficient and is assumed to be 10.5 W/(m$^2 \cdot$ °C). The thermal parameters for all the tissues in the head model have been directly obtained from previous studies (DeMarco et al., 2003; Bernardi et al., 2003).

In order to validate the thermal method and model used, *in vivo* experiments conducted with dogs were simulated, and experimental and computational results were compared. The experiment comprised of mechanically holding a heater probe (1.4 × 1.4 × 1.0 mm in size) dissipating 500 mW in the vitreous cavity of the eye of the dog for 2 h (Gosalia et al., 2004; Piyathaisere et al., 2003). The experimental set up included thermocouples to measure the temperature rise at different locations in the vitreous cavity and the retina during this period. Figure 17.5 shows the comparison between the experimentally observed and the simulated results for temperature rise at the retina and the vitreous cavity. The uncertainty in the exact locations of the thermocouples during the actual experiment is the likely cause of the small difference between simulated and experimental results.

17.5.1 Heat and the Telemetry System

As mentioned in the preceding paragraphs, the wireless telemetry system can be a source of thermal rise since it causes deposition of electromagnetic (EM) power in the head and eye tissues. Using the FDTD technique, the deposited EM power can be quantified in terms of the specific absorption rate (SAR) and several studies have quantified the thermal effects in the human head and eye tissues based on the evaluated SAR using the bio-heat equation (DeMarco et al., 2003; Bernardi et al., 1998, 2000; Hirata et al., 2000). SAR is expressed as $\sigma \vec{E}_2/(2\rho)$ for conductivity σ, electric field $\vec{E}$, and mass density ρ at each cell (x, y, z) in the computational model. In the radiofrequency range, the IEEE/ANSI (IEEE standard safety levels, 1999) safety limit for peak 1-g EM power deposition is 1.6 W/kg for the general population (the reader is encouraged to refer to the standard for a detailed description of maximum permissible exposure [MPE], SAR, and effect of the frequency for EM safety considerations). In general, if the EM power deposition remains well within this limit, the thermal effects induced will be negligible. Therefore, it is necessary to quantify the EM power deposition in the head tissues due to the wireless telemetry link to establish if there could be potential hazards. As an example and to illustrate the procedure, we have used a circular coil of approximately 37 mm diameter modeled at a distance of 20 mm from the eye and excited by a 2 A current at the center operation frequency of 10 MHz. Computed peak 1-g SAR observed in the head model due to such an excitation was 0.02 W/kg. At this currently estimated operating current level for the wireless telemetry link, the SAR values do not exceed the IEEE safety limits for power absorption (IEEE Standard exposure to RF, 1999). Thus, it can be reasonably concluded that the contribution of SAR to the final temperature elevation would be negligible compared to the rise in temperature due to power dissipation in the implanted chip. In these cases, the power dissipation due to the implanted chip and coil alone can be considered as the extraneous heat source (besides the natural metabolism of the eye).

However, it should be noted that this will not always be the case. The peak 1-g SAR value directly depends upon the wireless link employed for supplying power and data to the implanted device, the geometrical characteristics of the wireless devices, the frequency of operation, their placement with respect to the human body, and their power level. In general, one must evaluate the SAR to ensure that it is within guidelines and determine whether such SAR could result in a thermal increase and therefore would need to be included in the bio-heat equation.

17.5.2   Power Dissipation of Implanted Electronics

In order to compute the thermal elevation due to implanted electronics, the implanted chip was modeled in the three-dimensional head model. The chip was modeled to have a composite thermal conductivity $K = 60$ J/(m sec °C) and encapsulated in a 0.5-mm thick layer of insulation ($K = 60$ J/[m sec °C]). These values of thermal conductivity are very high compared to the values of the tissues in the human head (Gosalia et al., 2004).

When an actual prosthesis is implanted, there are several parametric options that can be explored to minimize the thermal elevation in the surrounding tissues. In order to characterize these options, several thermal simulations were performed with the chip modeled with different sizes, placed at different locations (within the eyeball) and also dissipating different amounts of power in order to gain an insight into the best possible configuration (from the point of view of least thermal elevation) for an implant in the eye.

As an example of the impact of the location of the implanted microchip on the temperature increase, we considered two locations for positioning the implanted unit within the eyeball of the patient. In the first case, the lens can be removed and the implanted chip hinged between the ciliary muscles of the eye (referred to as the anterior position). The other considered position is in the middle of the vitreous cavity parallel to the axis of the eyeball (referred to as the center position). Both these cases were characterized computationally. The implanted chip was modeled at both these locations and thermal simulations were performed to study the variation in temperature increase in different human head tissues as a function of the implant location.

For both the above cases, the size of the implanted chip was kept constant at $4 \times 4 \times 0.5$ mm and was allowed to dissipate 12.4 mW (anticipated worst case power dissipation from an implanted current stimulator chip driving a 16 electrode array positioned on the retina). The power density for each cell of the model of the chip was calculated from the total power dissipated (12.4 mW) and was kept uniform throughout the total volume of the chip (it should be noted that uniform power dissipation is a further simplification since such an implanted device could, in effect, exhibit nonuniform "hot-spots"). It was observed that within 26 min of actual stimulation time (because of the extremely small time step in the FDTD simulations, the actual simulation time was significantly higher), the thermal elevation profiles in the tissues reached to within 5 to 7% of their final values. Since this provided a good indication of the approximate thermal rise, all the simulations were performed for approximately 26 min (physical time).

The maximum temperature increase for both chip positions was observed on the surface of the insulating layer. In both cases, the maximum thermal increase was approximately 0.82°C. In the first case where the chip was placed in the anterior position, the temperature of the ciliary muscles rose by 0.36°C as compared to 0.19°C when the chip was placed in the center position. In the vitreous cavity, temperature rise was 0.26°C for the chip placed in center of the eye while the anterior chip raised its temperature by 0.16°C (Gosalia et al., 2004).

A chip placed in the anterior chamber of the eye raised the temperature of the retina by less than half the amount that a chip placed in the center did (0.05 °C by anterior chip as compared to 0.12°C by a center chip) (Gosalia et al., 2004). In these simulations, it was observed that the vitreous cavity was acting as a heat sink since the rise in temperature of tissues beyond the eyeball is very small. A graphic comparison of the thermal elevation observed for the anterior and the center placed chips is provided in Figure 17.6. The anterior position is certainly preferable for the implanted unit in order to minimize the temperature rise in the vitreous cavity and on the retina.

A similar analysis can be performed to compute the impact of the size of the implant and dissipated power on the temperature increase in the tissue (Gosalia et al., 2004). It is worth pointing out, however, that power dissipation of the implanted microchip is probably the most significant parameter among all to be considered.

Two cases were considered in this example: in the first case, the chip dissipated 12.4 mW and in the second case, it dissipated 49.6 mW. For both of these cases, the size of the chip was

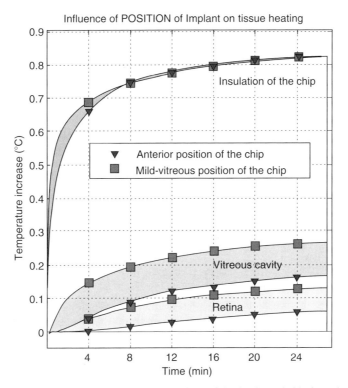

Figure 17.6 Thermal rise observed due to different locations of the implanted chip (anterior and center of the eyeball). (From Gosalia K, Weiland J, Humayun M, and Lazzi G. *IEEE Transactions on Biomedical Engineering*, 51(8): 1469–1477, 2004. With permission.)

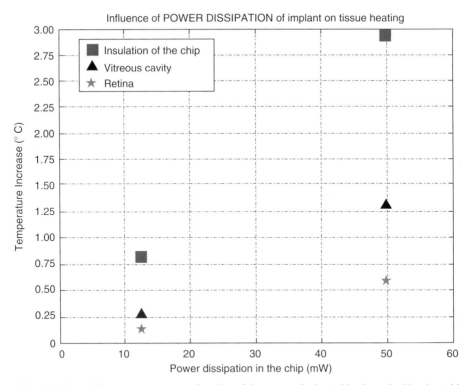

Figure 17.7   Variation of the temperature as a function of the power dissipated by the retinal implant chip.

$4 \times 4 \times 0.5$ mm and it was placed in the center of the eyeball. Power density was again kept uniform throughout the chip. The computation was performed for 26 min of simulated physical time.

Figure 17.7 graphically compares the temperature increase observed on the insulation, in the vitreous cavity and on the retina for both cases. From the thermal elevation results, it is observed that increasing the power dissipation by a factor of 4 does not necessarily lead to a rise in the temperature by the same factor. In the majority of tissues, a temperature rise by a factor of around 3.5 to 5 is observed for a four times increase in the power dissipation in the implant.

This preliminary investigation provided a qualitative and quantitative estimate of the thermal influence of such an implanted prosthetic system in the eye. Also, in the actual system, the various parametric variations can be optimized to yield the least harmful configuration from the point of view of thermal damage to the tissues of the eye of head. Several efforts are currently underway to accurately quantify the contribution of each aspect of such a prosthetic configuration to the eventual thermal and electromagnetic influence on the human tissues.

17.6 FUTURE IMPLICATIONS

A retinal prosthesis will form several interfaces with the eye including thermal, electrical, and mechanical. All of these interfaces must be considered simultaneously during the design of a safe and effective retinal prosthesis. For example, it may be possible to reduce the thermal concerns by using a larger electrode that consumes less power. However, such an electrode may stimulate a large area of the retina and not allow fine resolution vision. Many other optimization problems are presented by such a complex interaction. Therefore, future designs may well need to use automated optimization algorithms to yield the most effective device.

While future implants will depend on the continued advances in technology, the success of these implants (i.e., helping the blind see) will be jeopardized if we do not understand the neurobiology of the electrically stimulated visual system (Weiland and Humayun, 2003). The sense of vision is enormously complex and the nervous system has the ability to remodel in response to new stimuli. The development of prototypes that can be permanently implanted in research animals now gives us the ability to study these effects by applying advanced microscopy and tissue labeling methods developed in neuroscience basic research. While these studies are absolutely necessary and will yield valuable information, human implant studies are the only way to verify the effectiveness of the devices. Therefore, a multifaceted effort including technology development, biological research, and strict monitored, limited human tests is needed to advance the current artificial vision devices from proof-of-principle to accepted clinical treatment for blindness.

17.7 SUMMARY

The work in visual prostheses has come a long way from the days of laboratory research and the initial volunteer experiments. Today, we have a few patients implanted with the actual device; these devices have shown no major side-effect or complication related to surgery. Some of these patients have shown encouraging responses. Artificial visual stimulus is being tried at various levels, from the retina all the way to the cortex. Each type of implant has its own advantages and problems. The implant has to be not only biocompatible, but also be able to avoid damage from corrosion in the biological spaces the device will be implanted in. Long term damage from electrical current is an issue, as is the issue with the type of vision generated by the blind patients through these implants.

There are several challenges involved and issues to be considered during the design and development of a retinal prosthetic system, which can restore a limited form of vision. The electrical considerations of the prosthetic system (size and shape of electrodes, magnitude of current injection, size and shape of the implanted unit and its power dissipation, frequency, and strength of the wireless telemetry link) are closely coupled with safety considerations of the entire system (maximum allowable current densities and thermal elevation). These issues have to be resolved to realize a safe and effective retinal prosthesis system or any other implantable neuro-stimulator with a large number of channels. Several electromagnetic methods and computational techniques are being utilized to investigate the electrical performance characteristics of a prosthetic implant. The impedance (or admittance) method coupled with the multiresolution meshing scheme (to represent the intricate details of the retinal tissues — with a 5 μm resolution) appears very promising for characterizing the current spread in the retinal layers for given current stimulation and electrode array parameters. The computational implementation of the bio-heat equation through the FDTD method has been utilized to characterize the thermal elevation in the eye and head tissues due to the operation of the wireless telemetry link and power dissipation of the implant. Both these numerical techniques employ a very high spatial resolution and anatomically accurate model of the human head and eye. Tissues are represented by their dielectric and thermal properties as required for the specific computational investigation. Using these methods, it is possible to optimize the performance of an implantable neurostimulator such as the epiretinal prosthesis system with respect to effectiveness of stimulation and power dissipation.

REFERENCES

Adair ER and Petersen RC. Biological effects of radiofrequency/microwave radiation. *IEEE Transactions on Microwave Theory and Techniques*, 50:953–962, March 2002.

Armitage DW, LeVeen HH, and Pethig R. Radiofrequency-induced hyperthermia: computer simulation of specific absorption rate distributions using realistic anatomical models. *Physics in Medicine and Biology*, 28:31–42, 1983.

Bak M, Girvin JP, Hambrecht FT, Kufta CV, Loeb GE, and Schnidt EM. Visual sensations produced by intracortical microstimulation of the human occipital cortex. *Medical and Biological Engineering and Computing*, 28:257–259, 1990.

Bernardi P, Cavagnaro M, Pisa S, and Piuzzi E. SAR distribution and temperature increases in an anatomical model of the human eye exposed to the field radiated by the user antenna in a wireless LAN. *IEEE Transactions on Microwave Theory and Techniques*, 46:2074–2082, December 1998.

Bernardi P, Cavagnaro M, Pisa S, and Piuzzi E. Specific absorption rate and temperature increases in the head of a cellular phone user. *IEEE Transactions on Microwave Theory and Techniques*, 48:1118–1126, July 2000.

Bernardi P, Cavagnaro M, Pisa S, and Piuzzi E. Specific absorption rate and temperature elevation in a subject exposed in the far field of radio-frequency sources operating in the 10–900 MHz range. *IEEE Transactions on Biomedical Engineering*, 50:295–304, March 2003.

Berson EL, Rosner B, Sandberg MA, Hayes KC, Nicholson BW, Weigel-DiFranco C, and Willett WA. A randomized trial of vitamin A and vitamin E supplementation for retinitis pigmentosa (see comments). *Archives of Ophthalmology*, 111:761–772, 1993.

Brabyn JA. New developments in mobility and orientation aids for the blind. *IEEE Transactions on Biomedical Engineering*, 29:285–289, 1982.

Button J and Putnam T. Visual responses to cortical stimulation in the blind. *Journal of Iowa State Medical Society*, 52:17–21, 1962.

Brindley GS. The number of information channels needed for efficient reading. *Journal of Physiology*, 177:44, 1965.

Brindley GS and Lewin WS. The sensations produced by electrical stimulation of the visual cortex. *Journal of Physiology*, 196:479–493, 1968.

Brindley G and Rushton D. Implanted stimulators of the visual cortex as visual prosthetic devices. Transactions of the American Academy of Ophthalmology and Otolaryngology, 78:OP741–OP745, 1974.

Cha K, Horch KW, and Normann RA. Simulation of a phosphene-based visual field: visual acuity in a pixelized vision system. *Annals of Biomedical Engineering*, 20:439–449, 1992a.

Cha K, Horch KW, and Normann RA. Mobility performance with a pixelized visual system. *Vision Research*, 32:1367–1372, 1992b.

Cha K, Horch KW, Normann RA, and Boman DK. Reading speed with a pixelized vision system. *Journal of Optical Society of America*, 9:673–677, 1992c.

Chen SJ, Humayun MS, and Weiland JD. Electrical stimulation of the mouse retina: a study of electrically elicited visual cortical responses. The Association of Research in Vision and Ophthalmology Annual Meeting, 1999; Fort Lauderdale. Abstract 3886 S736.

Chiang YP, Bassi JL, and Javitt JC. Federal budgetary costs of blindness. *Milbank Q*, 70:319–340, 1992.

Chow AY and Chow VY. Subretinal electrical stimulation of the rabbit retina. *Neuroscience Letters*, 225:13–16, 1997.

Chow AY and Peachey NS. The subretinal microphotodiode array retinal prosthesis (letter; comment). *Ophthalmic Research*, 30:195–198, 1998.

Crapper DR and Noell WK. Retinal excitation and inhibition from direct electrical stimulation. *Journal of Neurophysiology*, 6:924–947, 1963.

Curcio CA, Medeiros NE, and Millican CL. Photoreceptor loss in age-related macular degenration. *Investigative Ophthalmology Visual Science*, 37:1236–1249, 1996.

Dawson TW, De Moerloose J, and Stuchly MA. Comparison of magnetically induced ELF fields in humans computed by FDTD and scalar potential FD codes. *Applied Computational Electromagnetics Society*, 11:63–71, 1996.

DeMarco SC, Clements M, Vichienchom K, Liu W, Humayun M, de Juan EJ, Weiland J, and Greenberg RJ. An epi-retinal visual prosthesis implementation. Annual Conference BMES, vol. 1, p. 475, 1999.

DeMarco SC, Lazzi G, Liu W, Weiland JD, and Humayun MS. Computed SAR and thermal elevation in a 0.25 mm 2-D model of the humal eye and head in response to an implanted retinal stimulator: parts 1 and 2. *IEEE Transactions on Antennas and Propagation*, 51(9):2274–2286, 2003.

Dobelle WH and Mladejovsky MG. Phosphenes produced by electrical stimulation of human occipital cortex, and their application to development of a prosthesis for the blind. *Journal of Physiology*, 243:553–576, 1974.

Dobelle WH, Mladejovsky MG, Evans JR, Roberts TS, and Girvin JP. "Braille" reading by a blind volunteer by visual cortex stimulation. *Nature*, 259:111–112, 1976.

Dougherty SH and Simmons RL. Infections in bionic man: the pathobiology of infections in prosthetic devices — part II. *Current Problems in Surgery*, 19:265–319, 1982.

Eberdt M., Brown PK, and Lazzi G. Two-dimensional SPICE-linked multiresolution impedance method for low-frequency electromagnetic interactions. *IEEE Transactions on Biomedical Engineering*, 50, 881–889, July 2003.

Eckmiller R. Learning retina implants with epiretinal contacts. *Ophthalmic Research*, 29:281–289, 1997.

Foerster O. Beitrage zur pathophysiologie der sehban und der spehsphare. *Journal of Psychology and Neurology* (Lpz), 39:435–463, 1929.

Gandhi OP, DeFord JF, and Kanai H. Impedance method for calculation of power deposition patterns in magnetically induced hyperthermia. *IEEE Transactions on Biomedical Engineering*, 31:644–651, 1984.

Gosalia K, Weiland J, Humayun M, and Lazzi G. Thermal elevation in the human eye and head due to the operation of a retinal prosthesis. *IEEE Transactions on Biomedical Engineering*, 51(8):1469–1477, 2004.

Greenberg RJ, Velte TJ, Humayun MS, Scarlatis GN, and de Juan E Jr. A computational model of electrical stimulation of the retinal ganglion cell. *IEEE Transactions on Biomedical Engineering*, 46:505–514, 1999.

Guenther E, Troger B, Schlosshauer B, and Zrenner E. Long term survival of retinal cell cultures on retinal implant materials. *Vision Research*, 39:3988–3994, 1999.

Hirata A, Matsuyama SI, and Shiozawa T. Temperature rises in the human eye exposed to EM waves in the frequency range 0.6–6 GHz. *IEEE Transactions on Electromagnetic Compatibility*, 42:386–393, 2000.

Humayun MS, de Juan E Jr, Dagnelie G, Greenberg RJ, Propst RH, and Phillips DH. Visual perception elicited by electrical stimulation of retina in blind humans. *Archives of Ophthalmology*, 114:40–46, 1996.

Humayun MS, Prince M, de Juan EJ, Barron Y, Moskowitz M, Klock IB, and Milam AH. Morphometric analysis of the extramacular retina from postmortem eyes with retinitis pigmentosa. *Investigative Ophthalmology and Visual Science*, 40:143–148, 1999a.

Humayun MS, de Juan EJ, Weiland JD, Dagnelie G, Katona S, Greenberg RJ, and Suzuki S. Pattern electrical stimulation of the human retina. *Vision Research*, 39:2569–2576, 1999b.

IEEE Standard for Safety Levels With Respect to Human Exposure to Radio Frequency Electromagnetic Fields. 3 kHz to 300 GHz. IEEE Standard C95.1, 1999.

Jones KE and Normann RA. An advanced demultiplexing system for physiological stimulation. *IEEE Transactions on Biomedical Engineering*, 44:1210–1220, 1997.

Kanda H, Morimoto T, Fujikado T, Tano Y, Fukuda Y, and Sawail H. Electrophysiological studies of the feasibility of suprachoroidal–transretinal stimulation for artificial vision in normal and RCS rats. *Investigative Ophthalmology and Visual Science*, 45:560–566, 2004.

Karny H. Clinical and physiological aspects of the cortical visual prosthesis. *Surveys of Ophthalmology*, 20:47–58, 1975.

Kim A, Sadda S, Pearlman J, Humayun M, de Juan EJ, Melia M, and Green WR. Morphometric analysis of the macula in eyes with disciform age-related macular degeneration. *Retina*, 22: 471–477, 2002.

Lederman RJ, Noell WK. Optic nerve population responses to transretinal electrical stimulation. *Vision Research*, 9:1041–1052, 1969.

Liu W, Vichienchom K, Clements M, DeMarco SC, Hughes C, McGucken E, Humayun MS, de Juan EJ, Weiland JD, and Greenberg RJ. A neurostimulus chip with telemetry unit for retinal prosthetic device, *IEEE Journal of Solid-State Circuits*, 35:1487–1497, October 2000.

Lowenstein J, Rizzo JF, Shahin M, Coury A. Novel retinal adhesive used to attach electrode array to retina. The Association for Research in Vision and Ophthalmology Annual Meeting, 1999; Fort Lauderdale. Abstract 3874.

Majji AB, Humayun MS, Weiland JD, Suzuki S, D'Anna SA, and de Juan E Jr. Long term histological and electrophysiological results of an inactive epiretinal electrode array implantation in dogs. *Investigative Ophthalmology and Visual Science*, 40:2073–2081, 1999.

Margalit E, Fujii G, Lai J, Gupta P, Chen S, Shyu J, Piyathaisere DV, Weiland JD, de Juan E Jr., and Humayun MS. Bioadhesives for intraocular use. *Retina*, 20:469–477, 2000.

Margalit E, Dagnelie G, Weiland JD, de Juan E, and Humayun MS. Can vision be restored by electrical stimulation? In: Horch KW and Dhillon GS (eds) (Chapter 7.5) *Nueroprostheses: Theory and Practice.* Singapore: World Scientific, 1067–1102, 2004.

Maynard EM, Nordhausen CT, and Normann RA. The Utah intracortical electrode array: a recording structure for potential brain–computer interfaces. *Electroencephalography and Clinical Neurophysiology,* 102:228–239, 1997.

Maynard EM. Visual prostheses. *Annual Review of Biomedical Engineering,* 3:145–168, 2001.

Nordhausen CT, Maynard EM, and Normann RA. Single unit recording capabilities of a 100 microelectrode array. *Brain Research,* 726:129–140, 1996.

Necati O. *Heat Transfer: A Basic Approach.* New York: McGraw Hill, 1985.

Normann RA, Maynard EM, Rousche PJ, and Warren DJ. A neural interface for a cortical vision prosthesis. *Vision Research,* 39:2577–2587, 1999.

Oehmichen M. Inflammatory cells in the central nervous system: an integrating concept based on recent research in pathology, immunology and forensic medicine. *Progress in Neuropathology,* 5:277–335, 1983.

Peterman MC, Mehenti NZ, Bilbao KV, Lee CJ, Leng T, Noolandi J, Bent SF, Blumenkranz MS, and Fishman HA. The artificial synapse chip: a flexible retinal interface based on directed retinal cell growth and neurotransmitter stimulation. *Artificial Organs,* 27(11):975–985, 2003.

Peyman G, Chow AY, Liang C, Chow VY, Perlman JI, and Peachey NS. Subretinal semiconductor micro-photodiode array. *Ophthalmic Surgical Lasers,* 29:234–241, 1998.

Piyathaisere DV, Margalit E, Chen SJ, Shyu JS, D'Anna SA, Weiland JD, Grebe RR, Grebe L, Fujii G, Kim SY, Greenberg RJ, de Juan EJ, and Humayun MS. Heat effects on the retina. *Ophthalmology Surgery and Lasers Imaging,* 34(2):114–120, 2003.

Pollen DA. Responses of single neurons to electrical stimulation of the surface of the visual cortex. *Brain, Behavior and Evolution,* 14:67–86, 1977.

Potts AM and Inoue J. The electrically evoked response (EER) of the visual system II. Effect of adaptation and retinitis pigmentosa. *Investigative Ophthalmology,* 8:605–612, 1969.

Potts AM and Inoue J. The electrically evoked response of the visual system (EER) III. Further consideration to the origin of the EER. *Investigative Ophthalmology,* 9:814–819, 1970.

Potts AM, Inoue J, and Buffum D. The electrically evoked response of the visual system (EER). *Investigative Ophthalmology,* 7:269–278, 1968.

Ranck JB Jr. Which elements are excited in electrical stimulation of mammalian central nervous system: a review. *Brain Research,* 98:417–440, 1975.

Rita P, Kaczmarek ME, Tyler ME, and Garcia-Lara J. Form perception with a 49-point electrotactile stimulus array on the tongue: a technical note. *Journal of Rehabilitation Research and Development,* 35:427–430, 1998.

Rizzo J. and Wyatt J. Prospects for a visual prosthesis. *Neuroscientist,* 3:251–262, 1997.

Rizzo J, Wyatt J, Loewenstein J, and Kelly S. Acute intraocular retinal stimulation in normal and blind humans. The Association of Research in Vision and Ophthalmology Annual Meeting, 2000; Fort Lauderdale. Abstract 532 S102.

Rocha G, Baines MG, and Deschenes J. The immunology of the eye and its systemic interactions. *Critical Reviews of Immunology,* 12:81–100, 1992.

Rucker L and Lossinsky A. Percutaneous Connectors. 30th Neural Prosthesis Workshop, NINDS, NINCD, NIH, October 1999.

Santos A, Humayun MS, de Juan EJ, Greenberg RJ, Marsh MJ, Klock IB, and Milam AH. Preservation of the inner retina in retinitis pigmentosa. A morphometric analysis. *Archives of Ophthalmology,* 115:511–515, 1997.

Schmidt EM, Bak MJ, Hambrecht FT, Kufta CV, O"Rourke DK, and Vallabhanath P. Feasibility of a visual prosthesis for the blind on intracortical microstimulation of the visual cortex. *Brain,* 119(Pt 2):507–522, 1996.

Shimazu K, Miyake Y, and Watanabe S. Retinal ganglion cell response properties in the transcorneal electrically evoked response of the visual system. *Vision Research,* 39:2251–2260, 1999.

Stone JL, Barlow WE, Humayun MS, de Juan EJ, and Milam AH. Morphometirc analysis of macular photoreceptors and ganglion cells in retinas with retinitis pigmentosa. *Archives of Ophthalmology,* 110:1634–1639, 1992.

Sullivan D. *Electromagnetic Simulation Using the FDTD Method*. New York: IEEE Press, 2000.

Suzuki S, Humayun MS, de Juan E Jr, Weiland JD, and Barron YA. A comparison of electrical stimulation threshold in normal mouse retina vs. different aged retinal degenerate (rd) mouse retina. The Association of Research in Vision and Ophthalmology Annual Meeting, 1999; Fort Lauderdale. Abstract 3886 S735.

Tassicker. (US 2760483). 1956. Ref Type: Patent.

The National Library of Medicine. The Visible Human Project. (Online). Available: *http://www.nlm.nih.gov/ research/visible/visible_human.html*, 2000.

Uematsu S, Chapanis N, Gucer G, Konigsmark B, and Walker AE. Electrical stimulation of the celebral visual system in man. *Confinia Neurologica*, 36:113–124, 1974.

Veraart C, Wanet-Defalque MC, Gerard B, Vanlierde A, and Delbeke J. Pattern recognition with the optic nerve visual prosthesis. *Artificial Organs*, 27(11):996–1004, 2003.

Walter P, Szurman P, Vobig M, Berk H, Ludtke-Handjery HC, Richter H, Mittermayer C, Heimann K, and Sellhaus B. Successful long term implantation of electrically inactive epiretinal microelectrode arrays in rabbits. *Retina*, 19:546–552, 1999.

Wang J and Fujiwara O. FDTD computation of temperature rise in the human head for portable telephones. *IEEE Transactions on Microwave Theory and Techniques*, 47:1528–1534, August 1999.

Weiland JD, Humayun MS. Suzuki S, D'Anna SA, and de Juan E Jr. Electrically evoked response (EER) from the visual cortex in normal and retinal degenerate dog. The Association of Research in Vision and Ophthalmology Annual Meeting, 1999; Fort Lauderdale. Abstract 4125 S783.

Weiland JD and Humayun MS. Past, present and future of artificial vision. *Artificial Organs*, 27(11):961–962, 2003.

Yagi T and Watanabe M. A computional study on an electrode array in a hybrid retinal implant. *Proceedings of 1998 IEEE International Joint Conference on Neural Networks*, 780–783, 1998.

Zrenner E, Miliczek KD, Gabel VP, Graf HG, Guenther E, Haemmerle H, Hoefflinger B, Kohler K, Nisch W, Schubert M, Stett A, and Weiss S. The development of subretinal microphotodiodes for replacement of degenerated photoreceptors. *Ophthalmic Research*, 29(5): 269–280, 1997.

Zrenner E, Stett A, Weiss S, Aramant RB, Guenther E, Kohler S, Miliczek KD, Seiler MJ, and Hammerle H. Can subretinal microphotodiodes successfully replace degenerated photoreceptors? *Vision Research*, 39:2555–2567, 1999.

Zrenner E. Will retinal implants restore vision? *Science*, 295:1022–1025, 2002.

18

Artificial Support and Replacement of Human Organs

Pramod Bonde

CONTENTS

18.1   INTRODUCTION

Heart disease is a leading cause of death and contributes to 29% of total deaths in USA (Anderson and Smith, 2003). About five million people suffer from heart failure each year with additional 500,000 being diagnosed new every year (AHA, 2003). Approximately 1,000,000 will die within 2 years of their diagnosis. Heart transplant is the only definitive therapy for these patients (Baumgartner et al., 2002). Respiratory failure accounts for the fourth leading cause of death followed by kidney and liver failure (Anderson and Smith, 2003). The current gold standard for treating organ failure is transplantation (UNOS, 2003). There are strict criteria for patients to be accepted as suitable candidates for transplantation and in 2002, there were close to 80,000 patients in the USA on the waiting list to receive organ transplantation (UNOS, 2003). During the same year 24,000 received a transplant, with a majority (18,000) receiving them from deceased donors. The latter accounted mostly for kidney and liver transplantation. In 2002, approximately 14,000 patients had kidney transplants and in the same period 5,000 liver transplants were performed (UNOS, 2003). Each year approximately 3,000 heart transplants are performed (AHA, 2003; Baumgartner et al., 2002; UNOS, 2003). As pointed out earlier, the strict criteria for organ transplantation mean that many patients do not have the option of organ transplantation, in addition, as mentioned above, a significant number of patients die waiting for a transplant due to the mismatch in supply and demand of the organs (Baumgartner et al., 2002; UNOS, 2003).

   The only alternative for these patients today is the supportive management offered by artificial organ systems. The design and development of the most of the artificial organ systems can be traced to the 1950s and 1960s (Cooley et al., 1969; Gibbon, 1954; Gottschalk and Fellner, 1997; Kolff, 2002). The subsequent modifications were added later on as the experience with these systems increased. The substitution of organ function by artificial organs represents one of the most remarkable achievements in the 20th century (Lysaght and Reyes, 2001). It is currently estimated that close to 20 million people worldwide derive benefit of prolonging the organ function and quality of life with the use of some kind of artificial medical implant (Lysaght and Hazlehurst, 2004; Malchesky, 2001). Artificial organ supports constitute a part of this population. It represents a financial spending of 350 billion per year on organ replacement therapy and is likely to increase in the future as the population grows old in the next few decades (Lysaght and Reyes, 2001; Lysaght and Hazlehurst, 2004; Malchesky, 2001).

18.2   HISTORICAL PERSPECTIVE

The history of organ replacement can be traced to human origins. This happened when the primitive man took support of a wooden stick to support an injured limb. However, the replacement or mimicking of the internal human organs had to wait until after the industrial revolution, which brought about the technical expertise combined with newer insights and understanding of human anatomy and functioning. First such attempts were primarily to sustain the isolated organ function outside the body by perfusion. LeGalliois (1813) first proposed the idea of mechanically supporting the circulation. In 1885, Von Frey and Gruber built a perfusion apparatus to sustain organ function outside the body (Zimmer, 2001).

   Alexis Carrel contributed monumental work in the perfusion studies and cell and organ cultures in addition to some original work on organ transplantation at the beginning of the last century (Zimmer, 2001). His work on the heart and vessels led him to the problem of biocompatibility of materials (Malinin, 1996). A death of a close relative of Charles Lindbergh was the reason behind the unexpected and unique collaboration between these two to develop a perfusion apparatus (Bing, 1987; Malinin, 1996). The original dream of Charles Lindbergh to bypass the function of the heart and lungs to correct heart defects had to wait another 30 years, when Gibbon developed a heart–lung machine (Gibbon, 1954).

At the same time, Willem Kolff, who saw a young patient dying of kidney failure, reasoned that if urea can be removed from the blood, then that can prevent patients from dying. Using a simple sausage tubing made of cellophane he was able to remove urea from the blood; this lead to the development of the artificial kidney or what we call today, the hemodialysis machine (Kolff, 2002). A chance observation of blue blood turning red during the early experiments with rotating drum kidney led to the development of disc oxygenators. This was later helpful in devising the oxygenators in the heart–lung machine, and ultimately led to the development of modern artificial lung, what is known as extracorporeal membrane oxygenation (ECMO) (Wolfson, 2003). Further improvements in the artificial kidney led to the modern capillary membrane based hemodialysis machines (Gottschalk and Fellner, 1997).

The work of Gott and Daggett was important in understanding the biocompatibility issues in heart valve implants. They designed one of the first bileaflet heart valves, with a graphite–benzalkonium–heparin coating, and later proved the extraordinarily low thrombogeneicity with pyrolytic carbon (Gott et al., 2003). This has been the primary component of valve implants for the last 35 years.

The story of development of artificial human organs is both fascinating and remarkable. Fascinating because, it made possible things which could only be dreamed of before. And remarkable in the unique collaboration that developed between doctors, engineers, scientists, and physicists from diverse disciplines that led to the development of various organ support systems and replacement options. From the highs of achievements in 1950s and 1960s to the recent ugly lawsuits concerning patents for artificial support systems, artificial organ development has witnessed both public curiosity and skepticism with equal measure.

We will be reviewing the relevant historical landmark later in this chapter when we look at the individual organ replacement systems. I have tried to keep the language as simple as possible, avoiding medical jargon to aid easier understanding by nonmedical readers. It is impossible to cover all the technical and medical details of all the artificial organs and organ replacement systems, but I have made every effort to provide a glimpse of this fascinating field. In a true sense of an artificial organ, currently the heart is the only organ which can be replaced as an artificial implant in the human body after removing the native heart, and as such I have focused on the current available artificial heart and assist devices in more details. Other artificial medical implants have been covered in corresponding chapters.

18.3 ARTIFICIAL KIDNEY

We have come a long way from the simple construct of sausage skin, a type of cellophane tubing to remove toxins and harmful waste products (Kolff, 2002). The earlier advances consisted of an artificial kidney made at Johns Hopkins by Abel and colleagues in 1913 using colloidon and hirudin anticoagulant. It took another 15 years before the modification of the Hopkins kidney was used by Hass in Germany to perform first clinical hemodialysis (Vienken et al., 1999). With the use of a rotating drum kidney, developed by Willem Kolff in 1943, the modern era of hemodialysis truly began (Gottschalk and Fellner, 1997). The advances in artificial kidney development were halted due to the Second World War. Soon after the end of Second World War, unprecedented technological developments made what was essentially an experimental therapy into a routine clinical tool in treating kidney failure (Gottschalk and Fellner, 1997; Vienken et al., 1999).

The modern dialyzers consist of semipermeable membranes which are configured into a hollow fiber design. These membranes are essentially cellulose derived or noncellulose synthetic polymers. High flux membranes have a higher ultra filtration coefficient which facilitates higher clearance of the solutes during fluid removal. The technical and clinical aspects of the myriad of these devices available are beyond the scope of this chapter.

Although hemodialysis revolutionized the treatment of kidney failure, it is far from perfect in mimicking the functions of the kidney. Patients need to be hooked to the machine for prolonged periods and therefore limit their mobility. Besides removing the toxic waste and maintaining the electrolyte and water balance, the human kidney plays an important role in terms of endocrine and metabolic activities. To solve this problem, and mimic the functioning of a normal human kidney, developments are underway to develop a bio-artificial kidney which incorporates tubular cells in the hollow fibers (Moussy, 2000). Cells are grown as confluent monolayers along the inner surface of these hollow fibers; the membrane acts as a scaffold and allows the cells to carry out the important metabolic and endocrine activities (Humes et al., 1997; Nikolovski et al., 1999). Another novel aspect of these cell-seeded hollow fibers is that the cells are not exposed to the patient's blood and hence do not develop an immune response (Humes et al., 1997). Early results of these systems are encouraging.

18.4   ARTIFICIAL LIVER

The liver plays an important role in the detoxification, synthesis, and digestion in the body. Currently, liver transplantation is the only viable and satisfactory option for liver failure (UNOS, 2003; van de Kerkhove et al., 2004). But the paucity and mismatch of demand and supply of available donors is a major impediment for widespread application of this therapy. The liver has a tremendous capacity to regenerate and if given adequate time to rest, the liver has the capability of regrowing the damaged cells and can potentially recover. Currently, support systems function as a bridge and try to exploit this regenerative capacity of the damaged liver until recovery or transplantation. Attempts to replace the function of the liver are complex and currently are in their infancy. Several earlier attempts to use hemodialysis to remove undesirable toxic products did not meet with success (van de Kerkhove et al., 2004). Several other modalities like hemofiltration, hemodiafiltration, and hemodiabsorption were not particularly attractive (van de Kerkhove et al., 2004). One of the reasons is that these systems replace only one or two of the myriad functions undertaken by the liver. However, a few of the promising techniques include the Molecular Adsorbents Recirculating System (MARS), Artificial Liver Support Systems, and Albumin Dialysis System (Jalan et al., 2004; Mullin et al., 2004; van de Kerkhove et al., 2004). These are based on detoxification of water soluble and protein bound toxins in dialysis (Boyle et al., 2004). But all of these systems share the common disadvantage of inability to synthesize and produce liver specific factors and proteins.

The above limitations have turned attention to options of biologically mimicking organ function by using liver cells from animal and human origin (Kobayashi et al., 2003; Liu et al., 2004a–c). Theoretically, they can carry out detoxification, metabolic function, and synthesize important proteins. Earlier attempts involved using cross-circulation with animal livers or liver-tissue pre-parations (van de Kerkhove et al., 2004). Liver cells can be used in suspended, attached, or encapsulated fashion with the aid of a semipermeable membrane akin to a bio-artificial kidney. These are collectively called bio-artificial liver systems (Demetriou et al., 2004; Fruhauf et al., 2004; Kobayashi et al., 2003; Liu et al., 2004a–c). Currently, there are few systems available which have undergone even limited human trials (Demetriou et al., 2004). They include the Extracorporeal Liver Assist System (ELAD), which uses a transformed hepatocyte cell line (Figure 18.1). Other systems such as the HepatAssist System, the TECA-hybrid artificial liver support system, the bio-artificial liver support system, the radial flow bioreactor, the liver support system, the AMC-bio-artificial liver, and the bio-artificial hepatic support system, all use porcine derived hepatocyte cells (Demetriou et al., 2004; van de Kerkhove et al., 2004). However, there are concerns about using tumor derived or transformed cells due to their potential to develop cancer. On the other hand, porcine cells pose the risk of exposing the human body to animal tissue thus setting up an immune response and the added risk of transporting infections from animals to humans. The widespread

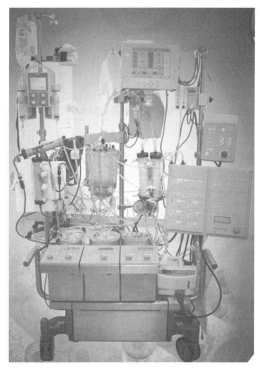

Figure 18.1 (See color insert following page 302) ELAD artificial liver system. (Courtesy of Vital Therapies Inc, San Diego, CA.)

clinical application of such systems is currently limited, although some of the bio-artificial liver support systems have shown favorable clinical outcomes (Demetriou et al., 2004; van de Kerkhove et al., 2004).

18.5 HEART AND LUNG MACHINE

A 20-year quest by John Gibbon realized the dream of building an artificial heart–lung machine, which in turn allowed the field of open heart surgery to bloom (Gibbon, 1954). The day was May 6, 1953, when this device was first used to repair a hole in the upper chambers of the heart. Since then, the machine has undergone several changes (Boettcher et al., 2003) from the initial disc and screen oxygenators to De Wall bubble oxygenators and finally to the membrane oxygenators (Cook, 2004). The modern heart–lung machine consists essentially of a venous reservoir which drains venous blood from the vena cava system. The blood is then pumped through a membrane oxygenator and subsequently pumped back into the aorta to support the circulation. There is a heat exchanger incorporated in the circuit. Over the years various sensors and safety features have been added to this system, although the basic design has remained the same for the last few decades (Boettcher et al., 2003).

The conventional bypass machine requires considerable priming fluid which can lead to significant hemodilution. This in turn can have adverse effects on the functioning of the cellular components of the blood. The large surface area initiates an immune systemic response. The latest efforts have been to miniaturize the heart–lung machine (Boettcher et al., 2003; Remadi et al., 2004; von Segesser et al., 2003).

The widespread use of the heart–lung machine has provided an opportunity for its use in heart failure. The earlier attempts were to use the heart–lung machine for extended time to allow the heart to recover (FC Spencer, 1959). But they had inherent problems associated with damage to the blood due to the heart–lung machine. This necessitated a development of artificial ventricular assist devices and total artificial heart to sustain the function of the heart (Deng et al., 2001; Portner, 2001). The application of the heart–lung machine also led to efforts at constructing an artificial lung, which we will review prior to the ventricular assist devices and artificial heart.

18.6 ARTIFICIAL LUNG

The human lungs are essential in the oxygenation of blood. An artificial ventilator can supply the oxygen in a controlled fashion to allow oxygenation in the lungs, and is used extensively during surgeries and sometimes for prolonged ventilation. But patients with diseased lungs and failing heart need extraordinarily high oxygen content and pressures to be maintained by the ventilators; this itself can cause fibrosis and destroy the lungs over time. The solution for these patients with potentially reversible lung and cardiac failure is to achieve oxygenation of blood without exposing the lungs to high ventilatory pressures and potentially toxic oxygen levels; this is particularly important in children and neonates (Hansell, 2003; Lawson et al., 2004; Wolfson, 2003).

In 1955, Clowes and associates reported oxygen diffusion across plastic films, which led to the foundation for the later development of artificial lungs, more commonly called Extracorporeal Membrane Oxygenation (ECMO) (Clowes et al., 1955). One of the authors of this original publication later reported the use of a silicone for membrane oxygenation; this is five times more efficient in carbon dioxide permeability compared to oxygen (Kolobow and Bowman, 1963). The use of a silicone membrane allows a much smaller force for carbon dioxide transfer. The ECMO interposes a semipermeable membrane of silicone between blood and oxygen, thus aiding gas transfer. A traditional ECMO circuit drains the deoxygenated blood from the right side of heart, and it is then pumped through a membrane oxygenator which allows the gas exchange to take place (Cook, 2004; Hansell, 2003). The blood is then rewarmed and returned to the left side of the heart. The transfer from and to the heart can be done by cannulating peripheral vessels such as the internal jugular and carotid artery. The concerns about manipulating and ligating the carotid artery led to a venovenous ECMO, in which a double lumen catheter does the job of taking blood from and returning to the heart (Hansell, 2003).

In spite of earlier discouraging results, ECMO has proven to be a very useful tool in treating the neonatal population needing cardio-respiratory support (Bartlett et al., 1976; Cook, 2004; Hansell, 2003; Petrou and Edwards, 2004). Its value in treating adults is currently limited. However, ECMO, in a true sense does not replace the lungs but allows them to rest and recover, and hence is a temporary substitute. Other systems such as hollow fiber systems have been used both clinically and preclinically with good results. Microporous hollow fiber oxygenators are widely applied and are popular for short-term cardiopulmonary bypass. Hollow fiber nonporous oxygenators are mostly employed for long-term extracorporeal circulatory support. Unlike other organs there is no reliable method available for bridging patients waiting for lung transplantation. Recent developments in fluid dynamics have allowed development of low-resistance membrane oxygenators (Figure 18.2). One such system relies on the pumping capacity of the right ventricle to sustain an artificial lung oxygenator (Figure 18.3). Initial animal studies have been encouraging (Lick et al., 2001; Zwischenberger et al., 2001). How this paracorporeal lung device will influence the treatment of acute lung failure in the clinical setting is yet to be explored.

18.7 VENTRICULAR ASSIST DEVICES

The quest to support the function of the heart or to temporarily support it commenced soon after the introduction of the Gibbon screen oxygenator (Gibbon, 1954). Earlier attempts employed the heart–lung machine itself to support patients for extended period (FC Spencer, 1959). The first ventricular

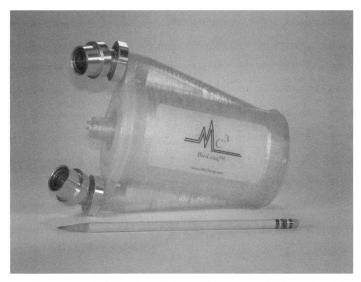

Figure 18.2   The small BioLung artificial implantable lung. (Courtesy of MC3 Corp, Ann Arbor, MI.)

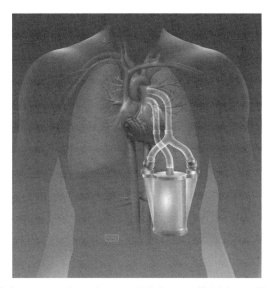

Figure 18.3   The site of the proposed attachment of BioLung artificial lung. (Courtesy of MC3 Corp, Ann Arbor, MI.)

device was developed by DeBakey et al. in 1964 (DeBakey, 1971; Hall et al., 1964). Cooley made an attempt to support the heart by one of the earlier artificial hearts as a bridge to transplant (Cooley et al., 1969). Oyer was first to successfully implant the Novacor device (Pierce, 1988). DeVries and colleagues successfully implanted the Jarvik-7 model in Barney Clark (DeVries et al., 1984). Pulsatility has been the main difference between the different devices, as one tries to mimic nature by producing a beat with every ejection of the pump. These mechanical systems consist of the pusher plate activated devices or compression of collapsible sacs by pneumatic power (Deng et al., 2001; Portner, 2001). Nonpulsatile devices are essentially motor driven centrifugal pumps (Portner, 2001). Here we review some of the clinically used devices. All that the ventricular assist devices do is to bypass the native heart; they do not replace the heart. The heart is kept in place, and the devices merely bypass the blood flow. The devices can either be connected with tubings to and from the

Figure 18.4   The BioMedicus centrifugal pump.

heart with the pump lying outside the body (centrifugal pumps and paracorporeal devices) or are implantable in the body (intracorporeal). On the other hand, artificial hearts provide total replacement of the heart, which is excised.

In the last few years, mechanical circulatory support has provided clinically relevant solutions in the form of bridge to recovery, bridge to transplantation, and as a definitive therapy (Rose et al., 2001). The *rematch* trial conclusively demonstrated benefit with reduction of the mortality by 48% in patients treated with devices versus those who received maximal medical management (Rose et al., 2001). Patients in this study were ineligible for heart transplantation.

18.7.1 Centrifugal Pumps

Centrifugal pumps do not have valves or multiple moving or occluding parts; blood is pumped by rotating blades or by use of impellers (Curtis et al., 1999). They are able to provide high flow rates with low pressure rises. There are various devices available; for example, BIO-PUMP (Medtronic, Inc., Minneapolis, MN), St Jude pump (Bard Cardiopulmonary Division, Haverhill, MA), Carmeda Bio-Pump (Medtronic, Inc., Minneapolis, MN), Sarns pump (3M Healthcare, Inc., Ann Arbor, MI), Nikkiso pump (Nikkiso Pumps America, Inc., Plumsteadville, PA) (Curtis et al., 1994, 1996, 1999; Magovern, 1993; Noon et al., 1995). The BioMedicus pump is depicted in Figure 18.4. One of the disadvantages of centrifugal pumps is that they can only be used for short-term support of hours to days (Hoy et al., 2000).

18.7.2 Paracorporeal Devices

Paracorporeal devices are placed outside the body and support each ventricle separately. For example, the Abiomed BVS 500 is a pneumatically driven, asynchronous, pulsatile, polycarbonate housed dual chamber pump (Dekkers et al., 2001; Wassenberg, 2000). During systole, compressed air enters the ventricular chamber and compresses the polyurethane bladder. Another paracorporeal device used clinically is the Thoratec VAD (Figure 18.5), which is also pneumatically driven (Farrar, 2000; Farrar et al., 2002).

18.7.3 Intracorporeal Devices

These devices have the pump mechanism implanted in the body with power and driveline, being connected to an external console. The HeartMate device is an implantable, pulsatile, pneumatically

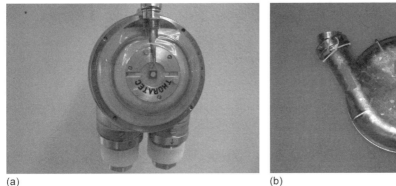

(a)   (b)

Figure 18.5   (a) Thoratec paracorporeal VAD and (b) HeartMate device.

actuated (IP) device (Frazier, 1994). The unit is made from sintered titanium and houses a flexible, textured polyurethane diaphragm and operated by a pusher plate mechanism. The newer electrically vented model is HeartMate VE (Figure 18.6) and which was introduced in 1991 (Frazier et al., 2001). This latter unit is composed of textured titanium, incorporating titanium microspheres; the polyurethane diaphragm specially treated with textured polyurethane, encouraging deposition of fibrin–collagen matrix, and is more biocompatible with reduced need for anticoagulation (Morgan et al., 2004; Rose et al., 2001). Novacor LVAS (World Heart Corp., Ottawa, Canada) incorporates a dual pusher-plate sac type pump, with a smooth blood contacting surface (Di Bella et al., 2000; Robbins et al., 2001). The system uses a high efficiency linear motor, with a pulsed solenoid energy converter with a two armature assembly (Figure 18.7). This system requires no gears, cams, or intermediate hydraulic conversion; theoretically, it has advantage of low mechanical failure. The advantage of the LionHeart LVAS (Arrow International, Reading, Pennsylvania) is that it is completely implantable; it has a brushless motor which actuates a pusher plate using a roller and screw mechanism (El-Banayosy et al., 2003; Mehta et al., 2001). Energy transfer through the chest wall is achieved using radiofrequency induction (Figure 18.8).

18.7.4 Newer Rotary Axial Pumps

In order to overcome the disadvantages offered by the various paracorporeal and intracorporeal devices such as external drivelines and the large consoles, the radial axial flow devices were devised (Kung and Hart, 1997; Nose, 1998; Okada et al., 1997; Wu et al., 1999). These devices have only one moving part and hence are smaller in size with less need for energy (Mesana, 2004; Nose et al., 2000). There are, however, some inherent problems associated with the simplicity of the design. For example, the absence of a valve can result in significant backflow with device failure, which can be detrimental in an already diseased heart. There are also concerns about the very high speeds at which axial pumps rotate and some aspects regarding bio-compatibility are still open to question (Mesana, 2004). The three systems which are clinically used include: (1) DeBakey VAD (MicroMed Technology, Inc, Houston, Texas), a small 30-mm diameter device, 76 mm in length, which weighs 95 gm (Figure 18.9) and is made of titanium (Goldstein, 2003; Noon et al., 2001); (2) The Flowmaker (Jarvik 2000), (Jarvik Heart, Inc., New York, NY) titanium based pump, which is 25 mm in diameter, 51 mm in length and weighs 90 gm (Frazier et al., 2004; Kaplon et al., 1996); and (3) HeartMate II (Thoratec Corp., Pleasanton, CA), which has a diameter of 40 mm, length of 70 mm and weighs 176 gm (Figure 18.10) (Burke et al., 2001; Griffith et al., 2001).

All these devices work on similar principle and are composed of an impeller and inducer assembly with blood lubricated bearings. In spite of their perceived shortcomings, these devices offer the advantages of small size and potential for a destination therapy for heart failure.

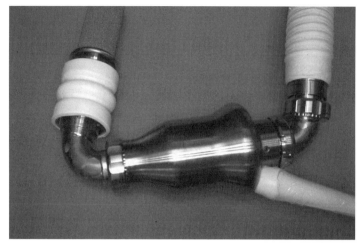

Figure 18.6 (See color insert following page 302)   HeartMate II rotary axial pump device.

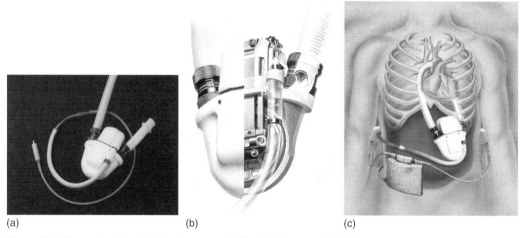

(a)   (b)   (c)

Figure 18.7 (See color insert following page 302)   (a) Novacor VAD, (b) cross-section of Novacor, and (c) diagrammatic representation of Novacor. (With permission from World Heart Corporation, Ottawa, Canada.)

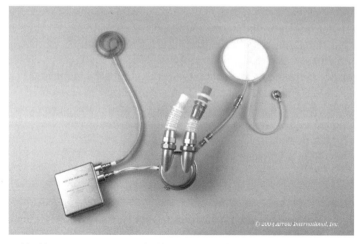

Figure 18.8   Arrow LionHeart support system. (With permission from Arrow International, Reading, PA.)

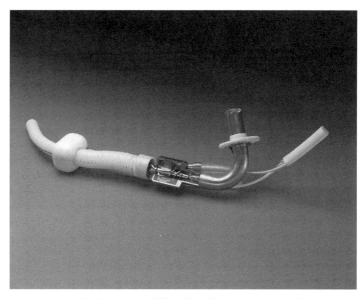

Figure 18.9 (See color insert following page 302) MicroMed DeBakey ventricular assist device. (With permission from MicroMed Technology, Inc, Huston, TX.)

Some of the newer magnetically levitated centrifugal pumps like HeartQuest VAD (MedQuest Products, Inc.) (Chen et al., 2002), DuraHeart LVAS (Terumo Corp.) (Nojiri, 2002) and CorAide LVAS (Arrow Intl.) (Figure 18.11) (Doi et al., 2004) offer the similar advantages as above and are in the early stages of clinical investigation.

18.8 TOTAL ARTIFICIAL HEART

The era of total artificial heart replacement began in 1958 with the successful sustenance of circulation of a dog for 90 min by Akustu and Kolff (1958). This was followed by the implantation of the Liotta heart by Cooley in 1969 (Cooley et al., 1969) and the Akutsu III heart in 1981 (Frazier

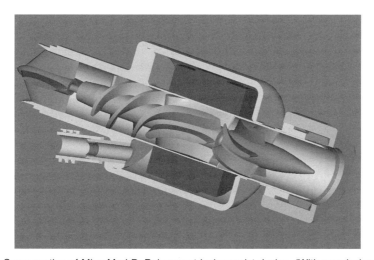

Figure 18.10 Cross-section of MicroMed DeBakey ventricular assist device. (With permission from MicroMed Technology, Inc, Huston, TX.)

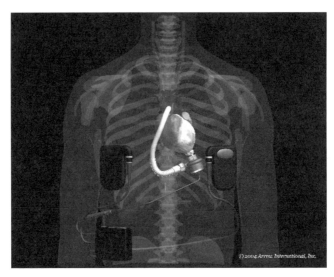

Figure 18.11   The CorAide device for ventricular support. (With permission from Arrow International, Reading, PA.)

et al., 1982); both had unfavorable outcome. It was in 1982 that the Jarvik-7 total artificial heart (TAH) was implanted in Barney Clark by DeVries and colleagues (DeVries et al., 1984). This operation attracted much public attention as a result of significant coverage in the lay media. The New York Times went as far as calling this a "Dracula" of medical therapy. The following devices are some of the clinically available TAH which have been used.

18.8.1   AbioCor Total Artificial Heart (ABIOMED, Inc, Denver, CO)

This device is composed of a high efficiency miniature centrifugal pump situated between the two artificial ventricles; this system is implanted after removing the native heart. A two position switching valve is used to alternate the direction of the hydraulic flow between the pumping chambers resulting in alternate contraction of the chambers (Dowling et al., 2001, 2003). In addition, there is a balance chamber which balances the volume between right and left sided chambers. This device is in the initial clinical trial stage and at the time this publication went to press, 15 patients have undergone device implantation.

18.8.2   CardioWest TAH (SynCardia Systems, Inc, Tucson, AZ)

This device is a precursor of the Jarvik-7. It is a pneumatically driven biventricular pulsatile device (Figure 18.12) (Leprince et al., 2003). Blood collects in to a polyurethane blood sac and is compressed by the air (Copeland, 2000; Copeland et al., 2001). There are two artificial ventricles which are connected to the native atria. An external console delivers the compressed air.

18.8.3   Penn State TAH (ABIOMED, Inc, Denver, CO)

This device is an electromechanical TAH and consists of a titanium rigid case with blood sacs and energy converters. The energy converter is a brushless DC electric motor. The actuation moves dual pusher plates to both sides by a roller screw. It has low associated vibrations and minimal noise (Weiss et al., 1999). It has a stroke volume of about 64 ml and can reach a maximum output of about 8 l/min. This device is in the preclinical testing stage.

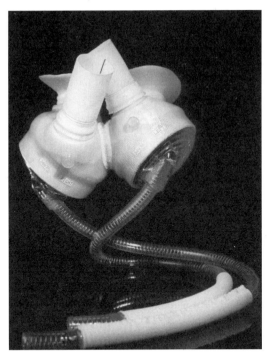

Figure 18.12 (See color insert following page 302)   The SynCardia CardioWest total artificial heart. (With permission from SynCardia Systems, Inc., Tucson, AZ.)

18.9 TOTAL JOINT REPLACEMENTS

Attempts have been made since decades to find a cure for the treatment of arthritis which affects mainly the hip and knee joints. Earlier treatments included fusion of joint spaces (arthrodesis), removing some part of bone (osteotomy), disrupting the nerve supply and cleaning (débridement) of the joint spaces to remove the irregular coverings over the bone in the joints. In an attempt to smoothen the joint spaces, calcium deposits, parts of cartilage, and extraneous bony spurs were removed. Attempts were made to cover the joint spaces with fat, muscle, fascia, or metals like gold, magnesium, and zinc. These strategies had limited success, and they proved short lived at the best. Smith Peterson in 1925 in Boston proposed using a glass molding of the ball of the hip joint to reduce friction (Neff, 1954). However, it was soon apparent that glass could not withstand the stress of the normal body weight in the erect posture. Other materials like plastics and stainless steel were tried, and steel was proven to be biocompatible. Other material used was cobalt–chromium to allow a resistant free joint movement; this met with some mixed success. Austin Moore and Frederick Thompson independently demonstrated the feasibility of replacing the entire ball of the hip, but this only addressed the problems of the arthritic femoral head, since the hip socket (acetabulum) was not replaced.

In 1938, Jean Judet and his brother used acrylic material to replace the hip surfaces; although unsuccessful initially due to the limited adhesiveness offered by acrylic, this was the precursor for using the dental fast setting acrylic in joint replacement (Lukes and Merckelbach, 1958). John Charnley is credited with using polyethylene to create the hip joint socket and used to achieve a smooth surface between the metal ball component and the new socket, which was cemented using polymethyl acrylate (Mallory, 2004). This essentially established the field of total hip replacement (Wroblewski, 2004). Around 100,000 hip replacements are performed annually in the United States. The aspect of breakdown in the cement fixation has been the area of intense research in the recent years, and implants with textured surfaces have been developed in an effort to allow bone to grow into the implant, this theoretically has the advantage of allowing much stronger biological cementing. One of the long-term problems after hip replacement is loosening of the components, which can result in bone loss and pain. This restricts the use of total hip replacement among younger patients. This happens due to very small plastic particles produced by the wearing of the cup. Recently metal on metal joints have regained popularity and are particularly suited for the hip joint replacement in middle age patients since it gives a much longer lasting results compared to the other hip replacements (Dorr et al., 2000).

Parallel developments allowed the development of total knee replacement. Initial attempts were to replace the joint cavity with hinges which can cover the joint space to reduce friction. But problems with loosening and infection frequently occurred. Frank Gunston developed a metal on plastic knee replacement joint in 1968 (Gunston, 1971). A three component knee-joint prosthesis was proposed by John Insall in 1972 which covered the femur, tibia, and the patella, and were held in place using cement (Ranawat et al., 1975). This has resulted in the development of the modern knee-joint prosthesis. Currently more than 150,000 knee-joint replacements are undertaken in United States alone (Noble et al., 2005). Similar to the hip prosthesis, attempts have been underway in recent years to achieve a cementless joint replacement, using biological ability to glue these components together by allowing new bone growth in the roughened surfaces of these devices, which then can give strength and eliminate the need for artificial gluing materials that could come loose.

18.10   BIO-ARTIFICIAL PANCREAS

Long standing diabetes mellitus (types I and II) results due to the inability of the pancreas to secrete insulin. Therapy has been focused at administering the insulin exogenously to achieve acceptable blood sugar levels, however, it is often difficult to manage. Transplantation of the isolated islet cells (which secrete insulin) although promising is limited due to the associated need for immunosuppression and limited organ supply.

Devices such as microencapsulated islets (small diameter spherical chamber), and microencapsulated islets (including hollow fiber, disk-shaped diffusion chambers and Millipore cellulose membranes) have been proposed (Lanza et al., 1992; Lim and Sun, 1980; Reach et al., 1981; Sullivan et al., 1991). Advancements in glucose sensing and insulin sensing technology have allowed developing automated closed loop insulin delivery systems that can deliver insulin in a more physiologic way. One such system currently undergoing clinical trials is a diffusion chamber for a bio-artificial endocrine pancreas (Bio-AEP), which is constructed by placing pancreatic islet cells, trapped in a scaffold; this is sandwiched between semipermeable membranes, and shielded by silicone (Hirotani et al., 1999). Although some of the results achieved in animal studies have been difficult to reproduce in large animal models, this therapy holds promise for the future treatment of diabetes mellitus.

18.11   VISUAL PROSTHESIS (ARTIFICIAL EYE)

The understanding of the mammalian visual system has given impetus for conceptualizing an artificial visual prosthesis that can be used in the profoundly blind. The goal of these systems is to produce a visual perception to allow activities like reading, recognizing shapes and faces, negotiating complex spaces, and giving the perception of light surroundings. This is dealt with in greater detail in Chapters 11 and 17.

18.12   ARTIFICIAL SKIN SUBSTITUTES

Successful application of skin substitutes has been applied widely in the clinical field for a few decades now. The development of skin substitutes or artificial skin began with growing sheets of cells in culture media and has progressed to developing complex structures with bi-layered skin that mimics the human skin. A deeper dermal element is constructed using synthetic epidermis. Currently there are three approaches used for manufacturing artificial skin, the gel approach where cells are grown in a gel of extracellular material like collagen; the scaffold approach where porous scaffolds created from collagen or synthetic material are used to allow cells to be seeded subsequently (Jones et al., 2002); the third approach entails, self-assembly, it is still in animal testing stage and has to await clinical application.

Some of the artificial skin substitutes available are, Alloderm® introduced in market in 1992 and is based on treating fresh cadaver skin in which the epidermal layer is removed and cellular components are destroyed (Bello et al., 2001). The freeze drying of this skin substitute renders it immunologically inert and hence is not rejected by the recipient (Losee et al., 2005; Terino, 2001). Integra™ approved in 1996 by FDA is another skin substitute available commercially and is made from cellular collagen and glycosaminoglycans matrix (Winfrey et al., 1999). The dermal component is made of collagen and the epidermal element is substituted by synthetic silicon. Dermagraft® is an allogenic dermal substitute, it comprises of a scaffold of polyglactin seeded with allogenic fibroblasts (Eaglstein, 1998). This is now used to treat skin ulcers and burn wounds. Another allogenic frozen dermal substitute is TransCyte®, which is used as a temporary replacement for wounds and burns (Noordenbos et al., 1999). It is created by seeding fibroblasts into a scaffold made from nylon mesh and silicone sheet. Bilayered substitutes are composed of allogenic keratinocytes seeded on a nonporous collagen gel and covered with a bovine collagen scaffold containing fibroblasts (OrCel®). They offer the more biologically mimicking skin substitute (Still et al., 2003).

18.13   ARTIFICIAL BLOOD

Inadequate oxygen delivery to the tissues is common sequelae when significant blood loss occurs due to trauma or surgery. This is commonly treated in clinical practice by administering donated human blood. However, the availability of donors and the risk of transmission of infections limit this approach. Fatal reactions can occur due to a mismatch or presence of antibodies in the blood of the recipient; in addition, repeated blood transfusions can depress the immune function in the host. This is one of the reasons why an artificial blood substitute is highly desirable since it can avoid these complications. Two main approaches are used for achieving an artificial blood substitute, bio-artificial oxygen carriers and totally synthetic oxygen carriers. Bio-artificial oxygen carriers are hemoglobin-based oxygen carriers and use human, animal, or recombinant hemoglobin. Synthetic oxygen carriers use metal chelates that mimic the hemoglobin's oxygen binding capacity. Artificial fluorinated organic compounds can physically dissolve large amounts of oxygen, perflurocarbon-based oxygen carriers are commonly employed for this purpose. However, in a strict sense they constitute oxygen carrier substitutes and not blood substitutes since they lack the coagulation factors and immune cells fighting infection that are essential in aiding coagulation and clot formation and fighting infection, which can be vital in the patients receiving these therapies.

Examples of bio-artificial oxygen carriers include modified human or animal hemoglobin-based carriers, stabilized hemoglobin tetramers, polymerized hemoglobin, conjugated hemoglobin, and liposome encapsulated hemoglobin. Other carriers also include recombinant hemoglobin or from transgenic studies. Synthetic oxygen carriers include lipid–heme vesicles, hemoglobin aquasoms, and perflurocarbon based carriers. More detailed review is presented elsewhere (Kim and Greenburg, 2004).

18.14   OTHER SUBSTITUTES

The last few decades have seen an explosive growth in the development of various implants such as pacemakers, stents for the arteries, cochlear implants (Rubinstein, 2004) to improve hearing, apheresis, small joints for the fingers and other joints, etc., the list is quite long and a brief review like this is unable to cover these areas in detail. Another field that is currently undergoing intense research is the field of xenotransplantation. Theoretically, this should allow transplantation of organs from animals to humans; however, there are several issues which need to be addressed include the risk of transmission of animal diseases to humans, the altered immune response that may accompany the species specific difference (Hammer, 2004; Schmidt et al., 2004).

18.15   LIMITATIONS OF THE CURRENT ORGAN REPLACEMENT SYSTEMS

In spite of the significant advances made in the development of the artificial organs, some common problems plague all the systems. Biocompatibility (Hernandez et al., 2004; Jalan et al., 2004) is still a major problem, necessitating heparinization to avoid thrombosis. The use of heparin, to combat thrombosis, puts the patient at risk of bleeding-associated complications (Boyle et al., 2004; Minami et al., 2000; Rose et al., 2001). The organ systems do not truly replace the organs except in the case of total artificial heart. Most of the systems work on the principle of passive transport as in artificial kidney, artificial liver, and lung and hence fail to mimic the physiological functions of these individual organs.

Most of these systems expose the body to increased infection risk due to the various lines and ports used for access (Rose et al., 2001; Tobin and Bambauer, 2003). This risk of infection can be serious in an already sick group of patients (El-Banayosy et al., 2001; Minami et al., 2000). Other limitations include nonphysiological support; for example, the organ support in case of kidneys need not be continuous as is the case of normal kidney which carries out the work 24 h a day, and this can disturb the delicate physiological balance necessary for the optimum biological functioning. Mobility is restricted in all types of the support devices.

The issue of energy supply is very important in the case of artificial ventricular assist devices and the artificial heart; these devices need to work continuously and lack of back-up systems can be catastrophic (Portner, 2001). Mechanical failure is an important issue if long term support is envisaged.

18.15.1   Impact of Other Technologies

Technological advances are rapidly taking place around us and it is natural that these will significantly affect future organ support systems. The current organ replacement systems were designed in the 1960s and 1970s; it is a natural evolutionary step that new technology will replace the older systems.

In concluding this section, we will take a glimpse at current developmental research in related fields and how it will impact the future of organ replacement systems.

18.15.1.1   Tissue Engineering

Tissue engineering is a science that uses living cells combined with biomaterials for diagnostic and therapeutic purposes. This involves generation of cells, tissues, and complex organoid structures in the laboratory to replace natural organ function partially or completely (Fuchs et al., 2001). Application of tissue engineering has resulted in the development of bio-artificial kidney (Aebischer et al., 1987), liver (Chamuleau, 2002; Kulig et al., 2004), tissue-engineered heart valves (Hoerstrup et al., 2000a,b; Stock et al., 2002) and generation of myocardial cells to treat heart failure (Thompson et al., 2003). From the initial euphoria in 1990 to disappointment in 2004, tissue engineering has been put to test; a number of products have not shown benefit in clinical trials, that in turn is reflected in the lack of market interest in these products (Lysaght and Reyes, 2001; Lysaght and Hazlehurst, 2004).

Tissue engineering has the necessary potential of seeding appropriate scaffolding with cells of interest as in the case of tubule cells used in artificial kidney (Fey-Lamprecht et al., 2003; Humes, 2000; Ozgen et al., 2004). As our understanding increases in terms of cell growth characteristics in relation to biomaterials, we are likely to move towards bio-artificial organ replacement systems. Normal organs, however, are composed of many different cell types with complex messaging and interactions. Using a single cell type may not necessarily guarantee adequate functioning of such systems. The importance of developing appropriate scaffolds for the blood vessels to grow can be key to future development of solid organs (Kaihara et al., 2000; MacNeill et al., 2002). The current systems use altered cancerous cells or cells from animal origin, which raises the likelihood of risk of cancerous transformation and transmission of animal originated diseases (van de Kerkhove et al., 2004). However, using adult stem cells from the patients' own bone marrow may be the solution which will be more widely applied in the future.

18.15.1.2 Stem Cell Technology

Stem cells are the precursor cells from which any type of cell differentiation is possible (Jain, 2002). There are two types of stem cell sources that can be used, one from the embryonic stage and another from the adult stem cells within the bone marrow. Stem cells from the embryonic stage offer the characteristic of differentiating into any possible cell type (Kakinuma et al., 2003; Sukhikh and Shtil, 2002); but recent findings, however, of increasing plasticity shown by the human hematopoietic stem cells to differentiate into different cell types has led to interest in developing them as a cell therapy for organ failure (Liu et al., 2004a–c; Schuster et al., 2004; Strom et al., 2004; Yokoo et al., 2003).

18.15.1.3 Impact of Understanding the Human Genome

The human genome sequence now has been decoded (Venter et al., 2001). This offers the potential of synthetic DNA which can create proteins of interest. Theoretically, this can be used to develop synthetic organ systems and conceivably a complete organism. However, there are several limitations to this concept since we still do not have the insight into the function and role of all the human genes. Early indications suggest a possibility of tailor-made treatment based on the individual patient's genomic characteristics; how this will apply to the treatment and replacement of organ systems remains to be fully explored.

18.15.1.4 Microelectromechanical Systems

Microdevices have been applied for certain diagnostic, therapeutic, and selected surgical procedures (Evans et al., 2003; Polla et al., 2000; Richards Grayson et al., 2004). Microelectromechanical systems (MEMS) employ the same manufacturing methods as silicone chips for computer industry. They can be a useful tool for rapid screening of diseases, measurement of blood levels of hormones and drugs, targeted drug delivery, and novel micro-stimulators in neurosciences (Evans et al., 2003; Huang et al., 2002; Liu et al., 2004a–c; Polla et al., 2000; Roy et al., 2001).

What makes MEMS more promising is the building of small rotors capable of running on miniscule energy (Epstein and Senturia, 1997; Miki et al., 2003). These have enormous potential to provide the energy source for organ replacement systems. In addition, they can provide the capability to detect the minute changes in hormones and endorphins on which the response of the organ support system can be tailored.

18.15.2 Nanotechnology and Biomimetics

Living organisms are the perfect example of the advanced nanotechnological manufacturing by nature. What could be more interesting than trying to build artificial organs from the beginning and mimicking nature?

Advances in the development of artificial organs to date have relied mainly on supplanting the function of an organ with an alternative process. As in the case of heart, it is the pumping mechanism, in lungs the oxygenation of the blood, and in liver and kidney, the removal of harmful wastes. But biological organs play even more complex and dynamic role in the physiological mileu in terms of metabolic and other organ function. The biofeedback in these natural organs takes place at nano-dimensions, which the present replacement systems are unable to mimic precisely enough to bring about changes in the functionality of the devices.

Nanotechnology can provide molecularly manipulated nanostructured materials which will mimic the natural surfaces. Sensing and control can be achieved in these systems using microelectronics and novel interface technologies (Lee et al., 2004). Drug delivery systems at nanoscale can maintain the function of normal cells (Prokop, 2001). Molecular self-assembly can simulate the surface geometry by polymeric patterning; since this has immense importance in the behavior of the individual cell and cell to cell communication, adhesion and migration (Chaikof et al., 2002; Hilt, 2004). Current cell and tissue culture systems fail to mimic the natural processes that provide extracellular matrix. Extracellular matrix plays an important role in the repair processes and thus influences cell behavior and survival. Scaffolds at a micro-level can be created using nanotechnology, and can incorporate the extracellular matrix containing glycosaminoglycans and glycoproteins supporting cell growth and proliferation (Bouhadir et al., 2001; Chaikof et al., 2002).

The advances in nanotechnology allow us to synthesize novel materials, fabricate them in two or three-dimensional forms as scaffolds and allow the growth of new cells and ultimately whole organs (Chaikof et al., 2002; Karlsson et al., 2004; Moldovan and Ferrari, 2002).

The National Institute of Health (NIH) has taken a big initiative in funding nanomedicine-related research and development. The NIH roadmap aims to have applications in drug delivery, cell repair, anticancer methodologies, and biomachines that could remove and replace a damaged cell or tissue. The biggest advantage of nanotechnology will be in understanding the organ function at minute levels and creating bio-engineered cells and tissues capable of replacing human organs.

Structural and functional creation of artificial organs using nanotechnology will need precise understanding of the structure and function of the organ; the current knowledge of anatomical structures can greatly help in this regard. This can allow bioengineers to create exact scaffolds for the blood vessels and cells to grow. The issue of energy source can only be solved, however, if micro-machines are built which can derive energy from oxygen, glucose and other substances that are easily available in the body.

18.16 SUMMARY

The current emphasis on replacement by mechanical systems is already profoundly affected by newer technologies. In the future, bio-compatible surfaces will be designed keeping in mind the precise interactions at atomic and molecular levels rather than the trial-and-error approach that was adopted several decades ago. These newer technologies will definitely have an impact on future artificial medical implants, be it artificial heart valves, vascular conduits, or artificial organ systems.

Design and technology will certainly move to center stage in the coming years. Unique problems will be posed for the today's scientists, physicians, and engineers, who are slow to adjust to collaborative research. Current funding structure is limited in supporting such collaborations; the cost of such design and manufacturing will be prohibitive for one group or individual organizations to sustain. Answers to these problems will hopefully be addressed in the future federal funding mechanism as outlined in the initiative by NIH on nanotechnology.

One of the questions that is frequently debated is whether future organ replacement technology will involve miniaturizing the current systems or building newer organ replacement systems from scratch. As outlined above, current organ replacement systems have several disadvantages which will be difficult to overcome even if they are miniaturized. Miniaturization will certainly play an important role in devising therapeutic interventions such as drug delivery. Devising organ replacement systems from scratch will help address the current problems of biocompatibility and better mimic the organ function at cellular level. This will involve creating novel anatomical models of scaffoldings which are biocompatible and bioactive to allow cell growth and differentiation so that complex organs can be developed. Such new organ systems will need to produce energy from oxygen, glucose, and other substances freely available in the blood and be self-sufficient.

How far we are from the reality of buying off-the-shelf artificial organs? May be in next 10 years? As the pace of developments in the fields of nanotechnology, tissue engineering, and others is accelerating, the reality of having a self-sustaining artificial organ replacement system is a possible reality in the upcoming years.

REFERENCES

Aebischer, P, Ip, TK, Panol, G, et al. The bioartificial kidney: progress towards an ultrafiltration device with renal epithelial cells processing, *Life Support Syst*, 5, 2, 1987, 159–68.

AHA. *Heart Disease and Stroke Statistics: 2004 Update*, American Heart Association, Dallas, TX, 2003.

Akustu, T and Kolff, WJ. Permanent substitute for valves and hearts, *Trans Am Soc Artif Intern Organs*, 4, 1958.

Anderson, R and Smith, B. Deaths: leading causes for 2001, *National Vital Statistics Report*, 52, 9, 2003, National Center for Health Statistics, Hyattsville, Maryland.

Bartlett, RH, Gazzaniga, AB, Jefferies, MR, et al. Extracorporeal membrane oxygenation (ECMO) cardiopulmonary support in infancy, *Trans Am Soc Artif Intern Organs*, 22, 1976, 80–93.

Baumgartner, W, Reitz, B, Kasper, E, et al. *Heart and Lung Transplantation*, second edition, WB Saunders, Philadelphia, PA, 2002.

Bello, YM, Falabella, AF and Eaglstein, WH. Tissue-engineered skin. Current status in wound healing, *Am J Clin Dermatol*, 2, 5, 2001, 305–13.

Bing, RJ. Lindbergh and the biological sciences (a personal reminiscence), *Tex Heart Inst J*, 14, 3, 1987, 230–7.

Boettcher, W, Merkle, F and Weitkemper, HH. History of extracorporeal circulation: the invention and modification of blood pumps, *J Extra Corpor Technol*, 35, 3, 2003, 184–91.

Bouhadir, KH and Mooney, DJ. Promoting angiogenesis in engineered tissues, *J Drug Target*, 9, 6, 2001, 397–406.

Boyle, M, Kurtovic, J, Bihari, D, et al. Equipment review: the molecular adsorbents recirculating system (MARS(R)), *Crit Care*, 8, 4, 2004, 280–6.

Burke, DJ, Burke, E, Parsaie, F, et al. The Heartmate II: design and development of a fully sealed axial flow left ventricular assist system, *Artif Organs*, 25, 5, 2001, 380–5.

Chaikof, EL, Matthew, H, Kohn, J, et al. Biomaterials and scaffolds in reparative medicine, *Ann N Y Acad Sci*, 961, 2002, 96–105.

Chamuleau, RA. Bioartificial liver support anno 2001, *Metab Brain Dis*, 17, 4, 2002, 485–91.

Chen, C, Paden, B, Antaki, J, et al. A magnetic suspension theory and its application to the HeartQuest ventricular assist device, *Artif Organs*, 26, 11, 2002, 947–51.

Clowes, GJ, Hopkins, A and Kolobow, T. Oxygen diffusion through plastic films, *TASAIO*, 1, 1955, 23–4.

Cook, LN. Update on extracorporeal membrane oxygenation, *Paediatr Respir Rev*, 5 Suppl A, 2004, S329–37.

Cooley, DA, Liotta, D, Hallman, GL, et al. Orthotopic cardiac prosthesis for two-staged cardiac replacement, *Am J Cardiol*, 24, 5, 1969, 723–30.

Copeland, JG. Mechanical assist device; my choice: the CardioWest total artificial heart, *Transplant Proc*, 32, 7, 2000, 1523–4.

Copeland, JG, III, Smith, RG, Arabia, FA, et al. Comparison of the CardioWest total artificial heart, the Novacor left ventricular assist system and the Thoratec ventricular assist system in bridge to transplantation, *Ann Thorac Surg*, 71, 90030, 2001, 92S–97.

Curtis, JJ, Boley, TM, Walls, JT, et al. Frequency of seal disruption with the sarns centrifugal pump in postcardiotomy circulatory assist, *Artif Organs*, 18, 3, 1994, 235–7.

Curtis, JJ, Wagner-Mann, CC, Mann, F, et al. Subchronic use of the St. Jude centrifugal pump as a mechanical assist device in calves, *Artif Organs*, 20, 6, 1996, 662–5.

Curtis, JJ, Walls, JT, Wagner-Mann, CC, et al. Centrifugal pumps: description of devices and surgical techniques, *Ann Thoracic Surg*, 68, 2, 1999, 666–71.

DeBakey, ME. Left ventricular bypass pump for cardiac assistance 1: clinical experience, *Am J Cardiol*, 27, 1, 1971, 3–11.

Dekkers, RJ, FitzGerald, DJ and Couper, GS. Five-year clinical experience with Abiomed BVS 5000 as a ventricular assist device for cardiac failure, *Perfusion*, 16, 1, 2001, 13–8.

Demetriou, AA, Brown, RS, Jr., Busuttil, RW, et al. Prospective, randomized, multicenter, controlled trial of a bioartificial liver in treating acute liver failure, *Ann Surg*, 239, 5, 2004, 660–7; discussion 667–70.

Deng, L, El-Banayosy, A, et al. Mechanical circulatory support for advanced heart failure: effect of patient selection on outcome, *Circulation*, 103, 2, 2001, 231–7.

DeVries, W, Anderson, J, Joyce, L, et al. Clinical use of the total artificial heart, *N Engl J Med*, 310, 5, 1984, 273–8.

Di Bella, I, Pagani, F, Banfi, C, et al. Results with the Novacor assist system and evaluation of long-term assistance, *Eur J Cardiothorac Surg*, 18, 1, 2000, 112–6.

Doi, K, Golding, LA, Massiello, AL, et al. Preclinical readiness testing of the arrow international CorAide left ventricular assist system, *Ann Thorac Surg*, 77, 6, 2004, 2103–10.

Dorr, LD, Wan, Z, Longjohn, DB, et al. Total hip arthroplasty with use of the Metasul metal-on-metal articulation. Four to seven-year results, *J Bone Joint Surg Am*, 82, 6, 2000, 789–98.

Dowling, RD, Etoch, SW, Stevens, KA, et al. Current status of the AbioCor implantable replacement heart, *Ann Thorac Surg*, 71, 90030, 2001, 147S–149.

Dowling, RD, Gray, LA, Jr., Etoch, SW, et al. The AbioCor implantable replacement heart, *Ann Thorac Surg*, 75, 6 Suppl, 2003, S93–9.

Eaglstein, WH. Dermagraft treatment of diabetic ulcers, *J Dermatol*, 25, 12, 1998, 803–4.

El-Banayosy, A, Korfer, R, Arusoglu, L, et al. Device and patient management in a bridge-to-transplant setting, *Ann Thorac Surg*, 71, 90030, 2001, 98S–102.

El-Banayosy, A, Arusoglu, L, Kizner, L, et al. Preliminary experience with the LionHeart left ventricular assist device in patients with end-stage heart failure, *Ann Thorac Surg*, 75, 5, 2003, 1469–75.

Epstein, AH and Senturia, SD. Macro power from micro machinery, *Science*, 276, 5316, 1997, 1211.

Evans, M, Sewter, C and Hill, E. An encoded particle array tool for multiplex bioassays, *Assay Drug Dev Technol*, 1, 1 Pt 2, 2003, 199–207.

Farrar, DJ. The thoratec ventricular assist device: a paracorporeal pump for treating acute and chronic heart failure, *Semin Thorac Cardiovasc Surg*, 12, 3, 2000, 243–50.

Farrar, DJ, Holman, WR, McBride, LR, et al. Long-term follow-up of Thoratec ventricular assist device bridge-to-recovery patients successfully removed from support after recovery of ventricular function, *J Heart Lung Transplant*, 21, 5, 2002, 516–21.

FC Spencer, BE, Trinkle, JK, et al. Assisted circulation for cardiac failure following intracardiac surgery with cardiorespiratory bypass, *J Thorac Cardiovasc Surg*, 49, 56, 1959.

Fey-Lamprecht, F, Albrecht, W, Groth, T, et al. Morphological studies on the culture of kidney epithelial cells in a fiber-in-fiber bioreactor design with hollow fiber membranes, *J Biomed Mater Res*, 65A, 2, 2003, 144–57.

Frazier, O. First use of an untethered, vented electric left ventricular assist device for long-term support (published erratum appears in Circulation 1995 June 15, 91, 12, 3026), *Circulation*, 89, 6, 1994, 2908–14.

Frazier, OH, Akustu, T and Cooley, DA. Total artificial heart (TAH) utilization in man, *Trans Am Soc Artif Intern Organs*, 23, 1982.

Frazier, OH, Rose, EA, Oz, MC, et al. Multicenter clinical evaluation of the HeartMate vented electric left ventricular assist system in patients awaiting heart transplantation, *J Thorac Cardiovasc Surg*, 122, 6, 2001, 1186–95.

Frazier, OH, Shah, NA, Myers, TJ, et al. Use of the Flowmaker (Jarvik 2000) left ventricular assist device for destination therapy and bridging to transplantation, *Cardiology*, 101, 1–3, 2004, 111–6.

Fruhauf, NR, Oldhafer, KJ, Holtje, M, et al. A bioartificial liver support system using primary hepatocytes: a preclinical study in a new porcine hepatectomy model, *Surgery*, 136, 1, 2004, 47–56.

Fuchs, JR, Nasseri, BA and Vacanti, JP. Tissue engineering: a 21st century solution to surgical reconstruction, *Ann Thorac Surg*, 72, 2, 2001, 577–91.

Gibbon, J. Application of a mechanical heart and lung apparatus to cardiac surgery, *Minn Med*, 37, 1954, 171.

Goldstein, DJ. Worldwide experience with the MicroMed DeBakey ventricular assist device(R) as a bridge to transplantation, *Circulation*, 108, 90101, 2003, 272II–277.

Gott, VL, Alejo, DE and Cameron, DE. Mechanical heart valves: 50 years of evolution, *Ann Thorac Surg*, 76, 6, 2003, S2230–9.

Gottschalk, CW and Fellner, SK. History of the science of dialysis, *Am J Nephrol*, 17, 3–4, 1997, 289–98.

Griffith, BP, Kormos, RL, Borovetz, HS, et al. HeartMate II left ventricular assist system: from concept to first clinical use, *Ann Thorac Surg*, 71, 90030, 2001, 116S–120.

Gunston, FH. Polycentric knee arthroplasty. Prosthetic simulation of normal knee movement, *J Bone Joint Surg Br*, 53, 2, 1971, 272–7.

Hall, CW, Liotta, D, Henly, WS, et al. Development of artificial intrathoracic circulatory pumps*1, *2, *Am J Surg*, 108, 5, 1964, 685–92.

Hammer, C. Xenotransplantation — will it bring the solution to organ shortage? *Ann Transplant*, 9, 1, 2004, 7–10.

Hansell, DR. Extracorporeal membrane oxygenation for perinatal and pediatric patients, *Respir Care*, 48, 4, 2003, 352–62; discussion 363–6.

Hernandez, MR, Galan, AM, Cases, A, et al. Biocompatibility of cellulosic and synthetic membranes assessed by leukocyte activation, *Am J Nephrol*, 24, 2, 2004, 235–41.

Hilt, JZ. Nanotechnology and biomimetic methods in therapeutics: molecular scale control with some help from nature, *Adv Drug Deliv Rev*, 56, 11, 2004, 1533–6.

Hirotani, S, Eda, R, Kawabata, T, et al. Bioartificial endocrine pancreas (Bio-AEP) for treatment of diabetes: effect of implantation of Bio-AEP on the pancreas, *Cell Transplant*, 8, 4, 1999, 399–404.

Hoerstrup, SP, Sodian, R, Daebritz, S, et al. Functional living trileaflet heart valves grown *in vitro*, *Circulation*, 102, 19 Suppl 3, 2000a, III44–9.

Hoerstrup, SP, Sodian, R, Sperling, JS, et al. New pulsatile bioreactor for *in vitro* formation of tissue engineered heart valves, *Tissue Eng*, 6, 1, 2000b, 75–9.

Hoy, FBY, Mueller, DK, Geiss, DM, et al. Bridge to recovery for postcardiotomy failure: is there still a role for centrifugal pumps? *Ann Thoracic Surg*, 70, 4, 2000, 1259–63.

Huang, Y, Mather, EL, Bell, JL, et al. MEMS-based sample preparation for molecular diagnostics, *Anal Bioanal Chem*, 372, 1, 2002, 49–65.

Humes, HD. Bioartificial kidney for full renal replacement therapy, *Semin Nephrol*, 20, 1, 2000, 71–82.

Humes, HD, MacKay, SM, Funke, AJ, et al. The bioartificial renal tubule assist device to enhance CRRT in acute renal failure, *Am J Kidney Dis*, 30, 5 Suppl 4, 1997, S28–31.

Jain, KK. Stem cell technologies in regenerative medicine, *Expert Opin Biol Ther*, 2, 7, 2002, 771–3.

Jalan, R, Sen, S and Williams, R. Prospects for extracorporeal liver support, *Gut*, 53, 6, 2004, 890–8.

Jones, I, Currie, L and Martin, R. A guide to biological skin substitutes, *Br J Plast Surg*, 55, 3, 2002, 185–93.

Kaihara, S, Borenstein, J, Koka, R, et al. Silicon micromachining to tissue engineer branched vascular channels for liver fabrication, *Tissue Eng*, 6, 2, 2000, 105–17.

Kakinuma, S, Tanaka, Y, Chinzei, R, et al. Human umbilical cord blood as a source of transplantable hepatic progenitor cells, *Stem Cells*, 21, 2, 2003, 217–27.

Kaplon, RJ, Oz, MC, Kwiatkowski, PA, et al. Miniature axial flow pump for ventricular assistance in children and small adults, *J Thorac Cardiovasc Surg*, 111, 1, 1996, 13–8.

Karlsson, M, Davidson, M, Karlsson, R, et al. Biomimetic nanoscale reactors and networks, *Annu Rev Phys Chem*, 55, 2004, 613–49.

Kim, HW and Greenburg, AG. Artificial oxygen carriers as red blood cell substitutes: a selected review and current status, *Artif Organs*, 28, 9, 2004, 813–28.

Kobayashi, N, Okitsu, T, Nakaji, S, et al. Hybrid bioartificial liver: establishing a reversibly immortalized human hepatocyte line and developing a bioartificial liver for practical use, *J Artif Organs*, 6, 4, 2003, 236–44.

Kolff, WJ. Lasker Clinical Medical Research Award. The artificial kidney and its effect on the development of other artificial organs, *Nat Med*, 8, 10, 2002, 1063–5.

Kolobow, T and Bowman, R. Construction and evaluation of an alveolar membrane heart lung, *Trans Am Soc Artif Intern Organs*, 9, 1963, 238–45.

Kulig, KM and Vacanti, JP. Hepatic tissue engineering, *Transpl Immunol*, 12, 3–4, 2004, 303–10.

Kung, RT and Hart, RM. Design considerations for bearingless rotary pumps, *Artif Organs*, 21, 7, 1997, 645–50.

Lanza, RP, Borland, KM, Lodge, P, et al. Treatment of severely diabetic pancreatectomized dogs using a diffusion-based hybrid pancreas, *Diabetes*, 41, 7, 1992, 886–9.

Lawson, DS, Walczak, R, Lawson, AF, et al. North American neonatal extracorporeal membrane oxygenation (ECMO) devices: 2002 survey results, *J Extra Corpor Technol*, 36, 1, 2004, 16–21.

Lee, SC, Bhalerao, K and Ferrari, M. Object-oriented design tools for supramolecular devices and biomedical nanotechnology, *Ann N Y Acad Sci*, 1013, 2004, 110–23.

LeGallois, C. *Experiences on the Principle of Life*, Thomas, Philadelphia, 1813.

Leprince, P, Bonnet, N, Rama, A, et al. Bridge to transplantation with the Jarvik–7 (CardioWest) total artificial heart: a single-center 15-year experience, *J Heart Lung Transplant*, 22, 12, 2003, 1296–303.

Lick, SD, Zwischenberger, JB, Alpard, SK, et al. Development of an ambulatory artificial lung in an ovine survival model, *Asaio J*, 47, 5, 2001, 486–91.

Lim, F and Sun, AM. Microencapsulated islets as bioartificial endocrine pancreas, *Science*, 210, 4472, 1980, 908–10.

Liu, JP, Gluud, LL, Als-Nielsen, B, et al. Artificial and bioartificial support systems for liver failure, *Cochrane Database Syst Rev*, 1, 2004a, CD003628.

Liu, J, Hu, Q, Wang, Z, et al. Autologous stem cell transplantation for myocardial repair, *Am J Physiol Heart Circ Physiol*, 287, 2, 2004b, H501–11.

Liu, R, Wang, X and Zhou, Z. Application of MEMS microneedles array in biomedicine, *Sheng Wu Yi Xue Gong Cheng Xue Za Zhi*, 21, 3, 2004c, 482–5.

Losee, JE, Fox, I, Hua, LB, et al. Transfusion-free pediatric burn surgery: techniques and strategies, *Ann Plast Surg*, 54, 2, 2005, 165–71.

Lukes, J and Merckelbach, FM. [Experiences with arthroplasty of the hip joint according to the method of the Judet brothers]. Arthroplastics of the hip joint; consideration of the Judet and Smith-Petersen surgical methods, *Acta Chir Orthop Traumatol Cech*, 25, 2, 1958, 127–33.

Lysaght, MJ and Hazlehurst, AL. Tissue engineering: the end of the beginning, *Tissue Eng*, 10, 1–2, 2004, 309–20.

Lysaght, MJ and Reyes, J. The growth of tissue engineering, *Tissue Eng*, 7, 5, 2001, 485–93.

MacNeill, BD, Pomerantseva, I, Lowe, HC, et al. Toward a new blood vessel, *Vasc Med*, 7, 3, 2002, 241–6.

Magovern, GJ, Jr. The biopump and postoperative circulatory support, *Ann Thoracic Surg*, 55, 1, 1993, 245–9.

Malchesky, PS. Artificial organs and vanishing boundaries, *Artif Organs*, 25, 2, 2001, 75–88.

Malinin, TI. Remembering Alexis Carrel and Charles A. Lindbergh, *Tex Heart Inst J*, 23, 1, 1996, 28–35.

Mallory, TH. John Charnley remembered: regaining our bearings, *Orthopedics*, 27, 9, 2004, 921–2.

Mehta, SM, Pae, WE, Jr., Rosenberg, G, et al. The LionHeart LVD–2000: a completely implanted left ventricular assist device for chronic circulatory support, *Ann Thorac Surg*, 71, 90030, 2001, 156S–161.

Mesana, TG. Rotary blood pumps for cardiac assistance: a "must?" *Artif Organs*, 28, 2, 2004, 218–25.

Miki, N, Teo, CJ, Ho, LC, et al. Enhancement of rotordynamic performance of high-speed micro-rotors for power MEMS applications by precision deep reactive ion etching, *Sensors and Actuators A: Physical*, 104, 3, 2003, 263–7.

Minami, K, El-Banayosy, A, Sezai, A, et al. Morbidity and outcome after mechanical ventricular support using Thoratec, Novacor, and HeartMate for bridging to heart transplantation, *Artif Organs*, 24, 6, 2000, 421–6.

Moldovan, NI and Ferrari, M. Prospects for microtechnology and nanotechnology in bioengineering of replacement microvessels, *Arch Pathol Lab Med*, 126, 3, 2002, 320–4.

Morgan, JA, John, R, Rao, V, et al. Bridging to transplant with the HeartMate left ventricular assist device: the Columbia Presbyterian 12-year experience, *J Thoracic Cardiovasc Surg*, 127, 5, 2004, 1309–16.

Moussy, Y. Bioartificial kidney. I. Theoretical analysis of convective flow in hollow fiber modules: application to a bioartificial hemofilter, *Biotechnol Bioeng*, 68, 2, 2000, 142–52.

Mullin, EJ, Metcalfe, MS and Maddern, GJ. Artificial liver support: potential to retard regeneration? *Arch Surg*, 139, 6, 2004, 670–7.

Neff, G. Hip arthroplasty according to Smith-Peterson or Judet, *Helv Chir Acta*, 21, 5–6, 1954, 380–3.

Nikolovski, J, Gulari, E and Humes, HD. Design engineering of a bioartificial renal tubule cell therapy device, *Cell Transplant*, 8, 4, 1999, 351–64.

Noble, PC, Gordon, MJ, Weiss, JM, et al. Does total knee replacement restore normal knee function? *Clin Orthop*, 431, 2005, 157–65.

Nojiri, C. Left ventricular assist system with a magnetically levitated impeller technology, *Nippon Geka Gakkai Zasshi*, 103, 9, 2002, 607–10.

Noon, GP, Ball, JW, Jr. and Papaconstantinou, HT. Clinical experience with BioMedicus centrifugal ventricular support in 172 patients, *Artif Organs*, 19, 7, 1995, 756–60.

Noon, GP, Morley, DL, Irwin, S, et al. Clinical experience with the MicroMed DeBakey ventricular assist device, *Ann Thorac Surg*, 71, 90030, 2001, 133S–138.

Noordenbos, J, Dore, C and Hansbrough, JF. Safety and efficacy of TransCyte for the treatment of partial-thickness burns, *J Burn Care Rehabil*, 20, 4, 1999, 275–81.

Nose, Y. Design and development strategy for the rotary blood pump, *Artif Organs*, 22, 6, 1998, 438–46.

Nose, Y, Yoshikawa, M, Murabayashi, S, et al. Development of rotary blood pump technology: past, present, and future, *Artif Organs*, 24, 6, 2000, 412–20.

Okada, Y, Ueno, S, Ohishi, T, et al. Magnetically levitated motor for rotary blood pumps, *Artif Organs*, 21, 7, 1997, 739–45.

Ozgen, N, Terashima, M, Aung, T, et al. Evaluation of long-term transport ability of a bioartificial renal tubule device using LLC-PK1 cells, *Nephrol Dial Transplant*, 19, 9, 2004, 2198–207.

Petrou, S and Edwards, L. Cost effectiveness analysis of neonatal extracorporeal membrane oxygenation based on four year results from the UK Collaborative ECMO Trial, *Arch Dis Child Fetal Neonatal Ed*, 89, 3, 2004, F263–8.

Pierce, WS. Permanent heart substitution: better solutions lie ahead, *Jama*, 259, 6, 1988, 891.

Polla, DL, Erdman, AG, Robbins, WP, et al. Microdevices in medicine, *Annu Rev Biomed Eng*, 2, 2000, 551–76.

Portner, PM. Permanent Mechanical Circulatory Assistance, *Heart and Lung Transplantation*, second edition, 2001.

Prokop, A. Bioartificial organs in the twenty-first century: nanobiological devices, *Ann N Y Acad Sci*, 944, 2001, 472–90.

Ranawat, CS, Insall, J and Shine, J. Duo-condylar knee replacement, *Curr Pract Orthop Surg*, 6, 1975, 28–35.

Reach, G, Poussier, P, Sausse, A, et al. Functional evaluation of a bioartificial pancreas using isolated islets perifused with blood ultrafiltrate, *Diabetes*, 30, 4, 1981, 296–301.

Remadi, JP, Marticho, P, Butoi, I, et al. Clinical experience with the mini-extracorporeal circulation system: an evolution or a revolution? *Ann Thorac Surg*, 77, 6, 2004, 2172–5; discussion 2176.

Richards Grayson, AC, Scheidt Shawgo, R, Li, Y, et al. Electronic MEMS for triggered delivery, *Adv Drug Deliv Rev*, 56, 2, 2004, 173–84.

Robbins, RC, Kown, MH, Portner, PM, et al. The totally implantable Novacor left ventricular assist system, *Ann Thorac Surg*, 71, 90030, 2001, 162S–165.

Rose, EA, Gelijns, AC, Moskowitz, AJ, et al. Long-term mechanical left ventricular assist device for end-stage heart failure, *N Engl J Med*, 345, 20, 2001, 1435–43.

Roy, S, Ferrara, LA, Fleischman, AJ, et al. Microelectromechanical systems and neurosurgery: a new era in a new millennium, *Neurosurgery*, 49, 4, 2001, 779–97; discussion 797–8.

Rubinstein, JT. How cochlear implants encode speech, *Curr Opin Otolaryngol Head Neck Surg*, 12, 5, 2004, 444–8.

Schmidt, P, Andersson, G, Blomberg, J, et al. Possible transmission of zoonoses in xenotransplantation: porcine endogenous retroviruses (PERVs) from an immunological point of view, *Acta Vet Scand Suppl*, 99, 2004, 27–34.

Schuster, MD, Kocher, AA, Seki, T, et al. Myocardial neovascularization by bone marrow angioblasts results in cardiomyocyte regeneration, *Am J Physiol Heart Circ Physiol*, 287, 2, 2004, H525–32.

Still, J, Glat, P, Silverstein, P, et al. The use of a collagen sponge/living cell composite material to treat donor sites in burn patients, *Burns*, 29, 8, 2003, 837–41.

Stock, UA, Vacanti, JP, Mayer, JE, Jr., et al. Tissue engineering of heart valves — current aspects, *Thorac Cardiovasc Surg*, 50, 3, 2002, 184–93.

Strom, TB, Field, LJ and Ruediger, M. Allogeneic stem cell-derived "repair unit" therapy and the barriers to clinical deployment, *J Am Soc Nephrol*, 15, 5, 2004, 1133–9.

Sukhikh, GT and Shtil, AA. Transplantation of embryonic hepatocytes. Experimental substantiation of a new approach to the therapy of liver failure, *Bull Exp Biol Med*, 134, 6, 2002, 519–24.

Sullivan, SJ, Maki, T, Borland, KM, et al. Biohybrid artificial pancreas: long-term implantation studies in diabetic, pancreatectomized dogs, *Science*, 252, 5006, 1991, 718–21.

Terino, EO. Alloderm acellular dermal graft: applications in aesthetic soft-tissue augmentation, *Clin Plast Surg*, 28, 1, 2001, 83–99.

Thompson, CA, Nasseri, BA, Makower, J, et al. Percutaneous transvenous cellular cardiomyoplasty. A novel nonsurgical approach for myocardial cell transplantation, *J Am Coll Cardiol*, 41, 11, 2003, 1964–71.

Tobin, EJ and Bambauer, R. Silver coating of dialysis catheters to reduce bacterial colonization and infection, *Ther Apher Dial*, 7, 6, 2003, 504–9.

UNOS. 2003 Annual Report of the U.S. Organ Procurement and Transplantation Network and the Scientific Registry of Transplant Recipients: Transplant Data 1993–2002, *Department of Health and Human Services, Health Resources and Services Administration, Office of Special Programs, Division of Transplantation, Rockville, MD*, 2003, United Network for Organ Sharing, Richmond, VA; University Renal Research and Education Association, Ann Arbor, MI.

van de Kerkhove, MP, Hoekstra, R, Chamuleau, RA, et al. Clinical application of bioartificial liver support systems, *Ann Surg*, 240, 2, 2004, 216–30.

Venter, JC, Adams, MD, Myers, EW, et al. The sequence of the human genome, *Science*, 291, 5507, 2001, 1304–51.

Vienken, J, Diamantoglou, M, Henne, W, et al. Artificial dialysis membranes: from concept to large scale production, *Am J Nephrol*, 19, 2, 1999, 355–62.

von Segesser, LK, Tozzi, P, Mallbiabrrena, I, et al. Miniaturization in cardiopulmonary bypass, *Perfusion*, 18, 4, 2003, 219–24.

Wassenberg, PA. The Abiomed BVS 5000 biventricular support system, *Perfusion*, 15, 4, 2000, 369–71.

Weiss, WJ, Rosenberg, G, Snyder, AJ, et al. Steady state hemodynamic and energetic characterization of the Penn State/3M Health Care Total Artificial Heart, *Asaio J*, 45, 3, 1999, 189–93.

Winfrey, ME, Cochran, M and Hegarty, MT. A new technology in burn therapy: INTEGRA artificial skin, *Dimens Crit Care Nurs*, 18, 1, 1999, 14–20.

Wolfson, PJ. The development and use of extracorporeal membrane oxygenation in neonates, *Ann Thorac Surg*, 76, 6, 2003, S2224–9.

Wroblewski, BM. Total hip arthroplasty: results and consequences, *Ann R Coll Surg Engl*, 86, 6, 2004, 439–41.

Wu, ZJ, Antaki, JF, Burgreen, GW, et al. Fluid dynamic characterization of operating conditions for continuous flow blood pumps, *Asaio J*, 45, 5, 1999, 442–9.

Yokoo, T, Sakurai, K, Ohashi, T, et al. Stem cell gene therapy for chronic renal failure, *Curr Gene Ther*, 3, 5, 2003, 387–94.

Zimmer, HG. Perfusion of isolated organs and the first heart–lung machine, *Can J Cardiol*, 17, 9, 2001, 963–9.

Zwischenberger, JB, Anderson, CM, Cook, KE, et al. Development of an implantable artificial lung: challenges and progress, *Asaio J*, 47, 4, 2001, 316-20.

Nastic Structures: The Enacting and Mimicking of Plant Movements

Rainer Stahlberg and Minoru Taya

CONTENTS

19.1 INTRODUCTION

In the preface to his 1992 popular book *Exploring Biomechanics*, R. McNeill Alexander reflected that "one of the liveliest and most fascinating branches of biomechanics is the study of animal locomotions." In this chapter, we hope to show that nastic movements in plants are just as intriguing and reveal design principles that are uniquely fitted to the sessile lifestyle of plants and to the many challenges encountered by human engineering. This expectation may be surprising if we confine our considerations to the *free movement of entire individuals (locomotion)*. Higher plants are rooted and migrate only as seeds dispersed by wind and animals. Only the climbing seedlings of one higher plant — the parasitic dodder in the genus *Cuscuta* — have shown that they can actively change their individual location by abandoning and regrowing root and root-like haustoria.

However, translocation of individuals occurs frequently in some developmental stages of lower plants and in motile single-celled and multicellular algae. Botanists call this type of individual movement taxis or tactic movement. Depending on the type of activating stimuli, such as light, ions, sugars, or hormones, and electric fields we find directed translocations in the form of phototactic, chemotactic, or even galvanotactic movements. The direction of the response to the triggering stimulus is indicated by a positive or negative sign, for example, a single green alga moving *towards* a light source carry out a *positive* phototactic move whereas light avoidance is referred to as *negative* phototaxis. In higher plants, we find positive chemotaxis in fern sperms, which respond to a malate signal from unfertilized archegonia. Similarly, pollen tubes change their direction of growth towards a calcium or hormone signal from unfertilized ovules. Whether the last example should be referred to as a tactic movement is disputable, since only the internal part of the original organism — the pollen grain — is moving.

The locomotion of entire large organisms can be viewed as a zoological specialization that derived from organ movements. It was useful only for the evolution of animals to coordinate these organ movements with a purpose to translocate the entire organism as a unit. Primarily, both animals and plants show only autonomous movements of their organs whether these are leaves, flower parts, tendrils, legs, wings, or fins. In plants, there are two types of *organ movements*; tropisms and nastic movements. Tropisms are based on an induced difference in the irreversible expansion between two flanks of a growing plant organ that cause the organ to bend and adopt a new direction of growth. Accordingly, tropisms are growth responses that respond to the direction of the triggering stimuli. Stimuli, such as light, gravity, hormones, mechanical stimulation, and electric fields induce directed responses in the form of phototropic, gravitropic, chemotropic, seismotropic, and galvanotropic curvatures. A well-known example for positive gravitropism is the growth of the main root towards the earth's center of gravity. Negative gravitropism is the response that leads a growing organ away from the earth's center of gravity and appears, for example, in the vertical reerection of trampled or fallen plants. Since tropistic changes in growth direction are caused by different expansion rates on the opposite flanks of plant organs, they resemble monomorph and bimorph actuators. Contrasting to tropisms are organ movements that are independent from the direction of the inducing stimulus, for example, raising and lowering of leaves, folding and unfolding of leaves, opening and closure of the Venus flytrap. This second type of organ movement is called *nastic*. Like before, with tactic and tropistic movements we can also specify photonastic, seismonastic, and chemonastic movements. Nastic movements are often reversible and defined by joint-like structures that confine their mobility options. Although these characteristics are shared by many human-made machines, we will find nastic motors and structures to be uniquely arranged.

In engineering terms, plants are adaptive (smart) structures with remarkable capabilities that were developed and perfected over millions of years of evolution in constantly changing and increasingly complex environments. It is therefore smart to study, understand, mimic, and modify nature's time-tested principles and mechanisms. Life originated in water, and nastic structures as well as their motors are optimized to use the potentialities of this unique solvent. In addition to ATP-dependent molecular motors, contracting and inflating molecules, such as the P-protein in the phloem conduits, plants rely heavily on three hydration motors (osmotic, colloidal, and fibrous) that figure as the major workhorses for nastic plant movements. After examining the three major types of plant motors, we will review selected examples for their action in simple and then more complex nastic structures.

19.2 MOTORS IN NATURE'S NASTIC DESIGNS

When exploring the designs of moving things, one usually starts with the force-generating units or motors. Animal locomotion involves molecular motors called muscles, which consist of long fiber cells with the ability to contract. With an almost ubiquitous force of contraction of about 30 g mm^{-2}, muscles develop their strength in pull but not in push. It follows that they must operate any reversible joint in antagonistic pairs attached to different locations of the relevant bone levers.

To increase speed, muscles can slowly tension elastic polymers like resilin and elastin that can release the stored energy much more rapidly (Alexander, 1992). To increase muscle strength, animals can only increase the cross-sectional area by bundling the fibers in thicker and longer fascicles. Only engineering is able to amplify force beyond this natural limit by combining elementary machines (lever, wedge, inclined plane, pulley, wheel, and screw) in devices like cranking wheels, compound pulleys, and hydraulic amplifiers.

Although plants possess similar molecular motors as animals (cytoplasmic streaming and pollen tube movement use actin-based myosin motors, while cytokinetic chromosome movements use microtubule-based kinesin motors), these remain confined to movements inside cells and do not play a causal role in the macroscopic movements of plant organs (Asada and Collings, 1997; Shimmen et al., 2000). Only animals have adopted molecular motors to drive their macro-movements by an energy-consuming cell contraction. In these movements, motor elements (muscles) are external to the moving parts (bones). Plants use hydration motors to drive their macro-movements and here the motors are an inconspicuous, integrated part of the structure. Plant cells differ from those of animals in their ability to undergo multifold expansions by the ability to incorporate large amounts of water into a special organelle, the vacuole. Consequently, nastic motors extract work from a change in hydration pressure. Their operation differs not only from muscles but also from related technical, that is, hydraulic designs and is for this reason alone worthwhile reviewing. Moreover, plant motors generally power autonomous movements that are locally controlled and operate without remote control through central nervous systems or computers.

19.2.1 Osmotic Motors

Hydraulic motors and actuators work on the basis of a change in hydrostatic pressure. In animals and human-made designs this is achieved by muscles and mechanical pumps that subject water to pressure. Plant osmotic motors are different. Rather than compressing the water directly, plants generate hydrostatic pressure by injecting solutes into a confined space that must be surrounded by a selective membrane that retains the solutes but allows water to permeate freely into this space. Osmosis therefore requires two components: a semipermeable membrane inside to concentrate the solutes and a restraining, but elastic and expandable wall outside to prevent the compartment from bursting when water is taken up during the hydration of these solutes. The hydration of the solutes generates hydrostatic pressure inside the osmotic compartments. All plants use osmosis to pump and concentrate water-binding electrolytes and nonelectrolytes into the inside of their cells and in particular into the vacuole, a membrane-surrounded compartment specifically designed for storing solutes and water. Osmotically operating plant cells allow the build-up of internal pressures far exceeding that of car tires.

The accurateness of this principle was first demonstrated by a model ingeniously devised by Traube; an artificial cell consisting of a porous clay cylinder covered with a copper ferrocyanide membrane permeable only for the small water molecules. The combination of this device with a manometer allowed the experimental determination of osmotic pressure values for a variety of concentrations and solutes (Pfeffer, 1873). The model allowed Pfeffer to predict the existence and properties of membranes too thin to be visible in the light microscopes of his time. Pfeffer's osmometer was the first truly man-made osmotic motor, one of the earliest biomimetic designs and instrumental for a breakthrough in the biology of ion and water transport.

The internal cell pressure of plant cells can be determined with external solutions of an equal or higher osmotic pressure that draw water from the cells, relieve the internal pressure so that the cell membranes are no longer closely pressed against the cell wall but separate from it; a process called plasmolysis. Internal cell pressures can reach up to 5 MPa in storage roots of sugar beets and in the shoots of some halophytic and xerophytic desert plants (Walter, 1953). As in human-made inflatable structures (e.g., sleeping bags) pressurization of the cells leads to the expansion as well as stiffening and hydrostatic stabilization of the cells, tissues, and entire structures (e.g., Niklas,

1992). To prevent an explosive rupture of the membranes, the huge pressures have to be counter-balanced by strong but flexible cell walls made of cellulose and other polymers. However, when considering mobile, nastic structures, the walls must have also the ability to yield to the pressure and so to allow plastic and elastic expansion. Not all cell expansions are driven by the pressure of vacuolar osmotic motors though. The 15-fold volume increase in root cap cells occurs without the formation and expansion of a large vacuole and — at least in part — is likely to be driven by a colloid motor (Juniper and Clowes, 1965).

The osmotic motor is the most common of the three types of hydration motors used by plants. Some plant species can also use osmosis to pump water and solutes into the upper shoot, an action that becomes necessary when a highly humid air prevents normal ion transport by transpiration. Pumping ions into the xylem vessels of the lower end of the root cylinder, roots generate a local pressure increase of up to 0.1 MPa sufficient to push a water column 9 to 10 m above the ground. This so-called root pressure is generated by only some plant species and becomes apparent in such phenomena like guttation (the appearance of droplets at the leaf periphery of grasses and broad-leafed plants) and the so-called bleeding of decapitated stumps (e.g., Stahlberg and Cosgrove, 1997).

Osmotic motors have the disadvantages that they depend on the intactness of a very thin, fragile membrane that also must be permeable to the small water molecules alone. Freezing and subsequent thawing destroy these membranes and with it all osmosis-based mechanisms. The same failure occurs when the ionic solutions are so severely dehydrated that they crystallize. The shrinking of osmotically operating vacuoles that often occupy more than 90% of the cell volume leads to harmful structural deformations of tissues (exceptions are discussed in Section 19.2.2.4). Some plants, for example, in the genus *Selaginella,* can repeatedly dry and rehydrate without structural damage. They avoid critical cell deformations during severe dehydration by using vacuoles of smaller size that are filled with tannin colloids instead of ions. Upon dehydration these colloids undergo minimal volume changes (Walter, 1956). Nature itself points here to the interesting alternative of replacing crystallizing small molecules with larger-sized colloids.

19.2.2 Colloid-Based Motors

Colloids are hydrating particles with a size ranging from 5 nm to 0.5 μm. Most colloids do not form true solutions but suspensions that are not completely transparent and show light diffraction (Tyndall effects) and other optical effects not found in true solutions. Many natural macromolecules, such as starch, pectins, latex, nucleic acids, and proteins, fit this definition. Due to their larger particle size, colloidal solutions cannot be as concentrated as osmotic solutions and their osmotic effect is therefore considerably smaller. This is demonstrated in the following comparison. A 10% (w/v) solution of glucose has an osmotic pressure of 1.35 MPa whereas a 10% solution of the colloidal bovine serum albumin (BSA, a soluble protein) has only 3.2 kPa, that is, it is osmotically almost three orders of magnitude less effective (Levitt, 1969).

For the purpose of constructing a colloid-based motor, there are primarily three desirable characteristics of colloids: (i) the potential expandability, that is, the volume change they undergo per volume absorbed water; (ii) the reversibility of the volume change; (iii) a low hydraulic capacitance, defined as the volume change per unit applied pressure (Meidner and Sheriff, 1976). A high force development per volume change is desirable for any motor and leads to the practical question for both nature and human engineers of whether colloid-based motors can equal the generated pressures of osmotically operating systems like vacuoles filled with ionic solutes. Natural macromolecules bind water to different degrees, for example, 1 g of starch binds 0.8 g water. Due to high particle size and molecular weight, colloid hydration looks more impressive if we express it as the binding of 30,000 to 100,000 water molecules per molecule gelatin (Walter, 1957). This high degree of hydration is not osmotic but due to the presence of adsorptive forces (called adhesion or imbibition) that can equal and exceed the pressure of osmotic systems by reaching values of up to

100 MPa. Due to internal adsorption, the volume increase of colloidal systems is smaller than the volume the adsorbed water occupies in its free form. Note also that the response of colloid-based hydration motors is not limited to water alone but can also be strongly affected by the concentration of Ca^{2+}, ATP, temperature, and pH (Pollack, 2004). This is one reason to consider colloid-based hydration motors as more promising for biomimetic designs than osmotic motors.

Water-soluble proteins (albumins and globulins) from seeds are well-hydrating colloids that bind 1.3 g of water per gram. Although proteins have been studied for a long time, a recently isolated slime or P-protein from beans generated a lot of interest because of its large, calcium-dependent expansion (and contraction) that led to a rapid plugging of pores near the wound sites (Knoblauch et al., 2003; Mavroidis and Dubey, 2003). Many other interesting colloids in plants remain unidentified; they often contain both carbohydrate and protein polymers and are summed under the name of mucilage (e.g., Wainwright, 1995). Mucilage is known to swell perceptibly in water and can be precipitated (dehydrated) with alcohol. The water-holding capacity of mucilage is known and has been used in the herbal sciences where mucilage coating soothes inflamed mucous membranes and keeps them hydrated (emollients) during coughs and bronchial infections. Large amounts of mucilage are found outside and inside cells from plants in highly dehydrating environments, for example, cacti and halophytes (Englmaier, 1987). Interesting are recent findings of mucilage in the water-transporting xylem vessels of plants. The adhesion of water molecules to capillary walls filled with mucilage could become stronger than their cohesion in the column and so localized mucilage secretion may be a way to segment the long and heavy water column that otherwise would stretch uninterruptedly from the soil to the tip of the plant and require a pull of more than 0.01 MPa per meter to be moved (Zimmermann et al., 1994). An increase in xylem-bound mucilage was found after mechanical wounding and might also have a blocking function for pathogens (Crews et al., 2003). Having a confirmed role in water storage and volume changes in desert, salt, and normal plants, we suggest that mucilages are worth testing as potential motor material.

19.2.2.1 Macroscopic Swelling Bodies

One of the few investigated colloid-based plant motors are seeds, in particular pea seeds (Kuhne and Kausch, 1965; Larson, 1968; see also Figure 19.1). The authors showed that the first phase of hydration is not associated with the development or enlargement of vacuoles and is hence not an osmotic but simple colloid-based hydration process. The hydration is easy to follow since the seeds are big and their expansion can be measured macroscopically. The imbibition pressure of hydrating pea seeds is very high. For centuries botany students in Europe have been impressed by their

Figure 19.1 Dry (left) and hydrated (right) natural swelling bodies in the form of split pea seeds (A, shells have been removed) and corn grits (B, grits are particles of broken seed endosperm). The final volume increase is about 150% in split pea seeds and 200% in corn seed particles. Note that the smaller grit particles hydrate much faster than the larger peas.

teacher's demonstration of how hydrating pea or bean seeds easily and effectively crack glass bottles and animal skulls by what must be considered a slow, controlled, and silent demolition (Brauner and Bukatsch, 1980). Although consisting of many smaller colloidal subunits (storage cells) pea seeds are large, compact bodies that hydrate and expand as one unit. A hydration motor operating with such compact swelling bodies rather than solute molecules has a huge advantage over an osmotic design. It no longer needs a fragile membrane with fine pores but can operate with a sturdy and inexpensivemetallic or ceramic sieve with pores small enough to retain the dehydrated swelling bodies.

However, pea seeds are designed to hydrate slowly and for that reason less than ideal motor material. Two factors keep their rate of hydration low: (i) the mechanical resistance of seed shells to expansion and (ii) the limited accessibility of the internal seed space to water (Larson, 1968). To become technically attractive one needs swelling bodies with an increased access for water or other solvents. The access to internal parts of a swelling body depends on the rate of diffusion and this parallels the surface to volume ratio or *specific surface* (Levitt, 1969). The specific surface S_v is the total area of all surfaces (S) of a body divided by its volume (V):

$$S_v = S/V.$$

In the case of a sphere (pea) with radius r this equation simplifies to $S_v = 3/r$ and shows that with increasing radius of round swelling bodies their S_v and with it accessibility to water decreases. Large storage organs, such as potato tubers, fruits, and seeds, have very densely packed cells and with no internal surfaces; that is, their accessibility or exchange of matter is restricted. The swelling rate of such compact bodies can be increased by a reduced size. It is for this reason that the breaking of seeds into smaller particles (grits) accelerates their hydration process (Figure 19.1).

Alternatively, hydration can be sped up by using particles with a more elaborate internal surface as in plant organs like leaves where large intercellular airspaces provide an internal surface that is 10 to 32 times larger than their external surface (Turill, 1936). Exchange rates for leaves are many orders of magnitude higher than for seeds. Rapidly swelling bodies can be produced by packing colloid particles inside a porous and elastic shell. Such rapid swelling bodies would allow to construct a simple and robust sieve-based colloid motor as shown in Figure 19.2.

A last consideration for the efficiency of colloid hydration is based on the fact that the diffusion of water molecules is by 2 orders of magnitude faster in the gas phase than in liquid. The use of water-saturated air rather than water in the motor shown in Figure 19.2 would considerably increase its

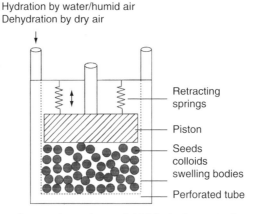

Figure 19.2 Schematic view of a membrane-free colloidal hydration motor-based on macroscopic swelling bodies like seeds and biomimetic colloid clusters. Water or water-saturated air (fog) initiates the power stroke by expanding the swelling bodies. Dry air reverses the expansion with springs pushing the piston back into the initial position.

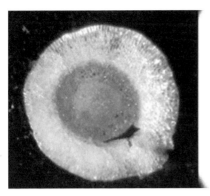 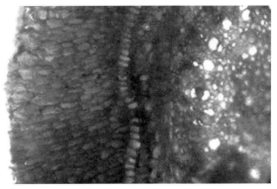

Figure 19.3   Cross section through a 3-mm-wide aerial root of a tropical orchid (left panel) shows the living central part of the root (gray) and the nonliving hygroscopic fibrous layer, called velamen (white), surrounding it. Right side picture shows velamen in higher magnification.

efficiency. We can combine this fact with the finding that quite a few plants developed hygroscopic tissues that are able to extract water from the air. The list includes not only lichens, mosses, and tree ferns but also higher plants like the gigantic mammoth trees (*Sequoiadendron*) and rootless epiphytic orchids and bromeliads, including the so-called Spanish moss in the genus *Tillandsia*. Their hygroscopic tissues contain colloidal and matrix substances that hydrate easily and reversibly, characteristics of prime interest for motor material. Another example is velamen, the dead tissue that covers the aerial roots of epiphytic orchids (Figure 19.3). Although the hygroscopic swelling of this tissue has no apparent mechanical function, it functions to transfer the humidity of air to the internal root tissue and the plant. Even among plants, this is an incredible achievement. It is useful to study the hydration of such materials in more detail to understand the aggressive mechanisms of such hygroscopic structures. In the case of velamen, the hydration material consists of easily accessible, hygroscopic, and dead cell walls; that is, the material is fibrous.

19.2.3   Fibrous Motors

Fibrous motors are based on the adhesive absorption of water in internal capillary spaces of parallel-arranged fibers. Fibers are used as reinforcing elements in both technical and biological designs; their role as hydration motors identifies them as multifunctional materials. And yet this multifunctionality has often been overlooked. With the exception of blotting paper and the wooden stone splitters of ancient times, hydration-dependent form changes of wood have been mostly seen as an annoyance rather than an opportunity.

The most common natural fibrous material is cellulose. Cellulose is a polymer with a clearly defined hierarchical structure. Cellulose fibers are made of long chains of glucose molecules twisted together in a micellar bundle. These bundles are often found in a parallel arrangement to increase the breaking strength of the material. Unlike isometrically expanding colloids, parallel fibers swell in a diametric fashion, that is, they expand only in the two directions that are perpendicular to the direction of the fibers (Figure 19.4).

Diametric expansion occurs when water fills the inter-fibrillar spaces where it pushes the fibers apart without altering their length (Figure 19.4 and Figure 19.5). The space between the microfibril bundles resembles small-sized capillaries and is large enough for water to move and be adsorbed (Figure 19.5; Frey-Wyssling, 1959; Robards, 1974). The large internal or specific surface (S_v) of cellulose bundles and other fibers accounts for their rapid hydration. Fibrous hydration occurs in the walls outside the living cell and therefore does not contribute to internal cell (turgor) pressure. The obvious lack of operating osmotic motors in dead tissues led to an early acceptance of operating fibrous or wall-based hydration motors in nonliving tissues. Although this has not been

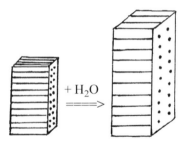

Figure 19.4 A fibrous hydration motor working as a hydration lift. The unequal or diametric expansion of hydrating fibrous motors is confined to the two directions perpendicular to the parallel-oriented fibers. Ancient engineers used similar wooden motors to break rocks and walls.

demonstrated yet, there is no obvious reason that would exclude fibrous motors from having a role in some living tissues as well. Fibrous wall materials also line the plant's water transport tubes (called xylem vessels) and are known for their high adhesion for water molecules. This high adhesion results in a considerable capillary pull of water. The high affinity of wall materials is technically used in blotting paper, which can absorb a considerable amount of water without a considerable volume change (due to the random arrangement of the cellulose fibers).

At some cells the same capillary spaces can be rapidly filled with the polyphenolic resin lignin instead of water (Brett and Waldron, 1996). This process turns the water-permeable, hydrating–swelling net of cellulose microfibrils into an inexpandable, water-repelling composite called wood (Figure 19.5). Another way to create a layer with little hydration-dependent area change is a random arrangement of the fibers (Figure 19.6a). By combining swelling, modestly swelling, and non-hydrating cellulose layers, plants generate a various monomorph actuators (Figure 19.6). Botanists were among the first who studied and correctly explained the movements of dead fibrous hydration motors as actuators consisting of two (or sometimes more) layers with different hydration-mediated expansion characteristics. The encountered plant movements were modeled with paper laminates in the form of two joined paper strips (Figure 19.6), each with a different orientation of the fibers (Jost, 1933). The simple combination of a modestly or nonswelling layer (random orientation of the fibers) and a swelling layer with parallel fibers generates a simple bending when the laminate is hydrated (Figure 19.6a). A combination of two swelling layers with a perpendicular orientation of their parallel aligned fibers adds a rolling movement to the simple bending (Figure 19.6b). Finally, with a 45° angle between the parallel oriented fibers of the two strips, hydration subjects the laminate to twisting and torsion (Figure 19.6c).

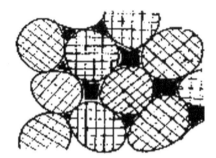

Figure 19.5 Cross section of a cellulose fiber. With about 10 nm diameter, the spaces (black) between the microfibril bundles (x) are large enough for the adhesive absorption of water molecules, a process that increases the diameter of the fiber. The spaces resemble fine capillaries and follow similar rules. As an alternative to water, they can be immediately filled with hydrophobic, water-proofing lignin generating a sturdy, no longer swelling resin–fiber composite.

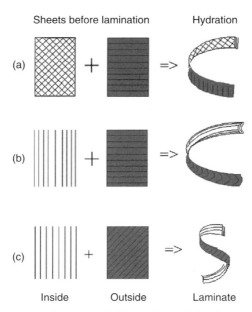

Sheets before lamination   Hydration

(a)

(b)

(c)

Inside   Outside   Laminate

Figure 19.6 Laminate paper models of monomorph (a) and bimorph (b, c) hydraulic actuators. The joining of a swelling and a nonswelling layer leads to a structure that will show simple bending upon hydration (a). When joining two swelling layers with their fibers and swelling direction at a 90° angle, hydration causes rolling and bending in the same structure (b). A 45° angle between the fibers of the two layers (c) leads to torsion and a twisted structure.

One should not forget that these paper models differ from nature's designs in that they are glued together when they are flat and dehydrated and that they bend and twist when hydrated. The opposite is the case in plant designs (see about seed containers in the following), which develop as straight structures when hydrated and green, while tension and torsion arise with the severe dehydration of the ripening seeds. This difference aside, these models are extremely useful to mimic and hence fully comprehend existing natural fibrous motors. They also elegantly demonstrate the countless possibilities for multifunctional biomimetic designs where fibers can serve both as structural elements and as motors. One can safely predict that the biomimetic use of fibrous hydration will not stop with the development of blotting paper that is lignin-free and features a loose and random orientation of the fibers. Bio-inspired design of fibers with larger relative volume increase and force development than the original cellulose fibers are a useful step on the way.

One example for the use of fibrous motors in plants is the opening of seed containers like legume pods to release the seeds. The walls of the seed pods are made of two attached fibrous hydration layers that have an angle of $\leq 90°$ to each other. Dehydration of the maturing walls leads to a torsional twist that rips the pod apart spreading the seeds in different directions (Figure 19.7). Although dehydration-induced torsion of fibrous motors was used for an "advanced version of medieval catapults," called the ballista, few human-made designs use fibrous torsion today.

19.3 NASTIC STRUCTURES IN PLANTS

Nastic movements are strictly defined as organ responses lacking directional input from the causal stimuli (see Introduction). In spite of this classic limitation to organs, one cannot avoid realizing that similarly responding structures exist at any structural level from protein complexes to subcellular swelling bodies (e.g., P-proteins, mucilage, and colloid vacuoles), dead cell walls (Figure 19.5), one or more living cells (guard cell of stomatal complex, leaf-rolling bulliform cells), organs (trap closure, shoot bending) to entire organisms (e.g., emergence of mescal cactus

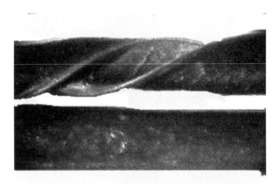

Figure 19.7   Dehydrated and twisted (top) and smooth hydrated halves (bottom) from the same seed pod of *Lathyrus japonicus,* a legume. The wall is made of two hydration layers with the fibers pointing in different directions (angle of ≤90°). When dehydrating (top), this torsion motor rips the pod apart and releases the seeds.

from the soil). In spite of their different complexity, all these structures show movements that are reversible, independent from the direction of the triggering stimuli, and powered by some type of hydration motors. While the definition could be expanded to suborgan levels of structure, it is general enough to include two different types of nastic structures: (i) autonomous structures that are driven by their own hydration motor and (ii) nonautonomous structures that depend on external energy sources to work. Nonautonomous nastic structures are most frequently found in flowers where they work as strictly mechanical machines that force pollen on unsuspecting visitors. It is the transmitted energy of the weight and impetus of a landing insect (legume flowers) or bird (*Strelitzia reginae* or bird-of-paradise flowers) that causes the movement of stamens (legume flowers) or the opening of a protective envelope around the stamens (*S. reginae*). Details of these structures can be found in many reviews and monographs on plant reproduction (e.g., Meeuse and Morris, 1984).

The following part focuses on examples for the even more intriguing, autonomously working nastic structures that contain their own motors and hence depend on external factors only for stimulation. Such autonomous nastic devices consist of a motor and structures that modify and direct the forces it generates. Therefore, nastic structures could be well defined as asymmetric or heterogeneous elastic restraints of hydration motors. Whether they surround osmotic, colloidal or fibrous motors, the unequal restraints are almost always realized by heterogeneous depositions of the cell wall material and in particular by a different direction of the deposited layers of cellulose microfibrils. Depending on the complexity of the nastic structures, the asymmetry can be located between the opposite walls of one cell (e.g., guard cell), between cell layers (Venus flytrap) or between internal and external tissues (growing stems). Nastic cells or organs are characterized by asymmetric capacities for elastic expansion that result in asymmetric expansions in the form of curved responses. The following part of the chapter describes the functioning of nastic structures through different levels of complexity and provides useful examples of how plants use their motors.

19.3.1   Hydrostat Motor Cells — Source and Location of Movements

Most plant movements and growth are driven by *osmotic motors*. These motors are located in either young cells or specialized cells that feature thin, expandable walls and a large, osmotically pressurized central vacuole (Figure 19.8). If not part of specialized units (guard cells of stomatal pores, pulvinus cells of bendable petioles, etc.) these young and undifferentiated cells are constituents of a tissue called parenchyma. The behavior of these cells compares well to pressure-dependent mechanical softness and rigidity of human-made inflatables and they are referred to sometimes as

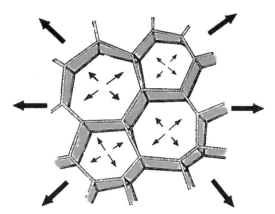

Figure 19.8 A hydrostat network of thin-walled, elastic and expandable parenchyma cell walls internally pressurized by vacuoles. This kind of tissue is soft when dehydrated and rigid when fully pressurized (inflated) and provides the basis for most reversible (cell walls undergo elastic expansion) and irreversible (cell walls yield to stress and are plastically altered) volume changes in plant tissues. Volume changes of these thin-walled cellular networks are the motor behind many nastic movements.

hydrostats (Niklas, 1989, 1992), sometimes as motor cells (Pfeffer, 1873). The term motor cell refers to the ability of cells to rapidly expand or shrink their vacuoles, cell volume, and surface area. The term hydrostat refers to the characteristic of cells and cell complexes (tissues) to undergo striking changes in their mechanical properties in dependence on the degree of water uptake and inflation. The soft walls and plasticity of hydrostat tissues turns into a very rigid structure upon full hydration or inflation, which generates a high internal pressure and exposes the cell walls to high tension and rigidity. The most common and basic function of such hydrostats is (i) to constitute the expanding, growing parts of the plant and (ii) to provide rigidity, stability, and expandability to these parts. Niklas (1992) calculated that only cells with a wall thickness considerably less than 20% of the cell radius will exhibit hydrostat and motor behavior. Reversibly expanding motor cells need to have walls with sufficient elasticity, that is, the walls should stretch while building an elastic tension that returns the cell to its original size when the initiating pressure is relieved.

A fitting example for a hydrostat motor tissue is the pith parenchyma in the core of younger sunflower stems (Figure 19.9). When well-hydrated, the parenchyma cells of the pith exert radial pressure on the peripheral cell layers of the stem and give it rigidity and straightness (Kutschera, 1989). Dehydration of the pith cells softens and bends (wilts) the upper, younger part of the stem. Observation of stem slices shows that it is the lack of pressure from the shrinking pith cells that causes wilted bending of the stem (Figure 19.9). In older, no longer growing parts of the stem, pith cells die and fill with air (sectioned pith appearance turns to white) before disappearing altogether (stem becomes hollow). Meanwhile the older stem is reinforced through lignin depositions in the peripheral ring of vascular bundles that turns their walls into a stable fiber–resin composite. The older stem is no longer a hydrostatically stabilized structure.

19.3.2 From Isotropic Cell Pressure to Anisotropic Cell Expansion

The form of cell expansions is always determined by a combination of motors and nonexpandable materials surrounding them. While fiber motors are inherently structural by showing anisotropic expansions, the common vacuole-based osmotic and colloid motors are unstructured in the sense that they exert equal pressure in all directions. To convert the isotropic vacuolar pressure into a anisometric cell expansion, plant cells use anisotropic depositions in their cell walls. The original cell wall is made of cellulose and similar fibers, polyelectrolyte gels like pectin (made of galacturonic acid monomers that keep an unbalanced carboxyl group after polymerization), as well as

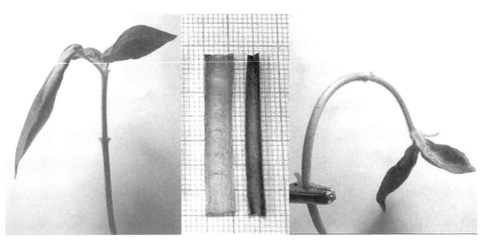

Figure 19.9 (See color insert following page 302)   A stable, erect sunflower stem (left panel) depends on the pressure of internal, easily expandable hydrostat tissue (pith = transparent cells in center panel) that tensions the stronger-walled surface layers of the stem giving them rigidity and stability. The well-known limp shape of wilting young plant stems (right panel) occurs when the internal, thin-walled pith tissue dehydrates, shrinks (center panel) and ceases to exert radial pressure on the surface layer and keeps it under tension. Left panel shows fully hydrated sunflower stem and right panel a dehydrated stem; the center panel demonstrates volume reduction in pith during dehydration of a segment slice. Pith parenchyma acts as a hydrostat motor that provides herbaceous stems with stability and the driving force for expansion.

other colloid materials (for detailed reviews see Vincent, 1982; Taiz, 1984; Cleland et al., 1990; Brett and Waldron, 1996).

Cellulose fibrils wrap the internal cell motor in either a single or many subsequently deposited layers creating either a primary or secondary walls. If there are many layers and the microfibril directions between these separate layers are diverse enough, the wall will provide multidirectional support and the cell has no directional preference for expansion. Older cells with secondary walls resist the vacuolar pressure equally well in all directions. The same principle applies to the multidirectional strength of plywood (Figure 19.10). However, if a cell is young or part of a nastic structure, it has only a thin primary wall with one cellulose layer with one dominant direction of the microfibrils. Such anisotropic walls provide young cells with a preferred direction for their expansion. Young tubular stem cells most frequently show preference for a transverse (radial) orientation of the microfibrils or for a low pitch helix. Since such microfibril arrangements favor axial and restrict diametrical extension, the cells are bound to elongate and develop into a narrow cylindrical shape.

During the growth of plant stem cells have most of their microfibrils around them in rings or helices with increasing pitch. The feature of helical microfibril bundles, particularly prominent in the collenchyma cells of vascular tissues and tendrils, provides stems with an astonishingly high resistance to tension, high strain, and breaking strength they are famous for (Wainwright, 1980; Vincent, 1982). Both osmotically driven expansion and external pull affect an increasingly vertical reorientation of the cellulose microfibrils that finally terminates cell expansion (Robards, 1974). Additional processes like the deposition of multiple cellulose layers or the resin lignin, as well as the disappearance of expansion-catalyzing enzymes, complete the termination of cell expansion. Some older cells may also lose their liquid content and turn into dead wood or cork cells. The stability of such cells no longer depends on internal pressure but exclusively on the static stability of their walls (Gibson and Ashby, 1982). Even in dead wood cells it is the spiral arrangement of the most prominent S2 cell wall layer that resists tension and deflects cracks and makes wood ten times more resistant to fracture than plain fiber–resin composites (Gordon and Jerominidis, 1980).

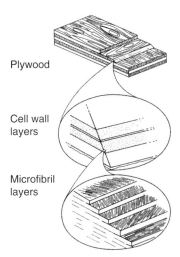

Figure 19.10   If the microfibril directions between the separate cell wall layers are diverse enough, the wall will provide multidirectional support and the cell has no directional preference for expansion. This principle is also the biomimetic basis for the design of plywood. (From Brett C, Waldron K (1996) *Physiology and Biochemistry of Plant Cell Walls*. Chapman & Hall, London. With permission of CRC Press.)

19.3.3   Guard Cells or How to Make a Pore

Although directional deposition of microfibrils provides different wall resistance to diametric and axial cell expansion of stem cells, the wall resistance does not differ on both sides of the elongating cell axis and therefore such cells have a directed but linear expansion. Guard cells, however, are an example where asymmetrical wall resistance on the flanks of the cell axis lead to an unequal expansion of the flanks and a curved shape of the cell. Since these cells are used to create openings in a surface layer, guard cells always occur in pairs. Their asymmetric expansion leads to the opening of a stomatal pore within the inexpandable flanks of two cells (Figure 19.11). A stomatal pore is a reversible, nastic structure that operates as a valve in the leaf surface to balance the photosynthetic intake of carbon dioxide (to be assimilated into sugars) with the loss of water exiting the leaf in a process of forced evaporation that generates negative pressure.

Plants use the opening of these pores to (i) cool the leaf surface, (ii) access the carbon dioxide in the air, and (iii) as the jets of a capillary vacuum pump that through evaporation generates (absolute) negative pressures far beyond the capacity of human-made designs and lifts water and ions in the capillary tubes of the xylem. Different models have been developed to explain the asymmetric expansion of guard cells (Figure 19.12). All models agree that the basis for this behavior is an asymmetry of cell wall characteristics but they differ in the specifics of this structural requirement. The oldest model claims that this kind of expansion is due to the conspicuousthickening of the inner walls adjacent to the stomatal pore. Later models deny the role of wall thickening and explain the asymmetric expansion of the guard cells and the resulting pore formation by either the symmetry of guard cells (Cooke et al., 1977) or more convincingly with the arrangement of the radial microfibril bundles that are tightly wound near the ends and looser near the center of the guard cells (Aylor et al., 1973). Still others (including the authors of this chapter) think that it is not so much the thickening of the inner walls but the enigmatic structure that is behind the thickening — a ring made of cutin, a waterproofing resin polymer found as a common coating on the walls of most epidermal cells of leaves. Even after the guard cell walls have been digested by enzymatic mixtures, the cutin rings remain intact and float in the solution in the form of a gaping mouth. Aside from separating the inner walls and water-proofing the pore walls, the ring itself is the only direct physical connection between the two guard cells. Attached only to the inner

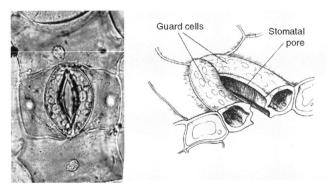

Figure 19.11 Photographic and schematic view (plus cross section) of a stomatal pore from a dicot plant (*Zebrina pendula*) consisting of two guard cells. K^+ ion uptake, for example, under light, into these cells increases the turgor pressure inside the cells, the impeded and asymmetric expansion of which opens the pore. (After Walton PD (1988) *Principles and Practices of Plant Science*, 1st Edition. Prentice-Hall, NJ. Reprinted with permission from Pearson Education, Inc., NJ.)

walls, it connects only the ends of the cells and resists all attempts of the guard cell to increase the vertical axis of the pore (Figure 19.11 and Figure 19.12). With the expansion of the inner walls prevented by the ring and an enlargement of the diameter prevented by radial arrangement of the microfibril bundles, the only expandable surface left is then the outer wall and any volume increase of the guard cell must now lead to cell bending and the formation of a pore.

Using hydrostatic devices like bicycle tubes, balloons, tapes, and ropes all three mechanical models were able to reproduce opening and closure of a pore and hence, supported the validity of each of the suggested explanations (Figure 19.12). With three alternative mechanical models simulating pore openings it seems possible that nature used more than just one mechanism — the principle of redundancy — to produce a highly reliable design.

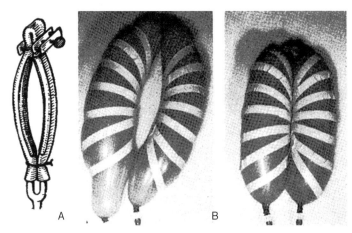

Figure 19.12 Two working mechanical models of stomatal expansion. Inflation of the two halves of a bicycle tire with equal pressure results in a pore formation when the inner flanks have been reinforced by gluing an additional tube layer to them. (Left panel; after Brauner L, Bukatsch F (1980) *Das Kliene Pflanzenphysiologische Praktikum©*. Gustav Fischer Verlag, Jena. Reprinted with permission of Elsevier GmbH, Spektrum Verlag, Heidelberg.) Pore formation also occurs when two balloons are connected at both ends and the balloon walls are spirally taped in a biomimetic simulation of the radial wall structure of guard cells. (From Aylor DE, Parlange J.-Y., Krikorian AD (1973) *American Journal of Botany* 60:163–171. Reprinted with permission from the *American Journal of Botany*.) A third working model would fix the vertical distance between the two connected ends of the balloons with a rope (not shown).

19.3.2 Leaf Motors

Although microstructural arrangements at the cell wall level suffice to generate functioning structures like stomatal valves, most nastic structures involve multiple cells and cell layers. We show here the functioning of three different nastic structures that serve (i) to rapidly bend (ii) to fold or unfold, and (iii) to position a leaf in either a horizontal or vertical direction.

19.3.2.1 *Venus Flytrap* (Dionea muscipula)

Venus flytraps have a leaf with two rapidly closing halves designed to trap insects needed to supplement their nitrogen supply. The closure of Venus flytraps is an example showing how nastic movement is organized at a more complex level than the cell pair of stomatal pores. Although there is still considerable uncertainty about the exact mechanism (for a recent version see Forterre et al., 2005), we can state with confidence that the operating principle also consists of an internal hydrostat or motor being asymmetrically restricted. Although the flytrap leaf seems to bend like a bimorph actuator, it is a trimorphic structure with exceptionally large motor cells in the center of a trap leaf (Figure 19.12). These motor cells are so intimately connected to the two adjacent, epidermal surface layers that they cannot be separated experimentally (Hodick and Sievers, 1988). The inner (upper or red) surface has sensory hairs and the digestive glands and with only 5% a strikingly lower extensibility than the outer (lower or green) surface featuring 20% or a five times higher extensibility. According to the different extensibilities, a sudden, rapid and powerful expansion of the internal motor cells leads to an equally rapid curving of the leaf halves towards the center and the closure of the trap (Figure 19.13). Forced to beat the rapid movements of the muscles and extremely rapid resilin-springs of its insect prey (Alexander, 1992) the flytrap developed one of the fastest nastic movement known in plants.

Figure 19.13 (See color insert following page 302)   Leaves of Venus flytraps have large, powerful motor cells in their center. The rapid expansion of these cells is modified by different extensibilities of the two adjacent surface (epidermis) layers. These anisotropic restraints turn a linear expansion into the rapid curving of the leaf and closure of the trap. This mechanism is triggered when the sensory hairs at the upper surface are repeatedly touched.

Figure 19.14   Leaf rolling in grass blades of *Agropyron elongatum,* the tall wheatgrass. This nastic movement is driven by leaf rollers — small hydration motors that consist of only a few large bulliform cell joints (left panel) placed at regular intervals along the upper surface of the leaf. The easy dehydration of these uncoated cell clusters contracts the upper surface and so rolls and even sharply folds the leaves of grasses in a rapid and reversible manner. Right panel: rapid rolling is visible in dehydrating leaf slices (sequential pictures taken with interval of 1 min).

19.3.2.2   Leaf Rolling or the Autonomous Unfolding of Surface Area

Similar, slower leaf-folding movements are carried out by many by other plants, not with the purpose to trap insects but to restrict transpirational water loss. Structural unfolding of rhododendron leaves, grass leaves, cacti stems, etc., occurs in response to favorable environmental conditions (in particular sufficient humidity or root water supply). Under adverse conditions many grasses, such as Kentucky bluegrass *Poa pratense*, tall wheatgrass *Agropyron elongatum, Tradescantia* species, as well as dicot leaves of some rhododendrons, roll their leaves to prevent excessive dehydration. In grasses this response is triggered by large, bulliform cells in the upper epidermis (Figure 19.14). Since these cells have no or only a very thin impregnating coat (cuticle) they lose water more rapidly than other epidermal cells, thereby shrinking the surface area of the upper leaf and rolling it around the midvein. This measure drastically reduces any further evaporation from the upper surface. A similar mechanism operates in dicot leaves of rhododendrons that roll in the opposite direction. The preferred dehydration and shrinking of the lower epidermis cells (all of them), as well as the shrinking of patches of very loosely arranged mesophyll cells in the blade center (cross sections not shown), cause a very rapid, reversible coiling of the blade around the midvein of *Rhododendron* leaves (Figure 19.15).

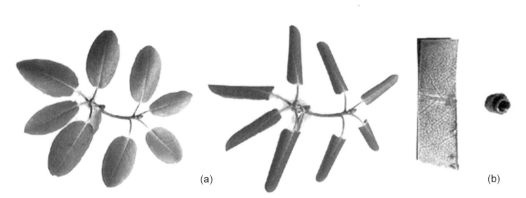

(a)   (b)

Figure 19.15   Rolling in dehydrated rhododendron blades differs from grass blades in that it is downwards and not driven by discrete bulliform cell joints. The preferred dehydration and shrinking of the lower epidermis, as well as patches of very loosely arranged mesophyll cells in the blade center (cross sections not shown), cause a downward curving of the blade around the midvein (a). A rectangular leaf section turns into a roll within a few minutes exposure to dry lab air (b).

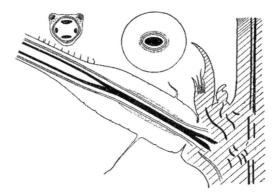

Figure 19.16   Longitudinal and cross sections of the swollen base (pulvinus) of a leaf stalk of *Mimosa pudica*. Cross sections (pictured above the stalk) show that a centralization of the vascular bundles allows bending of this organ in an almost joint-like fashion. Loss of ions from the hydrostat parenchyma cells below the central vascular tubing leads to simple and reversible buckling of the structure and drops the leaf into a vertical position. (From Fitting H, Harder R, Schumacher W, Firbas F, (1954) *Lehrbuch der Botanik fur Hochschulen*. Gustav Fischer Verlag, Stuttgart. Reprinted with permission of Specktrum Verlag GmbH, Heidelberg.)

19.3.2.3 Leaf "Muscles" or Pulvini

Leaves of *Phaseolus*, *Oxalis*, *Desmodium*, *Mimosa* and many other plants can undergo slow or sudden nastic "sleep movements," in which the drooping leaf stalk brings the leaf blade from a horizontal into a more vertical position. Although reasons for this movement are still unknown, the fact that several unrelated plant species independently developed similar designs, makes one safely conclude that it is worth the effort. The organ involved in elevating and lowering the antenna-like structure of the leaf blade is a so-called leaf muscle or pulvinus, a swollen joint-like structure at the basal part of a lever (the leaf stalk or petiole) that attaches the leaf to the stem. The functioning of this unusual crane rests solely with the hydrostat motor cells of the parenchyma below the central vascular bundle of the pulvinus (Figure 19.16). As these cells dehydrate and shrink the leaf drops rapidly. It rises slowly again when they regain a pressure large enough to overcome the gravitational force of the leaf blade. An antagonistic course of events in the cells on top of the vascular bundle of the petiole help these movements but is less critical for the mechanism (Satter and Morse, 1990). Intriguing is that so far no ion channel has been found that is rapid and large enough to account for the rapid loss of pressure and volume in the motor cells of the lower petiole of *Mimosa* and *Oxalis* species. It is instructive to note that the petiole construction (cross sections pictured above the leaf stalk in Figure 19.16) changes from a statically stable design supported by fibrous bundles in the periphery to a statically unstable structure (after the vascular support moved from the periphery to the center) that now has to be supported by the hydrostatic pressure of the parenchyma cells at the lower half of the pulvinus.

19.3.2.4 External Structures Allowing Large Volume Changes

The expansion and shrinking of entire plant bodies qualifies as a slow nastic movement. A truly remarkable example was first mentioned by Paturi (1976). He refers to the impressive adaptation of the mescal cactus *Lophophora williamsii*, which converts the dehydration-induced shrinking during the beginning of the dry season to reduce the shoot length and submerge below the desert floor. After just one seasonal rainfall the hydrating shoot reemerges by pushing the photosynthesizing apex out of its soil cover into the light and open air (Figure 19.17). Although smaller-sized volume oscillations occur also in fruits, leaves, and stems, they are most prominent in the strange, leafless

during dry season   after rainfall

Figure 19.17 Picture shows the almost completely buried shoot of a mescal cactus. Vertical shoot expansion after rainfall and shrinking in the dry period cause the appearance and disappearance of the photosynthetically active shoot apex in the ground — a very useful adaptation of this plant to a harsh environment, the Mesoamerican desert.

column-like shoots of xerophytic desert plants. The unique circumferential shape of these xerophytes reminds us of an accordion or the Eiffel tower (Figure 19.18). It is this structure that allows the ribbed xerophyte shoots to undergo extensive volume changes (shrinking and swelling) without exposing the most important surface to a parallel but highly dangerous change in their area. The external cell layers include both the shoot's only photosynthesizing cells and the mechanical support structure of the shoot in the form of an external, fibrous network.

These accordion-like, inflatable shoot structures allow substantial volume changes for water storage in the stem center (white area in Figure 19.18b), while an unchanged surface area is maintained by changing from a concave to a convex shape (Figure 19.18c). The unique adaptation of this stem design becomes apparent if we compare it with volume changes in herbal plants from moderate climate zones. Figure 19.9 shows clearly how a shrinking pith tissue in a sunflower stem reduces the surface tension and with it the stability of the entire stem, which turns limp showing the familiar structural collapse known as wilting. Another adaptive advantage of this pleated-column structure is that dehydration and shrinking increase the length of protruding ridges and so provide increased self-shading of the shoot surface in an environment that has little other shade to offer.

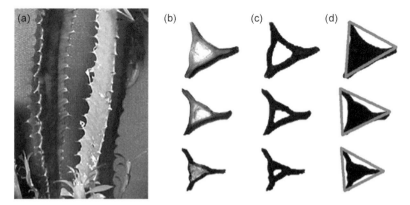

Figure 19.18 (See color insert following page 302)   The shoots of desert plants in the *Cactus* and *Euphorbia* families often adopt the shape of pleated columns (a) a structure which allows the photosynthesizing periphery to maintain its area in spite of massive volume losses in the shoot center. Cross sections of a pleated-column shoot show increasing dehydration from upper to lower pictures (b) with a considerable volume loss of the shoot center (c) and remarkable constancy in the area of a surface that changes its shape from a convex to a concave outline.

19.4 BIO-INSPIRED MATERIALS FOR BIOMIMETIC ACTION — CONCLUSIONS

Plants are adaptive machines that have developed and perfected their structures over millions of years of evolution in constantly changing and increasingly complex environments. With their autonomous and locally controlled responses plant movements do not require remote control through a central nervous system or computer. Life originated in water and nastic structures as well as their motors are optimized to use the potentialities of this unique solvent. In addition to ATP-dependent molecular motors, contracting or inflating molecules, such as the P-protein of the phloem conduits, we presented here three types of hydration motors (osmotic, colloidal, and fibrous) that figure as the major workhorses for nastic plant movements. Most plant movements and growth are driven by osmotic motors, and several examples illustrate how the isometric, internal cell pressure is transmitted into anisometric movement patterns. Colloid motors are known to be used by plants to overcome the formidable resistance of multilayered seed shells to start germination. Fibrous motors are known to be involved in the opening of seed containers.

A striking particularity of nastic structures is an efficient and inconspicuous design that integrates two or more functions (like motor and valve or sensor, motor and lever) in one smoothly operating unit. The knowledge gained from studying nastic structures is key input for designing human-made adaptive structures and smart materials and includes novel and increasingly complex multifunctional materials like fiber-reinforced bio-composites, pressure-sensing, humidity-sensing, sugar (or other chemical)-sensing fiber-reinforced hydrogels, nastic-inspired combinations of actuators, and force-guiding structures like matrixes with pressurized body inclusions, as well as arrayed actuators combining memory shape alloys, photomechanical films, electro-active polymers, and sensing hydrogels in one multifunctional unit. Even the most modest, selective presentation of such bio-inspired materials and designs is far beyond the scope of this chapter but good examples can be found in some chapters of this book and other publications as well (e.g., Taya, 2003). In the past, some passive plant structures have served as the basis for a few successful phytomimetic designs of buildings (Paxton's Crystal Palace in London following rib construction of floating leaves from *Victoria amazonica*), attachments [VELCRO fastener after burdock or *Arctium* seed pods (Paturi, 1976; Benyus, 1997; Vincent, 1997; Vogel, 1998)], and airplanes [wings of glider and motor planes with extremely reduced stalling features were designed after *Zanonia* seeds (Etrich, 1915)]. The next generation of phytomimetic designs is more likely to be inspired by active plant structures like those reviewed here. Powerful and energy-efficient, auto-sensing and autonomously adjusting, multifunctional, exhaust-free and silent actuators, tools, motors, as well as new materials from leak-free storage of liquids to the silent demolition of buildings, can be based on the integrated workings of nastic structures. By reviewing the major guiding principles and selected examples of natural nastic structures, this chapter is meant to inspire, stimulate, and broaden their current and future technical application.

REFERENCES

Alexander RM (1992) *Exploring Biomechanics: Animals in Motion*. Scientific American Library, New York.

Asada T, Collings D (1997) Molecular motors in higher plants. *Trends in Plant Science* 2:29–36.

Aylor DE, Parlange J-Y, Krikorian AD (1973) Stomatal mechanics. *American Journal of Botany* 60:163–171.

Benyus JM (1997) *Biomimicry — Innovation Inspired by Nature*. William Morrow and Company, New York.

Brauner L, Bukatsch F (1980) *Das kleine pflanzenphysiologische Praktikum*. Gustav Fischer Verlag, Jena.

Brett C, Waldron K (1996) *Physiology and Biochemistry of Plant Cell Walls*. Chapman & Hall, London.

Cleland RE, Virk SS, Taylor D, Bjorkman T (1990) Calcium, cell walls and growth. In *Calcium in Plant Growth and Development* (Leonard RT and Hepler PK (eds)). American Society of Plant Physiologists, Rockville, USA, pp. 9–16.

Cooke JR, Rand RH, Mang HA, Debaerdemaeker JG (1977) A non-linear finite element analysis of stomatal guard cells. American Society of Agricultural Engineers paper 77–5511, St Joseph, USA.

Crews LC, McCully ME, Canny MJ (2003) Mucilage production by wounded xylem tissue of maize roots — time course and stimulus. *Functional Plant Biology* 30:755–766.

Englmaier P (1987) Carbohydrate metabolism of salt-tolerant fructan grasses as exemplified with *Puccinellia peisonis*. *Biochemie Physiologie Pflanzen* 182:65–182.

Etrich I (1915) *Die Taube — Memoiren des Flugpioniers Dr.-Ing. h.c. Igo Etrich*. Econ Verlag, Vienna.

Fitting H, Harder R, Schumacher W, Firbas F (1954) *Lehrbuch der Botanik fur Hochschulen*. Gustav Fischer Verlag, Stuttgart.

Forterre Y, Skotheim JM, Dunals J, Mahadevan L (2005) How the Venus flytrap snaps. *Nature* 433:421–425.

Frey-Wyssling A (1959) *Die Pflanzliche Zellwand*. Springer Verlag, Berlin.

Gibson LJ, Ashby MF (1982) The mechanics of three-dimensional cellular materials. *Proceedings of the Royal Society of London A* 383:43–59.

Gordon JE, Jerominidis G (1980) Wood — a natural polymer. *Philosophical Transactions of the Royal Society London A* 294:545–550.

Hodick D, Sievers A (1988) On the mechanism of trap closure of Venus flytrap (*Dionea muscipula* Ellis). *Planta* 174:8–18.

Knoblauch M, Noll G, Muller T, Prufer D, Schneider-Huther I, Scharner D, Van Bel AJE, Peters W S (2003) ATP-independent contractile proteins from plants. *Nature Materials* 2:573–574.

Kuhne L, Kausch W (1965) Uber das Quellungsmaximum der Kotyledonen und Keimachsen von *Pisum sativum* L. *Planta* 65:27–41.

Jost R (1933) *Reizerscheinungen der Pflanzen*. Thieme Verlag, Leipzig.

Juniper BE, Clowes FAL (1965) Cytoplasmic organelles and cell growth in root caps. *Nature* 208:864–865.

Larson LA (1968) The effect soaking pea seeds with or without seed coats has on seedling growth. *Plant Physiology* 43:255–259.

Levitt J (1969) *Introduction to Plant Physiology*. CV Mosby Company, St Louis, MO.

Mavroidis C, Dubey A (2003): Biomimetics: from pulses to motors. *Nature Materials* 2:573–574.

Meidner H, Sheriff DW (1976) *Water and Plants*. Wiley, New York.

Meeuse B, Morris S (1984) *The Sex Life of Flowers*. Oxford Scientific Films Ltd, New York.

Niklas KJ (1989) Mechanical behavior of plant tissues as inferred from the theory of pressurized cellular solids. *American Journal of Botany* 76:929–937.

Niklas KJ (1992) *Plant Biomechanics. An Engineering Approach to Plant Form and Function*. The University of Chicago Press, Chicago, IL.

Paturi FR (1976) *Nature, Mother of Invention. The Engineering of Plant Life*. Thames and Hudson, London.

Pfeffer W. (1873) *Physiologische Untersuchungen*. Thieme Verlag, Leipzig.

Pollack GH (2004) Cells, gels and mechanics. In *Design and Nature II. Comparing Design in Nature with Science and Engineering* (Collins MW, Brebbia CA (eds)). WIT press, Southampton, UK, pp. 433–442.

Robards AW (1974) *Dynamic Aspects of Plant Ultrastructure*. McGraw Hill, London.

Satter RL, Morse MJ (1990) Light-modulated, circadian rhythmic leaf movements in nyctinastic legumes. In *The Pulvinus: Motor Organ for Leaf Movement* (Satter RL, Gorton HL, Vogelmann TC (eds)). American Society of Plant Physiologists, Rockville, USA, pp. 10–24.

Shimmen T, Ridge RW, Lambiris I, Plazinski J, Yokota E, Williamson RE (2000) Plant myosins. *Protoplasma* 214:1–10.

Stahlberg R, Cosgrove DJ (1997) Mannitol inhibits the growth of intact cucumber but not pea seedlings by mechanically collapsing the root pressure. *Plant Cell Environment* 20:1135–1144.

Taiz L (1984) Plant cell expansion: regulation of cell wall mechanical properties. *Annual Reviews of Plant Physiology* 35:585–657.

Taya M (2003) Design of bio-inspired active materials (invited paper). *Proceedings of the SPIE Symposium on Electroactive Polymers and Devices* (Y. Bar-Cohen, (ed)). San Diego, CA, vol. 5051, pp. 54–65.

Turill FM (1936) The area of the internal exposed surface of dicotyledon leaves. *American Journal of Botany* 23:255–264.

Vincent JFV (1982) *Structural Biomaterials*. Wiley, New York.

Vincent JFV (1997) Stealing ideas form nature (Trueman Wood lecture). *RSA Journal* August/September: 36–42.

Vogel S (1998) *Cats' Paws and Catapults: Mechanical Worlds of Nature and People*. Norton and Company, New York.

Wainwright SA (1980) Adaptive materials: a view from the organism. In *The Mechanical Properties of Biological Materials*. Cambridge University Press, Cambridge, MA, pp. 437–453.

Wainwright SA (1995) What can we learn from soft biomaterials and structures? In *Biomimetics: Design and Processing of Materials*. AIP Press, New York, pp. 1–12.

Walter H (1950) *Grundlagen des Pflanzenlebens und ihre Bedeutung für den Menschen,* Verlag Eugen Ulmer, Stuttgart.

Walton PD (1988) *Principles and Practices of Plant Science*, 1st Edition. Prentice Hall, Englewood Cliffs, NJ.

Zimmermann U, Zhu J, Meinzer F, Goldstein G, Schneider H, Zimmermann G, Benkert R, Thurmer F, Melcher P, Webb D, Haase A (1994) High molecular weight organic compounds in the xylem sap of mangroves: implications for long-distance water-transport. *Botanica Acta* 7:218–229.

Biomimetics: Reality, Challenges, and Outlook

Yoseph Bar-Cohen

CONTENTS

20.1 INTRODUCTION

After 3.8 billion years of evolution, nature has learned how to use minimum resources to achieve maximal performance and come up with numerous lasting solutions (Gordon, 1976). Recognizing that nature's capability continues to be significantly ahead of many of our technologies, humans have always sought to mimic nature. The field of study pertaining to this, which is also called biomimetics, bionics, or biogenesis, has reached impressive levels. It includes imitating some of the human thinking process in computers by mimicking such human characteristics as making decisions and operating autonomously. Biology offers a great model for the development of mechanical tools, computational algorithms, effective materials, as well as novel mechanisms and information technology. Some of the commercial implementations of the progress in biomimetics can be seen in toy stores, where toys seem and behave like living creatures (e.g., dogs, cats, birds, and frogs). More serious benefits of biomimetics include the development of prosthetic implants that appear very much like they are of biological origin, and sensory aiding mechanisms that are interfaced to the brain to assist in hearing, seeing, or controlling instruments. As described and discussed throughout this book, the topic of biomimetics is very broad and covers many disciplines, with applications and implications for numerous areas of our life.

Robotics is one biomimetic area in which advances are continually being made. The movie industry has created a vision of robots that are human-like at a level significantly far beyond what is currently feasible. However, even though it will be a long time before such robotic capabilities become a reality, there are already numerous examples of accomplishments (Bar-Cohen and Breazeal, 2003). Initially, robots were not well received because they were considered too bulky and too expensive, requiring major amount of work to employ, maintain, modify, and upgrade. Solving these problems by making robots more biomimetic became feasible when powerful lightweight microprocessors were introduced. These improvements included high computation speed, very large memory, wireless communication with a wide bandwidth, effective control algorithms, miniature position indicators using Global Positioning Satellites (GPS), and powerful software tools including artificial intelligence techniques. Advancements in computers and control methodologies led to the development of sophisticated robots with a significant expansion of the capability to emulate biological systems. Autonomous robots were developed and they have successfully demonstrated their ability to perform many human- and animal-like functions. Such robots offer superior capabilities to operate in harsh or hazardous environments that are too dangerous for humans. Progress in intelligent biomimetic robots is expected to impact many aspects of our lives, especially in performing tasks that are too risky to execute by humans, or too expensive to employ humans (e.g., operate as movie actors). These robots may also be used in tasks that combine the advantages of biological creatures in a hybrid form, which are far beyond any known system or creature, including operating in multiple environments (flying, walking, swimming, digging, etc.).

This book has focused on aspects that are related to biology which have inspired artificial applications and technologies. Many inventions have been based on concepts that have had their roots in biology. However, since natural inventions are not recorded in a form that one can identify in engineering terms, the inventions that were produced by humans may have been coincidently similar, subconsciously inspired, or their origin in nature may not have been well documented. In this chapter, the author makes an attempt to summarize the current status of biomimetics, its challenges, and its outlook for the future.

20.2 BIOLOGY AS A MODEL

Nature has an enormous pool of inventions that passed the harsh test of practicality and durability in a changing environment. In order to harness the most from nature's inventions it is critical to bridge the gap between the fields of biology and engineering. This bridging effort can be a key to turning nature's inventions into engineering capabilities, tools, and mechanisms. In order to approach nature in engineering terms it is necessary to sort biological capabilities along technological categories using a top-down structure or vice versa. Namely, one can take each aspect of the biologically identified characteristics and seek an analogy in terms of an artificial technology. The emergence of nano-technologies, miniature, highly capable and fast microprocessors, effective power storage, large compact and fast access memory, wireless communication and so on are making the mimicking of nature capabilities significantly more feasible. One reason for this is both natural and artificial structures depend on the same fundamental units of atoms and molecules. Generally, biological terms can be examined and documented analogously to engineering categories as shown in Table 20.1.

Some of nature's capabilities can inspire new mechanisms, devices, and robots. Examples include the beaver's engineering capability to build dams, and the woodpecker's ability to impact wood while suppressing the effect from damaging its brain. Another inspiring capability is the ability of numerous creatures to operate with multiple mobility options including flying, digging, swimming, walking, hopping, running, climbing, and crawling. Increasingly, biologically inspired capabilities are becoming practical including collision avoidance using whiskers or sonars,

Table 20.1 Characteristic Similarities of Biology and Engineering Systems

Biology	Engineering	Bioengineering, Biomimetics, Bionics, and Biomechanics
Body	System	Systems with multifunctional materials and structures are developed emulating the capability of biological systems
Skeleton and bones	Structure and support struts	Support structures are part of every human made system. Further, exoskeletons are developed to augment the operation of humans for medical, military, and other applications (Chapter 6)
Brain	Computer	Advances in computers are being made modeling and emulating the operation of the human brain, for example, the adaptation of the association approach of memory search in the brain to make faster data access (Chapters 3 to 5)
Nervous system	Electric systems and neural networks	Our nervous system is somewhat analogous to electrical systems, especially when it is incorporated with neural networks. The connections of elements in both systems are based on significantly different characteristics
Intelligence	Artificial intelligence	There are numerous aspects of artificial intelligence that have been inspired by biology including: Augmented Perception, Augmented Reality, Autonomous Systems, Computational Intelligence, Expert Systems, Fuzzy Logic, Intelligent Control, Learning and Reasoning Systems, Machine Consciousness, Neural Networks, Path Planning, Programming, Task Planning, Simulation, Symbolic Models, etc. (Chapters 3 to 5)
Senses	Sensors	Computer vision, artificial vision, acoustic and ultrasonic technology, radar, and other proximity detectors all have direct biological analogies. However, at their best, the capability of the human-made sensors is nowhere near as good as biosensors (Chapters 11 and 17)
Muscles	Actuators	Electroactive polymers are artificial actuators with very close functional similarity to natural muscles (Chapters 2, 9, and 10)
Electrochemical power generation	Rechargeable batteries	The use of biological materials, namely, carbohydrates, fats, and sugars to produce power will offer mechanical systems with enormous advantages
DNA	Computer code	Efforts are being made to develop artificial equivalent of DNA (Chapters 7 and 8)

controlled camouflage, and materials self-healing. One of the challenging capabilities will be to create reconfigurable systems that match or exceed the butterfly life stages that include egg, caterpillar, cocoon, and butterfly. Other challenges include making miniature devices that can fly with enormous maneuvering capability like a dragonfly; adhere to smooth and rough walls like a gecko; camouflage by adapting itself to the texture, patterns, and shape of the surrounding environment like a chameleon, or reconfigure its body to travel through very narrow tubes like an octopus. Further challenges also include processing complex 3D images in real time; recycling mobility power for highly efficient operation and locomotion; self-replication; self-growing using resources from the surrounding terrain; chemical generation and storage of energy; and many such capabilities for which biology offers a model for science and engineering inspiration. While many aspects of biology are still beyond our understanding, significant progress has been made.

Biological designs and processes follow the template that is written in the organisms' DNA, which defines the building blocks of all living organisms. This archival storage of construction codes of all organisms' is stored in the nucleus of all living cells and it consists of strands of nucleic acids: guanine, adenine, thymine, and cytosine. These four nucleic acids are assembled as long sentences of biological laws and they guide the function of living cells through a simple universal process. Information contained in the DNA is transcribed in the nucleus by RNA polymerase and sent out of the nucleus as messenger RNA that is translated at the ribosomes into amino acids, the building blocks of proteins. Proteins are the foundation of all life: from cellular to organism levels and they play a central role in the manifestation of populations, ecosystems, and global dynamics. Designers of human-made systems are seeking to produce sequence-specific polymers that constitute proteins in order to make products and services that meet the needs of humans and the demand of consumers. Cloning the DNA allows you to produce synthetic life while adapting nature's principles allows you to create artificial life and biomimetic tools and capabilities.

20.3 CHARACTERISTICS OF BIOLOGICALLY INSPIRED MECHANISMS

There are many characteristics that identify a biomimetic mechanism and some of the important ones include the ability to operate autonomously in complex environments, perform multifunctional tasks and adaptability to unplanned and unpredictable changes. Making mechanisms with such characteristics dramatically increases the possible capabilities and can reach levels that can be as good or superior to humans or animals. This may include operating for 24 h a day without a break or operating in conditions that pose health risks to humans. Benefits from such capabilities can include performance of security monitoring and surveillance, search and rescue operations, chemical, biological, and nuclear hazardous operations, immediate corrective and warning actions as well as others that are only limited by our imagination. Some of the biologically inspired capabilities that are/can be implemented into effective mechanisms include:

- **Multifunctional materials and structures (Chapters 12 and 14):** Biological systems use materials and structures in an effective configuration and functionality incorporating sensor and actuation to operate and react as needed. Using multifunctional materials and structures allows nature to maximize the use of the available resources at minimum mass (Rao, 2003). An example is our bones, which support our body weight and provide the necessary body stiffness while operating as our "factory" for blood that is produced in the bone marrow. Another example is the feathers in birds, which are used for flying as well as for thermal insulation and the control of heat dissipation. Mimicking multi-functionality capabilities, system are made to operate more effectively in robots provided with ability to grasp and manipulate objects and with mobility of appendages or sub-appendages (hands, fingers, claws, wings). Some of the concerns with regard to the application of multiple functionality is the associated design difficulties where there is a need to simultaneously satisfy many constraints. Design changes in one part of the system affect many other parts.
- **High strength configurations:** The geometry of birds' eggs have quite interesting characteristics. On the one hand, they are amazingly strong from the outside, so a bird can warm its eggs by sitting on them till the chicks hatch. On the other hand, they are easily breakable from the inside, so the chicks can break the shell with their beak once they are ready to emerge into the outside world.
- **Just-in-time manufacturing:** Producing as needed and at the time of the need is widely used in biology and such examples include the making of the web by spiders or the production of the toxic chemicals by snakes. Such a capability is increasingly adapted by industry as a method of lowering the cost of operation. Many industries are now manufacturing their products in small quantities as needed to meet consumers demand right at the assembly line. Thus, industry is able to cope with the changing demand and decline or rise in orders for its products.
- **Deployable structures:** The leaves of most plants are folded or rolled while still inside the bud. The way they unfold to emerge into to a fully open leaf can inspire deployable structures for space, including gossamer structures such as solar sails and antennae as well as terrestrial applications such as tents and other covering structures (Guest and Pellegrino, 1994; Unda et al., 1994; Kobayashi et al., 1998).
- **Hammering without vibration back-propagation:** The woodpecker (Picidae family) has the amazing capability to tap and drill holes in solid wood in search of insects and other prey (Bock, 1999). One example is the Northern Flicker (*Colaptes auratus*), which is a member of the woodpecker family, shown in Figure 20.1. The brain of the woodpecker is protected from damage as there is very little space between it and the skull preventing rotation during impact. Some woodpecker species have modified joints between certain bones in the skull and upper jaw, as well as muscles which contract to absorb the shock of the hammering. A strong neck, tail-feather muscles, and a chisel-like bill are other hammering adaptations in some species. This ability to absorb the shocks and prevent damage to the bird brain or cause disorientation could inspire a

Figure 20.1   A view of the Northern Flicker (*Colaptes auratus*) which belongs to the woodpecker family. (Courtesy of Ulf T. Runesson, Faculty of Forestry and the Forest Environment, Lakehead University, Ontario, Canada: www.borealforest.org.)

mechanism for protecting operators of jackhammers. The vibrations generated by the jackhammer back-propagate into the hand and body of the operator. These vibrations can cause severe damage including the pulling out the teeth from the operator mouth. Mimicking the shock-absorbing mechanism of the woodpecker beak may offer an effective approach to suppressing back-propagated vibrations from the jackhammer.

- **Nanostructures (Chapters 7 and 8):**   Biology consists of complex nanostructures that allow many capabilities that are far beyond current human capabilities. Recent developments in nano- and micro-fabrication, as well as self-assembly techniques, are driving the development of new functional materials and unique coatings that mimic biomaterials. For controlled adhesion, efforts are underway to mimic the geckos and their setae. These setae, which are microscopic hairs on the bottom of their feet, use van der Waals forces to run fast on smooth surfaces such as glass (Autumn and Peattie, 2003). Further, there are efforts to produce the biomimetic equivalence of cells as described in Chapters 1 and 15.

- **Behavior and cooperative operation (Chapters 3, 4, 5, and 16):**   Biologically inspired systems need to autonomously recognize and navigate in various environments, perform critical tasks that include terrain following, target location and tracking, and cooperative tasks such as hive and swarm behavior. Such activity requires the incorporation of principles that are derived from biological behaviors of social groups. Ants serve as a model for accomplishing tasks that are much bigger than an individual.

- **Mimicking aerodynamic performance:**   The development of aerodynamic structures and systems was inspired by birds and the shape of wind-dispersed seeds. Trees disperse their seeds to great distances using various aerodynamic principles that allow them to use the wind. The propelling capability of seeds has inspired designs of futuristic missions with spacecraft that could soft land on atmospheric planets such as Mars. Adapting this design may offer a better alternative than parachutes, with a better capability to steer towards selected sites. In recent years, increasing efforts have been made to develop miniature flying vehicles, especially since the terror attack in September 11, 2001. Micro-air-vehicle (MAV) with wing spans of several centimeters has been developed using a propeller, and efforts are currently underway to produce even smaller MAV units (http://uav.wff.nasa.gov/).

- **Mobility (Chapter 6):** Mobility is a characteristic of animals that involves multi-functionality, energy efficiency (not necessarily optimized), and autonomous locomotion. Animals can operate in multiple terrains, performing various locomotion functions and combinations, including walking, crawling, climbing (trees, cliffs, or walls), jumping and leaping, swimming, flying, grasping, digging, and manipulating objects. Integration of such locomotion functions into a hybrid mechanism would potentially enable mobile transitions between air, land, and water. Making robots with such capabilities will far exceed any biological equivalence.

- **Attaching to steep walls and upside down from a ceiling:** As shown in Chapter 1, the swallow is capable of attaching itself to walls by carrying its body weight on its fingernails. The gecko is capable of controlled adherence to rough and soft surfaces. Mimicking this capability, a gecko tape was made by microfabrication of dense arrays of flexible plastic pillars, the geometry of which was optimized to ensure their collective adhesion (Geim et al., 2003). This approach showed a way to manufacture self-cleaning, reattachable dry adhesives, although problems related to the gecko tapes durability and mass production are yet to be resolved. Generally, controlled adhesion is a capability that is sought by roboticists to adapt into robotic devices. A four-legged robot, named Steep Terrain Access Robot (STAR) (Badescu et al., 2005), is being developed at Jet Propulsion Laboratory (JPL) and is designed to climb rocks and steep cliffs using an ultrasonic/sonic anchor that uses low axial force to anchor the legs (Bar-Cohen and Sherrit 2003). This robot is shown in Figure 20.2.

- **Autonomous locomotion:** Inspiration from biology led to the introduction of robots and systems that operate autonomously with self-learning capability (Chapters 3, 4, and 6). Such a capability to operate without real-time control by a human operator is critical to the National Aeronautics and Space Administration (NASA) missions that are performed at distant extraterrestrial conditions where remote-control operation is not feasible. The distance of millions of miles from Earth to Mars causes a significant communication time delay, and necessitates an autonomous capability to assure the success of the NASA planetary exploration missions.

- **Sensors and feedback:** The integration of sensors into biomimetic systems is critical to their operation and it is necessary to provide closed-loop feedback to accomplish biologically inspired

Figure 20.2 (See color insert following page 302) A four-legged robot called Steep Terrain Access Rover (STAR) is under development at JPL. (Courtesy of Brett Kennedy, JPL.)

tasks. Nature uses many types of sensors, and some of them were already mimicked in artificial devices, including the collision-detection whiskers in automatic vacuum cleaners and mobile toys. Combining the input from the sensor and the control system is critical to the operation of the specific systems. The use of biologically inspired centralized and decentralized control architectures offers advantages in speed of operation and simplicity of the selected control architecture. The topic of vision as human sensing and its imitation were covered in Chapters 11 and 17 of this book.

- **Optimization tools and algorithms:** Various optimization tools have been developed using biological models. As described in Chapters 4 and 5, the simulation of natural selection and survival of the fittest, which is the key to the process of evolution, has been adapted mathematically in the form of genetic algorithm. To survive, individuals of any species must reproduce and regenerate and this requires new members of the population to be fit and adaptable to changing environmental conditions. Only the fittest individuals survive while the weak members perish or are killed by their natural enemies. Inherent to the genetic algorithm approach is, the definition of what features identify the fittest, where in nature, the definition keeps evolving with changing environmental conditions and across species. Unlike nature, in genetic algorithms the definition of the fittest is stability. By identifying the stable elements in a population, genetic algorithms allow for the ultimate achievement of an "ideal" population and this is a situation that is not paralleled in nature.
- **Machine–human interaction:** Intuitive interaction between human and machines is increasingly becoming an issue of attention of computer and instrument manufactures. As efforts are made to reach consumers outside the pool of high-tech individuals, it is increasingly critical to make human–machine interaction more users friendly. To address this need many computer monitors and input pads are equipped with touch screen capability. Systems with voice recognition are becoming a standard in information services that are provided over the phone. When calling your bank, airline, phone operator, and many businesses today you are greeted by a computer operator that interacts with you and understands your answers from a selected menu of choices. In parallel, efforts are underway to develop robots that can recognize body language and emotional expressions (sad, happy, etc.) and respond accordingly (Chapter 6; Bar-Cohen and Breazeal, 2003). Other forms of interaction that are emerging include direct control from the human brain to allow disabled individuals to operate independently.

20.4 TURNING SCIENCE FICTION INTO ENGINEERING REALITY

Biology is filled with solutions and inventions that has been the subject of mimicking and continues to offer enormous potential for human-made mechanisms, tools, and algorithms (Benyus, 1998). Some of the functions that are performed by creatures are far from becoming an engineering reality, such as the octopus' capability to travel through narrow passages significantly smaller than its body cross section. Making a robot that can camouflage itself as well as an octopus (Cott, 1938; Hanlon et al., 1999) and defend itself with multiple tentacles using numerous suction cups and poisonous needles offers enormous potential for homeland defense, but it is far from reality. Science-fiction movies and literature have created a level of expectation for the field of biomimetics and robotics that is far from reality, though these expectations offer creative ideas. Employing biologically inspired principles, mobility, sensing, and navigation are driving revolutionary capabilities in emerging robots. Development in biomimetics may lead to a day when intelligent robots could replace dogs, offering unmatched benefits in terms of capability and intellectual support. It may become possible to discuss with robots strategies for stock market investment, obtain advice about a personal problem, or possibly debate philosophical thoughts and politics. Also, one may be able to have the robot read books in any desired language, accent, or gender voice, and answer questions about unclear words or sentence in a book, as well as provide related information and background. The robot may be able to cheer you up, laugh when a funny situation occurs, smell and identify odors, as well as taste food and provide detailed nutrition and health information. Being fully autonomous, biomimetic robots would conduct self-diagnostics and go to the selected maintenance facility for periodic checkup and possibly repair themselves as needed. Rapid prototyping will enable fast development of this technology as improvements are introduced to the field. While many positive aspects may result from the development of such robots with humanistic capabilities and behavior, negative issues may arise that will require attention. Such issues may include owner liability in case of accident or "misbehavior" of the robot, as well as the potential use of robots for unlawful acts.

For many years, the beneficiaries of biologically inspired robot have been the entertainment industry, including toys and movies. Robots with biomimetic characteristics are becoming popular consumer products, reflecting the public fascination with the realistic capabilities that can be enabled in robots. Such products include robotic toys such as the Mattel's Miracle Moves Baby, which was created and developed in partnership with TOYinnovation, Inc. Miracle Moves Baby was introduced in 2001, and sold widely at its introduction. This doll wakes up the way a real baby would, yawns, appears tired, sucks her bottle and her thumb, giggles, burps, and is rocked to sleep in the most life-like manner.

Further, as the evolution in capability has increased it has reached the level that the more sophisticated and demanding fields as space science are considering biomimetic robots. At the Jet Propulsion Laboratory, which is part of the NASA, four and six-legged robots have been under development for future missions to Mars. Such robots include the Limbed Excursion Mobile Utility Robot (LEMUR) and the Steep Terrain Access Robot (STAR) (Badescu et al., 2005). These types of robots are developed to travel across rough terrain, acquire and analyze samples, and perform many other functions that are attributed to legged animals including walking, grasping, object manipulation, and wall climbing. Advances in this technology may potentially lead to future NASA missions, in which operations could resemble a plot from a movie or science-fiction book more than conventional mission operations. Equipped with multi-functional tools and multiple cameras, the new models of LEMUR are intended to inspect and maintain installations beyond humans' easy reach. This robot has six legs, each of which has interchangeable end-effectors as required to perform the required mission. The axi-symmetric layout is much like a starfish or octopus, with a panning camera system that allows omni-directional movement and manipulation operations.

Besides the possibility of robots that emulate human capabilities, science fiction also suggests humans with supernatural capabilities. A human being with bionic muscles is synonymous with superhuman characters in movies and TV series. Driven by bionic muscles, these characters are portrayed as capable of strength and speeds that are far superior to humans. The development of artificial muscles using electroactive polymers (EAP) materials has made the use of bionic muscles a potential reality. These materials can induce large strains (stretching, contracting or bending) in response to electrical stimulation (Bar-Cohen, 2004). EAP-based actuators may be used to eliminate the need for gears, bearings, and other components that complicate the construction of robots, reducing their costs, weight, size, and premature failures. Further, these materials can be used to make biomimetic robots that appear and behave more realistically. Robots are being introduced with increased capability and sophistication, including the ability to express emotions both verbally and facially as well as respond emotionally to such expressions. The first commercial product driven by EAP that emerged in 2002 is a Fish-Robot (Eamex, Japan) that swims without a motor or batteries. It uses EAP materials that simply bend upon stimulation. For power it uses inductive coils that are energized from the top and bottom of the fish tank. This toy represents a major milestone for the field, making a very realistic looking fish.

20.4.1 Simulators and Virtual Robots

For many years, the entertainment industry has been imitating living creatures using numerous forms that include puppets, cartoons, manikins, and others. Making animated movies is a well-established industry with an extensive heritage, where artists draw creatures that represent living animals, humans, or imagined creatures. These cartoon figures are made with biomimetic appearance and behavior but with capabilities that are only limited by the artist's imagination and creativity. Generally, such animated creatures do not have to obey the laws of physics, and they can perform unrealistic tasks that defy gravity and other forces of nature. However, cartoons can indicate future advances in biomimetic technology. While the operation of biomimetic robots could use some of the kinematic algorithms that are well developed by the animation industry, there are many issues that need to be addressed when making actual robots. These issues include control, stability, feedback, vibration suppression, effect of impact, power, mass, volume, obstacle avoidance, environmental conditions, workspace, and other real-world requirements. In order to address these issues without the costly process of making and testing real robots, one can use computer simulations, in which the laws of physics are accurately represented.

Computer simulation has become a critical development tool that can be used to test the behavior of simulated system and rapidly make modifications without the high cost of fabrication and testing. The analytical phase is followed by rapid prototyping and other procedures of accelerated software development. The development of computers and analytical tools, including numerical and logical models, has made possible a very powerful simulated representation of real-world activity. Such tools are used to investigate the performance of complex systems, and address such parameters as thermal, aerodynamic, mechanics, material behavior, and time-dependent effects. Also, electronic and mechanical issues of driving and operating the developed systems can be integrated into the simulation model and studied on the computer. Testing a real-world system can be prohibitively expensive, or even impossible for situations in which making changes can be very difficult and time consuming. Also, simulated testing to the point of failure can be repeated many times without serious consequences to the tested systems. An example includes the simulation of a car crash into a wall, in which safety engineers evaluate potential designs. The advantage is that it reduces the number of real cars that may need to be instrumented and sacrificed. Other examples of simulated systems include the response of an aircraft structure to bird strikes, and the effect of loads on mechanical systems and new products. Because of the complexity of products, their behavior cannot be perfectly modeled. Therefore, test products must still be physically built and tested to destruction.

20.4.2 Robots as an Integral Part of our Society

Making biomimetic robots requires attention to technical, philosophical, and social issues. Inspiration from science fiction sets expectations that will continually be bound by reality and the state of the art. Making biomimetic robots is the electro-mechanical analog of biological cloning. Being increasingly capable, the development of biomimetic robots, or the performance of artificial cloning, raises issues of concern with regard to questionable implementations. This issue may become a topic of public debate in years and may reach the level that is currently involved with the topics of fetal stem cells and human cloning. As biomimetic robots with human characteristics are becoming more an engineering reality, there may be a growing need to equip them with limited self-defense and controlled-termination. In parallel, there may be a rise in potential use of such robots for unlawful applications, and proper attention may be required by lawmakers to head off this possibility in order to assure that such robots are used for positive applications. As this need begins to rise, it will become more important to give serious attention to the laws of Asimov (1950) that he defined for robots. These laws address the human concern that robots may be designed to harm people. According to these laws, the desired status of robots is as slaves to humanity, where they are allowed to protect themselves only as long as no human is physically hurt. While these laws reflect the desire to see "peaceful" robots as productive support tools it might not be realistic to expect them to be designed only as Asimov's law-obedient robots. One would expect that some robots would be designed by various governments to perform military and law enforcement tasks that may involve violation of these laws.

One can expect a revolution in our lives as such robots are developed to the point that they become part of our daily activity. It would require the implementation of direct interaction schemes that include the ability of robots to express themselves both in body language and verbal expressions. Since different people view emotions and moods differently, users will need to have the capability of user-friendly programming of the robot's behavior, emotion, and mood. This may also be provided through self-learning and adaptive behavior just as kids learn that which is appropriate or acceptable and that which is not. Further, while computers will have superior capabilities over humans, there will be a need to assure the social order with the clear role of robots as the slaves in master–slave relation. A certain level of independence will need to be provided with a user-selected autonomous operation vs. fully programmable performance depending on the desired task. Also, robots will need to have selectable behavior specifications that define their desired personality. This personality may include friendliness, and "cool" operations with various algorithms of human interaction and behavior.

20.5   SMART STRUCTURES AND MATERIALS

The development of smart materials has been the objective of researchers and engineers for over three decades. Materials, systems, and structures are identified as smart if they can interact with the environment and have an ability to predict the required future actions and to respond to change in various ways. Adaptive capabilities have already been implemented in commercial materials. For instance, liquid crystals are used to indicate changes in temperature, and there are commercially available optical glasses that become dark with the increase in light intensity. To behave "smart" beyond the simple reactions to a specific condition, as sunglasses change their shade, it is necessary to provide systems with the ability to learn the required response to various stimulations from the environment and be capable of predicting future conditions and prepare to respond optimally.

It would be interesting to develop systems with individual characteristics that would be the result of learning from the environment in which they operate and exhibit relatively wide variety of shapes while still working properly. It may be feasible to define structures in terms of the ability to carry loads and the positions or places where it can hold or place objects. Thus, such systems would not need to be engineered to high tolerances, yet they will learn how to functionally deal with the design details. Such development will require taking advantage of the increasingly evolving nano-technology, where minute sensors will be integrated throughout the structure to provide information and feedback for smart control. Ultimately, such smart structures would need to design and construct themselves using resources from the environment or redistribute their structural materials to allow effective handling of large loads. Such an approach would enable producing lighter and safer structures that eliminate stress concentrations, perform optimally, and operate with long life duration. Structures will need to be designed with scalability in mind to allow adapting the technology to various aspects of our daily lives. An interesting distinction between biological structures with bones compared to robots is the fact that the biological elements are not rigidly connected. It would be a challenge for future roboticists to develop robots that have such a structural flexibility of being an integrated system while still able to carry loads, move rapidly, and perform all these functions that we recognize as biological.

20.6   IMPACT OF BIOMIMETICS ON NONENGINEERING FIELDS

Throughout the history of mankind, nature has been an inspiration to many nonengineering fields including entertainment, toys, and art with the results well documented in such artistic objects as paintings, statues, structures, and other artifacts. Engineering and art with biomimetic characteristics are increasingly being integrated in the construction of modern buildings and other structures.

Figure 20.3 (See color insert following page 302) The Singaporean giant "durians" building called the Esplanade Theater (left) has the shape of this fruit that is considered the king of fruits (right). (The photo on the right is the courtesy of Anand Krishna Asundi, Nanyang Technological University, Singapore.)

Examples of the influence of nature on construction are the architectural landmarks such as the Sydney Opera House in Australia and the Esplanade Theater in Singapore. The Sydney Opera House has an artistic configuration of sea shells. It was originally designed by the Danish architect Joern Utzonhas and opened in 1973. The construction of the Esplanade Theater in Singapore was influenced by the durian fruit (see Figure 20.3, right). In Singapore, this fruit is considered the king of fruits — it is sweet, spiky, and weighs about 1 to 2 kg. It is native to Malaysia and Indonesia, green to brown in color, oblong to round in shape, prickly with strong sharp thorns, emitting a strong, distinctive smell that puts most foreigners off. For this reason, in Singapore it is forbidden to carry it on public transportation such as aircraft and subway. The durian fruit inspired the construction of the Esplanade Theater that was opened in October 2002. This building consists of two domes having the shape of this fruit with durian-like spikes that are used as sun shields (see Figure 20.3, left). Due to its shape this theater is also known as the Durians Building.

Another example shown in Figure 20.4 is the beech (*Fagus sylvaticus*) leaf (Kobayashi et al., 1998), which serves as an inspiring model. The leaves of the beech emerge from the bud by unfolding its corrugated surface (Figure 20.4a). Interestingly, the leaf uses high angle folds to allow it to be folded more compactly within the bud, though this arrangement it requires more time to expand. This may be needed to allow the plant to optimize the timing of the leaf deployment with ecological and physiological conditions. An artistic object that mimicked this leaf is shown in Figure 20.4b. Called the "Leaf-Mat", it was created as a folding mat for children's play-time and it consists of a polypropylene base and felt.

Another area that mimics biology is economy. In nature, entities compete for energy, while in an economy they compete for money (Mattheick, 1994; Vincent, 2001). Plants compete to grow higher in order to gain more sunlight, while animals compete for territory, sex, and food. On the other hand, industry competes for customers to assure survival and growth. In business, if a company cannot survive competition in the changing environment of the marketplace, it goes bankrupt, analogous to death in biology. The use of subsidies to support small companies can be viewed as similar to a small plant that is supported with a stick or animal that is helped by its parents in its early stages. These animals and plants need to learn to operate independently; otherwise they will require support throughout their life and will never be able to handle the tough challenges of the real world.

As mentioned in Chapter 1, the use of terminologies that are biologically inspired makes communication of complex details easier to understand. Examples include the use of the terms male and female for plugs, erection of structures to describe their construction, head or tail for the location in a structure and many other such terms. Increased use of such terms can be highly beneficial to improvement in communication, training, and friendliness of users' manual for

(a)

(b)

Figure 20.4 (a) Unfolding of the common beech (*Fagus sylvaticus*) leaves as they open from the bud stage (left) to corrugated leaves (right). (These photos are a courtesy of the Royal Society and of Julian Vincent and they were taken by Biruta Kresling; both Julian and Biruta are from The University of Bath, England.) (From Kobayashi H., B. Kresling, and J. F. V. Vincent, The geometry of unfolding tree leaves, *Proceeding of the Royal Society, Series B*, vol. 265 (1998), pp. 147–154. http://www.bath.ac.uk/mech-eng/biomimetics/LeafGeometry.pdf. With permission.). (b) A biomimetic art called 'Leaf-Mat' mimics the folding leaves and it is a folding mat that is made of polypropylene deployment and felt. (Courtesy of Adi Marom, Landscape Products Co., Tokyo, Japan.)

operation of new instruments. The interface of machine and humans is becoming increasingly complicated, but the instructions for using them can be simplified by using biological terms and principles. One can make new instruments more intuitive if concepts from nature are used, making it easier to "figure out" how the instrument works, thereby reducing instructions or training.

20.7 HUMAN DEVIATION FROM NATURE MODELS

In order to ensure both the short-term existence and long-term species sustainability, all organisms must grow, maintain existence, feed, and reproduce. Generally, most organisms meet their basic life needs within the boundaries of the habitat in which they live. If they cannot compete in this habitat, then they must either adapt a different strategy, move to a different habitat in which they can compete, or die. In the short term, the adaptation capability of individual organisms helps species survive if followed by genetic modifications that sustain the long-term survival of the species. The specific characteristics of the adaptation are determined by the constraints of the environment and the genetic make up of the specific species.

For many thousands of years, humans lived harmoniously with nature and migrated periodically allowing for nature to recover from the damage to the specific habitat. As the human race advanced its capability, efforts have increasingly been made to deviate from the process of evolution. We have significantly extended our life, increased our survival rate, reduced our reproduction, and stopped migrating to allow recovery of our habitats. Also, we are using significant amount of energy, consuming oil at enormous levels, processing our food, polluting the environment with nondegradable chemicals, changing the temperatures around us, affecting the weather (e.g., inducing rain), blocking or diverting the path of rivers, bringing many species to extinction, destroying the ozone layer, and doing so many other things that affect our environment in nonreversible ways. The pollution that we have released into our environment has reached levels where every aspect of our life has been impacted including the air we breathe, the water we drink, and the food we eat. For example, there are some fish that are not recommended for consumption because of the levels of toxic chemicals in their system, including mercury, PCB, and others. In defiance of other organisms, we often adapt our environments to suit ourselves and even change those constraints that are supposed to make us adapt to the environment. Effectively, we are operating against the laws of biology and pushing the limits of our existence. However, we are also increasingly becoming aware of these facts and making greater efforts to live more in harmony with nature. There are examples now of our efforts to ensure our own sustainability with many success stories. These include increased use of biodegradable materials, making and operating mechanisms that consume energy more efficiently, recycling our resources, and protecting the ozone layer from our pollution and chemicals.

Biomimetics can provide an important guide in our efforts to live harmoniously with nature (Benyus, 1998). We can learn from plants how to use the Earth's pollution that is in the form of CO_2 to produce oxygen, which is also critical to human life. Also, plants pump water and minerals from the ground to great heights and use these as resource for growth and also as a source of energy that is completely Earth-friendly, that is, solar energy. Another example is nature's recycling of its resources where plants are eaten by plant-eaters, which in turn are consumed by predators whose bodies decompose to fertilize plants. The mimicking of this recycling process can be seen in the recycling of trash to produce recycled materials as well as energy and it is one of the human success stories.

20.8 PRESENT TECHNOLOGY, FUTURE POSSIBILITIES, AND POTENTIALS

The focus of this book has been on the making of technologies that one can label "artificial" as opposed to the ones that are known as "natural". Parallel to the efforts to mimic biology in engineering and science terms, there are also efforts to create synthetic systems that include making cells, tissues, and in future years, possibly organs. Although the latter is a form of mimicking nature it is outside the scope of this book.

Developing biomimetic mechanisms requires employing many disciplines, tools, and capabilities. It involves materials, actuators, sensors, structures, control, and autonomous operations. As described in this book, mimicking nature has immensely expanded the collection of tools that are available to us in performing tasks that were once considered science fiction. As technology evolves, increasing numbers of biologically inspired mechanisms and functions that emulate the capability of creatures and organisms are expected to emerge. The challenges to making such biomimetic technologies that are copied or adapted depend on the complexity that is involved. Many examples of biomimetic applications that are currently in use or expected to emerge in future were described and discussed in this book. Other examples may include marine vehicles that mimic shark skin by having low friction surface in water or use antifreeze proteins found in some marine creatures allowing them to sustain temperatures below freezing points.

One of the emerging areas of biomimetics is artificial muscles, a moniker for electroactive polymers (EAP). It offers enormous potential for many areas of our life. The easy capability to produce EAP materials in various shapes can be exploited to make future mechanisms and devices using such methods as stereolithography and ink-jet printing techniques. A polymer can be dissolved in a volatile solvent and ejected drop-by-drop onto various substrates. Such processing methods offer the potential of making systems and robots in full 3D details that include EAP actuators allowing rapid prototyping and quick mass production (Bar-Cohen, 2004). A possible vision for such technology can be the fabrication of insect-like robots that can be made to fly and pack themselves into a box, ready for shipping, once they are made. These miniature robots may help to inspect hard-to-reach areas of aircraft structures, where they can be launched to conduct the required inspection procedures and download information about the structure integrity. Other examples can be the rapid prototyping of robots with controlled characteristics that follow specific movie scripts and with the appearance and behavior of the desired artificial actors. The robots' appearance and behavior can be modified rapidly as needed for the evolving script, and when changes need to be made, the artificial actors can be rapidly produced with any desired modification. Using effective EAP actuators to mimic nature would immensely expand the collection and functionality of devices and mechanisms that are currently available. Important addition to this capability can be the application of telepresence combined with virtual reality using haptic interfaces (Mavroidis et al., 2004). While such capabilities are expected to significantly change future robots, additional effort is needed to develop robust and effective EAP-based actuators.

Considering the current limitations of artificial muscles and their capability to support biomimetic applications, the author posed a challenge to the worldwide science and engineering community to develop a robotic arm that is actuated by artificial muscles to win an arm-wrestling match against a human opponent (Figure 20.5) (Chapter 10; Bar-Cohen, 2004). The first competition was held on March 7, 2005 during the EAP Session of the SPIE's EAP Actuators and Devices (EAPAD) Conference, which is part of the Smart Structures and Materials Symposium. As described in Chapter 10, three EAP-actuated arms wrestled against a 17-year-old female student who won all three matches. Progress in making robotic arms that win a match against humans will lead to significant benefits, particularly in the medical area of effective prosthetics. A remarkable contribution would be to see a disabled person jogging to the grocery store using this technology.

Figure 20.5 (See color insert following page 302)   Grand challenge for the development of EAP-actuated robotics.

It would lead to exciting new generations of robots that can change our daily life; possibilities would include robots as a household assistant and intelligent companion replacing the dog as "man's best friend." Another important benefit that may be achieved with success in winning this challenge would be a milestone demonstration of the capability to produce superior biomimetic robots.

Availability of strong and robust artificial muscles may enable future years to produce biomimetic legged robots that can run as fast as a cheetah, carry mass like a horse, climb steep cliffs like a gecko, reconfigure its body like an octopus, fly like a bird, and dig tunnels like a gopher. This is an incredible vision of robots that can potentially be used in future exploration of planets in the universe leading to future NASA mission plans that may include a script for the robots operation that may follow science-fiction ideas. Hopefully, these robots will be able to operate autonomously, detecting water, various resources, and possibly even biological indicators of past or present life. They may even be able to construct facilities for future human habitats.

20.9 AREAS OF CONCERNS AND CHALLENGES TO BIOMIMETICS

Throughout this book there are descriptions and discussions of many examples of concepts, devices, and mechanisms that were mimicked or inspired by biology. One of the amazing capabilities of nature that were described include the spider's ability to create in room temperature and pressure, incredibly flat and strong web structures that are durable in outdoor conditions. The spider's web may have inspired the fishing net, the fabric of the clothing that we wear, and many other things that we use in our daily life. In some cases, the possibilities seen in nature have allowed us to make things with far superior capabilities. For example, human efforts to copy birds' wings in order to produce a flying machine led to very limited capabilities as was demonstrated in the late 1880s and 1890s by Horatio Phillips and Otto Lilienthal. Only after we mastered aerodynamic principles we managed to make flying machines far superior to birds. Aircraft capabilities are an incredible human success that far exceeded capabilities of any flying creature that ever existed. This includes flying significantly higher and faster, and carrying far more load as aircrafts have enormous volume as we can see in airports today. The only thing airplanes cannot do yet is perch on a wire (though microplanes now under development may end up doing just that).

Some human inventions that appear biomimetic may have not necessarily been the result of an actual adaptation of nature's ideas (Altshuller, 1988). The process of innovation and introduction of invention by humans as problem solvers can be difficult to trace. In some tools, nature may not have been the immediate model, and similarities may just be coincidental. Honeycombs, used in many aircraft structures (Gordon, 1976) may not have been directly inspired by the honeycombs made by the bees; however, it is still the same structure and the aircraft structure's name is the same as the product made by the bees. The potential of reinventing nature's innovation may be reduced if these inventions can be documented, not as biological observations but as engineering mechanisms and tools. Effectively, there is a need to establish a database and handbooks that logically catalog nature's capabilities, specifications, mechanisms, processes, tools, and functions in terms of principles, materials, dimensions, limitations, etc. Such documented information, which can be produced by biologists for use by engineers, may greatly help humans in making novel biomimetic inventions. Working towards such an objective, one can consider nature in technological terms, while possibly considering the use of a unified approach to describe biological inventions. Such documentation might help to accelerate advances in human-made technologies.

The December 2004 tsunami disaster caused over two hundred thousand casualties and led to million homeless people, where in contrast very few animals died. This fact suggests that humans have lost the ability to sense and be forewarned of such natural calamities. It is difficult to believe that such a sensing capability can be reacquired by humans and therefore alternative detection techniques are needed. Most countries do not have the required monitoring system due to the very high associated cost and the fact that such disaster may occur once in tens or hundreds of years. Adapting the sensing capability of animals would potentially lead to affordable detection technology.

Nature offers many capabilities that are unique to some species, and understanding the requirements for their adaptation can help us in many ways. Some of these capabilities are still mysteries that can offer enormous potential for humans. One may wonder about bears' ability to sleep for 6 months without urinating and poisoning its blood. For medical applications, learning the clues to this capability may help fight diabetes. The ability of the lizard to drop its tail as a decoy in case of danger, and grow it back without scars is another important model for the field of medicine. Adapting this capability can help heal the disabled and severely injured.

20.10 CONCLUSION

Over the 3.8 billion years of evolution, nature has come up with inventions that are great models for imitation and adaptation. Nature consists of a large pool of inventions although it has its own evolution drawbacks including that nature is irreversible, cannot be planned and has crevasses in its solution space. The field of biomimetics is multidisciplinary requiring the use of expertise from biology, engineering, computational and material sciences, robotics, neuroscience, biomechanics, and many other related fields. Further, several disciplines have emerged in recent years as a result of the effort to develop biomimetic systems. The technology requires the ability to produce scaleable mechanisms ranging from miniature — as small as nanometers scale — to giant sizes — as large as several meters. There are still numerous challenges, but the recent trends in the field of biomimetics — international cooperation, the greater visibility of this area of study and the surge in funding of related research projects — offer great potential.

Nature uses minimum resources to produce maximum results, and one of the characteristics of this aspect is the effective packing and deployment techniques that have been used by nature allowing organisms to be fitted for the environment in which they need to operate. As seen throughout this book, both plants and animals have used various techniques of packing where flowers and leaves grow from a highly packed structure in the bud. Further, animals are using appendages for locomotion that are configured in easy-to-deploy structures, which include the fins, legs and wings (Kresling, 2000). Beside the inspiration of effective robots, there are numerous other inventions and mechanisms that one can be inspired to develop using such packing techniques. One may consider deployable structures that can include tents and other large surface foldable structures as well as gossamer structures and deployable antennas for space applications. For commercial applications and user-friendly household products one may consider future improvements to such tools as the food mixer that has many parts and need to be assembled and disassembled each time the mixer is used. One may think of integrated parts that can be deployed like the wings of the bird. Another example can be vacuum cleaners that also consist of many parts which can be easily and rapidly deployed when needed and packed and stowed when not.

There are many areas where nature is superior, and one example is the ability to recognize patterns and objects. We can recognize people whom we have not seen for years and who may have grown quite older and changed significantly and we can do so even from some distance. Efforts have been done to develop such a technology of face recognition at airports as part of the US homeland defense technology. While significant success was observed in the early tests, the systems that were installed at airports for face recognition were removed. This has been the result of the many false positive indications that have been encountered.

One can find many examples in our daily life where human-made technology can be traced to nature's inventions that were mimicked or used as an inspiration. These include many aspects of science and engineering and learning how to do more will help humans even further. In this age of international terror and with the need for more innovative homeland security and defense tools one may want to examine nature's techniques and investigate the possibilities of learning more.

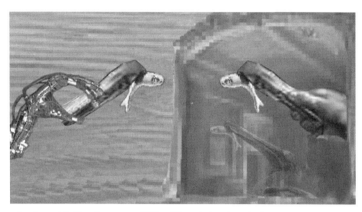

Figure 20.6 (See color insert following page 302)   The chain of evolution of our mimicking nature is drawn into the artificial world that we create. (Top graphics is the courtesy of David Hanson and Human Emulation Robotics, LLC. The bottom graphics is the modification that was made by Adi Marom, Research Artifacts Center Engineering, The University of Tokyo, Japan. The robotic arm in this figure was made by G. Whiteley, Sheffield Hallam U., UK, and photographed in the author's lab.)

We have learned to use various techniques of camouflage, shields and body armor as well as the stings and barbed arrows and there should be many more ideas that we can mimic.

As we learn from nature we are becoming better able to implement the adapted inventions into the world of artificial tools that we create and continue to improve. The cover page graphics (see the top panel of Figure 20.6) shows an illustration of human learning to make tools by watching nature. This idea can be turned into a vision for the future of the artificial world that we are continuing to create and improve as illustrated at the bottom panel of Figure 20.6. In this figure, an evolutionary chain of the inspiration of nature is drawn where technology that we learned from nature is implemented into robot's end-effectors including robotic arms, manipulators, and other biomimetic support fixtures.

For the question "what else can we learn?" it would be highly helpful to create a documented database that would examine biology from an engineering point of view and offer possibly a catalog of nature's inventions. This catalog needs to include the inventions that have already been used and possibly even offer different ways of looking at nature's innovations to enrich other fields that have not been benefited yet. This database can be documented in a format of webpages with hyperlinks that crossrefer related information. An example of a tool for documenting the database one can use the Wiki online database system (see for more information: http://wiki.org/wiki.cgi?WhatIsWiki).

There are many challenges to mimicking nature but the possibilities are endless. As long as we would not reach the level of having a chip grows from a micron size in a hibernated state to an active fully grown robot that is highly intelligent and autonomous, biomimetics will still be a useful source of inspiration for inventors. Success in developing and implementing nature's ideas will bring science fiction and imaginations to engineering reality. One of the great challenges to imitating biology is to create robots that mimic such creatures as octopus. This would mean having robots that are highly flexible and dexterous that operate intelligently and autonomously with the capability to crawl through very narrow strips, camouflage its body by matching the colors, shape and texture of the surrounding, be equipped with multiple tentacles and suction cups for gripping on objects, using ink as a smoke screen, see clearly without blind spots, and having many other capabilities and multifunctional components that can perform multiple tasks simultaneously.

The future of biomimetics is quite exciting but it is hard to predict what would be learned or mimicked next. One can envision in the years to come that many more tools and capabilities will emerge in every scale of our life from nano levels to macro and beyond. The benefits can be expected in such areas as medical, military, consumer products, and many others.

ACKNOWLEDGMENT

Research reported in this manuscript was partially conducted at the Jet Propulsion Laboratory (JPL), California Institute of Technology, under a contract with National Aeronautics and Space Administration (NASA).

REFERENCES

Altshuller G., *Creativity as an Exact Science*, Gordon and Breach, New York (1988).

Asimov I., *I Robot*, Fawcett Publications, Greenwhich, CT (1950).

Autumn K. and A. M. and Peattie, Mechanisms of adhesion in geckos, *Journal of Integrative and Comparative Biology*, vol. 42, no. 6 (2003), pp. 1081–1090.

Badescu M., X. Bao, Y. Bar-Cohen, Z. Chang, B. Kennedy, and S. Sherrit, Enhanced robotic walking mobility in geological analogues using extractable anchors, *Journal of Mechanical Design,* in preparation, (2005).

Bar-Cohen Y., *Electroactive Polymer (EAP) Actuators as Artificial Muscles — Reality, Potential and Challenges*, 2nd Edition, vol. PM136, SPIE Press, Bellingham, WA, March 2004, pp. 1–765, ISBN 0-8194-5297-1.

Bar-Cohen Y. and C. Breazeal (Eds), *Biologically-Inspired Intelligent Robots*, vol. PM122, SPIE Press, Bellingham, WA, May 2003, pp. 1–393, ISBN 0-8194-4872-9.

Bar-Cohen Y. and S. Sherrit, Self-Mountable and Extractable Ultrasonic/Sonic Anchor (U/S-Anchor), NASA New Technology Report, Docket No. 40827, December 9, 2003 (patent disclosure in preparation).

Benyus J. M., *Biomimicry: Innovation Inspired by Nature*, ISBN 0688160999, Harper Collins (Perennial Press), New York 1998, pp. 1–302.

Bock W. J., Functional and evolutionary morphology of woodpeckers, *The Ostrich*, vol. 70 (1999), pp. 23–31.

Cott H. B., Camouflage in nature and war, *Royal Engineers Journal*, (December 1938), pp. 501–517.

Geim A. K., S.V. Dubonos, I.V. Grigorieva, K.S. Novoselov, A.A. Zhukov, and S.Y. Shapoval, Microfabricated adhesive mimicking gecko foot-hair, *Nature Materials*, vol. 2, no. 7 (2003), pp. 461–463.

Gordon J. E. *The New Science of Strong Materials, or Why You Don't Fall Through the Floor*, 2nd Edition, Pelican-Penguin, London, 1976, pp. 1–287, ISBN: 0140209204.

Guest S. D. and S. Pellegrino, The folding of triangulated cylinders — I: Geometric considerations, *Journal of Applied Mechanics, ASME E*, vol. 61 (1994), pp. 773–777.

Hanlon R., Forsythe, J. Joneschild, and D. Crypsis, Conspicuousness, mimicry and polyphenism as antipredator defenses of foraging octopuses on Indo-Pacific coral reefs, with a method of quantifying crypsis from video tapes, *Biological Journal of the Linnean Society*, vol. 66 (1999), pp. 1–22.

Kobayashi H., B. Kresling, and J. F. V. Vincent, The geometry of unfolding tree leaves, *Proceeding of the Royal Society, Series B*, vol. 265 (1998), pp. 147–154. http://www.bath.ac.uk/mech-eng/biomimetics/LeafGeometry.pdf.

Kresling B., Coupled mechanisms in biological deployable structures, Pellegrino S. and S. D. Guest (Eds), *Proceedings of the IUTAM-IASS Symposium on Deployable Structures: Theory and Application*, Kluwer Academic Press, Dordrecht, The Netherlands, 2000, pp. 229–238.

Mattheick C., Design in nature, *Interdisciplinary Science Review*, vol. 19 (1994), pp. 298–314.

Mavroidis C., Y. Bar-Cohen, and M. Bouzit, Haptic interfacing via electrorheological fluids, Topic 7, Chapter 19, in Y. Bar-Cohen (Ed.), *Electroactive Polymer (EAP) Actuators as Artificial Muscles — Reality, Potential and Challenges*, 2nd Edition, vol. PM136, SPIE Press, 2004, pp. 659–685, ISBN 0-8194-5297-1.

Rao P. R., Biomimetics, *Sadhana*, vol. 28, Parts 3 and 4, June/August 2003, pp. 657–676.

Unda J., J. Weisz, J. Rivacoba, and I. R. Urfen, Family of deployable/retractable structures for space application, *Acta Astro*, 32 (1994), pp. 767–784.

Vincent J. F. V., Stealing ideas from nature, Chapter 3, in S. Pellegrino (Ed.), *Deployable structures*, Springer-Verlag, Vienna, 2001, pp. 51–58.

WEBSITES

http://www.amazon.com/exec/obidos/tg/detail/-/0688160999/ref = sib_rdr_dp/103-6613814-7815021?%5Fencoding = UTF8&no = 283155&me = ATVPDKIKX0DER&st = books

http://www.netinformations.com/Detailed/136232.html

http://www.rdg.ac.uk/biomim/projects.htm

http://www.ias.ac.in/sadhana/Pdf2003JunAug/Pe1106.pdf

Index